Robotics

Robotics

Science and Systems IV

edited by Oliver Brock, Jeff Trinkle, and Fabio Ramos

The MIT Press
Cambridge, Massachusetts
London, England

For information about special quantity discounts, please email special_sales@mitpress.mit.edu or write to Special Sales Department, The MIT Press, 55 Hayward Street, Cambridge, MA 02142.

Printed and bound in the United States of America.

Library of Congress Cataloging-in-Publication Data

Robotics: Science and Systems Conference (4th : 2008 : Swiss Federal Institute of Technology)
Robotics : science and systems IV / edited by Oliver Brock, Jeff Trinkle, and Fabio Ramos.
p. cm.
"Papers presented at Robotics: Science and Systems 2008, held at the Swiss Federal Institute of Technology (ETH) in Zurich, Switzerland, from June 25 to June 28, 2008"—Pref.

ISBN 978-0-262-51309-8 (pbk. : alk. paper)
1. Robotics—Congresses. I. Brock, Oliver. II. Trinkle, Jeffrey C. III. Ramos, Fabio. IV. Title.
TJ210.3.R6435 2009
629.8'92—dc22 2008054715

10 9 8 7 6 5 4 3 2 1

Contents

Preface

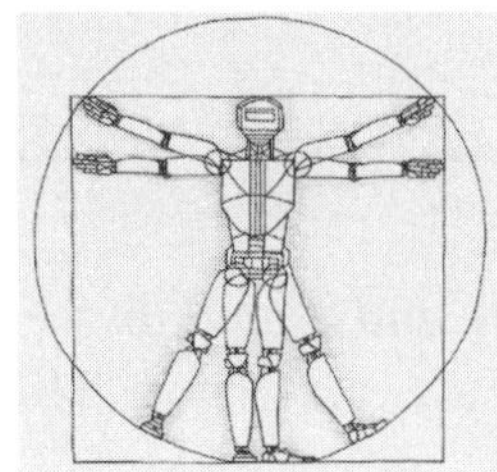

This volume contains the papers presented at *Robotics: Science and Systems 2008*, held at the Swiss Federal Institute of Technology (ETH) in Zurich Switzerland, from June 25 to June 28, 2008. This year's meeting brought together more than 280 researchers from Europe, Asia, North America, and Australia. The papers presented at the meeting and compiled here cover a wide range of topics in robotics spanning computer vision, mapping, terrain identification, distributed systems, localization, manipulation, collision avoidance, multibody dynamics, obstacle detection, micro-robotic systems, pursuit-evasion, grasping and manipulation, tracking, spatial kinematics, machine learning, sensor networks, and applications such as autonomous driving and design of manipulators for use in functional-MRI.

Following the RSS tradition, there were a number of invited talks: two Early Career Spotlight talks by rising stars in the robotics community, one banquet speaker, and six invited talks on topics on the fringes of robotics. The latter talks were chosen to give the audience new motivating perspectives on what is possible and on new ways to approach robotics problems.

- Prof. Armin Gruen of ETH delivered "Positioning Modeling and Navigation with Photogrammetric Techniques," in which he argued convincingly for the use of photogrammetric techniques in robotics and showed impressive results in which accuracy on the order of one part per million was obtained.

- Prof. Miguel Nicolelis of Duke University spoke about "Computing with Neural Ensembles." In this excellent talk, he discussed the functioning of neural ensembles in animals providing evidence that the ensemble of neurons that controls a given task is not fixed, thus raising questions about the design of bio-inspired robot controllers.

- Prof. Jean-Louis Deneubourg of the Universit Libre de Bruxelles shared his thoughts on "Shared Decision-Making in Mixed Societies of Animals and Robots." He provided deep insights into the "emergent" global behavior of large colonies of small animals capable of only simple local communication.

- Dr. Kevin O'Regan, Director of Research of the Laboratory of the Psychology of Perception at CNRS Paris discussed robot consciousness in his talk "How to Build Consciousness into a Robot: the Sensori-motor Approach." This was a highly engaging talk, in which it was conjectured that given today's computing power, it is only a matter of time before robots become self-aware, which gives rise to many complex questions of ethics.

- Prof. Howard Berg of Harvard University presented "Motile Behavior of E-coli: a Remarkable Robot." He showed detailed pictures of the inner workings of the remarkable nano-machines called "flagella," how they are constructed by bacteria, and how they provide propulsion and directional control.

- Prof. Toshio Yanagida of Osaka University described the engineering principles behind biological molecular motors in his talk "Mechanism Involved in Utilizing Thermal Fluctuations by Muscle Molecular Motor." He showed how molecular motors exploit thermal noise to achieve energy efficiency and talked about the implications for building artificial muscles.

- Prof. Rob Wood of Harvard University gave an engaging Early Career Spotlight presentation on his progress on meso-scale bio-inspired vehicles, that fly, walk, and swim. These would be useful in many many applications ranging from environmental monitoring to border surveillance. He covered aspects of design, fabrication, analysis, and control.

- Prof. Eric Klavins of the University of Washington was the second Early Career Spotlight speaker. He spoke about the motivations behind his Self-Organizing Systems Lab and a string of results. Along the way, he provided insights into the aspects of systems that lead to self-organization and robustness of global behaviors He showed the application of these ideas to a macro-scale testbed and discussed DNA applications.

- In an extremely engaging banquet talk, Prof. Luis von Ahn of Carnegie Mellon Institute focused on the unwitting use of humans to solve difficult problems while playing games on the Web. Among many other intriguing ideas, his talk, "Human Computation," described how the use of capchas to verify that a real human is interacting with a web application, is also providing free labor to translate old texts, not amenable to optical character recognition.

Thanks to the efforts of Workshop Chair Charlie Kemp of Georgia Tech., the fourth day of RSS 2008 was devoted to the following well-attended workshops:

- *Control of Locomotion: From Animals to Robots* organized by Auke Ijspeert, Paolo Dario, Sten Grillner;

- *Underwater Robotics ... at the Microscale* organized by Brad Nelson, Vijay Kumar, Sylvain Martel, Metin Sitti, Lixin Dong;

- *Topology and Minimalism in Robotics and Sensor Networks* organized by Robert Ghrist, Steve LaValle, George J. Pappas;

- *Teaching with Robots* organized by Chris Rogers, Pedro Lima, Roland Siegwart, Illah Nourbakhsh, Aaron Dollar;

- *Design and Control of Variable Impedance Actuators for Physical Interaction of Robots with Humans and their Environment* organized by Antonio Bicchi, Alin Albu Schaeffer, Bram Vanderborght;

- *Robot Manipulation: Intelligence in Human Environments* organized by Robert Platt, Sami Haddadin, Charlie Kemp, Lorenzo Natale, Neo Ee Sian;

- *Interactive Robot Learning* organized by Andrea Thomaz, Geert-Jan M. Kruijff, Henrik Jacobsson, Danijel Skocaj;
- *Inside Data Association* organized by Udo Frese, Jose Neira, Diedrich Wolter, Jorg Kurlbaum;
- *Quantitative Performance Evaluation of Navigation Solutions for Mobile Robots* organized by Raj Madhavan, Chris Scrapper, Alex Kleiner;
- *Experimental Methodology and Benchmarking in Robotics Research* organized by Fabio Bonsignorio, John Hallam, Angel P. del Pobil;
- *Advances in Simulation of Robot and Task Dynamics* organized by Evan Drumwright, Kurt Anderson, Roy Featherstone; and
- *Grand Challenges in Microrobotics and Microassembly* organized by Pierre Lambert, Stephane Regnier, Metin Sitti.

RSS 2008 was a success thanks to the efforts of many people. We gratefully acknowledge the enormous time spent by the area chairs, who each handled about 20 papers and flew to Zurich for the in-person area chair meeting. The area chairs were: Antonio Bicchi (University of Pisa), Karl Bohringer (University of Washington), Jaydev P. Desai (University of Maryland, College Park), Hugh Durrant-Whyte (University of Sydney), Shin-ichi Hirai (Ritsumeikan University), Atsushi Konno (Tohoku University), Kevin Lynch (Northwestern University), Yoky Matsuoka (University of Washington), Michael McCarthy (UC Irvine), Jose Neira (University of Zaragoza), Giuseppe Oriolo (University of Rome), Nick Roy (Massachusetts Institute of Technology), Thierry Simeon (LAAS-CNRS, Toulouse), Metin Sitti (Carnegie Mellon University), Frank van der Stappen (Utrecht University), and Kazuhito Yokoi (AIST). Together, their expertise covered an extraordinarily broad swath of the robotics landscape.

Paper reviewing was rigorous. All but two papers received four or more double-blind reviews (neither the authors nor the reviewers knew each others' identities) - that's over 650 reviews from 196 program committee members. After the reviews were completed, the program committee members and area chairs discussed reviews for each paper in an on-line forum within the conference management system. Then the authors were invited to rebut the reviews. Following the rebuttals, the program committee members reconsidered and finalized their reviews. Final acceptance decisions and presentation categories were made at the area chair meeting in Zurich, also attended by the program chair and general chair. Of the 163 submissions, 20 were selected for poster presentation and 20 were selected for oral presentation.

The local arrangement chairs were Roland Siegwart and Brad Nelson of ETH. They and their support staff did a fantastic job in organizing everything from the Travel Kit to the poster session at the faculty club to the alphorn player in the opening ceremony to the banquet in the restaurant overlooking the lake and the city. We particularly want to thank Cornelia Della Casa and Luciana Borsatti for sweating all the details with such grace and efficiency. We cannot thank them enough.

We sincerely thank Rudi Triebel and Ralf Kästner for taking the roles of general Webmaster for the Conference. Finally our thanks go out to Janosch Nikolic and Stefan Bertschi for dealing with all computer and wireless issues and to Markus Bhler and Dario Fenner for setting up the welcoming reception.

RSS 2008 had many sponsors; thanks to ABB, Evolution Robotics, Intel, Microsoft Research, and Willow Garage, for providing funds for general conference use and to the National Science Foundation and the Naval Research Lab for providing funds to support student travel. Last, but not least, we thank Springer for providing a $2500 prize for the best student paper award. The competition was fierce.

Finally, we would like to express our gratitude to the members of the robotics community who have adopted RSS and its philosophy, and who have made RSS 2008 an outstanding meeting by submitting manuscripts, participating in the review process, and by attending the Conference. We look forward to many more exciting meetings of RSS in the years to come.

Oliver Brock, University of Massachusetts Amherst
Jeff Trinkle, Rensselaer Polytechnic Institute
Fabio Ramos, University of Sydney

July 2008

Organizing Committee

General Chair	Oliver Brock, University of Massachusetts Amherst
Program Chair	Jeff Trinkle, Rensselaer Polytechnic Institute
Local Arrangement Chair	Brad Nelson, ETH Zurich
Local Arrangement Co-Chair	Roland Siegwart, ETH Zurich
Publicity Chair	Andrew Davison, Imperial College London
Publications Chair	Fabio Ramos, University of Sydney
Workshop Chair	Charlie Kemp, Georgia Tech
Web Masters	Rudolph Triebel, ETH Zurich Ralf Kaestner, ETH Zurich
Area Chairs	Antonio Bicchi, University of Pisa Jaydev P. Desai, University of Maryland, College Park Shin-ichi Hirai, Ritsumeikan University Kevin Lynch, Northwestern University Michael McCarthy, University of California Irvine Giuseppe Oriolo, University of Rome Thierry Simeon, LAAS-CNRS, Toulouse Frank van der Stappen, Utrecht University Karl Bohringer, University of Washington Hugh Durrant-Whyte, University of Sydney Atsushi Konno, Tohoku University Yoky Matsuoka, University of Washington Jose Neira, University of Zaragoza Nicholas Roy, Massachusetts Institute of Technology Metin Sitti, Carnegie Mellon University Kazuhito Yokoi, AIST
Conference Board	Oliver Brock, University of Massachusetts Amherst Wolfram Burgard, University of Freiburg Greg Dudek, McGill University Dieter Fox, University of Washington Lydia Kavraki, Rice University Eric Klavins, University of Washington Nicholas Roy, Massachusetts Institute of Technology Daniela Rus, Massachusetts Institute of Technology Sven Koenig, University of Southern California John Leonard, Massachusetts Institute of Technology Stefan Schaal, University of Southern California Gaurav Sukhatme, University of Southern California Sebastian Thrun, Stanford University

Program Committee

Agrawal, Sunil
Aiyama, Yasumichi
Albu-Schaeffer, Alin
Alterovitz, Ron
Amato, Nancy
Antonelli, Gianluca
Arai, Fumihito
Asano, Fumihiko
Bailey, Tim
Balaguer, Carlos
Balasubramanian, Ravi
Balkcom, Devin
Bamberg, Stacy
Bayazit, Burchan
Belta, Calin
Berard, Stephen
Berg, Jur van den
Berkelman, Peter
Bidaud, Philippe
Bleuler, Hannes
Bosse, Michael
Bretl, Timothy
Bruyninckx, Herman
Burgard, Wolfram
Caldwell, Darwin
Castellanos, Jose
Cavusoglu, M. Cenk
Chaumette, Francois
Cheah, Chien
Chella, Antonio
Chevallereau, Christine
Chirikjian, Greg
Chopra, Nikhil
Choset, Howie
Chung, Wan Kyun
Cortes, Jorge
Crandall, Jacob
Davison, Andrew
De Luca, Alessandro
Dissanayake, Gamini
Dong, Lixin
Doulgeri, Zoe
Dudek, Gregory
Egerstedt, Magnus
Emery-Montemerlo, Rosemary
Eustice, Ryan
Fatikow, Sergej
Ferreira, Antoine
Fitch, Robert
Folkesson, John
Frazzoli, Emilio
Freeman, Robert
Fukushima, Edwardo
Furukawa, Tomonari
Gauthier, Michael
Ge, Qiaode Jeffrey
Ghrist, Robert
Gienger, Michael
Gillespie, Brent
Gordon, Geoff
Gorman, Jason
Goswami, Ambarish
Grisetti, Giorgio
Guglielmelli, Eugenio
Gupta, Kamal
Han, Li
Harada, Kensuke
Hashimoto, Koichi
Henrich, Dominik
Higashimori, Mitsuru
Hirata, Yasuhisa
Hosoda, Koh
Hsu, David
Hu, Tie
Iagnemma, Karl
Ijspeert, Auke
Inamura, Tetsunari
Indiveri, Giovanni
Isler, Volkan
Isto, Pekka
Jensfelt, Patric
Jia, Yan-Bin
Julier, Simon
Kao, Imin
Kapoor, Ankur
Kavraki, Lydia
Kemp, Charlie
Kersting, Kristian
Kikuuwe, Ryo
Klavins, Eric
Koganezawa, Koichi
Kragic, Danica
Krovi, Venkat
Krut, Sebastien
Kuffner, James
Kurisu, Masamitsu
Lamiraux, Florent
Laschi, Cecilia
Lenarcic, Jadran
Leonard, John
Lien, Jyh-Ming
Lin, Ming
Lungarella, Max
Lutz, Philippe
Madhavan, Raj
Maeda, Yusuke
Maeyama, Shoichi
Martel, Sylvain
Martinez-Cantin, Ruben
Mascaro, Stephen
Matsumoto, Yoshio
Menciassi, Arianna
Minguez, Javier
Mitsuishi, Mamoru
Miura, Jun
Mochiyama, Hiromi
Moll, Mark
Montesano, Luis
Montiel, J.M
Morales, Marco
Morgansen, Kristi
Mori, Taketoshi
Morin, Pascal
Murphey, Todd
Nagatani, Keiji

Nakanishi, Jun
Nejat, Goldie
Nelson, Bradley
Newman, Paul
Nguyen, Binh
O'Kane, Jason
Odest, Jenkins
Ogata, Tetsuya
Oh, Paul
Olson, Edwin
Omata, Toru
Parviz, Babak
Patel, Rajni
Peine, William
Pennock, Gordon
Perez-Garcia, Alba
Peshkin, Michael
Peters, Jan
Pineau, Joelle
Plagemann, Christian
Posner, Ingmar
Pozidis, Haralampos
Prattichizzo, Domenico
Ramos, Fabio
Redon, Stephane
Regnier, Stephane
Rekleitis, Ioannis
Ribeiro, Isabel
Roberts, Jonathan
Rocco, Paolo
Roumeliotis, Stergios
Rudolph, Lee
Santos-Victor, Jose
Schechter, Evan
Scheding, Steve
Shibata, Tomohiro
Shimada, Nobutaka
Siegwart, Roland
Sim, Robert
Song, Dezhen
Song, Guang
Spong, Mark
Stachniss, Cyrill
Su, Haijun
Suarez, Raul
Sudsang, Attawith
Sugar, Thomas
Sugihara, Tomomichi
Sukhatme, Gaurav
Sun, Yu
Suzuki, Sho'ji
Svinin, Mikhail
Tahara, Kenji
Takahashi, Yasutake
Tan, Hong
Tardos, Juan D.
Tedrake, Russ
Tendick, Frank
Thrun, Sebastian
Umeda, Kazunori
Upcroft, Ben
Vanderborght, Bram
Villani, Luigi
Wada, Takahiro
Wakamatsu, Hidefumi
Williams, Brian
Williams, Stefan
Wimboeck, Thomas
Wisse, Martijn
Wollherr, Dirk
Xiao, Jing
Xiao, Jizhong
Yamane, Katsu
Yoon, Woo-Keun
Zavala-Rio, Arturo
Zefran, Milos
Zhou, Quan

Sponsors

The organizers of Robotics Science and Systems gratefully acknowledge the following conference sponsors:

United States National Science Foundation (NSF)

United States Naval Research Laboratory (NRL)

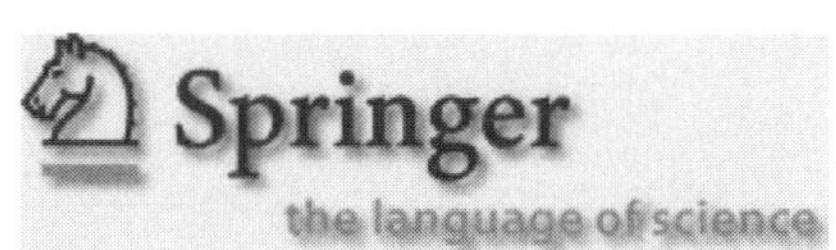

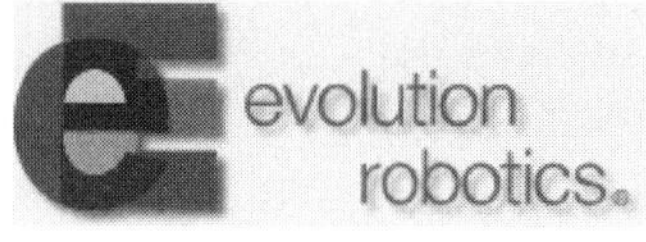

IEEE Robotics and Automation Society (RAS)

Association for the Advancement of Artificial Intelligence (AAAI)

International Foundation of Robotics Research (IFRR)

Multi-Sensor Lane Finding in Urban Road Networks

Albert S. Huang[†] David Moore[†] Matthew Antone[⋆] Edwin Olson[†] Seth Teller[†]

[†]MIT Computer Science and Artificial Intelligence Laboratory
Cambridge, MA 02139
{ashuang,dcm,eolson,teller}@mit.edu

[⋆]BAE Systems Advanced Information Technologies
Burlington, MA 01803
matthew.antone@baesystems.com

***Abstract*— This paper describes a system for detecting and estimating the properties of multiple travel lanes in an urban road network from calibrated video imagery and laser range data acquired by a moving vehicle. The system operates in several stages on multiple processors, fusing detected road markings, obstacles, and curbs into a stable non-parametric estimate of nearby travel lanes. The system incorporates elements of a provided piecewise-linear road network as a weak prior.**

Our method is notable in several respects: it estimates multiple travel lanes; it fuses asynchronous, heterogeneous sensor streams; it handles high-curvature roads; and it makes no assumption about the position or orientation of the vehicle with respect to the road.

We analyze the system's performance in the context of the 2007 DARPA Urban Challenge. With five cameras and thirteen lidars, it was incorporated into a closed-loop controller to successfully guide an autonomous vehicle through a 90 km urban course at speeds up to 40 km/h amidst moving traffic.

Fig. 1. Our system uses many asynchronous heterogeneous sensor streams to detect road paint and road edges (yellow) and estimate the centerlines of multiple travel lanes (cyan).

I. Introduction

The road systems of developed countries include millions of kilometers of paved roads, of which a large fraction include painted lane boundaries separating travel lanes from each other or from the road shoulder. For human drivers, these markings form important perceptual cues, making driving both safer and more time-efficient [12]. In mist, heavy fog or when a driver is blinded by the headlights of an oncoming car, lane markings may be the principal or only cue enabling the driver to advance safely. Moreover, public-safety officials use the number, color and type of lane markings to encode spatially-varying traffic rules, for example no-passing regions, opportunities for left turns across oncoming traffic, regions in which one may (or may not) change lanes, and preferred paths through complex intersections.

Even the most optimistic deployment scenario for autonomous vehicles assume the presence of massive numbers of human drivers for the next several decades. Given the centrality of lane markings to public safety, it is clear that they will continue to be maintained indefinitely. Thus autonomous vehicle researchers, as they design self-driving cars, may assume that lane markings will be encountered with significant frequency.

We define the lane finding problem as divining, from live sensor data and (when available) prior information, the presence of one or more travel lanes in the vehicle's vicinity, and the semantic, topological, and geometric properties of each lane. By semantic properties, we mean the lane's travel sense and the color (white, yellow) and type (single, double, solid, dashed) of each of its boundaries. By topological properties, we mean the connectivity of multiple lanes in regions where lanes merge, split, terminate, or start. The term geometric properties is used to denote the centerline location and lateral extent of the lane. This paper focuses on detecting lanes where they exist, and the determination of geometric information for each detected lane (Figure 1). We infer semantic and topological information in a limited sense, by matching detected lanes to edges in an annotated input digraph representing the road network.

Aspects of the lane finding problem have been studied for decades in the context of autonomous land vehicle development [5, 17] and driver-assistance technologies [8, 1, 2]. McCall and Trivedi provide an excellent survey [11]. Lane finding systems intended to support autonomous operation have typically focused on highway driving [5, 17], where roads have low-curvature and prominent lane markings, rather than on urban environments. Previous autonomous driving systems have exhibited limited autonomy in the sense that they required a human driver to "stage" the vehicle into a valid lane before enabling autonomous operation, and to take control whenever the system could not handle the required task, for example during highway entrance or exit maneuvers [17].

Driver-assistance technologies, by contrast, are intended as continuous adjuncts to human driving; one common class of such systems, lane departure warning (LDW) systems, is designed to alert the human driver to an imminent (unsignaled) lane departure [13, 10, 15]. These systems typically assume that a vehicle is in a highway driving situation and that a human driver is controlling the vehicle correctly, or nearly so. Highways exhibit lower curvature than lower-speed roads, and do not contain intersections. In vehicles with LDW systems, the human driver is responsible for selecting an appropriate travel lane, is assumed to spend the majority of driving time within such a lane, is responsible for identifying possible alternative travel lanes, and only occasionally changes into such a lane. Because LDW systems are essentially limited to providing cues that assist the driver in staying within the current lane, achieving fully automatic lane detection and tracking is not simply a matter of porting an LDW system into the front end of an autonomous vehicle.

Clearly, in order to exhibit safe, human-like driving, an autonomous vehicle must have good awareness of all other nearby travel lanes. In contrast to prior lane-keeping and LDW systems, this paper presents a lane finding system suitable for guiding a fully autonomous land vehicle through an urban road network. In particular, our system is distinct from previous efforts in several respects: it attempts to detect and classify all observable lanes, rather than just the single lane occupied by the vehicle; it operates in the presence of complex road geometry, static hazards and obstacles, and moving vehicles; and it uses prior information (in the form of a topological road network with sparse geometric information) when available.

The apparent difficulty of matching human performance on sensing and perception tasks has led some researchers to investigate the use of augmenting roadways with a *physical infrastructure* amenable to autonomous driving, such as magnetic markers embedded under the surface of the road [18]. While this approach has been demonstrated in limited settings, it has yet to achieve widespread adoption and faces a number of drawbacks. First, the cost of updating and maintaining hundreds of thousands of miles of roadway is highly prohibitive. Second, the danger of autonomous vehicles perceiving and acting upon a different infrastructure than human drivers do (magnets vs. visible markings) becomes very real when one is modified and the other is not, whether through accident, delay, or malicious behavior.

Advances in computer networking and data storage technology in recent years have brought the possibility of a *data infrastructure* within reach. In addition to semantic and topological information, such an infrastructure might also contain fine-grained road maps registered in a global reference frame; advocates of these maps argue that they could be used to guide autonomous vehicles. We propose that a data infrastructure is useful for topological information and sparse geometry, but reject relying upon it for dense geometric information.

While easier to maintain than a physical infrastructure, a data infrastructure with fine-grained road maps might still become "stale" with respect to actual visual road markings. Even for human drivers, mapping staleness, errors, and incompleteness have already been implicated in accidents in which drivers trusted too closely their satellite navigation systems, literally favoring them over the information from their own senses [3, 16]. Static fine-grained maps are clearly not sufficient for safe driving; to operate safely, in our view, an autonomous vehicle must be able to use local sensors to perceive and understand the environment.

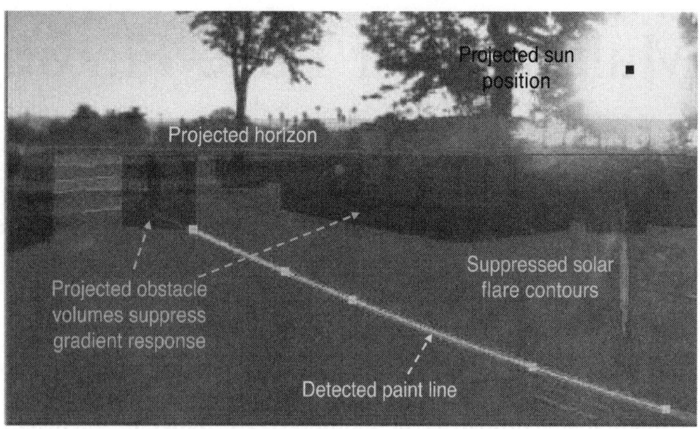

Fig. 2. Use of absolute camera calibration to project real-world quantities into the image.

The primary contributions of this paper are:

- A method for estimating multiple lanes of travel in a typical urban road network using only information from local sensors;
- A method for fusing these estimates with a weak prior, such as that derived from a topological road map with sparse metrical information;
- Methods for using monocular cameras to detect road markings; and
- Multi-sensor fusion algorithms combining information from video and lidar sensors.

We also describe our method's failure modes, and describe possible directions for future work.

II. Approach

Our approach to lane finding involves three stages. In the first stage, the system detects and localizes painted road markings in each video frame, using lidar data to reduce the false positive detection rate. A second stage processes the road paint detections along with lidar-detected curbs [6] to estimate the centerlines of nearby travel lanes. Finally, the detected centerlines output by the second stage are filtered, tracked, and fused with a weak prior to produce one or more non-parametric lane outputs.

Separation of the three stages provides simplicity, modularity, and scalability. Specifically, we are able to experiment with each stage independently of the others and easily substitute different algorithms for each stage. For example, we experimented with and ultimately used two separate algorithms in parallel for detecting road paint, both of which are described below. By introducing sensor-independent abstractions of environmental features, we are able to scale to many heterogeneous sensors.

A. Absolute Camera Calibration

Our road-paint detection algorithms assume that GPS and IMU navigation data are available of sufficient quality to

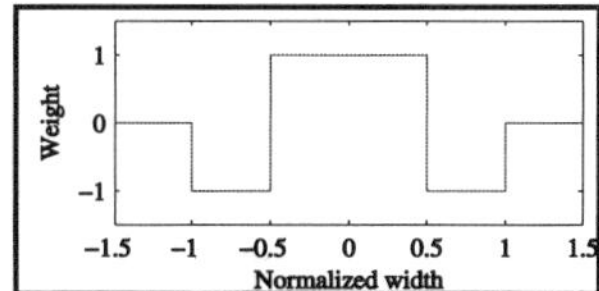

Fig. 3. The shape of the one-dimensional kernel used for matching road paint. By applying this kernel horizontally we detect vertical lines and vice versa. The kernel is scaled to the expected width of a line marking at a given image row and sampled according to the pixel grid.

correct for short-term variations in vehicle heading, pitch, and roll during image processing. In addition, the intrinsic (focal length, center, and distortion) and extrinsic (vehicle-relative pose) parameters of the cameras have been calibrated ahead of time. This "absolute calibration" allows preprocessing of the images in several ways (Figure 2):

- The horizon line is projected into each image frame. Only pixel rows below this line are considered for further processing.
- Our lidar-based obstacle detector supplies real-time information about the locations of obstructions in the vicinity of the vehicle [6]. These obstacles are projected into the image and their extents masked out as part of the paint detection algorithms, an important step in reducing false positives.
- The inertial data allows us to project the expected location of the ground plane into the image, providing a useful prior for the paint-detection algorithms.
- False paint detections caused by lens flare can be detected and rejected. Precise knowledge of the date, time, and Earth-relative vehicle pose allow computation of the solar ephemeris; line estimates that point toward the sun in image coordinates are then removed.

B. Road Paint Detection using Matched Filters

This section describes the first of two vision algorithms we use for detecting painted lines on the road. For each camera, we run a dedicated process that detects road paint and outputs a list of candidate line markings for each frame. These candidates are expressed as cubic hermite splines, which have the convenient property that the spline passes through each of the control points. Each frame is considered independently from the others; cross-frame tracking techniques could be used to improve the result.

The first step in our image processing pipeline is to construct matched one-dimensional filters, tuned to the expected width of a painted line marking at each row of the image. We consider two types of lines: Those that extend roughly away from the car towards the horizon and those that run transverse to the line of sight. The former is detected by a horizontal kernel; the latter by a vertical kernel. In both cases, each row of the image has its own kernel as computed by the projection of the expected ground plane into the image and the nominal painted line widths that such projection would imply. The shape of the kernel is shown in Figure 3.

(a) Original Image (b) Filtered Image

(c) Local maxima w/orientations (d) Spline fit

Fig. 4. Our first road paint detector: (a) The original image is (b) convolved with a matched filter at each row (horizontal filter shown here). (c) Local maxima in the filter response are enumerated and their dominant orientations computed. The figure depicts orientation by drawing the perpendiculars to each maximum. (d) Nearby maxima are connected into cubic hermite splines.

The kernel is sampled according to the pixel grid at each row, then convolved with that row to produce the output of the matched filter. As shown in Figure 4, this operation successfully discards most of the clutter in the scene and produces a strong response along line-like features. This is done separately for the vertical and horizontal kernels, giving two output images. We then compute a list of local maxima of the filter responses and a principal direction of the line at each maximum. This direction is computed using the dominant eigenvector of the Hessian in a small window around each maximum.

The system next connects nearby maxima into splines that represent continuous line markings. To do so, it randomly selects 100 "seed" maxima near the bottom of the image from the list of all maxima. For each seed, we set it as the first control point in a cubic hermite spline, and consider an annulus around the seed of radius 50 pixels and width 10 pixels. Each of the maxima within this annulus becomes a candidate for the spline's second control point and is assigned a score. The score is computed by sampling along the spline's length the value of the distance transform function of the list of maxima. The highest scored maximum is saved as the second spline control point. If no maximum has a score above a certain threshold, we reject the whole spline. We continue to "grow" the spline in the same fashion by considering additional annuli of successive control points and finish the spline when adding another control point results in a poor score. The maxima near finished splines are removed from the maxima list so that the same lines are not re-detected. After finishing searching the 100 seeds, the algorithm is complete. The splines are inverse-perspective mapped, intersected with the ground plane, discretized into piecewise linear curves, and transmitted for further processing by the centerline estimator.

C. Road Paint Detection using Symmetric Contours

A second road paint detection mechanism employed in our system relies on more traditional low-level image processing. In order to maximize frame throughput, and thus reduce the time between successive inputs to the lane fusion and tracking components, we designed the module to utilize fairly simple and easily-vectorized image operations.

The central observation behind this detector is that image features of interest – namely lines corresponding to road paint – typically consist of well-defined, elongated, continuous regions that are brighter than their surround. This characterization encompasses solid and dashed lane markings, stop lines, crosswalks, white and yellow paint on road pavements of various types, and markings seen through cast shadows across the road surface. Thus, our strategy is to first detect the potential boundaries of road paint using spatial gradient operators, and then estimate the desired line centers by searching for boundaries that enclose a brighter region; that is, boundary pairs which are proximal and roughly parallel in real-world space and whose local gradients point toward each other (Figure 5).

Fig. 5. Progression from original image through smoothed gradients (red), border contours (green), and symmetric contour pairs (yellow) to form a paint line candidate.

Three steps constitute the contour-based road line detector: low-level image processing to detect raw features; contour extraction to produce initial line candidates; and contour post-processing for smoothing and false alarm reduction. The first step applies local lowpass and derivative operators to produce the noise-suppressed direction and magnitude of the raw (grayscale) image's spatial gradients. The gradient magnitude is thresholded, and non-maximal suppression is performed in the gradient direction to produce a sparse feature mask.

Next, a connected components algorithm iteratively walks the feature mask to generate smooth contours of ordered points, broken at discontinuities in location and gradient direction. This results in a new image whose pixel values indicate the identities and positions of the detected contours, which in turn represent candidate road paint boundaries. In order to localize the centerlines between these boundaries, a second iterative walk is applied. At each boundary pixel p_i (traversed in contour order), the algorithm extends a virtual line in the direction of the gradient until it meets another contour at p_j. If the gradient of the second contour points in the opposite direction, then the midpoint between p_i and p_j is added to a growing centerline curve (Figure 2).

This step connects many short paint fragments, producing a smaller number of longer centerline candidates. The gradient constraint insures that each candidate is brighter than its surround. Since this candidate set may be corrupted by small line fragments and outliers, a series of higher-level post-processing operations is performed. We enforce global smoothness and curvature constraints by fitting parabolas to the curves and recursively breaking them at points of high deviation or spatial gaps. We then remove all curves shorter than a given threshold length (in pixels) to produce the final road paint lines. As with the first road paint detection algorithm, these are inverse-perspective mapped and intersected with the ground plane before further processing.

D. Lane Centerline Estimation

The second stage of lane finding estimates the geometry of nearby lanes using a weighted set of recent road paint and curb detections, both of which are represented as piecewise linear curves. To simplify the process, we estimate only lane centerlines, which we model as locally parabolic segments. While urban roads are not designed to be parabolic, this representation is generally accurate for stretches of road that lie within sensor range.

Lanes centerlines are estimated in two steps. First, a centerline evidence image D is constructed, where the value each pixel $D(\mathbf{p})$ of the image corresponds to the evidence that a point $\mathbf{p} = [p_x, p_y]$ in the local coordinate frame lies on the center of a lane. Second, parabolic segments are fit to the ridges in D and evaluated as lane centerline candidates.

1) Centerline Evidence Image: To construct D, road paint and curb detections are used to increase or decrease the values of pixels in the image, and are weighted according to their age (older detections are given less weight). The value of D at a pixel corresponding to the point $\mathbf{p}$ is computed as the weighted sum of the influences of each road paint and curb detection d_i at the point $\mathbf{p}$:

$$D(\mathbf{p}) = \sum_i e^{-a(d_i)\lambda} g(d_i, \mathbf{p})$$

where $a(d_i)$ denotes how much time has passed since d_i was received, λ is a decay constant, and $g(d_i, \mathbf{p})$ is the influence of d_i at $\mathbf{p}$. We chose $\lambda = 0.7$.

Before describing how the influence is determined, we make three observations. First, a lane is more likely to be centered $\frac{1}{2}$ lane width from a strip of road paint or a curb. Second, 88% of federally managed lanes in the U.S. are between 3.05 m and 3.66 m wide [14]. Third, a curb gives us different information about the presence of a lane than does road paint. From these observations and the characteristics of our road paint and curb detectors, we define two functions $f_{rp}(x)$ and $f_{cb}(x)$, where

x is the Euclidean distance from d_i to $\mathbf{p}$:

$$f_{rp}(x) = -e^{-\frac{x^2}{0.42}} + e^{-\frac{(x-1.83)^2}{0.14}} \quad (1)$$

$$f_{cb}(x) = -e^{-\frac{x^2}{0.42}}. \quad (2)$$

The functions f_{rp} and f_{cb} are intermediate functions used to compute the influence of road paint and curb detections, respectively, on D. f_{rp} is chosen to have a minimum at $x = 0$, and a maximum at one half lane width (1.83 m). f_{cb} is always negative, indicating that curb detections are used only to decrease the evidence for a lane centerline. We elected this policy due to our curb detector's occasional detection of curb-like features where no curbs were present. Let $\mathbf{c}$ indicate the closest point on d_i to $\mathbf{p}$. The actual influence of a detection is computed as:

$$g(d_i, \mathbf{p}) = \begin{cases} 0 & \text{if } \mathbf{c} \text{ is an endpoint of } d_i, \\ & \text{else} \\ f_{rp}(||\mathbf{p} - \mathbf{c}||) & \text{if } d_i \text{ is road paint, else} \\ f_{cb}(||\mathbf{p} - \mathbf{c}||) & \text{if } d_i \text{ is a curb} \end{cases}$$

This last condition is introduced because road paint and curbs are only observed in small sections. The effect is that a detection influences only those centerline evidence values immediately next to the detection, and not in front of or behind it.

In practice, D can be initialized once and incrementally updated by adding the influences of newly received detections and applying an exponential time decay at each update. Additionally, we improve the system's ability to detect lanes with dashed boundaries by injecting imaginary road paint detections connecting two separate road paint detections when they are physical proximate and collinear.

2) Parabola Fitting: Once the centerline evidence image D has been constructed, the set R of ridge points is identified by scanning D for points that are local maxima along either a row or a column, and also above a minimum threshold. Next, a random sample consensus (RANSAC) algorithm [7] is used to fit parabolic segments to the ridge points. At each RANSAC iteration, three ridge points are randomly selected for a three-point parabola fit. The directrix of the parabola is chosen to be the first principle component of the three points.

To determine the set of inliers for a parabola, we first compute its conic coefficient matrix $\mathbf{C}$ [9], and define the set of candidate inliers L to contain the ridge points within some algebraic distance α of $\mathbf{C}$.

$$L = \{\mathbf{p} \in R : \mathbf{p}^T\mathbf{C}\mathbf{p} < \alpha\}$$

For our experiments, we chose $\alpha = 1$. The parabola is then re-fit once to L using a linear least squares method, and a new set of candidate inliers is computed. Next, the candidate inliers are partitioned into connected components, where a ridge point is connected to all neighboring ridge points within a 1 m radius. The set of ridge points in the largest component is chosen as the set of actual inliers for the parabola. The purpose of this partitioning step is to ensure that a parabola cannot be fit across multiple ridges, and requires that an entire identified ridge be connected. Finally, a score for the entire parabola is computed.

Fig. 6. The second stage of our system constructs a centerline evidence image. Lane centerline candidates (blue) are identified by fitting parabolic segments to the ridges of the image. Front-center camera is shown in top left for context.

$$s = \sum_{\mathbf{p} \in L} \frac{1}{1 + \mathbf{p}^T\mathbf{C}\mathbf{p}}$$

The contribution of an inlier to the total parabola score is inversely related to the inlier's algebraic distance, with each inlier contributing a minimum amount to the score. The overall result is that parabolas with many very good inliers have the greatest score. If the score of a parabola is below some threshold, then it is discarded. Experimentation with different values resulted in us choosing a score threshold of 140.

After a number of RANSAC iterations (we found 200 to be sufficient), the parabola with greatest score is selected as a candidate lane centerline. Its inliers are removed from the set of ridge points, and all remaining parabolas are re-fit and re-scored using this reduced set of ridge points. The next best-scoring parabola is chosen, and this process is repeated to produce at most 5 candidate lane centerlines (Figure 6). Each candidate lane centerline is then discretized as a piecewise linear curve and transmitted to the lane tracker for further processing.

E. Lane Tracking

The primary purpose of the lane tracker is to maintain a stateful, smoothly time-varying estimate of the nearby lanes of travel. To do so, it uses both the candidate lane centerlines produced by the centerline estimator and an a-priori estimate derived from a road map.

In the context of the Urban Challenge, the road map was known as the Route Network Description File (RNDF). The RNDF can roughly be thought of as a directed graph, where each node is a waypoint in the center of a lane, and edges represent intersections and lanes of travel. Waypoints are given as GPS coordinates, can be separated by arbitrary distances, and a simple linear interpolation of connected waypoints may go off road, through trees and houses. For the purposes of our system, the RNDF was treated as a strong prior on the

(a) Two RNDF-derived lane centerline priors

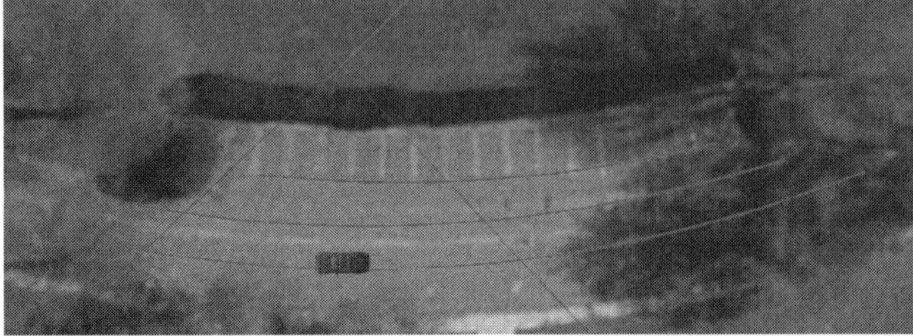

(b) Candidate lane centerlines estimated from sensor data

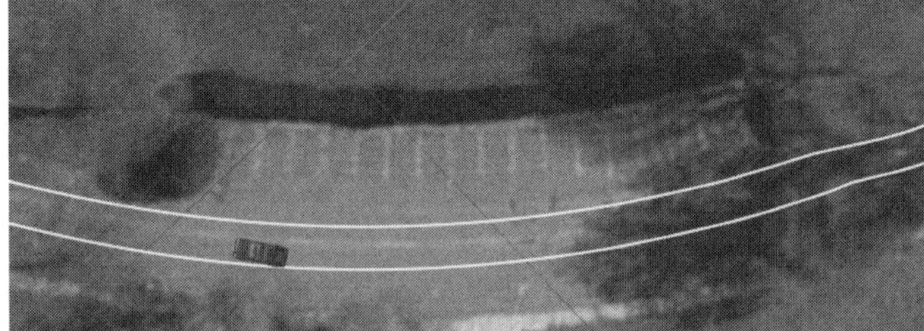

(c) Filtered and tracked lane centerlines

Fig. 7. (a) The RNDF provides weak a-priori lane centerline estimates (white) that may go off-road, through trees and bushes. (b) On-board sensors are used to detect obstacles, road paint, and curbs, which are in turn used to estimate lanes of travel, modeled as parabolic segments (blue). (c) The sensor-derived estimates are then filtered, tracked, and fused with the RNDF priors.

number and type of lanes, and a weak prior on their position and geometry.

As our vehicle travels, it constructs and maintains representations of all portions of all lanes within a fixed radius of 75 m. The centerline of each lane is modeled as a piecewise linear curve, with control points spaced approximately every 2 m. Each control point is given a scalar confidence value indicating the certainty of the lane tracker's estimate at that point. The lane tracker decays the confidence of a control point as the vehicle travels, and increases it either by detecting proximity to an RNDF waypoint or by updating control points with centerline estimates produced from the second stage.

As centerline candidates are generated, the lane tracker attempts to match each candidate with a tracked lane. If a matching is successful, then the candidate is used to update the lane estimate. To determine if a candidate c is a good match for a tracked lane l, the longest segment s_c of the candidate is identified such that every point on s_c is within some maximum distance τ to l. We then define the match score $m(c, l)$ as:

$$m(c, l) = \int_{s_c} 1 + \frac{\tau - d(s_c(x), l)}{\tau} dx$$

where $d(\mathbf{p}, l)$ is the distance from a point $\mathbf{p}$ to the lane l. Intuitively, if s_c is sufficiently long and close to this estimate, then it is considered a good match. We choose the matching function to rely only on the closest segment of the candidate, and not on the entire candidate, based on the premise that as the vehicle travels, the portions of a lane that it observes vary smoothly over time, and previously unobserved portions should not adversely affect the matching as long as sufficient overlap is observed elsewhere.

Once a centerline candidate has been matched to a tracked lane, it is used to update the lane estimates by mapping control points on the tracked lane to the centerline candidate, with an exponential moving average applied for temporal smoothing. At each update, the confidence values of control points updated from a matching are increased, and others are decreased. If the confidence value of a control point decreases below some threshold, then its position is discarded and recomputed as a linear interpolation of its closest surrounding confident control points. Figure 7 illustrates this process.

III. Urban Challenge Results

Often, the most difficult part of evaluating a lane detection and tracking system for autonomous vehicle operation lies in finding a suitable test environment. Legal, financial, and logistical constraints prove to be a significant hurdle in this process. We were fortunate to have the opportunity to conduct an extensive test in the 2007 DARPA Urban Challenge, which provided a large-scale real world environment with a wide variety of roads. Both the type and quality of roads varied significantly across the race, from well-marked urban streets, to steep unpaved dirt roads, to a 1.6 km stretch of highway. Throughout the duration of the race, approximately 50 human-driven and autonomous vehicles were simultaneously active, thus providing realistic traffic scenarios.

Our most significant result is that our lane detection and tracking system successfully guided our vehicle through a 90 km course in a single day, at speeds up to 40 km/h, with an average speed of 16 km/h. A post-race inspection of our log files revealed that at no time did our vehicle have a lane centerline estimate more than half a lane width off of the actual lane centerline, and at no time did it unintentionally enter or exit a lane of travel. In saying this, we note that the output of the lane tracking system was used directly to guide the navigation and motion planning systems. Specifically, if the lane tracking system produced an incorrect estimate, our vehicle would have traveled along that estimate, possibly into an oncoming traffic lane or off-road.

The first question that arises from these statements is that of determining how much our system relied on perceptually-derived lane estimates, and how much it relied on the prior knowledge of the road as given in the RNDF. To answer this, we examine the distance the vehicle traveled with high confidence visually-derived lane estimates, excluding control points where high confidence is a result of proximity to an RNDF waypoint.

Visual range (m)	Distance traveled (km)
≤ 0	30.3 (34.8%)
$1-10$	10.8 (12.4%)
$11-20$	24.6 (28.2%)
$21-30$	15.7 (18.0%)
$31-40$	4.2 (4.8%)
$41-50$	1.3 (1.5%)
>50	0.2 (0.2%)

TABLE I
DISTANCE TRAVELED WITH HIGH-CONFIDENCE VISUAL ESTIMATES IN CURRENT LANE OF TRAVEL.

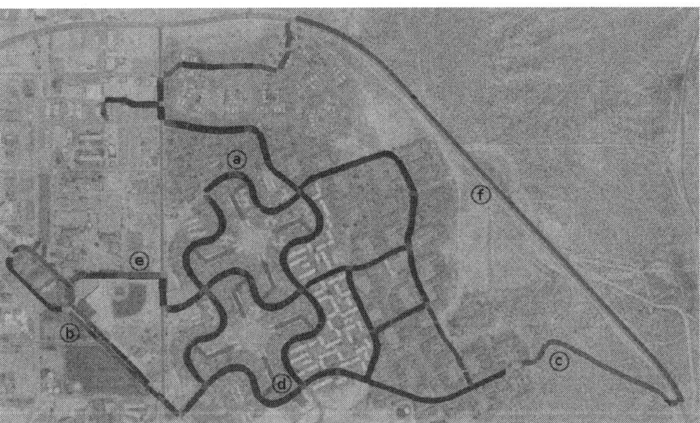

Fig. 8. Aerial view of the Urban Challenge race course in Victorville, CA. Autonomously traversed roads are colored blue in areas where the lane tracking system reported high confidence, and red in areas of low confidence. Some low-confidence cases are expected, such as at intersections and areas with no clear lane markings. Failure modes occurring at the circled letters are described in Fig. 9.

Fig. 9. Common failure cases. The most common failure was in areas with strong tree shadows, as seen in (a) and (b). Dirt roads, and those with faint or no road paint (c-e) were also common. In (f), a very wide lane and widely spaced dashed markings were a challenge due to our strong prior on lane width. In each of these failures, the system reported no confidence in its visual estimates.

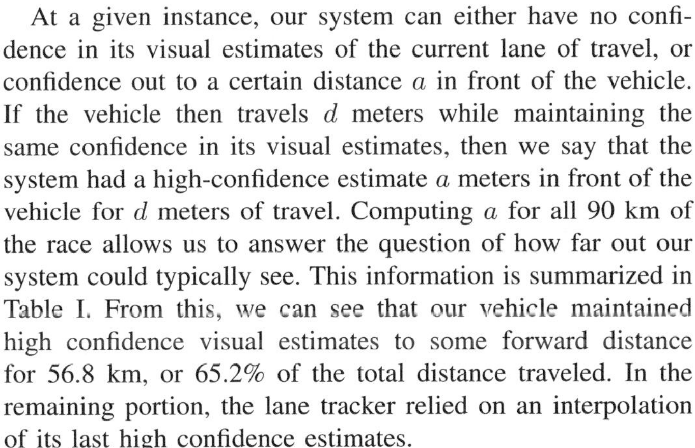

At a given instance, our system can either have no confidence in its visual estimates of the current lane of travel, or confidence out to a certain distance a in front of the vehicle. If the vehicle then travels d meters while maintaining the same confidence in its visual estimates, then we say that the system had a high-confidence estimate a meters in front of the vehicle for d meters of travel. Computing a for all 90 km of the race allows us to answer the question of how far out our system could typically see. This information is summarized in Table I. From this, we can see that our vehicle maintained high confidence visual estimates to some forward distance for 56.8 km, or 65.2% of the total distance traveled. In the remaining portion, the lane tracker relied on an interpolation of its last high confidence estimates.

A second way of assessing the system's performance is by examining its estimates as a function of location within the course. Figure 8 shows an aerial view of areas visited by our vehicle, colored according to whether or not the vehicle had a high confidence estimate at a given point. We note that our system had high confidence lane estimates throughout the majority of the high-curvature and urban portions of the course. Some of the low-confidence cases are expected, such as when the vehicle is traveling through intersections and roads with no discernible lane boundaries. In other cases, our system was unable to obtain a high confidence estimate whereas a human would have little trouble doing so.

Images from our logged camera images at typical failure cases are shown in Figure 9, and the locations at which these failures occurred are marked in Figure 8. A common failure mode was an inability to detect road paint in the presence of dramatic lighting variation such as that caused by cast tree shadows. However, we note that in all of these cases our system reported no confidence in its estimates and did not falsely estimate the presence of a lane.

Another significant failure occurred on the eastern part of the course, with a 0.5 km dirt road followed by a 1.6 km stretch of highway. Our vehicle traversed this path four times, for a total of 8.4 km. The highway was an unexpected failure, and the travel lane happened to be very wide. Its width did not fit the 3.66 m prior in the centerline estimator, which had trouble constructing a stable centerline evidence image. In addition, the dashed lane markings on the highway were spaced much further apart than dashed lane markings are on typical urban roads.

The final common failure mode occurred in areas with faint or no road paint, such as the dirt road and roads with well defined curbs but no paint markings. Since our system uses road paint as its primary information source, in the absence of road paint it is no surprise that no lane estimate ensues. Other environmental cues such as color and texture may be more useful [4].

The output of our system is used for high-speed motion planning; thus we would like for its estimates to remain relatively stable. Specifically, we desire that once the system produces a high confidence estimate, that the estimate does not change significantly. To assess the suitability of our system for this purpose, we can compute a *stability ratio* that measures how much its high confidence lane estimates change over time in the transverse direction.

Consider a circle of radius r centered at the current position of the rear axle. We can find the intersection $\mathbf{p}_0$ of this circle with the current lane estimate that extends ahead of the vehicle. When the lane estimate is updated at the next time step (10 Hz in this case) we can compute $\mathbf{p}_1$, the intersection of the same circle with the new lane estimate. We define the stability ratio as:

$$R = \frac{||\mathbf{p}_0 - \mathbf{p}_1||}{d_v} \tag{3}$$

where d_v is distance traveled by our vehicle in that time step.

Intuitively, the stability ratio is the ratio of the transverse movement of the lane estimate to the distance traveled by the car in that time, for some r. We can also compute an *average stability ratio* for some r by averaging the stability ratios for every time step of the vehicle's trip through the course (Figure 10). From this figure, we see that the average stability ratio remains small and relatively constant, but still nonzero, indicating that high-confidence lane estimates can be expected to shift slightly as the vehicle makes forward progress.

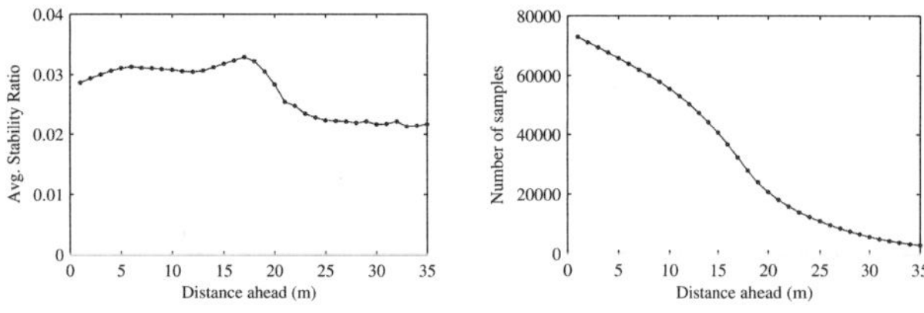

Fig. 10. (Left) The average stability ratio. (Right) The number of samples used to compute the stability ratio varies with r, as only control points with visually-derived high-confidence are used.

IV. Conclusion and Future Work

Our system attempts to extend the scope of lane detection and tracking for autonomous vehicles to the urban environment. We have presented a modular, scalable, perception-centric lane detection and tracking system that fuses asynchronous heterogeneous sensor streams with a weak prior to estimate multiple travel lanes in real-time. The system makes no assumptions about the position or orientation of the vehicle with respect to the road, enabling it to operate when changing lanes, at intersections, and when exiting driveways and parking lots. The vehicle using our system was, to our knowledge, the only vehicle in the final stage of the DARPA Urban Challenge to employ vision-based lane finding.

Despite these advances, the method is not yet suitable for real-world deployment. As with most vision-based systems, it is susceptible to strong lighting variations such as cast shadows. To address this, we are investigating the use of lidar intensity data for detecting road paint. Typical road paint has high infrared reflectivity and our preliminary lidar experiments are promising. Our highway experience in the race demonstrates the need to handle lanes with greater variance in width, which could be accomplished by first estimating lane width and then generating the centerline evidence image accordingly. Finally, since many roads do not use paint as boundary markers, we are extending our method to incorporate other environmental cues.

V. Acknowledgments

This work was conducted on the MIT 2007 DARPA Urban Challenge race vehicle. We give special thanks to Luke Fletcher, John Leonard, and Olivier Koch for numerous discussions and help given throughout the course of this work.

References

[1] Nicholas Apostoloff and Alex Zelinsky. Vision in and out of vehicles: Integrated driver and road scene monitoring. *International Journal of Robotics Research*, 23(4-5):513–538, Apr. 2004.

[2] Massimo Bertozzi and Alberto Broggi. GOLD: a parallel real-time stereo vision system for generic obstacle and lane detection. *IEEE Transactions on Image Processing*, 7(1):62–80, Jan. 1998.

[3] CNN. Doh! man follows GPS onto train tracks – when train coming. http://www.cnn.com/2008/US/01/03/gps.traincrash.ap/index.html, Jan. 2008.

[4] H. Dahlkamp, A. Kaehler, D. Stavens, S. Thrun, and G. Bradski. Self-supervised monocular road detection in desert terrain. In *Proceedings of Robotics: Science and Systems*, Philadelphia, USA, Aug. 2006.

[5] Ernst Dickmanns and Birger Mysliwetz. Recursive 3-d road and ego-state recognition. *IEEE Trans. Pattern Analysis and Machine Intelligence*, 14(2):199–213, Feb. 1992.

[6] John Leonard et al. A perception driven autonomous vehicle. *Journal of Field Robotics*, (to appear), 2008.

[7] Martin A. Fischler and Robert C. Bolles. Random sample consensus: A paradigm for model fitting with applications to image analysis and automated cartography. *Communications of the ACM*, 24(6):381–395, 1981.

[8] Luke Fletcher and Alexander Zelinsky. Context Sensitive Driver Assistance based on Gaze - Road Scene Correlation. In *International Symposium on Experimental Robotics*, Rio, Brazil, Jul. 2006.

[9] R. I. Hartley and A. Zisserman. *Multiple View Geometry in Computer Vision*. Cambridge University Press, ISBN: 0521623049, 2001.

[10] Iteris. http://www.iteris.com.

[11] Joel C. McCall and Mohan M. Trivedi. Video based lane estimation and tracking for driver assistance: Survey, system, and evaluation. *IEEE Transactions on Intelligent Transport Systems*, 7(1):20– 37, Mar. 2006.

[12] Ted R. Miller. Benefit-cost analysis of lane marking. *Public Roads*, 56(4):153–163, March 1993.

[13] Mobileye. http://www.mobileye.com.

[14] U.S. Department of Transportation, Federal Highway Administration, Office of Information Management. *Highway Statistics 2005*. U.S. Government Printing Office, Washington, D. C., 2005.

[15] D. Pomerleau and T. Jochem. Rapidly adapting machine vision for automated vehicle steering. *IEEE Expert*, 11(2):19–27, Apr. 1996.

[16] New York Times S. Lyall. Turn back. exit village. truck shortcut hitting barrier. http://www.nytimes.com/2007/12/04/world/europe/04gps.html, Dec. 2007.

[17] C. Thorpe, M. Hebert, T. Kanade, and S. Shafer. Vision and navigation for the Carnegie-Mellon Navlab. *IEEE Transactions on Pattern Anaysis and Machine Intelligence*, PAMI-10(3):362–373, May 1988.

[18] Wei-Bin Zhang. A roadway information system for vehicle guidance/-control. In *Vehicle Navigation and Information Systems*, volume 2, Oct. 1991.

Laser and Vision Based Outdoor Object Mapping

Bertrand Douillard
ARC Centre of Excellence
for Autonomous Systems
Australian Centre for Field Robotics
University of Sydney, Australia
Email: b.douillard@cas.edu.au

Dieter Fox
Dept. of Computer Science & Engineering
University of Washington
Seattle, WA, USA
Email: fox@cs.washington.edu

Fabio Ramos
ARC Centre of Excellence
for Autonomous Systems
Australian Centre for Field Robotics
University of Sydney, Australia
Email: f.ramos@cas.edu.au

Abstract— **Generating rich representations of environments can significantly improve the autonomy of mobile robotics. In this paper we introduce a novel approach to building object-type maps of outdoor environments. Our approach uses conditional random fields (CRF) to jointly classify laser returns in a 2D scan map into seven object types (car, wall, tree trunk, foliage, person, grass, and other). The spatial connectivity of the CRF model is determined via Delaunay triangulation of the laser map. Our model incorporates laser shape features, visual appearance features, structural information extracted from clusters of laser returns, and visual object detectors trained on image data sets available on the internet. The parameters of the CRF are trained from partially labeled laser and camera data collected by a car moving through an urban environment. Our approach achieves 91% accuracy in classifying objects observed along a 3 kilometer trajectory.**

I. Introduction

Generating rich representations of environments can bring another level of autonomy to mobile robotics. Over the last decade, much of the research in map building has focused on the simultaneous localization and mapping (SLAM) problem, *i.e.*, the problem of estimating the joint posterior distribution over the robot's location and the map of the environment. Research in this topic has produced various techniques that are able to build spatially consistent maps of large scale, cyclic environments [22].

More recently, several research groups extended SLAM approaches to generate maps that describe environments in terms of object types and places. Such representations can be extremely valuable, since they enable robots to perform high-level reasoning about their environments and the objects therein. For instance, in search and rescue tasks, a mobile robot that can reason about objects such as doors, and places such as rooms is able to coordinate with first responders in a much more natural way, being able to accept commands such as "Search the room behind the third door on the right of this hallway", and conveying information such as "There is a wounded person behind the desk in that room" [11]. As another example, consider autonomous vehicles navigating in urban areas. While the recent success of the DARPA Urban Challenge [5] demonstrates that it is possible to develop autonomous vehicles that can navigate safely in constrained settings, successful operation in more realistic, populated urban areas requires the ability to distinguish between objects such as cars, people, buildings, trees, and traffic lights.

In this paper we introduce a novel approach to building object type maps of outdoor environments. Our approach applies standard scan matching techniques to align 2D laser scans collected by a vehicle driving through urban environments. We use conditional random fields (CRF) to classify each laser return into the seven object types: car, wall, tree trunk, foliage, person, grass, and other. In contrast to previous work on outdoor object mapping [18], our model performs joint classification of the laser returns. This is done by connecting the nodes of the CRF based on a Delaunay triangulation of the laser data. An important aspect of CRFs is their ability to incorporate many features with arbitrary dependencies. Our model takes advantage of this ability by incorporating large sets of laser shape features and visual appearance features extracted from camera data. The parameters of our models are learned from partially labeled laser and camera data. We show that classification can be further improved by explicitly modeling within a CRF the information contained in the arrangement of clusters of returns. We also present results on the incorporation of visual object detectors trained on publicly available image data sets such as the LabelMe set [2].

We evaluate our technique on laser and camera data collected by a vehicle navigating through an urban environment. Tested using ten-fold cross validation, objects observed along a 3 kilometer long trajectory are identified with an accuracy of 91%.

This paper is organized as follows. Related work is discussed first, in Section II. In Section III, we introduce the probabilistic models underlying our mapping approach, followed by a description of features used for classification. Experimental results are presented in Section V. Finally, we conclude in Section VI.

II. Related Work

Object recognition is a long-standing problem in robotics and computer vision. Most of the approaches in computer vision aim at recognizing objects from single images. Classifiers are trained on labeled data and used to either classify images as containing or not an instance of the object, or to segment the object in the image. Examples are [8, 23, 25]. In robotics, the problem is different. Recognition can be performed in a sequence of images, in many cases combined with other sensor modalities. Alternatively, object recognition can be required on a full map, as addressed in this paper.

Within the robotics community, recent developments have created representations of the environment integrating more than one sensor modality. In [17], a 3D laser scanner and loop closure detection based on photometric information are brought together into the Simultaneous Localization and Mapping (SLAM) framework. This approach does not generate a semantic representation of the environment which can be obtained from the same multi-modal data using the approach proposed here.

In [20], a robust landmark representation is created by probabilistic compression of high-dimensional vectors containing laser and camera information. This representation is used in a SLAM system and updated on-line when a landmark is re-observed. However, it does not reason about landmark classes and therefore does not support the higher-level object detection described in this work.

Object recognition based on laser and video data has been demonstrated in [15]. Using a sum rule, this approach combines the outputs of two classifiers, each of them being assigned to the processing of one type of data. More recently, Posner and colleagues combine 3D laser range data with camera information to classify surface types such as brick, concrete, grass, or pavement in outdoor environments [18, 19]. The authors classify each laser scan return independently which can disregard important neighborhood information. As other researchers have shown, classification results can be improved by jointly classifying laser beams using techniques such as associative Markov networks [24] or conditional random fields [7].

In [3], a Markov Random Field is used to segment objects from 3D laser scans. The model is trained discriminatively using a max-margin objective function. The features used were simple geometric features capturing plane properties of groups of points. The authors considered four classes: ground, building, tree and shrubbery. Friedman and colleagues introduced Voronoi Random Fields, which generate semantic place maps of indoor environments by labeling the points on a Voronoi graph of a laser map using conditional random fields [10].

The key contribution of this paper is a methodology to build maps of objects in which accurate classification is achieved by exploiting the ability of CRFs to represent spatial correlations and to model the structural information contained in clusters of laser returns.

III. Mapping In Conditional Random Fields

To augment geometric maps with semantic information, we have developed three approaches corresponding to three different models. All these models are based on the framework provided by conditional random fields. Before describing how these models are built from laser and camera data, we provide background on learning and inference in conditional random fields.

A. Conditional Random Fields

Conditional random fields (CRF) are undirected graphical models developed for labeling sequence data [12]. CRFs directly model $p(\mathbf{x}|\mathbf{z})$, the *conditional* distribution over the hidden variables $\mathbf{x}$ given observations $\mathbf{z}$. In our framework, $\mathbf{x}$ is the set of object types to be estimated, a hidden state being instantiated for each laser return. The observations $\mathbf{z}$ correspond to shape and appearance features extracted from laser and vision data, respectively. A CRF can be formulated as follows:

$$p(\mathbf{x}|\mathbf{z}) = \frac{1}{\mathbf{Z}} \exp\left(w_A \sum_i A(x_i, \mathbf{z}) + w_I \sum_e I(x_e, x_{e'}, \mathbf{z}) \right) \quad (1)$$

Here, the term $1/\mathbf{Z}$ is a normalization factor. The functions A and I are the association and interaction potentials, respectively. In our framework, an association potential A is instantiated as a logitboost classifier [9] and estimates the object type of node x_i using the set of observations $\mathbf{z}$ but does not take into account information contained in the structure of the neighborhood. An interaction potential I is a function associated to each edge e of the CRF graph, where x_e and $x_{e'}$ are the nodes connected by edge e. Intuitively, interaction potentials measure the compatibility between neighboring nodes and act as smoothers by correlating the estimation across the network.

In our system, the first step of the CRF training is learning the logitboost classifier A which is performed as in [7]. The second step of the learning consists in finding optimal values for the set of weights w_A and w_I based on a labeled data set. Depending on the connectivity structure of the network to be trained, the system uses exact or approximate learning techniques. For non-cyclic networks, the systems uses a Maximum Likelihood approach since inference can be performed exactly. For networks containing cycles, the system uses the approximate version of this technique which is known as Maximum Pseudo-Likelihood learning [4].

Since the values of the local potential function A are obtained as the output of a logitboost classifier, our approach for training can be seen as an extension of boosting to structured classification tasks. As a result, this approach is very flexible and powerful. It not only learns the weights of the potentials, but also selects the subset of dimensions in the observation vectors $\mathbf{z}$ which are useful for classification [10, 13].

In this work, the maximum pseudo-likelihood learning is slightly extended in such a way that the labels of neighbor nodes are not required, allowing training to be performed on partially labeled data. This is achieved by optimizing the pseudo-likelihood written as:

$$\begin{aligned} pl(\mathbf{x}|\mathbf{z}) &= \prod_{i=1}^{N} p(x_i|\mathrm{MB(x_i)}, \mathrm{z}) \propto \prod_{i=1}^{\mathrm{N}} \exp(\mathrm{w_A A(x_i}, \mathbf{z})) \\ &\prod_{k \in \mathrm{MB(x_i)}} \exp(w_I I(x_i, x_k, \mathbf{z}) + w_A A(x_k, \mathbf{z})) \end{aligned}$$

where the last equation is obtained by breaking the exponential in Eq. 1 into two terms (the full derivation is not given here due to space constraints). N refers to the number of nodes in the network and $MB(x_i)$ is the Markov blanket of node x_i. The parameters to be adjusted to find the maximum value of the pseudo-likelihood are w_A and w_I. In this formulation, the usually required neighbor labels are replaced by the estimated distribution over the neighbor's label: $\exp(w_A A(x_k, \mathbf{z}))$. Via this formulation, the learning algorithm can use the unlabeled nodes in the neighborhood of each labeled node and be performed on partially labeled data.

Inference in CRFs estimates either the marginal distribution of each hidden variable $\mathbf{x}_i$ or the most likely configuration of all hidden variables $\mathbf{x}$ (*i.e.*, MAP estimation), based on their joint conditional probability (Eq. 1). We solve both tasks using belief propagation (BP) for non-cyclic networks. For cyclic networks, we use the approximate version of BP called Loopy Belief Propagation (loopy BP) [16].

B. From Laser Scans to Conditional Random Field

The input to our system is a collection of spatially aligned laser scans obtained by performing scan matching with the iterative closest point (ICP) algorithm [27] [1]. In this section, we present three types of CRFs which will be compared in order to better understand how to model the spatial correlations in a semantic map. We show how the three different models can be instantiated from aligned laser data and indicate which learning and inference techniques are used in each case. For these three networks, the hidden state for each node ranges over the seven object types: car, trunk, foliage, people, wall, grass, and other (any other object type).

1) Delaunay CRF: In this first type of network, each laser return is instantiated as one node in the CRF. The connections between the nodes are found using the Delaunay triangulation procedure [6] which efficiently finds a triangulation with non-overlapping edges. The system then removes links which are longer than a pre-defined threshold (50 cm in our application) since distant nodes are not likely to be strongly correlated. The resulting network is displayed as a set of blue edges in Fig. 2.

Since a Delaunay CRF contains cycles, training and inference are performed with maximum pseudo-likelihood and loopy BP, respectively.

2) Delaunay CRF with link selection: Generally speaking, structured classification as performed by CRFs is expected to improve on local classification since independence is not assumed, *i.e.*, neighborhood information is modelled through interaction potentials. However, as illustrated by the experimental results, the first type of CRF previously described does not improve on local classification. A too coarse modelling of the spatial correlations is responsible for this result. The term $\exp(w_I I(x_i, x_k, \mathbf{z}))$ of Eq. 1 is learnt in this first type of network as a constant matrix instantiated at each of the links. This gives the network a smoothing effect on top of the local classification. Since all the links are represented with the same matrix, only one type of node-to-node relationship is encoded, for example: neighbor nodes should have the same label. Such links are appropriate in very structured parts of the environment but may over-smooth in areas where the density of objects increases.

[1]In spatially more complex data sets containing loops, consistently aligned scans can be generated using various existing SLAM techniques [22]

In order to model more than one type of node-to-node relationships, the network is augmented with an additional node T for every pair of nodes $\{x_i, x_j\}$ as displayed in Fig. 1. The state of this node specifies which type of link is instantiated. For this second type of network, we consider two types of links encoding the following node-to-node relationships: (1) neighbor nodes have the same label, (2) neighbor nodes have a different label. Node T receives an observation S which is the output of a logitboost classifier learned to estimate whether node x_i and x_j are similar based on their respective local observation z_i and z_j. The observation S is a direct observation of the state of node T.

Since this second type of network contains loops, training and inference are also performed with maximum pseudo-likelihood and loopy BP, respectively.

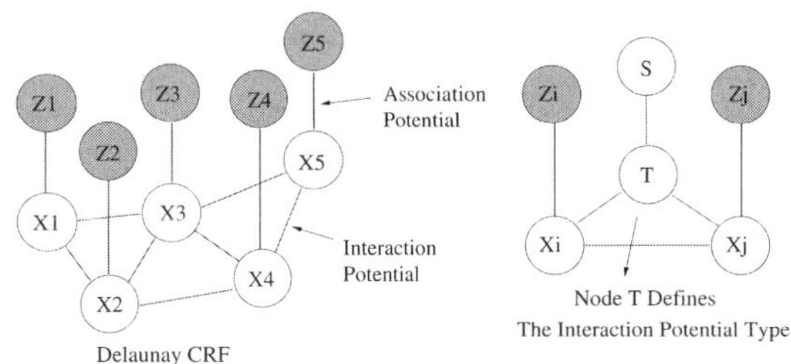

Fig. 1. Representation of the additional infrastructure required in a Delaunay CRF to perform link selection.

3) Tree based CRF: The previous two types of network contain cycles, which implies the use of approximate learning and inference algorithms. We now present a third type of network which is cycle free and does not require the use of approximate techniques. To design non-cyclic networks we start from the following observation: laser returns in a scan map are naturally organized into clusters. These clusters can be identified by analysising the connectivity of the Delaunay graph and finding its disconnected sub-components. Disconnected components appear when removing longer links of the original triangulation. In Fig. 2, the extracted clusters are indicated by green rectangles.

Once the clusters are identified, the nodes of a particular cluster are connected by a tree of depth one. A root node is instantiated for each cluster and each node in the cluster becomes a leaf node. The trees associated to the clusters in Fig. 2 are represented by green volumes. A tree-based CRF does not encode node-to-node smoothing but rather performs smoothing based on the identified clusters of laser returns.

The root node does not have an explicit state. It allows the instantiation of a network which does not contains cycles enabling learning and inference to be performed exactly. With this third type of network, the system uses a maximum likelihood approach for learning and belief propagation for inference. The possibility of using exact learning and inference is a strong advantage compared to the absence of theoretical results in terms of convergence of maximum pseudo-likelihood learning and loopy belief propagation.

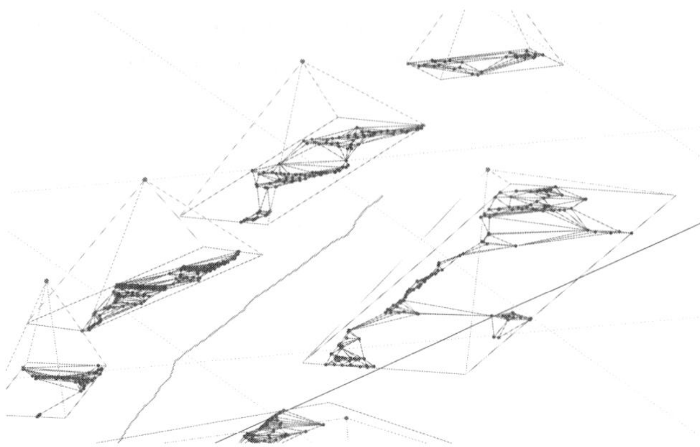

Fig. 2. Representation of a Tree based CRF in one region of a graph generated from data. The trajectory of the vehicle is displayed in orange. Laser returns are instantiated as nodes in the network and connected using the Delaunay triangulation. Nodes and edges are plotted in dark and light blue, respectively. Identified clusters are indicated by the green rectangles while root nodes are plotted in green. Root nodes are connected to all nodes in the cluster but for clarity this is represented by a rectangle enclosing the cluster.

IV. Features For Object Mapping

As formulated in Eq. 1, the computation of the posterior probability requires the set of observations $\mathbf{z}$. In this work, $\mathbf{z}$ consists of high-dimensional feature vectors $\mathbf{f}$ computed for each scan return. $\mathbf{f}$ results from the concatenation of three types of features, geometric features, visual features and features extracted from on-line datasets:

$$\mathbf{f} = [\mathbf{f}_{\text{geo}}, \mathbf{f}_{\text{visu}}, \mathbf{f}_{\text{www}}], \tag{2}$$

Geometric and visual features are first described. We then show how on-line labeled datasets freely available on the internet can provide additional binary features.

A. Geometric Features

Geometric features capture geometric properties of the objects in the laser returns. The feature vector computed for one scan return has a dimensionality of 231 and results from the concatenation of 38 different multi dimensional features. Due to limited space we only present a subset of these features below:

$$\mathbf{f}_{\text{geo}}(i, z_A) = [\mathbf{f}_{\text{dist}}, \mathbf{f}_{\text{angle}}, \mathbf{f}_{\text{oor}}, \mathbf{f}_{\text{cluster}}, \ldots], \tag{3}$$

where i indexes one of the returns in scan z_A.

$\mathbf{f}_{\text{dist}}$ or distance features are computed for each return $z_{A,i}$ in scan A as its distance to other points in scan A:

$$\mathbf{f}_{\text{dist}}(i, k, z_A) = \| z_{A,i} - z_{A,i+k} \|, \tag{4}$$

where k varies from -10 to $+10$.

$\mathbf{f}_{\text{angle}}$ or angle features are computed as angles formed by various configurations of neighbor returns:

$$\mathbf{f}_{\text{angle}}(i, k, l, z_A) = \| \angle (\overline{z_{A,i-k} z_{A,i}}, \overline{z_{A,i} z_{A,i+l}}) \|. \tag{5}$$

where k and l vary from -10 to $+10$. These two first types of features provide information about the local shape of the scan around return i.

$\mathbf{f}_{\text{oor}}$ or out of range features count the number of "out of range" beams between pairs of successive returns. These features allow the representation of open areas between valid beams of the laser scan.

$\mathbf{f}_{\text{cluster}}$ consists of various features computed to describe a cluster of laser returns. Cluster of returns within a single scan are extracted based on a simple distance criteria and characterized through the following quantities: geodesic length of the cluster, length of its two principal components, error generated by the fit of a spline to the cluster points. Note that two returns in the same cluster have the same $\mathbf{f}_{\text{cluster}}$ vector. The aim of the $\mathbf{f}_{\text{cluster}}$ features is to capture the organization of objects at the scale of one laser scan.

B. Visual Features

In addition to laser range scans, our system incorporates visual appearance by projecting the laser returns into camera images collected by a calibrated camera mounted on the vehicle, similar approach to [7, 19].

The CRF learned with a logitboost based algorithm can not only integrate geometric information but also any other type of data and, in particular, visual features extracted from monocular color images. As a consequence, the system extracts features in a region of interest (ROI) defined around the projection of each return into the corresponding image. The parameters required to carry out the projection are defined through the camera laser calibration procedure developed in [26]. The size of the ROI is changed depending on the range of the return. This provides a mechanism to deal with changes in scales across images. It was verified that the use of a size varying ROI improves classification accuracy by 4%.

The visual feature vector associated to each return has a dimensionality of 1239 and results from the concatenation of 51 multi-dimensional features computed in the ROI. Due to limited space, we only describe the most important of these features:

$$\mathbf{f}_{\text{visu}}(i) = [\mathbf{f}_{\text{pyr}}, \mathbf{f}_{\text{rgb}}, \mathbf{f}_{\text{hsv}}, \mathbf{f}_{\text{haar}}, \mathbf{f}_{\text{edges}}, \mathbf{f}_{\text{lines}}, \mathbf{f}_{\text{sift}}, \ldots], \tag{6}$$

where index i refers to the ROI associated to return i.

$\mathbf{f}_{\text{pyr}}$ returns texture information encoded as a vector containing the steerable pyramid [21] coefficients of ROI i as well as the minimum and the maximum of these coefficients. These extrema are useful to classify cars which from most point of views have a relatively low texture maxima due to their smooth surface.

$\mathbf{f}_{\text{rgb}}$ and $\mathbf{f}_{\text{hsv}}$ return a 3D histogram of the RGB and HSV data in ROI i.

$\mathbf{f}_{\text{haar}}$ returns Haar features of ROI i computed using the integral image approach proposed in [25].

$\mathbf{f}_{\text{edges}}$ uses a Canny edge detector to extract the number of pixels within ROI i recognized as belonging to an edge.

$\mathbf{f}_{\text{lines}}$ processes the whole image with the line detector [1] and extracts the number of lines intersecting ROI i as well as the maximum length of this subset of lines.

$\mathbf{f}_{\text{sift}}$ counts the number of Sift features [14] found in ROI i.

C. Using On-line Datasets

In our datasets, some of the classes such as the class people have no more than one hundred training samples. This can be detrimental to the accuracy of the classifier. To compensate for the lack of training data, we have used binary features computed with classifiers trained on on-line datasets. Across the web, large labeled datasets such as the LabelMe dataset [2] can be used to learn binary classifiers on large amount of training data. We used the LabelMe data to train binary object detectors for each of the ten classes: car, tree trunk, foliage, pedestrian, building, grass, road, pole, fence and road; and applied these detectors to our data to generate an additional binary feature vector $\mathbf{f}_{\text{www}}$ of dimensionality 10.

In addition to an algorithm which can be trained with partially labeled data, the use of on-line labeled data sets decrease the labelling effort. The results reported in Sec. V-B.4 with respect to the $\mathbf{f}_{\text{www}}$ features show the right trend while no significant improvement has been obtained yet. This part of the work is preliminary and aims at introducing the idea of generating additional features as output of classifiers trained on on-line datasets. We believe that understanding the requirements for features to be portable from standard datasets to a given robotics application is crucial for large-scale autonomy and this paper opens up this direction of research.

V. Experimental Results

A. Experimental Setup

Experiments were performed using outdoor data collected with a modified car traveling at 0 to 40 km/h along a 3km long trajectory. The car drove in a university campus which has structured areas with buildings, walls and cars, and unstructured areas with bush, trees and lawn fields. The overall dataset contains 4500 images representing 20 minutes of logging. Laser and vision data was acquired at a frequency of 4Hz. The laser sensor used belongs to the family of SICK devices and the camera was a high-resolution wide angle Hanvision camera.

The evaluation of the classifier was performed on a ten-fold cross validation setup which involves training each classifier on nine tenth of the trajectory and testing it on the remaining one tenth. These two operations are repeated ten times by changing the testing and training sets accordingly. The results presented below are averaged over the cross validation runs.

Each set of scans was converted into a probabilistic network as described in Sec. III-B. Training and testing sets were partly hand labeled to provide labels to the learning algorithm and a ground truth to evaluate classification accuracy.

The properties of the training and testing sets averaged over the ten tests are provided in Table I.

	Length vehicle trajectory	# scans total labeled	# nodes total labeled
Training set	2.6 km	3843 72	67612 5168
Testing set	290 m	427 8	7511 574

TABLE I

PROPERTIES OF THE TRAINING AND TESTING SETS

B. Classification Performance

This section presents the classification performances obtained with the three models presented in Sec. III-B. Results for local classification are first presented in order to provide a baseline for comparison.

1) Local Classification: A seven-class logitboost classifier is learned and instantiated at each node of the network as the association potential A (Eq. 1). Local classification, *i.e.*, classification which does not take neighborhood information into account is performed with the confusion matrix presented in Table II. This confusion matrix displays a strong diagonal which corresponds to an accuracy of 90.4%. A compact characterization of the confusion matrix is given by precision and recall values. These are presented in Table III. Averaged over the seven classes, the classifier achieves a precision of 89.0% and a recall of 98.1%.

Truth \ Inferred	Car	Trunk	Foliage	People	Wall	Grass	Other
Car	1967	1	7	10	3	0	48
Trunk	4	165	18	0	4	0	11
Foliage	25	18	1451	0	24	0	71
People	6	2	2	145	0	0	6
Wall	6	6	21	0	513	1	39
Grass	0	0	1	1	1	146	4
Other	54	5	123	3	24	0	811

TABLE II

LOCAL CLASSIFICATION: CONFUSION MATRIX

In %	Car	Trunk	Foliage	People	Wall	Grass	Other
Precision	96.6	81.7	91.3	90.1	87.5	95.4	79.5
Recall	97.9	99.3	96.4	99.7	98.5	99.9	95.4

TABLE III

LOCAL CLASSIFICATION: PRECISION AND RECALL

2) Delaunay CRF classification:

a) CRF without built-in link selection: the accuracy achieved by this first type of network is 90.3% providing no improvements on local classification. As developed in Sec. III-B.2, the modelling of the spatial correlation is too coarse since it contains only one type of link which cannot accurately model the relationships between neighbor nodes. As a consequence, the links end up representing the predominant relationship in the data. In our application the predominant neighborhood relationships are of the type "neighbor nodes possessing the same label". The resulting learned links enforce this "same-to-same" relationship across the network leading to over smooth estimates and explaining why this class of networks fails to improve on local classification. To verify that a better modelling of the CRF links improves the classification performance, we now presents results generated by the second proposed type of CRF, characterized by a built-in link selection process.

b) CRF with built-in link selection: the accuracy achieved by this second type of network is 91.4% which corresponds to 1.0% improvement in accuracy. Since the local accuracy is already high, the improvement brought by the network may be better appreciated when expressed as a reduction of the error rate of 10.4%. This result validates the claim that a set of link types encoding a variety of node-to-node relationships is required to exploit the spatial correlations in the laser map.

3) Tree based CRF classification: The two types of networks evaluated in the previous section contain cycles and require the use of approximate learning and inference techniques. The tree based CRFs presented in Sec. III-B.3 avoid these issues by allowing the use of exact learning and inference procedures.

This third type of network achieves an accuracy of 91.1% which is slightly below the accuracy given by a CRF with link selection while still improving on the CRF without link selection. However, the major improvement brought by this third type of network is in terms of computational time. Since the network has the complexity of a tree of depth one, learning and inference, in addition to being exact, can be implemented very efficiently. As displayed in Table IV, a tree based CRF is 80% faster at training and 90% faster at testing than a Delaunay CRF. Since both network types use as their association potential the seven classes logitboost classifier, they require the same features extracted from a scan and its associated image in 1.2 secs on average. As shown in Table I, the test set contains 7511 nodes on average which suggests that the tree based CRF approach is in its current state is very close to real time, feature extraction being the main bottleneck.

	Feature Extraction (per scan)	Learning (training set)	Inference (test set)
Delaunay CRF (with link selection)	1.2 secs	6.7 mins	1.5 mins
Tree based CRF	1.2 secs	1.5 mins	10.0 secs

TABLE IV
COMPUTATION TIMES

4) Using on-line data sets for training: Based on the LabelMe set, 10 binary object detectors are trained using the logitboost algorithm. The 10 classes considered are: car, tree trunk, foliage, pedestrian, building, grass, road, pole, fence and road. Since the LabelMe dataset contains vision data only, these binary classifiers are vision based detectors and, in order to use their output as additional features, we run them on the ROIs selected in each image of our urban dataset (the selection of these ROIs is performed as described in Sec. IV-B).

Within our urban dataset as well as within the LabelMe dataset, the size of the selected ROIs are not constant which requires designing the various vision features in such a way that the dimensionality of the vector $\mathbf{f}_{visu}$ is independent of the ROI size. Our approach consist in using features which are distributions (e.g. an histogram with a fixed number of bins) and whose dimensionality is constant (e.g. equal to the number of bins in the histogram). A larger ROI leads to a better sampled distribution (e.g. a larger number of samples in the histogram) and the actual feature dimensionality remains invariant.

The use of these additional $\mathbf{f}_{www}$ features slightly improves the local classification accuracy from 90.4% to 90.6%. We believe that there is no further increase in accuracy due to the fact that the lighting conditions in the two datasets differ significantly (our urban dataset contains images which are on average much darker than the ones in the LabelMe dataset). In the context of preliminary investigations, these results are encouraging and future tests will involve datasets with more similar lighting conditions.

C. Map of Objects

This section presents a visualization of the mapping results. It follows the lay out of Figure 3 in which the vehicle was travelling from right to left.

At the location of the first inset, the vehicle was going up a straight road with a fence on its left and right, and, from the foreground to the background, another fence, a car, a parking meter and bush. All these objects were correctly classified with the fences and the parking meter identified as other.

In the second inset, the vehicle was coming into a curve facing a parking lot and bush on the side of the road. Four returns misclassified as other can be seen in the background of the image. The class other regularly generated false positives which is possibly caused by the dominating number of training samples in this class. Various ways of re-weighting the training samples or balancing the training set were tried without significant improvements.

While reaching the third inset, a car driving in the opposite direction came into the field of view of our vehicle's sensors. The trace let by this car in the map appears in the magnified inset as a set of blue dots along side our vehicle's trajectory. Dynamic objects are not explicitly considered within this work. They are assumed to move at a speed which does not prevent ICP from performing accurate registration. In the campus areas where the data was obtained, this assumption has proven to be valid. In spite of a few miss-classifications in the bush on the left side of the road, the pedestrians on the side walk are correctly identified and the wall of the building is recognised.

Entering the fourth inset, our vehicle was facing a second car, scene which appears in the map as a blue trace intersecting our vehicle's trajectory. Apart from one miss-classified return on one of the pedestrians, and one miss-classified return on the tree in the right of the image, the inferred labels are accurate. Note that the first right return is correctly classified illustrating the accuracy of the model at the border between objects.

VI. CONCLUSIONS

This paper introduces a novel approach for object mapping in outdoor environments. Our technique applies conditional

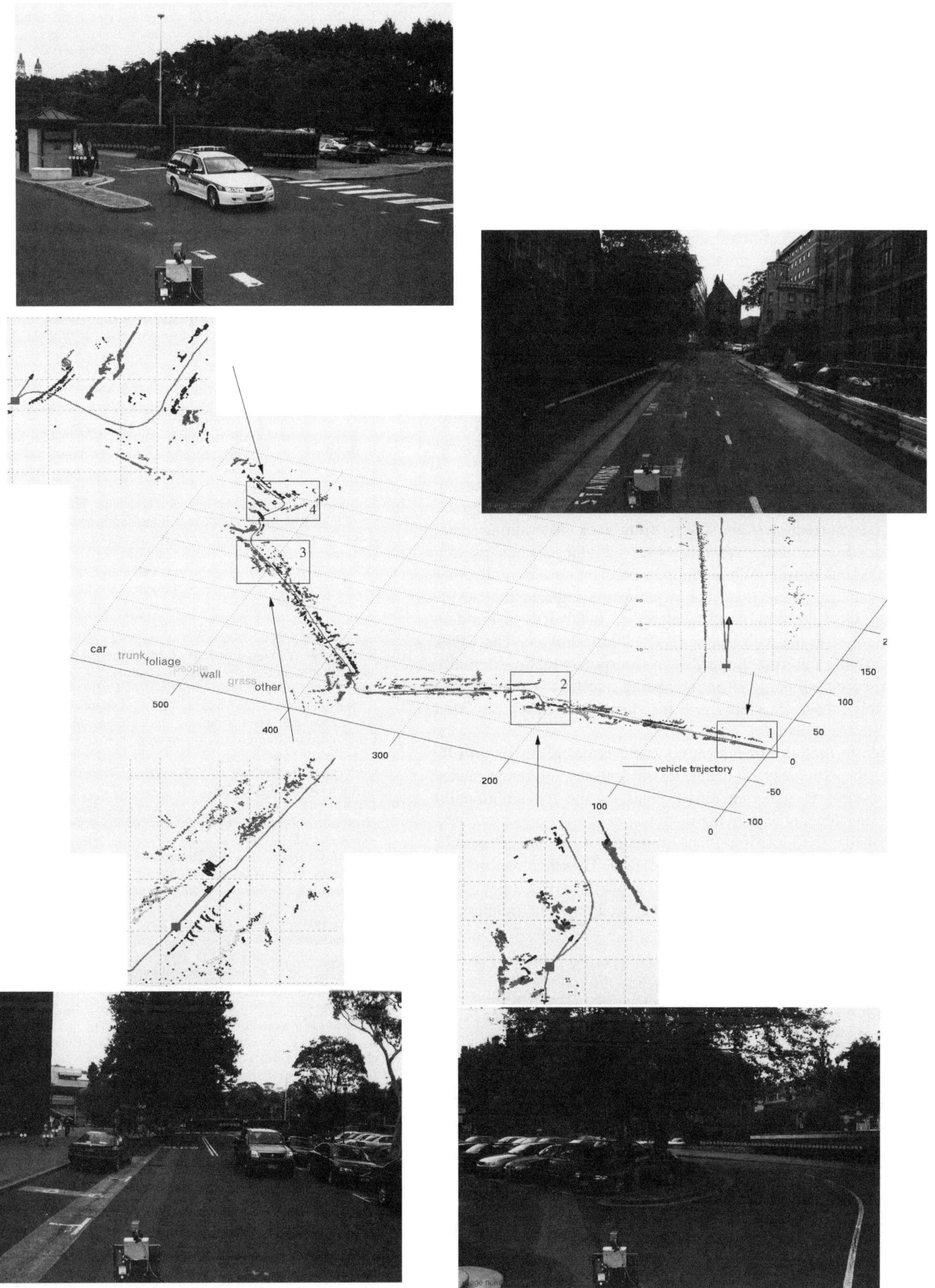

Fig. 3. Visualization of 750 meters long portion of the estimated map of objects with total length of 3km. The map was generated using the tree based CRF model. The legend is indicated in the bottom left part of the 2D plane. The color of the vehicle's trajectory is specified in the bottom right part of the same plane. The coordinate in the plane of the map are in meters. Each inset is magnified and associated to an image displayed with the inferred labels projected back onto the original returns. The location of the vehicle is shown in each magnified patch with a square and its orientation indicated by the arrow attached to it. The laser scanner mounted on the vehicle can be seen in the bottom part of each image.

random fields to label individual points in a 2D laser map annotated with camera data. We take advantage of CRFs' ability to handle dependent features by incorporating large sets of shape and appearance information extracted from laser scans and cameras. Spatial dependencies are modeled by connecting nodes in the CRF based on a Delaunay triangulation of the laser data. Label smoothing on the object level is achieved by three different graph structures based on a spatial segmentation of the laser data. Our approach learns both feature functions and model parameters using a combination of maximum likelihood and logitboost training on partially labeled data.

Experiments conducted on data collected along a 3km trajectory through an urban area indicate that our system achieves very good classification rates for object types such as car, trunk, foliage, people, wall, grass, and other. The approach achieves a reduction of the classification error of 10.4% with respect to a local approach solely integrating standard shape and appearance features. We also show how on-line datasets can be integrated by incorporating object detectors as additional features.

These results are extremely encouraging and the following aspects are promising directions for future work. The accuracy of our current system suffers from lack of training data, especially for more sparsely observed objects such as tree trunks and people. While this can be overcome by collecting and labeling more data, our experiments indicate that leveraging the large number of labeled (and unlabeled) vision data resources on the web is a more scalable technique. The CRFs underlying our system are able to incorporate many externally learned classifiers, and an interesting question is how to best combine these classifiers with the shape information provided by the laser data.

While the current mapping system is designed to run off-line, the efficiency of feature extraction and inference makes it possible to generate object maps on the fly, additionally labeling objects as moving or not.

Finally, the most important limitation of our current system is the reliance on 2D laser range data. However, we believe that our approach can also be applied to 3D laser data, which should greatly improve the accuracy and richness of the generated maps.

VII. ACKNOWLEDGMENTS

The authors would like to thank Roman Katz for numerous useful discussions, Juan Nieto, Jose Guivant, and Oliver Frank for helping with the dataset acquisition. This work is supported by the ARC Center of Excellence programme, the Australian Research Council (ARC), the New South Wales (NSW) State Government, the University of Sydney Visiting Collaborative Research Fellowship Scheme, and DARPA's ASSIST and CALO Programmes (contract numbers: NBCH-C-05-0137, SRI subcontract 27-000968).

REFERENCES

[1] Finding long, straight lines code. http://www.cs.uiuc.edu/homes/dhoiem/.

[2] Labelme data set. http://labelme.csail.mit.edu/.

[3] D. Anguelov, D. Koller, E. Parker, and S. Thrun. Detecting and modeling doors with mobile robots. In *Proc. of the IEEE International Conference on Robotics & Automation (ICRA)*, 2004.

[4] J. Besag. Statistical analysis of non-lattice data. *The Statistician*, 24, 1975.

[5] DARPA Urban Challenge. http://www.darpa.mil/grandchallenge/index.asp.

[6] M. De Berg, M. Van Kreveld, M. Overmars, and O. Schwarzkopf. Springer-Verlag, 2000. 2nd rev. ISBN: 3-540-65620-0.

[7] B. Douillard, D. Fox, and F. Ramos. A spatio-temporal probabilistic model for multi-sensor multi-class object recognition. In *Proc. of the International Symposium of Robotics Research (ISRR)*, 2007.

[8] L. Fei-Fei and P. Perona. A bayesian hierarchical model for learning natural scene categories. In *Proc. of the IEEE Computer Society Conference on Computer Vision and Pattern Recognition (CVPR)*, 2005.

[9] J. Friedman, T. Hastie, and R. Tibshirani. Additive logistic regression: A statistical view of boosting. *The Annals of Statistics*, 28(2), 2000.

[10] S. Friedman, D. Fox, and H. Pasula. Voronoi random fields: Extracting the topological structure of indoor environments via place labeling. In *Proc. of the International Joint Conference on Artificial Intelligence (IJCAI)*, 2007.

[11] V. Kumar, D. Rus, and S. Singh. Robot and sensor networks for first responders. *IEEE Pervasive Computing*, 3(4), 2004. Special Issue on Pervasive Computing for First Response.

[12] J. Lafferty, A. McCallum, and F. Pereira. Conditional random fields: Probabilistic models for segmenting and labeling sequence data. In *Proc. of the International Conference on Machine Learning (ICML)*, 2001.

[13] L. Liao, T. Choudhury, D. Fox, and H. Kautz. Training conditional random fields using virtual evidence boosting. In *Proc. of the International Joint Conference on Artificial Intelligence (IJCAI)*, 2007.

[14] D. Lowe. Discriminative image features from scale-invariant keypoints. *International Journal of Computer Vision*, 60(2), 2004.

[15] G. Monteiro, C. Premebida, P. Peixoto, and U. Nunes. Tracking and classification of dynamic obstacles using laser range finder and vision. In *Proc. of the IEEE/RSJ International Conference on Intelligent Robots and Systems (IROS)*, 2006.

[16] K. Murphy, Y. Weiss, and M. Jordan. Loopy belief propagation for approximate inference: An empirical study. In *Proc. of the Conference on Uncertainty in Artificial Intelligence (UAI)*, 1999.

[17] P. Newman, D. Cole, and K. Ho. Outdoor SLAM using visual appearance and laser ranging. In *Proc. of the IEEE International Conference on Robotics & Automation (ICRA)*, Orlando, USA, 2006.

[18] I. Posner, D. Schroeter, and P. M. Newman. Describing composite urban workspaces. In *Proc. of the IEEE International Conference on Robotics & Automation (ICRA)*, 2007.

[19] I. Posner, D. Schroeter, and P. M. Newman. Using scene similarity for place labeling. In *Proc. of the International Symposium on Experimental Robotics (ISER)*, 2007.

[20] F. Ramos, J. Nieto, and H.F. Durrant-Whyte. Recognising and modelling landmarks to close loops in outdoor slam. In *Proc. of the IEEE International Conference on Robotics & Automation (ICRA)*, 2007.

[21] E. Simoncelli and W. Freeman. The steerable pyramid: A flexible architecture for multi-scale derivative computation. In *Proc. of the International Conference on Image Processing*, 1995.

[22] S. Thrun, W. Burgard, and D. Fox. *Probabilistic Robotics*. MIT Press, Cambridge, MA, September 2005. ISBN 0-262-20162-3.

[23] A. Torralba, K. Murphy, and W. Freeman. Contextual models for object detection using boosted random fields. In *Advances in Neural Information Processing Systems (NIPS)*, 2004.

[24] R. Triebel, K. Kersting, and W. Burgard. Robust 3D scan point classification using associative Markov networks. In *Proc. of the IEEE International Conference on Robotics & Automation (ICRA)*, 2006.

[25] P. Viola and M. Jones. Robust real-time ob ject detection. In *International Journal of Computer Vision*, volume 57, page 2, 2004.

[26] Q. Zhang and R. Pless. Extrinsic calibration of a camera and laser range finder (improves camera calibration). In *Proc. of the IEEE/RSJ International Conference on Intelligent Robots and Systems (IROS)*, Sendai, Japan, 2004.

[27] Z. Zhang. Iterative point matching for registration of free-form curves and surfaces. *International Journal of Computer Vision*, 13(2):119–152, 1994.

Fast Probabilistic Labeling of City Maps

Ingmar Posner, Mark Cummins, and Paul Newman
Mobile Robotics Group,
Dept. Engineering Science
Oxford University
Oxford, UK
Email: {hip, mjc, pnewman}@robots.ox.ac.uk

***Abstract*— This paper introduces a probabilistic, two-stage classification framework for the semantic annotation of urban maps as provided by a mobile robot. During the first stage, local scene properties are considered using a probabilistic bag-of-words classifier. The second stage incorporates contextual information across a given scene via a Markov Random Field (MRF). Our approach is driven by data from an onboard camera and 3D laser scanner and uses a combination of appearance-based and geometric features. By framing the classification exercise probabilistically we are able to execute an information-theoretic bail-out policy when evaluating appearance-based class-conditional likelihoods. This efficiency, combined with low order MRFs resulting from our two-stage approach, allows us to generate scene labels at speeds suitable for online deployment and use. We demonstrate and analyze the performance of our technique on data gathered over almost 17 km of track through a city.**

I. INTRODUCTION

This paper addresses the fast labeling of mobile robot workspaces using a camera and a 3D laser scanner. We motivate this work by noting that, although contemporary online mapping and simultaneous localization techniques using lidar now produce compelling 3D geometric representations (a.k.a maps) of a mobile robot's workspace, these maps tend to be geometrically rich but semantically impoverished. Our work seeks to redress this shortcoming. Maps in the form of large unstructured point clouds are meaningful to human observers, but are of limited operational use to a robot. There is much to be gained by having the robot itself upgrade the map to include richer semantic information and to do so online. In particular, the semantics induced by online segmentation and labeling has an important impact on the action selection problem. For example, the identification of terrain types with estimates of their spatial extent has a clear impact on control. Similarly the identification of buildings and their entrances has a central role to play in mission execution and planning in urban settings.

In this paper we outline a probabilistic method which achieves fast labeling of 3D point clouds by using a combination of appearance and geometric features. In particular we use combined 3D range and image data to perform inference at two distinct levels. Firstly, over local scales, classification is based on the co-occurrence of appearance descriptors, which capture both visual and surface orientation information. We frame this classification problem in probabilistic terms, which allows the implementation of a principled "bail-out" policy to be invoked when evaluating class conditional likelihoods, resulting in very large computational savings. Secondly, at the scene-wide scale, we use a Markov Random Field (MRF) to model the expected relationships between patch labels and to thus incorporate the rich prior information common to many parts of our man-made environment. Our MRFs have a relatively low node-count, just one node for each scene patch, yielding rapid inference.

Fig. 1. Classification results for a typical urban scene: the original image **(top left)**; segments classified as 'pavement/tarmac' **(top right)**; segments classified as 'textured wall' **(bottom left)**; segments classified as 'vehicle' **(bottom right)**. The colour-coding is wrt. to ground-truth: green indicates a correct label; red indicates a false negative.

II. RELATED WORK

Recently there has been a surge in the literature regarding environment understanding within robotics, particularly as available sensory data becomes richer and the limitations of unannotated maps become more apparent. A variety of machine learning approaches to the problem have been explored, with more recent approaches utilizing contextual as well as local information to improve classification performance. In [1] the authors classify 2D laser data into types of indoor scenes using boosting. Contextual information was used explicitly in [2] by way of a model based on relational Markov networks to learn classifiers from segment-based representations of indoor workspaces. More recently [3] introduced an approach which takes into account spatial relationships between objects and object parts in 3D. 3D laser data were used in [4], where they were segmented to detect cars and classify terrain using Graph Cut applied to a Markov Random Field (MRF) formulation of the problem, an approach which was extended by [5].

Particularly relevant to the work presented here are papers which consider a combination of vision and laser data in an outdoor setting. [6] considers the task of pedestrian- and vehicle detection, using 2D laser data. In [7] a more sophisticated inference framework based on Conditional Random Fields was brought to bear on the vehicle detection problem, with preliminary results also reported for multi-class labelling. 3D laser data were combined with visual information in [8], which used support vector machines for classification but did not make use of contextual information.

The work presented here also leverages a combination of laser data with vision. Our main contribution lies in the definition of an efficient contextual inference framework, based on a graph over plane patches rather than over measurements (e.g. laser range data) directly. This yields substantial speed increases over previous approaches. As an integral part of this framework we further define a generative bag-of-words classifier and describe an efficient inference procedure for it. Finally, the work presented here further distinguishes itself from related work by combining information from two complimentary sensors – full 3D geometry and appearance. Thereby our approach gains the capacity of providing *more detailed* workspace descriptions such as the surface-type of building(s) encountered or the nature of ground traversed.

III. Classes And Features

The system described in this paper utilizes data from a calibrated combination of 3D laser scanner and monocular camera, both mounted on a mobile robot. Our basic processing pipeline is similar to that described in [8] – the major contribution of this paper is to extend the inference machinery. Briefly, incoming 3D laser data are segmented into local plane patches using a RANSAC procedure (see Figure 2). Plane patches are then sub-segmented into visually homogeneous areas using an off-the-shelf image segmentation algorithm [9]. The product of this feature extraction pipeline is a set of visually similar image patches which have 3D geometry attributes associated with them. Our classification framework proceeds by classifying each patch individually. The final stage then consideres scene-wide interactions between these local patches.

In contrast to much of the existing work in the area, we consider a relatively rich set of seven classes in three categories. Classes are listed in Table I, and comprise ground types, building types and two object categories. Labeling the environment into classes such as these is a useful step towards a number of autonomous tasks such as path following, location recognition and collision avoidance.

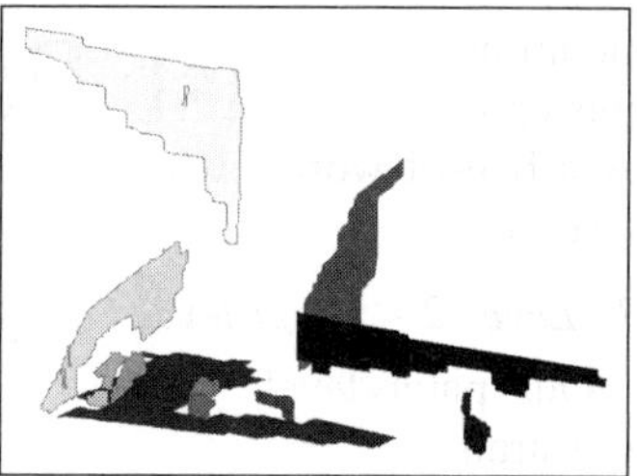

Fig. 2. An original 3D laser scan **(left)** and its approximation by planar patches as generated by the segmentation algorithm **(right)**.

Classification is performed on the basis of the features listed in Table II. These features are computed for all laser points in a patch, proivded that the points are visible in the camera image. Colour and texture features are computed over the 15x15 pixel local neighbourhood of each projected laser point.

IV. Generative Probabilistic Classification

The inference framework proposed in this paper is a multi-level approach based on successive combinations of lower-level features. At the lowest level, individual laser points are mapped to appearance-words based on the set of features described in Section III. The next level of the hierarchy pools information from multiple laser points by grouping them into patches based on boundaries in both the image and the point cloud. Each patch is then assigned a pdf over class membership by a bag of words classifier.

The highest level of the hierarchy takes account of spatial context by using an MRF defined over the set of patches. This improves local decisions by incorporating information from the gross geometric arrangement of classes in the scene.

A. Level 1 - Classification of Individual Laser Points

The lowest level input to our system is the collection of laser points in the scene. Each laser point is described by a feature vector, using the features described in Section III. Rather than deal with raw data directly, we adopt the *bag-of-words* representation [10], where the feature vectors are quantized with respect to a "vocabulary". The vocabulary is constructed by clustering all the feature vectors from a set of training data, using an incremental clustering algorithm. This yields a vocabulary of size $|v|$, the vocabulary size being determined by a user-specified threshold. The cluster centres then define the vocabulary. When the system has been trained,

TABLE I
Classes

Class	Description
Ground Type	
Pavement/Tarmac	Road, footpath.
Dirt Path	Mud, sand, gravel.
Grass	Grass.
Building Type	
Smooth Wall	Concrete, plaster, glass.
Textured Wall	Brickwork, stone.
Object	
Foliage	Bushes, tree canopy.
Vehicle	Car, van.

TABLE II
Features used for classification

Feature Descriptions	Dimensions
3D Geometry Orientation of surface normal of local plane	1
2D Geometry Location in image: mean of normalised x and y	2
Colour HSV: hue & sat. histograms in local neighbourhood	30
Texture HSV: hue & sat. variance in local neighbourhood	2

incoming sensory data is mapped to the approximate nearest cluster centre using a kd-tree. Each patch is then described by a bag-of-words, which is the input to the next level of the system.

B. Level 2 - Patch-level Classifier

Our patch-level classifier is inspired by the probabilistic appearance model introduced in [11] and the theory presented below is an extension of that work into a more general classification framework. Building on the output of the lower-level vector quantization step, an observation of a patch $\mathbf{z} = \{z_1, \ldots, z_{|v|}\}$ is a collection of binary variables where each z_i indicates the presence (or absence) of the ith word of the vocabulary within the patch. We would like to compute $p(\mathcal{C}|\mathbf{z})$, the distribution over the class labels given the observation, which can be computed according to Bayes rule:

$$p(\mathcal{C}^k|\mathbf{z}) = \frac{p(\mathbf{z}|\mathcal{C}^k)p(\mathcal{C}^k)}{p(\mathbf{z})} \qquad (1)$$

where $p(\mathbf{z}|\mathcal{C}^k)$ is the class-conditional observation likelihood, $p(\mathcal{C}^k)$ is the class prior and $p(\mathbf{z})$ normalizes the distribution.

C. Representing Classes

Given a vocabulary, individual classes are represented within the classification framework by a set of class-specific examples, which we call exemplars. Concretely, for each class k the model consists of n_k exemplars $\mathcal{C}^k = \{C_1^k, \ldots, C_{n_k}^k\}$ where C_i^k is the ith exemplar of class k. Exemplars themselves are defined in terms of a hidden "existence" variable e, each exemplar C_i^k being described by the set $\{p(e_1|C_i^k), \ldots, p(e_{|v|}|C_i^k)\}$. The term e_j is the event that a patch contains a property or artifact which, given a perfect sensor, would cause an observation of word z_j. However, we do not assume a perfect sensor — observations z are related to existence e via a sensor model which is specified by

$$\mathcal{D}: \quad \begin{cases} p(z_j = 1|e_j = 0), & \text{false positive probability.} \\ p(z_j = 0|e_j = 1), & \text{false negative probability.} \end{cases} \qquad (2)$$

with these values being a user-specified input. The reasons for introducing this extra layer of hidden variables, rather than modeling the exemplars as a density over observations directly, are twofold. Firstly, it provides a natural framework to incorporate data from multiple sensors, where each sensor has different (and possibly time-varying) error characteristics. Secondly, as outlined in the following section, it allows the calculation of $p(\mathbf{z}|\mathcal{C}^k)$ to blend local patch-level evidence with a global model of word co-occurrence.

D. Estimating the Observation Likelihood

The key step in computing the pdf over class labels as per Equation 1 is the evaluation of the conditional likelihood $p(\mathbf{z}|\mathcal{C}^k)$. This can be expanded as an integration across all the exemplars that are members of class k:

$$p(\mathbf{z}|\mathcal{C}^k) = \sum_{i=1}^{n_k} p(\mathbf{z}|C_i^k, \mathcal{C}^k)p(C_i^k|\mathcal{C}^k) \qquad (3)$$

where $\mathcal{C}^k$ is the class k, and C_i^k is an exemplar of the class. Given $p(\mathcal{C}^k|C_i^k) = 1$ (an assumption that none of the training data is mislabeled) and $p(C_i^k|\mathcal{C}^k) = \frac{1}{n_k}$ (all exemplars within a class are equally likely), this becomes

$$p(\mathbf{z}|\mathcal{C}^k) = \frac{1}{n_k}\sum_{i=1}^{n_k} p(\mathbf{z}|C_i^k) \qquad (4)$$

The likelihood with respect to the exemplar can now be expanded as:

$$p(\mathbf{z}|C_i^k) = p(z_1|z_2, ..., z_n, C_i^k)p(z_2|z_3, ..., z_n, C_i^k)...p(z_n|C_i^k) \qquad (5)$$

This expression cannot be tractably computed — it is infeasible to learn the high-order conditional dependencies between appearance words. We thus seek to approximate this expression by a simplified form which can be tractably computed and learned for available data. A popular choice in this situation is to make a Naive Bayes assumption — treating all variables z as independent. However, visual words tend to be far from independent, and it has been shown in similar contexts that learning a better approximation to their true distribution substantially improves performance [11]. The learning scheme we employ is the Chow Liu tree, which locates a tree-structured Bayesian network that approximates the true distribution [12]. Chow Liu trees are optimal within the class of tree-structured approximations, in the sense that they minimize the KL divergence between the approximate and true distributions. Because the approximation is tree-structured, its evaluation involves only first-order conditionals, which can be reliably estimated from practical quantities of training data. Additionally, Chow Liu trees have a simple learning algorithm that consists of computing a maximum spanning tree over the graph of pairwise mutual information between variables — this readily scales to very large numbers of variables.

We use the Chow Liu tree to model the fact that certain combinations of visual words tend to co-occur. It can be learnt from unlabeled training data across all classes, and approximates the distribution $p(\mathbf{z})$. To compute $p(\mathbf{z}|\mathcal{C}^k)$, the class-specific density, we find an expression that combines this global occurrence information with the class model outlined in section IV-C. Returning to Equation 5 and employing the Chow Liu approximation, we have

$$\begin{aligned} p(\mathbf{z}|C_i^k) &= p(z_1|z_2, .., z_n, C_i^k)p(z_2|z_3, .., z_n, C_i^k)..p(z_n|C_i^k) \\ &\approx p(z_r|C_i^k)\prod_{q=1}^{|v|} p(z_q|z_{p_q}, C_i^k) \end{aligned} \qquad (6)$$

where z_r is the root of the Chow Liu tree and z_{p_q} is the parent of z_q in the tree. Each term in Equation 6 can be further expanded as an integration over the state of the hidden variables in the exemplar appearance model, yielding

$$p(z_q|z_{p_q}, C_i^k) = \sum_{s_{e_q}\in\{0,1\}} p(z_q|e_q = s_{e_q}, z_{p_q}, C_i^k)p(e_q = s_{e_q}|z_{p_q}, C_i^k) \qquad (7)$$

which, assuming that sensor errors are independent of class and making the approximation $p(e_j|z_j) = p(e_j)\forall i \neq j$

becomes

$$p(z_q|z_{p_q}, C_i^k) = \sum_{s_{e_q} \in \{0,1\}} p(z_q|e_q = s_{e_q}, z_{p_q}) p(e_q = s_{e_q}|C_i^k) \tag{8}$$

further manipulation yields an expansion of the first term in the summation as

$$p(z_q = s_{z_q}|e_q = s_{e_q}, z_p = s_{z_p}) = \frac{a}{a+b} \tag{9}$$

where $s_{z_q}, s_{e_q}, s_{z_p} \in \{0, 1\}$ and

$$a = p(z_q = \overline{s_{z_q}}) p(z_q = s_{z_q}|e_q = s_{e_q}) p(z_q = s_{z_q}|z_p = s_{z_p})$$

$$b = p(z_q = s_{z_q}) p(z_q = \overline{s_{z_q}}|e_q = s_{e_q}) p(z_q = \overline{s_{z_q}}|z_p = s_{z_p})$$

which is now expressed entirely in terms of the known detector model and marginal and conditional observation probabilities. These can be estimated from training data. Thus we have a procedure for computing $p(\mathbf{z}|\mathcal{C}^k)$.

Returning to Equation 1, the prior $p(\mathcal{C}^k)$ can be learned simply from labeled training data, $p(\mathbf{z}|\mathcal{C}^k)$ we have discussed above, and to normalize the distribution we make the naive assumption that our set of classes fully partitions the world. Clearly this work would benefit from a background class, a change we plan to make in future versions of the system. The posterior distribution across classes, $p(\mathcal{C}^k|\mathbf{z})$, can now be computed for each patch.

E. Learning A Class Model

The final issue to address in relation to the patch-level classifier is the procedure for learning the class models described in section IV-C. Class models consist of a list of exemplars obtained from ground-truth (i.e. labeled) data. The term $p(e_q = 1|C_i^k)$ represents the probability that exemplar i of class k contained word q (this is a probability because our detector has false positives and false negatives). Given an observation labeled as this class, the properties of the exemplar can be estimated via

$$p(e_q = 1|C_i^k, \mathbf{z}) = \frac{p(\mathbf{z}|e_q = 1, C_i^k) p(e_q = 1|C_i^k)}{p(\mathbf{z}|C_i^k)} \tag{10}$$

where $p(\mathbf{z}|C_i^k)$ can be evaluated as described in the previous section and the prior term $p(e_q = 1|C_i^k)$ we initialize to the global marginal $p(e_q = 1)$.

F. Approximation Using Bounds

Computing the posterior over classes, $p(\mathcal{C}^k|\mathbf{z})$, requires an evaluation of the likelihood $p(\mathbf{z}|\mathcal{C}_j^k)$ for each of the exemplars in the training set. As the number of exemplars grows, this rapidly becomes the limiting computational cost of the inference procedure. This section outlines a principled approximation that accelerates this computation by more than an order of magnitude. The key observation is that while the posterior over classes depends on the summation over all exemplars (as per Equation 4), typically the value of the summation is dominated by a small number of exemplars, with the rest providing negligible contribution. By evaluating the exemplar likelihoods in parallel, those with negligible contribution can be identified and excluded before the computation is fully complete. This

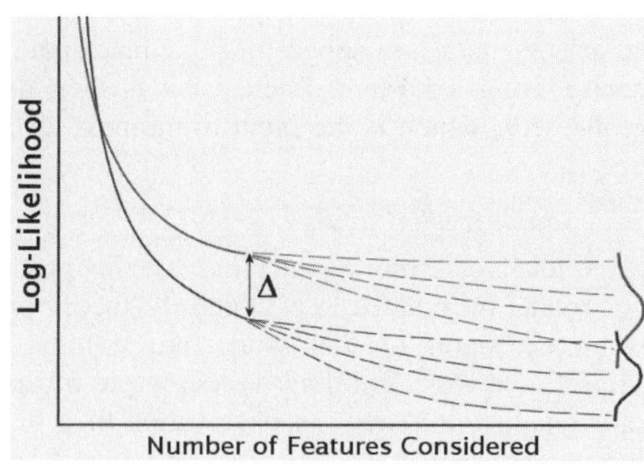

Fig. 3. Conceptual illustration of the bail-out test. After considering the first j words, the difference in log-likelihoods between two exemplars is Δ. Given some statistics about the remaining words, it is possible to compute a bound on the probability that the evaluation of the remaining words will cause one exemplar to overtake the other. If this probability is sufficiently small, the trailing exemplar can be discarded.

is a kind of preemption test, similar to procedures which have been outlined in other domains [13].

Recalling Equation 6, the log-likelihood of the current observation having been generated by exemplar i is given by

$$\ln(p(\mathbf{z}|C_i^k)) \approx \sum_{q=1}^{|v|} \ln(p(z_q|z_{p_q}, C_i^k)) \tag{11}$$

Now, define

$$d_q^i = \ln(p(z_q|z_{p_q}, C_i^k)) \tag{12}$$

and

$$D_j^i = \sum_{q=1}^{j} d_q^i = \sum_{q=1}^{j} \ln(p(z_q|z_{p_q}, C_i^k)) \tag{13}$$

where d_q^i is the log-likelihood of the ith exemplar given word q, and D_j^i is the log-likelihood of the ith exemplar after considering the first j words. At each step of the accelerated computation D_j^i is computed for all i, and incrementally increased j - that is, we are computing the log likelihoods of all exemplars in parallel, considering a greater proportion of the words at each step. After each step, a bail-out test is applied. This identifies and excludes from further computation those exemplars whose likelihood is *too far* behind the current leader. *Too far* can be quantified using concentration inequalities [14], which yield a bound on the probability that the discarded exemplar will catch up with the leader, given their current difference in log-likelihoods and some statistics about the properties of the words which remain to be evaluated.

Concretely, consider two exemplars a and b, whose log likelihood has been computed under the first j words, and whose current difference in log-likelihoods is Δ, as shown in Figure 3. Now, let X_j be the relative change in log likelihoods due to the evaluation of the jth word, and define

$$S_j = \sum_{q=j+1}^{|v|} X_q \tag{14}$$

so that S_j is that total relative change in log likelihoods due to all the words that remain to be evaluated. We are interested in $p(S_j > \Delta)$ – the probability that the evaluation of the remaining words will cause the trailing exemplar to *catch up*. If the probability is sufficiently small, the trailing hypothesis can be discarded. The key to our bail-out test is that a bound on the probability $p(S_j > \Delta)$ can be computed quickly, using concentration inequalities such as the Hoeffding or Bennett inequality [15]. These concentration inequalities are essentially specialized central limit theorems, bounding the form of the distribution S_j, given the statistics of the components X_j (which we can think of as distributions before their exact value has been computed). For the Hoeffding inequality, it is sufficient to know $\max(X_j)$ for each j, that is, the maximum relative change in log likelihood between any two exemplars due to the jth word. We can compute this statistic quickly - it is simply the difference in log likelihoods between the exemplars with highest and lowest probability of having generated word j, which we can keep track of with some simple book-keeping. Bennett's inequality additionally requires a bound on the variance of X_j, which can also be cheaply computed.

Applying the Bennett inequality, the form of the bound is

$$p(S > \Delta) < \exp\left(\frac{\sigma^2}{M^2}\cosh(f(\Delta)) - 1 - \frac{\Delta M}{\sigma^2} f(\Delta)\right) \quad (15)$$

where

$$f(\Delta) = \sinh^{-1}\left(\frac{\Delta M}{\sigma^2}\right) \quad (16)$$

and M and v are the maximum and variance values of the remaining features, such that

$$p\left(|X_q| < M\right) = 1, \forall q \in [j+1, |v|] \quad (17)$$

$$\sum_{q=j+1}^{|v|} E\left[X_q^2\right] < \sigma^2 \quad (18)$$

Typically we set our bail-out threshold $p(S > \Delta) < 10^{-6}$. The speed increase due to this bail-out test is data dependent — in our experiments it is typically a factor of 60 times faster than performing the full classification without bail-out test.

V. Markov Random Fields for Spatial Context

The estimation of the set of most likely values of a set of interdependent random variables from available data is a standard machine learning problem. Such context-dependent inference can be achieved using a family of graphical models known as Markov Random Fields (MRFs). An MRF models the joint probability distribution, $p(\mathbf{x}, \mathcal{Z})$, over the (hidden) states of the random variables, $\mathbf{x}$ and the available data, $\mathcal{Z}$. For pairwise MRFs, it is well known that this joint probability can be maximised by equivalently minimising an energy function incorporating a *unary* term modelling the data likelihood for each node and a *binary* term specifying the interaction potentials between neighbouring nodes over the set of possible values [16]. Under the assumption of every datum being equally likely (i.e. $p(\mathcal{Z})$ being uniform) a minimisation of this energy function is equivalent to finding the most likely configuration of labels given the observed data - i.e. a maximum a posteriori (MAP) estimate of $p(\mathbf{x}|\mathcal{Z})$. In the following we describe how an MRF can be applied in the context of our scene labelling endeavour. In particular, we outline how the model structure of an MRF is derived for each scene from the available data, how the model parameters are obtained and, finally, how a MAP estimate over $p(\mathbf{x}|\mathcal{Z})$ is achieved.

A. Model Structure

MRFs are a family of graphical models where the set of interdependent variables is modelled as a graph $G(\mathcal{V}, \mathcal{E})$, where $\mathcal{V}$ denotes the set of vertices and $\mathcal{E}$ denotes the set of edges connecting neighbouring nodes, respectively. In the context of our scene labelling problem, each vertex represents a patch as introduced in Section IV. Neighbourhood relations within each scene are established using the segmented image obtained in Section III using [9]. Of course, adjacency in an image implies, but does not guarantee, adjacency in the 3D scene. Therefore, in estimating adjacency from 2D information a trade-off is made between the ability of determining neighbourhood relations efficiently and the introduction of incorrect adjacencies due to the loss of depth information. In practice, we found the number of false adjacencies introduced by this approach to be negligible. Typical examples of graph structure extracted from scenes recorded by our mobile platform are shown in Figure 4.

It should be noted that the one-to-one correspondence between vertices and image patches implies that the number of nodes in the MRF for a particular frame is independent of the number of measurements taken of the scene. Thus, the abstraction away from individual measurements (e.g. laser range data) to the patch level decouples the complexity of our inference stage from the density of the underlying data. This provides a substantial advantage in terms of speed over related works [7, 4] where the complexity of the graphical models is directly proportional to the density of the underlying data.

B. Model Parameters

The specification of an energy function to be optimised provides a convenient and intuitive way of incorporating scene properties. Consider the set of labels, $\mathbf{x} \in \mathbb{Z}^{N_n}$, for a particular configuration of a graph with N_n nodes. Each node s has an observation vector, $\mathbf{z}^s$, associated with it (c.f. Section IV) and can be assigned one of N_c labels such that $x_s \in \{1, \ldots, N_c\}$. We specify the energy of any such configuration to be given by

$$E(\mathbf{x}|\theta, \lambda) = \lambda \sum_{s \in \mathcal{V}} \theta_s(x_s) + (1 - \lambda) \sum_{(s,t) \in \mathcal{E}} \theta_{st}(x_s, x_t) \quad (19)$$

where we adopt the notation of [17] in that θ defines the parameters of the energy: $\theta_s(\cdot)$ is a unary data penalty function; and $\theta_{st}(\cdot)$ is a pairwise interaction potential. λ represents a trade-off parameter which will be explained shortly. θ_s specifies the cost of assigning a given vertex any of the available labels. Intuitively, for a given node s, θ_s can be specified as a function of the posterior distribution over all classes for that node given the associated data, $p(\mathcal{C}|\mathbf{z}^s)$, as

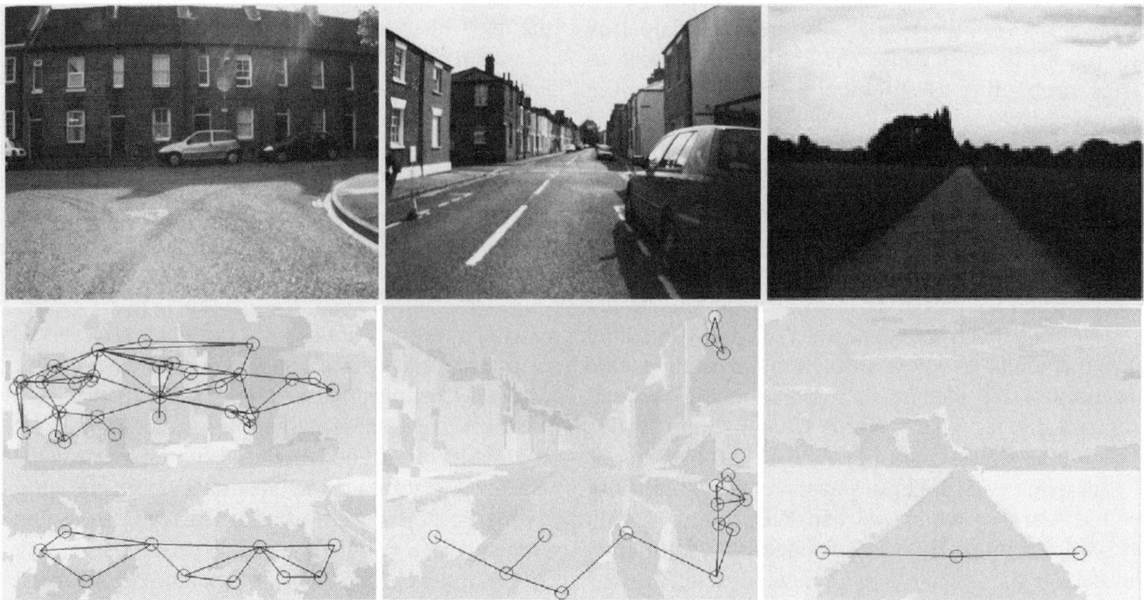

Fig. 4. Typical graphs extracted from urban scenes as recorded by our mobile robot. **Top:** the original scenes. **Bottom:** the corresponding segmented images with the extracted graph overlaid. Circles indicate nodes, lines indicate edges. For images patches which are not marked as nodes no reliable geometry estimates could be extracted from the laser data.

provided by the patch classifier introduced in Section IV. In particular, the penalty of assigning label k to node s can be expressed as

$$\theta_s(x_{sk}) = 1 - p(\mathcal{C}^k|\mathbf{z}^s) \tag{20}$$

The complement of $p(\mathcal{C}^k|\mathbf{z}^s)$ is used since θ_s refers to a penalty function which is to be minimised.

The pairwise potential θ_{st} encodes prior domain information in the form of penalties incurred by assigning specific labels to adjacent (i.e. connected) nodes. This is an intuitive formulation of the preference that nodes of certain labels are more likely to be connected to nodes of certain other labels. It follows that θ_{st} can be specified in terms of a square-symmetric matrix $\mathbf{\Phi}$ of size $N_c \times N_c$ such that

$$\theta_{st}(x_i, x_j) = 1 - \phi_{i,j} \tag{21}$$

where again the complement is used since a penalty function is specified. In this work we have chosen to specify $\mathbf{\Phi}$ such that, for two classes i and j,

$$\phi_{\mathbf{i},\mathbf{j}} = \frac{L_{i,j}}{L_i + L_j - Li, j} \tag{22}$$

Here $L_{i,j}$ denotes the total number of links connecting nodes of labels i and j, and L_i denotes the total number of links originating from nodes of label i. It follows that $\phi_{i,j} \leq 1 \; \forall(i,j)$. Appropriate values for both $L_{i,j}$ and L_i are obtained from a hand-labelled training set.

Finally, Equation 19 is a function of the trade-off parameter, λ, which provides control over the relative contributions of the unary and the binary terms to the overall energy. It is specified such that $\lambda \in [0,1]$. In this work λ is obtained by grid-search which selects a value that optimizes a measure of classifier performance on a set of labeled data. MAP estimation is performed using sequential tree-reweighted message passing (TRW-S) [17] because of its desirable convergence properties and speed.

VI. Results

We tested our algorithm using two extensive outdoor data sets spanning nearly 17 km of track gathered with an ATRV mobile platform. The system was equipped with a colour camera mounted on a pan-tilt unit and a custom-made 3D laser scanner consisting of a standard 2D SICK laser range finder (75 Hz, 180 range measurements per scan) mounted in a reciprocating cradle driven by a constant velocity motor. The camera records images to the left, the right and the front of the robot in a pre-defined pan-cycle triggered by vehicle odometry at 1.5 m intervals. The *Jericho* data set was recorded in a built-up area in Oxford over 13.2 km of track (16,000 images in total). The *Oxford Science Park* data set was recorded in the science park area in Oxford over 3.3 km of track (8,536 images in total). The two datasets were collected in different areas of the city, with only a very small overlap between the two regions.

The *Jericho* data set was used for training. The features from this set were used to learn the visual vocabulary and the Chow Liu tree. The class models were built from 1,055 patches which were segmented and labeled by hand. Automatically segmented versions of the same labeled data were used to learn the MRF binary potentials. An appropriate value for the sensor model used by our patch-level classifier was determined empirically as $p(z_i = 1|e_i = 1) = 0.35$ and $p(z_i = 0|e_i = 1) = 0$.

The *Jericho* data set is unsuitable for training the parameter λ since the patch-level classifier will correctly classify all patches in the training set, thus placing complete confidence in the unary potentials and leading to biased results. Therefore, λ was instead determined using an independent training set obtained by sampling randomly from the *Oxford Science Park* data. The sample comprised a quarter of the entire data set (55 of 220 frames). The parameter value was then determined by grid search over its range. Different values of λ lead to different classification results, thus to select a value we must define a measure of classifier performance which we wish to optimize. We present results for two different such 'tuning policies':

Tuning Policy 1. Define a per-class error function as

$$\mathbf{e} = 1 - \mathbf{p} \bullet \mathbf{r} \tag{23}$$

where $\mathbf{p}$ is the vector of class precision values, $\mathbf{r}$ is the vector of class recall values and $\bullet$ denotes the Hadamard product. Thus, classes with a low precision-recall product will have a large error. Tuning policy 1 selects λ so as to minimize $\|\mathbf{e}\|_2$. The intention here is to maximize the precision-recall product, with a bias toward improving the worst performing classes.

Tuning Policy 2. Maximize the number of true positives across all classes.

We evaluated the performance of the classifier using 3,938 patches from the *Oxford Science Park* data set, which were not involved in training λ and whose ground truth had been labeled by hand. Classification performance is summarized in Figure 5 and in Table III. A typical example is shown in Figure 1.

We present three sets of results, with confusion matrices visualized in Figure 5. 5(a) is based entirely on the output of the patch-level classifier, showing performance before MRF smoothing is applied. 5(b) shows the results incorporating the MRF tuned according to policy 1, and 5(c) the results from MRF policy 2. Prior to incorporating the MRF (5(a)), there is notable confusion between the *vehicle, foliage* and wall classes. Results incorporating the MRF (5(b),5(c)) show a visible improvement of the confusion matrix. Particularly noteworthy is improvement on the *vehicle* and *foliage* classes, where confusion with wall classes has been substantially reduced. The remaining confusion is primarily between closely related classes such as the two wall types.

Numerical measures of performance are presented in Table III. It should be noted that our test data is *unbalanced,* in the sense that there are many more instances of some classes than others, reflecting their relative frequency in the world. A consequence of this is that performance figures such as overall accuracy are not very informative, because they mostly represent classifier performance on the largest class. We chose not to balance the data because such an evaluation would be unrepresentative of classifier performance in the real world. We quote instead the per-class precision and recall. $F_{0.5}$ measures are also provided in order to provide a measure of overall classification performance per class for all policies.

The timing properties of our algorithm are outlined in Table IV. Run times are from a 2Ghz Pentium laptop. The mean total processing time was 3.9 seconds, which compares favourable to similar systems such as [7], where the authors quote 7 seconds to classify a single 2D laser scan.

TABLE IV

TIMING INFORMATION (IN MILLISECONDS).

Process	Mean (ms)	Max (ms)
Plane Segmentation	2000	2800
Feature Extraction	89	125
Feature Quantization	4	90
Image Segmentation	960	1130
Patch Classification	850	3480
MRF	2	9
Overall	3.9 seconds	7.6 seconds

VII. CONCLUSIONS

This paper has described and provided a detailed analysis of a two-stage approach to fast region labeling in 3D point-cloud maps of cities. The contributions of this work are two-fold: the first stage classifier is framed using a probabilistic bag-of-words approach, which provides for a principled bail out policy that greatly decreases the computational cost of evaluating likelihood terms. Further contribution lies in an efficient formulation of the MRF to integrate contextual information. In contrast to related approaches, the size of graph we use is small — indeed with just one node per region rather than one per laser range measurment. As a result, the overall per-scene compute time of this method is compelling: at 3.9 seconds (on average 5.6 times faster than our previous support-vector machine based approach [8]) it is suitable for online deployment.

The approach presented in this paper further provides several attractive features above and beyond our own previous work: the probabilistic nature of this approach enables a principled extraction of confidence estimates for classification results; the sensor model provides a mechanism to incorporate the notion that some of the robot's observations are more trustworthy than others; and finally, the class models can readily be updated online, allowing, in principle, for lifelong learning.

VIII. ACKNOWLEDGEMENTS

The authors would like to thank M. Pawan Kumar for many insightful conversations. The work reported here was funded by the Systems Engineering for Autonomous Systems (SEAS) Defence Technology Centre established by the UK Ministry of Defence.

REFERENCES

[1] O. Martínez-Mozos, R. Triebel, P. Jensfelt, A. Rottmann, and W. Burgard, "Supervised semantic labeling of places using information extracted from sensor data," *Robot. Auton. Syst.*, vol. 55, no. 5, pp. 391–402, 2007.

[2] B. Limketkai, L. Liao, and D. Fox, "Relational object maps for mobile robots." in *IJCAI*, L. P. Kaelbling and A. Saffiotti, Eds. Professional Book Center, 2005, pp. 1471–1476.

[3] A. Ranganathan and F. Dellaert, "Semantic modeling of places using objects," in *Proc. of Robotics: Science and Systems*, Atlanta, GA, USA, June 2007.

[4] D. Anguelov, B. Taskar, V. Chatalbashev, D. Koller, D. Gupta, G. Heitz, and A. Y. Ng, "Discriminative learning of Markov random fields for segmentation of 3D scan data." in *CVPR (2)*. IEEE Computer Society, 2005, pp. 169–176.

[5] R. Triebel, K. Kersting, and W. Burgard, "Robust 3D scan point classification using associative markov networks," in *"In Proceedings of the International Conference on Robotics and Automation(ICRA)"*, 2006.

[6] G. Monteiro, C. Premebida, P. Peixoto, and U. Nunes, "Tracking and Classification of Dynamic Obstacles Using Laser Range Finder and Vision," in *Workshop on "Safe Navigation in Open and Dynamic Environments - Autonomous Systems versus Driving Assistance Systems" at the IEEE/RSJ Intl. Conf. on Intelligent Robots and Systems (IROS)*, 2006.

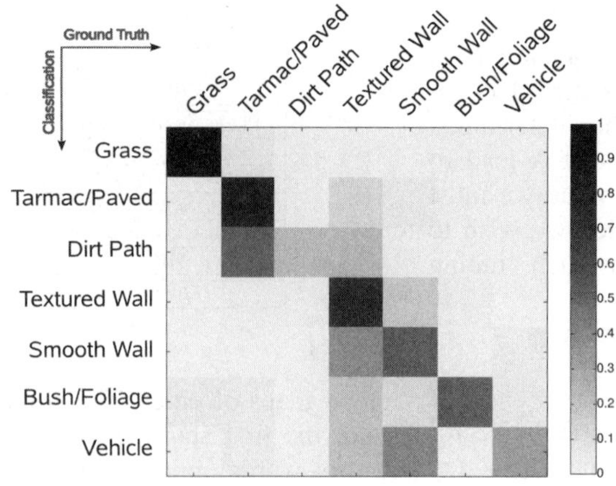

(a)

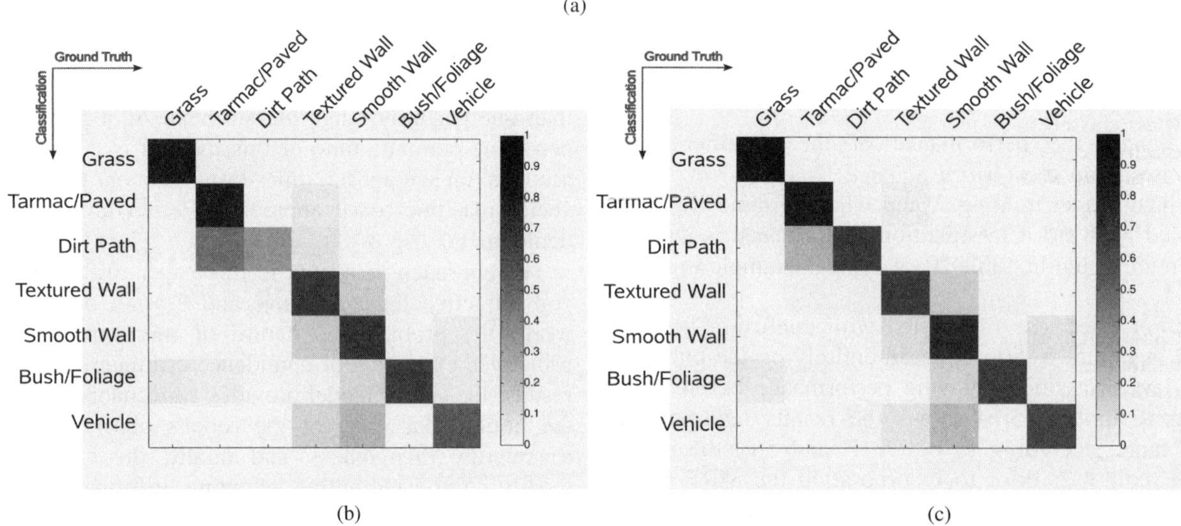

(b) (c)

Fig. 5. The confusion matrices resulting from an application of our classification framework to the (unbalanced) *Oxford Science Park* data set: **(a)** the output of the patch classification stage before MRF smoothing is applied; **(b)** the output after MRF smoothing obtained using tuning policy 1 and **(c)** the output after MRF smoothing obtained using tuning policy 2. Note that entries on the diagonals represent the *precision* with which the particular class is classified (cf. Table III). See text for more details.

TABLE III

DETAILED CLASSIFICATION RESULTS FOR THE OXFORD SCIENCE PARK DATA SET.

Class Details		Pre MRF			Post MRF (Tuning Policy 1)			Post MRF (Tuning Policy 2)		
Name	# Patches	Precision [%]	Recall [%]	$F_{0.5}$	Precision [%]	Recall [%]	$F_{0.5}$	Precision [%]	Recall [%]	$F_{0.5}$
Grass	74	90.0	73.0	86.0	95.1	52.7	81.9	100.0	25.7	63.3
Pavement/Tarmac	1078	79.0	84.8	80.1	85.4	92.1	86.7	87.4	91.6	88.2
Dirt Path	116	21.8	37.9	23.8	38.2	29.3	36.0	70.7	25.0	51.8
Textured Wall	1678	71.4	75.6	72.2	72.6	89.5	75.4	70.2	94.1	74.0
Smooth Wall	688	53.7	34.5	48.3	72.6	37.8	61.3	77.4	37.8	64.0
Bush/Foliage	161	52.7	49.1	51.9	67.0	44.1	60.7	69.5	45.3	62.8
Vehicle	143	32.2	34.3	32.6	55.4	43.4	52.5	63.8	25.9	49.3

[7] B. Douillard, D. Fox, and F. Ramos, "A Spatio-Temporal Probabilistic Model for Multi-Sensor Multi-Class Object Recognition," in *in Proc. 13th Intl. Symp. of Robotics Research (ISRR)*, 2007.

[8] I. Posner, D. Schröter, and P. M. Newman, "Describing composite urban workspaces," in *In Proc. Intl. Conf. on Robotics and Automation (ICRA)*, 2007.

[9] P. F. Felzenszwalb and D. P. Huttenlocher, "Efficient graph-based image segmentation," *Int. J. Comput. Vision*, vol. 59, no. 2, pp. 167–181, 2004.

[10] J. Sivic and A. Zisserman, "Video Google: A text retrieval approach to object matching in videos," in *Proceedings of the International Conference on Computer Vision*, Nice, France, October 2003.

[11] M. Cummins and P. Newman, "Probabilistic appearance based navigation and loop closing," in *Proc. IEEE International Conference on Robotics and Automation (ICRA'07)*, Rome, April 2007.

[12] C. Chow and C. Liu, "Approximating Discrete Probability Distributions with Dependence Trees," *IEEE Transactions on Information Theory*, vol. IT-14, no. 3, May 1968.

[13] O. Maron and A. W. Moore, "Hoeffding races: Accelerating model selection search for classification and function approximation," in *Advances in Neural Information Processing Systems*, 1994.

[14] S. Boucheron, G. Lugosi, and O. Bousquet, *Concentration Inequalities.* Heidelberg, Germany: Springer, 2004, vol. Lecture Notes in Artificial Intelligence 3176, pp. 208–240.

[15] G. Bennett, "Probability inequalities for the sum of independent random variables," *Journal of the American Statistical Association*, vol. 57, pp. 33–45, March 1962.

[16] S. Geman and D. Geman, "Stochastic relaxation, gibbs distributions, and the bayesian restoration of images," *IEEE Trans. on Pattern Analysis and Machine Intelligence*, vol. 6, no. 6, November 1984.

[17] V. Kolmogorov, "Convergent tree-reweighted message passing for energy minimization," *IEEE Trans. on Pattern Analysis and Machine Intelligence*, vol. 28, no. 10, pp. 1568–1583, 2006.

Clustering Sensor Data for Terrain Identification using a Windowless Algorithm

Philippe Giguere
Centre for Intelligent Machines
McGill University
Montreal, Quebec, Canada H3A 2A7
Email: philg@cim.mcgill.ca

Gregory Dudek
Centre for Intelligent Machines
McGill University
Montreal, Quebec, Canada H3A 2A7
Email: dudek@cim.mcgill.ca

***Abstract*— In this paper we are interested in autonomous systems that can automatically develop terrain classifiers without human interaction or feedback. A key issue is clustering of sensor data from the same terrain. In this context, we present a novel off-line windowless clustering algorithm exploiting time-dependency between samples. In terrain coverage, sets of sensory measurements are returned that are spatially, and hence temporally correlated. Our algorithm works by finding a set of parameter values for a user-specified classifier that minimize a cost function. This cost function is related to change in classifier probability outputs over time. The main advantage over other existing methods is its ability to cluster data for fast-switching systems that either have high process or observation noise, or complex distributions that cannot be properly characterized within the average duration of a state. The algorithm was evaluated using three different classifiers (linear separator, mixture of Gaussians and *k*-NEAREST NEIGHBOR), over both synthetic data sets and mobile robot contact feedback sensor data, with success.**

I. INTRODUCTION

Identifying the local terrain properties has recently become a problem of increasing interest and relevance. This has been proposed with both non-contact sensors, as well as using tactile feedback. This is because terrain properties directly affect navigability, odometry and localization performance. As part of our research, we are interested at using simple internal sensors such as accelerometers and actuator feedback information to help discover and identify terrain type. Real terrains can vary widely, contact forces vary with locomotion strategy (or gait, for a legged vehicle), and are difficult to model analytically. Therefore, the problem seems well suited to statistical data-driven approaches.

We approach the problem using unsupervised learning (clustering) of samples which represent sequences of consecutive measurement from the robot as it traverses the terrain, perhaps moving from one terrain type to another. Since those signals are generated through a physical system interacting with a continuous or piece-wise continuous terrain, time-dependency will be present between consecutive samples. The clustering algorithm we are proposing in this paper explicitly exploits this time-dependency. It is a single-stage batch method, eliminating the need for a moving time-window. The algorithm has been developed for noisy systems (i.e., overlapping clusters), as well as for systems that change state frequently (e.g., a robot traversing different terrain types in quick succession).

The paper is organized as follow. In Section II, we present an overview related work on the subject, pointing out some limitations with these methods. Our algorithm is then described in Section III, with theoretical justifications. In Section IV-A, Section IV-B and Section IV-C, we evaluate the performance of the algorithm on synthetic data with a linear separator classifier, a mixture of Gaussians classifier and *k*-NEAREST NEIGHBOR classifier, respectively. We compare our algorithm with a window-based method in Section IV-D. We then show in Section V how applying this method on data collected from a mobile robot enables robust terrain discovery and identification. This is followed by Section VI where we further point at differences between this algorithm and others.

II. RELATED WORK

Other techniques have been developed to exploit time dependencies for segmenting time-series or clustering data points. In Pawelzik *et al.* [1], an approximate probabilistic data segmentation algorithm is proposed on the segmentation algorithm on the assumption that the probability of having a transition within a sub-sequence is negligible, given a low switching rate between generating modes. Kohlmorgen *et al.* [3] present a method that uses mixture coefficients to analyze time-series generated by drifting or switching dynamics. In a similar fashion to the work we present here, these coefficients are found by minimizing an objective function that includes a squared difference between temporally adjacent mixture coefficients. The main argument behind their choice is that "solutions with a simple temporal structure are more likely than those with frequent change", an assumption we also exploit. In Kohlmorgen *et al.* [2], they present another segmentation method for time series. It is based again on minimizing a cost function that relates to the number of segments in the time series (hence minimizing the number of transitions), as well as minimizing the representation error of the prototype probability density function (*pdf*) of those segments. The distance metric used to compare *pdf*s is based on a L_2-Norm. Their simplified cost function has been designed to be computed efficiently using dynamic programming.

Lenser *et al.* [4] present both an off-line and on-line algorithm to segment time-series of sensor data for a mobile robot to detect changes in lighting conditions. The algorithm works by splitting the data into non-overlapping windows of fixed size, and then populating a tree structure such that similar regions are stored close to each other in the tree, forcing them to have common ancestor nodes. The tree is built leaf by leaf, resulting in an agglomerative hierarchical clustering. If the number of clusters is known, information in the tree structure can be used to group the data together and form clusters. The distance metric used to compare regions correspond to the absolute distance needed to move points from one distribution to match the other distribution.

For terrain identification using a vehicle, several techniques have been developed (Weiss *et al.* [5] [6], Brooks *et al.* [7], DuPont *et al.* [8], Sadhukan *et al.* [9]). Features are extracted from acceleration measurements (i.e., spectrum of acceleration, multiple moments, etc) and supervised learning is used to train a classifier, such as a support vector machine (Weiss *et al.* [5]) or a probabilistic neural network (DuPont *et al.* [8]). Another work by Lenser *et al.* [10], a non-parametric classifier for time-series is trained to identify states of a legged robot interacting with its environment. These techniques require part of the data to be manually labelled, and thus cannot be employed in the current context of unsupervised learning.

A. Limitations of Window-Based Clustering Algorithms

As long as a system is switching infrequently between states and the distributions are well separated, there will be enough data points within a window of time to properly describe these distributions. Algorithms such as Lenser *et al.* [4] or Kohlmorgen *et al.* [2] will be able to find a suitable *pdf* to describe the distributions or to detect changes, and the clustering or segmentation will be successful.

As the system switches state more frequently however, the maximum allowable size for a window will be reduced. This has to be done in order to keep the probability of having no transition in a window reasonably low. With this in mind, two particular cases become difficult:

- Noisy systems with closely-spaced distribution, resulting in significant overlap. According to Srebro *et al.* [11], the difficulty in properly clustering data from two normally distributed classes with means μ_a, μ_b, and identical standard deviation $\sigma_a = \sigma_b$ is related to the relative distance $\frac{|\mu_a - \mu_b|}{\sigma_a}$.
- Complex distribution that cannot be characterized within a small window size. In this case, there is enough information within a time-window to classify samples, but if the distributions are unknown, there is not enough information to decide whether the samples belong to the same cluster.

III. Approach

The algorithm works as follow. Given that we have:

- a data set $\vec{X}$ of T time-samples of feature vectors $\vec{x}_i$, $\vec{X} = \{\vec{x}_1, \vec{x}_2, ..., \vec{x}_T\}$ generated by a Markovian process with N_c states, with probability of exiting any state less than 50 *percent*,
- the sampled features $\vec{x}_i \in \vec{X}$ are representative, implying that *locally*, the distance between two samples is related to the probability that they belong to the same class,
- a classifier with parameters $\vec{\theta}$ used to estimate the probability $p(c_i|\vec{x}_t, \vec{\theta})$ that sample $\vec{x}_t$ belongs to class $c_i \subset C$, $|C| = N_c$,
- a classifier exploiting distance between data points $\vec{x}_i \in \vec{X}$ to compute probability estimates,
- a set of parameters $\vec{\theta}$ that is able to classify the data set $\vec{X}$ reasonably well.

The algorithm searches for the parameters $\vec{\theta}$ that minimize:

$$\arg\min_{\vec{\theta}} \sum_{i=1}^{N_c} \frac{\sum_{t=1}^{T-1} (p(c_i|\vec{x}_{t+1}, \vec{\theta}) - p(c_i|\vec{x}_t, \vec{\theta}))^2}{var(p(c_i|\vec{X}, \vec{\theta}))^2} \quad (1)$$

In our context of terrain identification, $\vec{X}$ represents a time-series of vehicle sensory information affected by the terrain e.g., acceleration measurements.

Roughly speaking, the cost in Eq. 1 tries to strike a balance between minimizing variations of classifier posterior probabilities *over time*, while simultaneously maintaining a wide distribution of posterior probabilities ($var(p(c_i|\vec{X}, \vec{\theta}))$). This is all normalized by the number of samples in each class (approximated as $var(p(c_i|\vec{X}, \vec{\theta}))$), thus preventing the algorithm from clustering all samples into a single class. A more thorough derivation of this cost function is provided in Section III-A.

An important feature of this algorithm is that it can employ either parametric or non-parametric classifiers. For example, if two classes that can be modelled with normal distributions, a linear separator is sufficient. On the other hand, a four-class problem requires a more complex classifier, such as mixture of Gaussians. If the shape of the distributions is unknown, *k*-Nearest Neighbor can be used.

A. Derivation of the Cost Function using Fisher Linear Discriminant

Let us assume that two classes, a and b, are normally distributed with means μ_a and μ_b with identical variance $\sigma_a = \sigma_b$. In *Linear Discriminant Analysis* (LDA), the data is projected on a vector $\vec{\omega}$ that maximizes the Fisher criterion $J(\omega)$:

$$J(\omega) = \frac{(\mu_{a\omega} - \mu_{b\omega})^2}{\sigma_{a\omega}^2 + \sigma_{b\omega}^2} \quad (2)$$

with $\mu_{a\omega}$, $\mu_{b\omega}$, $\sigma_{a\omega}$, $\sigma_{b\omega}$ being the means and variances of the data after projection onto $\vec{\omega}$. Data labels are required in order to compute Eq. 2. For an unlabelled data set $\vec{X}$ containing two normally distributed classes a and b, Eq. 2 can be approximated *if* the probability that consecutive samples belong to the same class is greater than 0.5. The within-class variance C_{var} projected on $\vec{\omega}$ is approximated by the average squared difference between consecutive time samples $\vec{x}_t$ projected on $\vec{\omega}$:

$$C_{var}(\vec{\omega}, \vec{X}) = \frac{1}{T-1} \sum_{t=0}^{T-1} (\vec{\omega} \cdot \vec{x}_{t+1} - \vec{\omega} \cdot \vec{x}_t)^2 \quad (3)$$

Its expected value is:

$$E\{C_{var}(\vec{\omega},\vec{X})\} = \sigma_{a\omega}^2 + \sigma_{b\omega}^2 + (\mu_{a\omega} - \mu_{b\omega})^2 P_{trans} \quad (4)$$

The between-class variance C_{dist} can be estimated by the variance of the projected data. Provided that each class has the same prior probability,

$$E\{C_{dist}(\vec{\omega},\vec{X})\} = E\{var(\vec{\omega}\cdot\vec{X})\} =$$

$$\frac{(\mu_{a\omega} - \mu_{b\omega})^2}{4} + \frac{\sigma_{a\omega}^2 + \sigma_{b\omega}^2}{2} \quad (5)$$

Dividing Eq. 5 by Eq. 4 and letting the probability of a transition $P_{trans} \rightarrow 0$, we get:

$$E\{\frac{C_{dist}(\vec{\omega},\vec{X})}{C_{var}(\vec{\omega},\vec{X})}\} \rightarrow \frac{1}{2} + \frac{(\mu_{a\omega} - \mu_{b\omega})^2}{4(\sigma_{a\omega}^2 + \sigma_{b\omega}^2)} \quad (6)$$

Minimizing the inverse of Eq. 6 corresponds to finding the Fisher criterion $J(\omega)$:

$$\arg\min_{\vec{\omega}} \frac{C_{var}(\vec{\omega},\vec{X})}{C_{dist}(\vec{\omega},\vec{X})} = \arg\max_{\vec{\omega}} \frac{C_{dist}(\vec{\omega},\vec{X})}{C_{var}(\vec{\omega},\vec{X})} \approx \arg\max_{\vec{\omega}} J(\vec{\omega}) \quad (7)$$

with

$$\frac{C_{var}(\vec{\omega},\vec{X})}{C_{dist}(\vec{\omega},\vec{X})} = \frac{\sum_{t=0}^{T-1}(\vec{\omega}\cdot\vec{x}_{t+1} - \vec{\omega}\cdot\vec{x}_t)^2}{var(\vec{\omega}\cdot\vec{X})} \quad (8)$$

The probability that sample $\vec{x}_t$ belongs to class c, for a linear separator classifier, can be expressed by a sigmoid function:

$$p(c|\vec{x}_t) = \frac{1}{1+e^{-kd}} \quad (9)$$

where d is the distance between $\vec{x}_t$ and the boundary decision, and k is a parameter that determines the "sharpness" of the sigmoid. Given a sufficiently small k and significant overlap of the clusters, most of the data will lie within the region $d \ll \frac{1}{k}$. Eq. 9 can then be approximated by:

$$p(c|\vec{x}_t) \approx \frac{d}{k} \quad (10)$$

and Eq. 8 approximated as:

$$\frac{C_{var}(\vec{\omega})}{C_{dist}(\vec{\omega})} \approx \frac{\sum_{t=0}^{T-1}(p(c|\vec{x}_{t+1},\vec{\theta}) - p(c|\vec{x}_t,\vec{\theta}))^2}{var(p(c|\vec{X},\vec{\theta}))} \quad (11)$$

Where $\vec{\theta}$ correspond to the linear separator parameters. Eq. 11 is then normalized by $p(c_a|\vec{X},\vec{\theta})p(c_b|\vec{X},\vec{\theta})$ to reflect the probability of leaving a given state c. This normalizing factor can be approximated by $var(p(c|\vec{X},\vec{\theta}))$, and the final cost function is:

$$E(\vec{X},\vec{\theta}) = \frac{\sum_{t=0}^{T-1}(p(c|\vec{x}_{t+1},\vec{\theta}) - p(c|\vec{x}_t,\vec{\theta}))^2}{var(p(c|\vec{X},\vec{\theta}))^2} \quad (12)$$

B. Optimization: Simulated Annealing

The landscape of the cost function being unknown, simulated annealing was used to find the classifier parameters $\vec{\theta}$ that minimize $E(\vec{X},\vec{\theta})$ described in Eq. Eq. 12. Although very slow, it is necessary due to the presence of local minima. For classifiers with few (less than 20) parameters, one parameter was randomly modified at each step. For *k*-NEAREST NEIGHBOR classifiers, modifications were made to a few points and a small random number of their neighbors. This strategy improved the speed of convergence. The cooling schedule was manually tuned, with longer schedules for more complex problems. Three random restart runs were used, to avoid being trapped in a deep local minimum.

IV. TESTING THE ALGORITHM ON SYNTHETIC DATA SETS

The algorithm was first evaluated using three different classifiers (linear separator, mixture of Gaussians, and *k*-NEAREST NEIGHBOR) on synthetic data sets. This was done to demonstrate the range of cases that can be handled, as our real robot data sets cannot cover some of those cases (e.g., complex-shaped distributions in Section IV-C.5). These sets were generated by sampling a distribution s times (our so-called *segment length*), then switching to the next distribution and drawing s samples again. This segment length determined the amount of temporal coherence present in the data. This process was repeated until a desired sequence length was reached. Fig. 1(b) shows a sequence of 12 samples drawn from two Gaussian distributions, with a segment of length 3. For the test cases presented in this section, the segment lengths were relatively short (between 3 and 5), to demonstrate how the algorithm is capable of handling signals generated from a system that changes state frequently. Results for these synthetic distributions are shown in the following subsections.

A. Linear Separator Classifier with Two Gaussian Distributions

For linear separators, we used two closely-spaced two-dimensional Gaussian distributions (Fig. 1(a)) with identical standard deviation $\sigma_{x\{1,2\}} = 0.863$, $\sigma_{y\{1,2\}} = 1$, and the distance between the means was 1. This simulated cases where features are extremely noisy. Without labels, the combination of the distributions is radially symmetric (Fig. 1(b)). The optimal Bayes classification rate for a single data point for these distributions was 70.7 *percent*, an indication of the difficulty of the problem. The linear separator was trained using Eq. 1, with probabilities computed from Eq. 9. A value of $k = 3$ was chosen, although empirically results were similar for a wide range of k values. 100 time sequences of 102 samples were randomly generated. The average classification success rate was 68.5 ± 6.8 *percent*, which is close to the Bayes classification rate. Fig. 2 shows an example of the classifier posterior probability over time after cost minimization.

B. Mixture of Gaussians Classifier with Three Gaussian Distributions

For *mixture of Gaussians* classifiers, three normal distributions in two dimensions (see Fig. 3) were used. The classi-

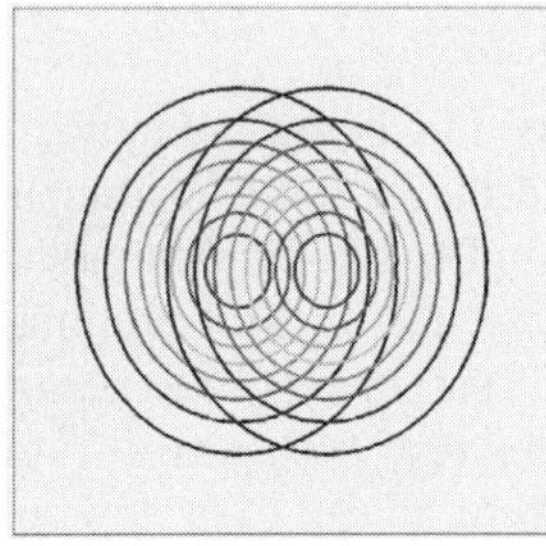

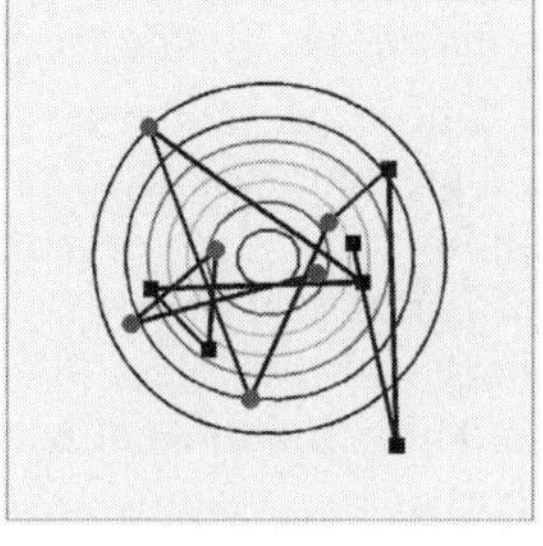

(a) Individual distributions (b) Combined Distributions

Fig. 1. Contour plot 1(a) of the two normal distributions used in testing the algorithm with a linear separator. Their combination 1(b) is resembles a single, radially-symmetric Gaussian distribution. A synthetic sequence of 12 data samples with segment length of 3 is also shown in 1(b), with a line drawn between consecutive samples.

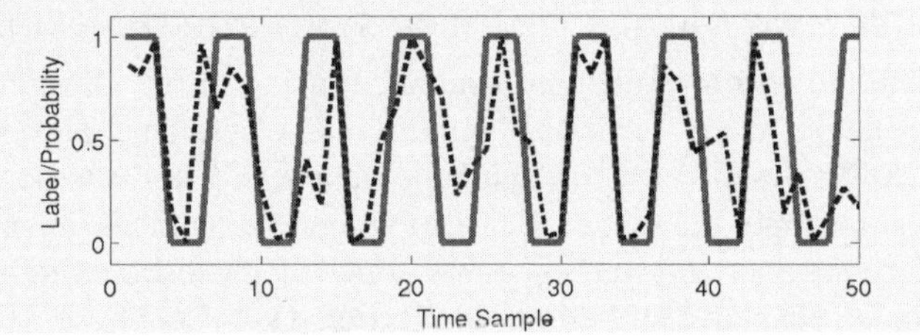

Fig. 2. Classifier posterior probability over time for the two classes drawn from Gaussians distributions depicted in Fig. 1(a), after optimization (dashed line). Ground truth is shown as solid green line. The segment length is 3. The first 50 samples are shown.

fier itself had 9 free parameters: 6 for the two-dimensional Gaussian locations, and 3 for the standard deviation (the Gaussians were radially-symmetric). The three distributions had covariance equal to:

$$\sigma_1 = \begin{pmatrix} .5 & 0 \\ 0 & 1 \end{pmatrix}, \sigma_2 = \begin{pmatrix} .8 & .1 \\ .1 & .6 \end{pmatrix}, \sigma_3 = \begin{pmatrix} .8 & -.1 \\ -.1 & .6 \end{pmatrix}$$

The distribution centers were located at a distance of 0.95 from the $(0, 0)$ location and at 0, 120 and 240 *deg* angles. The optimal Bayes classification rate for these distributions is 74 *percent*. 100 time sequences of 207 samples with segment length of 3 were generated. The average classification rate was 69.2 ± 9.3 *percent*.

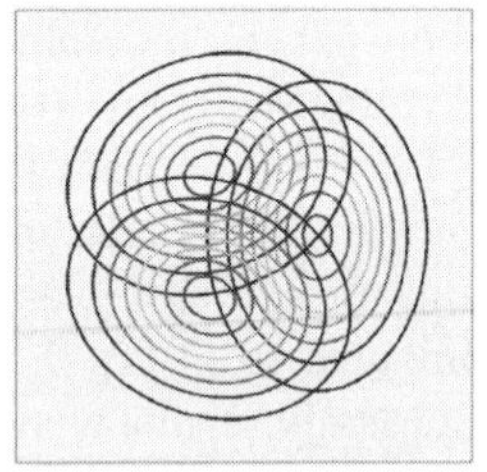

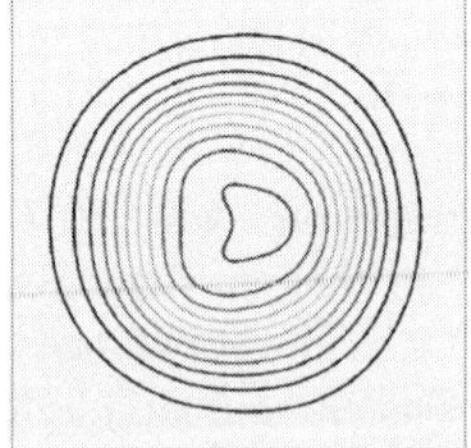

(a) Individual distributions (b) Combined Distributions

Fig. 3. Contour plot of the three normal distributions 3(a) and their sum 3(b) used to test the algorithm with a mixture of Gaussians classifier.

C. K-*Nearest-Neighbor Classifier with Uniform Distributions*

As an instance-based learning method, *k*-NEAREST NEIGHBOR [12] has the significant advantage of being able to represent complex distributions. A major drawback associated with this classifier is the large number of parameters (proportional to the number of points in the data set). This results in lengthy computation time in order to find the parameters that minimizes the cost function. Five different test cases (few data points, unequal number of samples per class, overlapping distributions, six-class distributions and complex-shaped distributions) were designed to test the performance of the algorithm using this classifier.

1) Few Data Points: For each test sequence, only 36 samples were drawn from 3 square, non-overlapping uniform distributions, with a segment length of 3. The classifier used $k = 10$ neighbors with a Gaussian kernel $\sigma = 0.8$. The mean classification success rate over 100 trials was 93.7 ± 6.1 *percent*. Fig. 4 shows these results in more detail.

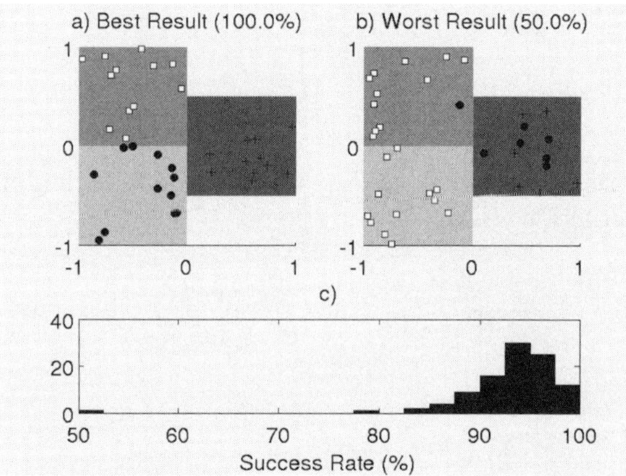

Fig. 4. Results of clustering method applied on a data set of 36 points drawn from three equal distributions. The best a) and worst b) results of clustering are shown, with distributions shown as background grey boxes. c) shows the histogram of success rates over 100 trials.

2) Unequal Number of Samples per Class: Three uniform rectangular distributions of equal density, but different area were sampled with a segment length is 3. In total, 84 samples were drawn from the smallest, 168 from the medium and 252 from the largest distribution. The classifier used $k = 20$ neighbors with a Gaussian kernel $\sigma = 0.8$. The average classification success rate over 80 trials was 87.8 ± 11.6 *percent*. Fig. 5 shows these results in more details.

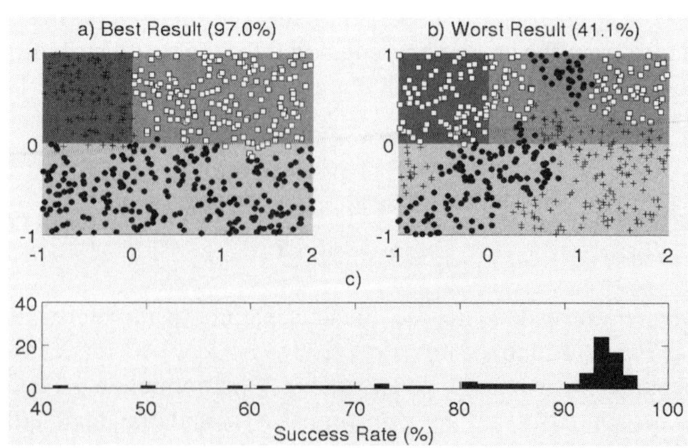

Fig. 5. Results of clustering method applied on a data set of 504 points, drawn from three distributions of different sizes. The best a) and worst b) results of clustering are shown, with distributions shown as background grey boxes. c) shows the histogram of success rates over 80 trials.

3) Overlapping Distributions: Two distributions with significant overlapping (40 *percent*) were used to generate the data. The overlapping regions were selected so non-overlapping regions within a class would be of different sizes. The classifier used $k = 20$ neighbors with a Gaussian kernel $\sigma = 0.8$. 57 test sequences of 504 points, with a segment length of 3 were randomly generated. The average classification success rate was $76.3 \pm .6.4$ *percent*, not too far from the maximum possible rate of 80 *percent*.

4) Six-Classes Distributions: Six square uniform distributions were used in these tests. The classifier used $k = 10$ neighbors with a Gaussian kernel $\sigma = 0.8$. 100 test sequences of 306 points, with a segment length of 3 were randomly generated, with detailed results shown in Fig. 6. The average classification rate was 90.0 ± 5.7 *percent*.

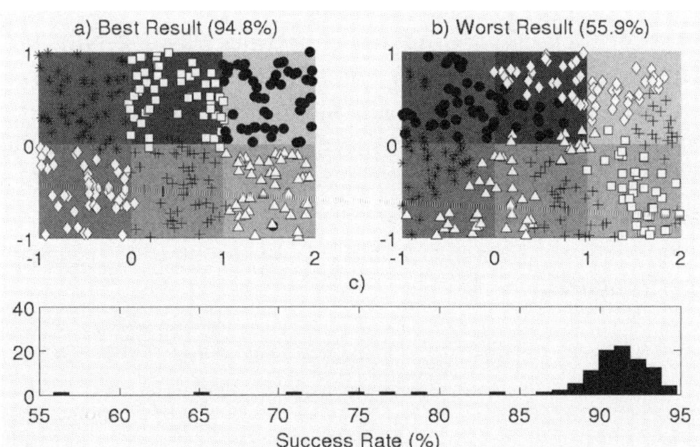

Fig. 6. Clustering results for a data set of 306 points drawn from 6 distributions of equal sizes, with segment length of 3. The best a) and worst b) results are shown, with distributions shown as background grey boxes. c) shows the histogram of success rates over 100 trials.

5) Complex-Shaped Distributions: Complex-shaped distributions were simulated using two, two-dimensional spirals. Data was generated according to the following equations:

$$x_1 = (tt + d_\perp) * cos(\phi + \phi_0), \quad x_2 = (tt + d_\perp) * sin(\phi + \phi_0) \tag{13}$$

with $\phi = d_{arc}\sqrt{rand\{0..1\}}$ the arc distance from the center, $d_\perp - N(0, \sigma_{SRnoise})$ a perpendicular, normally distributed distance from the arc, and ϕ_0 equal to 0 for the first distribution and π for the second. 5,000 data points were drawn for each trial, with segment length of 5 using $d_{arc} = 15$ and $\sigma_{SRnoise} = 0.9$ for the distributions. Fig. 7 shows time sequences of 10 and 50 samples from the test sequence used. One can see that the shape of the distributions cannot be inferred from short sequences, even when data labelling is provided.

The k-NEAREST NEIGHBOR classifier used $k = 40$ neighbors, with a Gaussian kernel of $\sigma = 0.4$. Fig. 8 shows classification success rate achieved for the 19 test cases generated. Fig. 9 shows a successful and unsuccessful case of clustering. If we exclude the 3 unsuccessful cases, considering them as being outliers due to the simulated annealing getting stuck in a local minimum, the average value of classification success was 92.6 ± 0.4 *percent*.

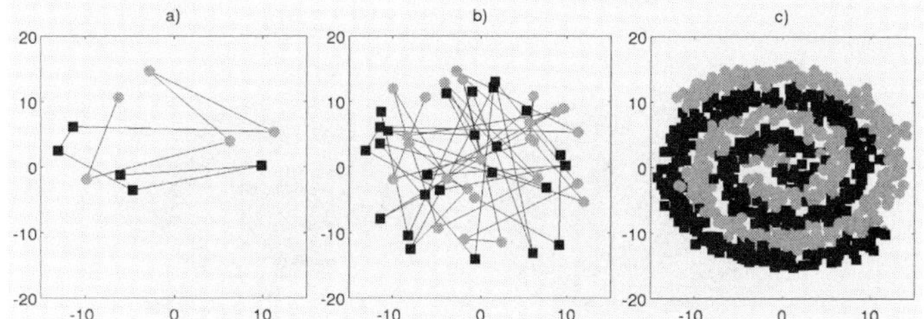

Fig. 7. Time sequence shown in feature space for a) ten samples and b) fifty samples drawn randomly from the distributions described in Eq. 13 and shown in c), with segment length of 5 samples. The spirals are not visible in a) and barely discernible in b).

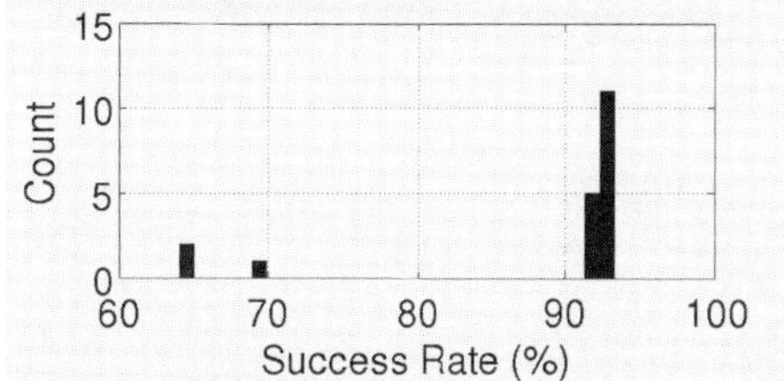

Fig. 8. Histogram showing the distribution of the 19 classification success rates after clustering. The majority of results were located around 92 *percent*, with three cases failing to completely identify the underlying structure.

D. Comparison with a Window-based Method

Performance of this algorithm was compared to the segmentation algorithm described in Lenser *et al.* [4]. The latter algorithm was run until only two clusters were left. The data used was generated from a simpler version of the two-spiral distributions, with $d_{arc} = 5$ and $\sigma_{SRnoise} = 0.9.$, with 250 samples per test case. A test case is shown in Fig. 10 a), without temporal information for clarity. For long segment lengths (over 60 samples), success rates are similar for both methods. As expected, Fig. 10 b) shows that shorter segment lengths negatively affect the window-based method, with larger windows being affected most. This can be explained by the merger of the two clusters through windows containing data from both distributions. The number of such windows is greater for a larger window size and a smaller segment length.

V. TESTING ALGORITHM ON ROBOT SENSOR DATA: AUTONOMOUS TERRAIN DISCOVERY

The vehicle used to collect various ground surfaces data (Fig. 11) [14] was a hexapod robot specifically designed for amphibious locomotion. It was previously shown in Giguere *et al.* [15] to be able to recognize terrain types using supervised learning methods. The robot was equipped with a 3-axis Inertial Measurement Unit (3DM-GX1^{TM}) from Microstrain. The sensors relevant for terrain identification are: 3 accelerometers, 3 rate gyroscopes, 6 leg angle encoders and 6 motor current estimators. Each sensor was sampled 23 times during a complete leg rotation, thus forming a 23-dimensions vector. Multiple vectors could be concatenated together to improve detection, forming an even higher dimensionality feature vector for each complete leg rotation. Dimensionality was subsequently reduced by applying Principal Component Analysis (PCA). In the following experiments, only the two first main components

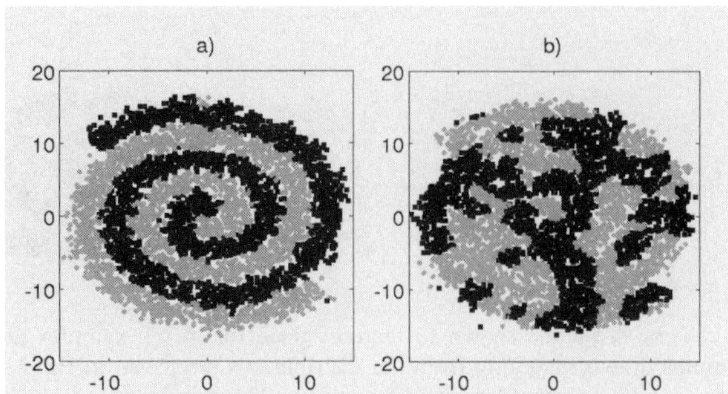

Fig. 9. Successful (92 *percent*) a) and unsuccessful b) clustering for a two-class problem of 5,000 points generated according to Eq. 13. The segment length was 5.

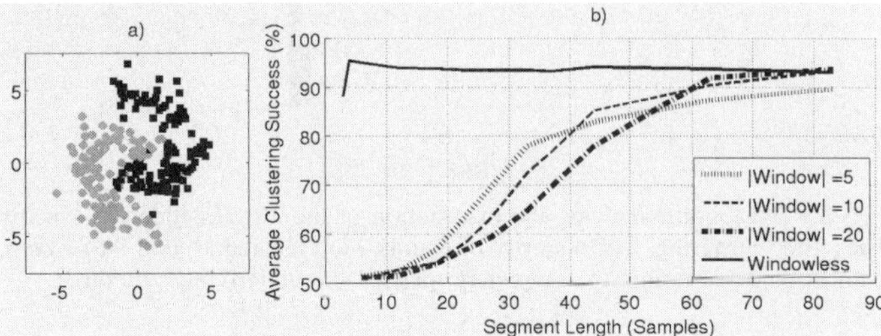

Fig. 10. Average clustering success rate for the windowless algorithm and the segmentation algorithm in Lenser *et al.* [4] for time-window sizes of 5, 10 and 15 samples. When transitions are infrequent (corresponding to segment length over 60), success rates are similar for both methods. For shorter segment lengths, the window-based method fails to identify the two clusters and instead simply merge them together.

were used. Even though some information is discarded, our previous results in [15] indicated that this was sufficient to distinguish between small numbers of terrains. If more discrimination is needed, other components can be added.

These experiments were restricted to level terrains, with small turning maneuvers. Changes in terrain slopes or complex robot maneuvers impact the dynamics of the robot, and consequently affects how a particular terrain is perceived by the sensors. Discarding data when the robot performs a complicated maneuver or when the slope of the terrain crosses a threshold would mitigate this issue. Terrains were selected to offer a variety of possible environments that might be encountered by an amphibious robot. They were also different enough that, from a locomotion point of view, they are distinct groups, and thus form classes on their own.

Fig. 11. The hexapod robot, shown equipped with semi-circle legs for land locomotion. The vehicle moves forward by constantly rotating legs in two groups of three, forming stable tripod configurations.

A. Overlapping Clusters with Noisy Data

The first data set was collected in an area covered with grass, with a section that had been recently tilled. The robot was manually driven in a straight line over the grass, crossing eventually to the tilled section. The robot was then turned around, and manually driven towards the grass. Eight transitions were collected in this manner. The problem was made more challenging by using only the pitch angular velocity vector. Using more sensor information would have reduced the noise and increased the relative separation between the two clusters.

Two classifiers were used in the clustering algorithm for this data set. The first one was a linear separator with a constant $k = 3.0$. Classification success after clustering was 78.6 *percent* (see Figs. 12 and 14(a)), a value certainly close to the Bayes error considering the overlap between the classes. The second classifier used for clustering was a k-NEAREST NEIGHBOR classifier with $k = 10$ and $\sigma = 1.0$ for the kernel. As expected, k-NEAREST NEIGHBOR had slightly inferior results, with a classification success rate of 73.9 *percent* (see Figs. 13 and 14(b)). The larger number of parameters (454 compared to 2 for the linear separator) makes this classifier prone to over-fit the data, potentially explaining the difference in performance.

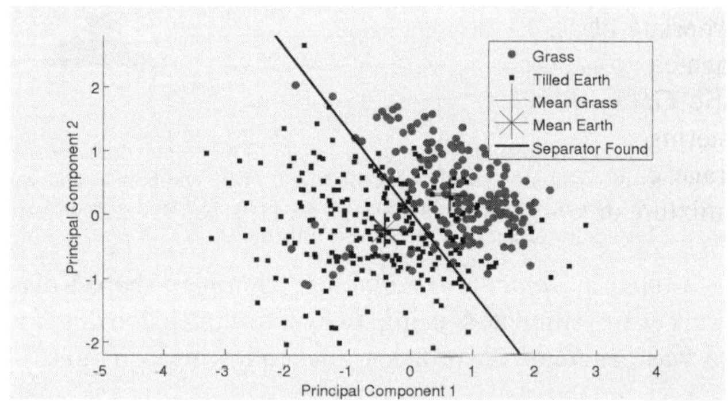

Fig. 12. Sensor data set collected for a robot walking on grass and tilled earth, with eight transitions present in the data. The solid line represents the separator found on the unlabelled data using the algorithm with a linear separator as classifier. Even though the clusters have significant overlap, the algorithm still managed to find a good solution. Notice how the separator is nearly perpendicular to a line joining the distribution centers, an indication that the solution is a close approximation to *LDA* on the labelled data.

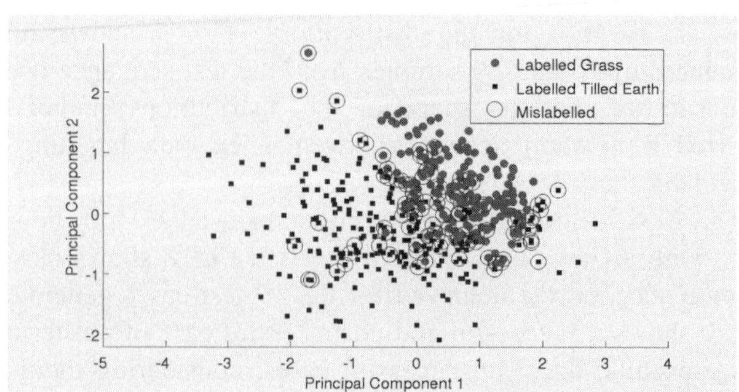

Fig. 13. Same data set as Fig. 12, clusterized with the algorithm using a k-NEAREST NEIGHBOR classifier. The circled data points are wrongly labelled. Most of them are located either at the boundary between the two distributions, or deep inside the other distribution.

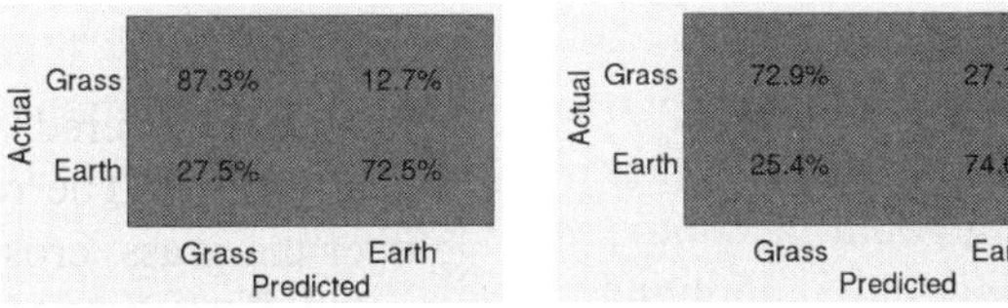

(a) Linear Separator (b) k-NEAREST NEIGHBOR

Fig. 14. Confusion matrix for classification of two-terrain data obtained after clustering using a) a linear separator and b) the k-NEAREST NEIGHBOR.

B. *Fast-Switching Semi-Synthetic Data*

Another data set was collected over five different terrains, and a high-dimensionality feature vector for each complete leg rotation was generated using 12 sensors (all 3 angular velocities, all 3 accelerations and all 6 motor currents). As in the previous case, only the two first principal components were used. Of all sensors, the motor currents were the most informative.

No rapid terrain changes were present in the original data set, so segments of data were selected according to the random state change sequence shown in Fig. 15. This artificially decreased the average segment length to a value of 6, with individual states having average segment lengths between 3.5 and 9.8. A classifier with a mixture of five radially-symmetric Gaussians was used in the clustering algorithm. Fig. 16 shows the labelling of the data set and Fig. 17 shows the confusion matrix. Overall, 91 *percent* of the data was grouped in the appropriate class. Even though some of the distributions are elongated (for example *linoleum*), the combination of symmetric Gaussians managed to capture the clusters. Standard clustering techniques struggled with these distributions, with average classification success rates of 68.2 and 63.0 *percent* for mixture of Gaussians and K-means clustering, respectively.

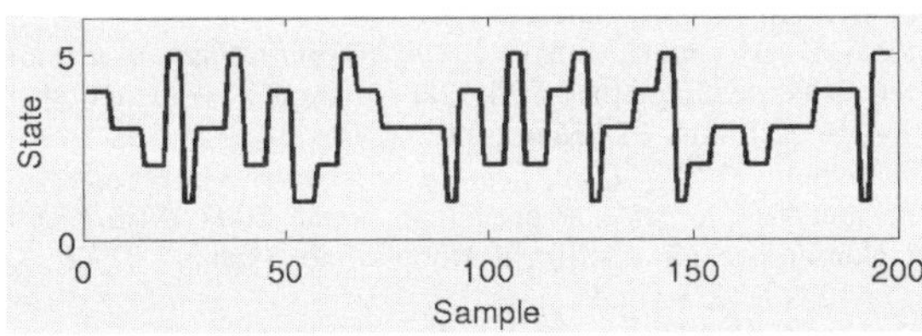

Fig. 15. State (from 1 to 5) sequence used to generate the time sequence of data in the clustering problem shown in Fig. 16. The shortest segment length is 3, and the average segment length is 6.

VI. DISCUSSION

A. *Comparing Cost Functions from Previous Works*

In [3], the cost function to be minimized is:

$$E(\Theta) = \sum_{t=1}^{T}(y_t - \sum_{s=1}^{N} p_{s,t} f_s(\vec{x}_t))^2 + C\sum_{t=1}^{T-1}\sum_{s=1}^{N}(p_{s,t+1} - p_{s,t})^2 \tag{14}$$

with $f_s(.)$ as a kernel estimator for data $\vec{x}_t$ for state s, $p_{s,t}$ as the mixing coefficient for state s at time t, $\theta = p_{s,t} : s = 1, ..., N; t = 1, ..., T$ as the set of coefficients to be found, and C as a regularization constant. In [2], the cost function is:

$$o(\vec{s}) = \sum_{t=W}^{T} d(p_{s(t)}(\vec{x}), p_t(\vec{x})) + Cn(\vec{s}) \tag{15}$$

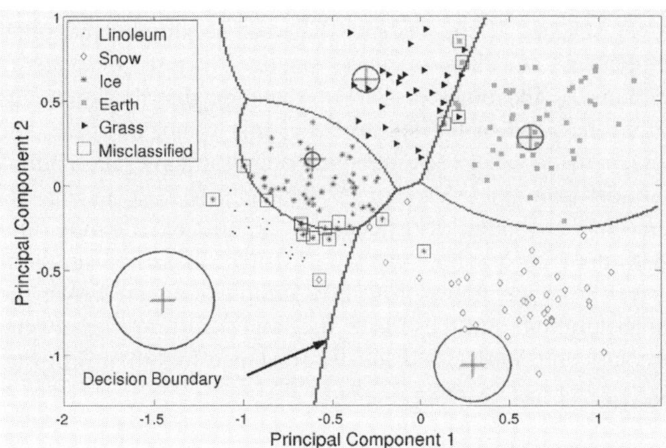

Fig. 16. Applying the clustering algorithm on a data set of five environments, with a mixture of five Gaussians as a classifier. Labels shown are the ground truth. The samples surrounded with a small square are wrongly labelled. The five red cross-hairs are the location of the Gaussians, the thick circles representing their standard deviation. The decision boundary is represented by the red curved lines.

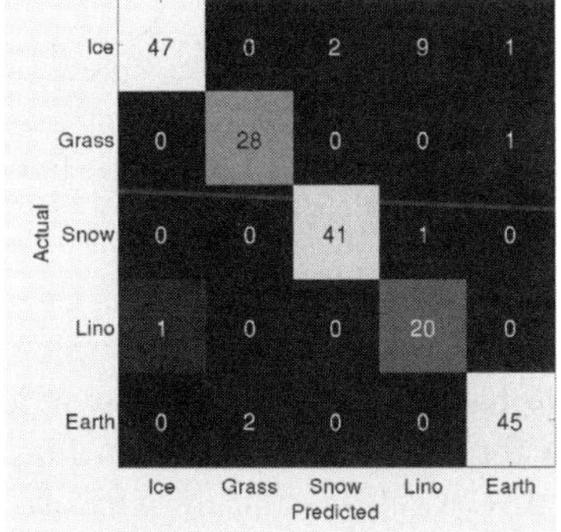

Fig. 17. Confusion matrix for the clustering shown in Fig. 16.

With $d(.,.)$ being a L_2-Norm distance metric function, $\vec{s}$ a state sequence, $p_t(\vec{x})$ the *pdf* estimate within the window at time t, $p_{s(t)}(\vec{x})$ the prototype *pdf* for state $s(t)$, $n(\vec{s})$ is the number of segments and C a regularization constant.

Eq. 14 and Eq. 15 can be separated in two parts: the first part minimizes representation errors, and the second part minimizes changes over time. A regularization constant C is needed to balance them, and selecting this constant is known to be a difficult problem. In contrast, our cost function (Eq. 12) does not take into account the classifier fit to the actual data $\vec{X}$. Instead we solely concentrate on the variation of the classifier output over time, thus completely eliminating the need for a regularization constant and associated stabilizer. This is a significant advantage. The fact that the classifier fit is not taken into account by our algorithm can be seen in the mixture of Gaussian result case shown in Fig. 16: the location of the Gaussians and their width (marked as circled red cross-hairs) do not match the position of the data they represent. Only the decision boundaries matter in our case. If the actual *pdf*s are required, a suiting representation can be fitted using the probability estimates found.

B. *Distances and Window Size for Complex Distributions*

Algorithms relying on distances between *pdf*s described within small windows such as the one used by Kohlmorgen *et* *al.* [2] succeed if the distance between a sample and other members of its class is much smaller than its distance to members of the other class. For complex distributions such as the one used in the spirals case, this is not the case: the average distance between intraclass points and interclass points is almost identical: 13.4 vs 13.9. This can be seen in Fig. 18, were the distribution of distances are almost overlapping. Our algorithm with a k-NEAREST NEIGHBOR classifier succeeds because it concentrates on nearby samples collected over the whole duration, and these have the largest difference between intraclass and interclass distances (distance less than 2 in Fig. 18).

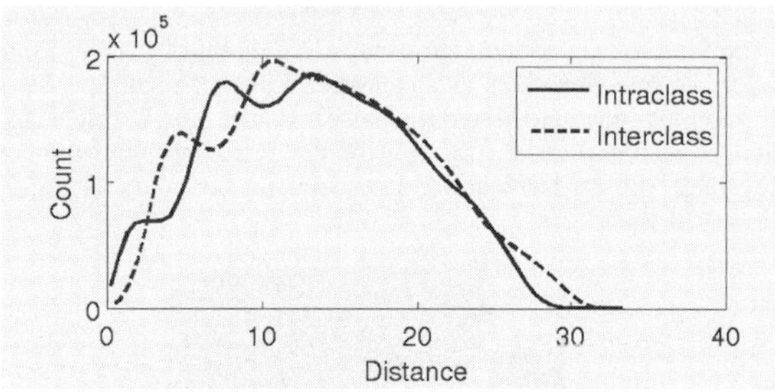

Fig. 18. Distribution of distances between samples belonging to the same class (intraclass) and samples belonging to different classes (interclass), for the spirals data set used in Fig. 9. The distributions almost completely overlap, except at shorter distances.

Moreover, to avoid transitions in their time-windows, these algorithms would have to limit their time-window sizes. Bearing in mind that for the spirals data set in Fig. 9 the segment length was 5, this would result in window sizes of 2 to 3 samples. These distributions cannot be approximated in a satisfactory manner with so few points. From simple visual inspection one can see that it requires, at a minimum, 10 to 20 points. An advanced dimensionality reduction technique such as *Isomap* [16] could be used to simplify this type of distribution. If data points bridging the gap between the spirals were present at regular intervals however, *Isomap* would be of little help.

VII. CONCLUSION AND FUTURE WORK

In this paper we presented a new clustering method based on minimizing a cost function related to the probability output of classifiers. Synthetic test cases were used to demonstrate its capabilities over a range of distribution types. The same method was employed on data collected with a walking robot, clustering its sensor data in terrain categories successfully.

As future work, we intend to quantify the impact of segment lengths, overlap and number of points in sequence on the clustering process. We are also looking at methods to estimate the number of classes present in the data set, so it would not need to be known beforehand. Since merging classes together does not affect the cost but splitting a valid class does, a rapid increase of cost-per-data point as we increase the number of clusters in the algorithm is a potential indicator that we have found the proper number of classes.

A major drawback of using simulated annealing to find the minimum cost is its prohibitive computing time, in the order of hours on a dual CPU 3.2 GHz Xeon machine for the most complex distributions. There are indications that other optimization techniques could be employed to drastically reduce the amount of computation required.

We are looking at applying this algorithm outside the scope of terrain discovery. Many segmentation problems exhibit continuity in their sets, such as texture segmentation (spatial continuity) or video (temporal continuity) segmentation. Feature ranking for continuous processes is another possible application, where the cost function could be employed to evaluate each feature individually.

REFERENCES

[1] Pawelzik, K., Kohlmorgen, J. and Muller, K.-R. (1996). *Annealed competition of experts for a segmentation and classification of switching dynamics*. Neural Computation, 8(2), 340–356.

[2] J. Kohlmorgen and S. Lemm. *An on-line method for segmentation and identification of non-stationary time series*. In NNSP 2001, pp. 113-122.

[3] J. Kohlmorgen, S. Lemm, G. Rätsch, and K.-R. Müller. *Analysis of nonstationary time series by mixtures of self-organizing predictors*. In Proceedings of IEEE Neural Networks for Signal Processing Workshop, pages 85-94, 2000.

[4] S. Lenser and M. Veloso. *Automatic detection and response to environmental change*. In Proceedings of the International Conference of Robotics and Automation, May 2003

[5] Weiss C., Fröhlich H., Zell A. *Vibration-based Terrain Classification Using Support Vector Machines*. In Proceedings of the IEEEE/RSJ International Conference on Intelligent Robots and Systems (IROS), October, 2006.

[6] Weiss C., Fechner N., Stark M., Zell A. *Comparison of Different Approaches to Vibration-based Terrain Classification*. in Proceedings of the 3rd European Conference on Mobile Robots (ECMR 2007), Freiburg, Germany, September 19-21, 2007, pp. 7-12.

[7] Brooks C., Iagnemma K., Dubowsky S. *Vibration-based Terrain Analysis for Mobile Robots*. In Proceedings of the 2005 IEEE International Conference on Robotics and Automation, 2005.

[8] Dupont, E. M., Moore, C. A., Collins, E. G., Jr., Coyle, E. *Frequency response method for terrain classification inautonomousground vehicles*. In Autonomous Robots, January, 2008.

[9] Sadhukan, D., Moore, C. (2003). *Online terrain estimation using internal sensors*. In Proceedings of the Florida conference on recent advances in robotics, Boca Raton, FL, May 2003.

[10] S. Lenser, M. Veloso, *Classification of Robotic Sensor Streams Using Non-Parametric Statistics*, Proceedings of the 2004 IEEE/RSJ International Conference on Intelligent Robots and Systems, IEEE, October, 2004, pp. 2719-2724 vol.3

[11] N. Srebro, G. Shakhnarovich, S. Roweis, *When is Clustering Hard?*. PASCAL Workshop on Statistics and Optimization of Clustering, July 2005.

[12] Cover, T.M. and Hart, P.E., 1967. *Nearest neighbor pattern classification*. IEEE Transactions on Information Theory, 13(1), 21-27.

[13] U. Saranli, M. Buehler, D.E Koditschek *RHex: A simple and highly mobile hexapod robot*, The International Journal of Robotics Research, vol. 20, no.7, pp.616-631, 2001.

[14] G. Dudek, M. Jenkin, C. Prahacs, A. Hogue, J. Sattar, P. Giguere, A. German, H. Liu, S. Saunderson, A. Ripsman, S. Simhon, L. A. Torres-Mendez, E. Milios, P. Zhang, I. Rekleitis *A Visually Guided Swimming Robot*, Proceedings of the 2005 IEEE/RSJ International Conference on Intelligent Robots and Systems, pp. 1749-1754, 2005.

[15] P. Giguere, G. Dudek, C. Prahacs, and S. Saunderson. *Environment Identification for a Running Robot Using Inertial and Actuator Cues*. In Proceedings of Robotics Science and System (RSS 2006), August, 2006.

[16] J. Tenenbaum, V. de Silva, J. Langford. *A Global Geometric Framework for Nonlinear Dimensionality Reduction*. Science 22 December 2000: Vol. 290. no. 5500, pp. 2319 - 2323.

Distributed Localization of Modular Robot Ensembles

Stanislav Funiak Michael P. Ashley-Rollman
and Seth Copen Goldstein
Carnegie Mellon University
Pittsburgh, PA 15213, USA
{sfuniak, mpa, seth}@cs.cmu.edu

Padmanabhan Pillai
and Jason D. Campbell
Intel Research Pittsburgh
Pittsburgh, PA 15213, USA
{padmanabhan.s.pillai, jason.d.campbell}@intel.com

Abstract— **Internal localization, the problem of estimating relative pose for each module (part) of a modular robot is a prerequisite for many shape control, locomotion, and actuation algorithms. In this paper, we propose a robust hierarchical approach that uses normalized cut to identify dense subregions with small mutual localization error, then progressively merges those subregions to localize the entire ensemble. Our method works well in both 2D and 3D, and requires neither exact measurements nor rigid inter-module connectors. Most of the computations in our method can be effectively distributed. The result is a robust algorithm that scales to large, non-homogeneous ensembles. We evaluate our algorithm in accurate 2D and 3D simulations of scenarios with up to 10,000 modules.**

I. Introduction

Large self-reconfigurable modular robots have received a growing interest from the robotics community. A self-reconfigurable modular robot (SRMR) is comprised of many discrete, physically connected modules which can be rearranged to adapt the robot's shape or capabilities to the task at hand. These robot ensembles have been proposed for various applications, such as product design and visualization [1], emergency search and rescue, and rapid prototyping [2, 3]. A fundamental task in such robot ensembles is *internal localization*, the establishment of relative pose amongst the robot's many individual components. Accurate internal localization is required for many tasks, including motion planning, mechanical stability, and control.

Internal localization for large robot ensembles presents a number of challenges. As systems scale to larger ensembles of smaller, finer grain modules, one can expect only limited capabilities at individual modules. In particular, modules only make noisy observations of their immediate neighbors, and do not have access to long distance measurements, such as global time-of-flight measurements, or external beacons. A lack of strong mechanical latches in small modules precludes mechanical constraints for accurate alignment and orientation.

This work is supported in part by the NSF under grants CNS 0428738 (ITR: Synthetic Reality) and NeTS-NOSS CNS-0625518, by the ONR under grant MURI N000140710747, by Intel Corporation, and by Carnegie Mellon University.

Although localization algorithms have been well studied in robotics, many of the existing approaches do not directly apply to large-scale internal localization. Constraint-based approaches [4, 3, 5] rely on strong prior assumptions about ensemble structure (e.g., lattices) or require exact observations to scale up to large ensembles. They are neither robust to noise nor well suited to irregular, non-lattice structures, common in some SRMRs. Local probabilistic approaches have been shown to be effective in localization of relatively small modular robots, such as PolyBot[6], but require assumptions of strong sensing, or robust mechanical latching to reduce errors in larger systems. Sparse approximation techniques [7, 8] used in simultaneous localization and mapping (SLAM), are effective in dealing with large amounts of noisy observations, but are difficult to apply to SRMRs, where modules are densely packed, forming grids and loops.

A more closely related problem is localization of wireless sensor networks. Here the nodes need to combine distance information about other nodes in order to accurately triangulate their positions. A standard formulation is to treat the distance information as weights of edges in a graph and obtain a Euclidean embedding, using methods such as regularized semidefinite programming relaxations [9, 10]. Internal localization can also be viewed as Euclidean embedding; however, only distances to immediate neighbors are known. As indicated by our experiments in Section VI, this restriction appears to impair the performance Euclidean embedding methods. Therefore, it is necessary to develop new techniques that are effective in this domain.

A key problem with applying incremental approaches like the ones seen in SLAM is that they can accumulate error, and this error takes a long time to resolve. In the case of a modular ensemble, the greatest error will tend to accumulate in a region with only a few inter-module observations, which we call a *weak region*. A substantial rotational uncertainty will be introduced in the partial solution, and will be magnified by subsequent additions. If we selectively incorporate the observations in the densely connected regions first, the partial solution will be constrained and the error will be substantially reduced. We use this intuition to formulate a hierarchical algorithm,

Fig. 1. Connectivity graph of ensemble with 8008 nodes, and resulting estimate of module positions; the results are accurate, subject to a rotation and translation of the coordinate space.

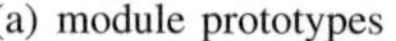
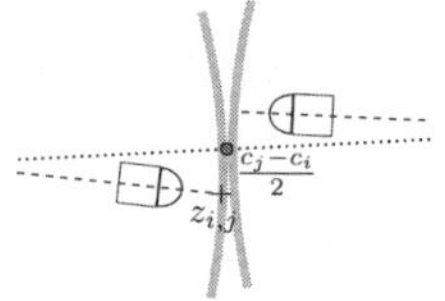

(a) module prototypes (b) sensor model

Fig. 2. (a) Sensor board from module prototype. (b) Sensor model, used in the paper. Each observation $z_{i,j}$ is represented as the location of the sensor, projected to the perimeter of the module. The circle indicates the midpoint of the two modules' centers. The model penalizes the module locations x_i and x_j, based on the distance between the midpoint and the observations $z_{i,j}$ and $z_{j,i}$.

where we recursively split the ensemble connectivity graph into well-connected components using normalized cut [11]. In order to keep the normalized cut computations tractable, we perform graph abstraction, analogous to over-segmentation in image segmentation [11].

A key challenge in internal localization is that observations are not stored centrally, and it is not feasible to collect the observations to a single node. This calls for a distributed approach, but the recursive nature of our algorithm and top-down partitioning make a distributed implementation difficult to achieve. We used a declarative, logical programming language called Meld to help address these issues and create an efficient distributed implementation of our algorithm.

We evaluate our algorithm on realistic 2D and 3D problems with up to 10,000 modules that accurately model unreliable observations and physical interactions among the modules. We demonstrate that the computational complexity of the approach is nearly linear in the size of the ensemble for a fixed ensemble structure, and outperforms methods from wireless sensor network localization based on classical multidimensional scaling [12] and semidefinite programming (SDP) relaxations [9, 10], as well as simpler incremental heuristics.

II. Localization of Modular Ensembles

We assume that the location of each module can be described by a small number of parameters, such as the coordinates of its center and orientation in space. In this paper, we focus primarily on circular and spherical modules in 2D and 3D space, respectively. Each module is equipped with sensors, such as infrared transmitters/receivers, that allow a pair of modules to detect when they are in close proximity. Such observations are inherently uncertain: two modules may be in sensing range, but not in physical contact, or a measurement can be made when sensors are not aligned. We do, however, assume that (i) the observations are symmetric (that is, whenever module i observes module j then module j also observes module i), and (ii) the modules know the identity of modules they sense (that is, we do not need to address the data association problem).

Figure 2(a) shows a current working prototype of a sensing subsystem fitting the properties described above. Each module shown has 8 IR transmitters and 16 IR receivers, oriented radially and spaced evenly around the circular perimeter. Note that for these modules, multipath interference, scattering, shadowing, and small dimensions effectively preclude techniques such as acoustic or radio time-of-flight-based localization.

III. Localization as Probabilistic Inference

In this section we define the probabilistic model that underlies our algorithm. We then discuss a simple incremental optimization method that motivates our approach.

A. Probabilistic Model With Attractive Potentials

We use a probabilistic model that describes the probability of a joint assignment of module locations $\mathbf{X} = (X_1, \ldots, X_N)$, given observations $\mathbf{Z}$ made by all modules in the ensemble. The location of each module i is represented by a vector, $X_i \triangleq (C_i, R_i)$, where C_i is the center of the module and R_i is its orientation (represented in 2D as an angle, and in 3D as a quaternion).

When two modules i and j are in the immediate neighborhood of each other, a pair of observations $(z_{i,j}, z_{j,i})$ is generated which represent the sensors at module i and j, respectively, that made the observation. We use a discrete model that captures whether two modules observe each other and with which sensors, but not the intensity of the readings. Also, for simplicity of notation, we assume that there is at most one pair of observations for every pair of modules, and we take $z_{i,j}$ to be the location of the sensor at module i, in module i's local reference frame (see Figure 2(b)). The model penalizes an observation $z_{i,j}$, based on how well it predicts the displacement between the two module centers.

$$\phi(x_i, x_j; z_{i,j}) \propto \exp\left\{-\frac{1}{2}\left\|r_i(z_{i,j}) - \frac{c_j - c_i}{2}\right\|_2^2\right\}. \tag{1}$$

Note that this model does not explicitly represent the constraint that the modules must not overlap; instead, we

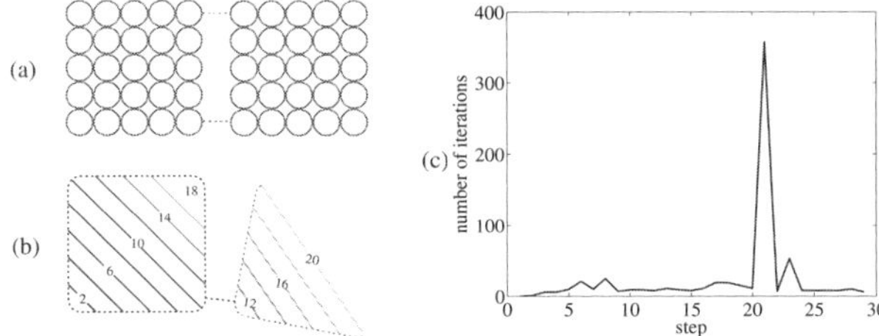

Fig. 3. (a) Ensemble, consisting of two tightly connected clusters. The clusters are connected by two pairs of observations. Within each component, modules make observations with all of their immediate neighbors, whereas the two components share only two observations, one at each side. (b) Intermediate result, obtained when incrementally conditioning on observations, starting from the lower left corner. The numbers indicate the order of conditioning. The solution accumulates substantial error; this error is not detected until loop closure, at step 21, and takes many iterations to resolve with conjugate gradient descent. (c) The number of iterations at each step to reach convergence.

have chosen to rely on the observations to obtain a non-overlapping solution. Alternatively, we could use a more accurate mode that captures properties of IR transmitters and receivers, such as quadratic decay and multi-modal response, but such a refinement is not key to the methods presented in this paper.

Combining the observation model (1) for each pair of neighboring modules i, j and instantiating the observations $z_{i,j}$ gives the likelihood of the joint state $\mathbf{x}$:

$$p(\mathbf{z}|\mathbf{x}) \propto \prod_{i,j} \phi(x_i, x_j; z_{i,j}). \tag{2}$$

For internal localization, we wish to compute the maximum likelihood estimate (MLE) of the location of all the modules, given all observations $\mathbf{z}$:

$$\mathbf{x}^* = \operatorname*{argmax}_{\mathbf{x}} p(\mathbf{z}|\mathbf{x}), \tag{3}$$

up to some global translation and rotation.

B. Computing the MLE Solution Incrementally

It is not easy to maximize the likelihood (2) directly, since the likelihood function is non-convex and high-dimensional. One approach is to compute the solution (3) incrementally, that is, compute the maximum likelihood estimate

$$\mathbf{x}_A^* = \arg\max p(\mathbf{z}_A|\mathbf{x}_A) = \arg\max \prod_{i,j \in A} \phi(x_i, x_j; z_{i,j})$$

for progressively larger connected sets of modules A. Here, $\mathbf{x}_A$ denotes the locations of the modules in A and $\mathbf{z}_A$ denotes all observations among the modules in A. At each step, the set $\mathbf{z}_A$ is expanded, incorporating modules connected to its perimeter, and then iteratively refining the position estimates with the correspondingly expanded set of observations.

Figure 3 illustrates the behavior such an incremental approach on a small ensemble with 200 modules that consists of two dense components. The observations are incorporate observations in breadth-first order, starting from the lower left corner. Figure 3(c) shows the running time of the algorithm at each step, expressed as the number of iterations of conjugate gradient descent until convergence. We see that while the number of iterations is typically small, it increases dramatically midway through the experiment when the observations close a loop, formed by the two square components. The computed solution accumulates error that takes a long time to resolve once the algorithm closes the loop.

IV. Guiding Localization With Normalized Cut

The experiment in Figure 3 points to an important drawback of an incremental maximum likelihood estimate (MLE) solution. Highly uncertain observations may be incorporated early, and errors magnified by subsequent module additions. With a simple MLE representation, this will remain undetected until a single observation closes the loop, at which point significant iterative computations are needed to shift the estimates back into low error bounds. However, if we were to first incorporate the observations within each dense cluster in Figure 3 and defer the observations that join the two clusters until later, the intermediate results would be more accurate, and would serve as a better starting solution for adding new observations. This suggests a hierarchical solution (Algorithm 1) that partitions the ensemble using clustering, recursively computes the estimates for each cluster, and uses the partial solutions to compute the globally optimal solution.

A. Determining an Effective Partition

Sparsely connected regions where only a few observations are made (*weak regions*) are one of the main sources of error and uncertainty as localization progresses. As illustrated in Figure 3, certain occurrences of weak regions can introduce a substantial rotational error – those that do not occur in pairs in 2D or triplets in 3D. An effective heuristic for identifying these weak regions is a cut on the connectivity graph of the ensemble: starting from a graph G, whose edges correspond to observations between modules, we seek to partition G, such that each component is well-connected and the inter-component observations are as few as possible. This criterion is effectively the one optimized in normalized cut [11]:

$$Ncut(A, B) = \frac{cut(A, B)}{assoc(A, V)} + \frac{cut(A, B)}{assoc(B, V)}, \tag{4}$$

where $Ncut(A, B)$ is the cut value, $cut(A, B)$ is the number of observations between module sets A and B, and $assoc(A, V)$ is the number of observations between the modules A and all modules in the graph. Minimizing (4) yields a partition of G into two components A, B. Our heuristic will first incorporate observations within the clusters A and B, and then the observations between A and B.[1]

Intuitively, normalized cut prefers partitions such that the number of observations between A and B is small, compared to all observations made by A and B. For example, in Figure 3a, the vertical cut that separates the two well-connected components has value $Ncut = O(\frac{1}{N})$, where N is the number of modules, whereas the value of the horizontal cut is $Ncut = O(\frac{1}{\sqrt{N}})$. Indeed, we see that normalized cut strongly discriminates these two orderings and yields the correct ordering.

B. Summary of the Algorithm

Our proposed approach is summarized in Algorithm 1. The algorithm starts by computing the normalized cut (A, B) for the connectivity graph G. By applying the localization procedure recursively, the algorithm computes a partial solution for the modules A, conditioned on all observations among A (in the algorithm description, G_A denotes the subgraph induced by A) and similarly for the modules B. We then use the partial solutions $\mathbf{x}_A^*$ and $\mathbf{x}_B^*$ to initialize the search for the optimum solution for the entire graph: we transform the observations between A and B, $\mathbf{z}_{A,B} \triangleq \{z_{i,j} : i \in A, j \in B\}$, into the global coordinate frame, using the module locations given by the partial solution $\mathbf{x}_A^*$; similarly for modules B. This procedure yields two sets of points $\mathbf{p} = \{p_i\}$ and $\mathbf{q} = \{q_i\}$, such that p_i and q_i are locations of matching observations in the global coordinate frame. Recall that the likelihood is maximal when sensors are in close proximity; thus, an effective initialization is to hold the relative locations of modules fixed within each cluster A, B, and compute the optimal rigid body transform between the clusters:

$$\arg \min_{R \in SO(d), t \in \mathbb{R}^d} \sum_k \|p_k - (Rq_k + t)\|_2^2, \quad (5)$$

where R is the rotation matrix (in 2D or 3D) and t is the translation vector. The optimal rigid body alignment (5) can be computed with closed-form solution in time linear in the number of observations between A and B [13]. This procedure yields an initial estimate of the locations of all modules, $\mathbf{x}_V^0$. The initial estimate is then refined using iterative methods, such as conjugate gradient descent or a quasi-Newton method.

[1] We selected binary partition, since as discussed below, the solutions between two clusters can be merged very efficiently.

Algorithm 1 $NormCutLocalize(G, V)$

if V is sufficiently small **then**
  compute $\arg\max p(\mathbf{x}_V|\mathbf{z}_V)$ using local heuristics
else
  Compute the normalized cut $(A, B) = NormCut(G)$
  $\mathbf{x}_A^* \Leftarrow NormCutLocalize(G_A, A)$
  $\mathbf{x}_B^* \Leftarrow NormCutLocalize(G_B, B)$
  $\mathbf{p} \Leftarrow$ transform the observations $\mathbf{z}_{A,B}$ into the coordinate frame, given by $\mathbf{x}_A^*$.
  $\mathbf{q} \Leftarrow$ transform the observations $\mathbf{z}_{B,A}$ into the coordinate frame, given by $\mathbf{x}_B^*$.
  Compute the optimal rigid alignment R, t:

$$\arg \min_{R \in SO(d), t \in \mathbb{R}^d} \sum_k \|p_k - (Rq_k + t)\|_2^2,$$

  Let $\mathbf{x}_V^0 = (\mathbf{x}_A^*, R\mathbf{x}_B^* + t)$.
  $\mathbf{x}_V^* \Leftarrow \arg\max p(\mathbf{x}_V|\mathbf{z}_V)$, starting from $\mathbf{x}_V^0$

C. Scaling Up the Solution

While the normalized cut formulation yields an effective sequence in which observations should be incorporated, computing the exact normalized cut is costly and dominates other operations. Specifically, the cost of the rigid alignment is linear in the total number of observations, whereas the complexity of computing a *single* normalized cut is $O(|V|^{3/2})$, where $|V|$ is the number of nodes [11]. A standard method to decrease the computational complexity is to compute an abstraction of the graph, using a simpler clustering algorithm, such as k-means [11]. In particular, in image segmentation, this procedure amounts to computing an over-segmentation of the image. The normalized cut is then computed on a smaller graph G', where each node of G' corresponds to a cluster of nodes in the original graph G.

Compared to other clustering tasks, the clustering task in localization is simpler in two ways. First, unlike in applications, such as image segmentation, where shifting the cut can adversely affect the visual quality of the segmentation, the clustering here is only used as a heuristic, and offsetting the cut does not substantially decrease the quality of location estimates. Furthermore, since the connectivity graph G has unit edge weights, the cut value itself increases at most linearly (in 2D) or quadratically (in 3D) in the number of hops away from the optimal cut. Therefore, we have found that it is often sufficient to partition the graph greedily into a fixed number of components. As discussed in Section VI, using as few as twenty components yields accurate solutions (the actual number of needed components will depend on the amount of uncertainty in the ensemble).

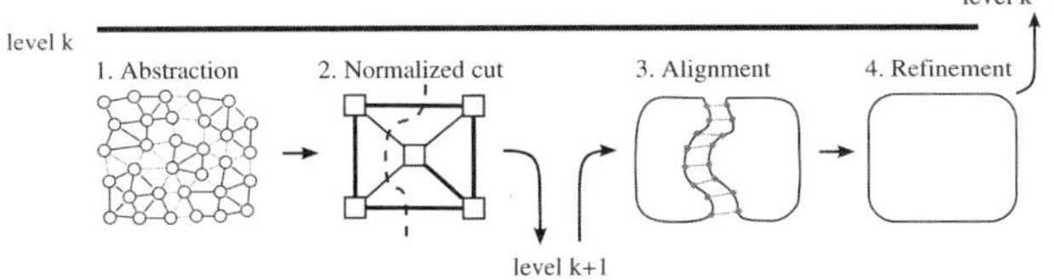

Fig. 4. Control flow for level k of the distributed implementation

V. Distributed Localization

While centralized localization in a self-reconfigurable modular robot is useful, a distributed localization is much more appealing, since it can significantly reduce the communication cost, enables online control, and avoids a centralized point of failure. In this section, we propose a distributed version of Algorithm 1 that uses a combination of data aggregation techniques and local refinement steps to compute each module's own location. In combination with a declarative programming language [14], we obtain a fully executable distributed solution.

A. Localization through Aggregation and Dissemination

Our distributed solution mirrors the operations of the centralized algorithm. Figure 4 summarizes the control flow for one level of the algorithm execution. The first three steps – graph abstraction, normalized cut, and rigid body alignment – use data aggregation techniques and perform key steps of the computation at the group leader. Note that the gradient of the log-likelihood (2) decomposes linearly over the nodes of the cluster and their neighbors. Therefore, the iterative refinement step can be performed locally, without global coordination.

In order to compute the graph abstraction, we use a random sampling strategy that partitions the graph into a set of connected subgraphs, centered around randomly chosen leaders. Each node elects itself as a leader with a small probability, and the nodes greedily join the nearest leader (as measured by the hop-count). The description of the abstracted graph is then aggregated to a single node that computes the normalized cut using a standard centralized implementation. Since we need to perform normalized cut only on small graph abstractions, a centralized implementation is sufficient. Alternatively, one could use a distributed algorithm based on decentralized power iteration. [15]

A key step in Algorithm 1 is computing the optimal rigid body alignment between two sides of the partition. While, at first glance, it is not clear how to distribute this step, a closer look at the method in [13] reveals that the method only depends on the first- and second-order statistics of the points $\{(p_i, q_i)\}$ in Equation 5. These statistics can be aggregated from the boundary towards the group leader. The leader then computes the optimal transform and disseminates the result. Since the aggregated information depends only on the dimensionality of the aligned points (2 or 3), rather than their count, the communication cost of aggregating and disseminating the optimal transform is small.

B. Declarative Implementation using Meld

The distributed algorithm described in the previous section presents a number of implementation challenges. Unlike simple message-passing style inference algorithms found commonly in literature [16], our localization approach uses multiple aggregation and dissemination steps that require both local and non-local communication across the ensemble and operate in asynchrony. These steps rely on a number of data structures, used to represent the graph, the location, and the rigid body transform statistics. Due to the recursive nature of the algorithm, the implementation may need to maintain parallel data structures for all of the concurrently active levels. These challenges make it tedious to implement the algorithm in a standard message-passing framework. In this section, we briefly outline our implementation that uses Meld [14], a logical, declarative, high-level programming language for modular robots.

Meld is a declarative language with syntax similar to Prolog. A Meld program consists of rules that specify sufficient preconditions to derive new facts from existing ones. A key benefit of Meld is that it lets the programmer focus on the logical, information processing aspects of an algorithm, while automatically taking care of the mechanics of distributed programming, such as communication. For example, a simple distributed spanning tree algorithm can be specified in two rules: a rule that determines the root of the tree, and a rule that lets a node join a tree that extends to one of its neighbors. In a similar manner, Meld simplifies implementation of other distributed data structures.

We found that many features of Meld fit well with the needs of our algorithm, but also exposed some drawbacks of our approach. The language let us naturally represent the graph abstraction process and aggregate and disseminate sufficient statistics for the rigid alignment. The results of different phases were easily chained together. Furthermore, when intermodule connections were lost or network layout changed, Meld was able to automatically recover, recomputing the relevant portions of these distributed data structures and rerunning parts of the localization algorithm. On the downside, Meld's declarative programming model made it more difficult to express certain imperative sequences and loops. More fundamentally, a change in or removal of one fact (for example, the origin of the coordinate system) may trigger the removal and subsequent rederivation of a large number of other facts. This drawback is inherent to our localization algorithm and is subject to ongoing research.

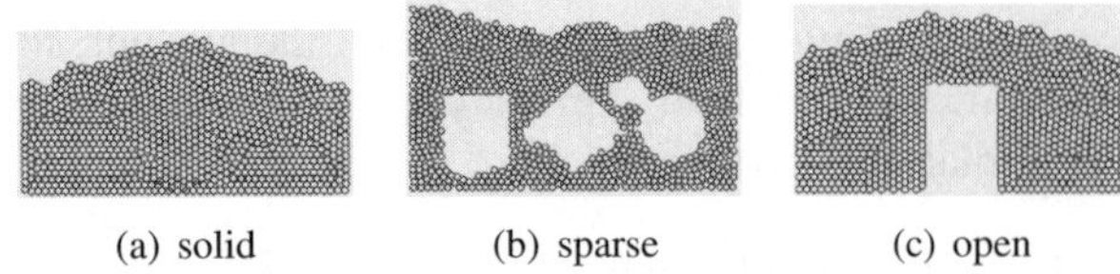

(a) solid (b) sparse (c) open

Fig. 5. Scenarios used in our experiments. The scenarios were generated by settling randomly inserted modules in a gravitational field.

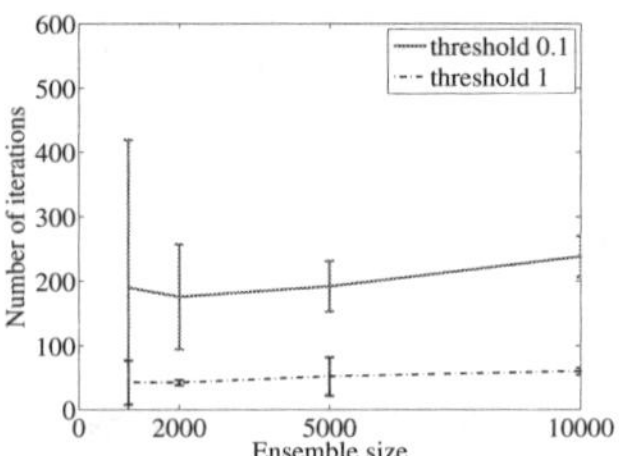

Fig. 6. The average number of iterations per module for the triple scenario as the number of modules grows. At different scales, the ensemble retains its shape and the proportions. With a threshold of 1.0, the average location RMS error was 1.26; with a threshold of 0.1, the average location RMS error was 0.80.

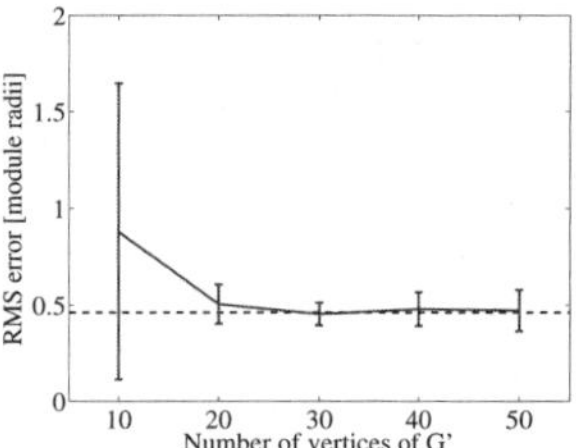

Fig. 7. The location RMS error for the triple scenario with 2000 modules, when using the normalized cut approximation in Section IV-C. The horizontal dashed line indicates the fidelity of the solution, obtained with exact normalized cut.

VI. Experimental Results

In this section, we present experimental results that illustrate both the centralized and the distributed aspects of our solution. We generated input scenarios with a C++ simulator [17] that models IR sensing and physical interactions between the modules. Each module in the simulation had 12 IR transceivers (colocated emitter/detector pairs), whose IR response was modeled according to an inverse-square law, similar to the model in [18]. The threshold for detecting observations between a pair of neighboring nodes was set to 20 per cent of the peak intensity. At this setting, a sensor can report a connection even if the modules are not in a physical contact and if the transmitter and the receiver are not perfectly aligned.

A. Scenarios

We constructed both 2D and 3D test ensembles. The 2D ensembles were generated by randomly settling simulated spheres under a simulated gravity field into a fixed container of the desired overall shape. The results were configurations with realistic, irregular, largely amorphous structures. Several of these configurations, illustrated in Figure 5 mimic planar slices of a 3D shape capture scenario [4]. Each shape in Figure 5 was instantiated ten times, with different initial velocities and locations of the modules, allowing us to average results across repeated runs using configurations very similar in overall shape but where module connectivity and spacing varies. The 3D test ensembles were generated by rasterizing 3D outlines (defined by OBJ files from various shape libraries) into a designated target lattice, either hexagonal-close-packed or cubic.

B. Scalability

In the first experiment, we evaluated the performance of the proposed method as the number of modules in an ensemble increases. We selected the structured triple scenario in Figure 5(b) and formed a set of progressively larger ensembles. At each scale, the ensemble retains the same overall shape and the proportions, but the number of modules that form the shape increases. We then run Algorithm 1 such that, at each level of the hierarchy, the estimate $\mathbf{x}_A^*$ reaches a fixed level of accuracy, as measured by the norm of the gradient of the likelihood function at $\mathbf{x}_A^*$. This procedure ensures that each estimate $\mathbf{x}_A^*$ is sufficiently accurate, before it is used at the higher level. Figure 6 shows the average number of iterations in preconditioned conjugate gradient descent as a function of the number of modules. We see that the number of iterations, needed to attain the same accuracy (as measured by the gradient norm), increases very slowly.

C. Sensitivity to Abstraction

In the second experiment, we evaluated the sensitivity of the proposed localization method to errors, introduced by performing normalized cut on the abstracted, rather than the original connectivity graph. We took the structured scenario in Figure 5(b) with 2000 modules. Figure 7 shows the root mean square (RMS) error as we vary the number of nodes in the abstraction of the connectivity graph (since we controlled the maximum diameter of clusters, rather than their count, the displayed node count is approximate). In order to account for the overlap, introduced by the objective (1), we uniformly scale the locations of the modules, so that the average spacing equals the module diameter. Then, using the ground truth locations of the modules, we compute an optimal rigid alignment and report the error for the aligned solution. We see that the performance of the proposed localization method is insensitive to abstraction errors: with 20 or more nodes in the abstracted graph, the approach yields a sufficiently small RMS error. These results suggest that small graph abstractions provide meaningful results and can be analyzed at each leader node centrally.

D. Performance in 3D

In the third set of experiments we extended the algorithm to three dimensions, using a quaternion representation for each catom's 3D orientation rather than the scalar orientation parameter used in the 2D case. Since the quaternion may become unnormalized in the process of gradient descent computations, we implicitly

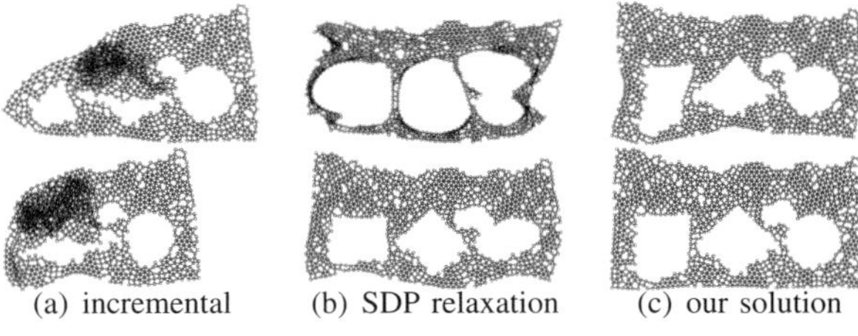

(a) incremental (b) SDP relaxation (c) our solution

Fig. 8. Example results using three algorithms on the triple scenarios. Lower images reflect results after additional iterative refinement steps.

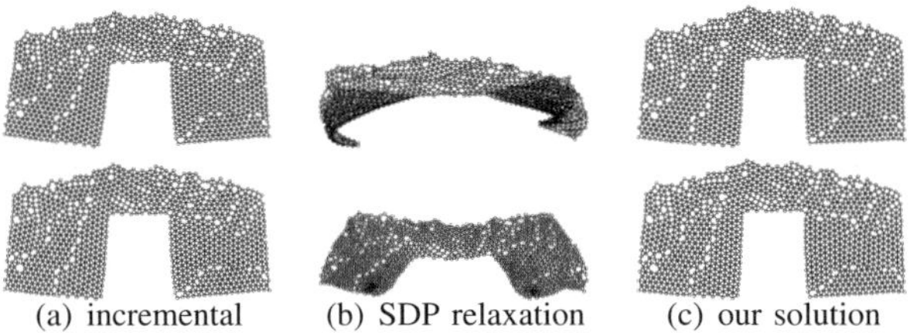

(a) incremental (b) SDP relaxation (c) our solution

Fig. 9. Example results using three algorithms on the open scenarios. Lower images reflect results after additional iterative refinement steps.

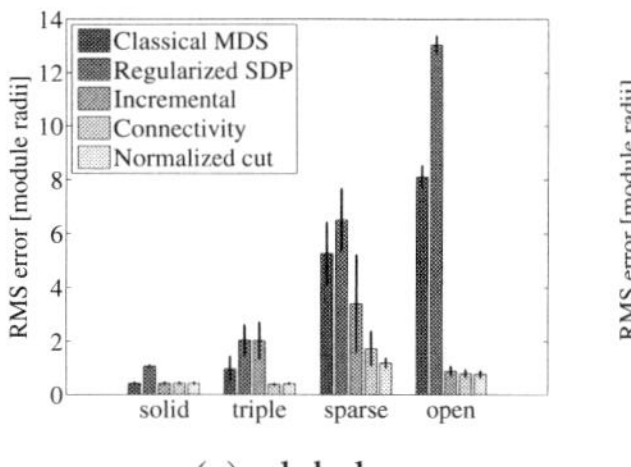

(a) global

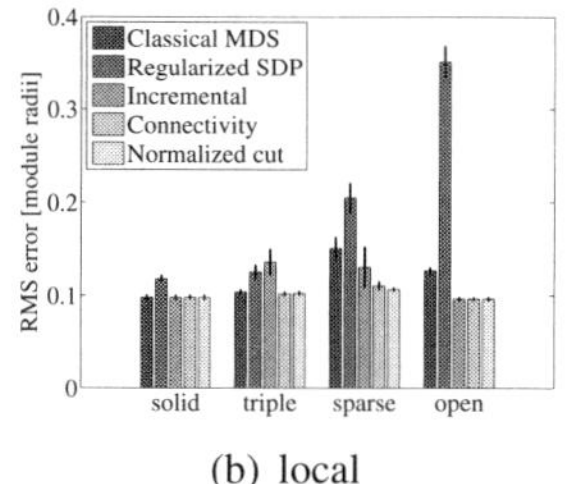

(b) local

Fig. 10. RMS error of the location estimates. (a) Global RMS error, averaged over all modules. (b) RMS error of modules relative to their neighbors. Here, we greedily partition the ensemble into connected regions with diameter of 6 modules or less and compute the RMS error using the optimal rigid alignment for each region.

normalize the quaternion in the observation model (1) and in the corresponding gradient.

We generated lattice-bound test ensembles from various 3D outlines. (The lattice-bound property of these ensembles was due to limitations of our ensemble-generation code.) We simulated spherical modules each with 50 sensors scattered across their surfaces. Despite the larger number of sensors when compared to the 2D case, the available angle constraints in the 3D test cases were generally much weaker. We found that our algorithm was capable of determining positions in these 3D tests very accurately, within 1 module radius. Figure 1 shows an example of the results obtained on a large 3D scenario with 8008 modules.

E. Comparison with Prior Work

In the fourth set of experiments, we compared the performance of the proposed algorithm to Euclidean embedding methods, used in wireless sensor network localization, as well as simpler incremental heuristics. Figures 8 and 9 show examples of how incremental and semidefinite programming approaches qualitatively perform poorly compared to our algorithm, even after applying significant number of iterative refinement steps. SDP in particular suffers from overestimation of distances, and artifacts due to projection to a 2D space from a manifold in a higher dimensional space.

Quantitatively, we evaluated the following methods: (i) classical multi-dimensional scaling [12], (ii) the inequality formulation of regularized semidefinite programming [9], (iii) the simpler incremental approach, discussed in Section III-B, (iv) a simple hierarchical approach that merges pairs of clusters bottom-up, in the order given by their algebraic connectivity, and (v) the proposed method, using exact normalized cut. We perform repeated experiments on the scenarios in Figure 5 with 1000 modules. The initial solution, obtained by each method is refined with 300 iterations of preconditioned conjugate gradient descent.

Figure 10 shows the average RMS error for each scenario. We see that approaches, based on Euclidean embedding (classical MDS, regularized SDP) generally do not perform very well in this setting, especially for the sparse version of the triple scenario and the large open-loop scenario. For classical multi-dimensional scaling, the error results from approximating true distances with hop-count; for regularized SDP, the errors come either from the SDP relaxation or the underlying solver. The incremental and simple hierarchical approaches perform better, but are outperformed by our normalized cut formulation on the scenarios with non-homogeneous structure (triple, sparse). It is worth noting that the Euclidean embedding methods are substantially more computationally expensive: an optimized implementation of a state-of-the-art SDP relaxation method [10] takes 5-10 minutes to run on an input with 5000 nodes, whereas the Matlab implementation of our hierarchical algorithm runs in less than a minute.

F. Distributed Results

Finally, we evaluate the message complexity of the distributed implementation. Figure 11 shows the average number of messages sent by each module as a function of the ensemble size. The figure confirms that the number of messages per module required by the algorithm increases only logarithmically in the total number of modules in the ensemble. Thus, the implementation scales to large ensembles. Figure 12 shows the split of the messages among different components of the algorithm for two ensemble sizes. Interestingly, the messages sent are dominated by the gradient descent. In particular, the extra work done to determine the normalized cut is small compared to the cost of iterative refinement.

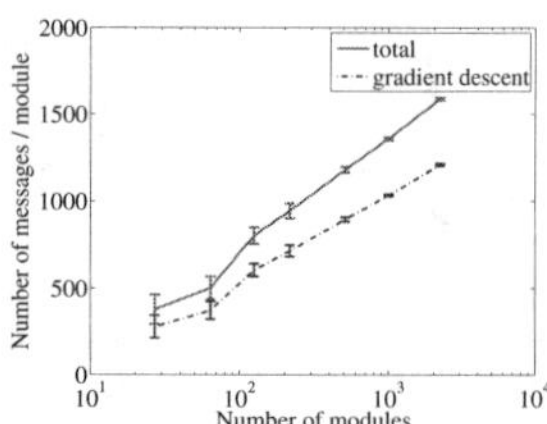

Fig. 11. The number of messages per module as a function of the ensemble size.

Procedure / Test case	$5 \times 5 \times 5$	$10 \times 10 \times 10$
Neighbor detection	0.5%	0.3%
Graph abstraction	7.7%	7.3%
Normalized cut	6.4%	6.5%
Rigid alignment	9.7%	9.5 %
Gradient descent	75.8%	76.3 %

Fig. 12. The relative number of messages sent by each component.

VII. Discussion and Future Work

In this paper, we examine large-scale localization in modular robot ensembles using uncertain, local observations. We formulate internal localization as a probabilistic inference problem and introduce a novel approach which hinges on selection of an effective ordering of observations using a normalized cut criterion. In combination with closed-form solutions for rigid alignment and simple graph abstraction scheme, this approach leads to accurate, scalable solutions. We perform an extensive evaluation of our proposed approach on a test suite of realistic 2D and 3D configurations with up to 10,000 nodes and demonstrate that our approach outperforms both recent methods using Euclidean embedding and simpler heuristics. Finally, we describe a fully distributed implementation of our algorithm that computes the results by sending only a few messages between the nodes.

While this paper goes a long way towards robust internal localization, there are some questions left to be answered. For example, it would be interesting to examine the merits of fast iterative methods developed for SLAM, such as [19]. These methods may make it possible to quickly recover from small changes in the ensemble and cope with dynamic settings. More broadly, one may hope to combine the normalized cut heuristic with additional structure in the problem to obtain a fully probabilistic solution. Such an approach would recover not only a point estimate, but also its uncertainty. Addressing both these questions would lead to truly robust solution to internal localization in modular robotics and drive research in other fields.

Acknowledgment

The authors thank Rahul Sukthankar and Dhruv Batra for their valuable feedback. We also thank Casey Helfrich and Michael Ryan for building the DPRSim simulator which was used for many of our experiments. Finally, we thank Rahul Biswas for an implementation of the regularized SDP localization algorithm.

References

[1] S. C. Goldstein and T. Mowry, "Claytronics: A scalable basis for future robots," in *Robosphere*, Nov 2004.

[2] K. Gilpin, K. Kotay, and D. Rus, "Miche: Modular shape formation by self-dissasembly." in *ICRA*, 2007, pp. 2241–2247.

[3] P. White, V. Zykov, J. Bongard, and H. Lipson, "Three dimensional stochastic reconfiguration of modular robots," in *Proceedings of Robotics Science and Systems*, 2005.

[4] P. Pillai, J. Campbell, G. Kedia, S. Moudgal, and K. Sheth, "A 3D fax machine based on claytronics," in *IEEE/RSJ Int'l Conf. on Intelligent Robots and Systems*, Oct. 2006, pp. 4728–35.

[5] G. Reshko, "Localization techniques for synthetic reality," Master's thesis, Carnegie Mellon University, 2004.

[6] M. Yim, D. Duff, and K. Roufas, "PolyBot: a modular reconfigurable robot," in *Proceedings of IEEE Intl. Conf. on Robotics and Automation (ICRA)*, 2000.

[7] M. A. Paskin, "Thin junction tree filters for simultaneous localization and mapping," in *Proc. IJCAI*, 2003.

[8] U. Frese and L. Schroder, "Closing a million-landmarks loop," in *Intelligent Robots and Systems, 2006 IEEE/RSJ International Conference on*, Oct. 2006.

[9] P. Biswas, T.-C. Lian, T.-C. Wang, and Y. Ye, "Semidefinite programming based algorithms for sensor network localization," *ACM Transactions on Sensor Networks (TOSN)*, vol. 2, no. 2, pp. 188–220, May 2006.

[10] Z. Wang, S. Zheng, S. Boyd, and Y. Yez, "Further relaxations of the SDP approach to sensor network localization," Stanford University, Tech. Rep., 2006.

[11] J. Shi and J. Malik, "Normalized cuts and image segmentation," *IEEE Transactions on Pattern Analysis and Machine Intelligence*, vol. 22, no. 8, pp. 888–905, 2000.

[12] Y. Shang, W. Ruml, Y. Zhang, and M. P. J. Fromherz, "Localization from mere connectivity," in *Proceedings of the 4th ACM international symposium on Mobile ad hoc networking & computing*. ACM Press New York, NY, USA, 2003, pp. 201–212.

[13] S. Umeyama, "Least-squares estimation of transformation parameters between two point patterns," *IEEE Transactions on Pattern Analysis and Machine Intelligence*, vol. 13, no. 4, April 1991.

[14] M. Ashley-Rollman, S. C. Goldstein, P. Lee, T. Mowry, and P. Pillai, "Meld: A declarative approach to programming ensembles," in *Proceedings of IEEE/RSJ Conference on Intelligent Robots and Systems (IROS)*, 2007.

[15] P. Yang, R. A. Freeman, G. Gordon, K. Lynch, S. Srinivasa, and R. Sukthankar, "Decentralized estimation and control of graph connectivity in mobile sensor networks," in *American Control Conference*, June 2008.

[16] C. Crick and A. Pfeffer, "Loopy belief propagation as a basis for communication in sensor networks," in *Proceedings of the Nineteenth Conference on Uncertainty in Artificial Intelligence (UAI-2003)*, C. Meek and U. Kjærulff, Eds. San Francisco: Morgan Kaufmann Publishers, Inc., 2003.

[17] "Dprsim: The dynamic physical rendering simulator," http://www.pittsburgh.intel-research.net/dprweb/.

[18] K. D. Roufas, Y. Zhang, D. G. Duff, and M. H. Yim, "Six degree of freedom sensing for docking using IR LED emitters and receivers," in *Proceedings of International Symposium on Experimental Robotics VII*, 2000.

[19] G. Grisetti, S. Grzonka, C. Stachniss, P. Pfaff, and W. Burgard, "Efficient estimation of accurate maximum likelihood maps in 3d," in *Intelligent Robots and Systems, 2007. IROS 2007. IEEE/RSJ International Conference on*, 29 2007-Nov. 2 2007.

Controlling Shapes of Ensembles of Robots of Finite Size with Nonholonomic Constraints

Nathan Michael and Vijay Kumar
University of Pennsylvania
Philadelphia, Pennsylvania 19104-6228
Email: {nmichael, kumar}@grasp.upenn.edu

Abstract—**In this paper we focus on the construction of distributed formation control laws that permit the control of individual mobile ground robots in a formation to a desired distribution with minimal knowledge of the global state. As in previous work, we consider an abstraction of the team that is derived from a shape descriptor of the ensemble and the position and orientation of the ensemble. We consider the control of the abstract state with decentralized control laws which are independent of the number of agents. However, we incorporate an important departure from previous work by explicitly modeling the shape of the robot, the geometric, non-interpenetration constraints and nonholonomic, kinematic constraints. Further, we propose a motion planning technique to plan motions for ensembles of robots and a technique for the splitting and merging of groups and subgroups. We demonstrate the effectiveness of the algorithms on a team of differential drive robots in simulation and on real hardware.**

I. Introduction

Effective strategies for controlling large teams of robots in complex environments are becoming increasingly relevant as the development of pervasive embedded computing, sensing, and wireless communication enables the application of multi-agent systems to challenging tasks such as environmental monitoring [1], surveillance and reconnaissance for security and defense [2], and support for first responders in a search and rescue operation [3]. In such scenarios, it is necessary to apply control strategies that allow robots to adapt to different environments and execute complex tasks, while avoiding collisions. Further, robot controllers must be robust to permit robot failures or changes in the team size.

Several methodologies exist to control large teams of robots. One way of reducing the complexity of the controller is to require the team to conform to a geometric rigid virtual structure [4]. Most of the recent works on stabilization and control of virtual structures model formations using *formation graphs* [5]. The controllers guaranteeing local asymptotic stability of a given rigid formation can be derived using standard techniques such as input-output linearization [6], input-to-state stability [7], Lyapunov energy-type functions [8], and biologically-inspired artificial potential functions [9]. Virtual structures unnecessarily constrain the problem, making this approach inappropriate for tasks in complex environments. Additionally, graph formulations and leader-follower architectures require identification and ordering of robots, which makes the overall architecture sensitive to failures.

The problem of controlling the trajectory of the group and shape of a large team of point robots was studied in [10], [11]. The authors defined an abstraction of the team that has a product structure of the Euclidean group and a shape space, and is independent of the number of robots. The group captures the pose of an ellipsoid spanning the team with semi-axes given by the shape variables. The overall abstract description is invariant to robot permutations. In addition, the model and the formulation is invariant to left actions of the group. This description allows one to define and control the behavior of the abstract state or the abstract description of the team at a high level, with automatic generation of individual robot control laws based only on the feedback of this abstract state. However, the control laws do not account for the physical constraints of the robots and ignore inter-agent interactions.

Coverage control schemes proposed by [12], [13] and their variants have a similar flavor. They enable large groups of robots to use local information to distribute themselves so that a suitable integral over this distribution is maximized. However, this formulation does not lend itself to the control of the position and orientation of the overall team.

In this paper we focus on a basic problem, the control of the position and orientation of a formation of mobile robots and the adaptation of the shape to the environment. Two related problems, the planning of the shape and trajectory of the ensemble and the development of effective coordination strategies to split the team into subgroups and to merge two subgroups, are also considered. We view these problems and their solutions to be building blocks that can enable a robot team to navigate an environment, adapting to the constraints imposed by obstacles in the environment. In contrast to most previous work we model the physical shape of the robot and consider controllers that are guaranteed to avoid collisions between the robots. Although the approach assumes global observation of the abstract state, it is possible to develop an estimator for individual robots to estimate this abstract state [14]. However, in the context of our work, we emphasize the control of the team of robots and decouple the challenges of considering the estimation problem. Further, we believe that some element of centralization is essential to command a large team of robots.

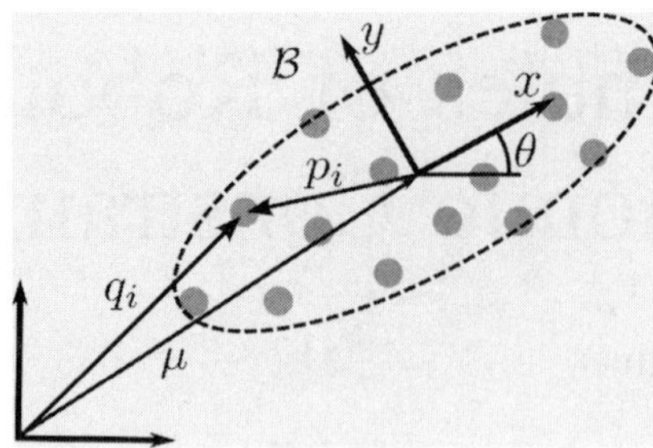

Fig. 1. The frame $\mathcal{B}$ fixed to the group of robots moves with respect to the inertial frame.

II. Background

This paper builds on the previous work in [10], [11], [15]. In these papers, an *abstraction* map is used to transform the high-dimensional state space into a smaller, tractable state space which captures only the position, orientation, and shape of the formation. The main advantages of this abstract representation are: (a) its dimension is independent of the number of robots in the team; and (b) it lends itself to planning in a lower dimensional space.

The state space of the N-robot system is constructed by creating N copies of Q_i, the state space of the ith robot:

$$Q = Q_1 \times Q_2 \times \ldots \times Q_N.$$

The abstract space, M, whose dimension is smaller and independent of the dimension of Q, is defined by a smooth, differentiable map

$$\phi : Q \to M, \quad \phi(q) = \boldsymbol{x}, \tag{1}$$

where ϕ is a mapping of the higher-dimensional state $q \in Q$ to the lower-dimensional abstract state $\boldsymbol{x} \in M$. In this paper, we consider kinematic robots in the plane (see [11] for a treatment of the three-dimensional case). Thus, q reflects the collection of the positions of the robots, $q = [q_1, \ldots, q_i, \ldots, q_N]^{\mathrm{T}}$ where $q_i \in \mathbb{R}^2$.

As in previous work, the shape is modeled by characterizing the distribution of robots about the mean position. The centroid of the group is given by:

$$\mu = \frac{1}{N}\sum_{i=1}^{N} q_i.$$

We can define a local frame, $\mathcal{B}$, whose origin is at the centroid, as shown in Fig. 1, by requiring the orientation to be such that the coordinates of the robots in this frame, $p_i = [x_i, y_i]$, satisfy

$$\sum_{i=1}^{N} x_i y_i = 0.$$

The distribution of robots in this local frame can be approximated by the inertia tensor (assuming uniform unit mass) or by a matrix of second moments:

$$\mathcal{I} = \sum_{i=1}^{N} p_i p_i^T = \begin{bmatrix} \mathcal{I}_{11} & 0 \\ 0 & \mathcal{I}_{22} \end{bmatrix}.$$

We define two shape variables proportional to the diagonal elements:

$$s_1 = \kappa \mathcal{I}_{11}, \quad s_2 = \kappa \mathcal{I}_{22},$$

where $\kappa \neq 0$. Choosing $\kappa = \frac{1}{N-1}$ gives the shape variables a geometric interpretation. They become the semi-major and semi-minor axes for a concentration ellipse for a group of robots whose coordinates in the plane are chosen to satisfy a normal distribution. Alternatively, the shape variables may be defined such that abstract shape is described by a bounding rectangle as discussed in [10].

The abstract description of the team of robots, $\boldsymbol{x}$, is given by the position and orientation of the team, g, and the shape s. In this paper, we take g to be the position and orientation of $\mathcal{B}$:

$$g = \begin{bmatrix} \cos\theta & -\sin\theta & \mu_1 \\ \sin\theta & \cos\theta & \mu_2 \\ 0 & 0 & 1 \end{bmatrix}$$

where $\mu = (\mu_1, \mu_2)$ are the components of the centroid in the inertial frame and the shape $s = (s_1, s_2)$. The map ϕ defined in this way can be easily shown to be a submersion [10].

The abstract space, M, is naturally decomposed into a shape space, S, and a Lie group, G, which in our case is $SE(2)$. Since ϕ is a submersion, it follows that there is a unique $\dot{x}$ for every $\dot{q}$ but not the other way around.

Using the natural kinetic energy metric on Q, it is possible to derive the optimal velocity (tangent vector) at any point $q \in Q$ for a desired $\dot{\boldsymbol{x}}$ at the corresponding point $\boldsymbol{x} = \phi(q) \in M$. It was shown in [10] that this input $u^\star$, for the system can be found by considering the time derivative of the transformation described by (1),

$$d\phi \dot{q} = \dot{\boldsymbol{x}}. \tag{2}$$

From (1), the definitions of (μ, θ, s_1, s_2), and algebraic simplification, the transformation $d\phi$ becomes

$$d\phi = \kappa \begin{bmatrix} \frac{1}{\kappa N} I_2 & \cdots & \frac{1}{\kappa N} I_2 \\ \frac{(q_1-\mu)^{\mathrm{T}}}{s_1 - s_2} H_3 & \cdots & \frac{(q_N-\mu)^{\mathrm{T}}}{s_1 - s_2} H_3 \\ (q_1 - \mu)^{\mathrm{T}} H_1 & \cdots & (q_N - \mu)^{\mathrm{T}} H_1 \\ (q_1 - \mu)^{\mathrm{T}} H_2 & \cdots & (q_N - \mu)^{\mathrm{T}} H_2 \end{bmatrix}. \tag{3}$$

Here H_1, H_2, and H_3 are defined by

$$H_1 = I_2 + R^2 E_2, \quad H_2 = I_2 - R^2 E_2, \quad H_3 = R^2 E_1,$$

where I_2 is the 2×2 identity matrix and

$$E_1 = \begin{bmatrix} 0 & 1 \\ 1 & 0 \end{bmatrix}, \quad E_2 = \begin{bmatrix} 1 & 0 \\ 0 & -1 \end{bmatrix}.$$

In [10] (and the extension to three-dimensions in [11]) it was shown that the minimum-energy solution satisfying (2) is obtained using the Moore-Penrose Inverse:

$$u^\star = d\phi^{\mathrm{T}} (d\phi d\phi^{\mathrm{T}})^{-1} \dot{\boldsymbol{x}}. \tag{4}$$

Further algebraic simplification of (4) using (3) results in the control law for each individual agent, $u_i = \dot{q}_i$,

$$\begin{aligned} u_i^\star = \dot{\mu} &+ \frac{s_1 - s_2}{s_1 + s_2} H_3 (q_i - \mu) \dot{\theta} \\ &+ \frac{1}{4 s_1} H_1 (q_i - \mu) \dot{s}_1 + \frac{1}{4 s_2} H_2 (q_i - \mu) \dot{s}_2. \end{aligned} \tag{5}$$

Although the description of shape variables is fairly simple, it is generalizable to include higher moments (beyond second order). However, the development of minimum-norm control inputs, such as (5), are harder with more general shapes.

In the next section, we will pursue a slightly different formulation by writing these equations in the moving frame $\mathcal{B}$.

III. Problem Formulation

A. Dynamics in the moving frame

At any point $\boldsymbol{x} = (g, s) \in M$ in the abstract space, the derivative can be written as:

$$\dot{\boldsymbol{x}} = \begin{bmatrix} \dot{g} \\ \dot{s} \end{bmatrix} = \begin{bmatrix} g & 0 \\ 0 & I_2 \end{bmatrix} \begin{bmatrix} \xi \\ \sigma \end{bmatrix}. \tag{6}$$

$\dot{\boldsymbol{x}} = (\dot{g}, \dot{s})$ is the time derivative of the abstract space in the inertial frame while $\zeta = (\xi, \sigma)$ is the time derivative in the moving frame $\mathcal{B}$, and

$$\Gamma = \begin{bmatrix} g & 0 \\ 0 & I_2 \end{bmatrix}$$

is a non-singular 5×5 transformation matrix. If ν_i is the robot velocity in the frame $\mathcal{B}$ so that $u_i = R\nu_i$,

$$\kappa \begin{bmatrix} \frac{I_2}{\kappa N} & \cdots & \frac{I_2}{\kappa N} \\ \frac{1}{s_1 - s_2} {p_1}^{\mathrm{T}} E_1 & \cdots & \frac{1}{s_1 - s_2} {p_N}^{\mathrm{T}} E_1 \\ {p_1}^{\mathrm{T}}(I_2 + E_2) & \cdots & {p_N}^{\mathrm{T}}(I_2 + E_2) \\ {p_1}^{\mathrm{T}}(I_2 - E_2) & \cdots & {p_N}^{\mathrm{T}}(I_2 - E_2) \end{bmatrix} \begin{bmatrix} \nu_1 \\ \vdots \\ \nu_N \end{bmatrix} = \begin{bmatrix} \xi \\ \sigma \end{bmatrix}.$$

The minimum-energy solution (5) can be written as:

$$\nu^\star = d\phi^{\mathrm{T}}(d\phi d\phi^{\mathrm{T}})^{-1}\zeta, \tag{7}$$

with the simplification:

$$\nu_i^\star = \begin{bmatrix} \xi_1 \\ \xi_2 \end{bmatrix} + \frac{s_1 - s_2}{s_1 + s_2} E_1 p_i \xi_3 + \frac{1}{4s_1}(I_2 + E_2)p_i\sigma_1 + \frac{1}{4s_2}(I_2 - E_2)p_i\sigma_2. \tag{8}$$

The control law defined by (8) does not consider inter-agent collisions or the spatial size of individual robots. Although [15] proposed an extension that resolved collisions, the strategy requires communication and negotiations during collisions, which adds unnecessary complexity. In the next subsection we address an approach in which collision avoidance is done without explicit arbitration.

B. Collision avoidance

The separation distance between the reference points on robots i and j is:

$$\delta_{ij} = \|p_i - p_j\|.$$

To avoid collisions between robots, we define a safe separation distance between two robots:

$$\epsilon = 2\rho + \epsilon_s, \tag{9}$$

where ρ is the radius of each robot and ϵ_s is a specified safety region.

We define the neighborhood $\mathcal{N}_i$ as the set of all robots sensed by or communicating with robot i such that i is able to gain knowledge of its neighbors' positions and velocities, $\{p_j, \nu_j\}, \forall j \in \mathcal{N}_i$. To ensure that the robots do not collide, we require that

$$(p_i - p_j) \cdot (\nu_i - \nu_j) \geq 0, \tag{10}$$

for all $j \in \mathcal{N}_i$ such that $\delta_{ij} \leq \epsilon$.

C. Asymptotic convergence to a desired abstract state

In the absence of collisions, the easiest way to guarantee convergence to a time-invariant abstract state $\boldsymbol{x}^{des}$ is to require the error $\tilde{\boldsymbol{x}} = (x^{des} - x)$ to converge exponentially to zero:

$$\dot{\boldsymbol{x}} = K\,\tilde{\boldsymbol{x}},$$

or equivalently,

$$\zeta = \Gamma^{-1} K\,\tilde{\boldsymbol{x}}, \tag{11}$$

where K is any positive-definite matrix, and use (7, 8) to obtain robot velocities that guarantee globally asymptotic convergence to any abstract state.

We next discuss the first contribution of this paper, where we propose a control law that guarantees convergence to an abstract state satisfying certain conditions, while guaranteeing safety (i.e., there are no inter-agent collisions).

IV. Control with Collision Avoidance

A. Monotonic convergence

We relax the requirement of exponential convergence to an abstract state and replace it with a slightly different notion of convergence in order to accommodate the safety constraints in (10). Specifically instead of insisting on the minimum-energy solution, (7), we find the solution closest to the minimum-energy solution satisfying the safety constraints.

First, we require that the error in the abstract state decrease *monotonically*:

$$\tilde{\boldsymbol{x}}^{\mathrm{T}} K \dot{\boldsymbol{x}} \geq 0. \tag{12}$$

Substituting (2) into (12), this inequality reduces to

$$\tilde{\boldsymbol{x}}^{\mathrm{T}}\, K\, \Gamma \begin{bmatrix} I_2 & \cdots & I_2 \\ \frac{1}{s_1 - s_2} {p_1}^{\mathrm{T}} E_1 & \cdots & \frac{1}{s_1 - s_2} {p_N}^{\mathrm{T}} E_1 \\ {p_1}^{\mathrm{T}}(I_2 + E_2) & \cdots & {p_N}^{\mathrm{T}}(I_2 + E_2) \\ {p_1}^{\mathrm{T}}(I_2 - E_2) & \cdots & {p_N}^{\mathrm{T}}(I_2 - E_2) \end{bmatrix} \begin{bmatrix} \nu_1 \\ \nu_2 \\ \cdots \\ \nu_N \end{bmatrix} \geq 0. \tag{13}$$

A sufficient condition to satisfy this *monotonic convergence condition* in (13) is that each robot select inputs that satisfy:

$$\tilde{\boldsymbol{x}}^{\mathrm{T}}\, K\, \Gamma \begin{bmatrix} I_2 \\ \frac{1}{s_1 - s_2} {p_i}^{\mathrm{T}} E_1 \\ {p_i}^{\mathrm{T}}(I_2 + E_2) \\ {p_i}^{\mathrm{T}}(I_2 - E_2) \end{bmatrix} \nu_i \geq 0. \tag{14}$$

If all robots choose controls satisfying (14), the error in the abstract state will monotonically decrease. It is useful to show that the minimum-energy control law (8) satisfies this inequality.

Proposition 1. *The minimum-energy control law (8) with ζ given by (11) satisfies the monotonic convergence condition (14).*

Proof: We define g_i and m_i such that

$$m_i = \left[I_2,\ \frac{s_1 - s_2}{s_1 + s_2} E_1 p_i,\ \frac{1}{4s_1}(I_2 + E_2) p_i,\ \frac{1}{4s_2}(I_2 - E_2) p_i \right]$$

$$g_i = \left[I_2,\ \frac{1}{s_1 - s_2} p_i^T E_1,\ p_i^{\mathrm{T}}(I_2 + E_2),\ p_i^{\mathrm{T}}(I_2 - E_2) \right]^{\mathrm{T}}.$$

Substituting (8) and (11) into the left hand side of (14) gives the quadratic form:

$$\tilde{\boldsymbol{x}}^{\mathrm{T}}\, K\, \Gamma\ [g_i\ m_i]\ \Gamma^{-1}\, K\, \tilde{\boldsymbol{x}}.$$

The 5×5 matrix $[g_i\ m_i]$, although asymmetric, can be shown to be positive semi-definite with the two non-zero eigenvalues to be given by:

$$\lambda_1 = 1 + \frac{\|p_i\|^2}{s_1 + s_2} \quad \text{and} \quad \lambda_2 = 1 + \frac{p_{i,x}^2}{s_1} + \frac{p_{i,y}^2}{s_2}.$$

Since K is chosen to be positive definite the inequality (14) is satisfied. ∎

B. A safe minimum-energy control law

In this subsection we derive a decentralized control law that selects a control input as close as possible to the minimum-energy controls while satisfying the monotonic convergence inequality and the safety constraints.

Proposition 2. *Equation (15) is a decentralized control law that selects a unique control input that has the smallest energy instantaneously while satisfying the monotonic convergence inequality and the safety constraints.*

$$\nu_i = \arg\min_{\hat{\nu}_i \in U} \|\nu_i^{\star} - \hat{\nu}_i\|^2, \quad s.t.\ (10, 14) \tag{15}$$

Proof: The constraints in (10, 14) provide the safety guarantees and the monotonic convergence condition. The function being minimized is the discrepancy from the minimum-energy input. Since the inequality constraints are linear in ν_i and the function being minimized is a positive-definite, quadratic function of ν_i, (15) is a convex, quadratic program with a unique solution. Further, since each robot only relies on its own state and knowledge of the error in the abstract state, it is a decentralized control law. ∎

Convergence properties of (15)

To investigate the global convergence properties, we introduce the Lyapunov function

$$V(q) = \frac{1}{2} \tilde{\boldsymbol{x}}^{\mathrm{T}} \tilde{\boldsymbol{x}}.$$

Since the solution of (15) must satisfy the inequality (14), we know that $\tilde{\boldsymbol{x}}^{\mathrm{T}} K \dot{\boldsymbol{x}} \geq 0$. If K is chosen to be diagonal with positive entries, this condition also implies $\tilde{\boldsymbol{x}}^{\mathrm{T}} \dot{\boldsymbol{x}} \geq 0$. In other words,

$$\dot{V}(q) = -\tilde{\boldsymbol{x}}^{\mathrm{T}} \dot{\boldsymbol{x}} \leq 0.$$

From [10], we know that q is bounded given that $\boldsymbol{x}$ is bounded and that $V(q) \to \infty$ as $\|q\| \to \infty$. Further, $V(q)$ is globally uniformly asymptotically stable. Therefore, from LaSalle's invariance principle, we know that the abstract state will converge to the largest invariant set given by $\tilde{\boldsymbol{x}}^{\mathrm{T}} \dot{\boldsymbol{x}} = 0$. From (2), we know that $\dot{\boldsymbol{x}} = 0$ only when $\nu = 0$. Thus the invariant set is characterized by the set of conditions that lead to the system of inequalities given by (10, 14) to have $\nu = 0$ as the only solution.

Proposition 3. *For any desired change in the abstract state $\tilde{\boldsymbol{x}}$, subject to the condition $\tilde{\boldsymbol{x}}_4 \geq 0$, $\tilde{\boldsymbol{x}}_5 \geq 0$, (i. e., a condition where the size of the shape of the formation is not decreasing), there is a non-zero solution to the inequalities (10, 14).*

Proof: Consider the solution given by the minimum-energy control law (8). In component form,

$$\nu_i^{\star} = \begin{bmatrix} \xi_1 + \frac{s_1 - s_2}{s_1 + s_2} y_i \xi_3 + \frac{x_i}{4s_1} \sigma_1 \\ \xi_2 + \frac{s_1 - s_2}{s_1 + s_2} x_i \xi_3 + \frac{y_i}{4s_2} \sigma_2 \end{bmatrix}.$$

It is easy to see that this satisfies the collision constraints (10) for every pair of robots $(i,\ j)$:

$$[(x_i - x_j)\ (y_i - y_j)] \begin{bmatrix} \nu_{i,x}^{\star} - \nu_{j,x}^{\star} \\ \nu_{i,y}^{\star} - \nu_{j,y}^{\star} \end{bmatrix} \geq 0.$$

As shown earlier in Proposition 1, (8) also satisfies the monotonic convergence inequality. ∎

Remark 1. *It is clear from the above proof that there are no guarantees when the shape in the abstract state is shrinking in area. If $\tilde{\boldsymbol{x}}_4 < 0$ or $\tilde{\boldsymbol{x}}_5 < 0$, there may not be a non-zero velocity vector ν that satisfies the inequalities (10, 14). It is only in this condition that the system will reach an equilibrium away from the desired abstract state.*

We next discuss the second contribution of this paper, where we propose an energy metric for motion planning of a deformable ellipse. Such a metric permits the computation of optimal motion plans in complex environments.

V. Motion Planning in the Abstract Space

The abstract representation of the team of robots permits the planning of motions that only require consideration of an abstract state space of fixed dimension, rather than one that scales with the number of robots. In this section we consider the problem of generating reference trajectories in the abstract space.

We start by defining a metric on M. On $SE(2)$, we can define a Riemannian metric as a bi-linear form derived by an inner product on $se(2)$. Given two twists $\{\xi_1,\ \xi_2\} \in se(2)$, we can define [16]:

$$< \xi_1, \xi_2 >= \xi_1^T\, W\, \xi_2,$$

where W is a positive definite matrix. At an arbitrary element $g \in SE(2)$, the inner product between two velocities or tangent vectors $\dot{g}_1,\ \dot{g}_2$ is obtained by left translation:

$$< \dot{g}_1, \dot{g}_2 >_g = < g^{-1}\dot{g}_1, g^{-1}\dot{g}_2 >_e,$$

where $g^{-1}\dot{g}_i$ are tangent vectors at the identity element e (the 3×3 identity homogeneous transformation) and therefore lie in $se(2)$. A metric defined in this way is a left-invariant Riemannian metric. Following [17], we can use the inertia tensor of a rigid body and its kinetic energy to define W_g:

$$W_g = \begin{bmatrix} mI_2 & 0 \\ 0 & \mathcal{I}_{11}+\mathcal{I}_{22} \end{bmatrix},$$

in the body-fixed coordinate system $\mathcal{B}$.

The above treatment was for a rigid shape. However, since $M = G \times S$ is a product space, we treat the shape space independently. We assume a constant metric $W_s = \alpha I_2$ to model the "cost" in changing the shape. Thus the rate of change of the abstract shape in $\mathcal{B}$ given by ζ has the norm:

$$\|\zeta\| = \frac{1}{2}\zeta^T \begin{bmatrix} W_g & 0 \\ 0 & W_s \end{bmatrix} \zeta, \tag{16}$$

which is well-defined everywhere on M.

Realistically one must also model the potential energy associated with deforming the shape. The simplest approach to creating an abstract model for potential energy storage is to think of the expansion or contraction as a reversible, adiabatic process in which no energy is lost. Compression results in an increase of internal energy which can then be recovered during expansion. It is well known that in such processes the pressure p and the volume v are related by the ratio of specific heats γ by the equation:

$$pv^\gamma = constant$$

and the work done to effect a change in volume from v_1 to v_2, and therefore an increase in internal energy, is given by:

$$\Delta V = k\left(\frac{1}{v_2^{\gamma-1}} - \frac{1}{v_1^{\gamma-1}}\right),$$

where k is a constant. In the plane, we can use the area of the ellipse (with a unit depth) instead of volume, which we know to be $\pi\sqrt{s_1 s_2}$. We define a reference shape $s^0(N)$ as a circular shape for N robots with zero potential energy. It is intuitively clear that the radius of a zero-energy circular shape, $r_0(N)$, must increase with N. In this paper, we take $r_0(N) = \frac{N\epsilon}{2}$. The potential energy associated with any shape $s \in S$ is given by:

$$V(s) = \beta\left(\frac{1}{(s_1 s_2)^{\frac{\gamma-1}{2}}} - \frac{1}{(r_0^2(N))^{\gamma-1}}\right), \tag{17}$$

where β is a constant. Thus, the total energy associated with any motion at any configuration is given by:

$$E(g,s,\zeta) = \frac{1}{2}\zeta^T \begin{bmatrix} W_g & 0 \\ 0 & W_s \end{bmatrix} \zeta + V(s). \tag{18}$$

Equation (18) gives us a principled approach to determine the cost of changes in configuration using a kinetic and potential energy with two constants α and β. It also allows us to formulate trajectory generation and motion planning problems as problems of finding geodesics.

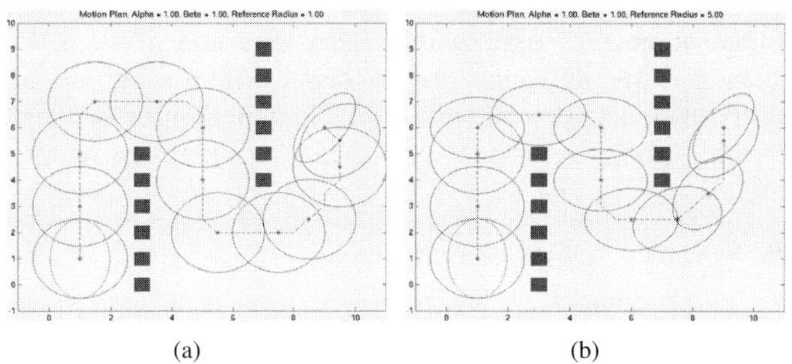

(a) (b)

Fig. 2. Two motion plans depict the effects of changing the radius of zero-energy, r_0, from $r_0 = 1$ (Fig. 2(a)) to $r_0 = 5$ (Fig. 2(b)). Although $\alpha = \beta = 1$ in both cases and the initial (red) and final (blue) abstract states are consistent, two different optimal motion plans are computed (using a Bellman-Ford search in a discretized abstract space). Obstacles are shown in magenta.

It is possible to split the space and derive the trajectories for $\mu \in \mathbb{R}^2$ and $(\theta, s_1, s_2) \in \mathbb{R}^2 \times SO(2)$ separately in open environments because of the product structure of the metric (18). Instead, we propose the design of motion plans using a discretization of the abstract space with constraints defined by obstacles and admissible abstract states. Many discrete optimal planning algorithms exist that permit the application of the energy metric defined by (18) to solve the for the minimum-energy path through open or cluttered environments [18].

VI. Splitting and Merging of Groups of Robots

When an ensemble of robots are forced to squeeze through constrained spaces, the only allowable shapes have small areas and correspond to large values of internal energy $V(s)$ as seen from (17). Therefore we define a threshold so that when $V(s) > V_{max}$ the group splits into two subgroups with shapes identical to the original shape but each with half the number of robots. This allows the group to reset its energy to a lower level by reducing the number of constraints, while remaining in the same configuration. Alternatively, a supervisory agent can decide when to divide the robots into subgroups. The formulation of Sect. IV can be applied to multiple subgroups with the new abstraction manifold $M = M_1 \times M_2$, where $M_i = G_i \times S_i$. The only additional mechanism that is required is a protocol for each robot to determine to which subgroup it belongs. Clearly the motion plan for each subgroup, $\boldsymbol{x}_i^{des}(t)$, can be generated and this desired trajectory can be broadcast to the group. With the feedback of the abstract state, $\boldsymbol{x}_i$, each robot can compute its own controls with the knowledge of the specific subgroup i to which it belongs. We propose two different event-triggered techniques for splitting and merging teams of robots in a decentralized and distributed manner.

A. Market-based auctioning method

Our first approach is based on the market-based auctioning method proposed in [19] which is guaranteed to converge in polynomial time but requires communication between the robots. It allows a team of robots to divide into subgroups by defining an auction determined by the desired abstract subgroup states $\{\boldsymbol{x}_1^{des}, \ldots, \boldsymbol{x}_k^{des}\}$, and the maximum number of agents allowed in each subgroup $\{n_1, \ldots, n_k\}$, where k

is the number of desired subgroups. The end result of the auction is that all agents are associated with a subgroup and distributed in agreement with the maximum number of agents in each group. To ensure correctness, the algorithm requires that $N = \sum_{i=1}^{k} n_i$.

B. Stochastic policy for splitting

As the number of robots grows, it is beneficial to use a mean-field model to model the distribution of robots and develop stochastic switching rules that guarantee the desired ensemble properties [20]. As $N \to \infty$, the ensemble properties of the group of robots using the stochastic switching rules converge to the desired properties. In other words, if each robot uses a probability distribution to select one subgroup versus the other, the ensemble properties of the group can be inferred from this probability distribution. From a practical standpoint, a split between k subgroups with $\{n_1, \ldots, n_k\}$ robots can be implemented approximately by each robot preferring group i with probability $p_i = \frac{n_i}{N}$.

C. Merging of groups

The merging of separate groups is trivial in the proposed framework using the controller (15). The redefinition of $\tilde{\boldsymbol{x}}$ to account for the desired merged abstract state results in a single group, while accounting for inter-agent collision avoidance.

VII. Simulation and Experimental Results

The remainder of the paper is dedicated to verifying the effectiveness of the control algorithm with collision avoidance presented in Sect. IV. We begin by discussing implementation details relevant to the analysis and experiment discussion that follows.

A. Implementation Details

The control algorithm was implemented in *C++* using the open-source robotics software *Player*, part of the *Player/Stage/Gazebo* project [21]. The *Player* server enables network communications between multiple robots. *Gazebo* is a three-dimensional simulation environment incorporating a dynamics engine and collision detection. *Player* also permits integration with *Gazebo*, allowing the same code base to be used in both simulation and experimentation on the real hardware.

The algorithm was tested in simulation via *Gazebo* and on an experimental infrastructure consisting of a team of small differential-drive robots, an indoor tracking system for ground-truth purposes, and a computer infrastructure to support wireless communication and data logging. Accurate models of the robots were created for use in simulation to emulate the real robots. Further, the asynchronous and distributed nature of the hardware was emulated by creating separate execution threads for each agent where all inter-agent communication was accomplished through the *Player* server.

Although robot dynamics play a significant role when considering inter-agent interactions, we are able to ignore these effects due to the fact that the robot platforms use stepper motors which permit "instantaneous" changes in velocity for sufficiently small magnitudes given the mass of the robot. We ensure during simulation and experimentation that the control velocities respect these thresholds.

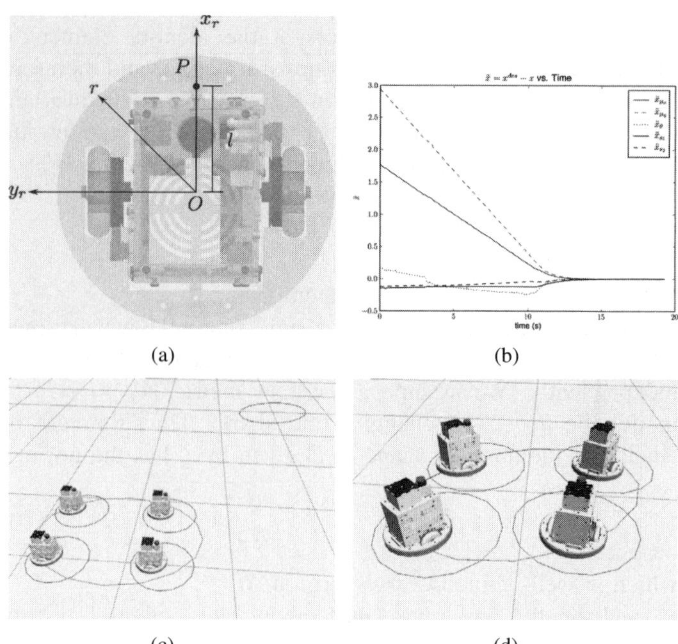

(a) (b) (c) (d)

Fig. 3. A top view of the robot model used in both simulation and experimentation showing the body-fixed coordinate system. P is a reference point on the robot whose position is regulated by the vector fields (Fig 3(a)). The convergence of a team of four robots in simulation to $\boldsymbol{x}^{des} = x$ (Fig. 3(b)). The team of robots controlling to the desired abstract state $\boldsymbol{x}^{des} = \{2, 2, 0.5, 0.2, 0.1\}$ (Figs. 3(c)–3(d)). The abstract states $\boldsymbol{x}$ and $\boldsymbol{x}^{des}$ are shown in green and black, respectively. The ϵ radius safety region is shown in blue, which for the differential drive robots is defined with respect to the feedback linearization point.

In both simulation and experimentation, the abstract state $\boldsymbol{x}$ (assessed using simulation data or the tracking system) and the desired abstract state $\boldsymbol{x}^{des}$ were broadcast to the robots. As the control law updates, each robot broadcasts its current pose and velocity while listening for the pose and velocity of its neighbors. The optimization defined by (15) is solved using the quadratic program routines provided by the open-source *Computational Geometry Algorithms Library* [22].

We consider a simple model of a point robot with coordinates (x, y) in the world coordinate system. On the differential-drive robot in Fig. 3(a), these are the coordinates of a reference point P on the robot which is offset from the axle by a distance l. Its velocities in the inertial frame and moving frame $\mathcal{B}$ are u_i and ν_i, respectively. The velocity of the reference point P can be converted into linear and angular velocities for the robot through the equations below:

$$\begin{bmatrix} v \\ \omega \end{bmatrix} = \begin{bmatrix} \cos\theta & \sin\theta \\ -\frac{\sin\theta}{l} & \frac{\cos\theta}{l} \end{bmatrix} \begin{bmatrix} \dot{x} \\ \dot{y} \end{bmatrix}.$$

It is well-known that if a robot's reference point is at the point P, and if r is the radius of a circle circumscribing the robot, all points on the robot lie within a circle of radius $r + l$ centered at P. In other words, if the reference point tracks a trajectory

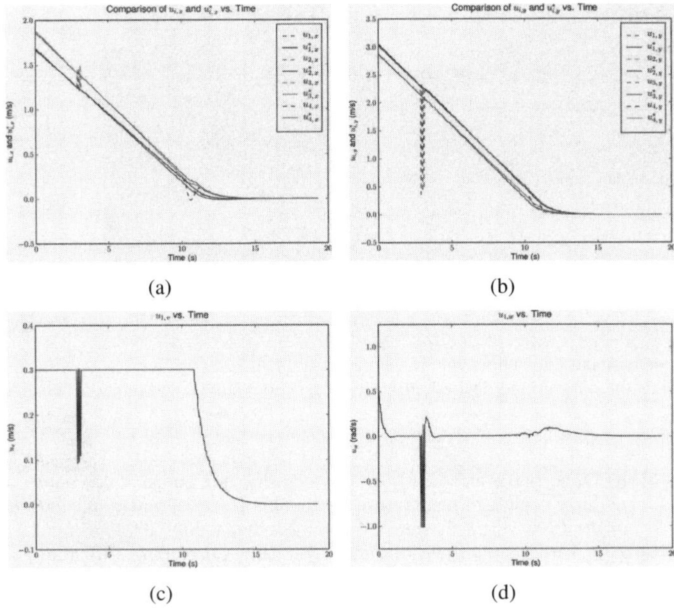

Fig. 4. The optimal control, $u_i^\star$, and u_i from (15) defined in the robot's local frame (Figs. 4(a)–4(b)). In general, the optimal control law is the solution to (15). However, during inter-agent interactions the resulting control varies from the optimal solution. The linear and angular velocities resulting from feedback linearization (Figs. 4(c)–4(d)).

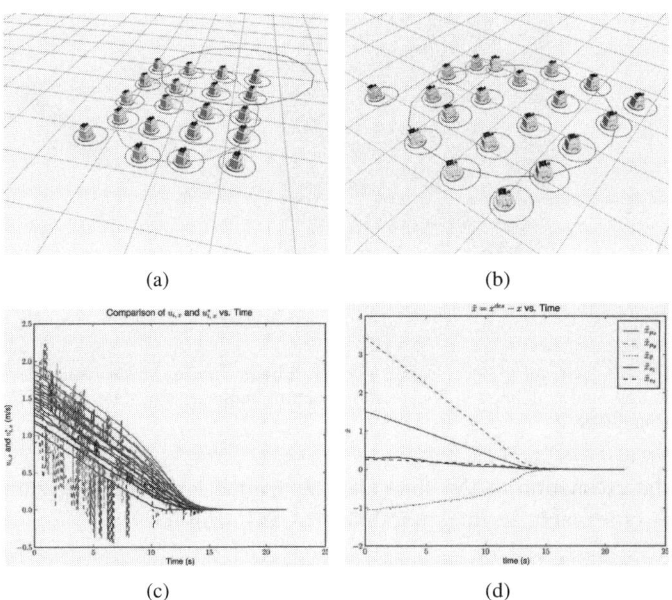

Fig. 5. A team of twenty robots control to $\boldsymbol{x}^{des} = \{3,\, 3,\, 0.5,\, 1.5,\, 1\}$. A red ϵ safety region indicates a state where $\delta_{ij} < \epsilon$, but in this case $\dot{\boldsymbol{x}} = 0$ (Fig. 5(b)). A comparison of $u_{i,x}$ from (15) and the optimal control, $u_{i,x}^\star$ defined in the robot's local frame (Fig. 5(c)). A comparison of $u_{i,y}$ to $u_{i,y}^\star$ results in similar control variations. Although there are numerous interactions between the agents, as depicted by the many variations of $u_{i,x}$ from $u_{i,x}^\star$, the system converges to the desired shape (Fig. 5(d)).

$(x_d(t), y_d(t))$, the physical confines of the robot are within a circle of radius $r + l$ of this trajectory. This allows us to use a geometric abstraction of a circular robot at P with a radius $\rho = r + l$. By ensuring that the reference points of adjacent robots are at least ϵ away as in (9), we guarantee that there are no collisions for the real robot.

B. Simulation

In this section we present two representative trials of the control law for collision avoidance: a basic system with limited interactions, and a large team of robots with numerous interactions. In both cases, we let $K = \text{diag}(1,\, 1,\, 0.8,\, 0.8,\, 0.8)$, $r = 0.15\,\text{m}$, $\epsilon_s = 0.1\,\text{m}$, $l = 0.1\,\text{m}$, and the maximum linear and angular velocities are $0.3\,\text{m/s}$ and $1\,\text{rad/s}$, respectively.

1) Controlling four robots with limited interactions: A small team of four stationary differential-drive ground robots were sent a desired abstract state (see Fig. 3). The convergence of the controller to the desired abstract state is shown in Fig. 3(b). A comparison of the resulting optimal control law (5) and the collision avoidance control law (15) in the separate kinematic controller dimensions is depicted in Figs. 4(a)–4(b). Indeed, as stated in Proposition 1, in general the optimal control law, $u_i^\star$, is applied. However, during collisions, the control varies from the optimal control to account for the inter-agent interactions. The linear and angular velocities of the first agent resulting from feedback linearization and velocity saturation are shown in Figs. 4(c)–4(d). These interactions are apparent after feedback linearization.

2) Controlling a team of twenty robots: The team of twenty robots depicted in Fig. 5 demonstrates the effectiveness of the controller in large groups to ensure collision avoidance while controlling to a desired shape. From Fig. 5(c) it is clear that numerous interactions occur between the agents and yet the system converges to the desired shape. In general, agents will control to a separation distance of ϵ since each agent is individually computing the collision constraints (10). However, it is possible that robots remain within the ϵ safety region of other robots at the time of convergence (see Fig. 5(b)) but do not collide since the solution to (15) when $\tilde{\boldsymbol{x}} = 0$ is $u_i = 0$.

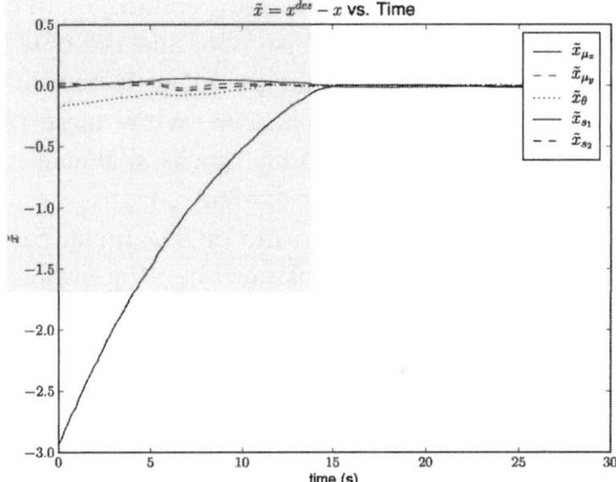

Fig. 6. The convergence of the team of seven robots in experimentation to $\boldsymbol{x}^{des}$. The trial represents a merging scenario where the robots were distributed in distinct groups separated by several meters.

C. Experiments

The control algorithm was verified on a team of seven nonholonomic robots. A trial run is depicted in Fig. 7. The algorithm was tested with several trials over a variety of desired states and scenarios with performance consistent with the convergence shown in Fig. 6. The noise associated with

(a) (b) (c) (d)

Fig. 7. A team of seven robots control through a series of abstract states in a corridor passing maneuver. A visualization of the experiment (following the representation defined in Fig. 3) corresponding to the system state in Fig. 7(a) depicts the current and desired abstract states, robot states, and corridor location (Fig. 7(b)).

the estimation of the abstract state via the localization system is observable in the smoothness of $\tilde{x}$ in Fig. 6.

VIII. Conclusion and Future Work

We presented solutions to the problem of planning and controlling the position, orientation, and shape of a formation of a team of robots. In contrast to most previous work, we model the physical shape of the robot and consider controllers that are guaranteed to avoid collisions between the robots. We also derive software abstractions that lend themselves to implementation on real platforms and to experimentation. In addition, we present a metric for the planning of deformable shapes and trajectories of the ensemble and the development of effective coordination strategies to split the team into subgroups and to merge subgroups. We view these problems and their solutions to be building blocks that can enable a robot team to navigate an environment adapting to the constraints imposed by obstacles in the environment. Simulation and experimental results demonstrate the effectiveness of the control algorithm when applied to nonholonomic robots.

As a direction of future research, we are interested in pursuing time-parameterized trajectories in the abstract space. We have also assumed the existence of a global observer in past work and are currently considering distributed estimation algorithms to relax this requirement.

Acknowledgment

We gratefully acknowledge partial support from NSF grants IIS-0427313, IIP-0742304, and IIS-0413138, ARO grant W911NF-05-1-0219, and ONR grant N00014-07-1-0829.

References

[1] G. S. Sukhatme, A. Dhariwal, B. Zhang, C. Oberg, B. Stauffer, and D. A. Caron, "Design and development of a wireless robotic networked aquatic microbial observing system," *Environmental Engineering Science*, vol. 24, no. 2, pp. 205–215, Mar. 2007.

[2] P. E. Rybski, S. A. Stoeter, M. D. Erickson, M. Gini, D. F. Hougen, and N. Papanikolopoulos, "A team of robotic agents for surveillance," in *Proc. of Int. Conf. on Autonomous Agents*, Barcelona, Spain, June 2000, pp. 9–16.

[3] V. Kumar, D. Rus, and S. Singh, "Robot and sensor networks for first responders," *IEEE Pervasive Computing*, vol. 3, no. 4, pp. 24–33, Oct. 2004.

[4] T. Eren, P. N. Belhumeur, and A. S. Morse, "Closing ranks in vehicle formations based on rigidity," in *Proc. of the IEEE Conf. on Decision and Control*, vol. 3, Las Vegas, NV, Dec. 2002, pp. 2959–2964.

[5] R. Olfati-Saber and R. M. Murray, "Distributed cooperative control of multiple vehicle formations using structural potential functions," in *Proc. of the IFAC World Congress*, Barcelona, Spain, July 2002.

[6] J. P. Desai, J. Ostrowski, and V. Kumar, "Controlling formations of multiple mobile robots," in *Proc. of the IEEE Int. Conf. on Robotics and Automation*, vol. 4, Leuven, Belgium, May 1998, pp. 2864–2869.

[7] H. Tanner, G. J. Pappas, and V. Kumar, "Input to state stability on formation graphs," in *Proc. of the IEEE Int. Conf. on Robotics and Automation*, Las Vegas, NV, Dec. 2002, pp. 2439–2444.

[8] M. Egerstedt and X. Hu, "Formation constrained multi-agent control," *IEEE Transactions on Robotics and Automation*, vol. 17, no. 6, pp. 947–951, Dec. 2001.

[9] P. Ogren, E. Fiorelli, and N. Leonard, "Formations with a mission: stable coordination of vehicle group maneuvers," in *Proc. of Int. Symposium on Mathematical Theory Networks and Systems*, Notre Dame, IN, Aug. 2002.

[10] C. Belta and V. Kumar, "Abstraction and control for groups of robots," *IEEE Transactions on Robotics*, vol. 20, no. 5, pp. 865–875, Oct. 2004.

[11] N. Michael, C. Belta, and V. Kumar, "Controlling three dimensional swarms of robots," in *Proc. of the IEEE Int. Conf. on Robotics and Automation*, Orlando, FL, May 2006, pp. 964–969.

[12] J. Cortes, S. Martinez, T. Karatas, and F. Bullo, "Coverage control for mobile sensing networks," *IEEE Transactions on Robotics and Automation*, vol. 20, no. 2, pp. 243–255, Apr. 2004.

[13] M. Schwager, J. McLurkin, and D. Rus, "Distributed coverage control with sensory feedback for networked robots," in *Robotics: Science and Systems*, Philadelphia, PA, Aug. 2006.

[14] R. A. Freeman, P. Yang, and K. M. Lynch, "Distributed estimation and control of swarm formation statistics," in *Proc. of the American Control Conf.*, Minneapolis, MN, June 2006, pp. 749–755.

[15] N. Michael, J. Fink, and V. Kumar, "Controlling a team of ground robots via an aerial robot," in *Proc. of the IEEE/RSJ Int. Conf. on Intelligent Robots and Systems*, San Diego, CA, Oct. 2007, pp. 965–970.

[16] R. Murray, Z. Li, and S. Sastry, *A Mathematical Introduction to Robotic Manipulation*. Florida: CRC Press, 1993.

[17] M. Zefran, V. Kumar, and C. B. Croke, "On the generation of smooth three-dimensional rigid body motions," *IEEE Transactions on Robotics and Automation*, vol. 14, no. 4, pp. 576–589, Aug. 1998.

[18] S. M. LaValle, *Planning Algorithms*. Cambridge, U.K.: Cambridge University Press, 2006.

[19] N. Michael, M. M. Zavlanos, V. Kumar, and G. J. Pappas, "Distributed multi-robot task assignment and formation control," in *Proc. of the IEEE Int. Conf. on Robotics and Automation*, Pasadena, CA, May 2008, pp. 128–133.

[20] A. Martinoli, K. Easton, and W. Agassounon, "Modeling swarm robotic systems: A case study in collaborative distributed manipulation," *The Int. Journal of Robotics Research*, vol. 23, no. 4-5, pp. 415–436, Apr. 2004.

[21] B. P. Gerkey, R. T. Vaughan, and A. Howard, "The player/stage project: Tools for multi-robot and distributed sensor systems," in *Proc. of the Int. Conf. on Advanced Robotics*, Coimbra, Portugal, June 2003, pp. 317–323.

[22] "CGAL, Computational Geometry Algorithms Library," http://www.cgal.org.

Stochastic Recruitment: A Limited-Feedback Control Policy for Large Ensemble Systems

Lael Odhner and Harry Asada

Abstract—**This paper is about stochastic recruitment, a control architecture for centrally controlling the ensemble behavior of many identical agents, in a manner similar to motor recruitment in skeletal muscles. Each agent has a finite set of behaviors, or states, which can be switched based on a broadcast command. By switching randomly between states with a centrally determined probability, it is possible to designate the number of agents in each state. This paper covers stochastic recruitment policies for the case when little or no feedback is available from the system. Feed-forward control policies based on rate equilibria are presented, with an analysis of the performance trade-offs inherent in the problem. Minimal feedback control laws are also discussed, and a policy is presented which minimizes the expected convergence time of the system given only the ability to halt the system when the desired output has been achieved.**

I. Introduction

Finding methods for centrally controlling the collective behavior of many identical agents is an active research topic across diverse communities, including robotics, bioengineering, and control. Many different systems can be modeled as a swarm of identical agents, such as bacteria, cells, robots in a swarm, or computers in a network. In this paper, we will explore control of a class of finite state or hybrid state agents that we believe to be applicable to a variety of these problems. In particular, we wish to regulate the ensemble distribution of many agents over their discrete states, or the fraction of agents having a particular discrete state. The discrete states in question could be tasks performed by swarm robots, cell migration behaviors, or switched modes governing the time evolution of some continuous state variable. One of the key components of the these regulatory mechanisms is stochasticity. In biological systems, stochasticity is a ubiquitous phenomenon that can be observed in mechanochemical cell and molecular behaviors. For example, angiogenesis cell migration and tissue development are often treated as directed random walks, where stochastic behaviors play key roles for developing multi-cellular structures that meet morphological and functional requirements [1], [2]. Natural systems are built upon random processes, and biological systems in particular exploit the stochastic nature of their building blocks. In control engineering, randomness has been treated as unwanted behavior that should be filtered out or avoided. Except for communication technology and some of the system identification techniques, randomness has not yet been fully exploited as a useful concept. In general, we assume that the transition between states can be modeled as a Markov state transition graph whose state transition probabilities are determined by some input given to all agents in the system. The authors are actively working on applying this control framework to the problems of endothelial cell migration and artificial muscle actuators. Figure 1 illustrates the basic system architecture.

The authors are with the Mechanical Engineering Department at the Massachusetts Institute of Technology, Cambridge, MA, USA, 02139 {lael, asada}@mit.edu

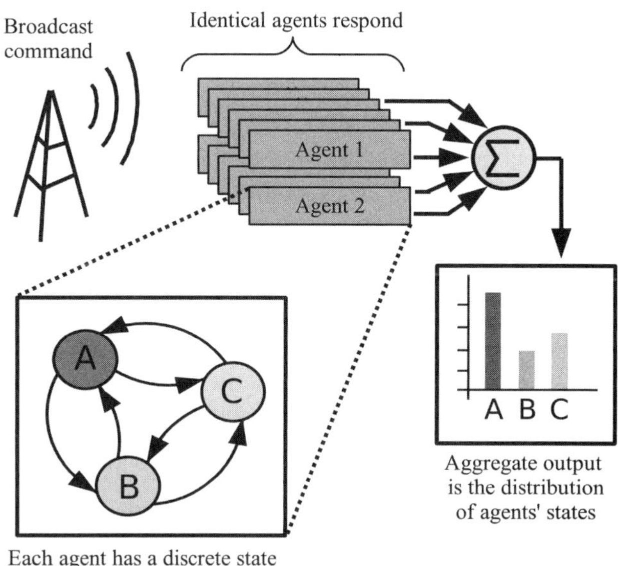

Fig. 1. Many systems in nature can be thought of as an ensemble of functionally similar discrete-state agents that each respond independently to some global stimulus. The ensemble output behavior of many such agents is the number of agents in each discrete state.

In this paper, we discuss the performance of feed-forward control policies for controlling the number of agents in each discrete state. If rich information usable for feedback control is available, then many control techniques that presume full-state knowledge can be applied. The authors have shown that simple feedback policies can be formulated for controlling the ensemble state distribution based on linear feedback laws [3], expectation-based control laws [4], and dynamic programming [5]. However, the output measurements from real systems often provide only uncertain estimates of system state distribution. For example, a robot swarm might have limited communication with a central control station. An artificial muscle composed of many individual sub-units may (and usually will) exhibit some output delay or hysteresis, so that some estimating filter must be used to predict the individual units' states. Uncertainty of this form will lead to increased uncertainty in the closed-loop behavior of the system.We will show that we can formulate open-loop control policies that cause the state distribution to reach a stochastic equilibrium that is close to any desired state distribution, having well-characterized variance. The performance of policies depending on minimal or no feedback provide a good bounding case guidelines for examining the performance of closed-loop laws. Depending on the degree of uncertainty in the state distribution estimated from measurements, it may actually be advantageous to use an open-loop policy instead of a closed-loop policy.

In this paper a stochastic cellular system with limited output feedback will be formulated in its simplest form. Basic convergence properties and control policy will be discussed to gain insights into collective behaviors of stochastic cellular systems. The framing of the problem is based on cell and systems biology and motivated by needs in robotics and control. The paper, however, does not aim to apply the results directly to a specific area; instead the objective is to better understand the possible regulatory mechanism that may govern a large population of cellular agents.

II. Introducing Stochastic Recruitment

In an attempt to investigate a regulatory mechanism that works with limited or no feedback, we look at biological control systems, extract key features from there, and formulate a simple, abstract problem for detailed analysis. One striking difference from traditional engineered systems is that biological cells are living in a wet environment, where signals propagate through diffusion. Stimuli to the process pervasively affect all the cells involved in the wet environment. It is not likely that each cell receives a specific control signal from a central controller. Rather, the control signal, if it exists, is broadcast in nature.

In response to stimuli, an important property of biological systems is that each cell's behavior is stochastic in nature. Although receiving the same stimuli, the cell's response is randomly chosen, conditioned on the environmental factors. Endothelial cell migration, for example, each cell's movement is a random process, switching directions stochastically [1]. Furthermore, the state transition probabilities often vary depending on the stimuli that individual cells receive. It is known that biochemical kinetics highly depend on temperature and other factors, resulting in changes to state transition probabilities. In the case of endothelial cell migration, state transition probabilities are modulated by the stimuli each agent receives as well as the conditions of the wet environment to which it is exposed [2].

These stochastic behaviors of biological cells suggest a new control methodology for engineered systems; a central controller broadcasts signals that modulate state transition probabilities to be used at individual agents, or directly broadcasts state transition probabilities across the cell population. Whichever the case, the central controller does not dictate each agent to take a specific transition, but only specifies the probabilities of transition. The actual control action is up to each cellular agent, but the collective behavior may be effectively controlled with the central controller. This broadcast control protocol will be useful for engineered systems, since it requires very little bandwidth, and allows each cellular agent to be completely anonymous.

The authors have developed several types of stochastic broadcast control [3], [5], [6]. The major difference is that the previous control architecture exploits feedback of aggregate output, while this paper addresses control issues with no feedback or limited feedback. We call this control "Stochastic Recruitment", which will be formally described next.

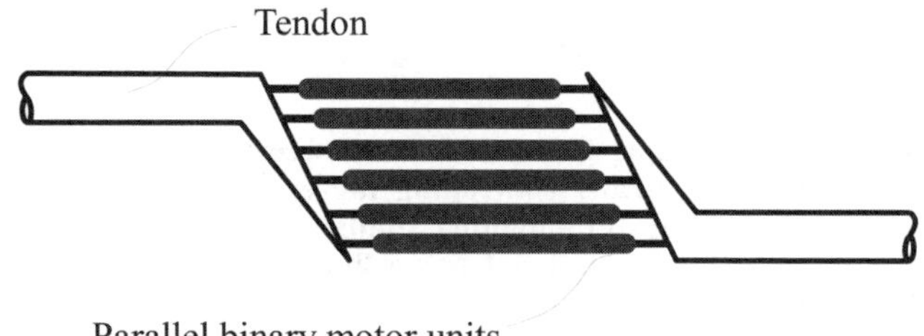

Fig. 2. Muscles are composed of many small motor units, which are either relaxed or activated, producing force. The net power and stiffness of the muscle depend on the number of recruited motor units.

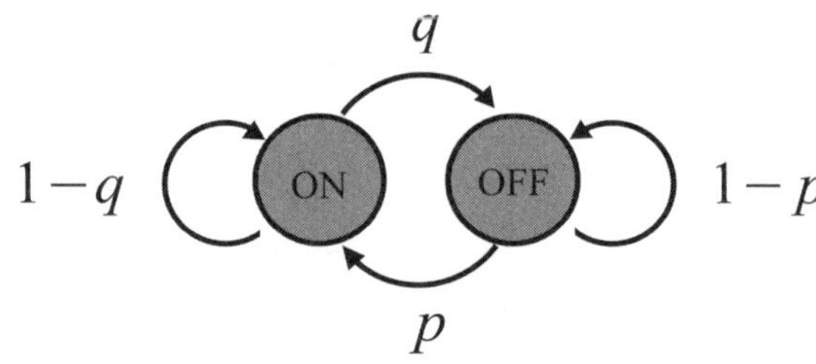

Fig. 3. A small automaton having two states, ON and OFF, can be commanded to transition between states with probabilities p and q.

The concept of recruitment is best explained by examining the function of skeletal muscle. A muscle is not a homogenous collection of individual cells; rather it is organized as a network of many small motor units, which respond autonomously to stimulus from the spinal column, as shown in Figure 2. A command from the nervous system produces a varied level of force by affecting the percentage of motor units which contract. [7]. Each motor unit has a threshold for response to the nervous excitation, so that varying the excitation being sent to the muscle acts to modulate the force produced [8]. This architecture can be replicated in engineered systems composed of a similar arrangement of subunits, like an artificial muscle [3]. A central controller broadcasts some excitation to all of the subunits, which change their behavior at a rate based on the excitation. The authors have been using two-state agents of this kind to control shape memory alloy muscle actuators made up of many binary units [5]. Collectively, N small actuators produce binary outputs which are modulated so that varied force and displacement are produced as a function of the number of ON (force-producing) agents, N_t^{on}. The probability per unit time of state transitions between the ON and OFF states are described by parameters p and q, as shown in Figure 3.

A. The Dynamics of Randomized Recruitment

A key appealing feature of this state transition framework is its similarity to statistical models for many real-world phenomena, including the dynamics of gene regulation [9], chemical reactions [10], and even swarms of insects [11]. It is well understood how kinetic chemical reaction models yield well-described transient and steady-state behaviors. These dynamics can be exploited in engineered systems, since an artificial system can be designed to arbitrarily vary the state transition probabilities of its agents. To demonstrate this, we will consider the simplest recruitment problem, a system made up of N small subsystems or agents having just two states, ON and OFF. The control task is to recruit a specific number of agents, N^{ref}, into the ON state. These states could represent two different modalities or behaviors exhibited by each individual agent. To keep track of the state evolution of the system, we will introduce a discrete distribution variable, x_t, describing the probability with which each agent is ON at time t,

$$x_t = P(state = ON) \tag{1}$$

Of course, using x_t as a sufficient statistic for predicting system behavior does not guarantee that N_t^{ON} can be known exactly. Instead, the likelihood that k units are on is calculated as a function of x_t using a binomial distribution,

$$P(N_t^{on} = k|x_t) = \left(\begin{array}{c} N \\ k \end{array} \right) x_t^k (1 - x_t)^{N-k} \tag{2}$$

There are multiple reasons why using x_t makes more sense than considering N_t^{on}, if little feedback information is available. First and foremost, it is quite simple to predict the future behavior of x given an initial condition and a broadcast command. In the absence of other information, such as a measured ensemble output, this will provide a good guess of long-term behavior. Second, as N becomes large, the central limit theorem will guarantee that N_t^{on} will approach its expected value,

$$E(N_t^{on}|x_t) = Nx_t \tag{3}$$

The evolution of x_t can be written as a recursive sequence based on the Markov graph parameters p and q,

$$x_{t+1} = (1-q)x_t + p(1-x_t) = (1-p-q)x_t + p \tag{4}$$

The time evolution of x_t can also be written as a deterministic sequence which satisfies (4),

$$x_{t+1} = \frac{p}{p+q} + \left(x_0 - \frac{p}{p+q}\right)(1-p-q)^t \tag{5}$$

Here x_0 is the initial likelihood that an arbitrarily selected unit is in the ON state. Some physical meaning can be gleaned from (5). The time-independent term of the sequence corresponds to the fraction of ON agents at steady state. This is equal to the probabilistic rate at which agents transition from OFF to ON, normalized by the sum of all transition rates between states,

$$x_{ss} = \frac{p}{p+q} \tag{6}$$

The rate at which x_t exponentially approaches x_{ss} from an initial condition x_0 is is represented in the transient term,

$$\lambda = 1 - p - q \tag{7}$$

These observations are important in formulating an open-loop control policy.

III. A NO-KNOWLEDGE CONTROL POLICY

In previous work, the authors formulated closed-loop control laws by finding the state transition graph parameters that minimize the expected future error,

$$E(N_{t+1}^{on}|p, q, N_t^{on}) = N^{ref} \tag{8}$$

If the central controller has no knowledge of the number of agents in each state, then the control policy must produce feed-forward dynamics that move the state distribution toward the desired goal. Equation (6) demonstrated that the time evolution of x_t has a steady-state component. Instead of choosing p and q to minimize the one-step ahead error conditioned on the present state distribution, p and q could be chosen so that some desired number of cells N^{ref} is expected in the steady state according to (3),

$$E(N_{ss}^{on}|p, q) = Nx_{ss} = N\frac{p}{p+q} = N^{ref} \tag{9}$$

Many policies satisfy this constraint for any given N^{ref}. For example, setting $p = 0.1$ and $q = 0.1$ will drive the agents to a 50% likelihood of being in either state. So will setting $p = q = 0.3$. In order to distinguish between these cases, a scaling analysis can be used to weigh the performance tradeoffs between these policies.

A. *Convergence Rate and Steady State Distribution are Independent.*

The control policy parameters p and q can be rewritten as βp_0 and βq_0, where $p_0 + q_0 = 1$ and β is a scaling factor that varies subject to the constraint that $0 < p < 1$ and $0 < q < 1$. When the transition probabilities are scaled in this way, The steady-state distribution x_{ss} is independent of β,

$$x_{ss} = \frac{\beta p_0}{\beta(p_0 + q_0)} = \frac{p_0}{p_0 + q_0} \tag{10}$$

However, the rate of convergence still depends on β,

$$1 - p - q = 1 - \beta(p_0 + q_0) = 1 - \beta \tag{11}$$

This means that β is a free parameter with which the convergence time can be arbitrarily varied while still satisfying the condition imposed in (9). In the most extreme case, β is chosen to be 1, so that $\lambda = 0$. In this case, x_t converges to $p/(p+q)$ after only one time interval. Figure 4 shows x_t converging to the same steady-state behavior from the same initial conditions, for several values of β.

B. Accuracy Varies Only as x_{ss} and N.

The other point of concern for this control system is the accuracy of the control system once it reached steady state. Specifying x_{ss} does not guarantee that the number of recruited cells N_{ss}^{on} will converge. Instead, the distribution from (2) will have some variance. When expressed as a variance normalized by the total number of units, this yields a measure of accuracy for recruitment,

$$Var\left(N_{ss}^{on}/N|x_{ss}\right) = x_{ss}(1-x_{ss})N = \frac{pq}{(p+q)^2N} \quad (12)$$

The β scaling argument from (10) and (11) can be applied to the variance calculation. The numerator and denominator of (12) both vary by a factor of β^2, so the variance is independent of the rate at which the actuator converges to its steady state probability distribution,

$$\frac{\beta^2 p_0 q_0}{\beta^2(p_0+q_0)^2N} = \frac{p_0 q_0}{(p_0+q_0)^2N} \quad (13)$$

This is an important observation; it means that nothing is to be gained by taking "baby steps", that is, selecting very small values of p and q in hopes of improving the accuracy of recruitment in exchange for a slower rate of response. The only parameter that can be varied to improve accuracy is N.

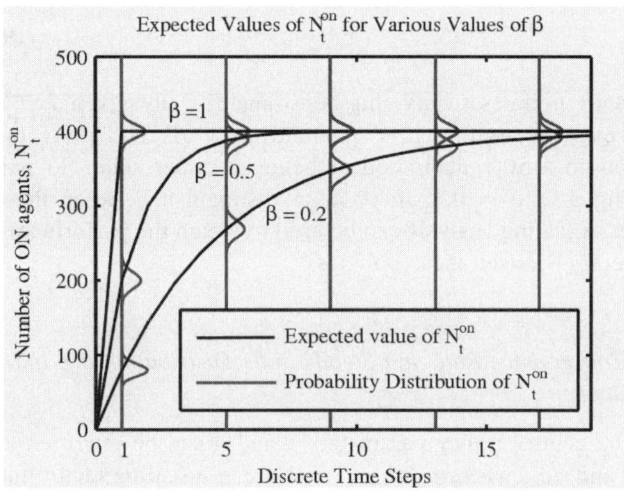

Fig. 4. The expected value of N_t^{on} and probability distribution of N_t^{on} at several points in time are shown for $N = 500$, $p_0 = 0.8$, $q_0 = 0.2$, and $\beta = 0.2$, 0.5 and 1. This plot illustrates the fact that the variance of N_t^{on} is independent of the rate of convergence. All three cases approach the same probability distribution in N_t^{on}.

C. The Number of Transitions per Unit Time

In a physical system, there is often a significant energy cost associated with switching agents from one behavior to another. For example, a mobile robot switching between patrolling two different areas will expend energy in driving from place to place. A shape memory alloy actuator has significant latent heat associated with the phase transition used for actuation, so spurious phase transitions are costly. As a consequence, it may be useful to consider the expected number of state transitions per unit time when formulating a control policy. The expected number of transitions can be calculated conditioned on x_t, p and q,

$$E(N_t^{trans}|x_t, p, q) = N(qx_t + p(1-x_t)) \quad (14)$$

In the steady state (9) can be substituted in, so that (14) is a function of N, p and q,

$$E(N_{ss}^{trans}|p, q) = \frac{2Npq}{p+q} \quad (15)$$

Using the scaling argument again, (15) can be rewritten in terms of βp_0 and βq_0. This implies that an increase in β implies more expected transitions per unit time in the steady state,

$$E(N_{ss}^{trans}|\beta p_0, \beta q_0) = \frac{\beta^2 2Np_0q_0}{\beta(p_0+q_0)} = \beta(2Np_0q_0) \quad (16)$$

The value of β minimizing the number of expected transitions is, naturally, 0, corresponding to the control policy that allows no random transitions between state.

D. Generalization to Many States

Both closed-loop and open-loop control laws of this form are not limited to the two-state case. In general, the centrally-specified state transition probabilities for a k-state graph could be represented as a matrix $\mathbf{M}$,

$$\mathbf{M} = \begin{bmatrix} p_{11} & p_{21} & \cdots & p_{k1} \\ p_{12} & p_{22} & \cdots & p_{k2} \\ \vdots & \vdots & \ddots & \vdots \\ p_{1k} & p_{2k} & \cdots & p_{kk} \end{bmatrix} \quad (17)$$

Each column of $\mathbf{M}$ must add up to 1. These $k^2 - k$ independent parameters could be chosen to minimize the expected error conditioned on the present knowledge for the number of agents in states $N_{t+1}^1 \ldots N_{t+1}^{k-1}$, as was done in (8),

$$E(N_{t+1}^i|\mathbf{M}, N_t^1, \ldots N_t^k) = N^{i,ref} \quad (18)$$

Similarly, (9) can be extended to multiple states so that the steady-state expected number of agents in each state is equal to the reference,

$$E(N_{ss}^i|\mathbf{M}) = Nv_1(\mathbf{M}) = N^{i,ref} \quad (19)$$

Here $v_1(\mathbf{M})$ is the eigenvector of $\mathbf{M}$ having eigenvalue 1. The second-largest eigenvalue will dominate the rate at which the feed-forward policy converges corresponding to the rate found in (7). Solving for these two conditions (the steady-state distribution and the rate of convergence) can be done algebraically or numerically. The other performance criteria also translate nicely to the multi-state case. The variance of the number of agents in state i is still determined only by the steady state probability of an agent being in that state, $v_{1i}(\mathbf{M})$, and by the total number of agents, N,

$$Var\left(N_{ss}^i|\mathbf{M}\right) = v_{1i}(\mathbf{M})(1 - v_{1i}(\mathbf{M}))N \quad (20)$$

Finally, the expected number of transitioning agents from (15) can be written for the k-state case in terms of the steady-state distribution,

$$E(N_{ss}^{trans}) = N\sum_{i=1}^{k} v_{1i}(1-p_{ii}) \tag{21}$$

The trade-offs for the multi-state case are not as easy to quantify nicely. However, the general observation is that as the second-largest eigenvalue of **M** decreases, the convergence time decreases and the expected number of transitions increases.

E. The One-Shot Policy

So far we have observed that it is impossible to minimize both convergence time and the number of transitions at steady state. This performance trade-offs in the no-knowledge, constant policy recruitment problem can be addressed by varying p and q with time. Suppose that we want to recruit, as accurately as possible, a specific number of agents, given no knowledge of N_t^{on} or x_0, in minimal time with minimal steady-state cost. It was demonstrated above that the accuracy of recruitment depends only on N and x_{ss}. As a consequence, the accuracy obtained by the $\beta = 1$ policy will not improve for more than one round of state transitions. This motivates the one-shot, time-varying recruitment policy:

1) Compute p and q that satisfies the steady state condition from (9), and sets the second-largest eigenvalue λ equal to 0.
2) Broadcast p and q to all agents.
3) After one round of stochastic state transitions, broadcast a command to all agents to set $p = 0$ and $q = 0$, so that all transitions cease.

This policy will guarantee that the agents get as close to the desired distribution as possible, subject to the variance of N_{ss}^{on} determined by (13).

IV. A Minimal Knowledge Feedback Policy

Often, it may be useful to consider the case in which the central controller has limited knowledge of the number of ON agents. Let y_t be a Boolean measurement which lets the controller know if the current distribution has reached the desired distribution, or is close enough. One definition of this could be when exactly the desired number of agents are in the ON state,

$$y_t = \begin{cases} true, & N_t^{on} = N^{ref} \\ false, & N_t^{on} \neq N^{ref} \end{cases} \tag{22}$$

The policy space to be searched is all policies for which a constant command (p, q) is broadcast, until the minimal feedback measurements determine that the desired state has been reached. At this point, all state transitions are commanded to cease, by setting $p = q = 0$. The broadcast command is assumed to be constant because, in the absence of additional information, there is no good reason for changing the command. This problem will be posed as a stochastic shortest path problem. In this framework, the cost function to be minimized in formulating a policy is the expected time that the system takes to converge to the desired state or output. We will formulate the cost J to be minimized as a function of the initial probability that agents are ON, x_0, given the additional knowledge that the system has not converged at time $t = 0$.

$$J = 1 + E\left[\sum_{t=1}^{\infty} g(y_t)\middle|\, x_0\right] \tag{23}$$

Here $g(y_t)$ is the cost per stage of the system, equal to 1 when the system has not yet reached the target, and 0 when the system is already there:

$$g(y_t) = \begin{cases} 1, & y_t = false \\ 0, & , y_t = true \end{cases} \tag{24}$$

At each point in time, Bellman's equation can be used to express the truncated cost J_t recursively forward in time,

$$J_t = E\left[g(y_t)|\, x_t\right] + E\left[J_{t+1}|\, x_t\right] \tag{25}$$

Keep in mind that the sequence $\{x_0, x_1, x_2, ..., x_t, ...\}$ is determined by (5), as long as the broadcast command (p, q) is constant. The stochastic nature of the cost function arises from the uncertainty about when the exact number of desired ON agents has been reached. After this point, the cost function becomes zero. The practical result of this is that the expectation of the cost-to-go will be equal to the probability of not converging at time t multiplied by the expected cost-to-go assuming that the desired state has not yet been reached. The likelihood of reaching the desired state can be described by thinking of the number of ON agents as a Bernoulli variable (the sum of many binary random variables),

$$H_t = P(y_t|x_t) = \binom{N}{N^{ref}} x_t^{N^{ref}}(1-x_t)^{N-N^{ref}} \tag{26}$$

Similarly, the likelihood of not reaching the desired state will be defined as $\bar{H}_t$,

$$\bar{H}_t = P(\bar{y}_t|x_t) = 1 - H_t \tag{27}$$

Using this, the expectation in (25) can be evaluated,

$$J_t = \bar{H}_t + \bar{H}_t J_{t+1} = \bar{H}_t[1 + J_{t+1}] \tag{28}$$

The benefit of rewriting the series representation of J_t recursively is that each term in the series carries a multiplicative term from the previous time step,

$$J_t = \bar{H}_t + \bar{H}_t\bar{H}_{t+1} + \bar{H}_t\bar{H}_{t+1}\bar{H}_{t+2} + ... + \prod_{k=t}^{m-1} H_k J_m$$

The recursive form of this expression can be used to derive the optimal policy. In order to demonstrate that a policy locally minimizes this cost function, we must first find a policy for which the gradient of J with respect to the policy parameters is zero, indicating that the policy lies at an extremum of J in parameter space. As in the no-knowledge case, nothing is known about the number of ON agents until the target distribution is achieved, so a policy with constant p and q must be pursued until y_t is true. The optimal policy to pursue is not hard to imagine; Essentially, it is to re-run the one-shot policy again and again until y_t is true. This can be proved analytically.

$$(p, q) = \left(\frac{N^{ref}}{N}, \ 1 - \frac{N^{ref}}{N} \right) \tag{29}$$

The first condition is that the gradient of J with respect to p and q is zero, for any initial probability x_0. J can be expanded recursively forward as a recursive series from any point in time using (28),

$$\frac{\partial J_t}{\partial p} = \frac{\partial \bar{H}_t}{\partial p}[1 + J_{t+1}] + \bar{H}_t \frac{\partial J_{t+1}}{\partial p} \tag{30}$$

The sign of each term in this series is determined by the sign of the partial derivative of $\bar{H}_t$ at each point in time. The other terms in the expression are probabilities, which are positive, or truncated cost functions, which must also be positive. The derivative of H with respect to x_t reduces to an expression in terms of H_t,

$$\frac{\partial H_t}{\partial x_t} = H_t \frac{N^{ref} - N x_t}{x_t(1 - x_t)} \tag{31}$$

The partial derivative of $\bar{H}_t$ with respect to p is:

$$\frac{\partial \bar{H}_t}{\partial p} = H_t \frac{N x_t - N^{ref}}{x_t(1 - x_t)} \frac{\partial x_t}{\partial p} \tag{32}$$

One multiplicative term within this expression, $N x_t - N^{ref}$, is particularly interesting. Using the allegedly optimal policy from (29), the value of x_t can be found for all t using (5):

$$x_t = \begin{cases} x_0, & t = 0 \\ N^{ref}/N, & t > 0 \end{cases} \tag{33}$$

For all $t > 0$, $N x_t - N^{ref}$ is equal to $N \cdot N^{ref}/N - N^{ref} = 0$. Consequently $\partial \bar{H}_t / \partial p = 0$ for $t > 0$, and by extension every term in the series defining $\partial J_t / \partial p$. Furthermore, this line of reasoning also works to show that $\partial J_t / \partial q = 0$, because each term of the series defining it will also contain a factor of $N^{ref} - N x_t$. The task remaining is to show that J is increasing as the policy parameters deviate from the optimal policy. It will not be shown here, but it is straightforward to demonstrate that J increases with even infinitesimal perturbations in p_0. As Figure 5, shows, however, the increase in J as β varies is very gradual, and second derivative tests do not suffice. One approach is to show that the gradient of J at some finite distance away from the critical point is always pointing away from the optimal policy. This is shown by making sure that the inner product between the vector distance to the optimal policy and the gradient at that point is positive:

$$(\Delta p, \Delta q) \cdot \nabla J_t = Z_t = \frac{H_t[1 + J_{t+1}]}{x_t(1 - x_t)} G_t + \bar{H}_t z_{t+1} \tag{34}$$

The sub-expression G_t determines the sign of each term in the series,

$$G_t = \left(N x_t - N^{ref}\right) \left(\Delta p \frac{\partial x_t}{\partial p} + \Delta q \frac{\partial x_t}{\partial q} \right) \tag{35}$$

As above, the only terms in the series defining the inner product which can be negative are within G_t. G_t can be evaluated for $p = \beta N^{ref}/N$ and $q = \beta(1 - N^{ref}/N)$, to obtain the expression:

$$G_t = N t (x_0 - p_0)^2 (1 - \beta)^{2t} \tag{36}$$

This is greater than zero for $\beta \neq 1$, so the cost function is also increasing in that direction. G_t can be shown to be positive for any small perturbation in p and q, but the expressions involved are lengthy and will not be repeated here.

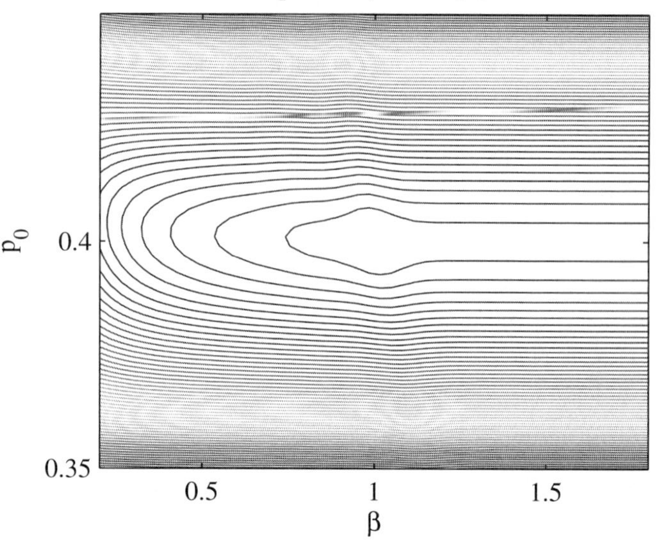

Fig. 5. The contour lines of the cost function for $N = 500$, $N^{ref} = 200$ show the shallowness of the minimum with respect to the scaling factor β. The cost steeply increases as p_0 is varied for a fixed N^{ref}. In contrast, changes in β result in only modest increases in J.

V. Computational Results

A swarm of 500 two-state agents was simulated with the goal of recruiting 200 agents into the ON state. The results illustrate the behaviors described above. Simple feed-forward control policies were computed, one setting $\beta = 0.1$ and the other$\beta = 1$. As predicted, the number of ON agents approaches 200 in both cases. The recruitment behavior of the $\beta = 0.1$ policy seems less random; however, this is the result of the auto-covariance of the system, not a difference in the actual variance of N_t^{on}. The cumulative distribution of values of N_t^{on} in the steady-state regime of both simulations shows that the overall distribution of both processes is nearly identical, as shown in Fig. 7 using the cumulative distribution function of N_t^{on} for each policy. However, the number of state transitions per unit time, shown in red on Figure 6, is much lower for the $\beta = 0.1$ policy, as predicted by (15).

The one-shot policy was simulated using the same reference as the constant policies, $N^{ref} = 200$. Figure 8 shows the result of 20 simulations. This control policy produces constant steady-state errors, unlike the constant policies. However, the distribution of these errors is exactly equal to those plotted in Fig. 7. Figure 9 demonstrates this using the result of 1000 simulations.

The minimal feedback policy introduced in Section IV was simulated for $N^{ref} = 200$, showing the typical convergence behavior. Figure 10 shows a simulated result, as well as the expected the expected convergence time to the desired distribution, 27.5 time intervals. The authors' previous work found that the expected convergence time under these same conditions, but with full knowledge of N_t^{on} was about 4.5 time intervals [5]. This gives a sense scale when considering the value of state information for feedback in recruitment problems.

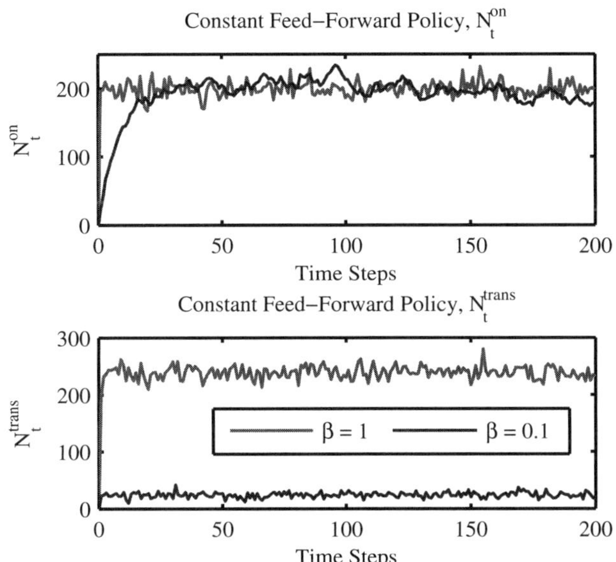

Fig. 6. A simulation of the constant policy recruitment algorithm. The number of ON cells for $N^{ref} = 200$ is plotted (blue) for $\beta = 0.1$ and $\beta = 1$, with the number of transitions per unit time (red). As predicted, reducing β by a factor of 10 reduces the number of transitions per unit time by a factor of 10.

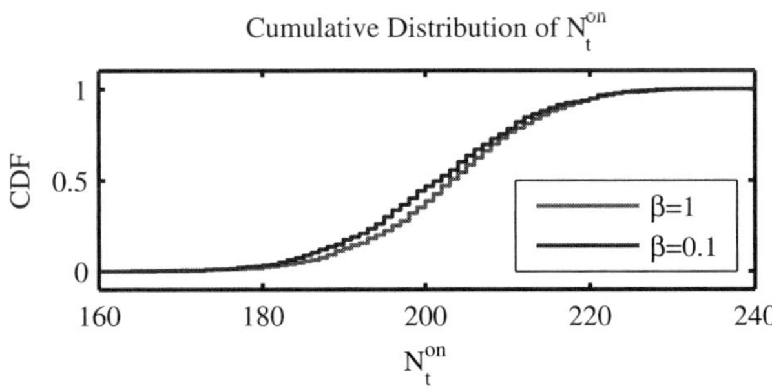

Fig. 7. The cumulative distribution of N_t^{on} in steady state, for the two simulations shown in Fig. 6. Despite the difference in convergence rate and smoothness of the time evolution, any value of β produces the same cumulative distribution.

VI. Conclusion

This paper has demonstrated that "recruiting" finite state agents into specified states in specified numbers is possible

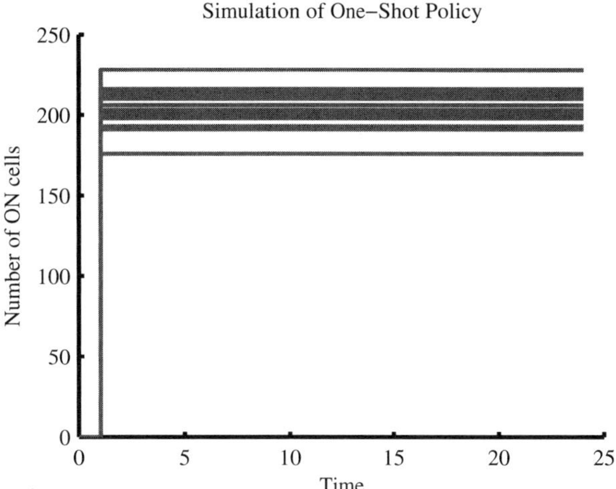

Fig. 8. Twenty simulation runs of the one-shot policy. This plot illustrates the trade-off made in this controller. In exchange for a steady-state error having the same variance as the constant policy, the agents make no state transitions at steady state.

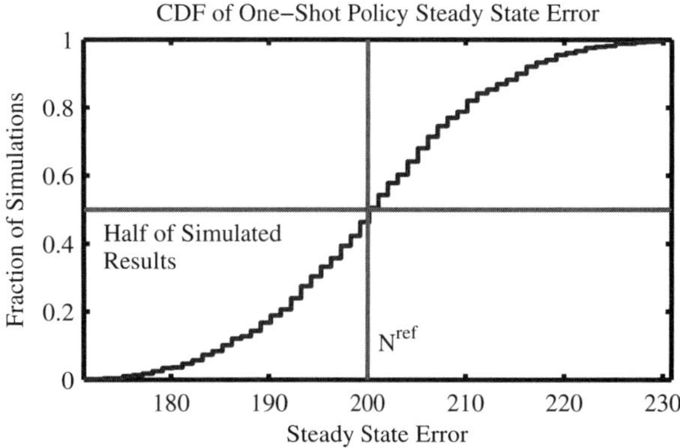

Fig. 9. A cumulative distribution function for the steady-state errors in the one-shot distribution. This distribution will approach exactly the same binomial limit distribution as the constant control policies.

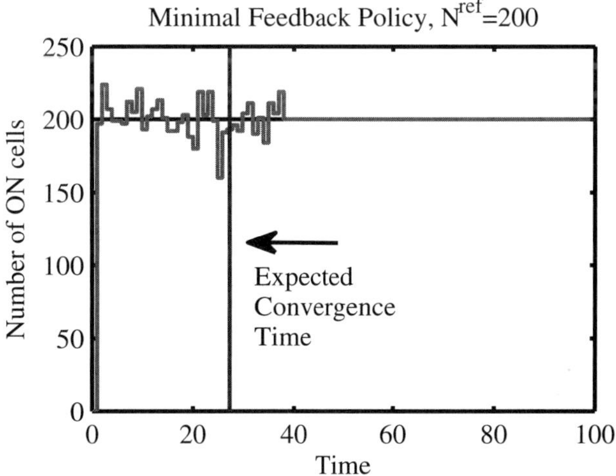

Fig. 10. The minimal feedback policy consists of using the $\beta = 1$ constant policy until some measurement confirms that the desired number of agents are recruited; The central controller then commands all agents to cease any state transitions.

using stable, feed-forward control policies as long as the controller can find inputs that satisfy a few eigenvector and eigenvalue constraints on the state evolution equation. Unlike the previously described closed-loop policies, these feed-forward control laws cannot cause the system to converge. However, the random distribution of agents among states will be tightly grouped about the desired distribution with a well-characterized variance. Because the variance is well-characterized and relatively independent of the other performance measures for the system, it is straightforward to assess whether a closed-loop law or a feed-forward law provides greater accuracy simply by comparing the variance in the response of the closed-loop policy with uncertain information to the variance of the feed-forward law. This analysis is restricted to very limited knowledge of the state of agents in the swarm; Past work on recruitment policies having full knowledge of N_t^{on} has yielded good results in finding numerically optimal control policies. Finding policies for systems in which partial knowledge is obtainable in the form of an estimating filter, or for systems with a limited range of control inputs both remain interesting future directions.

References

[1] N. Mantzaris, S. Webb, and H. Othmer, "Mathematical modeling of tumor-induced angiogenesis", *Mathematical Biology*, Vol. 49, pp.111-187, 2004

[2] S. McDougall, R.A. Anderson, and M. Chaplain, "Mathematical modeling of dynamic adaptive tumor-induced angiogenesis: Clinical implications and therapeutic targeting strategies", *Journal of Theoretical Biology*, Elsevier, 2006.

[3] J. Ueda, L. Odhner, S. Kim, H. Asada, "Distributed Stochastic Control of MEMS-PZT Cellular Actuators with Broadcast Feedback," *in 2006 IEEE-RAS International Conference on Biomedical Robotics and Biomechatronics*, 20-22 February 2006, p. 272-277.

[4] L.Odhner, J. Ueda, H. Asada, "Feedback Control of Stochastic Cellular Actuators," in *The 10th International Symposium on Experimental Robotics*, 2006, p. 481-490.

[5] L. Odhner, J. Ueda, H. Asada, "Stochastic Optimal Control Laws for Cellular Artificial Muscles," in *2007 International Conference on Robotics and Automation*, 1-14 April 2007, p. 1554-1559.

[6] J. Ueda, L. Odhner, H. Asada, "Broadcast Feedback of Stochastic Cellular Actuators Inspired by Biological Muscle Control", *International J. of Robotics Research*, v.26, n.11-12, pp.1251-1265

[7] Zajac, Felix, "Muscle and Tendon: Properties, Models, Scaling and Application to Biomechanics and Motor Control", *Crit. Reviews in Biomedical Engineering*, v.17, n.4, 1989, pp. 359-411

[8] E. Henneman, G. Somjen, and D.O. Carpenter, "Functional significance of cell size in spinal motoneurons," *J Neurophysiol*, 1965, v. 28, p. 560-580.

[9] A. Julius, A. Halasz, V. Kumar, G. Pappas, "Controlling biological systems: the lactose regulation system of Escherichia coli," *in 2007 American Control Conference*, 9-13 July 2007 p. 1305-1310.

[10] D. Gillespie, "Exact Stochastic Simulation of Coupled Chemical Reactions," *The Journal of Physical Chemistry*, Vol, 8 1, No. 25, 1977

[11] S. Berman, A. Halasz, V. Kumar, S. Pratt, "Bio-Inspired Group Behaviors for the Deployment of a Swarm of Robots to Multiple Destinations," *in 2007 International Conference on Robotics and Automation*, 1-14 April 2007, p. 2318-2323.

Prior Data and Kernel Conditional Random Fields for Obstacle Detection

Carlos Vallespi
National Robotics Engineering Center
Carnegie Mellon University
Pittsburgh, Pennsylvania 15201
Email: cvalles@ri.cmu.edu

Anthony (Tony) Stentz
National Robotics Engineering Center
Carnegie Mellon University
Pittsburgh, Pennsylvania 15201
Email: axs@ri.cmu.edu

***Abstract*— We consider the task of training an obstacle detection (OD) system based on a monocular color camera using minimal supervision. We train it to match the performance of a system that uses a laser rangefinder to estimate the presence of obstacles by size and shape. However, the lack of range data in the image cannot be compensated by the extraction of local features alone. Thus, we investigate contextual techniques based on Conditional Random Fields (CRFs) that can exploit the global context of the image, and we compare them to a conventional learning approach. Furthermore, we describe a procedure for introducing prior data in the OD system to increase its performance in "familiar" terrains. Finally, we perform experiments using sequences of images taken from a vehicle for autonomous vehicle navigation applications.**

I. Introduction

Obstacle detection (OD) is important in many mobile robot applications and autonomous vehicles. The most successful OD systems rely on range information to detect obstacles by size and shape. Among all the range sensors, laser rangefinders are the most popular and widely used range sensors, due to their quality of data.

Unfortunately, laser rangefinders contain mobile parts in their design which makes them complex and expensive. Furthermore, they may require extra hardware to scan the scene and a precise calibration. In contrast, color cameras are mass-produced and are comparatively inexpensive. However, the lack of range data makes the OD problem more challenging. Local features alone are not enough to extract enough information to detect obstacles reliably. We alleviate this problem by exploiting the contextual information in the image. For example, since we cannot measure the shape of the rock directly, we learn that rocks are gray objects with certain texture properties and surrounded by brown dirt.

This paper investigates several contextual techniques based on Conditional Random Fields (CRFs), which have been successfully used in the past for classification and segmentation tasks, and allow to exploit contextual features in the image. One of the CRF models presented in this paper uses a log-linear model which imposes a linear combination of the input features. Then, we apply the "kernel trick" to this model to allow the use of different kernels with the hope that, in the projected space, classes become linearly separable.

1: A John Deere tractor (4710 series) was used for some of the experiments in agricultural applications. It was equipped with cameras, positioning sensors, a computer and a SICK laser for perception.

Moreover, we present an algorithm to improve the OD system performance in agricultural environments and, in general, in applications where the system re-visits the same areas. In this case, we build a database of hand -or automatically- classified images of the terrain and integrate them in a contextual model to improve the obstacle detection process.

The contextual methods presented in this paper are compared to conventional non-contextual learning approaches. We use a OD system equipped with a SICK laser, a color camera, positioning sensors (IMU, GPS and wheel encoders for speed measurements) to extract the precise location of the obstacles (see fig. 1). Then, all the algorithms are trained to match the performance of this OD system.

This paper is structured as follows. The next section gives a short overview of prior work in this area. In section III we introduce the basic CRF model. Section IV describes our procedure to apply the kernel trick to a log-linear CRF model. In section V we show an algorithm to make use of prior data.

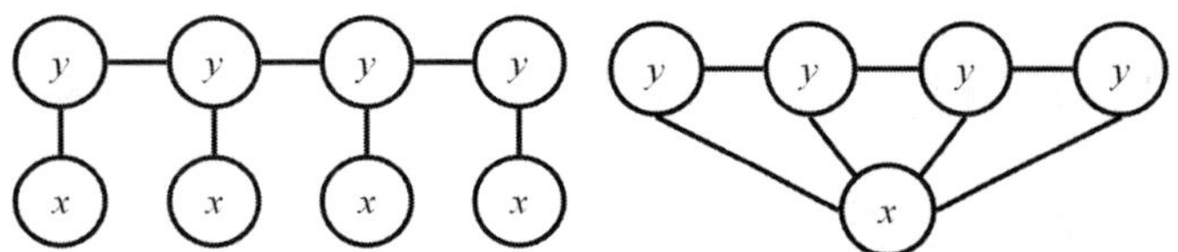

2: **Left:** MRF graph structure. **Right:** CRF graph structure.

Finally, we show the effectiveness of our algorithm in a batch of experiments in section VII.

II. Related Work

In the obstacle detection and avoidance field, Dima presented an algorithm which combined the information provided by different sensors in the image frame [3]. The data from each sensor was transformed to image coordinates, thus each part of the image contained features from the image plus features from other sensors (i.e. infrared, LADAR, etc). Whereas the algorithm enabled the fusion of heterogeneous sensors, the classifier that was used assumed independence among all the parts in the image.

Many cues needed for obstacle detection are contextual information, and Markov Random Fields (MRFs) are widely used machine learning tools to exploit this information. However, in the MRF framework, the observed data is assumed to be conditionally independent which can be very restrictive in some applications (see fig. 2). Unlike MRFs, CRFs model directly the conditional distribution. Thus, the relations between the input variables do not need to be explicitly represented. In the past, their main limitation was the use of slow training algorithm (such as iterative scaling -IIS-); however, recent advances in CRF theory have found efficient algorithms for parameter learning and inference in general CRF graphs [15].

Log-linear CRFs have been successfully used in the past for image labeling. In this form, CRFs allow for a parameter estimation guaranteed to find the global optimum due to the convex property of their conditional likelihood function. For instance, CRFs have been used for detection of man-made structures in natural images [7]. CRFs have been used for object detection and recognition given its parts in images [11]. They have been also used for object segmentation tasks in images [14], and with occlusion handling [17]. Also, this idea has been extended to segmentation in video sequences [16]. Saxena el al. [12] uses a discriminative MRF model to estimate the depth using a single still camera, which could be used as the input to an obstacle detection algorithm.

A linear relation between features and random variables in the CRF model has been widely used, but in some cases this can be a restrictive constraint. To overcome this limitation, an extension to the CRF model to allow the use of custom kernels was proposed by [8]. In this paper, we present an alternative algorithm. In general, the possibility of changing the kernel allows the model to adapt better to a specific problem resulting in better performance.

III. Conditional Random Field model

A. CRF model

A Conditional Random Field (CRF) is an undirected graphical model in which edges represent conditional dependencies between random variables at the nodes. The distribution of each random variable y_i is conditioned on an input sequence $\mathbf{x}$. The conditional dependency of the random variables on $\mathbf{x}$ is defined by using feature functions with some associated weights. Together, they can be used to determine the probability of each y_i. Dependencies among the input variables $\mathbf{x}$ do not need to be represented because the model is conditional, affording the use of complex and rich features of the input. Thus, CRFs are discriminative models, that is, they model $p(\mathbf{y}|\mathbf{x})$[1].

In a general way, to model the conditional probability distribution of a sequence of labels $\mathbf{y}$ given the observations $\mathbf{x}$, $p(\mathbf{y}|\mathbf{x})$ takes the form shown in (1):

$$p(\mathbf{y}|\mathbf{x}) = \frac{1}{\mathbf{Z}(\mathbf{x})} \prod_{c \in \mathbf{C}} \Psi_c(v_c), \tag{1}$$

where $\Psi_c(v_\mathbf{c})$ is a potential function that depends on the variables in a cluster c (defined as v_c). $\mathbf{Z}(\mathbf{x})$ is called the partition function, and it is a normalization factor to make sure that $\sum_{y_i} p(y_i|\mathbf{x}) = 1$. It depends on the data, therefore it takes different values as the input ($\mathbf{x}$) changes.

In this paper, we use a log-linear model for the CRF. Thus, Ψ are potentials of the form shown in (2):

$$\Psi_c(v_\mathbf{c}) = \exp\left\{\sum_{\mathbf{k}} \lambda_\mathbf{k} f_k(\mathbf{x_c}, \mathbf{y_c})\right\}, \tag{2}$$

where $\mathbf{x_c}, \mathbf{y_c} \in v_\mathbf{c}$ and f_k is feature k function over $\mathbf{x}, \mathbf{y}$.

In the image labeling task, the CRF model we use is a lattice, forming an undirected graph $G = (\mathbf{V}, \mathbf{E})$. $\mathbf{V}$ are the nodes or vertices and $\mathbf{E}$ are the edges. Every node and every edge contain a potential function that operates on a subset of the random variables present in G. Thus, we define the conditional probability distribution of the CRF as shown in (3):

$$\begin{aligned} p(\mathbf{y}|\mathbf{x}) &= \tfrac{1}{\mathbf{Z}(\mathbf{x})} \textstyle\prod_{i \in \mathbf{V}} \Psi(x_i, y_i) \prod_{(i,j) \in \mathbf{E}} \Psi(y_i, y_j, x_i, x_j) \\ Z(\mathbf{x}) &= \textstyle\sum_y \prod_{i \in \mathbf{V}} \Psi(x_i, y_i) \prod_{(i,j) \in \mathbf{E}} \Psi(y_i, y_j, x_i, x_j), \end{aligned} \tag{3}$$

where, functions Ψ are of the form:

$$\begin{aligned} \Psi(x_i, y_i) &= \exp\{\textstyle\sum_k \mu_{k y_i} g_k(x_i)\} \\ \Psi(y_i, y_j, x_i, x_j) &= \exp\{\textstyle\sum_k \lambda_{k y_i y_j} f_k(x_i, x_j)\} \end{aligned} \tag{4}$$

We set $\mu \in \Re^{K \times L}$ and $\lambda \in \Re^{K \times L \times L}$. L is the number of different labels or classes and K is the number of features.

[1]Bold letters denote an array of elements or variables. Functions are represented by non-bold letters followed by parentheses, i.e. $p(\mathbf{x})$. All non-bold no-function letters represent variables. In the CRF context, unless stated differently, $\mathbf{x}$ denotes data and $\mathbf{y}$ denotes labels. Sub-indexes denote elements of the array, i.e. x_i denotes the data at the i-th node. y_i denotes the label at the i-th node. $\{y_i, y_j\}$ denotes a pair of labels of $\mathbf{y}$ at nodes i, j. $\mathbf{y} = \{y_i, y_j\}$ represents all pair of labels of adjacent nodes of $\mathbf{y}$ equal to the pair $\{y_i, y_j\}$. $\mathbf{y}^m$ denotes the m-th sequence of $\mathbf{y}$ labels.

Unlike other representations found in the literature, we chose functions g_k and f_k which only depend on the data, and not on the labels. Because of this representation, the weights are the ones that depend on the labels. Thus, to account for the different classes, $L-1$ hyper-planes are needed.

The total number of node weights is $(L-1)K$ and it is equivalent to use $L \times K$ node weights and set $\forall k = 1..K \ \ \mu_{kL} = 0$. We chose the edge weights to be the absolute value of the difference of features in adjacent nodes. Hence, the total number of edge weights in this representation is $L \times L \times K$ because we need as many weights as node features and combinations of pairs of labels. However, we restrict $\lambda_{kls} = -\lambda_{kll} \forall l \neq s$, which reduces the number of edge weights to $L \times K$.

It is interesting to note that the model defined in (3) contains a logistic regression classifier in each node. Simply, by setting the edge weights to 0 (i.e., define $\forall i,j \in \mathbf{E} \ \ \lambda_{ky_iy_j} = 0$) it is easy to see that every node contains a multi-class logistic regression classifier.

In summary, the set of parameters for the CRF in our representation is the union of the node weights and the edge weights ($\phi = \{\mu_{1...K,1...L-1}, \lambda_{1...K,1...L}\}$), giving a total of $K \times (2L-1)$ parameters.

B. Inference in 2D CRF

It is worth noting that inference problems like marginalization and maximization are NP-hard to solve exactly and approximately (at least for relative error) in lattice graphical models, and in general, for most of the graph structures. In a CRF graph model, maximization is to find the most likely sequence of labels $\mathbf{y}$ given an input $\mathbf{x}$, that is:

$$\mathbf{y}^{max} = \underset{\mathbf{y}}{\text{argmax}} \ p(\mathbf{y}|\mathbf{x}, \phi) \tag{5}$$

However, finding $\mathbf{y}^{max}$ exactly is infeasible in practical cases for 2D CRFs. A brute force algorithm would need to explore all possible labelings, which in a binary CRF of size 24×32 would be 1.5×10^{231}. In theory, the marginalization problem for graphical models with loops is #P-complete and maximization is NP-complete. Thus, approximate inference is used to solve these problems.

There are several methods in the literature for approximate inference in graphs (i.e. maxent -although only works for binary labels-, variational methods, Monte Carlo methods, Belief Propagation (BP), etc...). We used BP for approximate inference, because it gives good results in practice [9] and provides solution to the marginalization and maximization problems.

C. Maximum Likelihood parameter learning

In our work, we use the MLE principle to learn the parameters ϕ such that the regularized negative log-likelihood is minimized. The algorithm assumes that we are given set of i.i.d. labeled images $(\mathbf{X}^m, \mathbf{Y}^m) \in \mathbf{M}$. Regularization is added in the form of a Gaussian centered at 0 over the parameters to avoid over-fitting. The negative regularized log-likelihood for a CRF model is given by (6):

$$nll(\phi) = -\sum_{m \in \mathbf{M}} \log\{p(\mathbf{y}|\mathbf{x}^m, \phi)\} + \frac{\lambda}{2}\phi^T\phi \tag{6}$$

Ignoring the regularization term, the derivatives of the log-likelihood over the parameters ϕ yield to equations in (7). The first term is the value of the features under the empirical distribution. The second term, which arises from the derivative of $\log Z(\mathbf{x})$, is the expectation of the features under the model distribution. Equation (7) shows the derivatives corresponding to μ and λ, respectively:

$$\begin{aligned} \frac{\partial nll(\phi)}{\partial \mu_{kl}} &= -E_{\tilde{p}(y=l,\mathbf{x})}[g_k] + E_{p(y=l|\mathbf{x};\phi)\tilde{p}(\mathbf{x})}[g_k] \\ \frac{\partial nll(\phi)}{\partial \lambda_{kls}} &= -E_{\tilde{p}(\mathbf{y}=\{l,s\},x)}[f_k] + E_{p(\mathbf{y}=\{l,s\})|\mathbf{x};\phi)\tilde{p}(\mathbf{x})}[f_k] \end{aligned} \tag{7}$$

Unfortunately, there is no analytic solution to this equation (setting the gradient to 0 and solving for λ does not always yield to a closed-form solution). Thus, an iterative algorithm is needed in order to approximate the optimal solution. Note that the function $nll(\phi)$ is concave, which follows from the convexity of functions of the form $g(\mathbf{x}) = \log \sum_i \exp x_i$. This property shows that every local optimum is also a global optimum. Adding *L2* regularization to the NLL ensures that $\log l(\phi)$ is strictly concave, which implies that it has exactly one global optimum.

Thanks to this property, methods like steepest descent can be used, although they may require many iterations to converge making them slow. Newton's method can converge much faster because it takes into account the curvature of the likelihood. However, computing the Hessian can be expensive, too, since it is quadratic in the size of the parameters. An intermediate solution for this problem is to use quasi-Newton methods, such as the Broyden-Fletcher-Goldfarb-Shanno (BFGS) method [1, 4, 6, 13], in which the Hessian is updated by analyzing successive gradient vectors. BFGS is the algorithm that we use in this work for optimization (through `fminunc` in Matlab), and, as it is discussed in [15], it provides a rapid convergence to the optimal solution.

IV. Kernel Conditional Random Fields

A. "Kernel Trick" and Logistic Regression

Logistic Regression (LR) is a discriminative linear classifier that estimates the $p(y|x)$ by using a linear combination of features of x. The conditional likelihood for LR is defined in (8):

$$p(y|x) = \frac{\exp\{\sum_k \mu_{ky} g_k(x)\}}{1 + \sum_{y=1}^{y=L-1} \exp\{\sum_k \mu_{ky} g_k(x)\}} \tag{8}$$

By applying the "kernel trick", one can convert a linear classifier algorithm into a non-linear one by using a non-linear function to map the original observations into a higher-dimensional space; this makes a linear classification in the new space equivalent to non-linear classification in the original space (see fig. 3). We apply the Mercer's theorem, which states that any continuous, symmetric, positive semi-definite kernel

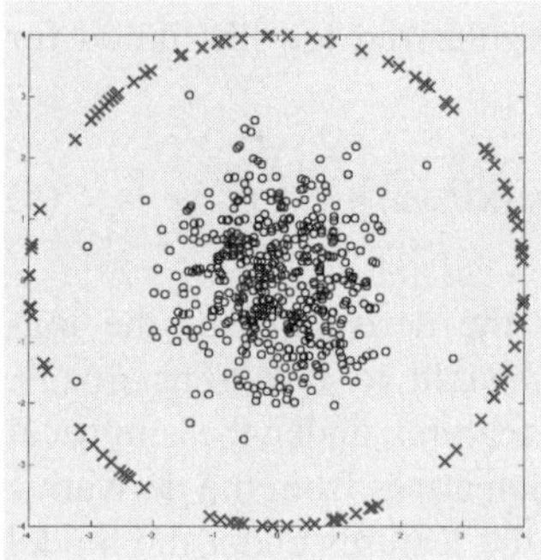

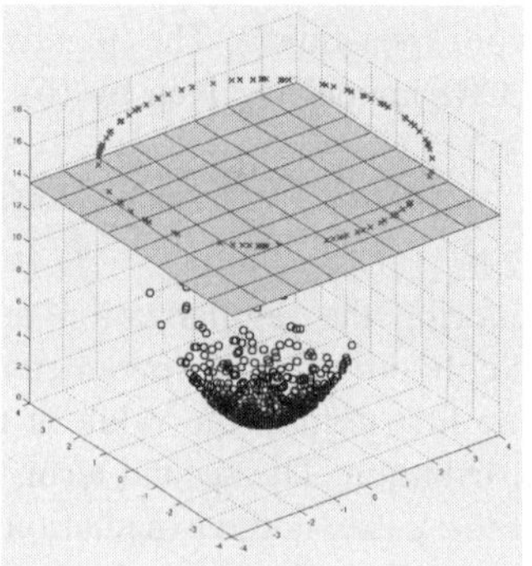

3: Example of a dataset not linearly separable in the original dimension space (on the **left**), but it becomes separable after using a quadratic kernel (on the **right**), augmenting the dimensionality of the input space by 1.

function $K(x_i, x_j)$ can be expressed as a dot product in a high-dimensional space, to equation 8:

$$p(y|x) = \frac{\exp\left\{\sum_{x_i \in S} \alpha_{iy} K(x, x_i)\right\}}{1 + \sum_{y=1}^{L-1} \exp\left\{\sum_{x_i \in S} \alpha_{iy} K(x, x_i)\right\}}, \quad (9)$$

where S is the space of vectors that span the kernel. As with LR, the log-likelihood is a convex function, and it is possible to compute the gradient and the Hessian of the log-likelihood, making suitable Newton-Raphson methods for rapid optimization. However, there are several performance penalties by using this method:

- Computing the kernel matrix can be computationally expensive ($O(N^2)$, where N is the number of vectors).
- A $N \times N$ matrix must be inverted at each iteration of Newton-Raphson method, increasing the computational cost of the training algorithm to the order of $O(N^3)$.
- In practice, most (if not all) α_i have non-zero values, which increases the cost of classifying samples (each new sample needs to be projected into all the x_i in the kernel).

These problems can be solved by fixing S to use a small subset of x_i. However, we need an algorithm to determine which and how many x_i should be in S.

In this paper, we will use the method proposed by [18]. The sub-model S found by IVM algorithm is an approximation to the full model found by KLR. The algorithm starts with an empty set of vectors for S. Then, at every iteration the vector that minimizes de NLL the most is added to S. The vectors in the kernel space S are called import vectors. The algorithm stops when the NLL does not decrease after some number of iterations. A toy example is shown in fig. 4.

B. *"Kernel trick" for CRFs (K-CRFs)*

K-CRFs were originally introduced in [8], and they allow the use of implicit feature spaces through Mercer kernels. Our approach differs mainly in the way we find the kernel space. In our algorithm, we use the IVM algorithm as described in section IV-A to find the vectors x_i that will span the kernel space S in the node potentials. Then, we compute the g_k projected into the kernel space found by the IVM algorithm

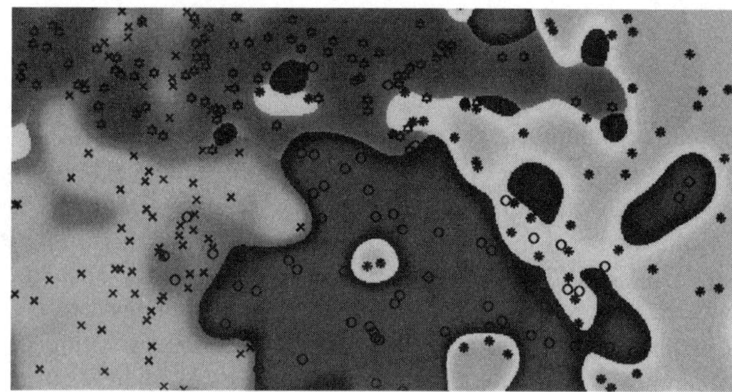

4: Example of boundaries obtained by a multi-class IVM classifier, 36 import vectors and a a Gaussian kernel ($\sigma = 0.1$).

in order to extract the set of node features that will be used by KCRF. Finally, we use the algorithm described in section III-C for parameter learning till its convergence.

In this work, we experiment with Gaussian kernels because they often provide good performance [2]. When they are used, the corresponding feature space is a Hilbert space of infinite dimension. However, the regularization used for parameter learning avoids the infinite dimension to spoil the results. In this paper, we refer to the Gaussian kernels the ones that take the form in (10):

$$K(x_i, x_j) = e^{-\frac{\|x_i - x_j\|^2}{2\sigma^2}}, \quad (10)$$

where x_i, x_j are input vectors.

V. Prior data and K-CRFs (PK-CRF)

In this section, we introduce an algorithm to take advantage of situations where prior labeled data is available (i.e., agricultural applications where vehicles revisit the same areas multiple times). If we label the data that corresponds to the working area once, we may be able to use these labelings to improve future labeling performances. Thus, in cases where the input image is similar to one in the prior data set, we may be able to re-use the prior labels. The parts of the image which experience changes (i.e. illumination, pose, new/missing objects, etc) may need new evaluations.

The algorithm described in this section assumes that the input data is tagged with pose (which does not need to be exact), and that the ground plane is mostly flat. For this task, we build a database with the prior labeled images tagged with pose. Then, for every new input image, the closest image in pose is recovered from the database and aligned with the input image (see section VI). At this point, we can refer each part/region in one image to the other. Thus, if we know the ground truth for the labels of one of the images, we can use this information in the image that we are trying to label.

In practice, there are differences between the input image and the reference image, even after the alignment, due to errors in the alignment process, errors in pose, possible changes in the environment, moving obstacles, etc. However, we let the CRF to decide for us whether the label in the reference image should be used.

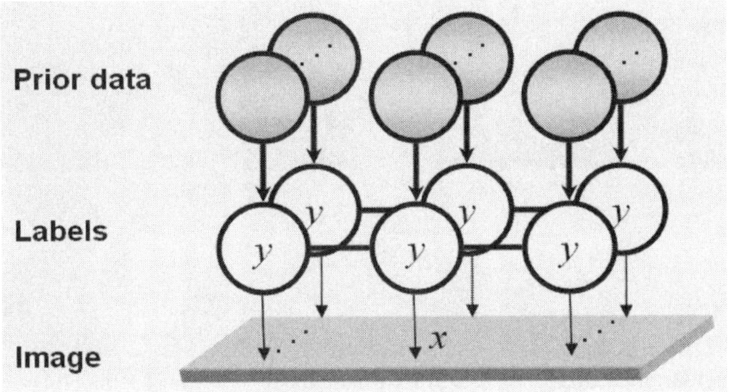

5: PK-CRF model. White nodes represent random variables.

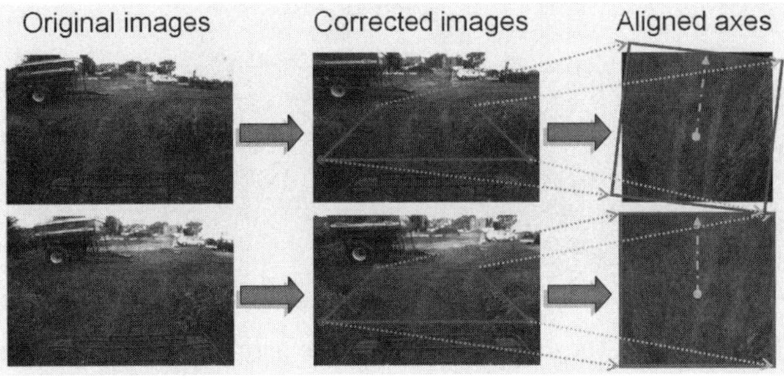

6: Axes alignment for two images. Images are rectified, transformed to a top-down view, and rotated to align their axes.

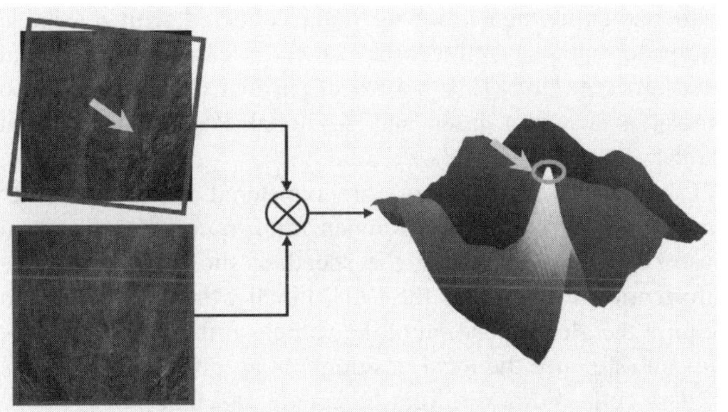

7: Normalized correlation between two images.

We added the prior data into the CRF model in the form of binary features for the nodes and real features for the edge. These features incorporate labeling information of the reference image to the input image. The node features have a value of 1 if the hypothesis about the label in the input image is the same as the reference image, and 0 otherwise. The edge features contain information about the differences between the region in the input image and the region in the reference image (similar to the edge features in the CRF). We refer to these features as h_k.

Each of these features is multiplied by a weight that depends on the class ($\gamma_n l$). Putting together this new features with the CRF model, we get the following node potentials:

$$\Psi(x_i, y_i) = \exp\left\{\sum_k^K \mu_{ky_i} g_k(x_i) + \sum_n^N \gamma_{ny_i} h_n(x_i)\right\} \quad (11)$$

Where, K is the number of node features and N is the number of prior features for each node. Fig. 5 shows a graphical representation of the model. Discrete random variables are connected to the image features and also connected to the features and label information from the reference image. Some of the prior features may be missing in the cases where no match is found for a patch in the reference image.

By adding these features we expect the CRF model to learn that if the node of the image and the node in the reference image are similar, then there should be a bias towards using the same label in the input image as the one used in the reference image. However, if the regions differ, then the information coming from the image taken in the past should be discarded and a full evaluation of the region would be required. We called this model Prior K-CRF (PK-CRF).

VI. Image Registration

We use an algorithm for aligning two images with different poses based in [10]. We work under the assumption that our test environment has a planar surface. Thus, we use an homography to transform the image to an orthonormal view (top-down in our case) by means of four fixed reference points in the ground. This transformation works for the ground (if it is planar) but does not work for trees or other objects which are not in the same plane which will often experience several distortion (see bushes in rightmost images in fig. 6).

Once both images are transformed in these coordinates, the optical axes are parallel. We can use the yaw angle to rotate one of the images and align the axes of the two images as shown in fig. 6. Due to differences in the actual X, Y, Z coordinates where each image was taken, the resulting images may not align, yet. However, after these transformations, the camera axes are parallel and we can compensate for these differences in pose by a simple translation. We use normalized correlation to compute the relative translation of one image w.r.t. the other (see fig. 7 for details). Finally, we can relate each patch from the reference image to each one in the input image (see fig. 8).

In our experiments, images with differences in position smaller than 3m, and differences in yaw smaller to 10 degrees were successfully registered. Beyond these numbers, this method may fail to successfully recover the correct translation for every patch in one image to match the patches in the reference image. Hence, it is important to find images in the database which are very close in pose to the image that we are labeling.

VII. Experiments

In this section, we show that the use of contextual models improves the performance for obstacle detection tasks from images. We compare the different contextual models presented in this paper (CRF, KCRF, PK-CRF), a logistic regression classifier (LR) and the Import Vector Machine (IVM).

8: **Left:** Flow for every 16×16 pixels patch to match patches in the right image. **Right:** Reference image.

(a) Original images (b) CRF labelings (c) LR labelings

9: Red boxes denote detections. **Top row:** Orchard detection. **Bottom row:** People detection.

A. Data acquisition platform and features

In our experiments, we use data collected with a vehicle equipped with several sensors: a color camera, an Inertial Measurement Unit (IMU), a wheel encoder (for vehicle speed input), a scanning laser and a Global Positioning System (GPS).

Camera and laser sensors are registered w.r.t. each other and the vehicle. We use a Kalman filter to compute the local pose of the vehicle using the speed of the vehicle and the information collected by the IMU. Finally, the GPS is used to acquire the global position of the vehicle at the start time, thus we can reference the local position among different sequences of data collected at different times (we call them logs).

Every image is divided into a grid of patches of 16×16, and we extract feature information from each patch independently, as described in [3]. The features extracted contain the mean and standard deviation for the U,V components in the LUV color space and texture information for a total of 28 features. Every feature was scaled to have 0 mean and standard deviation of 1.

The ground truth contains binary labels (obstacle/not obstacle) for every patch in each image and it is automatically computed using the laser data. The 3D data extracted from the laser is very accurate and it is used to get a good estimation of the ground. Once the ground is estimated, the detection of obstacles becomes very simple (i.e. any 3D point above the ground more than 0.5 m is an obstacle). We project the object location to the image frame and use that information for automatically getting the labels for every image patch in the grid.

LR classifier is trained using the 28 features + a constant feature to account for the bias. Similarly, we use the same features for the CRF node potentials and a total of 57 edge features (28×2(classes) $+ 1$(bias)). The edge features were computed as the Euclidean distance of two adjacent node features. Therefore, the total number of variables of the linear CRF was 86, which were successfully learned using the algorithm described in section III-C.

B. Comparison of LR and CRF

We compare the performance of LR and CRF (trained using the same features as described in section VII-A) for obstacle detection in an agricultural application. One of the tasks is to be able to drive a vehicle (i.e. tractor) through orchard tree lanes. The classifier must properly segment and detect the orchard tree lanes. We acquired two sequences of images and we used one for training the classifiers and the second one for producing the video: `http://www.cs.cmu.edu/~cvalles/videos/orchard.avi`

As can be seen in the video and in fig. 9 (top row), CRF provides a cleaner segmentation of the orchard tree lanes. Furthermore, the number of false positives produced by the CRF classifier is much lower. LR produces false positives continuously, some of them just in front of the vehicle, which would make the vehicle to stop. However, CRF does not produce any false positive in front of the vehicle while properly segmenting the orchard tree lanes throughout the video sequence. CRF significantly outperformed LR for detection and segmentation of people in our experiments as it is shown bottom row of fig. 9.

C. Comparison of LR, CRF, IVM, KCRF and PKCRF

We collected 8 sequences of data in a mostly flat and grassy environment, driving at 2m/s for about 4 minutes, with several obstacles scattered around. We logged 2 images per second. Two sequences were used for training and the other 6 were used for testing. Our definition of obstacle in this environment is anything that is above the ground more than a certain distance (i.e. 0.5m). Thus, in our gathered data, the obstacles are bushes, cones, trees, vehicles, fences, etc (see fig. 10). The data was collected in two different days, and some of the obstacles were placed in different positions. We followed similar trajectory for every log we took, just allowing deviations from the original path smaller than 5m.

CRF and LR were trained as described in section VII-A. We used a Gaussian kernel for IVM, KCRF and PKCRF classifiers, and we experimentally found the optimal kernel width to be $\sigma = 3$ by plotting the histogram of distances mapped by the kernel as proposed in [5]. The IVM algorithm found 119 import vectors. K-CRF used an extra 57 variables for the edge potentials (same ones as in the linear case) and 1 for bias, totaling 177 variables. In the case of PK-CRF, one of the sequences was used as reference to extract features described in section V, and was used as database for the test

10: Some snapshots of the environment in which data was collected. Note that an obstacle can be anything that may be a hazard for the vehicle (bushes, cones, other vehicles, trees, etc.)

TPR	LR	IVM	CRF	KCRF	PKCRF
0.95	61.6% ± 0.4	37.3% ± 0.7	37.0% ± 1.3	20.1% ± 2.8	13.5% ± 5.4
0.92	38.9% ± 0.5	16.6% ± 3.0	10.5% ± 6.1	4.3% ± 3.3	2.9% ± 2.4
0.90	24.3% ± 4.4	10.9% ± 2.2	4.1% ± 3.2	1.9% ± 1.6	1.6% ± 0.8
0.88	15.2% ± 3.5	7.2% ± 1.0	2.0% ± 1.6	1.2% ± 0.5	0.9% ± 0.3
0.85	9.5% ± 1.4	4.7% ± 0.5	1.0% ± 0.3	0.6% ± 0.2	0.4% ± 0.2
0.80	5.1% ± 0.6	2.8% ± 0.2	0.6% ± 0.1	0.3% ± 0.1	0.2% ± 0.1
0.75	3.3% ± 0.3	1.9% ± 0.1	0.4% ± 0.1	0.2% ± 0.1	0.1% ± 0.1

I: False Positive Rate (FPR) for a given True Positive Rate (TPR) for each algorithm evaluated in this paper.

experiments.

In order to compare the performance of the different algorithms, we considered a false positive an alarm from the classifier in an area of a 3×3 image patches that does not contain an obstacle. In this application, a false positive may cause the vehicle to stop for no apparent reason, degrading the performance of the autonomous vehicle. However, a false negative may be fatal. Hence, it is very important to achieve high obstacle detection rates when working at low false positive rates.

Table I shows the false positive rate of the various classifiers at different performance points. In this case, differences among the classifiers become apparent, specially at low false positive rates. The false positive rate (FPR) for a fixed true positive rate (TPR) is shown in table II. Whereas neither LR nor IVM could be used in practice because of its large FPR at any performance point in the table, contextual methods give enough performance boost to be considered.

At low FPRs, contextual methods perform several times better than non-contextual ones. For instance, at a fixed FPR of $1/250$, LR and IVM achieve obstacle detection rates of 30% and 43%, respectively (see table II). Contextual methods (CRF, KCRF and PKCRF) give performances over 75%. PKCRF gives a performance just shy of 85% at the same FPR.

Also, note that PKCRF achieves an obstacle detection

FPR	LR	IVM	CRF	KCRF	PKCRF
1/1000	4.5% ± 0.2	15.0% ± 1.3	57.1% ± 2.5	70.5% ± 3.3	75.3% ± 3.3
1/750	7.6% ± 0.50	21.2% ± 0.9	63.5% ± 2.8	73.2% ± 3.3	77.2% ± 3.3
1/500	12.7% ± 1.3	28.4% ± 1.5	68.5% ± 3.0	76.9% ± 3.0	80.0% ± 3.3
1/250	29.7% ± 1.5	43.3% ± 1.8	76.3% ± 2.7	82.1% ± 2.9	84.7% ± 3.0
1/100	50.2% ± 1.8	64.3% ± 2.2	84.7% ± 2.6	87.4% ± 2.8	88.2% ± 2.6
1/75	56.3% ± 2.0	70.2% ± 2.5	86.2% ± 2.6	88.7% ± 2.6	89.0% ± 2.7
1/50	66.46% ± 2.0	75.9% ± 2.6	88.1% ± 2.4	90.3% ± 2.5	91.0% ± 2.4
1/25	77.3% ± 2.3	83.6% ± 2.7	89.9% ± 2.4	91.8% ± 2.3	92.7% ± 2.1
1/10	85.5% ± 2.3	89.9% ± 2.3	91.9% ± 2.2	93.7% ± 1.9	94.4% ± 1.8

II: True Positive Rate (TPR) for a given False Positive Rate (FPR) for each algorithm evaluated in this paper.

LR	IVM	CRF	KCRF	PKCRF
> 100 img/s	∼ 20 img/s	∼ 10 img/s	∼ 3 img/s	∼ 1 img/s

III: Number of processed images per second in a Intel Core 2 Duo class machine.

rate of 75% generating a false positive every 1000 positive-classified patches. IVM and LR produce 1 false positive every 50 and 25 positive-classified patches to get the same obstacle detection rate. At this performance point, PKCRF performs 20 and 40 times better, respectively.

We briefly evaluated the computational complexity of these methods. We ran these experiments in a Intel Core 2 Duo class machine running at 2.0 Ghz. Code was not optimized to make use of the two cores, though. Table III shows the number of images per second that every classifier was able to process. This timing does not take into account the time needed to pre-process the image and extract its features.

We included the full Receiver Operator Characteristic (ROC) curves in fig. 11 and in table IV, we show the Area Under the Curve(AUC) for each of the methods. Note that the differences among the various algorithms shown in this table are small, and AUC is not significative enough to establish conclusions. However, in practice, non-contextual algorithms do not offer enough performance to be used in practice in these experiments, due to its bad obstacle detection rates at low false alarm rates.

Finally, we generated a video for one of the sequences with the outputs of each classifier discussed in the paper: `http://www.cs.cmu.edu/~cvalles/videos/test-obs.avi` In this sequence, the vehicle was manually driven and there were obstacles at different locations. The video shows red patches for detections, and a blinking red box around the image of the classifier that produces a false positive that would make the vehicle to stop. A snapshot of the video is shown in fig. 12 where the outputs of the different classifiers evaluated in this paper are displayed.

VIII. Conclusion

In this work, we presented an obstacle detection algorithm for autonomous vehicles using a monocular color camera. We extended a CRF model to allow the use of non-linear kernels and prior data. In the experimental section, we showed that the use of lattice models for image labeling tasks helps to obtain globally more accurate segmentations than classifiers that make locally independent decisions. Furthermore, we showed that Gaussian kernels work better than linear kernels

LR	IVM	CRF	KCRF	PKCRF
$91.9\% \pm 1.4$	$94.9\% \pm 1.1$	$95.7\% \pm 1.2$	$97.0\% \pm 0.9$	$97.4\% \pm 0.8$

IV: Area Under the Curve (AUC) for each algorithm.

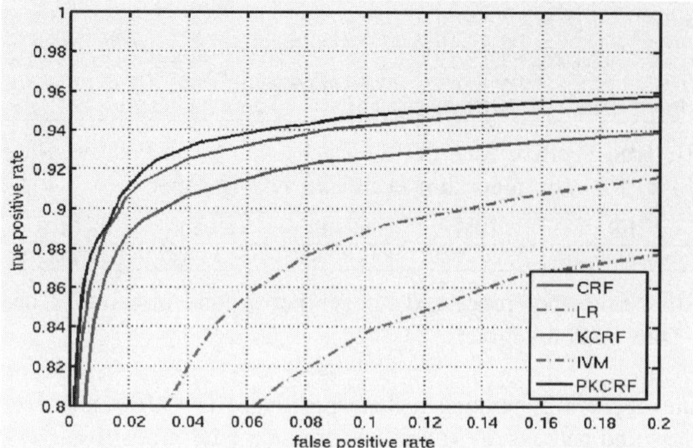

11: ROC curves for LR, IVM, CRF, KCRF and PKCRF.

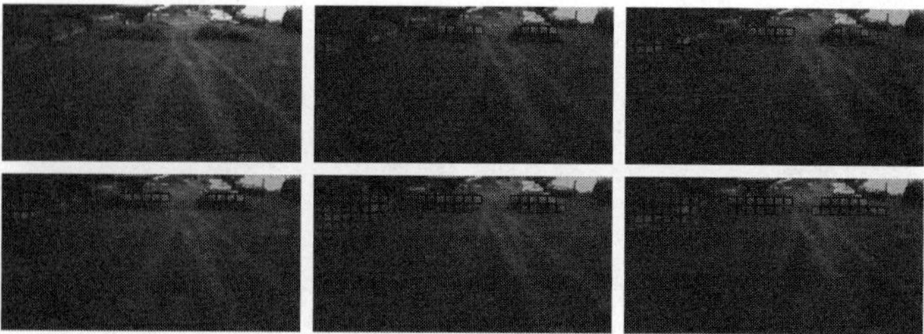

12: **From left to right and top to bottom:** comparison of original image, LR, IVM, CRF, KCRF and PKCRF for one of the test images. Obstacles are marked with red boxes.

in our obstacle detection experiments. Finally, prior data was introduced in the CRF model to produce better segmentations in "familiar" environments.

Even though the set of features we used for every image part was very limited (just color and texture features), the obtained results are promising towards building an obstacle detection system based only on a monocular color camera. Gaussian K-CRFs were the best performers when no prior data was available. However, PK-CRFs performed even better when prior data was available.

The algorithms proposed in this paper are not limited to monocular camera approaches. Future work includes the use of multiple camera solutions with different filters on them, the use of stereo features, experiments with different patch sizes (or non-uniform image divisions, such as super-pixels), the use of more node features, and extracting specific edge features for the CRF model. As one of the main issues when dealing with camera-only solutions is the exposure, High Dynamic Range (i.e. capture images at different exposures) could be used to address this problem.

In conclusion, CRFs provide a probabilistic framework with superior performance compared to classifiers that do not exploit context in image labeling applications. The model for CRFs is flexible enough to support different kernels, or the addition of a large variety of features. In our experiments, features that incorporate prior data helped to boost the obstacle detection performance, making PK-CRFs suitable for some robotic applications that use a monocular color camera.

ACKNOWLEDGMENTS

We would like to acknowledge the valuable support of Cristian Dima and Carl Wellington in the development of the infrastructure used for the experiments, and for their helpful comments and advice.

This work was supported by John Deere under contract 476169.

REFERENCES

[1] C. Broyden. The convergence of a class of double-rank minimization algorithms. *Journal of the Institute of Mathematics and Its Applications*, 6:76–90, 1970.

[2] C. J. C. Burges. A tutorial on support vector machines for pattern recognition. *Data Mining and Knowledge Discovery*, 2:121–167, 1998.

[3] C. Dima, N. Vandapel, and M. Hebert. Sensor and classifier fusion for outdoor obstacle detection: an application of data fusion to autonomous off road navigation. *The 32nd Applied Imagery Recognition Workshop (AIPR2003)*, October, 2003.

[4] R. Fletcher. A new approach to variable metric algorithms. *Computer Journal*, 13:317–322, 1970.

[5] D. Franois, V. Wertz, and M. Verleysen. About the locality of kernels in high-dimensional spaces. *ASMDA 2005, International Symposium on Applied Stochastic Models and Data Analysis*, pages 238–245, 2005.

[6] D. Goldfarb. A family of variable metric updates derived by variational means. *Mathematics of Computation*, 24:23–26, 1970.

[7] S. Kumar and M. Hebert. Discriminative fields for modeling spatial dependencies in natural images. *In Advances in Neural Information Processing Systems 16 (NIPS 2003)*, 2004.

[8] J. Lafferty, X. Zhu, and Y. Liu. Kernel conditional random fields: representation and clique selection. *In Proceedings of the Int. Conference on Machine Learning*, 2004.

[9] K. Murphy, Y. Weiss, and M. Jordan. Loopy belief propagation for approximate inference: An empirical study. *In proceedings of the 15th Annual Conference on Uncertainty in Artificial Intelligence (UAI-99)*, pages 467–47, 1999.

[10] K. Primdahl, I. Katz, O. Feinstein, Y. Mok, H. Dahlkamp, D. Stavens, M. Montemerlo, and S. Thrun. Change detection from multiple camera images extended to non-stationary camera. *In Proceedings of Field and Service Robotics (FSR05)*, 2005.

[11] A. Quattoni, M. Collins, and T. Darrel. Conditional random fields for object recognition. *In Advances in Neural Information Processing Systems 17 (NIPS 2004)*, 2005.

[12] A. Saxena, S. Chung, and A. Ng. 3-d depth reconstruction from a single still image. *Int. J. Comput. Vision*, 76(1):53–69, 2008.

[13] D. Shanno. Conditioning of quasi-newton methods for function minimization. *Mathematics of Computation*, 24:647–656, 1970.

[14] A. Torralba, K. Murphy, and W. Freeman. Contextual models for object detection using boosted random fields. *In Advances in Neural Information Processing Systems 17 (NIPS 2004)*, 2005.

[15] S. Vishwanathan, N. N. Schraudolph, M. W. Schmidt, and K. Murphy. Accelerated training of conditional random fields with stochastic meta-descent. *In Proceedings 23rd Intl. Conf. Machine Learning (ICML)*, pages 969–976, 2006.

[16] Y. Wang and Q. Ji. A dynamic conditional random field model for object segmentation in image sequences. *In IEEE Computer Society Conference on Computer Vision and Pattern Recognition (CVPR)*, 1, 2005.

[17] J. Winn and J. Shotton. The layout consistent random field for recognizing and segmenting partially occluded objects. *Computer Vision and Pattern Recognition, 2006 IEEE Computer Society Conference on*, 1:37–44, 2006.

[18] J. Zhu and T. Hastie. Kernel logistic regression and the import vector machine. *In Proceedings of Neural Information Processing Systems*, 2001.

SARSOP: Efficient Point-Based POMDP Planning by Approximating Optimally Reachable Belief Spaces

Hanna Kurniawati David Hsu Wee Sun Lee
Department of Computer Science, National University of Singapore
Singapore 117590, Singapore

Abstract— **Motion planning in uncertain and dynamic environments is an essential capability for autonomous robots. Partially observable Markov decision processes (POMDPs) provide a principled mathematical framework for solving such problems, but they are often avoided in robotics due to high computational complexity. Our goal is to create practical POMDP algorithms and software for common robotic tasks. To this end, we have developed a new point-based POMDP algorithm that exploits the notion of *optimally reachable belief spaces* to improve computational efficiency. In simulation, we successfully applied the algorithm to a set of common robotic tasks, including instances of coastal navigation, grasping, mobile robot exploration, and target tracking, all modeled as POMDPs with a large number of states. In most of the instances studied, our algorithm substantially outperformed one of the fastest existing point-based algorithms. A software package implementing our algorithm is available for download at `http://motion.comp.nus.edu.sg/projects/pomdp/pomdp.html`.**

I. Introduction

Partially observable Markov decision processes (POMDPs) [17] provide a principled mathematical framework for planning under uncertainty, an essential capability for robots operating in uncertain and dynamic environments. However, POMDPs are often avoided in robotics, because solving POMDPs exactly is computationally intractable [9]. Not long ago, the best algorithms could spend hours computing exact solutions to POMDPs with only a dozen states, which are woefully inadequate for modeling realistic robotic tasks. In recent years, point-based POMDP algorithms [5, 10, 16, 19, 20] have made impressive progress by computing good approximate solutions: POMDPs with hundreds of states have been solved in a matter of seconds (*e.g.*, [5, 16, 19]). These algorithms have the potential to make POMDPs practical for many applications in robotics and beyond.

Our goal is to create practical POMDP algorithms and software for common robotic tasks. To this end, we have developed a new point-based POMDP algorithm that exploits the notion of *optimally reachable belief spaces* to improve computational efficiency. In simulation, we successfully applied our algorithm to a set of common robotic tasks, including coastal navigation, grasping, mobile robot exploration, and target tracking, all modeled as POMDPs with a large number of states.

POMDP algorithms typically operate in a robot's *belief space*. A *belief* is a probability distribution over all possible robot states, and the set of all beliefs form the belief space. Intuitively, the difficulty of solving POMDPs is due to the "curse of dimensionality": in a discrete POMDP, the belief space $\mathcal{B}$ has dimensionality equal to $|\mathcal{S}|$, the number of robot states. The size of $\mathcal{B}$ thus grows exponentially with $|\mathcal{S}|$. Consider, for example, robot navigation in a simple planar environment modeled as a 10×10 grid. The resulting belief space is 100-dimensional!

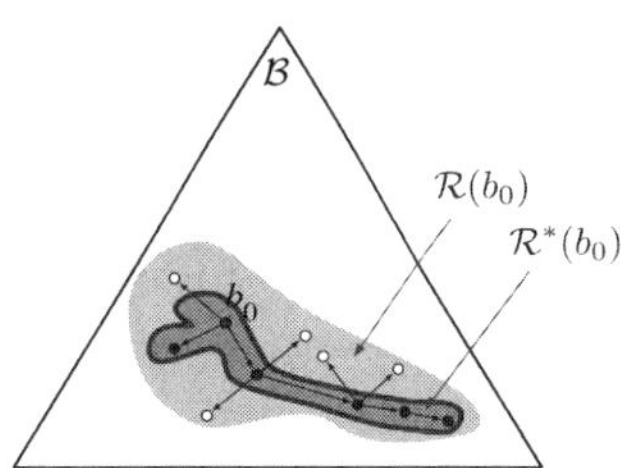

Fig. 1. Belief space $\mathcal{B}$, reachable space $\mathcal{R}(b_0)$, and optimally reachable space $\mathcal{R}^*(b_0)$. Note that $\mathcal{R}^*(b_0) \subseteq \mathcal{R}(b_0) \subseteq \mathcal{B}$.

To overcome this difficulty, one key idea of point-based POMDP algorithms is to *sample* a set of points from $\mathcal{B}$ and use it as an approximate representation of $\mathcal{B}$, instead of representing $\mathcal{B}$ exactly. Some early POMDP algorithms sample the entire belief space $\mathcal{B}$, using a uniform sampling distribution, such as a grid. However, it is difficult to sample a representative set of points from $\mathcal{B}$, due to its large size. More recent point-based algorithms sample only $\mathcal{R}(b_0)$, the subset of belief points reachable from a given initial point $b_0 \in \mathcal{B}$, under arbitrary sequences of actions (Fig. 1). It is generally believed that $\mathcal{R}(b_0)$ is much smaller than $\mathcal{B}$. Indeed, focusing on $\mathcal{R}(b_0)$ allows point-based algorithms to scale up to larger problems. To push further in this direction, we would like to sample near $\mathcal{R}^*(b_0)$, the subset of belief points reachable from b_0 under *optimal* sequences of actions, as $\mathcal{R}^*(b_0)$ is usually much smaller than $\mathcal{R}(b_0)$. Of course, the optimal sequences of actions constitute exactly the POMDP solution, which is unknown in advance. In fact, knowing $\mathcal{R}^*(b_0)$ is in some sense "equivalent" to knowing the POMDP solution (see Section III-A). So we need to approximate $\mathcal{R}^*(b_0)$.

The main idea of our algorithm is to compute successive approximations of $\mathcal{R}^*(b_0)$ and converge to it iteratively. Since $\mathcal{R}^*(b_0)$ is unknown in advance, the algorithm relies on heuristic exploration to sample $\mathcal{R}(b_0)$ and improves sampling over time through a simple on-line learning technique. It then uses a bounding technique to avoid sampling in regions that are unlikely to be optimal and focus sampling on the region near $\mathcal{R}^*(b_0)$, the subset of $\mathcal{B}$ most relevant to the POMDP solution. This leads to substantial gain in computational efficiency.

Focusing on $\mathcal{R}^*(b_0)$ also brings an indirect benefit. Under fairly general conditions, the solution to a POMDP can be represented as a convex, piecewise-linear *value function* [17].

We represent the value function as a set Γ of hyperplanes, each of which must dominate the rest at some sampled point. By pruning away sampled points that are suboptimal, *i.e.*, outside $\mathcal{R}^*(b_0)$, we can reduce the size of Γ, thus further improving computational efficiency.

II. Background

A. POMDPs

A POMDP models an agent taking a sequence of actions under uncertainty to maximize its reward. Formally it is specified as a tuple $(\mathcal{S}, \mathcal{A}, \mathcal{O}, T, Z, R, \gamma)$, where $\mathcal{S}$ is a set of states, $\mathcal{A}$ is a set of actions, and $\mathcal{O}$ is a set of observations.

In each time step, the agent lies in some state $s \in \mathcal{S}$; it takes some action $a \in \mathcal{A}$ and moves from s to a new state s'. Due to the uncertainty in action, the end state s' is modeled as a conditional probability function $T(s, a, s') = p(s'|s, a)$, which gives the probability that the agent lies in s', after taking action a in state s. The agent then makes an observation to gather information on its state. Due to the uncertainty in observation, the observation result $o \in \mathcal{O}$ is again modeled as a conditional probability function $Z(s, a, o) = p(o|s, a)$.

In each step, the agent receives a real-valued reward $R(s, a)$, if it takes action a in state s, and the goal of the agent is to maximize its expected total reward by choosing a suitable sequence of actions. For infinite-horizon POMDPs, the sequence of actions has infinite length. We specify a discount factor $\gamma \in [0, 1)$ so that the total reward is finite and the problem is well defined. In this case, the expected total reward is given by $\mathrm{E}\left[\sum_{t=0}^{\infty} \gamma^t R(s_t, a_t)\right]$, where s_t and a_t denote the agent's state and action at time t.

The solution to a POMDP is an *optimal policy* that maximizes the expected total reward. Normally, a policy is a mapping from the agent's state to a prescribed action. However, in a POMDP, the agent's state is partially observable and not known exactly. So we rely on the concept of beliefs. As described earlier, a belief is a probability distribution over $\mathcal{S}$. A POMDP policy $\pi\colon \mathcal{B} \to \mathcal{A}$ maps a belief $b \in \mathcal{B}$ to a prescribed action $a \in \mathcal{A}$.

A policy π induces a value function $V_\pi(b)$ that specifies the expected total reward of executing policy π starting from b. It is known that V^*, the value function associated with the optimal policy π^*, can be approximated arbitrarily closely by a convex, piecewise-linear function

$$V(b) = \max_{\alpha \in \Gamma} (\alpha \cdot b),$$

where Γ is a finite set of vectors called α-vectors, b is the discrete vector representation of a belief, and $\alpha \cdot b$ is the inner product of vectors α-vector and b. Each α-vector is associated with an action. The policy can be executed by selecting the action corresponding to the best α-vector at the current belief. So a policy can be represented as a set of α-vectors.

B. Related Work

POMDPs are a principled approach for planning and decision making under uncertainty [6, 17], but they are notoriously hard to solve [7, 9]. There have been significant efforts in developing approximation algorithms. See [1] for a recent survey.

Point-based algorithms have been particularly successful in computing approximate solutions to large POMDPs [2, 5, 10, 16, 19, 20]. Most of them use *value iteration* [13]. Exploiting the fact that the optimal value function must satisfy the Bellman equation [13], value iteration algorithms start with an initial policy represented as a value function V and perform backup operations on V by iterating on the Bellman equation until the iteration converges. One important idea shared by the point-based algorithms is to sample a representative set of points from the belief space $\mathcal{B}$ and compute an approximately optimal value function by performing backup operations over the sampled points rather than the entire $\mathcal{B}$. They differ in how they sample the belief space and perform backup operations. To improve computational efficiency, recent point-based algorithms sample only the reachable space $\mathcal{R}(b_0)$ from an initial belief point b_0.

PBVI [10] is the first point-based algorithm that demonstrated good performance on a large POMDP called Tag, which has 870 states. Later point-based algorithms improved the performance significantly on this and other larger POMDPs. To our knowledge, HSVI2 [19] so far has the best performance in general. HSVI2 uses heuristics to guide the sampling towards regions that help cut down the gap between the upper and lower bounds on the optimal value function. FSVI [16] is another point-based algorithm, which uses a Markov decision process (MDP) to guide the sampling. MDP-guided sampling is effective for some problems, but the performance degrades when uncertainty is high and long sequences of information-gathering actions are required.

Our algorithm is related to HSVI2 and FSVI, but it explicitly attempts to sample the optimally reachable space $\mathcal{R}^*(b_0)$ through learning-enhanced exploration and a bounding technique. Experimental results show that focusing on $\mathcal{R}^*(b_0)$ is a promising idea. An early version of our algorithm [5] exploits bounding in a limited way: bounds are compared locally at individual belief points to prune suboptimal actions. In contrast, the current algorithm sets up the bounds to reach a specified value function approximation level at b_0, thereby leveraging information globally to reduce the number of poor samples—those that are in $\mathcal{R}(b_0)$ but not in $\mathcal{R}^*(b_0)$.

One crucial reason for the computational intractability of POMDPs is the high dimensionality of $\mathcal{B}$. Low-dimensional approximations of $\mathcal{B}$ therefore improve computational efficiency greatly (*e.g.*, [12, 14]). These approaches are important, but beyond the scope of this paper.

III. SARSOP

We now describe our algorithm, SARSOP, which stands for Successive Approximations of the Reachable Space under Optimal Policies.

A. Optimally Reachable Spaces

A key idea of point-based POMDP algorithms is to sample a representative set of points from the belief space and

Algorithm 1 SARSOP.

Initialize the set Γ of α-vectors, representing the lower bound $\underline{V}$ on the optimal value function V^*. Initialize the upper bound $\overline{V}$ on V^*.
Insert the initial belief point b_0 as the root of the tree $T_{\mathcal{R}}$.
repeat
  SAMPLE($T_{\mathcal{R}}$, Γ).
  Choose a subset of nodes from $T_{\mathcal{R}}$. For each chosen node b, BACKUP($T_{\mathcal{R}}, \Gamma, b$).
  PRUNE($T_{\mathcal{R}}$, Γ).
until termination conditions are satisfied.
return Γ.

use it as an approximate representation of the space. For efficiency, most recent algorithms sample from $\mathcal{R}(b_0)$, the set of points reachable from a given point $b_0 \in \mathcal{B}$ under arbitrary sequences of actions. Theoretical analysis shows that approximate POMDPs solutions can be computed efficiently, when $\mathcal{R}(b_0)$ has a small *covering number* [4]. Informally, the δ-covering number $\mathcal{C}(\delta)$ of a set S is the minimum number of balls of radius δ needed to cover S. So it is a measure of the "volume" of S.

Theorem 1: For any $b_0 \in \mathcal{B}$, let $\mathcal{C}(\delta)$ be the δ-covering number of $\mathcal{R}(b_0)$. Given any constant $\epsilon > 0$, an approximation $V(b_0)$ of $V^*(b_0)$, with error $|V^*(b_0) - V(b_0)| \leq \epsilon$, can be found in time

$$O\left(\mathcal{C}\left(\frac{(1-\gamma)^2\epsilon}{4\gamma R_{\max}}\right)^2 \log_\gamma \frac{(1-\gamma)\epsilon}{2R_{\max}}\right).$$

However, for many realistic robotics tasks, the assumption of small $\mathcal{R}(b_0)$ may not hold. We would like our algorithm to do well when $\mathcal{R}(b_0)$ may be large, but $\mathcal{R}_{\pi^*}(b_0)$, the space reachable under an optimal policy π^*, is small. As $\mathcal{R}_{\pi^*}(b_0)$ is often much smaller than $\mathcal{R}(b_0)$, the assumption of small $\mathcal{R}_{\pi^*}(b_0)$ is more likely to hold. Unfortunately, this relaxed assumption is too weak, and the problem of computing approximate POMDP solutions remains hard, despite the assumption [4].

Theorem 2: Let b_0 be any point in $\mathcal{B}$ and π^* be an optimal policy. Given a constant $\epsilon > 0$, computing an approximation $V(b_0)$ of $V^*(b_0)$, with error $|V(b_0) - V^*(b_0)| \leq \epsilon|V^*(b_0)|$, is NP-hard, even if the covering number of $\mathcal{R}_{\pi^*}(b_0)$ is polynomial-sized.

On the other hand, if we are given a set of balls of radius δ that covers $\mathcal{R}_{\pi^*}(b_0)$, the problem becomes much easier [4]. We call the set C, which contains the centers of this set of balls, a δ*-cover* of $\mathcal{R}_{\pi^*}(b_0)$.

Theorem 3: For any $b_0 \in \mathcal{B}$ and any optimal policy π^*, given a proper δ-cover C of $\mathcal{R}_{\pi^*}(b_0)$ with $\delta = \frac{(1-\gamma)^2\epsilon}{2\gamma R_{\max}}$, an approximation $V(b_0)$ of $V^*(b_0)$, with error $|V^*(b_0) - V(b_0)| \leq \epsilon$, can be found in time

$$O\left(|C|^2 + |C| \log_\gamma \frac{(1-\gamma)\epsilon}{2R_{\max}}\right),$$

where $|C|$ is the size of C and $R_{\max} = \max_{s,a} |R(s,a)|$ is the maximum one-step reward.

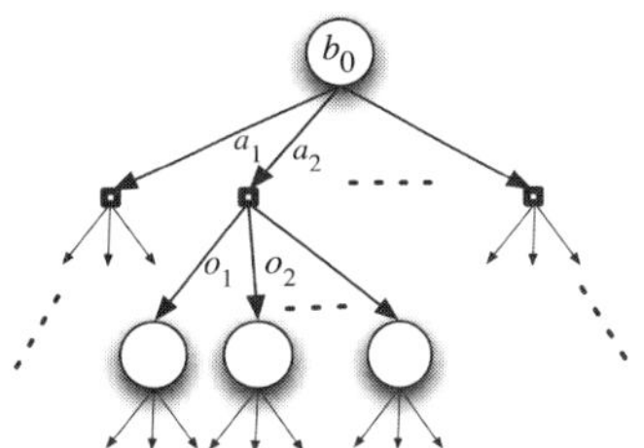

Fig. 2. The belief tree $T_{\mathcal{R}}$ rooted at b_0.

Together, Theorems 2 and 3 say that computing approximate POMDP solutions is hard, but the problem becomes much easier, if a proper δ-cover of $\mathcal{R}_{\pi^*}(b_0)$ is given. It follows that the key difficulty must lie in computing such a cover. Once the cover is obtained, we can find an approximate POMDP solution in time polynomial in the cover size. So, instead of following the common approaches of directly approximating V^* or searching for π^*, our SARSOP algorithm focuses on finding an approximate cover of $\mathcal{R}_{\pi^*}(b_0)$ through sampling.

Since there may be multiple optimal policies, SARSOP aims to sample $\mathcal{R}^*(b_0) = \bigcup_{\pi^*} \mathcal{R}_{\pi^*}(b_0)$, the union of all optimally reachable spaces.

In the following, to simplify the notations, we omit the argument b_0 in $\mathcal{R}(b_0)$ and $\mathcal{R}^*(b_0)$. It is understood that $\mathcal{R}$ and $\mathcal{R}^*$ are reachable from a given initial point b_0.

B. Overview of the Algorithm

SARSOP iterates over three main functions, SAMPLE, BACKUP, and PRUNE. A sketch is shown in Algorithm 1.

Like all point-based algorithm, SARSOP samples a set of points from the belief space. The sampled points form a tree $T_{\mathcal{R}}$ (Fig. 2). Each node of $T_{\mathcal{R}}$ represents a sampled point. As there is no confusion, we use the same symbol b to denote both a sampled point and its corresponding node in $T_{\mathcal{R}}$. The root of $T_{\mathcal{R}}$ is the initial belief point b_0. To sample a new point b', we pick a node b from $T_{\mathcal{R}}$ as well as an action $a \in \mathcal{A}$ and an observation $o \in \mathcal{O}$ according to suitable probability distributions or heuristics. We then compute b' using the formula

$$b'(s') = \tau(b,a,o) = \eta Z(s',a,o) \sum_s T(s,a,s')b(s),$$

where η is a normalization constant, and insert b' into $T_{\mathcal{R}}$ as a child of b. Clearly, every point sampled this way is reachable from b_0. If we apply all possible sequences of actions and observations, the set of nodes in $T_{\mathcal{R}}$ is exactly $\mathcal{R}$. The key is, of course, to avoid doing so and focus the sampling, instead, on $\mathcal{R}^*$.

To achieve this, SARSOP maintains both a lower bound $\underline{V}$ and an upper bound $\overline{V}$ on the optimal value function V^*. The set Γ of α-vectors represents a piecewise-linear approximation to V^* (Section II-A), and is also a lower bound when suitably initialized, using, *e.g.*, a fixed-action policy [1]. For the upper bound $\overline{V}$, SARSOP uses the sawtooth approximation [1]. The upper bound can be initialized in various ways, using the MDP or the Fast Informed Bound technique [1]. SARSOP uses the

Algorithm 2 Perform α-vector backup at a node b of $T_{\mathcal{R}}$.

```
BACKUP(T_R, Γ, b)
For all a ∈ A, o ∈ O, α_{a,o} ← argmax_{α∈Γ}(α · τ(b, a, o)).
For all a ∈ A, s ∈ S,
    α_a(s) ← R(s, a) + γ Σ_{o,s'} T(s, a, s')Z(s', a, o)α_{a,o}(s').
α' ← argmax_{a∈A}(α_a · b)
Insert α' into Γ.
```

upper and the lower bounds to bias sampling towards $\mathcal{R}^*$ ((see Section III-C).

Next, we perform backup at selected nodes in $T_{\mathcal{R}}$. A backup operation at a node b collates the information in the children of b and propagates it back to b. We perform the standard α-vector backup (Algorithm 2), with the value function approximation represented as a set Γ of α-vectors. The value function approximation at b, obtained from the α-vector backup, is the same as that from the Bellman backup. However, the Bellman backup propagates only the value, while the α-vector backup propagates the gradient of the value function approximation along with the value to obtain a global approximation over the entire belief space rather than a local approximation at b.

Invocation of SAMPLE and BACKUP generates new sampled points and α-vectors. However, not all of them are useful for constructing an optimal policy and are pruned to improve computational efficiency (see Section III-D).

SARSOP is an anytime algorithm that returns the best policy found within a pre-specified amount of time. It gradually reduces the gap ϵ between the upper and lower bounds on the value function at b_0, until it reaches either a pre-specified gap size or the time limit.

C. Sampling

The NP-hardness result described in Section III-A suggests that sampling from $\mathcal{R}^*$ is hard. We use heuristics and information gathered from earlier samples to guide the sampling and improve the sampling distribution over time. Furthermore, by using value function bounds, we try to avoid sampling in regions that are unlikely to be reachable under any optimal policy, *i.e.*, outside of $\mathcal{R}^*$. See Algorithm 3 for the pseudocode.

To sample new belief points, SARSOP sets a target gap size ϵ between the upper and lower bound at the root b_0 of $T_{\mathcal{R}}$ and traverses a single path down $T_{\mathcal{R}}$ by choosing at each node the action with the highest upper bound and the observation that makes the largest contribution to the gap at the root of $T_{\mathcal{R}}$. This is the same action and observation selection strategy used in HSVI2 [19]. The sampling path is terminated under suitable conditions. Together, the strategies for action and observation selection and the choice of termination conditions control the resulting sampling distribution.

One termination condition is to stop when the sampling path reaches a node whose gap between the upper and lower bounds is smaller than $\gamma^{-t}\epsilon$, where t is the depth of the node in $T_{\mathcal{R}}$ [19]. If each leaf of $T_{\mathcal{R}}$ has a gap smaller than $\gamma^{-t}\epsilon$, the gap at the root is guaranteed to be smaller than ϵ. This condition, although reasonable, is inadequate. As the target gap ϵ at the root gets smaller, the sampling path must traverse deeper down the tree. As we go down the tree, the set of points in $\mathcal{R}$ increases much faster than the set of points in $\mathcal{R}^*$, and it becomes increasingly difficult to sample from $\mathcal{R}^*$. To focus sampling near $\mathcal{R}^*$ and minimize sampling in $\mathcal{R}\backslash\mathcal{R}^*$, we would like to make the sampling path as shallow as possible while still achieving the target gap ϵ at the root of $T_{\mathcal{R}}$. A potential dilemma here is that some nodes with high expected rewards lie deep in the tree, and we must allow the sampling path to go deep enough in order to reach them.

a) Selective deep sampling: As each backup operation chooses the action that *maximizes* the expected reward, improvements in lower bounds are quickly propagated to the root when nodes with high expected rewards are found. This not only directly improves the policy but also provides information to stop sampling more quickly in regions that are likely outside $\mathcal{R}^*$. In contrast, upper bounds cannot be propagated beyond a node until the upper bounds for *all* the actions at the node are sufficiently improved. Finding the best action is not enough. Thus we give preference to lower bound improvements and continue down a sampling path beyond the node with a gap of $\gamma^{-t}\epsilon$, if we predict that doing so likely leads to improvement in the lower bound at the root.

To make such a predication, conceptually we predict the optimum value $V^*(b)$ at a node b and propagate the predicted value $\hat{V}$ up towards the root. If $\hat{V}$ improves the lower bound at the root, we expand b and then repeat the procedure at the next selected node down the sampling path. Otherwise, we proceed to check the gap termination criterion described in the next subsection.

To predict the optimal value $V^*(b)$, we use a simple learning technique. We cluster beliefs according to suitable features and use previously computed values of beliefs in the same cluster as b to predict the value of b. This allows us to learn which parts of the belief space is worth exploring. Currently, we use the initial upper bound and the entropy of b as the features and discretize the belief space into a finite number of bins according to these two features. The average value of beliefs in a bin is used as the prediction for the value of any new belief falling into the bin. If a bin is empty, the initial upper bound of the new belief is used as its predicted value.

To implement this idea efficiently, we do not actually propagate the predicted value $\hat{V}$ back to the root. Instead, we pass a lower-bound target level L down the sampling path. The predicted value $\hat{V}$ is checked against L. If $\hat{V}$ fails to meet the target L at a node b, the lower bound at b will not be propagated further up towards the root of $T_{\mathcal{R}}$.

Let us now consider how to pass the target L at a node b to a child node $b' = \tau(b, a, o)$. First, observe that value function information is propagated up from b' to b only if the action a that takes b to b' has higher value than all other actions at b. We thus calculate an intermediate target level L' for a, which is set to the maximum over L and the values of all the actions at b (Algorithm 3, lines 7–8). Next, observe that the lower bound on the value of action a is

$$\underline{Q}(b, a) = \sum_s R(s, a)b(s) + \gamma \sum_o p(o|b, a)\underline{V}(b').$$

Algorithm 3 Sampling near $\mathcal{R}^*$.

SAMPLE($T_\mathcal{R}$, Γ)

1: Set L to the current lower bound on the value function at the root b_0 of $T_\mathcal{R}$. Set U to $L + \epsilon$, where ϵ is the current target gap size at b_0.
2: SAMPLEPOINTS($T_\mathcal{R}$, Γ, b_0, L, U, ϵ, 1).

SAMPLEPOINTS($T_\mathcal{R}$, Γ, b, L, U, ϵ, t).

3: Let $\hat{V}$ be the predicted value of $V^*(b)$.
4: **if** $\hat{V} \leq L$ and $\overline{V}(b) \leq \max\{U, \underline{V}(b) + \epsilon\gamma^{-t}\}$ **then**
5:     **return**
6: **else**
7:     $\underline{Q} \leftarrow \max_a \underline{Q}(b, a)$.
8:     $L' \leftarrow \max\{L, \underline{Q}\}$.
9:     $U' \leftarrow \max\{U, \underline{Q} + \gamma^{-t}\epsilon\}$.
10:     $a' \leftarrow \arg\max_a \overline{Q}(b, a)$.
11:     $o' \leftarrow \arg\max_o p(o|b, a') \left(\overline{V}(\tau(b, a', o)) - \underline{V}(\tau(b, a', o))\right)$.
12:     Calculate L_t so that $L' = \sum_s R(s, a')b(s) + \gamma\left(p(o'|b, a')L_t + \sum_{o \neq o'} p(o|b, a')\underline{V}(\tau(b, a', o))\right)$.
13:     Calculate U_t so that $U' = \sum_s R(s, a')b(s) + \gamma\left(p(o'|b, a')U_t + \sum_{o \neq o'} p(o|b, a')\overline{V}(\tau(b, a', o))\right)$.
14:     $b' \leftarrow \tau(b, a', o')$.
15:     Insert b' into $T_\mathcal{R}$ as a child of b.
16:     SAMPLEPOINTS($T_\mathcal{R}$, Γ, b', L_t, U_t, ϵ, $t + 1$).

Hence the target level for b' is the value needed for $\underline{Q}(b, a)$ to achieve its target L' (Algorithm 3, line 12).

To guard against misleading predictions that result in unnecessarily deep samplings paths, we only continue down a sampling path until the gap between the upper and lower bounds is $\kappa\gamma^{-t}\epsilon$ for some $\kappa < 1$. The κ value is set to 0.5 in our current implementation.

b) The gap termination criterion: If our prediction shows no improvement of the lower bound at the root, we use the target gap size ϵ at the root to decide whether to terminate the sampling path and avoid sampling in regions unlikely to be in $\mathcal{R}^*$. As mentioned earlier, the straightforward way of achieving the target gap size ϵ between the upper and lower bounds at the root of $T_\mathcal{R}$ is to require a gap size $\gamma^{-t}\epsilon$ for all leaves of $T_\mathcal{R}$. However, it is in fact sufficient to ensure that the condition is satisfied somewhere along all the paths from the root to the leaves, rather than at the leaves themselves. This has the advantage of leveraging information globally from the other parts of $T_\mathcal{R}$ to terminate a sampling path as early as possible and thus improving computational efficiency.

To do this, we pass an upper-bound target level U down the sampling path as well. For a node b at depth t, we can terminate sampling if its upper bound is lower than $\underline{V}(b) + \epsilon\gamma^{-t}$ or the upper-bound target U passed down from parent of b. This termination criterion has the same effect as requiring all leaves to have a gap of no more than $\epsilon\gamma^{-t}$: if all leaves in $T_\mathcal{R}$ meet this termination criterion, the root b_0 achieves the target gap size of ϵ. The upper-bound target level U can be passed down a sampling path in a way similar to that for the lower-bound target level L. See Algorithm 3, lines 9 and 13.

The combination of selective deep sampling and the gap termination criterion leads to an effective sampling strategy that goes deep into $T_\mathcal{R}$ when need. This avoids unnecessarily sampling in $\mathcal{R}\backslash\mathcal{R}^*$ and gives a better approximation to $\mathcal{R}^*$.

D. Pruning

The efficiency of backup operations, which take up a significant fraction of the total computation time, depends significantly on the size of the set Γ of α-vectors. To improve computational efficiency, existing point-based algorithms usually prune an α-vector from Γ if it is dominated by others over the entire belief space $\mathcal{B}$. The notion of optimally reachable space suggests an alternative and more aggressive pruning technique: ideally, we want to prune an α-vector if it is dominated by other α-vectors over $\mathcal{R}^*$, rather than $\mathcal{B}$. Since $\mathcal{R}^*$ is potentially much smaller than $\mathcal{B}$, this may substantially reduce the size of Γ and improve the efficiency of the backup operations and thus the overall algorithm.

As $\mathcal{R}^*$ is not known in advance, we use B, the set of all sampled belief points contained in $T_\mathcal{R}$, as an approximation. To improve this approximation and to keep the size of B small, we prune from B those points that are provably suboptimal and do not lie in $\mathcal{R}^*$. For a node b in $T_\mathcal{R}$, if $\overline{Q}(b, a) < \underline{Q}(b, a')$ for two actions a and a', then we prune all the sampled points in the subtree resulting from taking action a at b, as an optimal policy will never take the action a at b and traverse the subtree underneath. It is possible that some pruned points may turn out to lie in $\mathcal{R}^*$, as there are other paths in $T_\mathcal{R}$ to reach them under an optimal policy. However, the benefits of keeping B small usually outweighs the loss in the approximation quality due to over-pruning. These points can also be eventually recovered from the other paths in $T_\mathcal{R}$.

Belief point pruning in turn enables more aggressive α-vector pruning. In SARSOP, an α-vector is pruned if it is dominated by others over B. A simple criterion for dominance is to say that for two α-vectors α_1 and α_2, α_1 dominates α_2 at a belief point b if $\alpha_1 \cdot b \geq \alpha_2 \cdot b$. However, this is not robust. The set B is a finitely sampled approximation of $\mathcal{R}^*$. Since SARSOP computes an approximately optimal policy over B only, the computed policy may choose an action that causes it to slightly veer off $\mathcal{R}^*$ and get into a region in which the value function approximation is poor. To address this issue, we impose the more stringent requirement of dominance over a δ-neighborhood: α_1 dominates α_2 at a belief point b if $\alpha_1 \cdot b' \geq \alpha_2 \cdot b'$ at every point b' whose distance to b is less than δ, for some fixed constant δ. We call this *δ-dominance*. We can check δ-dominance very quickly by computing the distance d from b to the intersection of the hyperplanes represented by α_1 and α_2 and making sure that $d \geq \delta$. In the implementation, the value of δ can be set adaptively according to the effectiveness of α-vector pruning. A similar idea for α-vector pruning, but without using the δ-neighborhood, is described in [15].

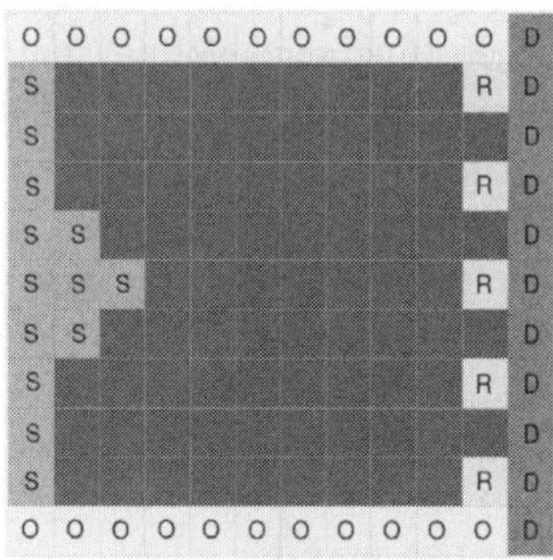

(*a*) Underwater Navigation, an instance of coastal navigation, shown on a reduced map with a 11×12 grid. "S" marks the possible initial positions for the robot. The robot is equally likely to start in any of these positions. "D" marks the destinations. "R" marks the rocks. "O" marks places that the robot can fully localize itself.

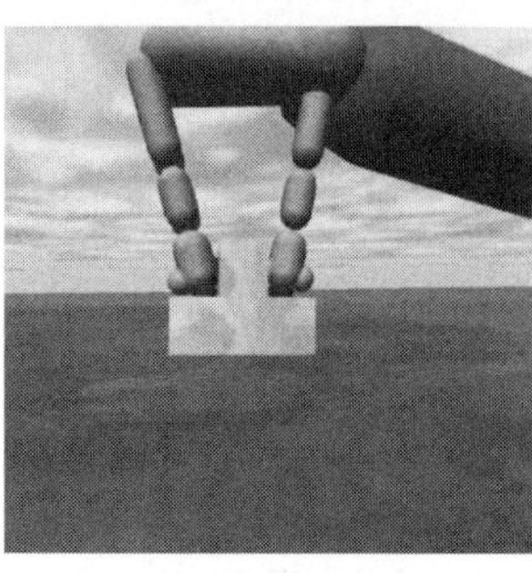

(*b*) Grasping. A fingered robot arm grasps a stepped block. Courtesy of L.P. Kaelbling and T. Lozano-Pérez.

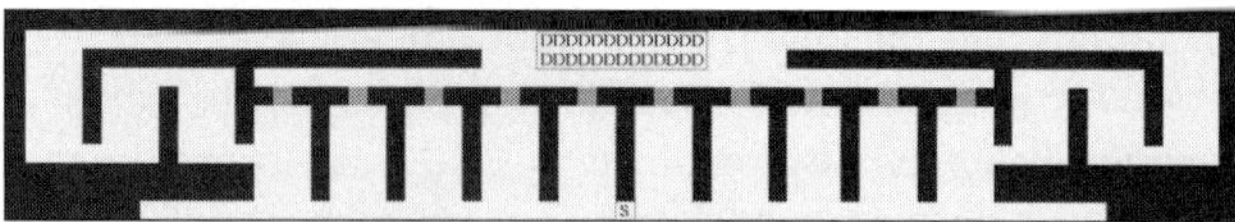

(*c*) Integrated Exploration. A robot navigates with an uncertain map. Areas shaded in black represent obstacles. Areas shaded in light gray represent (possibly damaged) bridges. "S" marks the start location for the robot. "D" marks destination locations.

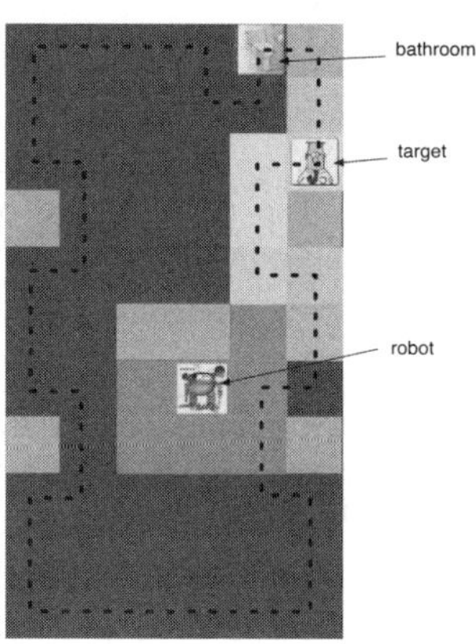

(*d*) Homecare. A robot follows a moving person, the target. The light blue areas indicate obstacles. The black dashed curve indicates the target's path. The green area around the robot indicates the the robot sensor' visibility region. The various shades of gray show the robot's belief of the current target position.

Fig. 3. Some common robotic tasks modeled as POMDPs.

IV. Experiments

We have successfully applied SARSOP to a set of distinct robotic tasks. In this section, we describe these tasks, the experimental setup, and the results.

A. Robotic Tasks Studied

Uncertainty arises in various ways in robotic systems. Suppose that the state of a robotic system is given by $(x_{\mathrm{r}}, x_{\mathrm{e}})$, where x_{r} represents the state of the robot and x_{e} represents the state of the environment. Inaccuracies in robot control and sensing are the typical causes for uncertainty in x_{r}. They are almost always present to some degree. Uncertainty in x_{e}, On the other hand, varies widely. We thus divide the robotic tasks studied here into three categories according to the uncertainty in x_{e}. In the first category, the environment is static and known with high accuracy. So uncertainty in x_{e} can be ignored, and we only need to consider uncertainty in x_{r} in planning the robot's actions. In the second category, the environment is static, but not known accurately. Thus, we must take into account the uncertainty in both x_{r} and x_{e} in planning. In the last category, the environment is not static and changes over time. We need a dynamic model of the environment and use it to plan actions for the robot to respond to changes in the environment.

a) Underwater Navigation: We start with an instance of the well known coastal navigation problem. An autonomous underwater vehicle (AUV) navigates in an environment modeled as a 51×52 grid map (Fig. 3*a*). The AUV needs to navigate from the left border of the map to the right border. It must avoid rocks scattered near the goals, as they may cause severe damages to the vehicle. In each step, the AUV can either stay in the current position or move to any of the four adjacent positions (directly above, below, left, and right). Due to poor visibility conditions, the AUV can only localize itself along the top or bottom borders, where there are beacon signals. The environment is static and known in advance. So this problem belongs to the first category.

Roughly, the optimal policy for the AUV is to move diagonally until it reaches the top or bottom border to localize itself. It can then safely pass through the rocks and get to the destinations on the right border. A feature of this problem is that heuristics assuming full observability (*e.g.*, an MDP policy) favor shorter horizontal paths rather than diagonal paths and thus often choose the wrong action.

b) Grasping: This problem was introduced in the work of Hsiao, Kaelbling, and Lozano-Pérez [3]. As a POMDP, this problem is similar to coastal navigation: the environment is static and known, but due to limited sensing capabilities, the robot has difficulty in determining its own state exactly. It needs to perform information-gathering actions to reduce the state uncertainty in order to reach the goal. However, as a robotic task, grasping has quite different physical characteristics. Here, a two-dimensional Cartesian robot arm with two fingers tries to grasp a stepped block on a table (Fig. 3*b*). It has only contact sensors at the tip and the sides of each finger to help determine the state. The robot performs compliant guarded moves (left, right, up, and down) and always maintains contact with the surface of the block or the boundary of the environment at the beginning and end of each move. The goal is to move the robot arm and have its two fingers straddle the block so that grasping is possible. More details on this problem can be found in [3].

c) Integrated Exploration: For some tasks, robots must traverse an area whose map is highly uncertain, for example, when robots perform SLAM tasks. In this situation, the robot must gather information to reduce map uncertainty, localize itself, and navigate to reach the goal. This is sometimes called *integrated exploration* [8]. When the environment is static, integrated exploration belongs to our second category. Unfortunately, despite a static environment, uncertainty in the environment map causes the number of states for x_e to grow exponentially. Recall further that increase in the number of states in turn causes the belief space size to grow exponentially. Currently, such doubly exponential growth is too difficult to manage, even for point-based POMDP algorithms.

Our problem here models a similar, but simplified scenario (Fig. 3*c*). In one step, the robot can move from its current location to one of the eight adjacent locations horizontally, vertically, and diagonally. The result of a move is uncertain. The robot can localize itself in several locations scattered around the environment. To reach the destination, the robot may follow one of the long routes along the far left and right sides of the environment or take a shortcut through one of the bridges (shaded in light gray in Fig. 3*c*). Due to flood damages, at most two bridges are still functional. The robot's goal is to reach the destination nodes as quickly as possible, using such an uncertain environment map. Even in this simplified setting, we still end up with more than 15,000 states.

d) Rock Sample: The Rock Sample problem first appeared in the work on HSVI [18]. In this problem, a rover explores an area modeled as a small grid and looks for rocks with scientific value. The rover always knows its own position exactly, as well as those of the rocks. However, it does not know which rocks are valuable. The rover can take noisy long-range sensor readings to gather information on the rocks. The accuracy of the sensor depends on the distance between the rover and the rocks. The rover can also sample a rock in the immediate vicinity. It receives a reward or a penalty, depending on whether the sampled rock is valuable.

In this problem, the environment is static, and a map with exact rock positions is available. However, the environment map that really matters is the one that marks the positions of *valuable* rocks. This map is unknown in advance. and the rover must infer this map from sensor readings. So this problem can be regarded as an instance of integrated exploration.

e) Tag: The Tag problem first appeared in the work on PBVI [10]. In Tag, the robot's goal is to follow a target that intentionally moves away. The robot and the target operate in a grid environment with 29 positions in total. In one step, they can either stay or move one of four adjacent positions (above, below, left, and right). The robot always knows its own position, but can observe the target's position only if they are in the same position. The robot pays a cost for each move and receives a reward every time it arrives in the same position as that of the target. Here, the environment changes over time due to the target motion. Thus the problem belongs to the third category.

f) Homecare: This problem models a robot following a person around at home for caretaking purposes (Fig. 3*d*). It is related to Tag, but involves a much larger number of states and more complex environment dynamics. Imagine that an elderly person moves around at home. His motion is non-deterministic: he follows a fixed path (marked as a black dashed curve in Fig. 3*d*), but in each time step, he may pause or proceed along the path with equal probabilities. Along the path, there is special location representing a bathroom, where the person may stay for an extended duration. The person has a call button to call the robot over for help. The call button stays on for some uncertain duration and then goes off. The robot gets a reward only if it arrives in time. The robot can observe the person's position when they are close enough. Clearly the robot should stay close to the person in order to track his position well and improve the chance of receiving rewards. At the same time, it also wants to minimize movement in order to reduce power consumption. POMDP provides a principled way to evaluate such trade-offs.

B. Experimental Setup

We applied SARSOP to the above tasks. For each task, we first performed long preliminary runs to determine approximately the reward level for the optimal policies and the amount of time needed to reach it. We then ran SARSOP for a maximum of two hours to reach this level and recorded the resulting policy. To estimate the expected total reward of the policy, we performed sufficiently large number of simulation runs until the variance in the estimated value was small. For comparison, we also ran HSVI2 on these tasks, following the same procedure. Both algorithms are implemented in C++. They were compiled with g++ v4.1.2. The experiments were performed on a PC with a 2.66GHz Intel processor and 2GB memory. For HSVI2, we used the newest software released by its original author, zmdp v1.1.3, which is a highly optimized implementation.

For SARSOP, the δ value for α-vector pruning was set at 1×10^{-2} for the two largest problems, Rock Sample and Homecare, and 1×10^{-4} for the rest. The performance of SARSOP is affected by the δ value, but not sensitive to it. One important consideration in the choice of δ is the dimensionality of the belief space involved, *i.e.*, the number of states. The rough guide that we have been using is 1×10^{-2} for POMDPs with about 10,000 states or more and 1×10^{-4} for those with substantially fewer states. We are currently implementing an adaptive technique to set δ automatically and will include it in the final software release.

C. Results

The results are shown in Table I. Column 2 of the table lists the estimated expected total rewards for the computed policies and the 95% confidence intervals. Column 3 lists the corresponding computation times.

For all six tasks, SARSOP obtained good approximate solutions within the two-hour limit. In five out of the six tasks, SARSOP substantially outperformed HSVI2, sometimes by several times. For two tasks (Integrated Exploration and Tag), HSVI2 was unable to reach a comparable reward level as that of SARSOP within the two-hour time limit. Thus, for these two tasks, we also report the reward level that HSVI2 was able to reach at the end of two hours (Table I).

TABLE I

PERFORMANCE COMPARISON.

	Reward	Time (s)
Underwater Navigation,		
$\|S\|$=2,653,$\|\mathcal{A}\|$=6,$\|\mathcal{O}\|$=103		
SARSOP	722.59 ± 1.30	72
HSVI2	721.45 ± 0.75	720
Grasping		
$\|S\|$=1,253,$\|\mathcal{A}\|$=6,$\|\mathcal{O}\|$=96		
SARSOP	320.00 ± 0.16	8
HSVI2	319.88 ± 0.14	60
Integrated Exploration		
$\|S\|$=15,517,$\|\mathcal{A}\|$=8,$\|\mathcal{O}\|$=1,015		
SARSOP	$(1.58 \pm 0.03) \times 10^6$	5,400
HSVI2	$(1.41 \pm 0.02) \times 10^6$	5,400
	$(1.43 \pm 0.02) \times 10^6$	7,200
Rock Sample (7,8)		
$\|S\|$=12,545,$\|\mathcal{A}\|$=13,$\|\mathcal{O}\|$=2		
SARSOP	21.27 ± 0.13	400
HSVI2	21.27 ± 0.09	250
Tag		
$\|S\|$=870,$\|\mathcal{A}\|$=5,$\|\mathcal{O}\|$=30		
SARSOP	-6.13 ± 0.12	6
HSVI2	-7.43 ± 0.11	6
	-6.40 ± 0.10	7,200
Homecare		
$\|S\|$=5,408,$\|\mathcal{A}\|$=9,$\|\mathcal{O}\|$=928		
SARSOP	16.86 ± 0.45	960
HSVI2	16.88 ± 0.37	2,880

On Rock Sample, SARSOP did not perform as well as HSVI2 for a very specific reason. HSVI2 implements an α-vector masking technique, which opportunistically computes only selected entries in the α-vectors. This technique is particularly beneficial here, because in Rock Sample, the robot position is fully observed, which substantially reduces the overall level of uncertainty involved. Furthermore, the remaining state variables that specify the status of rocks are independent, which also helps to improve the effectiveness of masking. Without masking, HSVI2 was only able to reach the reward level of 18.98 ± 0.09 after 400 seconds of computation time. This is worse than that of SARSOP. The effectiveness of masking degenerates for uncertain robot movements and noisy observations, which are the more common case in practice. For this reason, we currently do not incorporate masking in our implementation.

V. CONCLUSION

Point-based algorithms have greatly improved the speed of POMDP solution by sampling from the reachable space. This paper presents a new point-based algorithm, SARSOP, which exploits the notion of optimally reachable spaces to further improve computational efficiency. We applied SARSOP to a set of distinct robotic tasks, all modeled as POMDPs with a large number of states. SARSOP computed good approximate solutions to all of them in reasonable time. Further, it outperformed one of the fastest existing point-based algorithm in most of these tasks. These results indicate that approximating optimally reachable spaces through sampling is an interesting new angle to look at the problem. It has led to the more effective sampling and pruning strategies in SARSOP.

Along with other reports in literature [2, 3, 10, 11, 14, 19], our results indicate that with the advances in POMDP solution algorithms, the POMDP approach is gradually becoming practical for non-trivial robotic tasks. We have implemented SARSOP as a software package, and it is available for download at `http://motion.comp.nus.edu.sg/projects/pomdp/pomdp.html`.

Acknowledgements. We thank Yanzhu Du and Xan Huang for helping with the software implementation. This work is supported in part by AcRF grant R-252-000-327-112 from the Ministry of Education of Singapore.

REFERENCES

[1] M. Hauskrecht, "Value-function approximations for partially observable Markov decision processes," *J. Artificial Intelligence Research*, vol. 13, pp. 33–94, 2000.

[2] J. Hoey, A. von Bertoldi, P. Poupart, and A. Mihailidis, "Assisting persons with dementia during handwashing using a partially observable Markov decision process," in *Proc. Int. Conf. on Vision Systems*, 2007.

[3] K. Hsiao, L. Kaelbling, and T. Lozano-Pérez, "Grasping POMDPs," in *Proc. IEEE Int. Conf. on Robotics & Automation*, 2007, pp. 4485–4692.

[4] D. Hsu, W. Lee, and N. Rong, "What makes some POMDP problems easy to approximate?" in *Advances in Neural Information Processing Systems (NIPS)*, 2007.

[5] ——, "A point-based POMDP planner for target tracking," in *Proc. IEEE Int. Conf. on Robotics & Automation*, 2008, pp. 2644–2650.

[6] L. Kaelbling, M. Littman, and A. Cassandra, "Planning and acting in partially observable stochastic domains," *Artificial Intelligence*, vol. 101, no. 1–2, pp. 99–134, 1998.

[7] O. Madani, S. Hanks, and A. Condon, "On the undecidability of probabilistic planning and infinite-horizon partially observable Markov decision problems," in *Proc. Nat. Conf. on Artificial Intelligence*, 1999, pp. 541–548.

[8] A. Makarenko, S. Williams, F. Bourgault, and H. Durrant-Whyte, "An experiment in integrated exploration," in *Proc. IEEE/RSJ Int. Conf. on Intelligent Robots & Systems*, 2002.

[9] C. Papadimitriou and J. Tsisiklis, "The complexity of Markov decision processes," *Mathematics of Operations Research*, vol. 12, no. 3, pp. 441–450, 1987.

[10] J. Pineau, G. Gordon, and S. Thrun, "Point-based value iteration: An anytime algorithm for POMDPs," in *Proc. Int. Jnt. Conf. on Artificial Intelligence*, 2003, pp. 477–484.

[11] J. Pineau, M. Montemerlo, M. Pollack, N. Roy, and S. Thrun, "Towards robotic assistants in nursing homes: Challenges and results," *Robotics & Autonomous Systems*, vol. 42, no. 3–4, pp. 271–281, 2003.

[12] P. Poupart and C. Boutilier, "Value-directed compression of POMDPs," in *Advances in Neural Information Processing Systems (NIPS)*. The MIT Press, 2003, vol. 15, pp. 1547–1554.

[13] M. Puterman, *Markov Decision Processes: Discrete Stochastic Dynamic Programming*. John Wiley & Sons, 1994.

[14] N. Roy, G. Gordon, and S. Thrun, "Finding aproximate POMDP solutions through belief compression," *J. Artificial Intelligence Research*, vol. 23, pp. 1–40, 2005.

[15] G. Shani, R. Brafman, and S. Shimony, "Adaptation for changing stochastic environments through online POMDP policy learning," in *Proc. Eur. Conf. on Machine Learning*, 2005, pp. 61–70.

[16] ——, "Forward search value iteration for POMDPs," in *Proc. Int. Jnt. Conf. on Artificial Intelligence*, 2007.

[17] R. Smallwood and E. Sondik, "The optimal control of partially observable Markov processes over a finite horizon," *Operations Research*, vol. 21, pp. 1071–1088, 1973.

[18] T. Smith and R. Simmons, "Heuristic search value iteration for POMDPs," in *Proc. Uncertainty in Artificial Intelligence*, 2004, pp. 520–527.

[19] ——, "Point-based POMDP algorithms: Improved analysis and implementation," in *Proc. Uncertainty in Artificial Intelligence*, 2005.

[20] M. Spaan and N. Vlassis, "A point-based POMDP algorithm for robot planning," in *Proc. IEEE Int. Conf. on Robotics & Automation*, 2004.

Detection of Principal Directions in Unknown Environments for Autonomous Navigation

Dmitri Dolgov
AI & Robotics Lab
Toyota Research Institute
Ann Arbor, MI 48105
Email: ddolgov@ai.stanford.edu

Sebastian Thrun
Stanford Articial Intelligence Lab
Stanford University
Stanford CS 94305
Email: thrun@ai.stanford.edu

Abstract— **Autonomous navigation in unknown but well-structured environments (e.g., parking lots) is a common task for human drivers and an important goal for autonomous vehicles. In such environments, the vehicles must obey the standard conventions of driving (e.g., passing oncoming vehicles on the correct side), but often lack a map that can be used to guide path planning in an appropriate way. The robots must therefore rely on features of the environment to drive in a safe and predictable way. In this work, we focus on detecting one of such features, the principal directions of the environment.**

We propose a Markov-random-field (MRF) model for estimating the maximum-likelihood field of principal directions, given the local linear features extracted from the vehicle's sensor data, and show that the method leads to robust estimates of principal directions in complex real-life driving environments. We also demonstrate how the computed principal directions can be used to guide a path-planning algorithm, leading to the generation of significantly improved trajectories.

I. Introduction

Autonomous navigation in outdoor environments is an active area of research in robotics, with extensive work being done in two distinct areas: i) the creation of robotic vehicles capable of driving on streets and highways [7, 24, 25, 23], and ii) the development of robots for off-road navigation [14, 22, 4, 21, 1, 16]. In the former case of on-street driving, it is sensible and commonplace to assume that there is a map of the global road network available, and the robot must obey the standard rules of driving (drive on the appropriate side of the street, not straddle the lane boundaries, etc). In the other mode of off-road driving, a detailed map of the environment is typically not available to the robot *a priori*, but the robot is not confined by any rules of the road and is free to choose any drivable and safe path to its goal.

There is also a large middle ground between these two areas, consisting of well-structured environments where it is not reasonable to assume knowledge of a detailed map, but the robotic vehicle must nonetheless obey the common rules of driving. Such conditions arise, for example, in parking lots, shopping malls, and construction zones. In such areas—even in the absence of typical road markings and signs—human drivers are usually able to partition the space into drivable lanes and drive appropriately. For instance, when driving in a parking lot such as the one shown in Figure 1a, most people

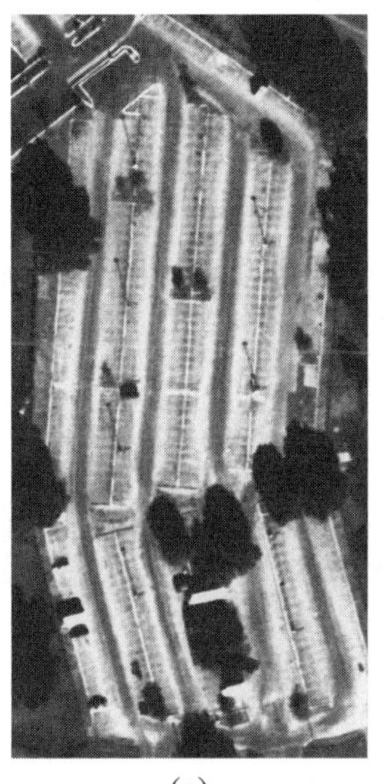

(a)

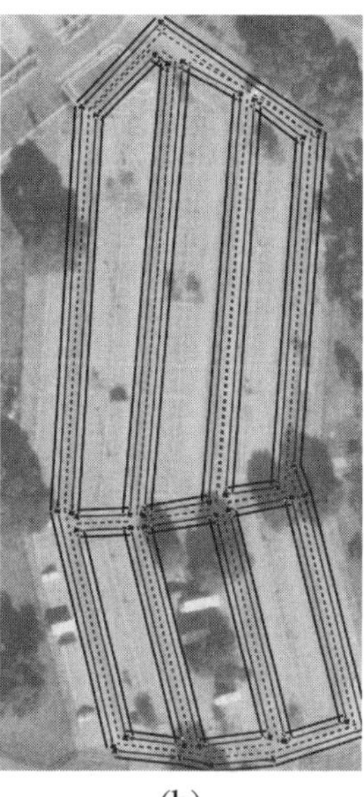

(b)

Fig. 1. Detecting structure of the driving environment. a) Aerial imagery of a parking lot; b) A typical structure imposed on the environment by human drivers.

will not have any difficulty detecting the main drivable lanes, similar to the ones shown in Figure 1b.

Similarly, a robot operating in such an environment needs to be able to—by using data from its on-board sensors—recognize features of the environment that will allow it to drive in accordance with common rules. Compliance to established rules and driving conventions is an especially important safety requirement if robots are to share the environment with human drivers, because it then becomes an issue of predictability.

In this work, we take a step towards this goal by considering one of the most basic features of environments: their principal directions. Knowing the principal directions of the environment is a prerequisite condition for implementing many high-level driving behaviors, such as driving on the correct side of the road (which is critical for collision avoidance) and avoiding diagonal paths across parking lots (which is often considered rude and unsafe).

Finding principal directions based purely on data from the robot's on-boards sensors is a challenging task due to the

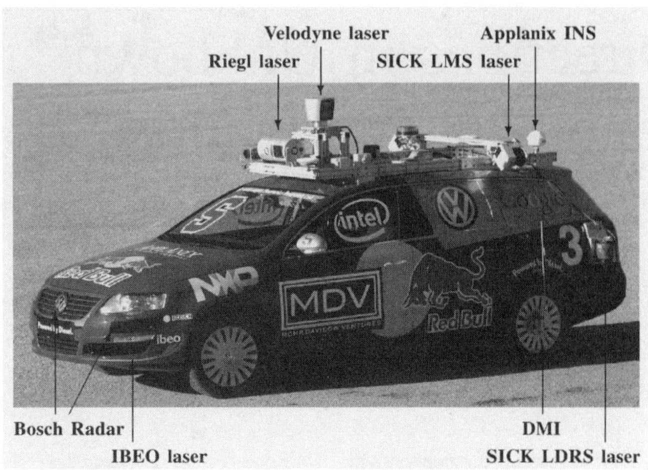

Fig. 2. Stanford Racing Team's robotic vehicle, Junior. The results presented in this work were obtained using the Applanix GPS+IMU system and the Velodyne 3D LIDAR.

following issues: i) Many environments have several principal directions that vary from point to point (e.g., Figure 1), ii) Sensor data is noisy and can give rise to conflicting local hypotheses regarding the global principal directions, and iii) Computation of principal directions has to be done efficiently to be useful for navigation.

We approach the problem of computing the field of principal directions in a probabilistic framework. Assuming that at every point in the two-dimensional space of the driving environment there is a main preferred direction of motion, we formulate a Markov-random-field (MRF) [6] model that allows us to infer a high-likelihood field of principal directions, corresponding to the observed evidence provided by the vehicle's on-board sensors. This approach is inspired by the extensive use of MRFs for image processing [18, 26].

The main steps of our approach are as follows. First, a map of the environment is computed from the robot's on-board sensors. In this work, we present results based on LIDAR data, but our approach can be applied to other sensor modalities, such as cameras (potentially even leading to improvements in accuracy). We then scan the resulting map for local linear features that provide evidence regarding the principal directions of the environment. Having thus obtained local evidence, we formulate an MRF whose nodes form a 2D grid and represent the global principal direction of the environment at the corresponding locations. A solution to the MRF is obtained using fast numeric-optimization techniques.

We also present results demonstrating the robust performance of the MRF approach on several real environments driven with our robotic vehicle (Figure 2), and show how the resulting field of principal directions can be used to guide path planning, leading to a significant improvement in generated trajectories.

II. Local Principal Directions

The core of our MRF-based approach is independent of the method used to obtain features of the environment that serve as local evidence. However, for completeness and continuity, in this section, we briefly describe the specifics of our method for obtaining evidence for the MRF.

We used a vehicle equipped with a 3D Velodyne LIDAR as the sole environmental sensor. Below, we outline the techniques we used to find lines in the environment, given the range data produced by the LIDAR. The main steps of the process are illustrated in Figure 3.

The Velodyne LIDAR outputs a 3D point cloud, as shown in Figure 3a. In the first step, we filter out the ground plane, integrate data over time, and project the points onto a plane, leading to a 2D obstacle map shown in Figure 3b. Any line segments in this 2D map (e.g., curbs, other cars) are evidence of the principal driving directions, and we can detect those local lines using common computer-vision approaches.

There are several good known techniques for detecting lines in images. We found that the following sequence of standard transforms leads to satisfactory results for out application. First, we smooth the data using a symmetric 2D Gaussian kernel (Figure 3c) and then apply a binary threshold (25%) to the result (leading to the data shown in Figure 3d). The effect of these two steps is to remove small noise from the data, smooth out jagged lines, and "fill in" small gaps within objects (such as cars or trees), thereby eliminating extra edges. We then apply the Canny algorithm [5] for edge detection, yielding the image shown in Figure 3e. Finally, we use a randomized Hough transform (RHT) [27] to find line segments in the data (Figure 3f).

III. Markov Random Field for Estimating Global Principal Directions

The lines obtained in the previous section can be viewed as local evidence regarding the principal directions of the environment. Our goal is therefore to estimate the maximum likelihood field of principal directions $\theta(x, y)$, which specifies the principal orientation at every point x, y in the region of interest. Note that most complex driving environments—e.g., Figure 1—do not have a single global orientation, but rather have several principal directions that differ from point to point.

We formulate a discrete version of the inference problem on a regular 2D grid, associating with each point $\langle x_i, y_i \rangle$ an MRF node with a scalar continuous value θ_i. Figure 4 illustrates our MRF construction. Each MRF node $\theta_i \in [0, \pi/2)$ has associated with it a set of evidence nodes $\alpha_{ik} \in [0, \pi/2)$, one for the angle of each line segment $k \in [1, K_i]$ that crosses the grid cell i.

All angles are normalized to $[0, \pi/2)$, because orthogonal lines (e.g., edges of a car) support the same hypothesis for the principal direction at a point.

Our MRF uses two sets of potentials associated with the nodes and edges of the graph in Figure 4. The first potential (Ψ) is defined on the nodes of the MRF and ties each θ_i to its local evidence α_{ik}. It is defined as follows:

$$\Psi(\theta, \alpha) = \sum_i \sum_{k=1}^{K_i} \lambda_{ik} \psi(\theta_i, \alpha_{ik}), \tag{1}$$

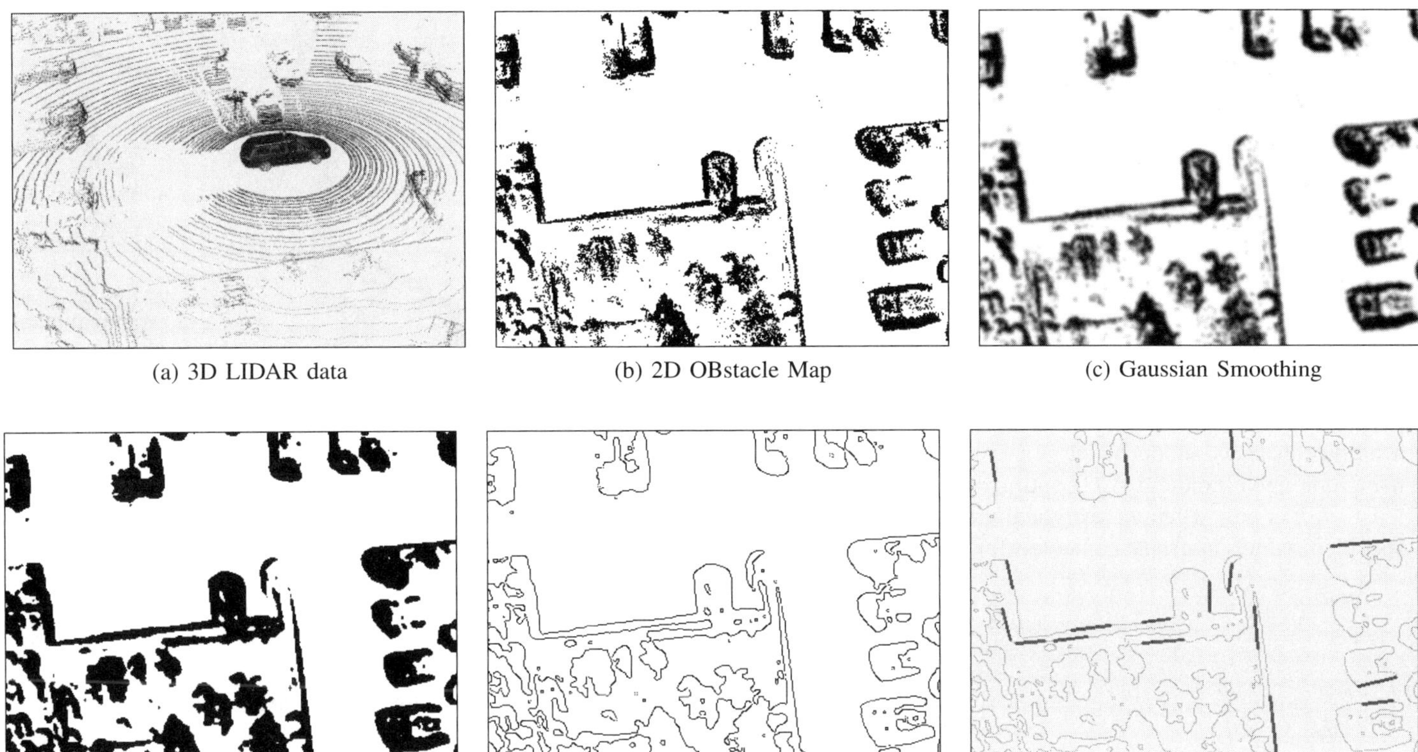

Fig. 3. Line detection in a parking lot: a) 3D LIDAR data; b) Obstacle map; c) After smoothing with a Gaussian kernel; d) After binary thresholding; e) After Canny edge detection; f) After Hough transform.

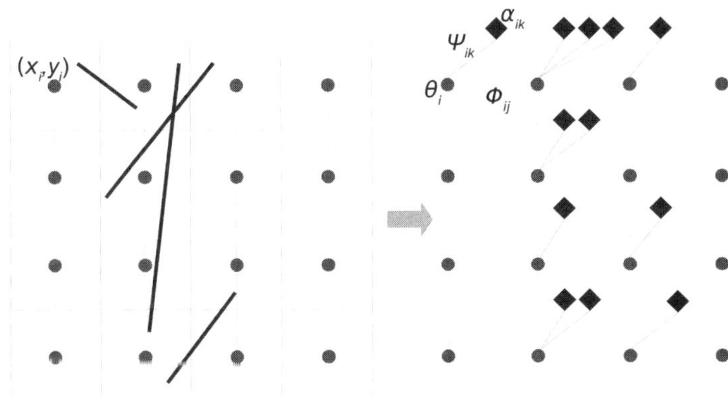

Fig. 4. MRF for inferring the field of principal directions of the environment. MRF variables θ_i are shown as red circles, α_{ik} are the input nodes, corresponding to the angles of the observed lines.

where λ_{ik} is the weight associated with line k, and ψ is a distance measure between two angles; both are defined below.

The second potential (Φ) is defined on the edges of the MRF and encodes a prior that enforces a smoothness relationship between the principal directions at neighboring nodes:

$$\Phi(\theta) = \frac{1}{2} \sum_{i} \sum_{j \in \mathcal{N}(i)} \phi(\theta_i, \theta_j), \tag{2}$$

where $\mathcal{N}(i)$ is the set of neighbors of node i, and ϕ is a distance measure between two angles (defined below).

There are many reasonable choices for the distance measures ψ and ϕ, as well as the weights λ_{ik}. We evaluated several options for each, and experimentally settled on the following. For the weights λ_{ik}, we used the length of the corresponding line segment (the longer the segment, the stronger the evidence). The choice of the distance measures ψ and ϕ is an interesting topic in itself: some distances favor smoother fields (e.g., L_2 norm), others (e.g., L_1 norm) have better discontinuity-preserving properties; see [9] for an applicable discussion of different norms in optimization. We empirically investigated several functions, and found the following to be a good choice of a norm for both evidence and smoothness potentials:

$$\psi(\beta, \gamma) = \phi(\beta, \gamma) = \sin^2 \Big(2(\beta - \gamma)\Big). \tag{3}$$

This measure behaves quadratically for small $\beta - \gamma$, and has natural wrap-around properties for our angles in the interval $[0, \pi/2)$.

Finally, the distribution of the MRF variables θ for a specific set of observed α is given by a Gibbs distribution:

$$\mathbb{P}(\theta|\alpha) = \frac{1}{Z} \exp\Big(-(w_\psi \Psi(\theta, \alpha) + w_\phi \Phi(\theta))\Big), \tag{4}$$

where w_ψ and w_ϕ are weights, and Z is a normalizer or the partition function.

Our goal is to find the maximum-likelihood field of principal directions θ, given the observed evidence α:

$$\theta^*_\alpha = \arg\max_\theta \mathbb{P}(\theta|\alpha), \tag{5}$$

or, in other words, find θ that minimizes the Gibbs energy $U = w_\psi \Psi + w_\phi \Phi$.

IV. Optimization

For computational reasons, it is infeasible to compute an exact maximum-likelihood solution to the MRF defined in the previous section for anything but the simplest problems (a typical MRF for a realistic environment will have several thousand to tens of thousands nodes). Similarly to the approach of Diebel and Thrun [8, 9], we therefore settle for a high-probability mode of the posterior, which we compute using conjugate-gradient (CG) optimization [12, 20].

CG works best when an analytical gradient of the energy is specified, which is easily computed for our MRF potentials:

$$\frac{\partial \Psi}{\partial \theta_i} = 4 \sum_k^{K_i} \lambda_{ik} \sin\Big(2(\theta_i - \alpha_{ik})\Big) \cos\Big(2(\theta_i - \alpha_{ik})\Big),$$

$$\frac{\partial \Phi}{\partial \theta_i} = -\frac{\partial \Phi}{\partial \theta_j} = 2 \sin\Big(2(\theta_i - \theta_j)\Big) \cos\Big(2(\theta_i - \theta_j)\Big).$$

Given the above potentials and the gradient, the implementation of conjugate gradient is standard [20]. The output of the optimization is a high-likelihood field $\theta(x_i, y_i)$ that corresponds to the observed lines α.

V. Results

In this section, we present results on the performance of our MRF-based approach to computing principal directions. The use of these directions in path-planning is discussed and evaluated in the next section.

We tested our algorithm using a vehicle equipped with an Applanix pose estimation GPS-IMU system and a Velodyne 3D LIDAR. Some representative examples of executing our method are shown in Figure 5. The left column shows the 2D obstacle map obtained from the vehicle's sensors. The center column shows the lines detected in the obstacles map using the method described in Section II; these lines serve as the input to our MRF. The right column of Figure 5 shows the resulting field of principal directions computed by conjugate-gradient optimization.

The top row of Figure 5 shows a nearly ideal scenario: a parking lot with two main orthogonal principal directions, which are easily computed by our algorithm. The second row shows data for another, more complex, parking lot. Notice the presence of trees and a second, differently oriented, parking lot in the bottom-left part of the map. Despite these challenges, the MRF computes a very good estimate of the preferred driving directions for this environment. The third row of Figure 5 demonstrates the ability of our approach to handle environments with gradually-changing orientations. Notice that the field of principal directions correctly follows the curved road segment. The fourth row of Figure 5 depicts another challenging situation with a curved street, an intersection,

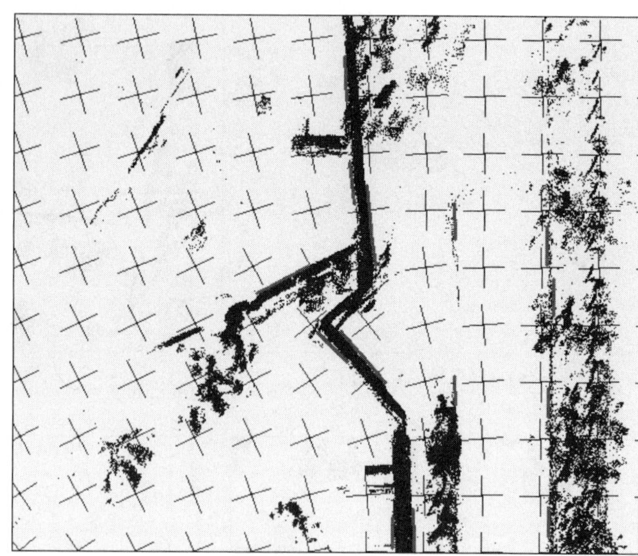

Fig. 6. The MRF computes the correct orientation for the drivable region of the parking lot, despite bad evidence provided by an adjacent building corner.

and an adjacent parking lot. It also highlights an interesting challenge for our method: parking lots with diagonally-parked cars (upper-right corner of the map). In this situation, the parked cars—whose sides are usually detected as lines—and features of the parking lots themselves (e.g., curbs) present conflicting evidence regarding the principal orientation of the environment.

Another complex situation is shown in Figure 6, where a building is located very close to a parking lot, but is oriented at a different angle, providing bad evidence for the MRF. However, as can be seen from the vector field in Figure 6, the MRF is able to compute the correct orientation for the drivable area, despite the fact that the curb separating the corner of the building from the parking lot is not detected.

Additional examples illustrating the computation of principal directions in real driving environments are shown in Figure 9.

Figure 7 shows the running time of our algorithm as performed on a 3Ghz Intel Core-2 PC. The data presented in Figure 7 is for the parking lot shown in the second row of Figure 5).

Figure 7a compares the running times of the main components of our approach. The timing results in Figure 7a are for experiments run with the following parameters: the obstacle grid was 260m$\times$260m with 15cm resolution, the MRF evidence grid was 260m$\times$260m with 5m resolution (resulting in an MRF with 2704 nodes). As can be seen from the data, the performance of our algorithm is usually dominated by the running time of the conjugate-gradient algorithm, and with the average computation time of under 300ms, the method is well-suited for use in online path planning (since it is usually not necessary to update the field of principal directions during every planning cycle).

Figure 7b shows how our MRF inference scales with the size of the MRF grid (with a 5-meter discretization of the MRF

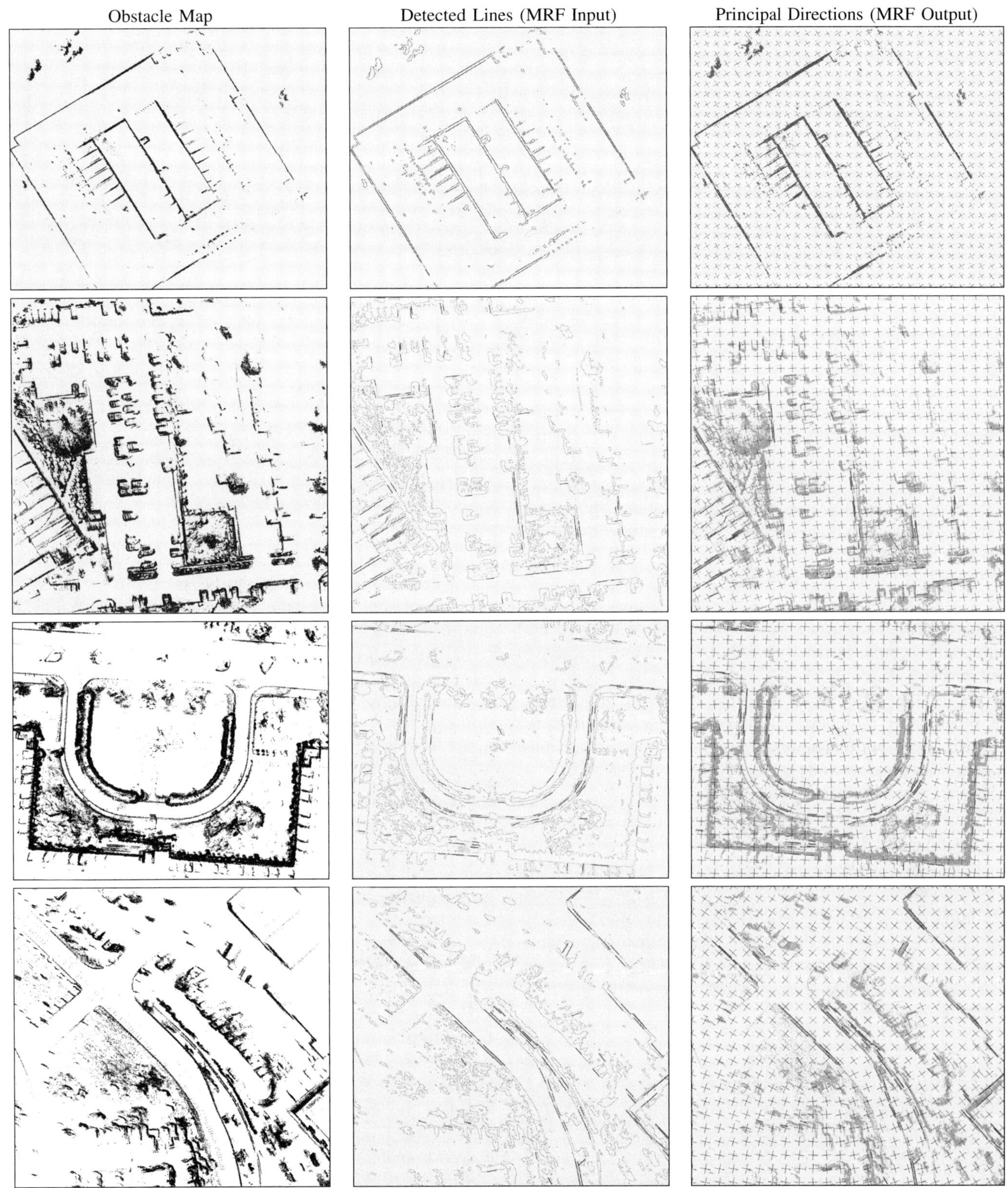

Fig. 5. Results of computing principal directions in several real driving environments.

grid). Figure 7c illustrates the scaling of the MRF inference as a function of the number of MRF variables θ_i (corresponding to MRF grids of size 40m to 300m, at 2.5m resolution).

We should note that a finer discretization of the MRF grid does not necessarily lead to better results. In fact, we found a resolution of around 5m to work best for typical driving areas.

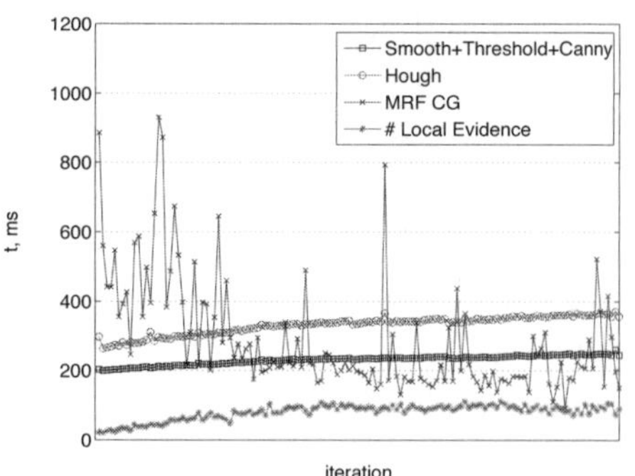

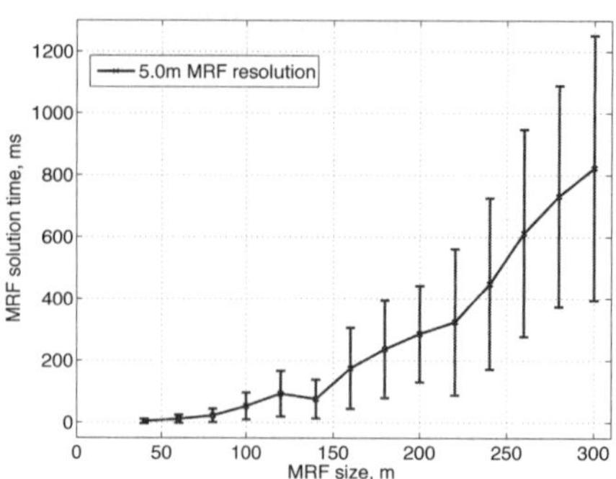

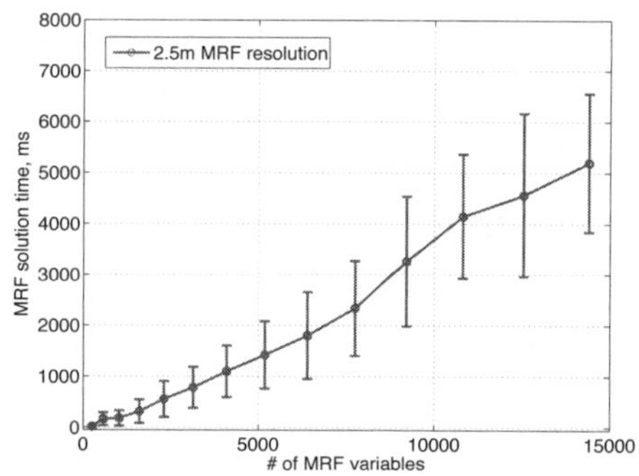

Fig. 7. (a) Running time of the main components of our method on a sequence of real obstacle maps: i) Time for map pre-processing and Canny edge detection, ii) Time for the Hough Tranform, iii) Time for the MRF conjugate-gradient optimization, and IV) Number of local linear features detected via the Hough transform. (b) MRF inference time as a function of the MRF grid size at a 5-meter resolution (number of MRF variables is quadratic in the size). (c) MRF inference time as a function of the number of MRF variables (grid cells) at a 2.5-meter discretization.

VI. Path Planning

There are several ways in which knowledge of principal directions can be fruitfully utilized in path planning, some of which were outlined in the Introduction. Below, we describe a path-planning algorithm that favors trajectories aligned with the environment, and illustrate the improvements this leads to, compared to a standard planner with a bias towards smooth trajectories with arbitrary headings.

In what follows we leave aside the problem of global planning, which in practice can be accomplished via several existing algorithms, such as continuous forward search via path sampling (e.g., [3, 13, 17, 19]) or discrete search (e.g., [10, 15, 11]), just to name a couple classes of algorithms. Assuming that a rough global plan has been computed, we show how the field of principal directions can be used in a path-smoothing phase of planning.

Given a global plan represented as a sequence $\langle x_l, y_l \rangle$, $l \in [1, N]$, we can formulate the smoothing problem as a continuous optimization on the coordinates of the vertices of the trajectory. For illustrative purposes, below we focus on two terms: smoothness and compliance with principal directions; other important aspects of a realistic path smoother (such as kinematic constraints of the vehicle and collision-avoidance terms) are omitted for brevity.

For a given trajectory $\{\langle x_l, y_l \rangle\}$, let us define the displacement vector at vertex l:

$$\Delta \mathbf{x}_l = \begin{pmatrix} x_l \\ y_l \end{pmatrix} - \begin{pmatrix} x_{l-1} y_{l-1} \\ , \end{pmatrix} \quad (6)$$

and the heading of the trajectory at vertex l:

$$\sigma_l = \tan^{-1}\left(\frac{y_{l+1} - y_l}{x_{l+1} - x_l}\right). \quad (7)$$

The smoothness of the trajectory and the bias towards driving along principal directions can then be expressed as:

$$f = w_{sm} \sum_{l=1}^{N-1} (\Delta \mathbf{x}_{l+1} - \Delta \mathbf{x}_l)^2 + w_{pd} \sum_{l=1}^{N} \mu\Big(\theta(x_l, y_l), \sigma_l\Big), \quad (8)$$

where $\theta(x_l, y_l)$ is the principal direction at the MRF node closest to $\langle x_l, y_l \rangle$; w_{sm} and w_{pd} are the weights on smoothness and the principal-direction bias, respectively; μ is a potential on the difference of the two angles, which can be defined in a variety of ways, for example:

$$\mu(\theta(x_l, y_l), \sigma_l) = 1 - \cos\Big(2(\theta(x_l, y_l) - \sigma_l)\Big). \quad (9)$$

This smoothing problem can be solved using several efficient numerical techniques. One approach is to again use the conjugate-gradient algorithm as in Section IV.

Figure 8 shows the output of different variants of path smoothing on the same problem. Figure 8a and Figure 8c shows the output of the smoother that minimizes the quadratic-curvature term in (8), subject to constraints on collision avoidance and kinematics of the car, but $w_{pm} = 0$ and principal directions are ignored. The driving style exemplified by such trajectories might in some cases be considered undesirable. Turning on the bias for aligning the trajectory with the principal directions leads to solutions shown in Figure 8b and Figure 8d; such compliance with the orientation of the environment can be useful in many driving situations.

VII. Discussion

We presented an MRF-based method for inferring principal directions of unknown environment using data from 3D LIDAR, and illustrated the usefulness of principal directions for path-planning in well-structured environments, such as parking lots. We evaluated the performance of our approach in real driving scenarios, demonstrating its ability to robustly estimate the field of principal directions in the presence of noise, conflicting local evidence, and regions with smoothly curved boundaries.

This work takes a step towards designing autonomous vehicles that can operate predictably in unknown environments and follow the standard conventions of human driving.

Still, the principal directions are a rather crude property of the environment, and it is necessary to detect more comprehensive features and use them in path planning to achieve predictable human-like driving. For instance, another high-level driving behavior that is challenging to implement is the

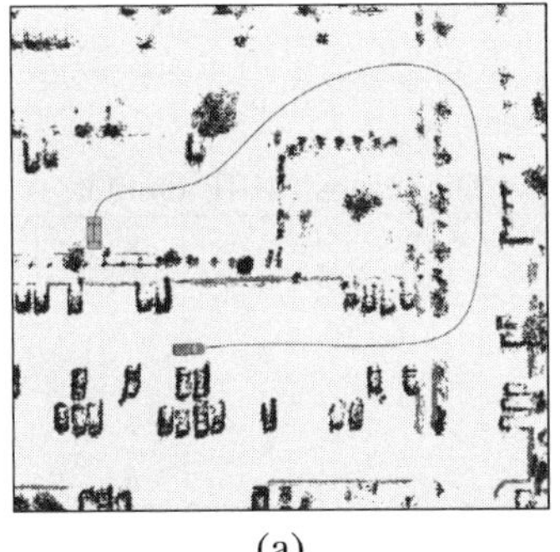
(a)

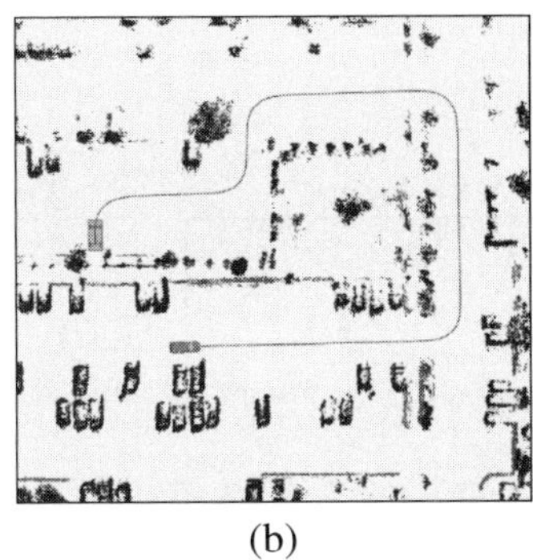
(b)

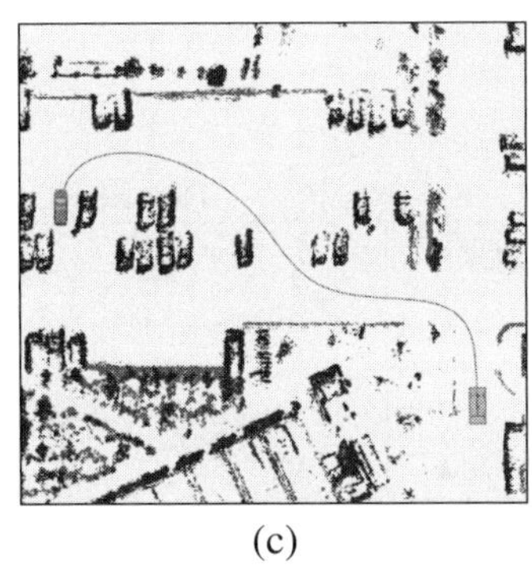
(c)

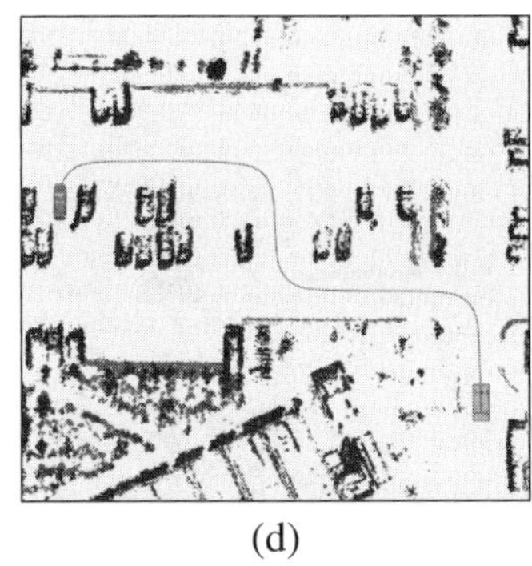
(d)

Fig. 8. Path planning using principal directions. a) and c) Principal directions are not used. b) and d) Planning favors trajectories that are aligned with principal directions of the environment.

adherence to the established convention of passing oncoming traffic on the correct side. A straightforward method of partitioning the space into "left" and "right" regions can be developed by using the Voronoi diagram [2] of the obstacle map and labeling points based on which side of the Voronoi edge they fall. However, a straightforward application of this method is brittle with respect to sensor noise, producing fragmented obstacle maps that can lead to highly irregular Voronoi diagrams. A more sophisticated method is needed for robust identification of drivable lanes in order for the robot to correctly handle situations involving other cars.

Another interesting thread of our current and future work lies along the direction of combining obstacle data with visual camera data as well as surface-reflectivity data from LIDARs for detecting and recognizing more advanced features in unknown environments (e.g., lane markings, curbs, other vehicles) and their use in path planning.

REFERENCES

[1] The 2005 DARPA Grand Challenge: The Great Robot Race. Springer, 2005.

[2] F. Aurenhammer. Voronoi diagrams – a survey of a fundamental geometric data structure. ACM Computing Surveys, 23(3):345–405, 1991.

[3] J. Barraquand, B. Langlois, and J.-C. Latombe. Robot motion planning with many degrees of freedom and dynamic constraints. In The fifth international symposium on Robotics research, pages 435–444, Cambridge, MA, USA, 1990. MIT Press.

[4] O. Brock and O. Khatib. High-speed navigation using the global dynamic window approach. In International Conference on Robotics and Automation (ICRA), 1999.

[5] J. Canny. A computational approach to edge detection. IEEE Trans. Pattern Anal. Mach. Intell., 8(6):679–698, November 1986.

[6] R. Chellappa and A. Jain. Markov Random Fields: Theory and Applications. Academic Press, Orlando, FL, 1993.

[7] E. D. Dickmanns and A. Zapp. Autonomous high speed road vehicle guidance by computer vision. Triennial World Congress of the International Federation of Automatic Control, 4:221–226, July 1987 1987.

[8] J. Diebel and S. Thrun. An application of markov random fields to range sensing. In Y. Weiss, B. Schölkopf, and J. Platt, editors, Advances in Neural Information Processing Systems 18, pages 291–298. MIT Press, Cambridge, MA, 2006.

[9] J. R. Diebel, S. Thrun, and M. Bruenig. A bayesian method for probable surface reconstruction and decimation. ACM Trans. Graph., 25(1):39–59, 2006.

[10] T. Ersson and X. Hu. Path planning and navigation of mobile robots in unknown environments. In IEEE International Conference on Intelligent Robots and Systems (IROS), 2001.

[11] D. Ferguson and A. Stentz. Field d*: An interpolation-based path planner and replanner. In Proceedings of the Int. Symp. on Robotics Research (ISRR), October 2005.

[12] M. R. Hestenes and E. Stiefel. Methods of conjugate gradients for solving linear systems. Journal of Research of the National Bureau of Standards, 49:409436, 1952.

[13] L. Kavraki, P. Svestka, J.-C. Latombe, and M. Overmars. Probabilistic roadmaps for path planning in high-dimensional configuration spaces. Technical Report CS-TR-94-1519, Department of Information and Computing Sciences, Utrecht University, 1994.

[14] A. Kelly. An intelligent predictive control approach to the high speed cross country autonomous navigation problem, 1995.

[15] S. Koenig and M. Likhachev. Improved fast replanning for robot navigation in unknown terrain. In IEEE Int. Conf. on Robotics and Automation (ICRA), 2002.

[16] S. Kolski, D. Ferguson, M. Bellino, and R. Siegwart. Autonomous driving in structured and unstructured environments. In IEEE Intelligent Vehicles Symposium, 2006.

[17] S. LaValle. Rapidly-exploring random trees: A new tool for path planning, 1998.

[18] S. Z. Li. Markov random field models in computer vision. In ECCV '94: Proceedings of the Third European Conference-Volume II on Computer Vision, pages 361–370, London, UK, 1994. Springer-Verlag.

[19] E. Plaku, L. Kavraki, and M. Vardi. Discrete search leading continuous exploration for kinodynamic motion planning. In Robotics: Science and Systems, June 2007.

[20] W. H. Press, S. A. Teukolsky, W. T. Vetterling, and B. P. Flannery. Numerical Recipes in C: The Art of Scientific Computing. Cambridge University Press, New York, NY, USA, 1992.

[21] S. Singh, R. Simmons, T. Smith, A. T. Stentz, V. Verma, A. Yahja, and K. Schwehr. Recent progress in local and global traversability for planetary rovers. In Proceedings of the IEEE International Conference on Robotics and Automation, 2000. IEEE, April 2000.

[22] A. T. Stentz and M. Hebert. A complete navigation system for goal acquisition in unknown environments. In Proceedings 1995 IEEE/RSJ International Conference On Intelligent Robotic Systems (IROS '95), volume 1, pages 425 – 432, August 1995.

[23] C. Thorpe, T. Jochem, and D. Pomerleau. Automated highway and the free agent demonstration. In Robotics Research – International Symposium, volume 8, 1998.

[24] S. Tsugawa, N. Watanabe, and H. Fujii. Super smart vehicle systemits concept and preliminary works. Vehicle Navigation and Information Systems, 2:269 –277, October 1991.

[25] P. Varaiya. Smart cars on smart roads: Problems of control. IEEE Transactions on Automatic Control, 38(2), February 1993.

[26] G. Winkler. Image Analysis, Random Fields and Markov Chain Monte Carlo Methods: A Mathematical Introduction (Stochastic Modelling and Applied Probability). Springer-Verlag New York, Inc., Secaucus, NJ, USA, 2006.

[27] L. Xu and E. Oja. Randomized hough transform (rht): basic mechanisms, algorithms, and computational complexities. CVGIP: Image Understanding, 57(2):131–154, 1993.

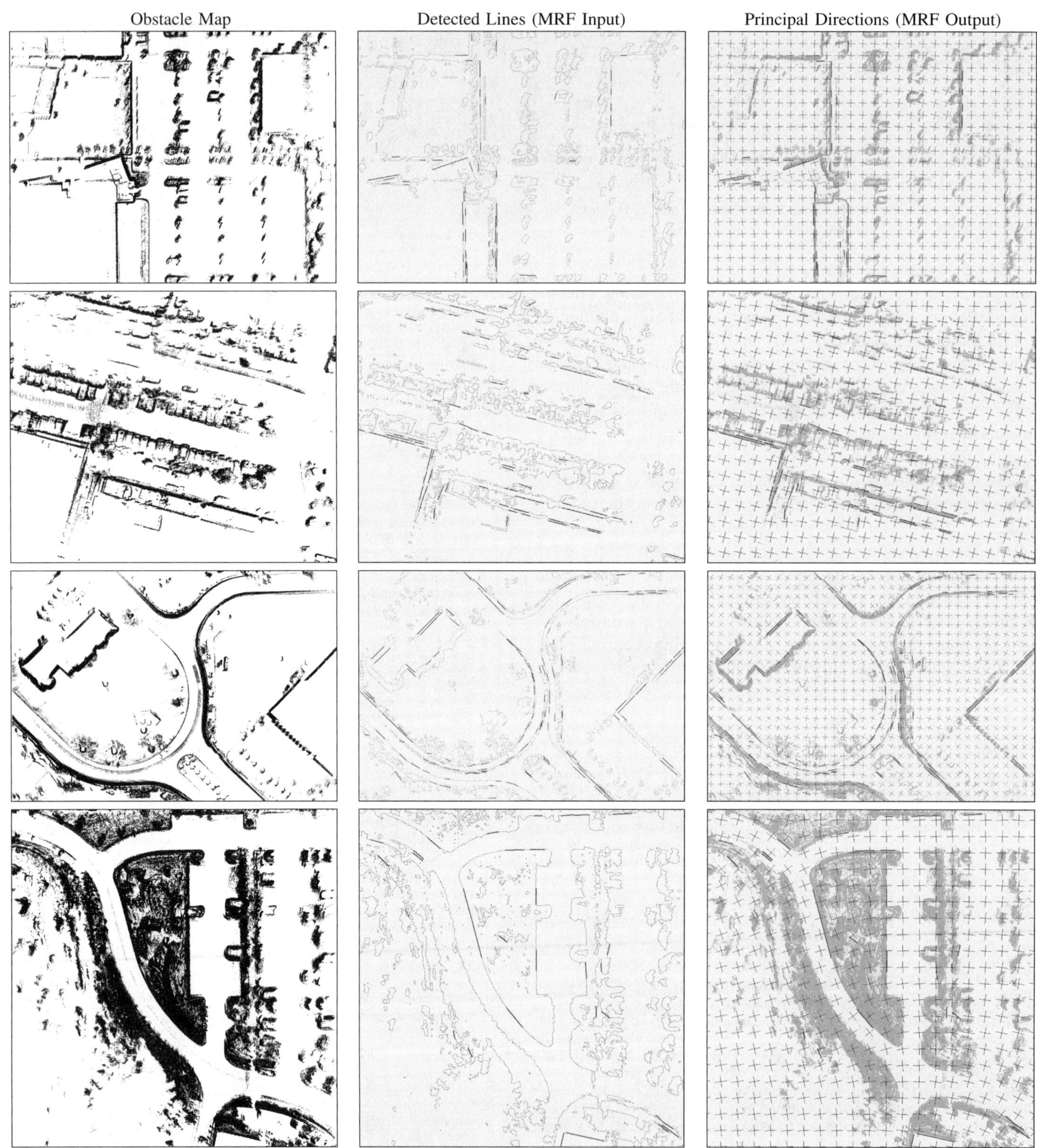

Fig. 9. Additional results of computing principal directions in several real driving environments.

Approximation Schemes for Two-Player Pursuit Evasion Games with Visibility Constraints

Sourabh Bhattacharya Seth Hutchinson
Department of Electrical and Computer Engineering
University of Illinois at Urbana Champaign
Urbana, Illinois
Email: {sbhattac, seth}@uiuc.edu

***Abstract*— In this paper, we consider the problem in which a mobile pursuer attempts to maintain visual contact with an evader as it moves through an environment containing obstacles. This surveillance problem is a variation of traditional pursuit-evasion games, with the additional condition that the pursuer immediately loses the game if at any time it loses sight of the evader. We present schemes to approximate the set of initial positions of the pursuer from which it might be able to track the evader.**

We first consider the case of an environment containing only polygonal obstacles. We prove that in this case the set of initial pursuer configurations from which it does not lose the game is bounded. Moreover, we provide polynomial time approximation schemes to bound this set. We then extend our results to the case of arbitrary obstacles with smooth boundaries.

I. Introduction

Target tracking is an interesting class of motion planning problems. It considers motion strategies for a mobile robot to track a moving target among obstacles. In case of an antagonistic target, the problem lies in the framework of pursuit-evasion which belongs to a special class of problems in game theory. The two players in the game are the pursuer and the evader. The goal of the pursuer is to maintain a line of sight to the evader that is not occluded by any obstacle. The goal of the evader is to escape the visibility region of the pursuer (and break this line of sight) at any instant of time.

This problem has some interesting applications. In security and surveillance systems, tracking strategies enable mobile sensors to monitor moving targets in cluttered environments. In home care settings, a tracking robot can follow elderly people and alert caregivers of emergencies. Target-tracking techniques in the presence of obstacles have been proposed for the graphic animation of digital actors, in order to select the successive viewpoints under which an actor is to be displayed as it moves in its environment [16]. In surgery, controllable cameras could keep a patient's organ or tissue under continuous observation, despite unpredictable motions of potentially obstructing people and instruments.

In this work, we address the problem of a single pursuer trying to maintain visibility of a single evader in a planar environment containing obstacles. The pursuer and the evader have bounded speeds. We address the following question: Given the initial position of the evader, what are the initial positions of the pursuer from which it can track successfully ? We use the term *decidable region* to refer to the set of initial positions of the pursuer at which the result of the game is known. Similarly, we use the term *undecidable region* to refer to the set of initial positions of the pursuer at which the result of the game is unknown.

The main contributions of this work are as follows. First; we prove that in an environment containing obstacles, the initial positions of the pursuer from which it can track the evader is bounded. Though this result is trivially true for a bounded workspace, for an unbounded workspace it is intriguing. Second; In this work, we provide polynomial-time approximation schemes to bound the set of initial positions of the pursuer from which it might be able to track successfully. If the initial position of the pursuer lies outside this region, the evader escapes. The size of the region depends on the geometry of the environment and the ratio of the maximum evader speed to the maximum pursuer speed. Third; we address the problem of target tracking in an environment containing non-polygonal obstacles. In the past, researchers [15] have addressed the problem of searching an evader in non-polygonal environments. However, we do not know of any prior work that addresses the problem of tracking an evader in non-polygonal environments. Fourth; although, we do not provide a complete solution to the *decidability* [5] of the tracking problem in general environments, we present partial solutions by providing polynomial time algorithms to bound the *undecidable region*.

The rest of the paper is organized as follows. Section II provides the related work. Section III presents the problem formulation. Section IV presents polynomial time approximation schemes to compute the *decidable region*. Section V extends the approximation schemes to environments containing non-polygonal obstacles. Section VI presents the conclusions and future research directions.

II. Related work

Some previous work has addressed the motion planning problem for maintaining visibility of a mobile evader. In [4], an algorithm is presented that operates by maximizing the probability of future visibility of the evader. In [14], algorithms are proposed for discrete-time representations of the system in deterministic and stochastic settings. The algorithms become computationally expensive as the number of stages of the game is increased. In [8], the authors take into account the positioning uncertainty of the robot pursuer. Game theory is proposed as a framework to formulate the tracking problem, and an approach is proposed that periodically commands the pursuer to move into a region that has no localization uncertainty in order to re-localize and better track the evader afterward.

In [5], the problem of tracking an evader around a single corner is addressed. The free workspace is partitioned according to the strategies used by the players to win the game. The authors have shown that the problem is completely *decidable* around a single corner. However, in reality, we seldom encounter environments having single corner. Hence the results about a single corner have limited application in real scenarios. In [18], the authors show that the problem of deciding whether or not the pursuer is able to maintain visibility of the evader in a general environment is at least NP-complete. This motivates the necessity to use randomized or approximation techniques to address the problem since any deterministic algorithm would be computationally inefficient.

Some variants of the tracking problem have also been addressed. [7] presents an off-line algorithm that maximizes the evader's minimum time to escape for an evader moving along a known path. In [9][3], a target tracking problem is analyzed for an unpredictable target and an observer lacking prior model of the environment. It computes a risk factor based on the current target position and generates a feedback control law to minimize it. [2] deals with the problem of *stealth target tracking* where a robot equipped with visual sensors tries to track a moving target among obstacles and, at the same time, remain hidden from the target. Obstacles impede both the tracker's motion and visibility, and also provide hiding places for the tracker. A tracking algorithm is proposed that applies a local greedy strategy and uses only local information from the tracker's visual sensors and assumes no prior knowledge of target tracking motion or a global map of the environment. In [19], the problem of target tracking has been analyzed at a fixed distance between the pursuer and evader. Optimal motion strategies are proposed for a pursuer and evader based on critical events.

Research has been done to track one or more evaders using multiple pursuers. [12] presents a method of tracking several evaders with multiple pursuers in an uncluttered environment. In [11] the problem of tracking multiple targets is addressed using a network of communicating robots and stationary sensors. A region-based approach is introduced which controls robot deployment at two levels, namely, a coarse deployment controller and a target-following controller.

III. Problem Formulation

In this paper we consider a mobile pursuer and evader on a plane. They are point robots and move with bounded speeds, $v_p(t)$ and $v_e(t)$. Therefore, $v_p(t) : [0, \infty) \rightarrow [0, \overline{v}_p]$ and $v_e(t) : [0, \infty) \rightarrow [0, \overline{v}_e]$. We use r to denote the ratio of the maximum speed of the evader to that of the pursuer$r = \frac{\overline{v}_e}{\overline{v}_p}$. The workspace contains obstacles that restrict pursuer and

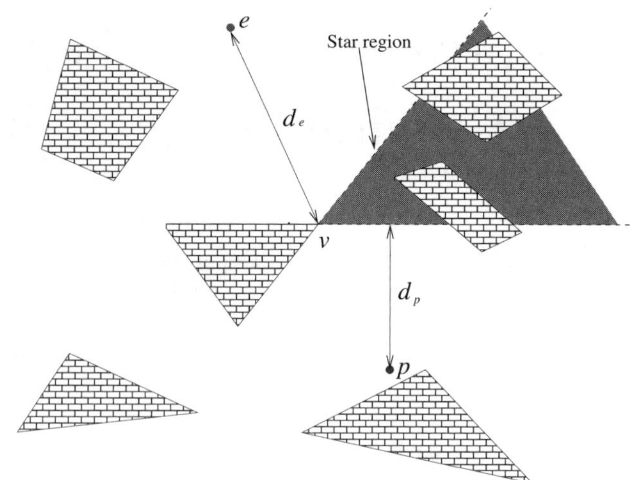

Fig. 1. *Star Region* associated with the vertex

evader motions and may occlude the pursuer's line of sight to the evader. The initial position of the pursuer and the evader is such that they are visible to each other. To prevent the evader from escaping, the pursuer must keep the evader in its visibility region. The visibility region of the pursuer is the set of points from which a line segment from the pursuer to that point does not intersect the obstacle region. The evader escapes if at *any* instant of time it can break the line of sight to the pursuer. Visibility extends uniformly in all directions and is only terminated by workspace obstacles (omnidirectional, unbounded visibility).

Now we present a sufficient condition of escape for an evader in general environments. We use it to prove some important results in the next section. The sufficient condition is based on the the concept of a *star region*. The *star region* associated with a vertex is defined as the region in the free workspace bounded by the lines supporting the vertex of the obstacle. The shaded region in Figure 1 shows the star region associated with the vertex *v*. The concept of *star region* is only applicable for a convex vertex(a vertex of angle less than π). Using the idea of the *star region*, a sufficient condition for escape for the evader can be stated as follows.

Sufficient Condition: If the time required by the pursuer to reach the star region associated with a vertex is greater than the time required by the evader to reach the vertex, the evader has a strategy to escape the pursuer's visibility region.

The sufficient condition arises from the fact that if the evader reaches the corner before the pursuer can reach the *star region* associated with the corner, the evader may escape from the side of the obstacle hidden from the pursuer. This is illustrated in figure 2. In the figure, the evader, e, is at the corner while the pursuer, p, is yet to reach the *star region* associated with the corner. If the pursuer approaches the *star region* from the left side as shown by the solid arrow, the evader can escape the visibility region of the pursuer by

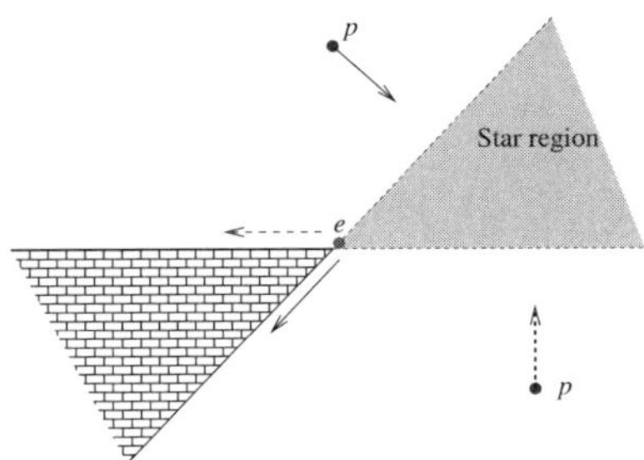

Fig. 2. Sufficient condition for escape

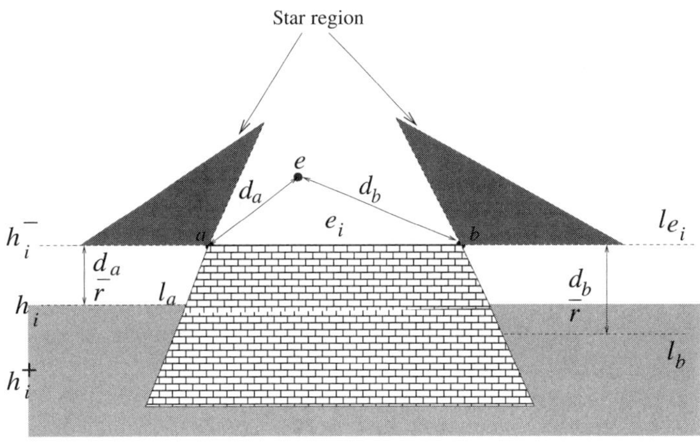

Fig. 3. Proof of Lemma 1

moving in the direction of the solid arrow. On the other hand, if the pursuer approaches the *star region* from the right side as shown by the dotted arrow, the evader can escape the visibility region of the pursuer by moving in the direction of the dotted arrow.

The relation between the time taken by the pursuer and evader can be expressed in terms of the distances traveled by the pursuer and the evader and their speeds. Referring to Figure 1, if d_e is the length of the shortest path of the evader from the corner, d_p is the length of the shortest path of the pursuer from the star region associated with the corner and r is the ratio of the maximum speed of the evader to that of the pursuer, the sufficient condition can also be expressed in the following way

SC: If $d_e < rd_p$, the evader wins the game.

For the sake of convenience, we refer to the sufficient condition as SC in the rest of the paper.

IV. Approximation schemes for Polygonal environment

In this section, we show that in any environment containing polygonal obstacles, the set of initial positions from which a pursuer can track the evader is bounded. First, we prove the statement for an environment containing a single convex polygonal obstacle. Then we extend the results to prove in case of a general polygonal environment. This leads to our first approximation scheme. Then we present two more approximation schemes to bound the set of initial positions of the pursuer from which it might be able to track the evader. The results presented in this section hold for unbounded as well as bounded environments.

Consider an evader, e, in an environment with a single convex polygonal obstacle having n sides. The edges of the polygonal obstacle are $e_1, e_2 \cdots e_n$. Every edge e_i is a line segment that lies on a line l_{e_i} in the plane. Let $r_e = (x_e, y_e)$ and $r_p = (x_p, y_p)$ denote the initial position of the evader and the pursuer respectively. Let $\{h_i\}_1^n$ denote a family of lines, each given by the equation $h_i(x, y, r_e, r) = 0$. The presence of the terms r_e and r in the equation imply that the equation of the line depends on the initial position of the evader and the speed ratio respectively. Each line h_i divides the plane into two half-spaces, namely, $h_i^+ = \{(x, y) \mid h_i(x, y, r_e, r) > 0\}$ and $h_i^- = \{(x, y) \mid h_i(x, y, r_e, r) < 0\}$. Now we use the SC to prove an important property related to the edges of the obstacle.

Lemma 1: For every edge e_i, there exists a line h_i parallel to e_i and a corresponding half-space h_i^+ such that the pursuer loses the game if $r_p \in h_i^+$.

Proof: Consider an edge e_i of a convex obstacle as shown in Figure 3. Since the obstacle is convex, it lies in one of the half-spaces generated by the line l_{e_i}. Without the loss of generality, let the obstacle lie in the half-space below the line l_{e_i}. Let d_a and d_b be the length of the shortest path of the evader from vertices a and b of the edge e_i respectively. Since the obstacle lies in the lower half-space of l_{e_i}, the *star region* associated with vertices a and b are in the upper half-space of l_{e_i} as shown by the green shaded region. Let l_a and l_b be the lines at a distance of $\frac{d_a}{r}$ and $\frac{d_b}{r}$ respectively, from the line l_{e_i}. If the pursuer lies at a distance d greater than $min(\frac{d_a}{r}, \frac{d_b}{r})$ below the line l_{e_i}, then the time taken by the pursuer to reach the line l_{e_i} is $t_p \geq \frac{d}{\bar{v}_p} \geq \frac{min(\frac{d_a}{r}, \frac{d_b}{r})}{\bar{v}_p}$. The minimum time required by the evader to reach corner a or b, whichever is nearer, is given by $t_e = \frac{min(d_a, d_b)}{\bar{v}_e}$. From the expressions of t_p and t_p we can see that $t_p > t_e$. Hence the pursuer will reach the nearer of the two corners before the evader reaches line l_{e_i}. Hence from SC, we conclude that if the pursuer lies below the line h_i parallel to e_i at a distance of $min(\frac{d_a}{r}, \frac{d_b}{r})$, then the evader wins the game by following the shortest path to the nearer of the two corners. In Figure 3, since $d_b > d_a$ the line h_i coincides with line l_a. ■

Given an edge e_i and the initial position of the evader, proof of Lemma 1 provides an algorithm to find the line h_i and the corresponding half-plane h_i^+ as long as the length of the shortest path of the evader to the corners of an edge is computable. For example, in the presence of other obstacles, the length of the shortest path of the evader to the corners can be obtained by Dijkstra's algorithm.

Now we present some geometrical constructions required to prove the next theorem. Refer to Figure 4. Consider a convex obstacle. Consider a point c strictly inside the obstacle. For

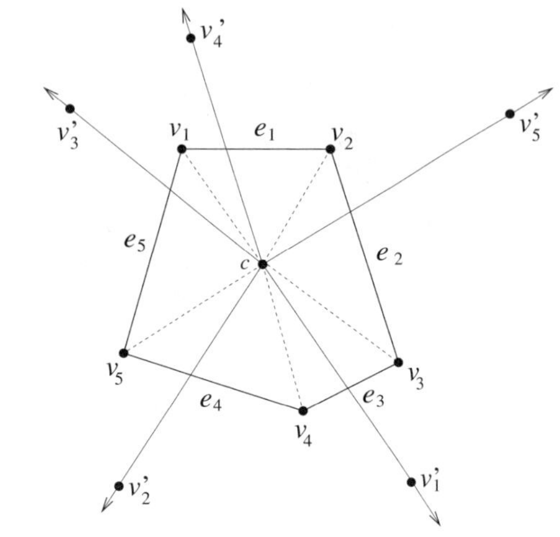

Fig. 4. A polygon and its sectors

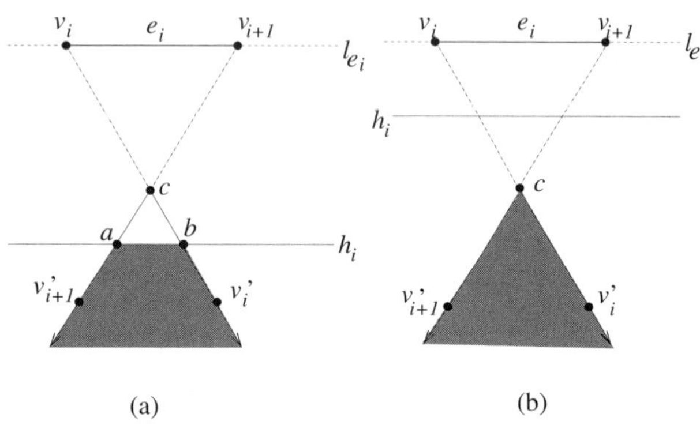

Fig. 5. Proof of theorem 1

each i, extend the line segment v_ic to infinity in the direction $\vec{v_ic}$ to form the ray cv_i'. Define the region bounded by rays cv_i' and cv_{i+1}' as *sector* $v_i'cv_{i+1}'$. The $sectors$ possess the following properties

1) Any two sectors are mutually disjoint.
2) The union of all the sectors is the entire plane.

We can extend the above idea to any n sided convex polygon. We use the construction to prove the following theorem.

Theorem 1: In an environment containing a single convex polygonal obstacle, given the initial position of the evader, the initial positions of the pursuer from which it can win the game is a bounded subset of the free workspace.
Proof: Refer to Figure 5. Consider an edge e_i of the convex obstacle with end points v_i and v_{i+1}. WLOG, the obstacle lies below l_{e_i}. Let c be a point strictly inside the convex polygon. Extend the line segments v_ic and $v_{i+1}c$ to form *sector* $v_i'cv_{i+1}'$. By Lemma 1, using the initial position of the evader, we can construct a line h_i parallel to e_i such that if the initial pursuer position lies below h_i, the evader wins the game. In case the line h_i intersects the *sector* $v_i'cv_{i+1}'$, as shown in Figure 5(a), the evader wins the game if the initial pursuer position lies in the shaded region. In case the line h_i does not intersect the *sector* $v_i'cv_{i+1}'$, as shown in Figure 5(b), the evader wins the game if the initial pursuer position lies anywhere in the sector. Hence for every sector, there is a region of finite area such that if the initial pursuer position lies in it then it might win the game. Every edge of the polygon has a corresponding sector associated with it. Since each sector has a region of finite area such that if the initial pursuer position lies in it, the pursuer might win the game, the union of all these regions is finite. Hence the proposition follows. Figure 6 shows the evader in an environment consisting of a hexagonal obstacle. The polygon in the center bounded by thick lines shows the region of possible pursuer win. ■

In the proof of theorem 1, we generate a bounded set for each convex polygonal obstacle such that the evader wins the game if the initial position of the pursuer lies outside this set. In a similar way, we can generate a bounded set for a non-convex obstacle. Given a non-convex obstacle, we construct its convex-hull. We can prove that Lemma 1 holds true for the convex-hull. Finally, we can use Theorem 1 to prove the existence of a bounded set. Due to limitations in space, the proof is omitted.

From the previous discussions, we conclude that any polygonal obstacle, convex or non-convex, restricts the set of initial positions from which the pursuer might win the game, to a bounded set. Moreover, given the initial position of the evader and the ratio of the maximum speed of the evader to the pursuer, the bounded set can be obtained from the geometry of the obstacle by the construction used in the proof of Theorem 1. For any polygon in the environment, let us call the bounded set generated by it, as the *B set*. If the initial position of the pursuer lies outside the *B set*, the evader wins the game. For an environment containing multiple polygonal obstacles, we can compute the intersection of all *B sets* generated by individual obstacles. Since each *B set* is bounded, the intersection is a **bounded set**. Moreover, the intersection has the property that if the initial position of the pursuer lies outside the intersection, the evader wins the game. This leads to the following theorem.

Theorem 2: Given the initial position of the evader, the set of initial positions from which the pursuer might win the game is bounded for an environment consisting of polygonal obstacles.
Proof: The bounded set referred in this theorem is the intersection of the *B sets* generated by the obstacles. If the initial pursuer position does not lie in the intersection it implies that it is not contained in all the *B sets*. Hence there exists at least one polygon in the environment for which the initial pursuer position does not lie in its *B set*. By Theorem 1, the evader has a winning strategy. Hence the theorem follows. ■

The intersection of the *B sets* generated by all the obstacles provides an approximation of the size of the *decidable regions*. For any initial position of the pursuer outside the intersection, the evader wins the game and hence the result is known. But we still do not know the result of the game for all initial position of the pursuer inside the intersection. However, we can find better approximation schemes and reduce the size of the region in which the result of the game is unknown. In the next subsection, we present one such approximation scheme.

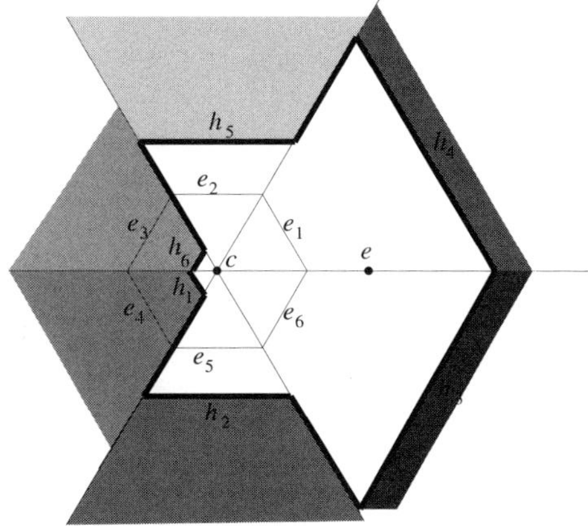

Fig. 6. *B set* for an environment consisting of a regular hexagonal obstacle and $r = 0.5$.

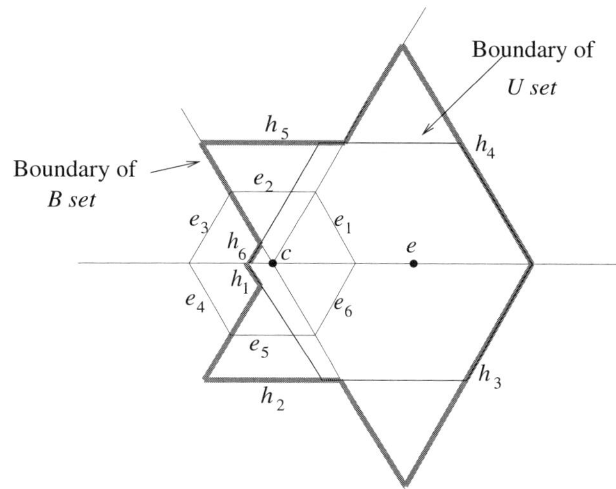

Fig. 7. *B set* and *U set* for an environment containing of a regular hexagonal obstacle and $r = 0.5$. The polygon bounded by thick lines is the *B set* and the polygon bounded by thin lines is the *U set*

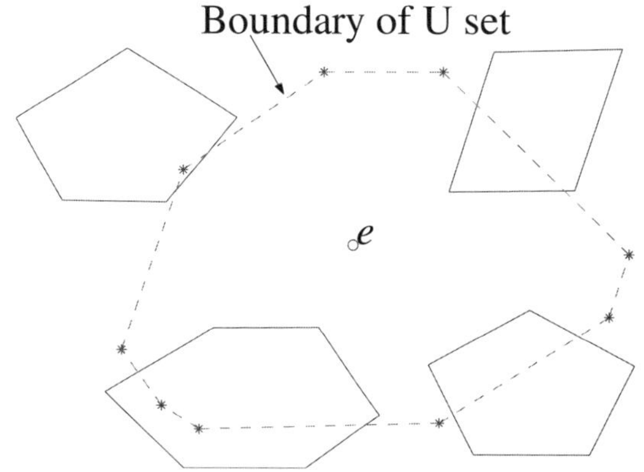

Fig. 8. *U set* for a general environment

A. U set

Now we present another approximation scheme that gives a tighter bound of the *undecidable region*. From Lemma 1, the evader wins the game if $r_p \in h_i^+$ for any edge. We can conclude that if $r_p \in \cup_{i=1}^n h_i^+$, the evader wins the game. Since $(\cup_{i=1}^n h_i^+)^c = \cap_{i=1}^n (h_i^+)^c = \cap_{i=1}^n h_i^-$, where S^c denotes the complement of set S, if r_p lies outside $\cap_{i=1}^n h_i^-$, the evader wins the game. Hence the set of initial positions from where the pursuer might win the game is contained in $\cap_{i=1}^n h_i^-$. We call $\cap_{i=1}^n h_i^-$ as the *U set*. An important point to note is that the intersection can be taken among any number of half-spaces. If the intersection is among the half-spaces generated by the edges of an obstacle, we call it the *U set* generated by the obstacle. If the intersection is among the half-spaces generated by all the edges in an environment, we call it the *U set* generated by the environment.

The next theorem proves that the *U set* generated by a single obstacle is a subset of the *B set* and hence a better approximation.

Theorem 3: For a given convex obstacle, the *U set* is a subset of the *B set* and hence bounded.

Proof: Consider a point q that does not lie in the *B set*. From the construction of the *B set*, q must belong to some half-plane h_j^+. If $q \in h_j^+$, then $q \notin h_j^- \implies q \notin \cap_{i=1}^n h_i^-$. This implies that the complement of the *B set* is a subset of the complement of the *U set*. This implies that the *U set* is a subset of the *B set*. ■

Figure 7 shows the *B set* and *U set* for an environment containing a regular hexagonal obstacle. In the appendix, we present a polynomial-time algorithm to compute the *U set* for an environment with polygonal obstacles. The overall time-complexity of this algorithm is $O(n^2 \log n)$ where n is the number of edges in the environment. Figure 8 shows the evader in a polygonal environment. The region enclosed by the dashed lines is the *U set* generated by the environment for the initial position of the evader. The *U set* for any environment having polygonal obstacles is a convex polygon with at most n sides[6]. Figure 9 shows the *U set* for an environment for various ratio of the maximum speed of the evader to that of the pursuer. In Figure 9, it can be seen that as the speed ratio between the evader and the pursuer increases, the size of the *U set* decreases. The size of the *U set* diminishes to zero at a critical speed ratio. At speed ratios higher than the critical ratio, the evader has a winning strategy for any initial position of the pursuer. Hence the problem becomes *decidable* [5] when the ratio of the maximum speeds is higher than a critical limit.

The next theorem provides a sufficient condition for escape of the evader in an environment containing obstacles using the *U set*.

Theorem 4 If the *U set* does not contain the initial position of either the pursuer or the evader, the evader wins the game.

Proof: To prove the theorem we need the following lemma.

Lemma 2: For $r \leq 1$, the evader lies inside the *U set*.

Proof: For $r \leq 1$, $\bar{v}_p \geq \bar{v}_e$. If the pursuer lies at the same position as the evader, its strategy to win is to maintain the same velocity as that of the evader. Hence if the pursuer and the evader have the same initial position, the pursuer can track the evader successfully. Since all the initial positions from

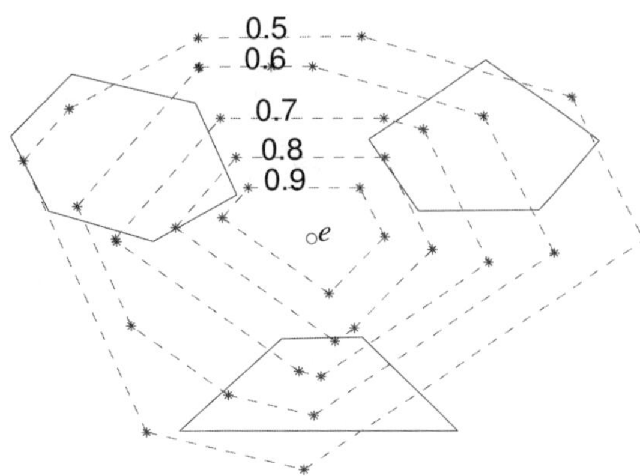

Fig. 9. *U set* for a various speed ratios of the evader to that of the pursuer

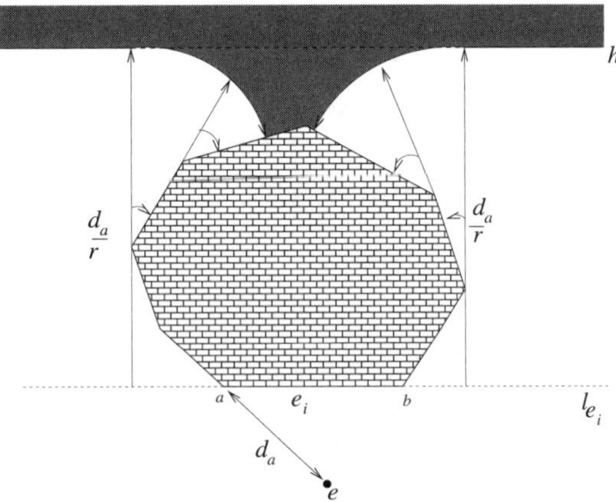

Fig. 10. A polygon in free space. The region shaded in red is obtained by using Lemma 1. The region shaded in green gets added by using a better approximation scheme.

which the pursuer can win the game must be contained inside the *U set*, the evader position must also be inside the *U set*. ■

Referring back to the proof of Theorem 4, by definition of the *U set*, if the pursuer lies outside the *U set*, it loses.

If the evader lies outside the *U set*, Lemma 2 implies $r > 1$. If $r > 1$, $\bar{v}_e > \bar{v}_p$. If $\bar{v}_e > \bar{v}_p$, the evader wins the game in any environment containing obstacles. Its winning strategy is to move on the convex hull of any obstacle. ■

B. Discussion

In the previous sections, we have provided a simple approximation scheme for computing the set of initial pursuer positions from which the evader can escape based on the intersection of a family of half-spaces. A slight modification to the proposed scheme leads to a better approximation. In the proof of Lemma 1, we presented an algorithm to find a half-space for every edge of the polygon such that if the initial position of the pursuer lies in the half-space, the evader wins the game. All the points in the half-space are at a distance greater than $\frac{d_a}{r}$ from l_{e_i}. By imposing the condition that the minimum distance of the desired set of points from l_{e_i} in the **free workspace** should be greater than $\frac{d_a}{r}$, we can include

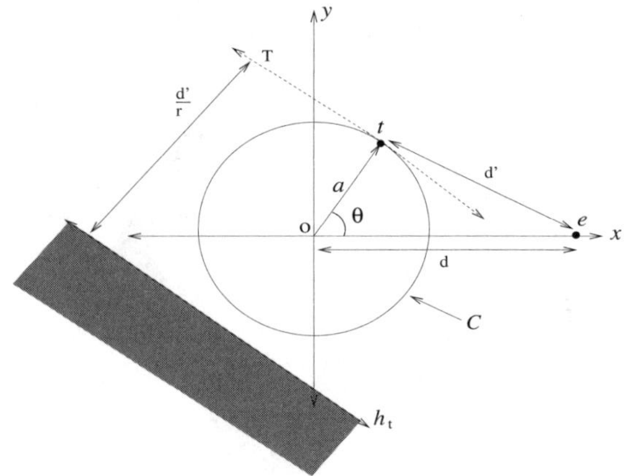

Fig. 11. A circular obstacle in free space

more points in the decidable regions as shown in Figure 10. The figure shows an obstacle in free space. From the proof of Lemma 1, we get the half space shaded in red. By adding the new condition, the region shaded in green gets included. When we repeat this for every edge, the set of initial positions from which the pursuer might win the game gets reduced and leads to a better approximation of the *decidable regions*. The boundary of the shaded region consists of straight lines and arc of circles. The boundary of the desired set is obtained by computing the intersections among a bunch of rays and arcs of circles generated by each edge. In this case a better approximation comes at the cost of expensive computation. We believe that better approximation schemes exist and one of our ongoing efforts is in the direction of obtaining computationally efficient approximation schemes.

None of the approximation schemes we have suggested so far restrict the initial position of the pursuer to be in the evader's visibility region. This condition can be imposed by taking an intersection of the output of the approximation algorithm with the *visibility polygon* at the evader's initial position. Efficient algorithms exist for computing the *visibility polygon* of a static point in an environment[10].

In the next section we extend the idea of *U set* to environments containing non-polygonal obstacles.

V. Approximation methods for non-polygonal obstacles

In this section we extend the approximation schemes presented in the previous section to non-polygonal obstacles. In order to illustrate the techniques required to handle non-polygonal obstacles, we compute an approximation for the initial positions of the pursuer from which the evader wins the game for the simple case of an evader in an environment containing a circular obstacle. Then we present the algorithm for any environment containing convex obstacles with smooth boundaries.

Figure 11 shows an evader, e, in an environment containing a circular obstacle of radius a in free space. The boundary of the obstacle is denoted by C. Let t be a point on C such that $\angle Ote = \theta$ and $\mid te \mid = d'$. T denotes the tangent to C at t. Let

h_t be a line at a distance of $\frac{d'}{r}$ from T on the same side of T as the obstacle. By Lemma 1, the evader wins the game if the pursuer lies in the half-space h_t^+, shown by the shaded region. The equation of line h_t is $y + x\cot\theta - (a - \frac{d'}{r})\csc\theta = 0$. For every point t on C, there exists a line h_t and the corresponding half-space h_t^+ such that if the initial position of the pursuer lies in h_t^+, the evader wins the game. Hence if the initial pursuer position lies in $\cup_{t\in C} h_t^+$, the evader wins the game$\Longrightarrow$ if the initial pursuer position lies outside $\cap_{t\in C} h_t^-$, the evader wins the game. Let us call $\cap_{t\in C} h_t^-$ as the *U set*.

Now we compute the boundary of the *U set*. Let $l(x, y, \theta)$ denote the family of lines h_t generated by all points t lying on C. Due to symmetry of the environment about the x-axis, the *U set* is symmetric about the x-axis. We present the construction of the boundary of the *U set* generated as θ increases from 0 to π. Let ∂U denote the boundary of the *U set*.

Theorem 6- ∂U is the envelope of the family of lines $l(x, y, \theta)$.

Proof: Consider any point q on ∂U. Since q belongs to the boundary of the *U set*, it belongs to some line, h_q, in the family $l(x, y, \theta)$. Either h_q is tangent to ∂U or else it intersects ∂U. In case it intersects ∂U, there is a neighborhood around q in which ∂U lies in both the half-spaces generated by h_q. This is not possible since one of the half-spaces generated by h_q has to be entirely outside the *U set*. Hence h_q is tangent to ∂U. Since q is any point on B, it implies that for all points q on ∂U, the tangent to ∂U at q belongs to the family $l(x, y, \theta)$. A curve satisfying this property is the envelope to the family of lines $l(x, y, \theta)$. Hence the proposition follows. ■

Using the *Envelope theorems* [20], the envelope of a family of lines $l(x, y, \theta)$ can be obtained by solving the following equations simultaneously

$$l(x, y, \theta) = y + x\cot\theta - (a - \frac{d'}{r})\csc\theta = 0 \quad (1)$$

$$\frac{\partial l}{\partial \theta} = 0 \quad (2)$$

d' as a function of θ is given by

$$d'(\theta) = \begin{cases} \sqrt{a^2 + d^2 - 2ad\cos\theta} & \text{if} \quad \theta \le \theta_0 \\ \sqrt{d^2 - a^2} + a(\theta - \theta_0) & \text{if} \quad \theta \ge \theta_0 \end{cases}$$

where $\theta_0 = \cos^{-1}\frac{a}{d}$.

The solution is

A. Case 1 ($\theta \le \theta_0$)

$$x = (a - \frac{\sqrt{a^2 + d^2 - 2ad\cos\theta}}{r})\cos\theta + \frac{ad\sin^2\theta}{r\sqrt{a^2 + d^2 - 2ad\cos\theta}}$$

$$y = (a - \frac{\sqrt{a^2 + d^2 - 2ad\cos\theta}}{r})\sin\theta - \frac{ad\sin\theta\cos\theta}{r\sqrt{a^2 + d^2 - 2ad\cos\theta}}$$

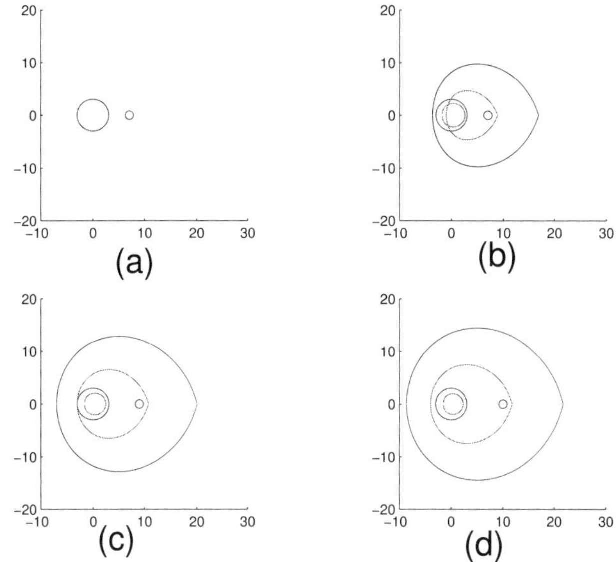

Fig. 12. (a) shows a circular obstacle with the initial position of the evader. The smaller circle is the evader. In (b),(c) and (d), $d = 5, 7$and 9 units respectively. In each of the figures (b), (c) and (d), the black boundary is for $r = 0.5$, the green boundary is $r = 1$ and the red boundary is for $r = 10$

B. Case 2 ($\pi \ge \theta \ge \theta_0$)

$$x = (a - \frac{\sqrt{d^2 - a^2} + a(\theta - \theta_0)}{r})\cos\theta + \frac{\sin\theta}{r}$$

$$y = (a - \frac{\sqrt{d^2 - a^2} + a(\theta - \theta_0)}{r})\sin\theta - \frac{\cos\theta}{r}$$

Since ∂U is symmetrical about the $x-$axis, the other half of ∂U is obtained by reflecting the above curves about the $x-$axis. Figure 12(a) shows an evader in an environment consisting of a disc-like obstacle. Figures 12(b),(c) and (d) show the boundary of the *U set* for varying distance between the evader and the obstacle. In each of these figures, the boundary of the *U set* is shown for three different values of r. We can see that for $r \le 1$, the evader lies inside the *U set* as given by Lemma 2.

The above procedure can be used to construct the *U set* for any convex obstacle with smooth boundary. Given the initial position of the evader, we present the procedure to construct the boundary of the *U set* for a obstacle with smooth boundary.

Consider an obstacle with smooth boundary given by the equation $f(x, y) = 0$. The procedure to generate the boundary of the *U set* is as follows

1) Given any point t on the boundary, compute the minimum distance of the point from the evader. Let it be d_t.
2) Find the equation of the line h_t at a distance of $\frac{d_t}{r}$ from the tangent to the obstacle at t.
3) Find the family $l(x, y, \theta)$ of lines generated by h_t as t varies along the boundary of the obstacle. θ is a parameter that defines t.

4) Compute the envelope of the family $l(x, y, \theta)$. This is the boundary of the *U set*. This is true since the proof of Theorem 5 does not depend on the shape of the obstacle.

VI. Conclusion and Future research

In this work we address the problem of target-tracking in general environments. We prove that in a general environment containing obstacles, given the initial position of the evader, the set of initial positions from which the pursuer might be able to track the evader is bounded. Moreover we provide an approximation algorithm to construct a convex polygonal region to bound that region. We provide a sufficient condition for escape of the evader in a general polygonal environment that depends on the geometry of the obstacles, the initial position of the evader and the ratio of the maximum speed of the evader to that of the pursuer. We extend the approximation schemes to obstacles with smooth boundaries.

Given the complete map of the environment, our results depend only on the initial position and the maximum speeds of the pursuer and evader. Hence our results hold for various settings of the problem such as an unpredictable or predictable evader [14] or localization uncertainties in the future positions of the players [8] or delay in pursuer's sensing abilities [17].

In the future, we would like to provide an algorithm to approximate the initial positions of the pursuer from which it can track the evader and also the strategies used by the pursuer to track successfully. We are using game-theory as a framework to provide feedback strategies for the pursuer to track successfully. We are also investigating the problem of target-tracking with multiple pursuers.

An interesting direction of future research would be to extend our results to the target-tracking in $\mathbb{R}^3$. Researchers have addressed the problem of target-tracking in $\mathbb{R}^3$ [1]. We believe that some of our results can be used in 3-d by considering polyhedrons as bounding sets instead of polygons. Another direction of future research would be to incorporate dynamics in the player's motion model.

References

[1] T. Bandyopadhyay, M.H. Ang Jr, and D. Hsu. Motion planning for 3-D target tracking among obstacles. *International Symposium on Robotics Research*, 2007.

[2] T. Bandyopadhyay, Y. Li, M.H. Ang Jr, and D. Hsu. Stealth Tracking of an Unpredictable Target among Obstacles. *Proceedings of the International Workshop on the Algorithmic Foundations of Robotics*, 2004.

[3] T. Bandyopadhyay, Y. Li, M.H. Ang Jr., and D Hsu. A Greedy Strategy for Tracking a locally Predicatable Target among Obstacles. *Robotics and Automation, Proceedings. ICRA'02. IEEE International Conference on*, pages 2342–2347, 2006.

[4] C. Becker, H. Gonzalez-Banos, J.C. Latombe, and C. Tomasi. An intelligent observer. *Proceedings of International Symposium on Experimental Robotics*, pages 94–99, 1995.

[5] Sourabh Bhattacharya, Salvatore Candido, and Seth Hutchinson. Motion strategies for surviellance. In *Robotics: Science and Systems - III*, 2007.

[6] M. de Berg, M. van Kreveld, M. Overmars, and O. Schwarzkopf. *Computational Geometry- Algorithms and Applicaions*. Springer-Verlag, Berlin Heidelberg, 1997.

[7] A. Efrat, HH Gonzalez-Banos, SG Kobourov, and L. Palaniappan. Optimal strategies to track and capture a predictable target. *Robotics and Automation, 2003. Proceedings. ICRA'03. IEEE International Conference on*, 3, 2003.

[8] P. Fabiani and J.C. Latombe. Tracking a partially predictable object with uncertainty and visibility constraints: a game-theoretic approach. Technical report, Technical report, Univeristy of Stanford, December 1998. http://underdog. stanford. edu/.(cited on page 76).

[9] HH Gonzalez-Banos, C.Y. Lee, and J.C. Latombe. Real-time combinatorial tracking of a target moving unpredictably among obstacles. *Robotics and Automation, 2002. Proceedings. ICRA'02. IEEE International Conference on*, 2, 2002.

[10] J. E. Goodman and J. O. Rourke. *Handbook of Discrete and Computational Geometry*. CRC Press, New York, 1997.

[11] B. Jung and G.S. Sukhatme. Tracking Targets Using Multiple Robots: The Effect of Environment Occlusion. *Autonomous Robots*, 13(3):191–205, 2002.

[12] Parker L. Algorithms for Multi-Robot Observation of Multiple Targets. *Journal on Autonomous Robots*, 12:231–255, 2002.

[13] J. P. Laumond, S. Sekhavat, and F. Lamiraux. *Guidelines in Nonholonomic motion planning for Mobile Robots*. Springer, 1998.

[14] S. M. LaValle, H. H. Gonzalez-Banos, C. Becker, and J. C. Latombe. Motion strategies for maintaining visibility of a moving target. In *Robotics and Automation, 1997. Proceedings., 1997 IEEE International Conference on*, volume 1, pages 731–736, Albuquerque, NM, USA, April 1997.

[15] S. M. LaValle and J. Hinrichsen. Visibility-based pursuit-evasion: The case of curved environments. *IEEE Transactions on Robotics and Automation*, 17(2):196–201, April 2001.

[16] T.Y. Li, J.M. Lien, S.Y. Chiu, and T.H. Yu. Automatically generating virtual guided tours. *Computer Animation Conference*, pages 99–106, 1997.

[17] R. Murrieta, A. Sarmiento, and S. Hutchinson. On the existence of a strategy to maintain a moving target within the sensing range of an observer reacting with delay. *Intelligent Robots and Systems, 2003.(IROS 2003). Proceedings. 2003 IEEE/RSJ International Conference on*, 2, 2003.

[18] R. Murrieta-Cid, R. Monroy, S. Hutchinson, and J. P. Laumond. A complexity result for the pursuit-evasion game of maintaining visibility of a moving evader. *Accepted in IEEE International Conference on Robotics and Automation*, 2008.

[19] R. Murrieta-Cid, T. Muppirala, A. Sarmiento, S. Bhattacharya, and S. Hutchinson. Surveillance strategies for a pursuer with finite sensor range. *International Journal of Robotics Research*, pages 1548–1553, 2007.

[20] E. Silberberg. The Viner-Wong Envelope Theorem. *Journal of Economic Education*, 30(1):75–79, 1999.

VII. Appendix

A. Algorithm for generating the U-set

Algorithm CONSTRUCTUSET(S,r,(x_e, y_e))
Input: A set S of disjoint polygonal obstacles, the evader position $r_e = (x_e, y_e)$, ratio of maximum evader speed to maximum pursuer speed r
Output: The coordinates of the vertices of the *U set*

1) For every edge e_i in the environment with end-points a_i, b_i
2) l_1 =DIJKSTRA(VG(S),r_e, a_i)
3) l_2 =DIJKSTRA(VG(S),r_e, b_i)
4) $d_{e_i} = \frac{\min(l_1,l_2)}{r}$
5) Find the equation of h_i using Lemma 1.
6) INTERSECTHALFPLANES($h_1^-, ...h_n^-$)

The subroutine VG(S), computes the visibility graph of the environment S. The subroutine DIJKSTRA(G,I,F) computes the least distance between nodes I and F in graph G. The subroutine INTERSECTHALFPLANES($h_1^-, ..., h_n^-$) computes the intersection of the half planes $h_1^-, ..., h_n^-$ [6]. The time complexity of the above algorithm is $O(n^2 \log n)$, where n is the number of edges in the environment.

A Numerically Robust LCP Solver for Simulating Articulated Rigid Bodies in Contact

Katsu Yamane
Dept. of Mechano-Informatics
University of Tokyo
Email: yamane@ynl.t.u-tokyo.ac.jp

Yoshihiko Nakamura
Dept. of Mechano-Informatics
University of Tokyo
Email: nakamura@ynl.t.u-tokyo.ac.jp

Abstract— **This paper presents a numerically robust algorithm for solving linear complementarity problems (LCPs), and applies it to simulation of frictional contacts of articulated rigid bodies each modeled as a general polygonal object. We first point out two problems of the popular pivot-based LCP solver called Lemke Algorithm and its extension with lexicographic ordering, due to numerical errors especially for ill-conditioned LCPs. Our new algorithm solves these problems by storing all pivot candidates and searching for a sequence of pivots that leads to a solution. An LCP-based contact dynamics formulation is combined with a forward dynamics algorithm for articulated rigid bodies to perform the whole simulation using a dynamic programming approach. Simulation examples using a humanoid robot show that the Lemke Algorithm (with or without lexicographic ordering) cannot solve complex contact problems, while our algorithm can successfully simulate such situations. We also demonstrate that the simulation results are qualitatively similar to those of hardware experiments.**

I. Introduction

Modeling collisions and contacts has been a long term research issue in both robotics and graphics. Most models can be categorized into penalty- and constraint-based methods. In penalty-based models, contact force at each contact point is modeled as the force exerted by a spring and damper. Although contact forces are easily computed from penetration depths and relative velocities, the approach suffers from numerical instability problem due to impulsive forces. This paper deals with the constraint-based approach, where we determine contact forces such that unilateral constraints on the post contact relative motion and force are satisfied.

Constraint-based approaches often employ linear complementarity problem (LCP) [1] to formulate the constraints. An n-dimensional LCP is to find a set of vectors $\boldsymbol{w} \in \boldsymbol{R}^n$ and $\boldsymbol{z} \in \boldsymbol{R}^n$ that satisfy

$$\boldsymbol{w} = \boldsymbol{M}\boldsymbol{z} + \boldsymbol{q} \quad (1)$$

$$\boldsymbol{w} \geq 0, \ \boldsymbol{z} \geq 0, \ \boldsymbol{w}^T\boldsymbol{z} = 0 \quad (2)$$

for a given square matrix $\boldsymbol{M} \in \boldsymbol{R}^{n\times n}$ and vector $\boldsymbol{q} \in \boldsymbol{R}^n$. In the rest of the paper, we shall denote condition (2) as

$$\boldsymbol{w} \geq 0 \perp \boldsymbol{z} \geq 0. \quad (3)$$

LCPs can be solved by either iterative or pivot-based approach. Iterative approaches (e.g. [2]) utilize the fact that the solution of an LCP is the equilibrium point of the associated quadratic cost function and employ numerical root-finding techniques such as Newton's method to find the equilibrium. Pivot-based approaches (such as Lemke Algorithm [3]), on the other hand, sequentially pivot a pair of elements of $\boldsymbol{w}$ and $\boldsymbol{z}$ according to specific rules until all elements of $\boldsymbol{q}$ of the pivoted equation become zero or positive. Once such pivot sequence is found, we can obtain the pivoted solution by setting $\boldsymbol{w} = \boldsymbol{q}$, $\boldsymbol{z} = 0$ and then moving the pivoted elements back to the original vectors.

Iterative approaches are generally easier to implement and numerically robust, although convergence is proven only for a limited class of $\boldsymbol{M}$. Pivot-based approaches are theoretically guaranteed to find a solution with finite number of trials (2^n) for general problems, and several systematic procedures are proposed to efficiently find a solution [1]. However, it is known that pivot-based approaches often suffer from numerical problem especially for large-scale and/or ill-conditioned problems.

There have been a body of research on developing efficient and robust methods for solving LCPs in the context of collision/contact modeling. Jourdan et al. [4] applied an iterative LCP solver similar to Gauss-Seidel algorithm to frictional contacts of rigid bodies and proved convergence in most practical cases. Förg et al. [5] utilized the sparsity of $\boldsymbol{M}$ to accelerate an iterative LCP solver. Stewart et al. [6] formulated frictional contacts between rigid bodies as an LCP and applied Lemke Algorithm. Lloyd [7] also utilized the structure of $\boldsymbol{M}$ in rigid-body contact model for reducing the computational cost for Lemke Algorithm. Guendelman et al. [8] combined a number of stabilization techniques to obtain visually plausible simulation results for highly complex scenes. All of these papers address contact dynamics between free rigid bodies, in which case $\boldsymbol{M}$ is generally sparse and the LCP is likely to be relatively easily solved by both iterative and pivot-based approaches.

In contrast, we are interested in modeling frictional contacts between articulated rigid bodies, each represented as a general polygonal object. A possible solution is to treat all links as free rigid bodies and extend the work described above to solve both unilateral (contact) and bilateral (joint) constraints at the same time. With such modeling strategy $\boldsymbol{M}$ would have similar structure as in free rigid-body case. This type of model has been employed in some dynamics engines such as [9]. However, integrating linear and angular accelerations of each rigid body independently occasionally breaks joint constraints, which should be corrected by applying heuristic recovering forces. Weinstein et al. [10] takes a different approach where joint constraints are maintained by sequentially applying impulses to joints, while contact constraints are handled as described in [8]. Although these models can generate visually plausible results for highly complex scenes, they are not suitable for applications that require physical precision.

Another approach is to combine a forward dynamics algorithm for articulated rigid bodies such as [11], [12] with an LCP-based contact formulation. Kokkevis [13] utilized Articulated-Body Algorithm [11] for computing the mass matrix in the LCP formulation and applied an iterative algorithm for solving the LCP. Gayle et al. [14] applied an adaptive forward dynamics algorithm [15] based on Divide-and-Conquer Algorithm (DCA) [12] to a collision and contact model based on Mirtich et al. [16]. Kry and Pai [17] derived an LCP-based contact formulation for simulating interactions between a rigid body and compliant fingers.

The general problem of the latter approach is that the associated LCP tends to be dense and ill-conditioned, in which case iterative methods do not guarantee convergence to a solution. In this paper, we pursue the application of Lemke Algorithm [3] to simulation of articulated rigid bodies under frictional contacts because of its potential generality, although its numerical robustness should be considerably improved to be practically applicable to complex problems. Lemke Algorithm has been successfully applied to contact problems of articulated rigid bodies in Kry and Pai [17], but they only consider one contact point per finger and hand dynamics is represented by finger compliance rather than its inertia.

The main contribution of this paper is improvement of Lemke Algorithm to deal with large-scale and ill-conditioned LCPs derived from frictional contacts between articulated rigid bodies of arbitrary geometry. The contact dynamics is formulated in a similar way as Stewart et al. [6], while the spatial mass matrix of free rigid bodies are replaced by inverse articulated-body inertias (IABI) [11] at all contact points. The contact model is combined with a forward dynamics algorithm called Assembly-Disassembly Algorithm (ADA) [18], which internally uses IABI for resolving the joint constraints and therefore fits well with our contact formulation.

A well-known extension of Lemke Algorithm is *lexicographic ordering* [1], [19] to solve *cycling* problem where the same pivot sequence is infinitely repeated when an inappropriate pivot choice is made. The problem is often encountered in ill-conditioned problems and the extension has been employed in [6], [7], [20].

Although lexicographic ordering can theoretically avoid cycling, we found that it is not enough for solving our contact problem under round-off errors. We will also point out another practical problem of numerical instability that, to our knowledge, has never been described in literature. Our solver addresses both of these problems. The basic idea of the method is to store all possible pivots at each pivot step and, in case a particular choice of pivot sequence resulted in an infinite pivot loop or numerical instability, track back the queue of possible pivots and try other possible pivots. In other words, the method searches for the best pivot sequence that leads to a solution of the LCP.

The rest of the paper is organized as follows. Section II reviews the Lemke Algorithm and point out the problems caused by round-off errors. In section III, we describe our numerically robust algorithm for solving LCPs. Section IV presents the LCP formulation of frictional contacts of articulated rigid bodies, along with several implementation issues. In Section V, we first show that the problems described in Section II actually happen in practical simulations using a simple example, and then demonstrate the robustness of the proposed solver by a number of simulation examples. Finally we conclude the paper in Section VI.

II. Lemke Algorithm

A. Algorithm Outline

We first show the outline of Lemke Algorithm [3] as explained in [7]. In general, pivot-based methods try to find a partition of Eq.(1):

$$\begin{pmatrix} \boldsymbol{w}_{\tilde{\alpha}} \\ \boldsymbol{w}_{\tilde{\beta}} \end{pmatrix} = \begin{pmatrix} \boldsymbol{M}_{\tilde{\alpha}\alpha} & \boldsymbol{M}_{\tilde{\alpha}\beta} \\ \boldsymbol{M}_{\tilde{\beta}\alpha} & \boldsymbol{M}_{\tilde{\beta}\beta} \end{pmatrix} \begin{pmatrix} \boldsymbol{z}_{\alpha} \\ \boldsymbol{z}_{\beta} \end{pmatrix} + \begin{pmatrix} \boldsymbol{q}_{\tilde{\alpha}} \\ \boldsymbol{q}_{\tilde{\beta}} \end{pmatrix} \quad (4)$$

such that the pivoted system

$$\begin{pmatrix} \boldsymbol{z}_{\alpha} \\ \boldsymbol{w}_{\tilde{\beta}} \end{pmatrix} = \boldsymbol{M}' \begin{pmatrix} \boldsymbol{w}_{\tilde{\alpha}} \\ \boldsymbol{z}_{\beta} \end{pmatrix} + \boldsymbol{q}' \quad (5)$$

satisfies the following conditions:

Condition 1: $\boldsymbol{w}_{\tilde{\alpha}}$ and $\boldsymbol{z}_{\alpha}$ contain the same set of indices, and

Condition 2: $\boldsymbol{q}' \geq 0$.

The vectors $(\boldsymbol{z}_{\alpha}^T \ \boldsymbol{w}_{\tilde{\beta}}^T)^T$ and $(\boldsymbol{w}_{\tilde{\alpha}}^T \boldsymbol{z}_{\beta}^T)^T$ are called *basic* and *non-basic* variables, respectively.

$\boldsymbol{M}'$ and $\boldsymbol{q}'$ are computed from the original matrix and vector as follows:

$$\begin{aligned} \boldsymbol{M}' &= \begin{pmatrix} \boldsymbol{M}_{\tilde{\alpha}\alpha}^{-1} & -\boldsymbol{M}_{\tilde{\alpha}\alpha}^{-1}\boldsymbol{M}_{\tilde{\alpha}\beta} \\ \boldsymbol{M}_{\tilde{\beta}\alpha}\boldsymbol{M}_{\tilde{\alpha}\alpha}^{-1} & \boldsymbol{M}_{\tilde{\beta}\beta} - \boldsymbol{M}_{\tilde{\beta}\alpha}\boldsymbol{M}_{\tilde{\alpha}\alpha}^{-1}\boldsymbol{M}_{\tilde{\alpha}\beta} \end{pmatrix} & (6) \\ \boldsymbol{q}' &= \begin{pmatrix} \boldsymbol{q}'_{\tilde{\alpha}} \\ \boldsymbol{q}'_{\tilde{\beta}} \end{pmatrix} = \begin{pmatrix} -\boldsymbol{M}_{\tilde{\alpha}\alpha}^{-1}\boldsymbol{q}_{\tilde{\alpha}} \\ \boldsymbol{q}_{\tilde{\beta}} - \boldsymbol{M}_{\tilde{\beta}\alpha}\boldsymbol{M}_{\tilde{\alpha}\alpha}^{-1}\boldsymbol{q}_{\tilde{\alpha}} \end{pmatrix}. & (7) \end{aligned}$$

Once such pivot is found, we can easily obtain the solution as $\boldsymbol{w}_{\tilde{\alpha}} = 0, \boldsymbol{w}_{\tilde{\beta}} = \boldsymbol{q}'_{\tilde{\beta}}, \boldsymbol{z}_{\alpha} = \boldsymbol{q}'_{\tilde{\alpha}}, \boldsymbol{z}_{\beta} = 0$.

Lemke Algorithm is one of the systematic methods to efficiently find an appropriate pivot. In Lemke Algorithm, we first introduce an auxiliary variable z_0 and modify the original LCP (1) as follows:

$$\boldsymbol{w} = \bar{\boldsymbol{M}} \begin{pmatrix} \boldsymbol{z} \\ z_0 \end{pmatrix} + \boldsymbol{q} \quad (8)$$

where

$$\begin{aligned} \bar{\boldsymbol{M}} &= \begin{pmatrix} \boldsymbol{M} & \boldsymbol{c} \end{pmatrix} \\ \boldsymbol{c} &= (1\,1\,\ldots\,1)^T. \end{aligned}$$

The solution of Eq.(8) can be found by the following steps:

Step 0 If $q \geq 0$, stop: $w = q, z = 0$ is the solution. Otherwise, obtain $r = \arg\min q_i/c_i$ and pivot z_0 with w_r. Compute $\bar{M}'$ and q', and set the *driving variable* $y_r = z_r$.

Step 1 Let m' denote the column vector of $\bar{M}'$ corresponding to y_r. If $m' \geq 0$, stop: there is no solution or this algorithm cannot solve the LCP. Otherwise, obtain $s = \arg\min\{-q_i/m'_i : m'_i \leq 0\}$ and let y_s denote the s-th element of the basic variables.

Step 2 Pivot y_s with y_r and update $\bar{M}'$ and q'. If $y_s = z_0$, stop: q' gives the solution. Otherwise set y_r to the complement of y_s and return to Step 1.

After Step 0, $q' \geq 0$ holds with the choice of r and the update rule Eq.(7). Similarly, all elements of subsequent q' are always equal to or larger than 0 with the choice of s in Step 1. The second condition above is therefore satisfied at every iteration. In Step 2, the first condition is met by setting the driving variable to the complement of the previously pivoted basic variable y_s, and by terminating when z_0 returns to a non-basic variable.

B. Problems of Lemke Algorithm

A well-known problem of Lemke Algorithm is that the minimum ratio test in Step 1 can result in tie, i.e. $-q_i/m'_i$ can take the same minimum value at multiple i's, in which case the LCP is said to be *degenerate*. This problem often occurs when the LCP includes redundant constraints, such as when there are more than three contact points between a pair of rigid bodies. Inappropriate choice of pivot in such cases can lead to cycling and should be avoided. It is known that an extension of Lemke Algorithm by *lexicographic ordering* [1], [19] (Lexicographic Lemke Algorithm) can resolve the tie by considering additional columns of $\bar{M}'$.

According to our experience, however, this solution still has a problem if the algorithm is implemented and executed on computers. In computer programs, exact tie of floating-point numbers almost never happens due to round-off errors even if two numbers are analytically equal. They would have very small difference and Step 1 would proceed without encountering a tie. However, the choice of pivot in such situations does not have any logical basis and, if the choice was inappropriate, the algorithm would fall into cycles. Alternatively, we could set some small threshold to determine if two values are equal. This approach however imposes another issue of choosing an appropriate threshold because if the threshold is too small the desirable pivot may be discarded due to numerical errors, and if too large even lexicographic ordering may not be able to discriminate the best pivot choice.

Another problem that, to our knowledge, has never been addressed in literature is that $M_{\bar{\alpha}\alpha}$ to be inverted in Eq.(6) may be close to singular with some of the pivots encountered during the process. In such cases, even if a solution is found, it may have large error with respect to the original equation (1). In contact simulation, this problem would result in physically unrealistic behavior such as penetration.

After presenting our new LCP solver and contact model in the following sections, we will demonstrate that these numerical problems actually occur in real simulation in Section V-A.

III. Robust Pivot-Based Solver for LCPs

The idea of our new algorithm is to store all the pivot candidates at every iteration of Steps 1 and 2, and return to them when cycling or numerical problem occurs. We store the i-th row as a pivot candidate at Step 1 if $m'_i < 0$ and the minimum element of q' after pivoting at row i is larger than a user-defined threshold. The threshold is usually chosen as a negative value with small absolute value to allow round-off errors. We define the cost function as a decreasing function of the minimum value of q' to prioritize pivot sequences with smaller errors.

In the algorithm, we construct a search tree composed of nodes each representing one pivot between a pair of basic and non-basic variables. The descendants of a node represent the possible pivots found in Step 1. The goal node is the one that includes z_0 in the pivot pair, and a successful sequence of pivots is reconstructed by tracing the ancestor nodes from the goal. We also construct a queue of nodes in which the nodes are sorted in the ascending order of the cost associated to each node.

Algorithm 1 shows the higher-level search algorithm, where

- Q is a queue of nodes,
- $Q.appendNode(x)$ adds a new node x to the queue,
- $Q.getBest()$ finds and pops the node with the smallest total cost,
- $x.isGoal()$ determines if node x is a goal by checking if z_0 is in the non-basic variables, and
- $Q.addDescendants(x)$ adds all possible descendant nodes of x to Q.

Details of $Q.addDescendants(x)$ is described in Algorithm 2, where

- $x.q\_min(i)$ computes the smallest element of q' after the i-th basic variable is pivoted,
- ϵ is a user-defined positive constant,
- $x.newNode(i)$ creates a new node representing a pivot of the i-th basic variable, and
- $x'.error()$ computes the norm of the current error $w_{x'} - Mz_{x'} - z_{0x'}c - q$ where $w_{x'}$, $z_{x'}$ and $z_{0x'}$ are the values of w, z and z_0 after performing the pivot x',
- e_{max} is a user-defined permissible numerical error,
- $Q.unique(x')$ returns true if Q does not include the same pivot set as x'.

After updating M' and q' in accordance with the current pivot set, the $addDescendants()$ function checks the minimum element of q' when the i-th basic variable is further pivoted (line 4). The minimum value would ideally be zero when $-q_i/m'_i$ is the minimum and negative otherwise. In our problem, however, there may be multiple i's that yield small negative q_{min} due to round-off errors as mentioned in the previous section. We keep such rows as pivot candidates if $q_{min} > -\epsilon$ ($\epsilon > 0$) (line 5). For each pivot candidate, we

Algorithm 1 Search Pivot Sequence

```
Require: an LCP
1: perform Step 0
2: create initial (dummy) node x_0
3: Q.appendNode(x_0)
4: while Q not empty do
5:    x ← Q.getBest()
6:    if x.isGoal() then
7:       return x
8:    end if
9:    Q.addDescendants(x)
10: end while
11: return NULL
```

Algorithm 2 $Q.addDescendants(x)$

```
1: update M̄' and q'
2: for i = 1, 2, ..., n do
3:    if m'_i < 0 then
4:       q_min ← x.q_min(i)
5:       if q_min > −ε then
6:          x' ← x.newNode(i)
7:          x'.totalCost ← x.totalCost + exp(−q_min)
8:          if x'.error() ≤ e_max and Q.unique(x') then
9:             Q.appendNode(x')
10:         end if
11:      end if
12:   end if
13: end for
```

verify that the error is smaller than the user-defined permissible error and that the same pivot set has never been visited before to avoid cycling (line 8).

The advantage of this method over directly comparing $-q_i/m_i'$ as in Lemke Algorithm is that $|q_{min}|$ has a clear physical meaning: either contact force or relative velocity in the normal direction, and therefore it is much easier to choose the threshold. The cost of each node is computed by $\exp(-q_{min})$ (line 7), which takes the maximum value $\exp(\epsilon)$ when $q_{min} = -\epsilon$. This cost penalizes the pivots with larger error, and as a result the optimal solution would be more physically reasonable.

Choosing ϵ and e_{max} is much easier than it would be with the threshold for determining tie in lexicographic ordering because we only have to make sure that it is sufficiently large not to drop correct pivots from the candidate list. Larger threshold degrades the speed because more candidates are kept in the queue, but it does not harm the result because the pivot sequence with the minimum cost is chosen anyway.

Updating $\boldsymbol{M}'$ and $\boldsymbol{q}'$ (line 1) based on Eqs. (6)(7) requires the inversion of a matrix of the size of pivot number, which can be computationally expensive for large problems. In fact, the update can be performed incrementally with less computational cost by using the $\boldsymbol{M}'$ and $\boldsymbol{q}'$ in the direct ancestor [7].

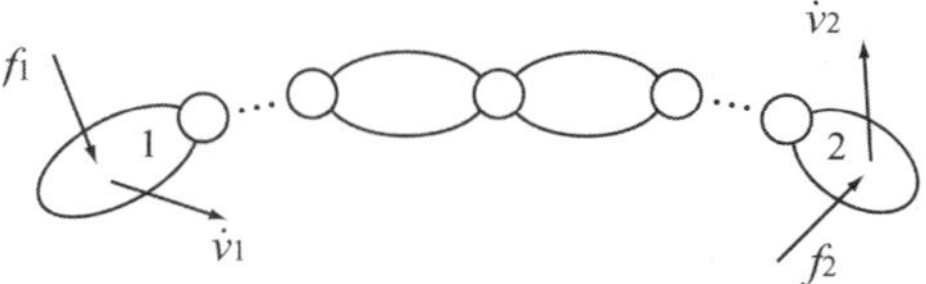

Fig. 1. Inverse articulated-body inertia.

IV. Contact Model for Articulated Rigid Bodies

A. IABI [11]

Inverse articulated-body inertia (IABI) is the inverse of the apparent inertia matrix of articulated bodies. This matrix describes the relationship between a force applied to a link called *handle* and resulting spatial acceleration, at current configuration. Furthermore, we can consider multiple handles, in which case we need m^2 IABIs to describe the relationship between forces and accelerations of m handles. In Fig. 1, for example, suppose links 1 and 2 are handles. The relationship between forces $\boldsymbol{f}_1, \boldsymbol{f}_2$ and accelerations $\dot{\boldsymbol{v}}_1, \dot{\boldsymbol{v}}_2$ is described using IABIs $\boldsymbol{\Phi}_{ij}$ $(i, j = 1, 2)$ as follows:

$$\begin{pmatrix} \dot{\boldsymbol{v}}_1 \\ \dot{\boldsymbol{v}}_2 \end{pmatrix} = \begin{pmatrix} \boldsymbol{\Phi}_{11} & \boldsymbol{\Phi}_{12} \\ \boldsymbol{\Phi}_{21} & \boldsymbol{\Phi}_{22} \end{pmatrix} \begin{pmatrix} \boldsymbol{f}_1 \\ \boldsymbol{f}_2 \end{pmatrix} + \begin{pmatrix} \boldsymbol{\phi}_1 \\ \boldsymbol{\phi}_2 \end{pmatrix} \tag{9}$$

where $\boldsymbol{\phi}_1$ and $\boldsymbol{\phi}_2$ are the bias accelerations of links 1 and 2, respectively. IABIs can be computed recursively as described in [12].

Forward dynamics algorithms such as DCA [12] and ADA [18] utilize IABI to describe the equation of motion of articulated bodies. We may be able to directly use IABIs in our contact model. The contact model of Kokkevis [13] is based on Articulated-Body Algorithm (ABA) [11], which uses articulated-body inertia (ABI) rather than IABI. In [13], IABIs are computed by applying unit test forces to the contact links and computing link accelerations by ABA. Gayle et al. [14] uses an extension of DCA as the basic forward dynamics engine; however, they apply a method similar to [13] to compute IABIs between contact links. In our implementation, we explicitly specify contact links as handles and let the forward dynamics algorithm compute IABIs between contact links.

B. LCP Formulation of Contacts

We apply the formulation in [7] to articulated rigid bodies, whose dynamics is represented by ABIs instead of spatial inertia matrix of rigid bodies.

Suppose N_L links are mutually in contact at N_C contact points. We first compute IABIs $\hat{\boldsymbol{\Phi}}_{ij}$ $(i, j = 1, 2, \ldots, N_L)$ and bias accelerations $\hat{\boldsymbol{\phi}}_i$ $(i = 1, 2, \ldots, N_L)$ of the N_L links using DCA or ADA. These IABIs describe the relationship between forces applied to contact links and their accelerations as

$$\dot{\hat{\boldsymbol{v}}} = \hat{\boldsymbol{\Phi}} \hat{\boldsymbol{f}} + \hat{\boldsymbol{\phi}} \tag{10}$$

where $\hat{\boldsymbol{v}} \in \boldsymbol{R}^{6N_L}$ and $\hat{\boldsymbol{f}} \in \boldsymbol{R}^{6N_L}$ are vectors composed of the spatial velocities and forces of all links respectively, and

$$\hat{\boldsymbol{\Phi}} = \begin{pmatrix} \hat{\boldsymbol{\Phi}}_{11} & \hat{\boldsymbol{\Phi}}_{12} & \dots & \hat{\boldsymbol{\Phi}}_{1N_L} \\ \hat{\boldsymbol{\Phi}}_{21} & \hat{\boldsymbol{\Phi}}_{22} & \dots & \hat{\boldsymbol{\Phi}}_{2N_L} \\ \vdots & \vdots & \ddots & \vdots \\ \hat{\boldsymbol{\Phi}}_{N_L 1} & \hat{\boldsymbol{\Phi}}_{N_L 2} & \dots & \hat{\boldsymbol{\Phi}}_{N_L N_L} \end{pmatrix}$$
$$\hat{\boldsymbol{\phi}} = \begin{pmatrix} \hat{\boldsymbol{\phi}}_1^T & \hat{\boldsymbol{\phi}}_2^T & \dots & \hat{\boldsymbol{\phi}}_{N_L}^T \end{pmatrix}^T.$$

Let $\boldsymbol{f} \in \boldsymbol{R}^{3N_C}$ and $\boldsymbol{v} \in \boldsymbol{R}^{3N_C}$ denote contact forces and relative velocities at contact points, respectively. The relationship between forces and velocities of links and contact points can be described by a Jacobian matrix $\boldsymbol{J}$ as

$$\boldsymbol{v} = \boldsymbol{J}\hat{\boldsymbol{v}} \tag{11}$$
$$\hat{\boldsymbol{f}} = \boldsymbol{J}^T \boldsymbol{f}. \tag{12}$$

Substituting Eqs.(11)(12) into Eq.(10) yields

$$\begin{aligned} \dot{\boldsymbol{v}} &= \boldsymbol{J}\hat{\boldsymbol{\Phi}}\boldsymbol{J}^T\boldsymbol{f} + \boldsymbol{J}\hat{\boldsymbol{\phi}} + \dot{\boldsymbol{J}}\hat{\boldsymbol{v}} \\ &= \boldsymbol{\Phi}\boldsymbol{f} + \boldsymbol{\phi} \qquad (13) \\ \boldsymbol{\Phi} &\triangleq \boldsymbol{J}\hat{\boldsymbol{\Phi}}\boldsymbol{J}^T \\ \boldsymbol{\phi} &\triangleq \boldsymbol{J}\hat{\boldsymbol{\phi}} + \dot{\boldsymbol{J}}\hat{\boldsymbol{v}} \end{aligned}$$

which represents the dynamics at contact points.

We then discretize the equation of motion Eq.(13). Let $\boldsymbol{v}^-$ and $\boldsymbol{v}^+$ denote relative velocities at contact points before and after the current integration. Assuming that we apply Euler integration with time step Δt, we can write $\boldsymbol{v}^+$ as

$$\boldsymbol{v}^+ = \bar{\boldsymbol{\Phi}}\boldsymbol{f} + \bar{\boldsymbol{\phi}} \tag{14}$$

where

$$\bar{\boldsymbol{\Phi}} = \Delta t \boldsymbol{\Phi} \tag{15}$$
$$\bar{\boldsymbol{\phi}} = \boldsymbol{v}^- + \Delta t \boldsymbol{\phi}. \tag{16}$$

We now derive an LCP formulation of unilateral constraints to model the contact, similar to the one used in [7]. The friction cone is approximated by an M-sided polyhedral cone. We also assume that each contact point has the same static and slip friction coefficients. Let $\boldsymbol{n}_i$ denote the normal vector at contact i, and $\boldsymbol{c}_{im}$ $(m = 1, 2, \dots, M)$ the normal vectors of the side faces of the cone projected onto the contact tangential plane and normalized.

We write the contact force at contact point i, $\boldsymbol{f}_i$, as a linear combination of $\boldsymbol{n}_i$ and $\boldsymbol{c}_{im}$ $(m = 1, 2, \dots, M)$ by the non-negative coefficients a_i and b_{ik} $(k = 1, 2, \dots, M)$, i.e.

$$\begin{aligned} \boldsymbol{f}_i &= a_i \boldsymbol{n}_i + \sum_{m=1}^{M} b_{im} \boldsymbol{c}_{im} \\ &= a_i \boldsymbol{n}_i + \boldsymbol{C}_i \boldsymbol{b}_i \qquad (17) \end{aligned}$$

where

$$\boldsymbol{C}_i = \begin{pmatrix} \boldsymbol{c}_{i1} & \boldsymbol{c}_{i2} & \dots & \boldsymbol{c}_{iM} \end{pmatrix} \in \boldsymbol{R}^{3\times M} \tag{18}$$
$$\boldsymbol{b}_i = \begin{pmatrix} b_{i1} & b_{i2} & \dots & b_{iM} \end{pmatrix}^T \in \boldsymbol{R}^{M}. \tag{19}$$

By combining Eq.(17) at all contact points, we obtain

$$\boldsymbol{f} = \boldsymbol{N}\boldsymbol{a} + \boldsymbol{C}\boldsymbol{b} \tag{20}$$

where

$$\begin{aligned} \boldsymbol{N} &= diag\{\boldsymbol{n}_i\} \in \boldsymbol{R}^{3N_C \times N_C} \\ \boldsymbol{a} &= \begin{pmatrix} a_1 & a_2 & \dots & a_{N_C} \end{pmatrix}^T \\ \boldsymbol{C} &= diag\{\boldsymbol{C}_i\} \in \boldsymbol{R}^{3N_C \times N_C M} \\ \boldsymbol{b} &= \begin{pmatrix} \boldsymbol{b}_1^T & \boldsymbol{b}_2^T & \dots & \boldsymbol{b}_{N_C}^T \end{pmatrix}^T \end{aligned}$$

and $diag\{*\}$ denotes a block diagonal matrix.

The linear complementarity condition for normal directions is described as

$$\boldsymbol{N}^T \boldsymbol{v}^+ \geq 0 \perp \boldsymbol{a} \geq 0. \tag{21}$$

The condition for the friction force and tangential velocity is described as

$$\boldsymbol{\mu}\boldsymbol{a} - \boldsymbol{E}\boldsymbol{b} \geq 0 \perp \boldsymbol{\lambda} \geq 0 \tag{22}$$
$$\boldsymbol{C}^T \boldsymbol{v}^+ + \boldsymbol{E}^T \boldsymbol{\lambda} \geq 0 \perp \boldsymbol{b} \geq 0 \tag{23}$$

where $\boldsymbol{\lambda} \in \boldsymbol{R}^{N_C}$ is a Lagrangian, $\boldsymbol{\mu}$ is a diagonal matrix composed of the friction coefficients at all contact points, and $\boldsymbol{E} \in \boldsymbol{R}^{N_C \times N_C M}$ is a constant block-diagonal matrix defined as

$$\boldsymbol{E} = diag\{\boldsymbol{1}\}, \ \boldsymbol{1} = \begin{pmatrix} 1 & \dots & 1 \end{pmatrix} \in \boldsymbol{R}^M. \tag{24}$$

Substituting Eqs.(14)(20) into Eqs.(22)(23), we obtain the whole LCP:

$$\begin{pmatrix} \boldsymbol{N}^T\bar{\boldsymbol{\Phi}}\boldsymbol{N} & \boldsymbol{N}^T\bar{\boldsymbol{\Phi}}\boldsymbol{C} & 0 \\ \boldsymbol{C}^T\bar{\boldsymbol{\Phi}}\boldsymbol{N} & \boldsymbol{C}^T\bar{\boldsymbol{\Phi}}\boldsymbol{C} & \boldsymbol{E}^T \\ \boldsymbol{\mu} & -\boldsymbol{E} & 0 \end{pmatrix} \begin{pmatrix} \boldsymbol{a} \\ \boldsymbol{b} \\ \boldsymbol{\lambda} \end{pmatrix} + \begin{pmatrix} \boldsymbol{N}^T\boldsymbol{\phi} \\ \boldsymbol{C}^T\boldsymbol{\phi} \\ 0 \end{pmatrix} = \begin{pmatrix} \boldsymbol{w}_a \\ \boldsymbol{w}_b \\ \boldsymbol{w}_\lambda \end{pmatrix} \tag{25}$$

$$\begin{pmatrix} \boldsymbol{a} \\ \boldsymbol{b} \\ \boldsymbol{\lambda} \end{pmatrix} \geq 0 \perp \begin{pmatrix} \boldsymbol{w}_a \\ \boldsymbol{w}_b \\ \boldsymbol{w}_\lambda \end{pmatrix} \geq 0. \tag{26}$$

C. Implementation Issues

As with many contact models, the most important factor in determining the computational cost is the number of contact points. Because our collision detection library handles general polygonal objects, many contact points are detected for complex objects. We accelerate the computation by removing unnecessary contact points such as those placed within some small distance δ from another point or inside the contact area. Note that there still may be redundant constraints even after removing some contact points. Although in some papers [16], [21] contact points with positive normal velocities are also removed, we found that ignoring these points can cause penetration because contact forces at other points may change their normal velocities negative.

We also applied a two-step solution to further accelerate the computation. We first solve the frictionless version of the contact problem by considering only the normal direction of each contact point to identify which points are likely to be active in the current contact state. We then solve the frictional problem from the initial guess that the normal directions of all active contact points are constrained (zero velocity and positive contact force), while others are unconstrained. We expect that this method greatly reduce the number of additional pivots required to reach a solution.

V. Results

The experiments presented in this section were executed on a workstation with a Pentium Xeon 3.8GHz processor. The code was written in C++ and the compiler was Microsoft Visual Studio .NET 2003 with optimization. We used 4-th order Runge-Kutta integration with 1 ms time step except when otherwise noted. Collision detection between general polygonal objects was performed by a library called PQP [22] with an extension to compute penetration depths and normal vectors [23]. The constants were set as $e_{max} = 1 \times 10^{-3}$, $\epsilon = 1 \times 10^{-3}$, $M = 8$ and $\delta = 1 \times 10^{-3}$ (m), and we employed the incremental update of $\boldsymbol{M}'$ and $\boldsymbol{q}'$ mentioned in the last paragraph of Section III. The search queue is implemented as a binary tree to efficiently find the node with minimum cost.

A. Comparison with Conventional Algorithm

We first compare our LCP solver with Lemke and Lexicographic Lemke algorithms. The example used here is a squat motion of a small 20-joint humanoid robot [24] on a horizontal flat floor. The simulated robot is under high-gain feedback mode, i.e. the joint angle, velocity and acceleration computed from the reference trajectory were directly applied to each joint. The geometry data of the links were extracted from the CAD model.

We used our own implementations of the conventional algorithms with the following details (refer to the algorithm outline in Section II-A):

- *Lemke Algorithm*—Always chooses the row with the minimum $-q_i/m'_i$ for pivot in Step 1. In case of tie (in the sense of floating-point numbers), the row with the minimum index is chosen.
- *Lexicographic Lemke Algorithm*—A row is regarded as tie if its $-q_i/m'_i$ is within a threshold Δ from the minimum, and the lexicographic ordering test is repeated until a unique minimum is identified using the same threshold. If multiple rows are in tie in the last test, the row with the minimum value is chosen. If z_0 is among the tie rows in any test, it is immediately chosen as the pivot variable and hence the algorithm terminates.

The implementations were tested using the examples in [1] that are known to be solvable by Lemke and/or Lexicographic Lemke algorithms.

There are the following three possible failure modes:

1) *no solution* is the case when $\boldsymbol{m}' \geq 0$ occurred in Step 1,
2) *cycle* is emitted when the same set of pivot was already found in one of the previous steps, and
3) *error* is emitted when the error of the solution is larger than e_{max}.

TABLE I

Comparison of the proposed solver, original Lemke Algorithm and Lexicographic Lemke Algorithm with four different thresholds Δ.

	proposed	*Lemke*	*Lexicographic Lemke*			
			0	10^{-8}	10^{-6}	10^{-4}
contacts	15.4	16.1	16.0	15.9	12.1	15.5
active	2.18	3.03	3.02	3.03	2.30	2.55
total frames	999	998	998	998	1000	1000
success	999	506	514	900	990	996
failure	0	492	484	98	10	4
no solution	0	52	27	17	3	3
cycle	0	157	147	18	2	0
error	0	283	310	63	5	1

If the algorithm failed in solving the LCP with friction and the frictionless LCP was successfully solved, the frictionless result is applied to prevent penetration, although it causes slipping. The number of frames where both LCPs failed was at most three and resulting penetration was practically negligible in all simulations.

The results are summarized in Table I. The first two rows represent the average numbers of detected contact points and those identified to be active respectively, and the next three rows represent the number of total frames with contact, successfully terminated frames, and failure frames respectively. The number of frames with each failure mode is shown in the last three rows. We also tested Lexicographic Lemke Algorithm with $\Delta = 10^{-3}$ but the robot fell down due to a slip caused by applying a frictionless solution.

The three algorithms behaved differently even for this relatively static motion. The results obviously show the advantage of our algorithm. Lemke Algorithm could solve only around half of the frames, and Lexicographic Lemke Algorithm did not help much either when the threshold is too small. $\Delta = 10^{-4}$ gave the best result in this example, but still had a few failure frames. In contrast, our algorithm successfully solved the LCP in all frames.

Note that 12 to 16 contact points were detected in average even with the contact point reduction described in Section IV-C, and only 2 to 3 of them were identified to be active. This is physically reasonable because three contact points are enough to constrain the motion of the robot in high-gain feedback mode, and the original contact points would yield highly ill-conditioned LCP. This is probably the reason for failures in conventional algorithms.

Fig. 2 compares real and simulated total vertical forces of left and right feet. The simulation was performed using the proposed algorithm, and a single force plate measured the total contact forces and moments at the feet. The simulated force exhibits qualitatively similar pattern to the real one, but the measured force shows more oscillation. This discrepancy is probably because the real hardware has some elasticity in the links, joints and controller. The constant offset is due to the error in the mass parameters estimated from the CAD model that does not include wires and screws.

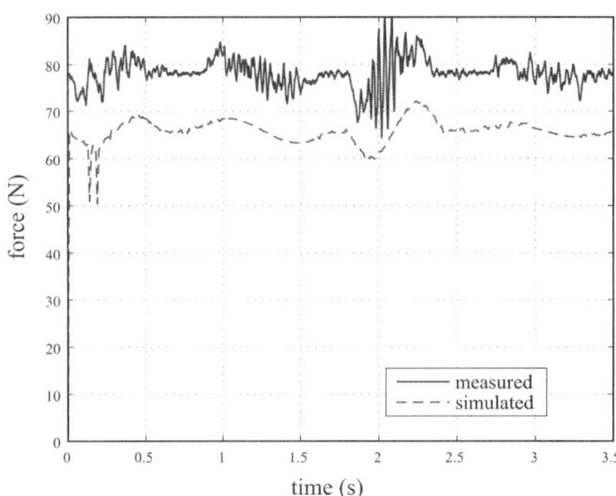

Fig. 2. Measured and simulated vertical forces for the humanoid motion.

TABLE II

COMPARISON OF SUCCESS RATE AND AVERAGE COMPUTATION TIME OF LEXICOGRAPHIC LEMKE AND PROPOSED ALGORITHMS FOR THE HUMANOID EXAMPLE.

	Lemke	*proposed*
contacts	6.89	7.19
active	2.02	1.70
total frames	9656	9670
success	9620	9670
failure	36	0
no solution	13	0
cycle	7	0
error	16	0
LCP solve time (ms)	0.375	0.182
simulation time (ms)	3.49	3.54
total pivots	71102	34335

B. Tap-Dancing of a Humanoid Robot

We perform another comparison including computation time using a tap-dancing motion of the same humanoid robot. This is a more challenging task because it includes frequent collisions as well as static contacts. We applied our solver and Lexicographic Lemke Algorithm. The best Δ was 10^{-10} in this case, which implies the necessity of finding threshold tailored to each task.

The result is shown in Table II, where "LCP solve time" indicates the time for solving the main frictional LCP. The time for solving the frictionless one is included in the total simulation time. Lexicographic Lemke Algorithm still failed to find a solution in about 0.4% of the frames with contact, while the proposed algorithm succeeded in all frames. Our LCP solver is also faster because lexicographic ordering requires more pivot computations than the search in our algorithm as shown in the "total pivots" row, although the total simulation time is longer because of the larger number of contact points that require preprocessing. A visual comparison of simulated and actual motions is shown in Fig. 3. Note the qualitatively similar behaviors such as yaw rotation.

TABLE III

COMPUTATION TIME FOR THE LONG AND CLOSED CHAIN EXAMPLES.

	hoist	*hoist* (10 ms)	*ring*	*net*
duration (s)	10	10	4	2
contacts	34.2	48.4	26.1	25.7
active	20.8	23.0	21.7	16.9
total frames	9357	935	3926	1818
failure	0	0	0	0
LCP solve time (ms)	62.0	96.5	35.0	9.72
simulation time (ms)	120	187	80.6	62.5

C. More Complex Scenarios

Figure 4 shows three simulation examples involving complex collisions and contacts of open and closed articulated rigid bodies. In the *hoist* example, a string-like object modeled as a 25-joint chain is subject to continuous contact with the rod and therefore takes long time for solving the LCP. We also varied the time step for integration. The simulation result with 10 ms timestep was similar to that with 1 ms and the total computation time was about 4.5 times shorter, although each step requires longer computation time because the penetration depth tends to be larger due to integration errors and more contact points are detected.

The *ring* example includes a number of contacts of non-convex objects. Each wire is composed of five spherical joints with two ring-shaped links at the ends. In the *net* example, five cylinders fall onto a net composed of four strings each modeled as a 16-joint chain with both ends fixed to the inertial frame, forming closed loops. Our algorithm still yields realistic results without failure.

VI. CONCLUSION

The conclusions of this paper are summarized by the following three points:

1) We pointed out two numerical issues of Lemke and Lexicographic Lemke algorithms, and proposed a robust algorithm for solving general LCPs. The main idea of the new algorithm is to store all pivot candidates at each step, and back trace the queue in case a numerical problem is found.
2) We modeled frictional contact of articulated rigid bodies as an LCP using inverse articulated-body inertia (IABI) [11], and applied above algorithm to solve the LCP. We combined the formulation with a forward dynamics algorithm called ADA [18] but it can also be combined with DCA [12].
3) Experimental results showed that our algorithm can robustly solve LCPs formulating contact dynamics of articulated bodies, which cannot be solved by Lexicographic Lemke Algorithm.

ACKNOWLEDGEMENTS

The authors thank Professor Tomomichi Sugihara (Kyushu University) and Mr. Kou Yamamoto (University of Tokyo) for their help in the humanoid hardware experment. This work was supported by a MEXT Special Coordination Fund for Promoting Science and Technology.

Fig. 3. Comparison of simulated (left) and actual (right) motions of humanoid tap-dancing.

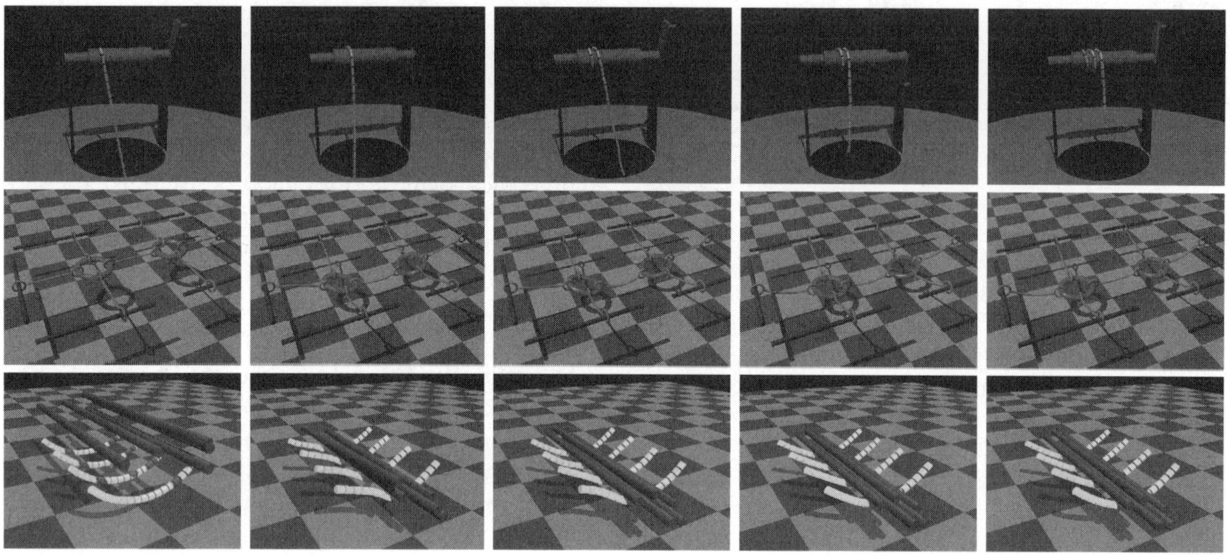

Fig. 4. Examples of contact simulation of articulated rigid bodies; from the top row: *hoist*, *ring*, and *net*.

REFERENCES

[1] K. Murty, *Linear Complementarity: Linear and Nonlinear Programming*. Lemgo, Germany: Heldermann-Verlag, 1988.

[2] M. Ferris and T. Munson, "Complementarity Problems in GAMS and the PATH Solver," University of Wisconsin–Madison, Tech. Rep. 98-12, September 1998.

[3] C. Lemke and J. Howson, "Equilibrium points of bimatrix games," *SIAM Journal on Applied Mathematics*, vol. 12, pp. 413–423, 1964.

[4] F. Jourdan, P. Alart, and M. Jean, "A Gauss-Seidel Like Algorithm to Solve Frictional Contact Problems," *Computer Methods in Applied Mechanics and Engineering*, vol. 155, pp. 31–47, 1998.

[5] M. Förg, F. Pfeiffer, and H. Ulbrich, "Simulation of unilateral constrained systems with many bodies," *Multibody System Dynamics*, vol. 14, pp. 137–154, 2005.

[6] D. Stewart and J. Trinkle, "An Implicit Time-Stepping Scheme for Rigid Body Dynamics with Coulomb Friction," in *Proceedings of IEEE International Conference on Robotics and Automation*, San Francisco, CA, May 2000, pp. 162–169.

[7] J. Lloyd, "Fast Implementation of Lemke's Algorithm for Rigid Body Contact Simulation," in *IEEE International Conference on Robotics and Automation*, Barcelona, Spain, April 2005, pp. 4538–4543.

[8] E. Guendelman, R. Bridson, and R. Fedkiw, "Nonconvex rigid bodies with stacking," *ACM Transactions on Graphics*, vol. 22, no. 3, pp. 871–878, 2003.

[9] R. Smith, "Open dynamics engine," http://www.ode.org/, 2007.

[10] R. Weinstein, J. Teran, and R. Fedkiw, "Dynamic simulation of articulated rigid bodies with contact and collision," *IEEE Transactions on Visualization and Computer Graphics*, vol. 12, no. 3, pp. 365–374, 2006.

[11] R. Featherstone, *Robot Dynamics Algorithm*. Boston, MA: Kluwer Academic Publishers, 1987.

[12] ——, "A Divide-and-Conquer Articulated-Body Algorithm for Parallel $O(\log(n))$ Calculation of Rigid-Body Dynamics. Part1: Basic Algorithm," *International Journal of Robotics Research*, vol. 18, no. 9, pp. 867–875, September 1999.

[13] E. Kokkevis, "Practical physics for articulated characters," in *Game Developers Conference*, 2004.

[14] R. Gayle, M. Lin, and D. Manocha, "Adaptie dynamics with efficient contact handling for articulated robots," in *Robotics: Science and Systems*, 2006, pp. 231–238.

[15] S. Redon, N. Galoppo, and M. Lin, "Adaptive dynamics of articulated bodies," *ACM Transactions on Graphics*, vol. 24, no. 3, pp. 936–945, 2005.

[16] B. Mirtich and J. Canny, "Impulse-based simulation of rigid bodies," in *Proceedings of Symposium on Interactive 3D Graphics*, Monterey, CA, 1995, pp. 181–188.

[17] P. Kry and D. Pai, "Interaction capture and synthesis," *ACM Transactions on Graphics*, vol. 25, no. 3, pp. 872–880, 2006.

[18] K. Yamane and Y. Nakamura, "Efficient Parallel Dynamics Computation of Human Figures," in *Proceedings of the IEEE International Conference on Robotics and Automation*, May 2002, pp. 530–537.

[19] R. Cottle and G. Dantzig, "Complementary pivot theory of mathematical programming," *Linear Algebra and Its Applications*, vol. 1, pp. 103–125, 1968.

[20] V. Acary and Pérignon, "An Introduction to Siconos," INRIA, Tech. Rep. 340, July 2007.

[21] D. Kaufman, T. Edmunds, and D. Pai, "Fast frictional dynamics for rigid bodies," *ACM Transactions on Graphics*, vol. 24, no. 3, pp. 946–956, 2005.

[22] UNC Gamma Group, "PQP—A Proximity Query Package," http://www.cs.unc.edu/~geom/SSV/.

[23] K. Yamane and Y. Nakamura, "Stable penalty-based model of frictional contacts," in *Proceedings of IEEE International Conference on Robotics and Automation*, Orlando, FL, May 2006, pp. 1904–1909.

[24] T. Sugihara, K. Yamamoto, and Y. Nakamura, "Architectural design of miniature anthropomorphic robots towards high-mobility," in *Proceedings of IEEE/RSJ International Conference on Intelligent Robots and Systems*, 2005, pp. 1083–1088.

Hybrid Motion Planning Using Minkowski Sums

Jyh-Ming Lien
jmlien@cs.gmu.edu
Department of Computer Science
George Mason University

Abstract—**Probabilistic and deterministic planners are two major approximate-based frameworks for solving motion planning problems. Both approaches have their own advantages and disadvantages. In this work, we provide an investigation to the following question: Is there a planner that can take the advantages from both probabilistic and deterministic planners? Our strategy to achieve this goal is to use the point-based *Minkowski sum* of the robot and the obstacles in workspace. Our experimental results show that our new method, called M-sum planner, which uses the geometric properties of Minkowski sum to solve motion planning problems, provides advantages over the existing probabilistic or deterministic planners. In particular, M-sum planner is significantly more efficient than the Probabilistic Roadmap Methods (PRMs) and its variants for problems that can be solved by reusing configurations.**

I. Introduction

In motion planning, we study the problem of finding a feasible path for a movable object to navigate in an environment with obstacles. Researchers have shown that any *complete* method that solves a general motion planning problem exactly will take time exponential to the complexity of the robot [1]. Approximate motion planners have since been intensively studied; see surveys by LaValle [2]. One of the most well known approximate planners is the *probabilistic* motion planners (e.g., PRMs [3] and its variants [4], [5], [6]). These planners are able to solve high dimensional problems that were not solvable before.

The success of the probabilistic motion planners is largely due to their simplicity and efficiency gained from sacrificing the completeness. As a consequence, when such a planner fails to find a solution, it cannot be certain whether a path exists or not. One of the most common reasons causing the failure of the PRM planners is the presence of *narrow passages* (the so called 'narrow passage problem'). Due to these problems, some recent work focused on developing *deterministic* approximate motion planners [7], [8], [9] that use more sophisticated geometric algorithms to approximate the obstacles in configuration space. These methods are provably less sensitive to narrow passages, thus providing stronger confidence on the path nonexistence problem. However, as far as we know, the motion planners in this category can only handle problems in low (≤ 4) dimensions and are in general more difficult to implement than PRMs.

Even with active research on probabilistic and deterministic motion planners, it is clear that the gap between these two approaches is still huge. Therefore, in order to bridge the gap, the question that we will investigate in this paper is:

- Is there a planner that can take the advantages from both probabilistic and deterministic planners?

The same question has also been raised by Hirsch and Halperin [10] although their focus is on a more specific problem: two-disc motion planning. By combining a complete planner for a single disc with a PRM strategy to coordinate two discs, their hybrid motion planner efficiently solves problems with narrow passages. In this paper, we adapt a totally different strategy and focus on more general problems. More specifically, we are interested in developing a planner that is simple and easily extensible to high dimensional space (the advantages from probabilistic planners) and remains efficient even with the presence of the narrow passages (the advantages from deterministic planners).

Our strategy in developing such a planner is to combine PRMs with the point-based *Minkowski sum* [11] of the robot and the obstacles in the workspace. Minkowski sum boundary is closely related to the concept of the "contact space" of translational robots in motion planning. We will discuss in detail regarding the definition of the Minkowski sum and its relationship to the contact space in Section II. For the rest of this section, we will provide an overview of our planner.

Our Approach. We investigate a method, called M-sum planner, that uses the Minkowski sum of the robot and the obstacles to facilitate the process of creating a roadmap. Similar to the probabilistic roadmap methods (PRMs) [3], [4], [5], [6], the roadmap constructed by M-sum planner represents the connectivity of the entire free space and can be used to solve motion planning queries. Due to this similarity we will focus on the process of building the roadmap only.

Intuitively, M-sum planner produces a set of n "shapes" of a robot by rotating or changing its joint angles. We treat each shape as one translational robot and compute the Minkowski sum of each shape and obstacles. The vertices of the Minkowski sum are then connected to form a small graph. There will be n such graphs constructed at the end of the process. Finally, we will merge these graphs into a global roadmap.

An important property of M-sum planner is that it is significantly more efficient than the Probabilistic Roadmap Methods (PRMs) and its variants for problems, e.g., Fig. 1, that can be solved by reusing configurations.

II. Related Work

In this section, we will discuss closely related work on PRMs and Minkowski sum.

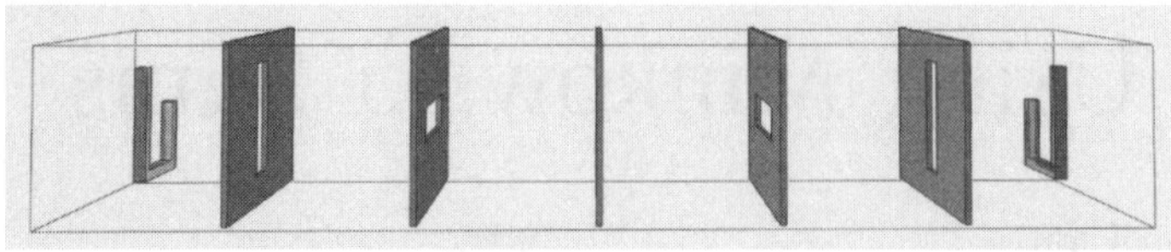

Fig. 1. A motion planning problem that can be solved more efficiently by M-sum planner. The workspace is composed of five parallel walls with horizontal and vertical windows. M-sum planner takes advantage of

A. Probabilistic Roadmap Methods

Probabilistic roadmap methods (PRMs) generally operate as follows (see, e.g., [3]). During a preprocessing phase, a set of configurations in the free space is generated by sampling configurations at random and retaining those that are valid. These nodes are then connected to create a roadmap by inserting edges between nodes if they can be connected by a simple and fast local planning method. This roadmap is then queried by first connecting the given start and goal configurations to the roadmap and then searching for a path in the roadmap connecting them.

An important shortcoming of PRMs is their poor performance on problems requiring paths that pass through narrow passages in the free space. This is a direct consequence of how the nodes are sampled. For example, using the traditional uniform sampling [3], any corridor of sufficiently small volume is unlikely to contain any sampled nodes whatsoever.

Effort has been made to modify the sampling strategy to increase the number of nodes sampled in narrow corridors. Intuitively, such narrow corridors may be characterized by their large surface area to volume ratio. For example, in [4], [12], nodes are sampled from the *contact space*, the set of configurations for which the robot is in contact with an obstacle. In [5], the sampling strategy samples pairs of nearby configurations that are separated by a Gaussian distance d. If one configuration is free and the other is in collision, then the free configuration is added to the roadmap. Otherwise, both configurations are discarded. The Gaussian sampler generates a higher density of nodes near C-obstacle boundaries. Following a similar strategy, the bridge test approach [6] samples two in-collision configurations separated by a Gaussian distance d and keeps their midpoint if it is free. In [13], preliminary configurations are generated by allowing the robot to penetrate the obstacles by a small amount. The areas near these nodes are then re-sampled to find nearby collision-free configurations. Work has also been proposed to address the narrow passage problem by analyzing the *workspace* properties. However, most of the work using this strategy [14], [15], [16] only focused on properties of the obstacles. On the contrary, our method considers both the robot and the obstacles.

B. Hybrid Motion Planners

Recently, several hybrid motion planners have been proposed [17], [18], [19], [20]. All these *meta-planners* focus on combining different PRMs using machine learning or statistics collected during sampling to discover when and where to apply certain sampling strategies.

Few hybrid methods attempt to combine deterministic and probabilistic planning strategies. Hirsch and Halperin's hybrid planner [10] studied two-disc motion planning. Zhang et al. [9] combines adaptive cell decomposition with PRMs but can only handle problems up to 4 DOF. Another 'hybrid' planner that connects 'slices' of configuration space into a global roadmap has also been discussed in the book by de Berg et al. [21, pp. 283–287] and by Lamiraux and Kavraki [22]. Each slice is computed by a complete planner for 2D translational robot using cell decomposition. Two slices are connected if the subdivisions from both slices overlaps. Unlike these methods that are limited in specific problems or in low dimensional space (≤ 4), our hybrid method can handle high dimensional problems.

C. Minkowski Sum

The Minkowski sum of two sets P and Q in $\mathbb{R}^d$ is defined as:

$$P \oplus Q = \{p + q \mid p \in P, q \in Q\}. \tag{1}$$

Typically, P and Q represent polygons in $\mathbb{R}^2$ or polyhedra in $\mathbb{R}^3$. Minkowski sum boundary is closely related to the concept of "contact space." Every point in the contact space represents a configuration that places the robot in contact with (but without colliding with) the obstacles. Given a translational robot P and obstacles Q, the contact space of P and Q can be represented as $\partial((-P) \oplus Q)$, where $-P = \{-p \mid p \in P\}$. In other words, if a point x is on the boundary of the Minkowski sum of two polyhedra P and Q, then the following condition must be true:

$$(-P^\circ + x) \cap Q^\circ = \emptyset \ , \tag{2}$$

where Q° is the open set of Q and $P + x$ denotes translating P to x.

Many methods have been proposed to compute Minkowski sum (see surveys in [23], [24], [25]). Ghosh [23] proposed a unified approach to handle 2-d or 3-d convex and non-convex objects by introducing negative shape and slope diagram representation. Slope diagram is closely related to *Gaussian map*, which has been used to implement very efficient Minkowski sum computation of convex objects by Fogel and Halperin [25]. Several other methods have been proposed to handle convex objects. Guibas and Seidel [26] proposed an output sensitive method to compute convolution curves, a super-set of the Minkowski sum boundaries. Kaul and Rossignac [27] proposed a linear time method to generate a set of Minkowski sum facets. Output sensitive methods that compute the Minkowski sum of polytopes in d-dimension have also been proposed by Gritzmann and Sturmfels [28] and Fukuda [29].

Because computing the Minkowski sum of convex polyhedra is easier, most methods that compute the Minkowski sum of non-convex polyhedra first compute the convex decomposition and then compute the union of the Minkowski sums of the convex components [30], [24]. Unfortunately, neither the convex decomposition nor the union of the Minkowski sums is trivial.

Peternell et al. [31] proposed a method to compute the Minkowski sum using points densely sampled from the solids, and compute local quadratic approximations of these points. However, their method only identifies the outer boundary of the Minkowski sum, i.e., no hole boundaries. This can be a serious problem in particular for motion planning. In this paper, we use the point-based Minkowski sum proposed by Lien [11] that does not have the undesirable issues above.

III. PRELIMINARY & OVERVIEW

In this section, we define notations that we will use throughout this paper. We will also give a more detailed overview of our method (M-sum planner) to end this section.

Separating translational and rotational motions. Given a configuration C, we separate the configuration into two components: Translational and rotational configurations. We represent the configuration as $C = \{T_C \times R_C\}$, where T_C and R_C represents the translational and rotational components of the configuration C, respectively. To ease our discussion, we use the notation $C(x)$ to denote the coordinate of a point x on the robot after the robot is placed at the configuration C. Similarly, we denote $T_C(x)$ (or $R_C(x)$) as the coordinates of a point x after the robot is translated (or rotated) by T_C (or R_C). By separating motions, we can handle translational and rotational motions differently. As we will see later, this separation provides many benefits in generating and connecting configurations.

$\mathcal{C}$-slice. Intuitively, a $\mathcal{C}$-slice is a slice (subspace) of the entire $\mathcal{C}$-space. All configurations in a $\mathcal{C}$-slice are generated from a "seed" configuration S and the Minkowski sum of the robot and the obstacles. That is a configuration C in a $\mathcal{C}$-slice must has the following form: $C = \{p \times R_S\}$, where p is the position of a point on the Minkowski sum surface and R_S is the rotational component of S. Therefore, a $\mathcal{C}$-slice resides only in a translational-subspace of the $\mathcal{C}$-space. An example of $\mathcal{C}$-slice is shown in Fig. 2.

Origins. A point x on the Minkowski sum surface is a combination of two points, which are called the origin of x. To simplify our discussion later, we define an operation $\mathcal{O}(x)$ to denote the origin of x.

> *Definition 3.1:* The origin $\mathcal{O}(x)$ of a point x on the Minkowski sum surface is a pair of points p and q from the robot and the obstacles, respectively, such that $x = p + q$.

Note that later in Definition 4.2, we will encounter another definition of origin for points on the surface of the robot or the obstacles.

Overview of M-sum planner. Essentially, we iteratively generate n $\mathcal{C}$-slices from n randomly selected seeds and then we connect $\mathcal{C}$-slices into a global roadmap. The main steps of M-sum planner include: Generate $\mathcal{C}$-slices (see Section IV), connect configurations in each $\mathcal{C}$-slice (see Section V-A), and connect configurations among $\mathcal{C}$-slices (see Section V-B).

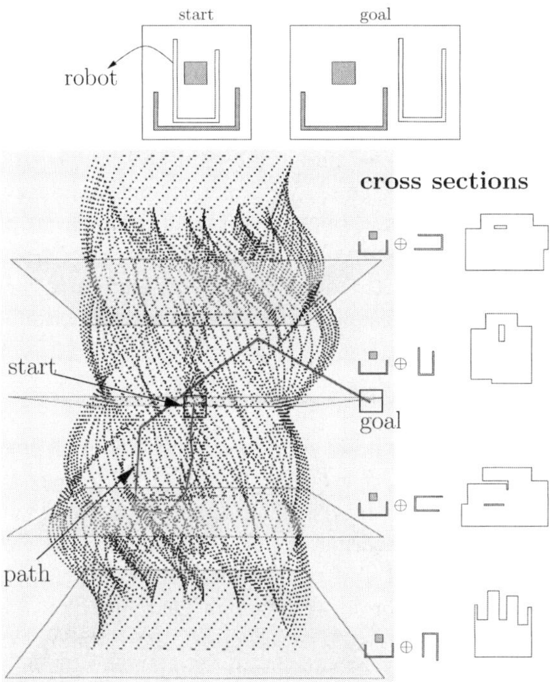

Fig. 2. A set of configurations sampled from the boundary of $\mathcal{C}$-obst using the Minkowski sum of the robot and the obstacles. Four $\mathcal{C}$-slices are highlighted in this figure.

IV. MINKOWSKI SUM ROADMAP

GENERATE CONFIGURATIONS

As mentioned earlier, we generate configurations by computing the vertex coordinates of the Minkowski sum of the obstacles and the robot whose rotation and joint angles are sampled at random. That is we sample a random configuration as our 'seed.' The translational information of the seed is discarded (i.e., set to zeros). Then the seed configuration is placed at the positions of the Minkowski sum vertices to generate *a family of configurations*, which we call a '$\mathcal{C}$-slice.' Note that our method does not depend on any kind of sampling strategy. For example, we can use a configuration generated by an obstacle-based PRM [4], [5], [6] as our seed.

Even though our strategy is straightforward, the difficulty of computing the Minkowski sum and its boundary remains unsolved. As we have seen in Section II, no existing methods can provide a robust and efficient method to compute the Minkowski sum boundary of polyhedra. Fortunately, using the recent work proposed by Lien [11], we can efficiently compute the Minkowski sum boundary if it is represented by points only. In the following section, we will provide a sketch of how to create such a point-based representation.

A. Generate points on the Minkowski sum boundary

Our goal is to produce a set of points that *cover* the boundary of the Minkowski sum of two given polyhedra, P and Q. More specifically, we will generate a point set S so that S is a d-covering of the Minkowski sum boundary, where d is a user-specified value. Intuitively, d controls the sampling density of a boundary. A smaller d will produce a denser approximation of the boundary.

Our approach is composed of three main steps. First, we sample two point sets from the input P and Q. Second, we generate the Minkowski sum of the point sets simply using the definition in Eqn. 1. Third, we separate the boundary points (both hole and external boundaries) from the internal points.

Step 1: Sample points. Let P and Q be two polyhedra. We generate two point sets from P and Q, denoted as S_P and S_Q. The point set S representing the Minkowski sum boundary of P and Q is simply

$$(S_P \oplus S_Q) \cap \partial(P \oplus Q) \ . \quad (3)$$

Because we want the point set S to cover the entire Minkowski sum boundary w.r.t. a user specified interval d, we have to make sure that the points S_P is a d_p-covering of ∂P and the points S_Q is a d_q-covering of ∂Q. It is our task to determine the values of d_p and d_q from the input d.

As shown in Theorem 4.1, we can guarantee that the final point set is at least a d-covering of the Minkowski sum boundary of P and Q by simply letting $d_p = d_q = d$. Moreover, since the boundaries of P and Q are known, we can easily generate S_P and S_Q that d-cover ∂P and ∂Q, respectively.

Theorem 4.1: [11] Let S_P and S_Q be two d-covering point sets sampled from two polyhedral surface ∂P and ∂Q and let $S_{P\oplus Q} = S_P \oplus S_Q$ and $S = S_{P\oplus Q} \cap \partial(P \oplus Q)$. Then, S must be a d-covering point set of $\partial(P \oplus Q)$.

Step 2: Compute the Minkowski sum. This step is straightforward. Using S_P and S_Q, we compute $S_{P\oplus Q}$ by simply following the Minkowski sum definition in Eqn. 1.

Step 3: Extract boundary points. In this final step, we separate (filter) points to two groups: Boundary points and inner points. Boundary points will be returned as our final answer and inner points will be discarded.

The first filter, named *normal filter* determines if a pair of sample points (from P and Q, resp.) is an inner point by examining their *origins* (defined later in Definition 4.2) and orientations. Kaul and Rossignac [27] have shown that a facet of the Minkowski sum boundary can only come from a facet of P and a vertex from Q (or vice versa) or from a new facet formed by two edges of P and Q if the facet, vertex and edges are properly oriented [27]. Our strategy is derived directly from their observation. Since our points are sampled from the polyhedral surface, we define the *origin* of a sample to ease our discussion.

Definition 4.2: The **origin of a sample** x, denoted as $\mathcal{O}(x)$, is a facet, an edge or a vertex of a polyhedron from which x is sampled.

Let p and q be a pair of points sampled from P and Q, respectively. We decide if $p+q$ is an inner point by checking the orientation of $\mathcal{O}(p)$ and $\mathcal{O}(q)$.

Consider the case when $\mathcal{O}(p)$ is a vertex and $\mathcal{O}(q)$ is a facet (or vice versa). We first define a supporting plane $\mathcal{P}$ at the point $p+q$ parallel to facet $\mathcal{O}(q)$. Then, we translate P by q so that vertex $\mathcal{O}(p)$ coincides with the point $p+q$. The point $p+q$ must be an inner point when the (open) half space defined by the plane $\mathcal{P}$ intersects at least one edge incident to the vertex $\mathcal{O}(p)$.

Now, consider the case when $\mathcal{O}(p)$ and $\mathcal{O}(q)$ are both edges. Similarly, we define a supporting plane $\mathcal{P}$ at point $p+q$ whose outward normal is the cross product of two vectors parallel to edges $\mathcal{O}(p)$ and $\mathcal{O}(q)$. Then, we translate P by q and Q by p so that edges $\mathcal{O}(p)$ and $\mathcal{O}(q)$ coincide with the plane $\mathcal{P}$. The point $p+q$ must be an inner point when the facets that incident to edges $\mathcal{O}(p)$ and $\mathcal{O}(p)$ are on the different sides of the plane $\mathcal{P}$.

When $\mathcal{O}(p)$ and $\mathcal{O}(q)$ are both vertices or when $\mathcal{O}(p)$ and $\mathcal{O}(q)$ are a vertex-edge pair, we can break them into several instances of the edge-edge and vertex-facet cases above.

This filter is efficient, but it alone *cannot* filter out all inner points. The second filter, named *CD filter* uses collision detection to separate boundary points from inner points. CD filter is computational more expensive but it provides an unambiguous decision. An example of the Minkowski sum generated by this point-based representation is shown in Fig. 3.

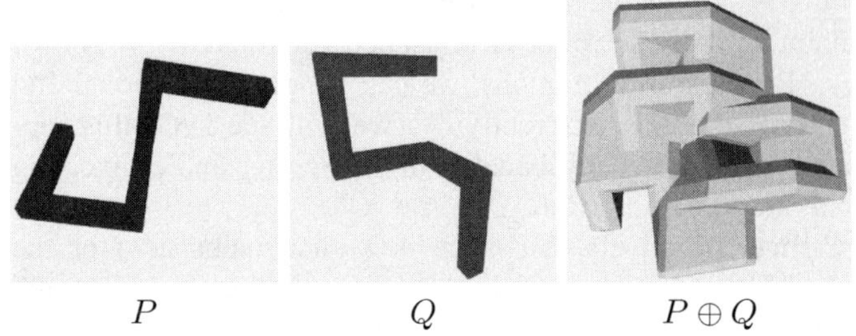

Fig. 3. This figure shows a 0.01-covering point set of the Minkowski sum boundary of two hook-like models. Note that all P, Q and $P \oplus Q$ are represented by densely sampled points.

V. MINKOWSKI SUM ROADMAP
CONNECT CONFIGURATIONS

Connecting configurations is usually the most expensive step in building a roadmap. In the following, we will show that, using some simple properties of the Minkowski sum, not only we can connect configurations more efficiently but also can increase the chance of connecting configurations. It is important to note that the new local planners proposed below are not applicable to samples generated by regular PRMs. To ease our discussion, we separate our approaches into connecting configurations within a $\mathcal{C}$-slice and among $\mathcal{C}$-slices.

A. *Connecting Configurations within A $\mathcal{C}$-slice*

Connecting configurations in a $\mathcal{C}$-slice can be done more efficiently than connecting configurations generated by random sampling. The reason for this is that we can quickly eliminate configurations that cannot be connected by simply examining the connectivity of the geometries (mesh) of the robot and the obstacles. We will make this claim more clearly next.

Let C_1 and C_2 be two configurations in the same $\mathcal{C}$-slice i.e., C_1 and C_2 have the same rotation and joint angles. Moreover, we can represent C_i as $p_i + q_i$ where p_i is a point from the robot P and q_i is a point from the obstacle Q. Now, we can only make a connection between C_1 and C_2 if C_1 and C_2 satisfy one of the following requirements:

- $q_1 = q_2$ and $\overline{p_1p_2}$ lies on a triangle of P.
- $p_1 = p_2$ and $\overline{q_1q_2}$ lies on a triangle of Q.
- Origins of p_1, p_2, q_1, q_2 are edges and $\mathcal{O}(p_1) = \mathcal{O}(p_2)$ and $\mathcal{O}(q_1) = \mathcal{O}(q_2)$.

These tests can be done in constant time. All we have to do is to keep these information during the construction of the Minkowski sum. If C_1 and C_2 do not satisfy all the requirements above, we will skip the pair. Otherwise, we will use collision detection of check of they are indeed connectable. The following theorem supports this approach.

Theorem 5.1: Only a pair of configurations that satisfy one of the criteria above can form a connection.

Proof: The boundary of the Minkowski sum of two polygons can only come the edges of the polygons. Therefore if two vertices on the Minkowski sum boundary are connected, then they must come from an edge of one of the polygons.

The boundary of the Minkowski sum of two polyhedra can only come the facets of the polyhedra or from the sweep area of two edges, one from each polyhedron, i.e., one edge from the robot and one edge from the obstacle. In both cases, if two vertices on the Minkowski sum boundary are connected, then they must come from a facet of the polyhedra or from a new facet that is generated by a pair of edges; each from one of the polyhedra. ∎

Note that if there are more than one obstacle in the workspace, the method mentioned above will not connect configurations that are generated from different obstacles. We still need the traditional approaches (e.g., k-closest) to connect the configurations between obstacles.

B. Connecting Configurations Among $\mathcal{C}$-slices

Connecting configurations among $\mathcal{C}$-slices is similar to connecting configurations among connected components of a roadmap in PRMs. Similar to connecting individual configurations, it is also more desirable to connect each connected component (CC) to its k-closest CCs (instead of to all CCs). However, in PRMs, there is no well defined distance metrics for CCs [32]. On the contrary, for M-sum planner, the distance between two $\mathcal{C}$-slices can simply be measured as the difference between their rotation and joint angles (of the seeds). Therefore connecting configurations among $\mathcal{C}$-slices can be handled more naturally for M-sum planner when we attempt to order $\mathcal{C}$-slices (or CCs) from near to far.

Moreover, following the same strategy of connecting configurations within a $\mathcal{C}$-slice, we are allowed to use the properties of Minkowski sum to increase the chance of connecting configurations from two $\mathcal{C}$-slices. In the rest of this section, we will proposed two local planners. The key characteristic of these local planners is that they connect configurations by 'walking' on the $\mathcal{C}$-obst boundary.

Connecting two configurations with the same origins. Given two configurations C_1 and C_2 that are generated from different $\mathcal{C}$-slices and have the origins (see Definition 3.1) from the same points p and q of the robot and the obstacle, respectively, i.e, $\mathcal{O}(C_1) = \mathcal{O}(C_2) = (p, q)$. Let $C_1 = S_1(p) + q$ and $C_2 = S_2(p) + q$, where S_i are the seed configurations of the $\mathcal{C}$-slice i.

When a straight-line local planner (or other simple local planners [33]) fails to connect C_1 and C_2, we can attempt to connect them as follows. First, we construct a new seed configuration $S_3 = \frac{S_1+S_2}{2}$, which is the mid point the two seed configurations S_1 and S_2. Next, we use C_3 to compute a new Minkowski sum point $C_3 = S_3(p) + q$. If C_3 is on the $\mathcal{C}$-obst surface, i.e., C_3 is collision free, then we recursively connect $\overline{C_1C_3}$ and $\overline{C_3C_2}$ in the same manner (because now C_1, C_2 and C_3 all have the same origin). In short, this local planner connects two configurations between two $\mathcal{C}$-slices by walking on the surface of $\mathcal{C}$-obst.

Connecting two configurations whose origins are connected in workspace. Given two configurations C_1 and C_2 that are generated from different $\mathcal{C}$-slices. Assuming C_1 and C_2 share one of the point in their origins. Let the shared point be a point q of the obstacle. That is $C_1 = S_1(p_1) + q$ and $C_2 = S_2(p_2) + q$, where p_1 and p_2 are two points on the robot and S_i is the seed configurations of the $\mathcal{C}$-slice i. Everything will be the same if the shared point is from the robot. Next, we will see that we can increase the chance of connecting C_1 and C_2 if p_1 and p_2 are from the same edge or the same triangle of the robot.

We first compute the midpoint $p_3 = \frac{p_1+p_2}{2}$. Then, we can split $\overline{C_1C_2}$ into three segments: $\overline{C_1(S_1(p_3)+q)}$, $\overline{(S_1(p_3)+q)(S_2(p_3)+q)}$, and $\overline{(S_2(p_3)+q)C_2}$. Observe that the first and the last segments connect two configurations in the same $\mathcal{C}$-slice, which is a problem that we have already handled in Section V-A and the second segment connects two configurations *with the same origin* in different $\mathcal{C}$-slices. This is exactly the problem that we have encountered earlier.

VI. Putting it All Together

Algorithm VI.1 summarizes all the methods we have discussed so far. The output of Algorithm VI.1 is a roadmap.

In the rest of this section, we will discuss the advantages and the limitations of the M-sum planner.

A. Advantages of M-sum planner

There are several important advantages of M-sum planner over the PRM planners. First, M-sum planner can connect the configurations more efficiently (see Sections V-A and V-B) using more powerful local planners, which are not applicable to the regular PRM samples. In addition, M-sum planner reuses configurations, including the "good" configurations that fit into the narrow passages. This property allows M-sum planner to solve problems more efficiently even in high dimensions.

Algorithm VI.1: M-SUM-ROADMAP(P, Q, n, k)

comment: P and Q are the robot and the obstacles, respectively

Initialize the roadmap $R \leftarrow \emptyset$
Initialize $\mathcal{C}$-slices $S \leftarrow \emptyset$
for $i \leftarrow 1$ **to** n
  do $\begin{cases} \text{Sample a configuration } C_i \text{ and set its translation to } 0 \\ S_i \leftarrow \text{points on } \partial(-C_i(P) \oplus Q) \\ S \leftarrow S \cup \{C_i, S_i\} \end{cases}$
$R \leftarrow S$
Sort S using the distance from a randomly picked $\mathcal{C}$-slice
for $i \leftarrow 1$ **to** n
  do $\begin{cases} R \leftarrow R \cup (\text{edges in } S_i) \\ \textbf{for } j \leftarrow (i - \frac{k}{2}) \textbf{ to } (i + \frac{k}{2}) \\ \quad \textbf{do } R \leftarrow R \cup (\text{edges between } S_i \text{ and } S_j) \end{cases}$

In our experiments, we see that M-sum planner outperforms PRMs regardless the dimensionality of the C-space.

Second, M-sum planner expresses different behaviors when different inputs are given. For example, given problems with the translational robot, M-sum planner automatically becomes a deterministic planner. These are the problems that can be solved significantly more efficiently by the deterministic planners than by the probabilistic planners, in particular when there are narrow passages. M-sum planner becomes a probabilistic planner for problems whose rotational motions dominate the $\mathcal{C}$-space.

Third, M-sum planner separates the translational and rotational motions. Configurations are first generated and connected using only translation. Then the configurations are connected into the final roadmap using only rotation. This strategy provides several advantages. For example, we can use a deterministic manner to generate translational portion of the configuration and use a probabilistic manner to generate rotational portion of the configuration. We can also use a distance metric for translation and use another distance metric for rotations and avoid the confusing of combining or weighting different distance metrics [33].

Finally, M-sum planner can generate samples that cover the surface of the C-space obstacles ($\mathcal{C}$-obst). This can be done by giving M-sum planner a small d (smaller than the length of the shortest edge in workspace). The consequences of this is that the possibility of generating configurations in narrow passages must be increased.

Theorem 6.1: Given a translational robot, the possibility of generating configurations using M-sum planner in narrow passages must be larger than that using the traditional PRM if the same number of configurations are generated.

Proof: Sketch. Because narrow corridors can be characterized by their large surface area to volume ratio, M-sum planner that generates samples that cover the surfaces of $\mathcal{C}$-obst must has higher probability of generating samples inside the corridors than PRM does. ■

Although Theorem 6.1 is theoretically interesting since no existing obstacle-based PRMs can guarantee this, practically speaking, a small d makes computation more expensive. Moreover, M-sum planner cannot guarantee to increase the sampling inside the narrow passages surrounded by $\mathcal{C}$-obst that is the result of robot's self-collision. This leads us to the limitations of the M-sum planner.

B. Limitations of M-sum planner

We envision M-sum planner provide a new framework to combine probabilistic and deterministic planners. Even though it does not provide a total solution to our question, M-sum planner provides a simple and efficient planner to solve a certain type of common motion planning problems. In this section, we discuss its limitations.

One of the limitations of M-sum planner is that the user need to decide the value of d. From the completeness perspective, a small d is desirable since it allows M-sum planner to tightly cover the $\mathcal{C}$-obst surfaces. From the efficiency perspective, a larger d (e.g., larger than the length of the longest edge of the robot and obstacles in the workspace) is desirable since fewer configurations are generated. In the optimal situation, only the vertices of the Minkowski sum boundary are included in the samples. However, it is well known that the Minkowski sum of the vertices of two polyhedra may not include all the vertices of the Minkowski sum of the polyhedra. Because this happens only in some rare cases (e.g., two grate-like shapes), in our experiments, we simply use a large d. Further research is required to determine the value of d from a given problem.

Another limitation of M-sum planner is that it cannot efficiently handle problems, such as the alpha puzzle or fixed-base robot arms, which require simultaneous translations and rotations or have no translational degrees of freedom. In these problems, reusing configurations will not be helpful and M-sum planner downgrades to a PRM planner.

VII. EXPERIMENTAL RESULTS

Implementing M-sum planner is straightforward. We developed software based on the proposed planner in C++. All experimental results are collected on an Intel CPU at 2.13 GHz with 3 GB of RAM. The software is available from our project webpage.

In this section, we compare M-sum planner to three PRM variants: PRM [3], Gaussian PRM [5], and Bridge-test PRM [6]. In our experiments, we use four workspaces shown in Fig. 4. These problems have robots with 3, 6, 8, 10 degrees of freedom, respectively. We study the efficiency of configuration generation and the efficiency of solving these four motion planning problems. The results are summarized in Tables I and II.

M-sum planner generates configurations near $\mathcal{C}$-obst more efficiently than PRMs do in all studied cases. In Table I, we collect the *configuration generation times* from the planners. It is clear that PRM is the most efficient method since it does not deliberately place or filter samples. M-sum

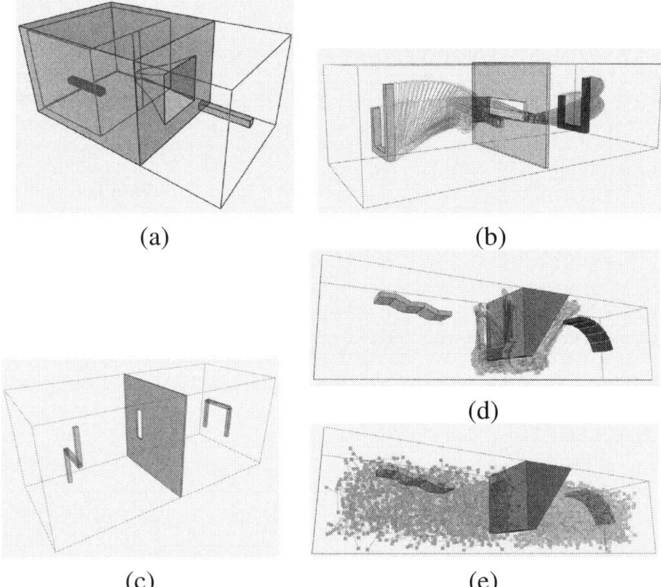

Fig. 4. (a) Bug trap environment. The robot (bug) is a translational robot in a 3D workspace. The width of the workspace is 23.5 (units). (b) A 3D free-flying rigid robot with 6 DOF. The width of the workspace is 10 (units) (c) A 8-DOF articulated robot. The width of the workspace is 30 (units). (d) A 10-DOF articulated robot. The width of the workspace is 40 (units). It also shows a roadmap with 3000 nodes generated by M-sum planner (e) A roadmap with 3000 nodes generated by Gaussian PRM (with $d = 0.1$) is shown.

TABLE I
COMPUTATION TIME TO GENERATE n CONFIGURATIONS

Environment	n	M-sum	PRM	Gaussian PRM	Bridge test PRM
Fig. 4(a)	50	0.04 s	0.02 s	0.20 s	105.05 s
Fig. 4(b)	200	0.25 s	0.02 s	0.74 s	53.77 s
Fig. 4(c)	1000	0.32 s	0.06 s	2.58 s	336.13 s
Fig. 4(d)	2000	14.54 s	0.30 s	23.50 s	8176.41 s

planner is the most efficient among the planners that attempt to generate configurations near the $\mathcal{C}$-obst. Figs. 4(d) and (e) show two roadmaps generated by M-sum planner and the Gaussian PRM, respectively. It is clear that configurations generated by M-sum planner are much denser around the boundary while the configurations generated by Gaussian PRM are more scattered. A reason of the scatteredness is because the Gaussian PRM not only samples configurations near the $\mathcal{C}$-obst generated from the workspace obstacle but also samples near the $\mathcal{C}$-obst generated from self-collisions. Another reason of the scatteredness is due to the Gaussian distance parameter d required by Gaussian PRM. Picking a good value of d used by both Gaussian and Bridge test PRMs is usually tricky and is problem dependent.

M-sum planner solves all studied cases more efficiently than PRMs do. In Table II, we study the *expected* computation time to solve these four problems. The expected solution time E_t is measured as:

$$E_t = \frac{t_x}{p},$$

TABLE II
EXPECTED SOLUTION TIME ($E_t = \frac{t_x}{p}$)

Environment	M-sum	PRM	Gaussian PRM	Bridge test PRM
Fig. 4(a)	0.3 s	21.9 s	14.9 s	513.4 s
Fig. 4(b)	0.4 s	104.0 s	10.9 s	1760.8 s
Fig. 4(c)	27.2 s	2002.6 s	82.0 s	7023.4 s
Fig. 4(d)	22.8 s	5073.1 s	3509.6 s	35735.1 s

(All PRMs connect a configuration to its $k = 20$ closest configurations. Gaussian and Bridge-test PRMs use $d = 0.1$ in all environments. Bridge-test PRM is *not* combined with uniform PRM.)

where t_x is the averaged running time over x runs using a given planner and p is the probability of successfully solving the problem from these x runs using the same planner. In all experiments, we set $x = 100$. One can view E_t as the time spent before the planner can find a solution (which may require several runs).

We observe from our experimental results that M-sum planner is the most efficient planner in all four environments. More precisely, in all four environments, M-sum planner is 40, 35, 3 and 150 times, respectively, faster than Gaussian PRMs, the best planner among the three PRMs.

It is clear that in the bug-trap environment M-sum planner is much more efficient than the all the other PRMs because M-sum planner essentially becomes a deterministic motion planner. In the U-shape robot environment (Fig. 4(b)), M-sum planner still outperforms PRM planners because once a configuration that fits into the hole in the wall is generated, M-sum planner will use this particular configuration to generate a family of configurations (i.e., $\mathcal{C}$-slice) around the hole and solves the problem.

Although we expect the performs of M-sum planner and PRMs become closer when the $\mathcal{C}$-space has higher dimensionality (> 6), M-sum planner still outperforms PRM planners in the cases with articulated robots (Figs. 4(c) and (d)). This is because M-sum planner has ability to *reuse* configurations including the "good" configurations, e.g., configurations that fit into the narrow passage. For example, in Fig. 4(c), the robot can fold into a triangle and fit into the hole, and in Fig. 4(d) the robot can make itself flat and slide through the bottom of the obstacle. M-sum planner picks up these promising configurations and generates more configurations from them ($\mathcal{C}$-slices). In PRMs, good configurations are generated and used only once.

VIII. CONCLUSION

We proposed a motion planner, called M-sum planner, that takes advantages from the probabilistic and the deterministic approximate motion planners. We have shown that Minkowski sum is the key of this hybrid planner. Using the properties of the Minkowski sum, we are able to generate configurations uniformly on the surface of the $\mathcal{C}$-obst and make more connections between configurations than PRMs do using more powerful local planners. In our experimental results we show that M-sum planner outperforms PRMs in all studied problems, even for problems with a 10 DOF robot.

Finally, we would like to conclude this paper by pointing out the similarity between M-sum planner and the ideas sketched in the 'Future Research' section in [10].

> One way to [solve the full rigid motion planning problem for polyhedron among polyhedra] is to use a "slicing" method, where we build a coarse gird (which fixes the rotational dofs of the robot) we construct an explicit representation of the free space (we call these representations complete cross-sections). We then use PRM techniques to connect between the complete cross-sections. How to effectively make these connections is a non-trivial challenge. — Hirsch and Halperin [10].

ACKNOWLEDGEMENT

The bug trap environment is modified from the Motion Planning Puzzles provided by Parasol Lab at Texas A&M University. The author thanks Roger Pearce and Nancy Amato for their help with this paper. The author also thanks anonymous reviewers for their valuable comments.

REFERENCES

[1] J. F. Canny, *The Complexity of Robot Motion Planning*. Cambridge, MA: MIT Press, 1988.

[2] S. M. LaValle, *Planning Algorithms*, 6th ed. Cambridge University Press, 2006.

[3] L. E. Kavraki, P. Svestka, J. C. Latombe, and M. H. Overmars, "Probabilistic roadmaps for path planning in high-dimensional configuration spaces," *IEEE Trans. Robot. Automat.*, vol. 12, no. 4, pp. 566–580, August 1996.

[4] N. M. Amato, O. B. Bayazit, L. K. Dale, C. V. Jones, and D. Vallejo, "OBPRM: An obstacle-based PRM for 3D workspaces," in *Robotics: The Algorithmic Perspective*. Natick, MA: A.K. Peters, 1998, pp. 155–168, proc. Third Workshop on Algorithmic Foundations of Robotics (WAFR), Houston, TX, 1998.

[5] V. Boor, M. H. Overmars, and A. F. van der Stappen, "The Gaussian sampling strategy for probabilistic roadmap planners," in *Proc. IEEE Int. Conf. Robot. Autom. (ICRA)*, vol. 2, May 1999, pp. 1018–1023.

[6] D. Hsu, T. Jiang, J. Reif, and Z. Sun, "Bridge test for sampling narrow passages with proabilistic roadmap planners," in *Proc. IEEE Int. Conf. Robot. Autom. (ICRA)*, 2003, pp. 4420–4426.

[7] G. Varadhan and D. Manocha, "Star-shaped roadmaps - a deterministic sampling approach for complete motion planning," in *Proc. Robotics: Sci. Sys. (RSS)*, 2005.

[8] L. Zhang, Y. Kim, and D. Manocha, "A simple path non-existence algorithm using c-obstacle query for low dof robots," in *Proc. Int. Workshop on Algorithmic Foundations of Robotics (WAFR)*, July 2006.

[9] L. Zhang, Y. J. Kim, and D. Manocha, "A hybrid approach for complete motion planning," in *IEEE/RSJ International Conference On Intelligent Robots and Systems (IROS)*, 2007.

[10] S. Hirsch and D. Halperin, "Hybrid motion planning: Coordinating two discs moving among polygonal obstacles in the plane," in *Proc. 5th Workshop on Algorithmic Foundations of Robotics (WAFR)*, Nice, 2002, pp. 225–241.

[11] J.-M. Lien, "Point-based minkowski sum boundary," in *PG '07: Proceedings of the 15th Pacific Conference on Computer Graphics and Applications*. Washington, DC, USA: IEEE Computer Society, 2007, pp. 261–270.

[12] N. M. Amato and Y. Wu, "A randomized roadmap method for path and manipulation planning," in *Proc. IEEE Int. Conf. Robot. Autom. (ICRA)*, 1996, pp. 113–120.

[13] D. Hsu, L. E. Kavraki, J.-C. Latombe, R. Motwani, and S. Sorkin, "On finding narrow passages with probabilistic roadmap planners," in *Proc. Int. Workshop on Algorithmic Foundations of Robotics (WAFR)*, 1998, pp. 141–153.

[14] C. Holleman and L. E. Kavraki, "A framework for using the workspace medial axis in prm planners," in *Proc. IEEE Int. Conf. Robot. Autom. (ICRA)*, 2000, pp. 1408–1413.

[15] H. Kurniawati and D. Hsu, "Workspace importance sampling for probabilistic roadmap planning," in *Proceedings. 2004 IEEE/RSJ International Conference on Intelligent Robots and Systems, 2004. (IROS 2004)*, 2004, pp. 1618–1623.

[16] J. P. van den Berg and M. H. Overmars, "Using workspace information as a guide to non-uniform sampling in probabilistic roadmap planners," *The International Journal of Robotics Research*, vol. 24, no. 12, pp. 1055–1071, 2005.

[17] M. Morales, L. Tapia, R. Pearce, S. Rodriguez, and N. M. Amato, "A machine learning approach for feature-sensitive motion planning," in *Proc. Int. Workshop on Algorithmic Foundations of Robotics (WAFR)*, Utrecht/Zeist, The Netherlands, July 2004, pp. 361–376.

[18] D. Hsu, G. Sánchez-Ante, and Z. Sun, "Hybrid PRM sampling with a cost-sensitive adaptive strategy," in *Proc. IEEE Int. Conf. Robot. Autom. (ICRA)*, 2005, pp. 3885–3891.

[19] B. Burns and O. Brock, "Toward optimal configuration space sampling," in *Proc. Robotics: Sci. Sys. (RSS)*, 2005.

[20] S. Rodriguez, S. Thomas, R. Pearce, and N. M. Amato., "(resampl): A region-sensitive adaptive motion planner," in *Proc. Int. Workshop on Algorithmic Foundations of Robotics (WAFR)*, July 2007.

[21] M. de Berg, M. van Kreveld, M. Overmars, and O. Schwarzkopf, *Computational Geometry: Algorithms and Applications*, 2nd ed. Berlin, Germany: Springer-Verlag, 2000.

[22] F. Lamiraux and L. Kavraki, "Planning paths for elastic objects under manipulation constraints," *The International Journal of Robotics Research*, vol. 20, no. 3, pp. 188–208, 2001.

[23] P. K. Ghosh, "A unified computational framework for Minkowski operations," *Computers and Graphics*, vol. 17, no. 4, pp. 357–378, 1993.

[24] G. Varadhan and D. Manocha, "Accurate Minkowski sum approximation of polyhedral models," *Graph. Models*, vol. 68, no. 4, pp. 343–355, 2006.

[25] E. Fogel and D. Halperin, "Exact and efficient construction of Minkowski sums of convex polyhedra with applications," in *Proc. 8th Wrkshp. Alg. Eng. Exper. (Alenex'06)*, 2006, pp. 3–15.

[26] L. J. Guibas and R. Seidel, "Computing convolutions by reciprocal search," *Discrete Comput. Geom.*, vol. 2, pp. 175–193, 1987.

[27] A. Kaul and J. Rossignac, "Solid-interpolating deformations: construction and animation of PIPs," in *Proc. Eurographics '91*, 1991, pp. 493–505.

[28] P. Gritzmann and B. Sturmfels, "Minkowski addition of polytopes: computational complexity and applications to Gröbner bases," *SIAM J. Discret. Math.*, vol. 6, no. 2, pp. 246–269, 1993.

[29] K. Fukuda, "From the zonotope construction to the minkowski addition of convex polytopes," *Journal of Symbolic Computation*, vol. 38, no. 4, pp. 1261–1272, 2004.

[30] T. Lozano-Pérez, "Spatial planning: A configuration space approach," *IEEE Trans. Comput.*, vol. C-32, pp. 108–120, 1983.

[31] M. Peternell, H. Pottmann, and T. Steiner, "Minkowski sum boundary surfaces of 3d-objects," Vienna Univ. of Technology, Tech. Rep., August 2005.

[32] D. Xie, M. A. Morales, R. Pearce, S. Thomas, J.-M. Lien, and N. M. Amato, "Incremental map generation (IMG)," in *Proc. Int. Workshop on Algorithmic Foundations of Robotics (WAFR)*, July 2006.

[33] N. M. Amato, O. B. Bayazit, L. K. Dale, C. V. Jones, and D. Vallejo, "Choosing good distance metrics and local planners for probabilistic roadmap methods," *IEEE Trans. Robot. Automat.*, vol. 16, no. 4, pp. 442–447, August 2000.

Bridging the Gap of Abstraction for Probabilistic Decision Making on a Multi-Modal Service Robot

Sven R. Schmidt-Rohr, Steffen Knoop, Martin Lösch, Rüdiger Dillmann

Abstract— **This paper proposes a decision making and control supervision system for a multi-modal service robot. With *partially observable Markov decision processes* (POMDPs) utilized for scenario level decision making, the robot is able to deal with uncertainty in both observation and environment dynamics and can balance multiple, conflicting goals. By using a flexible task sequencing system for fine grained robot component coordination, complex sub-activities, beyond the scope of current POMDP solutions, can be performed. The sequencer bridges the gap of abstraction between abstract POMDP models and the physical world concerning actions, and in the other direction multi-modal perception is filtered while preserving measurement uncertainty and model-soundness. A realistic scenario for an autonomous, anthropomorphic service robot, including the modalities of mobility, multi-modal human-robot interaction and object grasping, has been performed robustly by the system for several hours. The proposed filter-POMDP reasoner is compared with classic POMDP as well as MDP decision making and a baseline finite state machine controller on the physical service robot, and the experiments exhibit the characteristics of the different algorithms.**

I. Introduction

Service robots are meant to act autonomously and robustly in real world environments. Yet, observations of the physical world by robots are limited and noisy, thus the environment is partially observable. Also, the course of events in the real world is never completely deterministic but stochastic. Both aspects of uncertainty need to be regarded by decision making of an autonomous service robot.

In general, decision making of a multi-modal robot uses perceptions of multiple sensors together with background knowledge to choose one of the available actions which will contribute most likely to mission success. The chosen action is performed by coordinating available actuators.

This paper introduces a decision making and supervision system considering uncertainty, which utilizes *partially observable Markov decision processes* (POMDPs) for symbolic, scenario-level decisions. A main focus is bridging the gap of abstraction between symbolic POMDPs and multi-modal, real world perception as well as multiple actuators. Sensor information, including uncertainty, is filtered and fused into belief states. Abstract POMDP decisions are executed by processing sequential task programs to execute more complex and deterministic sub-tasks.

The presented approach is evaluated on a physical, autonomous, anthropomorphic service robot within a realistic waiter cup-serving scenario.

Supported by European Community's projects DexMART, COGNIRON
Institute of Computer Science and Engineering (CSE), University of Karlsruhe, Germany {srsr|knoop|loesch|dillmann} @ira.uka.de

II. State of the Art

Research has approached the challenge of building powerful reasoning systems for real world environments from two directions: construction of reasoning and supervision systems for robots in environments with assumed simplified properties like *fully observable*, *deterministic* or *discrete* on the one hand and the development of probabilistic decision theory coupled with algorithms for decision retrieval on the other hand.

A hierarchical approach for the design of reasoning and control systems for robots has proven to help coping with the complexity of task environments [1]. Three layer architectures are a very popular design [2]. The first layer performs low level processing of sensory data and reactive controlling of actuators. The second layer usually supervises an ongoing task on a symbolic level while the third layer handles the deliberative selection of abstract tasks. The supervisor and deliberative layer of most systems use classical planning which is not aimed at dealing with uncertainty.

Uncertainty in observation and environment dynamics can be handled by using probabilistic techniques. Probabilistic decision theory deals with reasoning of rational agents in the presence of uncertainty. A very promising framework within general probabilistic decision theory are partially observable Markov decision processes (POMDPs), especially the class of discrete, model based POMDPs. A POMDP is an abstract environment model for reasoning under uncertainty [3], [4]. A POMDP models a flow of events in discrete states and discrete time. A specific POMDP model is represented by the 8-tupel $(S, A, M, T, R, O, \gamma, b_0)$. S is a finite set of states, A is a discrete set of actions and M is a discrete set of measurements. The transition model $T(s', a, s)$ describes the probability of a transition from state s to s' when the agent has performed action a. The observation model $O(m, s)$ describes the probability of a measurement m when the intrinsic state is s. The reward model $R(s, a)$ defines the numeric reward given to the agent when being in state s and executing action a. The parameter γ controls the time discount factor for possible future events. The initial belief state is marked by b_0. As POMDPs handle partially observable environments, there exists only an indirect representation of the intrinsic state of the world. In POMDPs, the belief state, a discrete probability distribution over all states in a scenario model, forms this representation. At each time step, the belief state is updated by Bayesian forward-filtering.

A decision about which action is most favorable for the agent when executed next, can be retrieved from a policy which contains information about the most favorable action for any possible belief distribution. The policy incorporates balancing the probabilities of the course of events into the future with the accumulated reward which has to be maximized.

Computing a policy is computationally challenging and computing exact, optimal policies is intractable [5]. Approximate solutions as *point based value iteration* (PBVI) [6], discrete PERSEUS [7] or HSVI2 [8], however, are quite fast and yield good results for most mid-size scenarios.

For scenarios which can be modelled as *fully observable*, *Markov Decision Processes* (MDPs) can be used for decision making. In MDPs, the decision is derived from a policy function based on the known true world state.

POMDP decision making has already been applied to several different modalities in robotics, like autonomous navigation [9], dialog management [10] and grasping [11], however only for low level controlling of one modality at a time.

These recent investigations encourage to integrate abstract POMDP decision making into reasoning and supervision architectures for autonomous, multi-modal robots now and especially bridging the gap of abstraction between the physical world and discrete decision models. This paper presents such an approach and an evaluation in the following.

III. Approach: *filter*POMDP

When using real world sensory data, there are two possibilities to perform the belief update during scenario runtime when using POMDP decision making. The first is the classical approach, where a distinct observation is perceived and is then processed by using the observation and transition models of a POMDP in a Bayesian update step. The same models are used for calculating the policy. These models must be known a priori and are formulated as classical linear POMDP models. As a drawback, one cannot benefit from any dedicated models of uncertainty which might exist in the lower level algorithms.

The second approach uses dedicated Bayesian filtering methods for each sensor complex, which are then fused into a single belief state. The classic POMDP models are only used for calculating the policy with an approximation of the specific filtering methods. This *filter*POMDP approach has the advantage that the uncertainty, as determined by specific methods, is much more precise concerning the current situation than it could be delivered by a static linear observation model.

Therefore, the question about which approach to take depends on the tradeoff between uncertainty precision concerning the belief and policy precision concerning the process on which the belief is calculated. In mono-modal settings, e.g. mobile robot navigation settings as often used in conjunction with POMDPs, the classical approach is superior, because a general observation and transition model can be set up which models the behavior of both the self-localization process and the general POMDP well. In multi-modal settings, e.g. a service robot with mobility, manipulation, spoken dialogue and visual human activity recognition, the specific uncertainties are better derived from each sensor complex. In this case, the static observation model of the POMDP which is used to calculate the actual policy is in any case a simplification of the process taking place in the specific filters, while the transition model in the POMDP may be able to represent the predictions taking place in the filters.

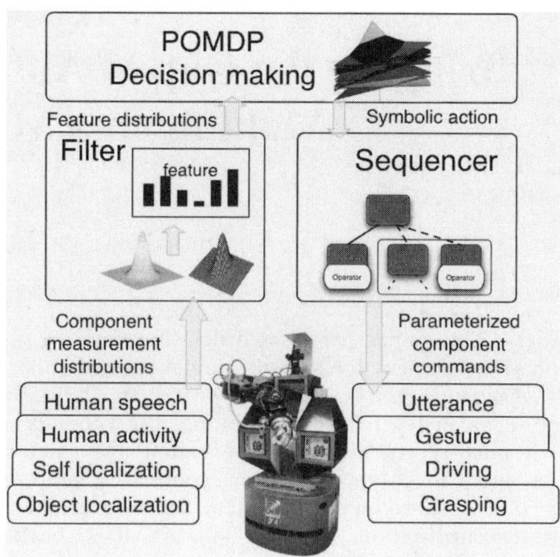

Fig. 1. The system architecture showing the three layer architecture. Low level component control for perception and actuation is at the bottom, the feature filter is on the left, the deliberative layer at the top and the sequencer at the right.

Discrete POMDP models, including all computationally tractable POMDP policy calculation algorithms, for real world scenarios are always approximations of the real dynamics. Thus, in real world scenarios, the policies are an approximation of an ideal policy in both approaches. In practice, in a multi-modal scenario, the approximate POMDP models and policy generated from it will be quite similar, while the belief reflects the true sensor uncertainties better in the second case.

This paper presents the *filter*POMDP approach and compares it in real settings with classic POMDP and MDP reasoning as well as a FSM controller on a multi-modal service robot system.

IV. System Architecture

Fig. 1 shows the general architecture of the presented approach. It is designed along the classical structure of a three-layer architecture, with low-level control on the lowest, measurement filtering and execution supervision on the middle and decision making on the top level.

A. Control layer

Three different capability domains exist on the control level: mobility, human-robot interaction and manipulation. Low level modules control the corresponding hardware directly, closely managing low level commands while processing sensor readings and delivering measurements, including uncertainty, to the layer above. The available capabilities

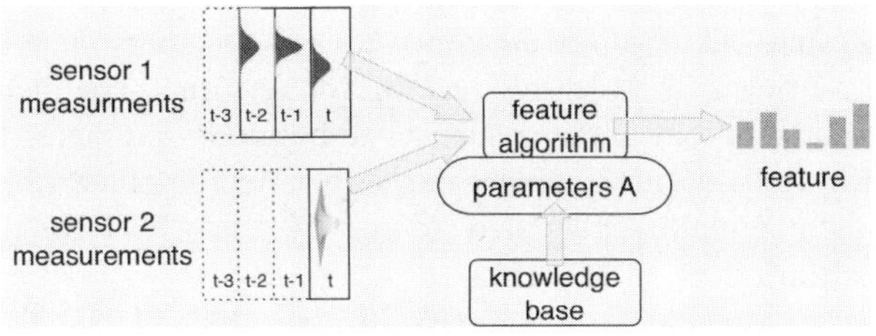

Fig. 2. Schematic computation of a feature.

in the mobility domain are driving and self-localization. Human-robot interaction is possible by speech output, speech recognition, robot arm/hand gestures and human body tracking with symbolic activity recognition by a state-of-the-art procedure [12], [13]. Manipulation capabilities include arm movement, hand grasping as well as force-torque measurements. All measurements delivered by low level modules are probability distributions - either continuous, parametric (e.g. self localization) or discrete, non-parametric (e.g. speech recognition). Actuator commands use a wide range of parameter types - e. g. symbolic utterances or numeric positions.

B. Filter

The filtering module handles processing of measurement data from low level modules. It is basically a modification of the Bayesian forward filter of the POMDP belief state. This forward filter has been sourced out from the main POMDP reasoning, split into a variable number of individual filters, each of which uses an algorithm specialized on a certain observation element and performs a discretization of the continuous data of the logical sensor at the same time. For some components, measurements already include Bayesian updates on the lowest layer, as e.g. for self localization.

Fig. 2 shows the filtering process of an exemplary feature. An individual feature filter takes the data of one or several low level modules, which can also include past observations and applies its specific algorithm, parameterized by the knowledge base. The result is a discrete probability distribution over a set of symbolic categories having their origin in the feature parameter set.

With a new observation, all features have to be updated and after the updates, there exists a new *feature state* which is a set of m discrete probability distributions p, defined over n_j sets of categories $c_{i,j}$ with $i \leq m, j \leq n_j$.

$$\textit{feature state} = \begin{matrix} p(c_{1,1}) & \dots & p(c_{1,n_1}) \\ \dots & \dots & \dots \\ p(c_{m,1}) & \dots & p(c_{m,n_m}) \end{matrix} \quad (1)$$

In the state model, each state in a POMDP scenario model is defined by at least one category from each feature. This mapping connects the abstract, symbolic state through the numerical descriptions of the feature categories to properties of the real world. Given both models, the space of all state models loaded from the knowledge base and the feature state created by feature extraction, the belief state can easily be calculated by calculating the probability b of each state s_k:

$$b(s_k) = \prod_{i=1}^{m} \sum_{j, c_{i,j} \in s_k} p(c_{i,j}) \quad (2)$$

This can also be seen as *fusing* all feature distributions into a single belief state by the means of state descriptions. The set $b(s_1)...b(s_k)$ is the probability distribution of the new belief state.

In the following some specific feature filters are presented.

1) Robot mobility: The low level self-localization module delivers a trivariate gaussian describing position and uncertainty of the robot. The *region occupation* feature filter computes the cumulative distribution function (CDF) of the robot's position distribution for a set of certain, predefined regions. As the orientation is neglected, the CDF over each region is computed over the bivariate by a state-of-the art numerical algorithm [14]. It currently works with rectangles, and larger, complex regions describing a single category can be constructed out of many, smaller cells:

$$p(r_{ij}) = \int_{x_{rp1}}^{x_{rp2}} \int_{y_{rp1}}^{y_{rp2}} N(x, y; \vec{\mu_{pos}}, \Sigma_{pos})\, dx\, dy \quad (3)$$

$$p(c_j) = \sum_{r_i \in c_j} p(r_i) \quad (4)$$

This approach is much more flexible than the usual primitive, one square per state, grid-based one. The resulting discrete probability distribution $p(c_1), ..., p(c_n)$ describes the probability of the robot being in the corresponding region. The regions are defined in the knowledge base.

2) Dialog and human activity filter: POMDPs are also very suitable for spoken dialog [10]. However, because no decision takes place at the filter level, the dialog filter uses a hidden Markov model (HMM), although enhanced with properties found only in MDPs. The dialog HMM model is constructed in the following way:

S is a set of abstract states in which the dialog can be, which reflects human intention. U is a set of possible utterances of the robot, and M is a set of possible human utterances, the robot can detect. $T(s', u, s)$ represents the transition model for each robot utterance, while $O(m, s)$ is the observation model mapping human utterances to states.

Because not any dialog step might consist of alternating human and robot utterances, *idle* utterances fill gaps. In perception, idle utterances are recognized as longer periods of no utterance by either human or robot. The transition model takes into account actions (utterances) of the robot which is not a characteristic in standard HMMs, but in MDPs. A dialog filter probability distribution is forward filtered by the following equation:

$$f'(s') = \alpha(\sum_m P(m)\, O(m, s'))\, (\sum_s T(s', u, s)\, f(s)) \quad (5)$$

With $P(m)$ being the probability for a specific utterance concerning the last detection as delivered by the speech

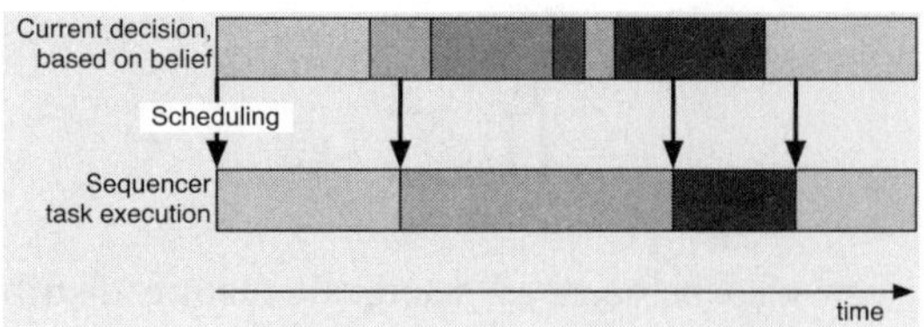

Fig. 3. Connection between reasoner and sequencer. The reasoner continually generates a decision, which is handed to the task sequencer whenever the previous task is finished.

recognition module which delivers discrete probability distributions over recognized utterances. The models for certain dialog scenarios are defined in the knowledge base.

The human activity filter works in the same way. The human activity recognition module delivers discrete probability distributions over a set of perceived, symbolic activities M. U is a set of specific actions of the robot in this case. The states S are a set of true activities.

C. Sequencer

The sequencer receives and processes the commands to execute symbolic programs which represent the actions selected by the reasoner. These symbolic programs represent basic actions for the reasoner. On the sequencing level, they are expanded into complex robot tasks, which are then triggered from the sequencer and executed within the control layer. The task is described as a hierarchical network of basic actions which is processed with a depth-first left-to-right search strategy. A detailed description of the task description called *Flexible Programs* can be found in [15].

The task set in the task database for the presented experiments comprises the tasks *DriveToPos*, *GraspObject*, *SpeakText*, *MonitorHumanActivity* and *PlaceObject*. By decoupling the atomic sensor and actuator controlling from abstract reasoning, it is possible to reduce the decision state space to computationally reasonable dimensionality. The reasoning system decides on the global task to be carried out, while the sequencer performs the actual subtasks that reach the associated goal.

The connection between reasoner and sequencer is depicted in fig. 3. The reasoning system continually generates a decision for a most promising action, based on the current belief state. This decision is passed to the sequencer for execution each time the previous task in the sequencer is finished. Execution of an action may take a non-fixed time span (e.g. *DriveToPos*), while a sensory filter update is quite fast (self-localization, speech recognition etc.). The feature filter runs at a frequency of 20 Hz and updates the belief many times during execution of an action. However, as each action is performed completely and the new action is chosen based on the latest belief, the discrete-step (PO)MDP principle is not violated, while the latest belief reflects the world more precisely.

D. Reasoner

The deliberative layer utilizes established algorithms for discrete decision making: it takes a concrete belief state or a state distribution as input, uses it to query a policy and retrieves a favorable action to perform. The action command is sent to the sequencer to be processed there. The policy for a scenario model can either be computed offline or incrementally calculated online by the PBVI anytime algorithm.

For the results and comparison presented in this paper, MDP and POMDP decision making algorithms are used on the reasoning level.

V. Model design

A fundamental and mostly unsolved question is how to obtain the observation and transition models for real-world scenarios. For model generation, we propose a rule-based approach.

The stochastic behavior of a single feature can often be described by parametric rules (e.g. taking the distance into account when modeling the uncertainty in driving around). Based on the definition of the states, actions and features of the scenario, we define parametric rules as Rule $= \{M, S, S', F, Op, P\}$ with M corresponding matrix, S corresponding origin state, S' target state, F sub state space, Op operation name and P the parameters. Wildcards can be used in M, S, S' which applies a rule to a row, column, matrix or the whole transition model.

The application of such rules shall be addressed by using an example of those actually used in our system. The mobile platform with topological navigation on a graph is more likely not to reach the goal when driving long sections and over many nodes. It is more likely to get stuck in the origin or close to the goal than in between, however most of the time it reaches a goal successfully. Applied to a POMDP transition model T this means that for all actions including a *Goto* command a_g, transition probabilities between states which represent different locations have to include the aforementioned characteristics. It can be achieved by two functions, one realizes getting stuck in the origin: $b(i,j,loc) = \begin{cases} p(\text{stuck}), & \text{if } s_i = s'_j \\ 0, & \text{otherwise} \end{cases}$ and one calculates the likelihood to end up at the goal $p(g)$ and less likely, depending on distance d, somewhere close before, with scaling $\phi()$:

$$b(i,j,loc) = \begin{cases} \max(0, p(g) - \phi(d)), & \text{if } s'_j \text{ on shortest} \\ & \text{path to goal from } s_i \\ 0, & \text{otherwise} \end{cases}$$

A few dozen rules can describe a quite complex scenario with non-uniform transition models containing several thousand probability entries.

VI. Experiments and results

A. Scenario design

The system was evaluated on an anthropomorphic robot acting in a typical service scenario where the mission is to fetch a cup for persons expressing their interest and bring it to a location they choose. The robot can move to several locations or wait for humans interested in interaction. In case a person interacts, the robot can find out if it shall fetch a cup or not. When finally holding the cup, it can interact with the person to find out where to bring the cup. The scenario design is open, not strictly sequential, thus the interaction can stop at any moment, as the human might leave. Therefore, the human behavior is modeled as stochastic - the intention of a person is not deterministically predictable. The duration of a scenario is not fixed, as the robot may wait for humans and serve cups indefinitely.

The scenario is realized as a POMDP model with specific states and actions, as well as dynamics of uncertainty. Complex, deterministic tasks are modeled as single actions (Flexible Programs) within the POMDP decision process, taking advantage of the sequencer. There are 11 actions, consisting of idleing, driving to different locations, human activity information gathering, utterances of the robot as well as pick and place actions. The state space is composed of the sub-spaces of self-localization, dialog and human activity. By combining redundancy in the state-space, the state-space can be reduced to 28 distinct and relevant states.

The observation model is derived from rules describing the uncertainty of self-localization, the speech recognizer and the human activity recognizer. The transition probabilities model the behavior of each action, e.g. glitches in moving and also stochastic human behavior in the scenario. Finally, the reward model contains the rules to define the mission motives of the robot. Actions (except idle) cost a small penalty while fetching the cup when desired as well as delivering it to the correct location give a reward.

The POMDP policy is generated from this model and used for decision making during the experiments.

B. Setup for experiments with a physical robot

To evaluate the presented *filter*POMDP approach on a physical robot, it had to be controlled exclusively and completely autonomously by the presented system, running on onboard-computers, with the sole input being the sensor domains as presented in sec. IV. For evaluation, the system was also provided with the ability to use the MDP as well as the classical POMDP approach for decision making while using the same models and in the POMDP case the same policy. In these cases, the observation probability distributions of the sensor complexes were discarded and instead the observation with the highest probability assumed as distinct measurement. Additionally in one experiment, the robot was controlled by an enhanced version of Flexible Programs exclusively, which include dynamic branching and recursive tree expansion, being able to model a complete finite state machine (FSM).

Because of the open-ended nature of a scenario, time is the only relevant measure of duration of an experiment when comparing gathered rewards. In case a method takes a lot of reassurance actions to determine a human intention, it will be able to fetch and bring a cup less often. Thus, all four experiments had exactly the same duration and starting condition. Concerning the behavior of interacting persons during experiments, it was assured that the behavior of the human corresponded on average to the probabilities in the transition model. Finally, true requests and actual robot behavior were recorded by a human supervisor.

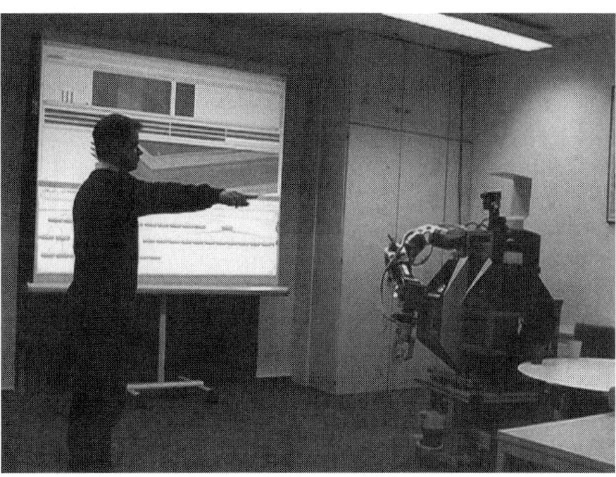

Fig. 4. The robot during experiments, after fetching a cup and awaiting a destination order. The camera-head contains a stereo-color-camera and a 3D-time-of-flight camera for human activity recognition. Speech recognition is performed using an onboard microphone. The platform uses a laser-scanner for self-localization. A live-visualisation of the belief state, a 3D cut of the 28D value function and the current Flexible Program as sent from the robot over wireless can be seen projected at the background.

C. Results

A representative set of four runs with our robot Albert 2 (see fig. 4), each lasting exactly 30 minutes, but controlled by different methods: FSM, MDP, classical POMDP and *filter*POMDP shall be further analyzed to compare the techniques. The following tables show these correlations between the action requested (Req.) and the actually performed behavior (Perf.) for the most important parts of the scenario. Desired behavior is shown on the main diagonal, the first column shows reassurance actions of all kinds, while other entries indicate bad behavior.

1) Finite state machine experiment:

Req. \ Perf.	Reassure	Fetch cup	Put to A	Put to B
Other	0	1	0	1
Fetch cup	3	4	0	0
Put to A	5	0	2	0
Put to B	7	0	0	2

While the state machine does not make many big mistakes, it is conservative and annoys the human with many reassurance questions. Thus, it is not able to perform many delivery actions in the given time, as it has no inherent risk assessment as POMDP methods.

2) MDP experiment:

Req. \ Perf.	Reassure	Fetch cup	Put to A	Put to B
Other	0	0	0	3
Fetch cup	0	8	0	0
Put to A	5	0	2	1
Put to B	2	0	0	2

First, it should be noted that the reassurance actions in this case are performed, when only one of the two human indicators (*say* and *point*) of the intention where to bring the cup, are present. The MDP has problems with contradicting *point* and *say* indicators because as soon as one of the two indicators is measured with highest probability - even if it is only slightly ahead of the other - the MDP will decide for the wrong location. In that case a POMDP still performs information gain actions.

3) Classical POMDP experiment:

Req. \ Perf.	Reassure	Fetch cup	Put to A	Put to B
Other	0	4	0	1
Fetch cup	1	5	0	0
Put to A	6	0	4	1
Put to B	3	0	0	3

In the POMDP, reassurance actions also include actions indicating typical POMDP information gain. Concerning the location to bring the cup to, the POMDP performs information gain until it is quite sure that both indicators (*say* and *point*) refer to the same location, thus it performs better than the MDP. However, it tends to bring the cup even when not requested, because of the reliance on a uniform distribution concerning prediction of human intention and the static observation model when calculating the belief.

4) filterPOMDP experiment:

Req. \ Perf.	Reassure	Fetch cup	Put to A	Put to B
Other	0	0	0	1
Fetch cup	1	9	0	0
Put to A	3	0	5	1
Put to B	1	0	0	2

The *filter*POMDP shows to match both the strong points of the MDP and the classical POMDP. When being requested to fetch the cup, the decision is made, based on the actual speech recognition uncertainty. In the case of deciding where to bring the cup, it also performs information gain, but not as often as the classical POMDP, because there is more precise knowledge about current uncertainty.

5) Comparison:

Type \ Method	FSM	MDP	POMDP	fPOMDP
Correct fetch/put	8	12	12	16
Incorrect fetch/put	2	4	6	2
Reassurance	15	7	10	5

The table shows the performance summary. The FSM is very conservative, as it has no dynamic risk assessment, the MDP has more problems with initial human interaction than the POMDP, thus it wastes time and the total number of fetch/put actions is slightly smaller.

D. Discussion

As shown by the results, exploiting available information about specific uncertainty of current perceptions can be beneficial for decision making by a multi-modal service robot. Static POMDP observation models only contain information about average uncertainty, not about the current one. Using available dedicated methods for determining this uncertainty is especially promising with multi-modal robots where complex perceptive components like speech recognition or human activity recognition exist. However, the POMDP model is still valid for policy generation as it has to use average observation uncertainties. This also holds for the transition model.

VII. Conclusion and Outlook

As shown, using the reasoning capabilities of POMDPs in real robot scenarios is promising as it leads to very robust reasoning in partially observable and stochastic domains. This work has presented a feature filter concept which incorporates sensor uncertainty directly, instead of using only an indirect, static observation model for obtaining belief states. The main remaining challenge is obtaining the transition and observation models for arbitrary scenarios. As there has been only very little research to learn models of real world domains for discrete, model based POMDPs so far [16], this should be pursued intensively now.

References

[1] R. Simmons, R. Goodwin, K. Z. Haigh, S. Koenig, and J. O'Sullivan, "A layered architecture for office delivery robots," in *Proceedings of the First International Conference on Autonomous Agents (Agents'97)*, W. L. Johnson and B. Hayes-Roth, Eds. New York: ACM Press, 5–8, 1997, pp. 245–252.

[2] E. Gat, "On three-layer architectures," *Artificial Intelligence and Mobile Robots. MIT/AAAI Press*, 1997.

[3] E. J. Sondik, "The optimal control of partially observable markov decision processes," Ph.D. dissertation, Stanford university, 1971.

[4] A. R. Cassandra, L. P. Kaelbling, and M. L. Littman, "Acting optimally in partially observable stochastic domains," in *In Proceedings of the Twelfth National Conference on Artificial Intelligence*, 1994.

[5] L. P. Kaelbling, M. L. Littman, and A. R. Cassandra, "Planning and acting in partially observable stochastic domains," *Artif. Intell.*, vol. 101, no. 1-2, pp. 99–134, 1998.

[6] J. Pineau, G. Gordon, and S. Thrun, "Point-based value iteration: An anytime algorithm for POMDPs," in *International Joint Conference on Artificial Intelligence (IJCAI)*, August 2003, pp. 1025 – 1032.

[7] M. Spaan and N. Vlassis, "Perseus: Randomized point-based value iteration for pomdps," *Journal of Artificial Intelligence Research*, vol. 24, pp. 195–220, 2005.

[8] T. Smith and R. Simmons, "Focused real-time dynamic programming for mdps: Squeezing more out of a heuristic," in *Nat. Conf. on Artificial Intelligence (AAAI)*, 2006.

[9] A. Foka and P. Trahanias, "Real-time hierarchical pomdps for autonomous robot navigation," *Robot. Auton. Syst.*, vol. 55, no. 7, pp. 561–571, 2007.

[10] J. D. Williams, P. Poupart, and S. Young, "Using factored partially observable markov decision processes with continuous observations for dialogue management," Cambridge University Engineering Department Technical Report: CUED/F-INFENG/TR.520, Tech. Rep., March 2005.

[11] K. Hsiao, L. P. Kaelbling, and T. Lozano-Pérez, "Grasping pomdps," in *ICRA*, 2007, pp. 4685–4692.

[12] S. Knoop, S. Vacek, and R. Dillmann, "Sensor fusion for 3d human body tracking with an articulated 3d body model," in *Proceedings of the 2006 IEEE International Conference on Robotics and Automation (ICRA)*, Orlando, Florida, 2006.

[13] M. Lösch, S. Schmidt-Rohr, S. Knoop, S. Vacek, and R. Dillmann, "Feature set selection and optimal classifier for human activity recognition," in *Robot and Human Interactive Communication 2007 (ROMAN 2007)*, 2007.

[14] A. Genz, "Numerical computation of rectangular bivariate and trivariate normal and t probabilities," *Statistics and Computing*, vol. 14, pp. 151–160, 2004.

[15] S. Knoop, S. R. Schmidt-Rohr, and R. Dillmann, "A Flexible Task Knowledge Representation for Service Robots," in *The 9th International Conference on Intelligent Autonomous Systems (IAS-9)*, 2006.

[16] R. Jaulmes, J. Pineau, and D. Precup, "Active learning in partially observable markov decision processes," in *ECML*, 2005.

Adaptive Body Scheme Models for Robust Robotic Manipulation

Jürgen Sturm Christian Plagemann Wolfram Burgard
Albert-Ludwigs-University Freiburg, Department for Computer Science, 79110 Freiburg, Germany
{sturm, plagem, burgard}@informatik.uni-freiburg.de

Abstract— **Truly autonomous systems require the ability to monitor and adapt their internal body scheme throughout their entire lifetime. In this paper, we present an approach allowing a robot to learn from scratch and maintain a generative model of its own physical body through self-observation with a single monocular camera. We represent the robot's internal model as a compact Bayesian network, consisting of local models that describe the physical relationships between neighboring body parts. We introduce a flexible Bayesian framework that allows to simultaneously select the maximum-likely network structure and to learn the underlying conditional density functions. Changes in the robot's physiology can be detected by identifying mismatches between model predictions and the self-perception. To quickly adapt the model to changed situations, we developed an efficient search heuristic that starts from the structure of the best explaining memorized network and then replaces local components where necessary. In experiments carried out with a real robot equipped with a 6-DOF manipulator as well as in simulation, we show that our system can quickly adapt to changes of the body physiology in full 3D space, in particular with limited visibility, noisy and partially missing observations, and without the need for proprioception.**

I. Introduction

Autonomous robots deployed in real world environments have to deal with situations in which components change their behavior or properties over time. Such changes can for example come from deformations of robot parts or material fatigue. Additionally, to make proper use of tools, a robot should be able to incorporate the tool into its own body scheme and to adapt the gained knowledge in situations in which the tool is grabbed differently. Finally, components of the robot might get exchanged or replaced by newer parts that no longer comply with the models engineered originally.

Kinematic models are widely used in practice, especially in the context of robotic manipulation [1, 2]. These models are generally derived analytically by an engineer [3] and usually rely heavily on prior knowledge about the robots' geometry and kinematic parameters. As robotic systems become more complex and versatile or are even delivered in a completely reconfigurable way, there is a growing demand for techniques allowing a robot to automatically learn body schemes with no or only minimal human intervention.

Clearly, such a capability would not only facilitate the deployment and calibration of new robotic systems but also allow for autonomous re-adaptation when the body scheme changes, e.g., through regular wear-and-tear over time. Furthermore, the

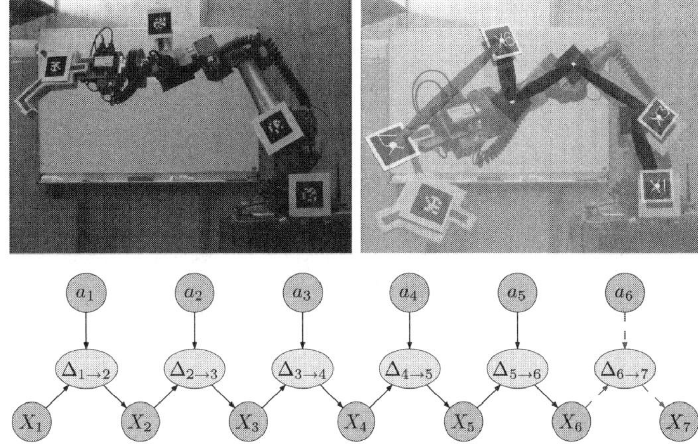

Fig. 1. **Upper left:** Our 6-DOF robotic manipulator arm learns and monitors its own body-scheme using an external monocular camera and visual markers. **Upper right:** After a different tool is placed in the robot's end-effector, the model predictions do not fit the current observations anymore. **Bottom:** The current body scheme linking action signals a_i and body parts X_j using local models $\Delta_{j\to k}$. Here, a mismatch between the internal model and recent self-observation has been detected at $\Delta_{6\to7}$.

ability to learn a body scheme is important in the context of tool use scenarios in which a robot has to identify the effects of its actions on the tool.

In this paper, we investigate how to equip autonomous robots with the ability to learn and adapt their own body schemes and kinematic models using exploratory actions and self-perception only. We propose an approach to learn a Bayesian network for the robot's kinematic structure including the forward and inverse models relating action commands and body pose. More precisely, we start with a fully connected network containing all perceivable body parts and available action signals, perform random "motor babbling," and iteratively reduce the network complexity by analyzing the perceived body motion. At the same time, we learn non-parametric regression models for all dependencies in the network, which can later be used to predict the body pose when no perception is available or to allow for gradient-based posture control.

One of the major advantages of the approach presented in this paper is that it addresses all of the following practical problems that frequently arise in robotic manipulation tasks in a single framework:

- **Prediction:** If both the structure and the CDFs of the

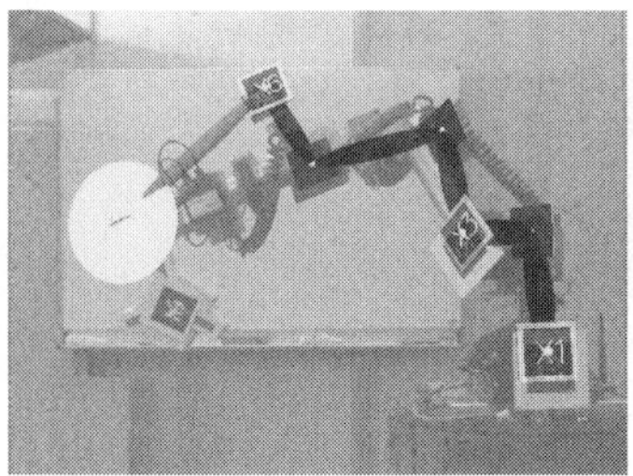

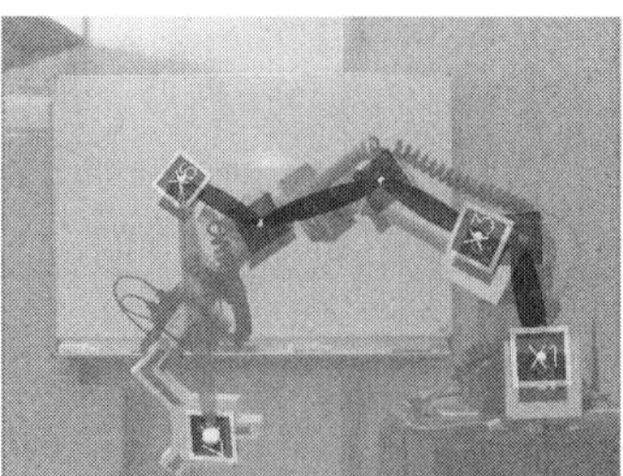

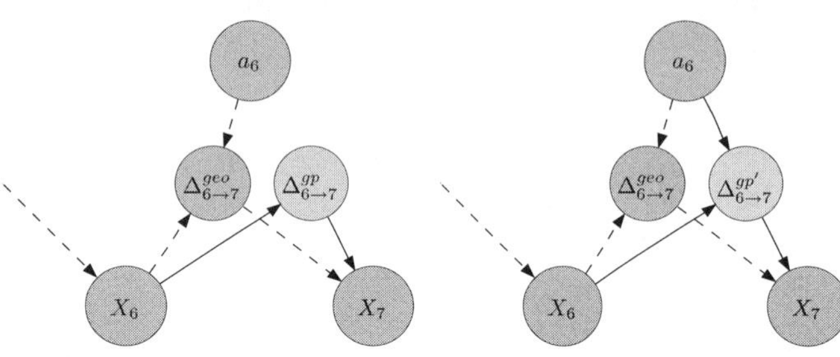

Fig. 2. Continued experiment from Figure 1. The robot samples a local model as replacement for the mismatching component $\Delta_{6\to7}$. **Left:** The first newly sampled model ($\Delta^{gp}_{6\to7}$) has high uncertainty, because of the missing dependency on action a_6. **Right:** The second sampled model ($\Delta^{gp'}_{6\to7}$) is a more suitable replacement for the mismatching component.

Bayesian network are known, the robot is able to predict for a given action command the expected resulting body configuration.

- **Control:** Conversely, given a target body pose, our approach is able to generate appropriate action commands that will lead to this pose.
- **Model testing:** Given both a prediction and an observation of the current body pose, the robot is able to estimate the accuracy of its own pose predictions. Model accuracy can, for example, be defined in terms of a distance metric or a likelihood function.
- **Learning:** Given a sequence of action signals and the corresponding body postures, the Bayesian network and its parameters can be learned from the data.
- **Discovering the network structure:** When the structure of the Bayesian network is unknown, the robot is able to build it from the available local models which are most consistent with the observed data.
- **Failure detection and model adaptation:** When the robot's physiology changes, e.g., when a joint gets blocked or is deformed, or a visual marker is changed, this is efficiently detected so that only the affected local models of the Bayesian network need to be replaced.

II. Related Work

The problem of learning kinematics of robots has been investigated heavily in the past. For example, Kolter and Ng [4] enable a quadruped robot to learn how to follow omnidirectional paths using dimensionality reduction techniques and based on simulations. Their key idea is to use the simulator for identifying a suitable subspace for policies and then to learn with the real robot only in this low-dimensional space. A similar direction has been explored by Dearden *et al.* [5], who applied dimensionality reduction techniques to unveil the underlying structure of the body scheme. Similar to this work, their approach is formulated as a model selection problem between different Bayesian networks. Another instance of approaches based on dimensionality reduction is the work by Grimes *et al.* [6] who applied the principal component analysis (PCA) in conjunction with Gaussian process regression for learning walking gaits on a humanoid robot.

In previous work [7], we have presented an approach to deal with the problem of learning a probabilistic self-model for a robotic manipulator. This approach, however, neither covered aspects of failure detection and life-long model revision nor did it address partial observability of model components. In this work, we give a more rigorous formulation of the the body-scheme learning framework, we significantly extend the model toward life-long adaptation and self monitoring, and we give experimental results in complex and realistic scenarios.

Yoshikawa *et al.* [8] used Hebbian networks to discover the body scheme from self-occlusion or self-touching sensations. Later, [9] learned classifiers for body/non-body discrimination from visual data. Other approaches used for example nearest-neighbor interpolation [10] or neural networks [11]. Recently, Ting *et al.* [12] developed a Bayesian parameter identification method for nonlinear dynamic systems, such as a robotic arm or a 7-DOF robotic head.

The approach presented in this paper is also related to the problem of self-calibration which can be understood as a subproblem of body scheme learning. When the kinematic model is known up to some parameters, they can in certain cases be efficiently estimated by maximizing the likelihood of the model given the data [13]. Genetic algorithms have been used by Bongard *et al.* [14] for parameter optimization when no closed form is available. To a certain extend, such methods can also be used to calibrate a robot that is temporarily using a tool [15]. In contrast to the work presented here, such approaches require a parameterized kinematic model of the robot.

To achieve continuous self-modeling, Bongard *et al.* [16] recently described a robotic system that continuously learns its own structure from actuation-sensation relationships. In three alternating phases (modeling, testing, prediction), their system generates new structure hypotheses using stochastic optimization, which are validated by generating actions and by analyzing the following sensory input. In a more general context, Bongard *et al.* [17] studied structure learning in arbitrary non-linear systems using similar mechanisms.

In contrast to all the approaches described above, we propose an algorithm that both learns the structure as well as functional mappings for the individual building blocks. Furthermore, our model is able to revise its structure and component models on-the-fly.

III. A Bayesian Framework for Robotic Body Schemes

A robotic body scheme describes the relationship between available action signals $\langle a_1, \ldots, a_m \rangle$, self-observations $\langle Y_1, \ldots, Y_n \rangle$, and the configurations of the robot's body parts $\langle X_1, \ldots, X_n \rangle$. In our concrete scenario, in which we consider the body scheme of a robotic manipulator arm in conjunction with a stationary, monocular camera, the action signals $a_i \in \mathbb{R}$ are real-valued variables corresponding to the joint angles. Whereas the $X_i \in \mathbb{R}^6$ encode the 6-dimensional poses (3D Cartesian position and 3D Euler angles) of the body parts w.r.t. a reference coordinate frame, the $Y_i \in \mathbb{R}^6$ are generally noisy and potentially missing observations of the body parts. Throughout this paper, we use capital letters to denote 6D pose variables to highlight that these also uniquely define homogeneous transformation matrices, which can be concatenated and inverted. Note that we do not assume direct feedback/proprioception telling the robot how well joint i has approached the requested target angle a_i.

Formally, we seek to learn the probability distribution

$$p(X_1, \ldots, X_n, Y_1, \ldots, Y_n \mid a_1, \ldots, a_m), \qquad (1)$$

which in this form is intractable for all but the simplest scenarios. To simplify the problem, it is typically assumed that each observation variable Y_i is independent from all other variables given the true configuration X_i of the corresponding body part and that they can thus be fully characterized by an observation model $p(Y_i \mid X_i)$. Furthermore, if the kinematic structure of the robot was known, a large number of pair-wise independencies between body parts and action signals could be assumed, which in turn would lead to the much simpler, factorized model

$$p(X_1, \ldots, X_n \mid a_1, \ldots, a_m) = \qquad (2)$$
$$\prod_i p(X_i \mid parents(X_i)) \cdot p(parents(X_i) \mid a_1, \ldots, a_m).$$

Here, $parents(X_i)$ denotes the set of locations of body parts, which are directly connected to body part i.

The main idea behind this work is to make the factorized structure of the problem explicit by introducing (hidden) transformation variables $\Delta_{i \to j} := X_i^{-1} X_j$ for all pairs of body parts (X_i, X_j) as well as their observed counterparts $Z_{i \to j} := Y_i^{-1} Y_j$. Here, we use the 6D pose vectors X and Y as their equivalent homogeneous transformation matrices, which means that $\Delta_{i \to j}$ reflects the (deterministic) relative transformation between body parts X_i and X_j. Figure 3 depicts a *local model*, which fully defines the relationship between any two body parts X_i and X_j and their dependent variables, if all other body parts are ignored.

Since local models are easily invertible ($\Delta_{i \to j}$ are homogeneous transformations), any set of $n-1$ local models which form a spanning tree over all n body parts defines a model for the whole kinematic structure.

In the following, we explain (1) how to continuously learn local models from data and (2) how to find the best spanning tree built from these local models that explains the whole robot. In this work, we consider the single best solution only and do not perform model averaging over possible alternative structures.

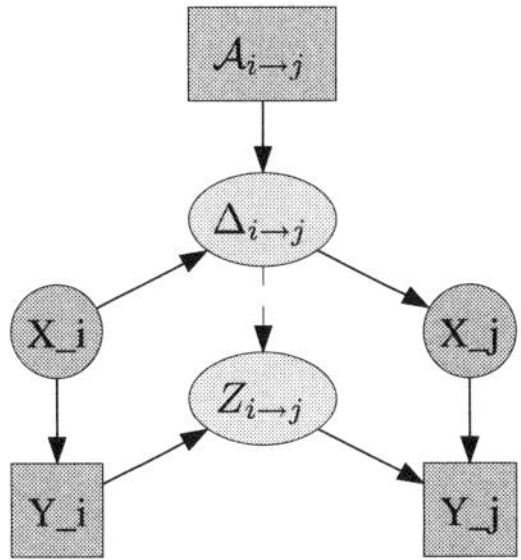

Fig. 3. Graphical model for two body parts X_i and X_j as well as their dependent variables. $\mathcal{A}$ denotes the set of independent action variables that cause a local transformation $\Delta_{i \to j}$. Y_i and Y_j are the observed part locations, and $Z_{i \to j}$ is their relative geometric transformation.

Please note that in theory, it would be straight-forward to keep multiple structure hypotheses and to average over them using Bayes' rule. Control under structure uncertainty is a slightly more difficult problem. One would have to average over all possible structures and assess the individual risks and gains for possible actions. Then, the one action sequence should be selected that maximizes the overall gain while keeping all possible risks low [18].

In practice, we found that considering the most-likely structure only is sufficient for most relevant tasks. Our approach is conservative in this respect since it requires a certain minimal model accuracy from all parts of the body scheme.

A. Local Models

The local kinematic models are the central concept in our body scheme framework. A *local model* $\mathcal{M}$ describes the geometric relationship $p_{\mathcal{M}}(Z_{i \to j} \mid \mathcal{A}_{i \to j})$ between two observed body parts Y_i and Y_j, given a subset of the action signal $\mathcal{A}_{i \to j} \subset \{a_1, \ldots, a_n\}$.

The probability distribution underlying a local model can be defined in various ways. If an analytic model of the robot exists from its specifications, it can be used directly to construct $p_{\mathcal{M}}(Z_{i \to j} \mid \mathcal{A}_{i \to j})$. The standard way to describe a geometric model for robot manipulators is in terms of the Denavit-Hartenberg parameters [1, 19]. When available, the advantages of these models are outstanding: they are exact and efficient in evaluation. In practice, however, such models need to be calibrated carefully and often require re-calibration after periods of use.

B. Learning Local Models from Noisy Observations

On the real robotic platform used in our experiments, the actions a_i correspond to the target angle requested from joint i and the observations Y_i are obtained by tracking visual markers in 3D space including their 3D orientation [20] (see the top right image of Figure 1). Note that the Y_i's are inherently noisy and that missing observations are common, for example in the case of (self-)occlusion.

The probability distribution $p_{\mathcal{M}}(Z_{i \to j} \mid \mathcal{A}_{i \to j})$ of a local model $\mathcal{M}$ can be learned from a sequence of observations $\mathcal{D} = \{(Z_{i \to j}, \mathcal{A}_{i \to j})\}_{1:t}$. If we assume Gaussian white noise with zero mean on the observations, the sensor model becomes $Y_i \sim X_i + \mathcal{N}(0, \sigma_{sensor})$. Note that we can connect the two body parts X_i and X_j in Figure 3 either by learning $p_{\mathcal{M}}(\Delta_{i \to j} \mid \mathcal{A}_{i \to j})$ or $p_{\mathcal{M}}(Z_{i \to j} \mid \mathcal{A}_{i \to j})$. The link $p(\Delta_{i \to j} \mid \mathcal{A}_{i \to j}) = p(X_i^{-1} X_j \mid \mathcal{A}_{i \to j})$ is noise-free. It, however, requires inference starting from Y_i and Y_j through both observation models via the indirect Bayesian pathway $Y_i \leftarrow X_i \rightarrow \Delta_{i \to j} \rightarrow X_j \rightarrow Y_j$. Thus, we propose to learn the model for $p_{\mathcal{M}}(Z_{i \to j} \mid \mathcal{A}_{i \to j}) = p_{\mathcal{M}}(Y_i^{-1} Y_j \mid \mathcal{A}_{i \to j})$ directly. As the noise distribution $p_{\mathcal{M}}(Z_{i \to j} \mid \Delta_{i \to j})$ is determined by integrating Gaussian random variables along $X_i \rightarrow Y_i \rightarrow Z_{i \to j} \rightarrow Y_j \rightarrow X_j$ it can nicely be approximated by a Gaussian [21].

The problem of learning the probability distribution now comes down to learning the function $f_{\mathcal{M}} : \mathbb{R}^{|\mathcal{A}_{i \to j}|} \rightarrow \mathbb{R}^6$, $\mathcal{A}_{i \to j} \mapsto Z_{i \to j}$, from the training data. A flexible model for solving such non-linear regression problems given noisy observations is the popular Gaussian process (GP) approach. The main feature of the Gaussian process framework is, that the observed data points are explicitly included the model and, thus, no parametric form of $f_{\mathcal{M}}$ needs to be specified. Data points can be added to the training set at any time, which facilitates incremental and online learning. Due to space constraints, we refer the interested reader to work by Rasmussen [22] for technical details about GP regression. For simplicity, we assume independence between all 12 free components of $f_{\mathcal{M}}(\mathcal{A}_{i \to j})$ and consider the functional mapping for each component separately. Due to this simplification, we cannot guarantee that the prediction corresponds to a valid, homogeneous transformation matrix. In practice, however, invalid transformations occur only rarely and they lie close to similar, valid transformations, such that a simple normalization step resolves the problem.

C. Learning a Factorized Full Body Model

We seek to find the best factorized model according to Eqn. 3 and, thus, require a suitable optimization criterion. Given a training set $\mathcal{D}$ of independent, time-indexed actions and their corresponding observations, $\mathcal{D} = \{(Y_i^t, Y_j^t, \mathcal{A}_{i \to j}^t)\}_{t=1}^T$, or, equivalently for our purposes, $\{(Z_{i \to j}^t, \mathcal{A}_{i \to j}^t)\}_{t=1}^T$, the data likelihood $p(\mathcal{D} \mid \mathcal{M})$ under a local model $\mathcal{M}$ can directly be computed from its probability distribution $p_{\mathcal{M}}(Z_{i \to j} \mid \mathcal{A}_{i \to j})$ as

$$p(\mathcal{D} \mid \mathcal{M}) = \prod_{k=1}^{t} p_{\mathcal{M}}(Z_{i \to j}^k \mid \mathcal{A}_{i \to j}^k) \; . \tag{3}$$

In practice, this product is highly sensitive to outliers, and makes the comparison of different classes of models difficult. We therefore developed an alternative model quality measure $q(\mathcal{D} \mid \mathcal{M})$ that is proportional to both the prediction accuracy and a penalty term for model complexity:

$$\log q(\mathcal{D} \mid \mathcal{M}) := \log(1/\epsilon_{pred}(\mathcal{D} \mid \mathcal{M})) + C(\mathcal{M}) \log \theta \tag{4}$$

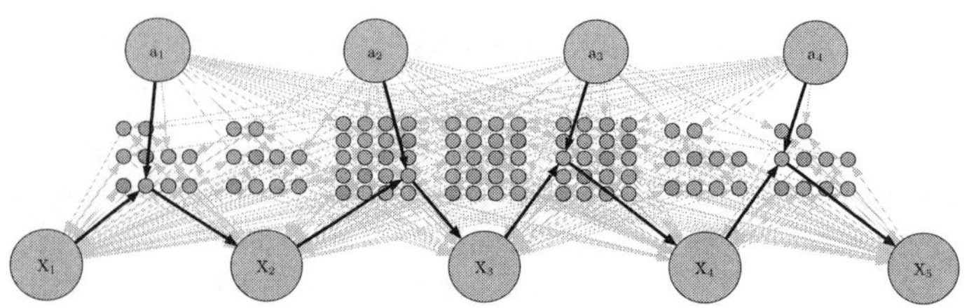

Fig. 4. In an early learning phase, the robot knows only little about its body structure, i.e., all possible local models need to be considered in parallel. From the subset of valid local models, a minimal spanning tree can be constructed which, in turn, forms a Bayesian network. This can subsequently be used as a body scheme for prediction and control.

where $C(\mathcal{M}) \in \mathbb{Z}$ is the complexity of model $\mathcal{M}$ and $\epsilon_{pred}(\mathcal{D} \mid \mathcal{M})$ is the prediction error defined as

$$\epsilon_{pred}(\mathcal{D} \mid \mathcal{M}) := \frac{1}{|\mathcal{D}|} \sum_{(Z_{i \to j}, \mathcal{A}_{i \to j}) \in \mathcal{D}} \epsilon_{pred}(Z_{i \to j} \mid \mathcal{A}_{i \to j}, \mathcal{M}) \tag{5}$$

with

$$\epsilon_{pred}(Z_{i \to j} \mid \mathcal{A}_{i \to j}, \mathcal{M}) := \int_Z \|Z_{i \to j} - Z\| \cdot \tag{6}$$

$$p_{\mathcal{M}}(Z' \mid \mathcal{A}_{i \to j}) \, dZ \; . \tag{7}$$

We define a local model $\mathcal{M}$ to be $valid_{\mathcal{M}}(\mathcal{D})$ given a set of observations, if and only if its observed prediction error is below some threshold θ, i.e., $\epsilon_{pred}(\mathcal{D}) < \theta$. Our experiments revealed that a good value for θ is 3σ, where σ is the standard deviation of the sensor model.

1) Bootstrapping: If no prior knowledge of the robot's body scheme exists, we initialize a fully connected network model (see Figure 4), resulting in a total set of $\sum_{k=0}^{m} \binom{n}{2} \binom{m}{k}$ local models. Given a set of self observations, the robot can determine the validity of the local models by evaluating Eq. 7. Certain ambiguities will, however, remain even after infinitely many training samples: if, for example, $p_{\mathcal{M}_1}(Z_{1 \to 2} \mid a_1)$ has been determined to be a valid local model, then $p_{\mathcal{M}_2}(Z_{1 \to 2} \mid a_1, a_2)$ will also be. Although $\mathcal{M}_1$ and $\mathcal{M}_2$ might not be distinguishable regarding prediction accuracy, these models differ significantly in terms of complexity and therefore in model quality $q(\mathcal{D} \mid \mathcal{M})$.

2) Finding the Network Topology: From the superset of all valid local models $\mathbb{M}_{valid} = \{\mathcal{M}_1, \ldots\}$, we seek to select the minimal subset $\mathbb{M} \subset \mathbb{M}_{valid}$ that covers all body part variables and simultaneously maximizes the overall model fit $q(\mathcal{D} \mid \mathbb{M}) := \prod_{\mathcal{M} \in \mathbb{M}} q(\mathcal{D} \mid \mathcal{M})$. It turns out that $\mathbb{M}$ can be found efficiently by computing the minimal spanning tree of $\mathbb{M}_{valid}$ taking the model quality measure of the individual local models as the cost function. Such a body spanning tree needs to cover all body parts $X_1, \ldots, X_n$ but not necessarily all action components of $a_1, \ldots, a_m$. Note that, in order to connect all n body poses in the Bayesian network, exactly $(n-1)$ local models need to be selected. This yields the astronomical number of $\#_{structures} = \binom{\#_{local\ models}}{n-1}$ possible network structures to be considered. In practice, however, simple search heuristics allow us to strongly focus the search on the relevant parts of the structure space. Recall that the quality measure $q(\mathcal{D} \mid \mathcal{M})$ for a local model is composed of the (data-dependent) prediction accuracy and a (data-independent) complexity penalty. If we consider two valid local models, i.e., with $\epsilon_{pred}(\mathcal{D} \mid \mathcal{M}_{1|2}) < \theta$, then by the definition of $q(\mathcal{D} \mid \mathcal{M})$, the quality of a model with lower complexity is always higher compared to a local model with higher complexity for any $\mathcal{D}$, i.e.,

$$C(\mathcal{M}_1) < C(\mathcal{M}_2) \Longleftrightarrow \forall \mathcal{D} : q(\mathcal{D} \mid \mathcal{M}_1) > q(\mathcal{D} \mid \mathcal{M}_2) \ .$$

Thus, it is sufficient to evaluate only the first k complexity layers of local models in $\mathbb{M}_{valid}$ until a minimal spanning tree is found for the first time. This spanning tree then corresponds to the global maximum of overall model quality.

D. Prediction and Control

The *kinematic forward model* is directly available by noting

$$\begin{aligned} p(Y_1, \ldots, Y_n \mid a_1, \ldots, a_m) \\ &= \prod_i p(Y_i \mid parents(Y_i)) p(parents(Y_i) \mid a_1, \ldots, a_m) \\ &= p(Y_{root}) \prod_{\mathcal{M} \in \mathbb{M}} p_{\mathcal{M}}(Z_{i \to j} \mid \mathcal{A}_{i \to j}) \ , \end{aligned} \tag{8}$$

where Y_{root} is the position of the robot's trunk, which is serving as the coordinate origin of all other body parts. In practice, instead of a probability distribution $p(Y_1, \ldots, Y_n \mid a_1, \ldots, a_m)$, we rather require the maximum likelihood (ML) estimate of the resulting body posture given an action signal. This can be computed efficiently by concatenating the geometric transformations of the individual mapping functions $f_{\mathcal{M}_i}$.

Although the *inverse kinematic model* can in principle be derived by applying the rules of Bayes,

$$\begin{aligned} p(X_1, \ldots, X_n \mid a_1, \ldots, a_m) \\ &= \frac{p(X_1, \ldots, X_n)}{p(a_1, \ldots, a_m)} p(a_1, \ldots, a_m \mid X_1, \ldots, X_n) \\ &\propto p(a_1, \ldots, a_m \mid X_1, \ldots, X_n), \end{aligned} \tag{9}$$

it is in general difficult to determine the maximum likelihood (ML) estimate for the action signal $a_1, \ldots, a_m$ that is supposed to generate a given target body posture $X_1, \ldots, X_n$. Since all individual functions $f_{\mathcal{M}_i}$ are continuous, and so is the ML posture estimate f of the forward kinematic model, we can compute the Jacobian $\nabla f(\mathbf{a})$ of the forward model as

$$\nabla f(\mathbf{a}) \quad = \quad \left[\frac{\partial f(\mathbf{a})}{\partial a_1}, \ldots, \frac{\partial f(\mathbf{a})}{\partial a_m} \right]^T . \tag{10}$$

A gradient descent algorithm can then be used to minimize $f(\mathbf{a})$ and thereby to iteratively control the manipulator to its target position [7].

E. Failure Awareness and Life-Long Model Adaptation

Until now, we have assumed that the robot's physiology remains unchanged during its whole life-time. It is clear, however, that in real-world applications, the robot will change in the course of time. This insight requires that the robot revises parts of its experience over time, allowing it to discriminate between earlier and more recent observations. This enables the robot to detect changes in its physiology by testing the validity of its local models at different points in time and at different temporal scales.

It might even be useful for the robot to maintain multiple body schemes at different time scales. Consider, for example, a robot that uses an accurate pre-programmed model over a long period of time, but simultaneously is able to create and use a short-term model that takes over as soon as the body structure of the robot changes occur (which could be as little as the displacement of one visual marker). From a formal point of view, time is simply another dimension in the model space which can be included in the definition of local models.

A *temporal local model* $\mathcal{M}^T$ describes the geometric relationship $p^T_{\mathcal{M}}(Z_{i \to j} \mid \mathcal{A}_{i \to j}, T)$ between two observed body parts Y_i and Y_j, given a subset of the action signal $\mathcal{A}_{i \to j} \subset \{a_1, \ldots, a_n\}$ and a particular time interval T.

However, the size of the learning problem in the bootstrapping case now grows exponentially in time yielding the immense upper bound of $\sum_{k=0}^{m} \binom{n}{2} \binom{m}{k} 2^{|T|}$ local models to be considered. As it would be practically infeasible to evaluate all of these local models even for small periods of time, three additional assumptions can be made such that an efficient algorithm for real-time application can be devised:

1) Changes in body physiology can be assumed to be relatively rare events.
2) Changes in physiology most probably happen incrementally.
3) Whatever local models were useful in the past, it is likely that similar (or maybe even the same) local models will be useful in the future.

Because of the first assumption it is not necessary to consider new local models as long as the current body scheme still yields a high prediction accuracy. Only when one of the local models of the current body scheme becomes invalid, incremental learning (assumption 2) has to be triggered. Then, according to assumption 3, it is reasonable to begin the search for new models that are similar to previously useful models. To incorporate these assumptions in the quality measure for local models, we first define the concept of relative complexity of a local model $\mathcal{M}_2$ given a previously used model $\mathcal{M}_1$ as

$$C(\mathcal{M}_2 \mid \mathcal{M}_1) := d(\mathcal{M}_2, \mathcal{M}_1),$$

where $d(\cdot, \cdot)$ is a (data-independent) similarity metric between two local models and $C(\mathcal{M}_2 \mid \mathcal{M}_1) \in \mathbb{Z}$. In practice, $d(\cdot, \cdot)$ can for example be defined as the ratio of shared nodes between two local models in the Bayesian network. The refined version of the model quality measure $q_2(\mathcal{D} \mid \mathcal{M}_1, \mathcal{M}_2)$ of some recent observations $\mathcal{D}$ given a newly sampled model $\mathcal{M}_2$ as a replacement for an invalidated previous model $\mathcal{M}_1$ can then be defined as

$$\begin{aligned} \log q_2(\mathcal{D} \mid \mathcal{M}_1, \mathcal{M}_2) &:= \log(1/error_{prediction}(\mathcal{D})) \\ &\quad + C(\mathcal{M}_2 \mid \mathcal{M}_1) \log \theta \\ &\quad + \log |T_{\mathcal{M}_2}| \ . \end{aligned} \tag{11}$$

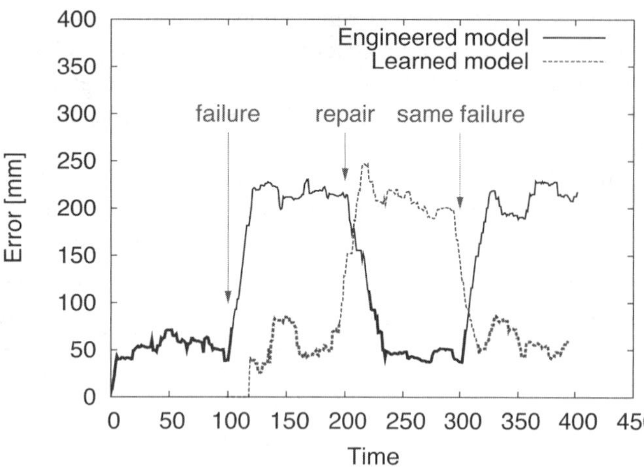

Fig. 5. At $t = 100$, a joint gets blocked, which causes the initial local model $p_{engineered}(Z_{6\to 7} \mid a_4)$ to produce substantially larger prediction errors. At $t = 126$, the robot samples a new local model $p_{learned}(\Delta 6 \to 7)$ as replacement.

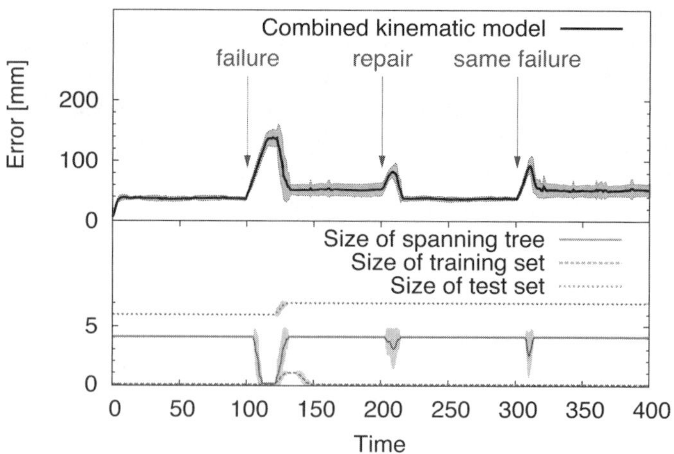

Fig. 6. The absolute prediction error of the combined kinematic model $p(Z_{1\to 7} \mid a_1, \ldots, a_4)$ of our 6-DOF manipulator. This model is composed of 6 individual local models of which one is replaced by a newly learned model at $t = 126$ (cmp. Figure 5). As can be seen from the plot, the prediction accuracy recovers quickly after each of the three external events.

Please note that, by construction, the quality measure of two local models with different relative complexity have no overlapping ranges in model quality independently of the observation data $\mathcal{D}$, i.e.,

$$C(\mathcal{M}_1 \mid \mathcal{M}_3) < C(\mathcal{M}_2 \mid \mathcal{M}_3)) \iff \forall \mathcal{D} : q(\mathcal{D} \mid \mathcal{M}_1) > q_2(\mathcal{D} \mid \mathcal{M}_2) \ . \quad (12)$$

It is, like in the static case, sufficient to sample and evaluate only the first k complexity layers of local models until a minimum spanning tree is found. By definition of the quality function, this minimum spanning tree is then by construction the global maximum of overall model quality.

IV. Experiments

We tested our approach in a series of experiments, both on a real robot and in simulation. The goal of our experiments was to verify that

1) physiological changes are detected confidently (blocked joints / deformations),
2) the body scheme is updated automatically without human intervention, and
3) the resulting body scheme can be used for accurate prediction and control.

The robot used to carry out the experiments is equipped with a 6-DOF manipulator composed of Schunk PowerCube modules. The total length of the manipulator is around 1.20m. With nominal noise values of ($\sigma_{joints} = 0.02°$), the reported joint positions of the encoders were considered to be sufficiently accurate to compute the ground truth positions of the body parts from the known geometrical properties of the robot. Visual perception was obtained by using a Sony DFW-SX900 FireWire-camera at a resolution of 1280x960 pixels. On top of the robot's joints, 7 black-and-white markers were attached (see Figure 1), that were detectable by the ARToolkit vision module [20]. Per image, the system perceives the unfiltered 6D poses of all detected markers. The standard deviation of the camera noise was measured to $\sigma_{markers} = 44$mm in 3D space, which is acceptable considering that the camera was located two meters apart from robot.

We averaged the prediction error over a test set of the latest $|\mathcal{D}_{testing}| = 15$ data samples. New local models were trained with $|\mathcal{D}_{training}| = 30$ succeeding training samples after the model was instantiated. In order for a local model to be valid, its translational and rotational error on the test set needed to be below a threshold of $\theta_{trans} = 3\sigma_{trans} = 150$mm and $\theta_{rot} = 3\sigma_{rot} = 45°$, with σ_{trans} and σ_{rot} as the standard deviation of the translational and rotational observation noise, respectively. New local models were only sampled when no valid spanning tree could be constructed for $|\mathcal{D}_{testing}|$ succeeding time steps, as this is the time it takes to replace most if not all (because of possibly missing observations) data samples of the test set. Note that otherwise it could happen that available local models cannot be selected because the test set temporarily consists of data samples partly observed just before and partly after a change in physiology.

A. Evaluation of Model Accuracy

To quantitatively evaluate the accuracy of the kinematic models learned from scratch as well as the convergence behavior of our learning approach, we generated random action sequences and analyzed the intermediate models using a 2-DOF robot of which the kinematic model is perfectly known.

Figure 7 gives the absolute errors of prediction and control after certain numbers of observations have been processed. For a reference, we also give the average observation noise, i.e. the absolute localization errors of the visual markers.

As can be seen from the diagram, the body scheme converges robustly within the first 10 observations. After about

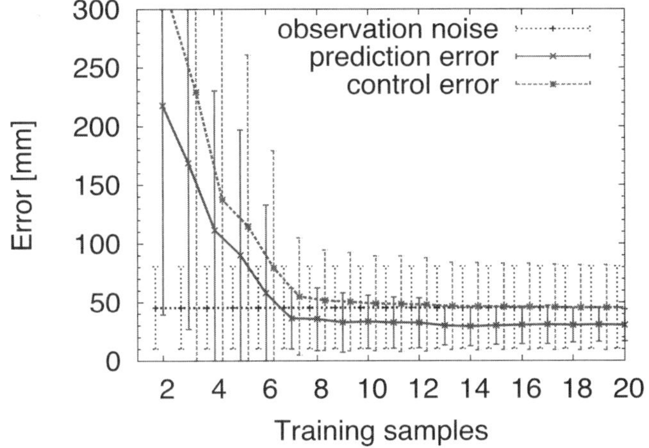

Fig. 7. Prediction and control errors for a kinematic model that is learned from scratch. Already after 7 samples, the average prediction error is lower than the average localization error of the visual markers.

15 training samples, the accuracy of the predicted body part positions even outperformed the accuracy of the direct observations. The latter is a remarkable result as it means that, although all local models are learned from noisy observations, the resulting model is able to blindly predict positions that are more accurate than immediate perception. The figure also gives the accuracy of the gradient-based control algorithm. Here, we used an additional marker for defining a target location for the robot's end effector. We learned the full body scheme model from scratch as in the previous experiment and used the gradient-based control algorithm to bring the end effector to the desired target location. The average positioning error is in the order of the perception noise (approx. 50mm, see Figure 7), i.e. slightly higher than the prediction error alone.

B. Scenario 1: Joint stuck

We generated a large sequence of random motor commands $\langle a_1, \ldots, a_m \rangle$. Before accepting a pose, we checked that the configuration would not cause any (self-)collisions, and that the markers of interest (X_6 and X_7) would potentially be visible on the camera image. This sequence was sent to the robot and after each motion command, the observed marker positions $\langle Y_1, \ldots, Y_n \rangle$ were recorded. In the rare case of a anticipated or a real (self-)collision during execution, the robot stopped and the sample was rejected. Careful analysis of the recorded data revealed that, on average, the individual markers were visible only in 86.8% of the time with the initial body layout. In a second run, we blocked the robot's end-effector joint a_4, such that it could not move, and again recorded a log-file. Note that we allow arbitrary 3D motion (just constrained by the geometry of the manipulator) and thus do not assume full visibility of the markers.

An automated test procedure was then used to evaluate the performance and robustness of our approach. For each of the 20 runs, a new data set was sampled from the recorded log-files, consisting of 4 blocks with $N = 100$ data samples each. The first and the third block were sampled from the initial body shape, while the second and the fourth block were sampled from the log-file where the joint got blocked.

Figure 5 shows the prediction error of the local models predicting the end-effector pose. As expected, the prediction error of the engineered local model increases significantly after the end-effector joint gets blocked at $t = 100$. After a few samples, the robot detects a mismatch in its internal model and starts to learn a new dynamic model (around $t = 130$), which quickly reaches the same accuracy as the original, engineered local model. At $t = 200$, the joint gets repaired (unblocked). Now the estimated error of the newly learned local model quickly increases while the estimated error of the engineered local model decreases rapidly towards its initial accuracy. Later, at $t = 300$, the joint gets blocked again in the same position, the accuracy of the previously learned local model increases significantly, and thus the robot can re-use this local model instead of having to learn a new one.

The results for 20 reruns of this experiment are given in Figure 6. The hand-tuned initial geometrical model evaluates to an averaged error at the end-effector of approx. 37mm. After the joint gets blocked at $t = 100$, the error in prediction increases rapidly. After $t = 115$, a single new local models gets sampled, which already is enough to bring down the overall error of the combined kinematic model to approximately 51mm. Training of the new local model is completed at around $t = 135$.

Later at $t = 200$, when the joint gets un-blocked, the error estimate of the combined kinematic model increases slightly, but returns much faster to its typical accuracy: switching back to an already known local model requires much fewer data samples than learning a new model (see Table I). At $t = 300$, the same quick adaption can be observed when the joint gets blocked again.

C. Scenario 2: Deformed limb

In a second experiment[1], we changed the end-effector limb length and orientation and applied the same evaluation procedure as in the previous subsection. This was accomplished by placing a tool with an attached marker in the robot's gripper at different locations (see Figure 1).

[1]A demonstration video of this experiment can be found on the internet at `http://www.informatik.uni-freiburg.de/~sturm/media/resources/public/zora-7dof-demo.avi`

TABLE I

EVALUATION OF THE RECOVERY TIME REQUIRED AFTER BEING EXPOSED TO DIFFERENT TYPES OF FAILURES. IN EACH OF THE 4×20 RUNS, FULL RECOVERY WAS AFTER EACH EVENT ROBUSTLY ACHIEVED.

Visibility rate	Failure type	Recovery time after failure	repair	same failure
91.9%	Joint blocked	16.50 ± 1.20	0.45 ± 0.86	0.65 ± 1.15
79.0%	Limb deformed	20.20 ±1.96	11.10 ± 0.83	12.10 ± 1.64

 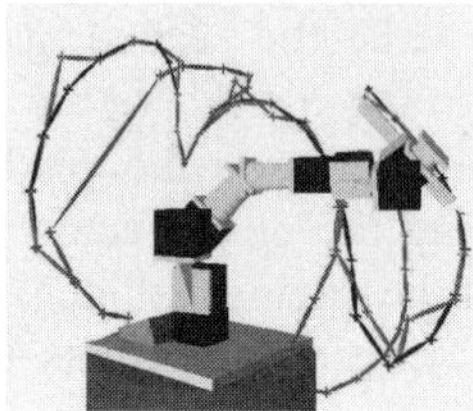

Fig. 8. The manipulator robot with a deformed limb has to follows the blue target trajectory. With a static body model, it suffers from strong derivation (red trajectory). By using our approach, the body scheme is dynamically adapted, and the trajectory is very well approached (green trajectory).

Although the overall result closely resembles the case of a blocked joint, there are a few interesting differences. After the tool gets displaced at $t = 100$, on average two local models need to be sampled because the first one is not sufficient.

Also note that it takes much more training samples for the GPs to learn and validate the underlying probability distribution $p(Z_{6 \to 7} \mid a_4)$ (see Table I). The prediction accuracy of the whole system closely resembles the levels as in the case of the blocked joint: On average, we measured after recovery an accuracy of 47mm.

D. Controlling a Deformed Robot

Finally, we ran a series of experiments to verify that dynamically maintained body schemes can be used for accurate positioning and control. The experiments were executed on a 4-DOF manipulator in simulation.

We defined a 3D trajectory consisting of 30 way-points that the manipulator should approach by inverse kinematics using its current body scheme, see Figure 8. When the initial geometric model was used to follow the trajectory by using the undamaged manipulator, a positioning accuracy of 7.03mm was measured. When the middle limb was deformed by $45°$, the manipulator with a static body scheme was significantly off course, leading to an average positioning accuracy of 189.35mm. With dynamic adaptation enabled, the precision settled at 15.24mm. This shows that dynamic model adaption enables a robot to maintain a high positioning accuracy after substantial changes to its body physiology.

V. Conclusion

In this paper, we presented a novel approach to life-long body scheme adaptation for a robotic manipulation system. Our central idea is to continuously learn a large set of local kinematic models using non-parametric regression and to search for the best arrangement of these models to represent the full system.

In experiments carried out with a real robot and in simulation, we demonstrated that our system is able to deal with missing and noisy observations, operates in full 3D space, and is able to perform relevant tasks like prediction, control, and online adaptation after failures. Challenging topics for further investigation include developing an active exploration strategy, learning from marker-less observations, point-like features, or range observations and learning for fully unobservable parts of the robot.

Acknowledgments

This work has partly been supported by the EU under FP6-004250-CoSy and by the German Ministry for Education and Research (BMBF) through the DESIRE project.

References

[1] J. J. Craig, *Introduction to Robotics: Mechanics and Control*. Boston, MA, USA: Addison-Wesley Longman Publishing Co., Inc., 1989.

[2] H. Choset, K. M. Lynch, S. Hutchinson, G. A. Kantor, W. Burgard, L. E. Kavraki, and S. Thrun, *Principles of Robot Motion: Theory, Algorithms, and Implementations*. MIT Press, June 2005.

[3] E. M. Rosales and Q. Gan, "Forward and inverse kinematics models for a 5-dof pioneer 2 robot arm," University of Essex, UK, Tech. Rep. CSM-413, 2004.

[4] J. Kolter and A. Ng, "Learning omnidirectional path following using dimensionality reduction," in *Proceedings of Robotics: Science and Systems*, Atlanta, GA, USA, June 2007.

[5] A. Dearden and Y. Demiris, "Learning forward models for robots," in *Proc. of the Int. Conf. on Artificial Intelligence (IJCAI)*, Edinburgh, Scotland, 2005, pp. 1440–1445.

[6] D. Grimes, R. Chalodhorn, and R. Rao, "Dynamic imitation in a humanoid robot through nonparametric probabilistic inference," in *Proc. of the Robotics: Science and Systems (RSS)*, Philadelphia, USA, 2006.

[7] J. Sturm, C. Plagemann, and W. Burgard, "Unsupervised body scheme learning through self-perception," in *Proc. of the IEEE Int. Conf. on Robotics & Automation (ICRA)*, Pasadena, CA, USA, 2008, to appear.

[8] Y. Yoshikawa, K. Hosoda, and M. Asada, "Binding tactile and visual sensations via unique association by cross-anchoring between double-touching and self-occlusion," in *Proc. of the International Workshop on Epigenetic Robotics*, Genoa, Italy, 2004, pp. 135–138.

[9] Y. Yoshikawa, Y. Tsuji, K. Hosoda, and M. Asada, "Is it my body? – body extraction from uninterpreted sensory data based on the invariance of multiple sensory attributes –," in *Proc. of the IEEE/RSJ Int. Conf. on Intelligent Robots and Systems (IROS)*, Sendai, Japan, 2004.

[10] P. Morasso and V. Sanguineti, "Self-organizing body-schema for motor planning," *Journal of Motor Behavior*, vol. 26, pp. 131–148, 1995.

[11] L. Natale, "Linking action to perception in a humanoid robot: A developmental approach to grasping," Ph.D. dissertation, Univerity of Genoa, Italy, May 2004.

[12] J. Ting, M. Mistry, J. Peters, S. Schaal, and J. Nakanishi, "A bayesian approach to nonlinear parameter identification for rigid body dynamics," in *Proceedings of Robotics: Science and Systems*, Philadelphia, USA, August 2006.

[13] N. Roy and S. Thrun, "Online self-calibration for mobile robots," in *Proc. of the IEEE Int. Conf. on Robotics & Automation (ICRA)*, 1999.

[14] J. C. Bongard, V. Zykov, and H. Lipson, "Automated synthesis of body schema using multiple sensor modalities," in *Proc. of the Int. Conf. on the Simulation and Synthesis of Living Systems (ALIFEX)*, 2006.

[15] C. Nabeshima, M. Lungarella, and Y. Kuniyoshi, "Timing-based model of body schema adaptation and its role in perception and tool use: A robot case study," in *Proc. of the IEEE International Conference on Development and Learning (ICDL2005)*, 2005, pp. 7–12.

[16] B. J., Z. V., and L. H., "Resilient machines through continuous self-modeling," *Science*, vol. 314., no. 5802, pp. 1118–1121, 2006.

[17] B. J. and L. H., "Automated reverse engineering of nonlinear dynamical systems," in *Proc. of the Nat. Academy of Science*, vol. 104, no. 24, 2007, p. 9943.

[18] C. Stachniss, D. Hahnel, W. Burgard, and G. Grisetti, "On actively closing loops in grid-based FastSLAM," *Advanced Robotics*, vol. 19, no. 10, pp. 1059–1080, 2005.

[19] L. Sciavicco and B. Siciliano, *Modelling and Control of Robot Manipulators*. Springer, January 2000.

[20] M. Fiala, "Artag, a fiducial marker system using digital techniques," National Research Council Canada, Tech. Rep., 2004.

[21] A. T. Ihler, E. B. Sudderth, W. T. Freeman, and A. S. Willsky, "Efficient multiscale sampling from products of gaussian mixtures." in *NIPS*, 2003.

[22] C. E. Rasmussen and C. K. Williams, *Gaussian Processes for Machine Learning*. Cambridge, Massachusetts: The MIT Press, 2006.

Bearing-Only Control Laws For Balanced Circular Formations of Ground Robots

Nima Moshtagh, Nathan Michael, Ali Jadbabaie, Kostas Daniilidis
GRASP Laboratory
University of Pennsylvania, Philadelphia, PA 19104
Email: {nima, nmichael, jadbabai, kostas} @grasp.upenn.edu

***Abstract*— For a group of constant-speed ground robots, a simple control law is designed to stabilize the motion of the group into a balanced circular formation using a consensus approach. It is shown that the measurements of the bearing angles between the robots are sufficient for reaching a balanced circular formation. We consider two different scenarios that the connectivity graph of the system is either a complete graph or a ring. Collision avoidance capabilities are added to the team members and the effectiveness of the control laws are demonstrated on a group of mobile robots.**

I. Introduction

Inspired by the social aggregation phenomena in birds and fish [1]–[3], researchers in robotics and control theory have been developing tools, methods and algorithms for distributed motion coordination of multi-vehicle systems. Two main collective motions that are observed in nature are *parallel motion* and *circular motion* [4]. One can interpret stabilizing the circular formation as an example of *activity consensus*, that is, individuals are "moving around" together. Stabilizing the parallel formation is another form of activity consensus in which individuals "move off" together [5].

The circular formation is a circular relative equilibrium in which all the agents travel around the same circle. This kind of behavior is observed in fish schooling, a well studied topic in ecology and evolutionary biology [3]. The *balanced* formation is an interesting family of equilibrium states where the agents are evenly spaced on a circular trajectory, and the geometric center of the agents is fixed. At the equilibrium, the relative headings and the relative distances of the agents determine the shape of the formation [6].

The primary contribution of this work is the presentation of a simple control law for achieving a balanced circular formation that only requires visual sensing such as bearing angles, *i.e.,* the input is in terms of quantities that do not require communication among nearest neighbors. In contrast with the work of Paley *et al.* [5], Sepulchre *et al.* [6], and Moshtagh *et al.* [7], where it is assumed that each agent has access to the values of its neighbors' positions and velocities, we design distributed control laws that use only visual clues from nearest neighbors to achieve motion coordination.

In [8]–[10] circular formations of a multi-vehicle system under cyclic pursuit is studied. The proposed strategy is distributed and relatively simple because each agent needs to measure the relative information from only one other agent. It is also shown that the formation equilibria of the multi-agent system are generalized polygons. In contrast to [8] our control law is a nonlinear function of the bearing angles and as a result our system converges to a different set of stable equilibria.

Verification of the theory through multi-robot experiments demonstrated the effectiveness of the bearing-only control law to achieve the desired formation. Of course in reality any formation control requires collision avoidance, and indeed collision avoidance cannot be done without range. In order to improve the experimental results, we provided inter-agent collision avoidance properties to the team members. What we show in this work is that the two tasks of formation-keeping and collision-avoidance can be done with decoupled additive terms in the control law, where the terms for keeping circular formation depends only on the bearing parameters.

The outline of the paper is as follows. First we provide some background information on graph theory and polygons that we are going to use throughout the paper. In Section III we derive the bearing-only controller that stabilizes a group of mobile agents into a balanced circular formation. In Section IV collision avoidance capabilities are added to the control laws. The derived controllers are tested on real robots and the experimental results are presented in Section V.

II. Background

In this section we briefly review a number of important concepts that we use in this paper.

A. Graph Theory

An (undirected) graph $\mathbb{G}$ consists of a vertex set, $\mathcal{V}$, and an edge set $\mathcal{E}$, where an edge is an unordered pair of distinct vertices in $\mathbb{G}$. If $x, y \in \mathcal{V}$, and $(x, y) \in \mathcal{E}$, then x and y are said to be adjacent, or neighbors and we denote this by writing $x \sim y$. The number of neighbors of each vertex is its valence. A path of length r from vertex x to vertex y is a sequence of $r+1$ distinct vertices starting with x and ending with y such that consecutive vertices are adjacent. If there is a path between any two vertices of a graph $\mathbb{G}$, then $\mathbb{G}$ is said to be connected.

The adjacency matrix $A(\mathbb{G}) = [a_{ij}]$ of an (undirected) graph $\mathbb{G}$ is a symmetric matrix with rows and columns indexed by the vertices of $\mathbb{G}$, such that $a_{ij} = 1$ if vertex i and vertex j are neighbors and $a_{ij} = 0$, otherwise. We also assume that $a_{ii} = 0$ for all i. The valence matrix, $D(\mathbb{G})$, of a graph $\mathbb{G}$

is a diagonal matrix with rows and columns indexed by $\mathcal{V}$, in which the (i,i)-entry is the valence of vertex i.

The symmetric singular matrix defined as:

$$L(\mathbb{G}) = D(\mathbb{G}) - A(\mathbb{G})$$

is called the Laplacian of $\mathbb{G}$. The Laplacian matrix captures many topological properties of the graph. The Laplacian L is a positive semidefinite M-matrix (a matrix whose off-diagonal entries are all nonpositive) and the algebraic multiplicity of its zero eigenvalue (*i.e.,* the dimension of its kernel) is equal to the number of connected components in the graph. The n-dimensional eigenvector associated with the zero eigenvalue is the vector of ones, $\mathbf{1}_n = [1, \ldots, 1]^T$.

Given an orientation of the edges of a graph, we can define the incidence matrix of the graph to be a matrix B with rows indexed by vertices and columns indexed by edges with entries of 1 representing the source of a directed edge and -1 representing the sink. The Laplacian matrix $L(\mathbb{G})$ of graph $\mathbb{G}$ is represented in terms of its incidence matrix as $L = BB^T$ independent of the orientation of the edges.

B. Regular Polygons

Let $d < n$ be a positive integer and define $p = n/d$. Let y_1 be a point on the unit circle. Let R_α be a clockwise rotation by the angle $\alpha = 2\pi/p$. The *generalized regular polygon* $\{p\}$ is given by the points $y_{i+1} = R_\alpha y_i$, and edges between points i and $i+1$ [11].

When $d = 1$ the polygon $\{p\}$ is called an ordinary regular polygon and its edges do not intersect. If $d > 1$ and n and d are coprime, then the edges intersect and the polygon is a *star*. If n and d have a common factor $l > 1$, then the polygon consists of l traversals of the same polygon with $\{n/l\}$ vertices and edges. If $d = n$ the polygon $\{n/n\}$ corresponds to all points at the same location. If $d = n/2$ (with n even), then the polygon consists of two end points and a line between them, with points corresponding to an even index on one end and points corresponding to an odd index on the other.

C. Kronecker Product

The Kronecker product, denoted by $\otimes$, is an operation on two matrices of arbitrary size resulting in a block matrix. If A is an $m \times n$ matrix and B is a $p \times q$ matrix, then the Kronecker product $A \otimes B$ is a $mp \times nq$ block matrix. If A, B, C and D are matrices of such size that one can form the matrix products AC and BD, then $(A \otimes B)(C \otimes D) = AC \otimes BD$. This is called the mixed-product property. Also the following property holds $(A \otimes B)^T = A^T \otimes B^T$.

III. Circular Formations of Planar Robots

Consider a group of n unit-speed planar agents. Each agent is capable of sensing information from its neighbors. The neighborhood set of agent i, $\mathcal{N}_i$, is the set of agents that can be "seen" by agent i. The precise meaning of "seeing" will be cleared later. The size of the neighborhood depends on the characteristics of the sensors. The neighboring relationship between agents can be conveniently described by a graph.

Definition 3.1 (Connectivity Graph): *The connectivity graph $\mathbb{G} = \{\mathcal{V}, \mathcal{E}\}$ is a graph consisting of:*

- *a set of vertices $\mathcal{V}$ indexed by the set of mobile agents;*
- *a set of edges $\mathcal{E} = \{(i,j) \mid i,j \in \mathcal{V}$, and $i \sim j\}$;*

The edge set $\mathcal{E}$ represents the links among the agents, and the neighborhood of agent i is defined by

$$\mathcal{N}_i \doteq \{j | i \sim j\} \subseteq \{1, \ldots, n\} \backslash \{i\}.$$

A circular formation is a circular relative equilibrium in which all the agents travel around the same circle. At the equilibrium, the relative headings and the relative distances of the agents determine the shape of the formation. We are interested in *balanced* circular formations as defined by:

Definition 3.2 (Balanced Circular Formation): *The set of equilibrium states where the agents are evenly spaced on a circular trajectory, and the geometric center of the agents is fixed is called the balanced circular formation.*

A. Kinematic Model for Mobile Robots

Let $\mathbf{r}_i$ represent the position of agent i, and $\mathbf{v}_i$ be its velocity vector. The dynamics of each unit-speed agent is given by:

$$\begin{aligned} \dot{\mathbf{r}}_i &= \mathbf{v}_i \\ \dot{\mathbf{v}}_i &= \omega_i \, \mathbf{v}_i^\perp \\ \dot{\mathbf{v}}_i^\perp &= -\omega_i \mathbf{v}_i \end{aligned} \tag{1}$$

where $\mathbf{v}_i^\perp$ is the unit vector perpendicular to the velocity vector $\mathbf{v}_i$. The orthogonal pair $\{\mathbf{v}_i, \mathbf{v}_i^\perp\}$ forms a body frame for agent i (See Figure 1). We represent the stack vector of all the velocities by $\mathbf{v} = [\mathbf{v}_1^T, \ldots, \mathbf{v}_n^T]^T \in \mathbb{R}^{2n \times 1}$.

The control input for each agent is the angular velocity ω_i. Since it is assumed that the agents move with constant unit speed, the force applied to each agent must be perpendicular to its velocity vector, *i.e.,* the force on each agent is a *gyroscopic force*, and it does not change its speed (and hence its kinetic energy). Thus, ω_i serves as a steering control [12] for each agent. In the following we design a *distributed* control law for achieving a balanced formation.

Let $\mathbf{c}_i$ represent the position of the center of the i-th circle with radius $1/\omega_o$, as shown in Figure 1, thus

$$\mathbf{c}_i = \mathbf{r}_i + (1/\omega_o)\mathbf{v}_i^\perp \ .$$

The shape controls for driving agents to a circular formation depend on the shape variables $\mathbf{v}_{ij} = \mathbf{v}_j - \mathbf{v}_i$ and $\mathbf{r}_{ij} = \mathbf{r}_j - \mathbf{r}_i$. The relative equilibria of the balanced formation are characterized by $\sum_{i=1}^n \mathbf{v}_i = 0$, and $\mathbf{c}_i = \mathbf{c}_o \in \mathbb{R}^2$ for all $i \in \{1, \ldots, n\}$, where $\mathbf{c}_o$ is the fixed geometric center of the agents.

The control input for each agent has two components:

$$\omega_i = \omega_o + u_i$$

The constant angular velocity ω_o takes the agents into a circular motion, and u_i puts the agents into a balanced formation. In

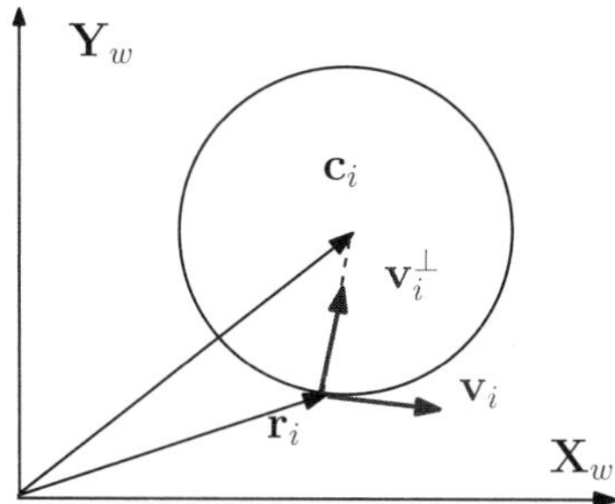

Fig. 1. Center of the circular trajectory is defined as $\mathbf{c}_i = \mathbf{r}_i + (1/\omega_0)\mathbf{v}_i^\perp$.

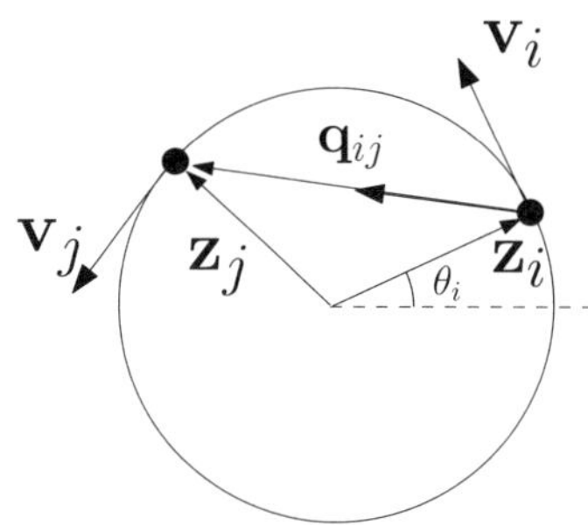

Fig. 2. By a change of coordinate $\mathbf{z}_i = \omega_o(\mathbf{r}_i - \mathbf{c}_i) = -\mathbf{v}_i^\perp$ the problem of generating circular motion in the plane reduces to the problem of balancing the agents on a circle.

order to design u_i we express the system in a *rotating frame*, which greatly simplifies the analysis. By a change of variable

$$\mathbf{z}_i = \omega_o(\mathbf{r}_i - \mathbf{c}_i) = -\mathbf{v}_i^\perp$$

the problem reduces to balancing the agents on a unit circle as shown in Figure 2. The new coordinate system is rotating with angular velocity ω_o. The dynamics in the rotating frame is given by

$$\begin{aligned} \dot{\mathbf{z}}_i &= \mathbf{v}_i u_i \\ \dot{\mathbf{v}}_i &= -\mathbf{z}_i u_i \ , \quad i = 1, \ldots, n \end{aligned} \tag{2}$$

The new position vector $\mathbf{z}_i$ is a unit vector, however its speed $|\dot{\mathbf{z}}_i|$ is not constant anymore, and it is proportional to u_i, which goes to zero as the group reaches a balanced formation.

Let us define $\mathbf{z}_{ij} = \mathbf{z}_j - \mathbf{z}_i$ and $\mathbf{q}_{ij} = \mathbf{z}_{ij}/|\mathbf{z}_{ij}|$ as the unit vector along the relative position vector $\mathbf{z}_{ij}$. We note that at the balanced equilibrium the velocity of each agent must be perpendicular to $\bar{\mathbf{q}}_i = \sum_{j\in\mathcal{N}_i} \mathbf{q}_{ij}$, which is a vector along the average of the relative position vectors that are incident to agent i. Thus, the quantity $< \mathbf{v}_i, \bar{\mathbf{q}}_i >$ vanishes at the balanced equilibrium. Hence we propose the following control law for the balanced formation:

$$u_i = -\kappa < \mathbf{v}_i, \bar{\mathbf{q}}_i >= -\kappa \sum_{j\in\mathcal{N}_i} < \mathbf{v}_i, \mathbf{q}_{ij} > \ , \ \kappa > 0. \tag{3}$$

B. Complete-Graph Topology

Suppose the underlying connectivity graph is a complete graph. We have the following theorem for reaching the balanced circular formation in a group of mobile planar agents with a complete-graph connectivity.

Theorem 3.3: *Consider a system of n agents with kinematics (2). Given a complete connectivity graph $\mathbb{G}$, and applying control law (3), the n-agent system (almost) globally asymptotically converges to a balanced circular formation as defined in Definition 3.2.*

Proof: Let us define vector $\mathbf{p}$ that points towards the geometric center of the group:

$$\mathbf{p} = \frac{1}{n}\sum_{i=1}^{n} \mathbf{z}_i = \frac{1}{n}\mathbf{1}^T\mathbf{z} \ , \quad \mathbf{1} = \mathbf{1}_n \otimes I_2 \in \mathbb{R}^{2n\times 2} \ .$$

The minimum $|\mathbf{p}| = 0$ is reached when the position vectors $\mathbf{z}_i$ are in a balanced position (splay state); and the maximum $|\mathbf{p}| = 1$ is reached when all the position vectors are aligned (state synchronized). Note that the balancing input (3) can be bounded above by a function of vector $\mathbf{p}$:

$$\begin{aligned} u_i &= -\kappa \sum_{j\in\mathcal{N}_i} < \frac{\mathbf{z}_{ij}}{|\mathbf{z}_{ij}|}, \mathbf{v}_i >= -\kappa \sum_{j=1}^{n} \frac{1}{|\mathbf{z}_{ij}|} < \mathbf{z}_j, \mathbf{v}_i > \\ &\leq -\frac{\kappa}{|\mathbf{z}|_{max}} \sum_{j=1}^{n} < \mathbf{z}_j, \mathbf{v}_i > \\ &= -\frac{n\kappa}{|\mathbf{z}|_{max}} < \mathbf{p}, \mathbf{v}_i > \end{aligned} \tag{4}$$

where $|\mathbf{z}|_{max} = \max\{|\mathbf{z}_{ij}|, \ (i,j) \in \mathcal{E}\}$, and we have used the fact that $\mathbf{v}_i \perp \mathbf{z}_i$.

Now consider the following Lyapunov function

$$w(\mathbf{z}) = \frac{n}{2}|\mathbf{p}|^2 = \frac{1}{2n}\mathbf{z}^T\mathbf{1}\mathbf{1}^T\mathbf{z} \tag{5}$$

which is minimized for the balanced formation. Given the gradient of $w(\mathbf{z})$:

$$\frac{\partial w(\mathbf{z})}{\partial \mathbf{z}_i} = \frac{1}{n}(\mathbf{1}\mathbf{1}^T\mathbf{z})_i = \frac{1}{n}\mathbf{1}^T\mathbf{z} = \mathbf{p}$$

the time derivative of $w(\mathbf{z})$ becomes

$$\begin{aligned} \dot{w}(\mathbf{z}) &= \sum_{i=1}^{n} < \frac{\partial w(\mathbf{z})}{\partial \mathbf{z}_i}, \dot{\mathbf{z}}_i >= \sum_{i=1}^{n} < \mathbf{p}, \mathbf{v}_i > u_i \\ &\leq -\frac{n\kappa}{|\mathbf{z}|_{max}} \sum_{i=1}^{n} < \mathbf{p}, \mathbf{v}_i >^2 \ \leq 0 \end{aligned} \tag{6}$$

where we have used (4).

A simple application of LaSalle's invariance principle over the configuration space which is an n-torus and therefore compact reveals that all trajectories starting in anywhere on the n-torus converge to the largest invariant sets in $E = \{\mathbf{z} \mid \dot{w}(\mathbf{z}) = \mathbf{0}\}$. This set is characterized by $< \mathbf{p}, \mathbf{v}_i >= 0$, for all $i \in \{1, \ldots, n\}$. Therefore the equilibria are given by either $\mathbf{p} = \mathbf{0}$, or $\mathbf{p} \perp \mathbf{v}_i$ for all $i \in \{1, \ldots, n\}$. $\mathbf{p} = \mathbf{0}$ is the global minimum of $w(\mathbf{z})$ and is asymptotically stable. At the equilibrium we have $u_i = 0$ for all $i \in \{1, \ldots, n\}$ and as a result the geometric center remains fixed because $\dot{\mathbf{p}} = \sum_i u_i \mathbf{v}_i = \mathbf{0}$.

The critical points given by $\mathbf{p} \perp \mathbf{v}_i$ correspond to a set of configurations that m agents are at antipodal position from the other $n - m$ agents, where $1 \leq m < n/2$. The instability of these equilibria is proved by showing that if we perturb the system at those equilibria, the system moves away from them and $w(\mathbf{z})$ will be decreasing. ■

Remark 3.4: The Laplacian matrix of a complete graph equals to $L_c = I_n - (1/n)\mathbf{1}_n\mathbf{1}_n^T$. Thus, one can see that minimizing $w(\mathbf{z})$ in (5) is equivalent to maximizing $\mathbf{z}^T\bar{L}_c\mathbf{z}$ with $\bar{L}_c = L_c \otimes I_2$. The maximum is achieved when all the agents are evenly spaced around the circle.

C. Ring Topology

Next we consider the situation that the connectivity graph has a ring topology. We denote this graph with $\mathbb{G}^{ring}$. We have the following theorem for the balanced circular formations of a group of mobile agents with ring topology.

Theorem 3.5: *Consider a system of n agents with kinematics (2). Suppose the connectivity graph has the ring topology $\mathbb{G}^{ring}$ and each agent applies the balancing control law (3). Let ϕ_o be the angle to which the relative headings converge, then if $\phi_o \in (\pi/2, 3\pi/2)$ the balanced equilibrium is locally exponentially stable.*

Proof: Let L_r be the Laplacian matrix of a graph with a ring topology, and $\bar{L}_r = L_r \otimes I_2$. Input (3) can be written in terms of the Laplacian of the connectivity graph:

$$\begin{aligned} u_i &= \kappa \sum_{j \in \mathcal{N}_i} \frac{1}{|\mathbf{z}_{ij}|} < \mathbf{z}_i - \mathbf{z}_j, \mathbf{v}_i > \\ &\geq \frac{\kappa}{|\mathbf{z}|_{max}} \sum_{j \in \mathcal{N}_i} < \mathbf{z}_i - \mathbf{z}_j, \mathbf{v}_i > \\ &= \frac{\kappa}{|\mathbf{z}|_{max}} < (\bar{L}_r\mathbf{z})_i, \mathbf{v}_i >, \quad \kappa > 0 \end{aligned} \tag{7}$$

where $(\bar{L}_r\mathbf{z})_i \in \mathbb{R}^2$ is the subvector of $\bar{L}_r\mathbf{z}$ associated with the ith agent. Now consider the function

$$s(\mathbf{z}) = \frac{1}{2}\mathbf{z}^T\bar{L}_r\mathbf{z}$$

that is maximized for the balanced formation, and this maximum exists because $s(\mathbf{z})$ is bounded from above. Using the dynamics (2) and input (3) we have that

$$\begin{aligned} \dot{s}(\mathbf{z}) &= \sum_{i=1}^{n} < \frac{\partial s(\mathbf{z})}{\partial \mathbf{z}_i}, \dot{\mathbf{z}}_i >= \sum_{i=1}^{n} < (\bar{L}_r\mathbf{z})_i, \mathbf{v}_i > u_i \\ &\geq \frac{}{|\mathbf{z}|_{max}} \sum_{i=1}^{n} < (\bar{L}_r\mathbf{z})_i, \mathbf{v}_i >^2 \geq 0 \end{aligned} \tag{8}$$

Thus $s(\mathbf{z})$ monotonically increases along the trajectories of system (2) with input (3), and converges to equilibria corresponding to

$$< (\bar{L}_r\mathbf{z})_i, \mathbf{v}_i >= 0, \quad \forall i \in \{1, \ldots, n\} \ . \tag{9}$$

Let us characterize the set of equilibria given by (9). We represent the unit vector $\mathbf{z}_i$ in the rotating frame by $\mathbf{z}_i = [\cos\theta_i \ \sin\theta_i]^T$. Then $\mathbf{v}_i = [-\sin\theta_i \ \cos\theta_i]^T$, and (9) is equivalent to

$$\sum_{j \in \mathcal{N}_i} \sin(\theta_i - \theta_j) = 0, \quad \forall i \in \{1, \ldots, n\} \ . \tag{10}$$

Let $\boldsymbol{\theta} = [\theta_1, \ldots, \theta_n]^T$. Then (10) becomes

$$B \sin(B^T\boldsymbol{\theta}) = 0 \ , \tag{11}$$

where $B \in \mathbb{R}^{n \times e}$ is the incidence matrix of $\mathbb{G}^{ring}$, where $e = |\mathcal{E}|$. For $\mathbb{G}^{ring}$, $n = e$ and B is a circulant matrix that satisfies $B\mathbf{1}_e = \mathbf{0}$. Let $\boldsymbol{\phi} = B^T\boldsymbol{\theta}$. Then the equilibria of system (11) are characterized by

$$\begin{aligned} \sin\boldsymbol{\phi} &= \alpha\mathbf{1}_e & (12) \\ \mathbf{1}_e^T\boldsymbol{\phi} &= m\pi \ . & (13) \end{aligned}$$

Vector $\boldsymbol{\phi}$ satisfies equation (12) iff $\phi_k = \{\phi_o, \pi - \phi_o\}$ for all $k \in \{1, \ldots, e\}$ and $\phi_o \in (0, 2\pi)$. Equation (13) is satisfied if $\phi_o = (m/e)\pi$ for $m \in \mathbb{N}$.

Next we prove the (local) exponential stability of the relative equilibria, *i.e.,* the balanced state. For the proof of the exponential stability of the equilibrium $\boldsymbol{\phi} = \phi_o\mathbf{1}_e$ we consider the linearization of system (11) about ϕ_o. The Jacobian of system $\dot{\boldsymbol{\theta}} = \kappa B \sin(B^T\boldsymbol{\theta})$ at the equilibrium is

$$J = \kappa B \text{diag}(\cos\phi_o)B^T = \kappa\cos\phi_o BB^T$$

where $\text{diag}(\cos\phi_o)$ is an $e \times e$ matrix with $\cos\phi_o$ as its diagonal elements. Since $\kappa > 0$, the linearized system $\dot{\boldsymbol{\theta}} = J\boldsymbol{\theta}$ is exponentially stable if $\phi_o \in (\pi/2, 3\pi/2)$.

As a result at the equilibrium the final configuration for $\mathbb{G}^{ring}$ is either a star polygon (for n odd), or a line (for n even) with odd-indexed agents on one side and even-indexed agents on the other side. This can be seen by noting that for a $\{n/d\}$ polygon, the angle between the connected nodes is $2\pi d/n$. Thus, the stable equilibria given by $\phi_o \in (\pi/2, 3\pi/2)$ correspond to polygons with $d \in (n/4, 3n/4)$. ■

For example, for $n = 5$, the stable polygons are $\{5/3\}$ and $\{5/4\}$ which are the same polygons with reverse ordering of the nodes. Simulations suggest that the largest region of attraction for n even belongs to a polygon $\{n/d\}$ with $d = n/2$, and a *star* polygon $\{n/d\}$ with $d = (n \pm 1)/2$ for n odd. These results are observed in experiments with real robots as demonstrated in Section V.

D. Bearing-based control law

In this section, we write input (3) in terms of a parameter that is measurable using a simple visual system. Similar attempt was done in [13] to obtain vision-based control laws for flocking of a group of nonholonomic agents. Let $\mathbf{r}_i = [x_i \ y_i]^T$ be the location of agent i in a fixed world frame, and $\mathbf{v}_i = [\dot{x}_i \ \dot{y}_i]^T$ be its velocity vector. The heading or orientation of agent i is then given by

$$\theta_i = \text{atan2}(\dot{y}_i, \dot{x}_i) \ . \tag{14}$$

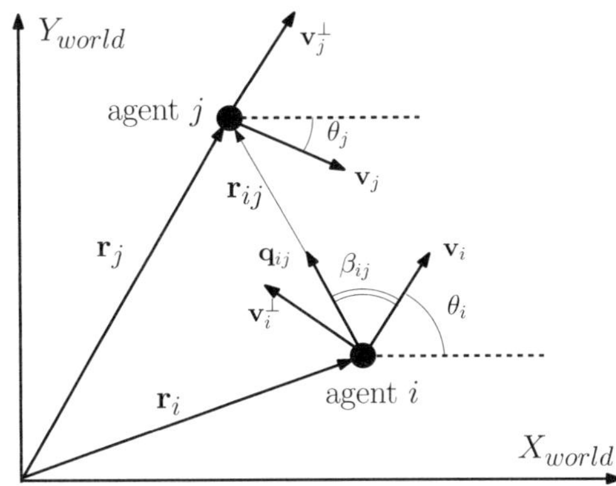

Fig. 3. Bearing angle β_{ij} is measured as the angle between the velocity vector (along body x-axis) and vector $\mathbf{r}_{ij}$, which connects the two neighboring agents.

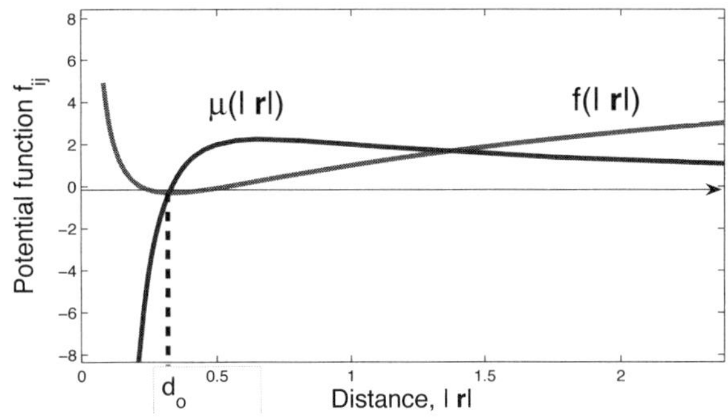

Fig. 4. Artificial potential function f_{ij}, and the norm of its gradient μ_{ij}.

Given the above definitions, dynamic model (1) becomes the unicycle model:

$$\begin{aligned} \dot{x}_i &= v \cos\theta_i \\ \dot{y}_i &= v \sin\theta_i \\ \dot{\theta}_i &= \omega_i \end{aligned} \tag{15}$$

where ω_i is the angular velocity of agent i, and v is the constant linear velocity (assuming $v = 1$ in this section). Let β_{ij} be the relative angle between agents i and j as measured in the local coordinate frame of agent i. The bearing angle β_{ij} is defined as (see Figure 3):

$$\beta_{ij} \doteq \text{atan2}(y_i - y_j, x_i - x_j) - \theta_i \ . \tag{16}$$

The only visual parameter that is required for generating a balanced circular formation is the *bearing angle*, β_{ij}. It is remarkable that we can generate interesting global patterns using only a single measurement of the bearing angle. Note that the inner product of two vectors is independent from the coordinate system in which they are expressed. Thus, given $\mathbf{v}_i = [1\ 0]^T$ and $\mathbf{q}_{ij} = [\cos\beta_{ij}\ \sin\beta_{ij}]^T$ in the body frame of agent i, the control input for a balanced circular formation can be written as:

$$\omega_i = \omega_o - \kappa \sum_{j \in \mathcal{N}_i} < \mathbf{v}_i, \mathbf{q}_{ij} > = \omega_o - \kappa \sum_{j \in \mathcal{N}_i} \cos\beta_{ij} \ , \tag{17}$$

where $\kappa > 0$. Input (17) is the desired bearing-only control input.

IV. Balanced Circular Formation with Collision Avoidance

The central contribution of this work is providing a simple bearing-only control law for reaching a balanced circular formation. Of course in reality any formation control requires collision avoidance, and indeed collision avoidance cannot be done without range. What we show here is that the two tasks can be done with decoupled additive terms in the control law, where the term for circular formation depends only on bearing.

To ensure collision avoidance and cohesion of the formation, an inter-agent potential function [14], [15] is defined. A control law from this artificial potential function results in simple steering behaviors known as *separation* and *cohesion* that govern how each agent maneuvers based on the relative position of its neighbors. The global minimum of this function is where all the agents are at the desired distances.

It was shown in [15] that only if the underlying proximity graph is a spanning tree, the formation stabilizes at a state where the potential function is at the global minimum, and all the agents are at the desired distances. Whereas, in the general case, the multi-agent system reaches a stable state where the potential energy of the system is minimized (a local minimum). Next we formally define the notion of potential function used in this paper.

The potential function $f_{ij}(|\mathbf{r}_{ij}|)$ is a symmetric function of the distance $|\mathbf{r}_{ij}| = l_{ij}$ between agents i and j, and is defined as follows [15]:

Definition 4.1 (Potential Function): *Potential f_{ij} is a differentiable, nonnegative function of the distance $|\mathbf{r}_{ij}|$ between agents i and j such that,*

- *$f_{ij} \to \infty$ as $|\mathbf{r}_{ij}| \to 0$.*
- *f_{ij} attains its unique minimum when agents i and j are located at a desired distance.*

This definition ensures that minimization of the inter-agent potential functions leads to the desired cohesion and separation in the group. Agent i's total potential is given by

$$f_i = \sum_{j \in \mathcal{N}_i} f_{ij}(|\mathbf{r}_{ij}|) \ . \tag{18}$$

The requirements for f_{ij} given in Definition 4.1 support a large class of functions. Similar potential functions as the following are used in both [15] and [16]:

$$f_{ij} = \frac{d_0}{|\mathbf{r}_{ij}|} + \log|\mathbf{r}_{ij}| \ ,$$

where d_0 is the desired distance between the pair (i, j). This choice of f_{ij} provides an *attractive* force when an agent is moving away from the group, and a *repulsive* force when two agents get too close to each other. The gradient of this function is given by

$$\nabla_{\mathbf{r}_{ij}} f_{ij} = \frac{\mathbf{r}_{ij}}{|\mathbf{r}_{ij}|}\left(\frac{1}{|\mathbf{r}_{ij}|} - \frac{d_0}{|\mathbf{r}_{ij}|^2}\right) = \mu(|\mathbf{r}_{ij}|)\mathbf{q}_{ij} = \mu_{ij}\mathbf{q}_{ij} \tag{19}$$

where $\mathbf{q}_{ij}$ is the unit-length bearing vector between agent i and its neighbor j. See Figure 4 for the plots of the potential function f_{ij}, and the norm of its gradient $\mu_{ij} = |\nabla_{\mathbf{r}_{ij}} f_{ij}|$.

The control input for balanced formations must have an additional components α_i that controls the spacing between the agents. α_i steers the agents to avoid collisions or pull them together if they are separating too far apart. For the inserted force to be gyroscopic, it must be perpendicular to the velocity vector $\mathbf{v}_i$ and along $\mathbf{v}_i^{\perp}$. The force is proportional to the negative gradient of the potential function f_i. Thus, as a result the spacing control must have the form

$$\alpha_i = -\kappa_s < \mathbf{v}_i^{\perp}, \nabla_{\mathbf{r}_i} f_i >, \quad \kappa_s > 0 \ . \tag{20}$$

Note that since $\mathbf{r}_{ij} = \mathbf{r}_j - \mathbf{r}_i$ we have

$$\nabla_{\mathbf{r}_i} f_i = -\nabla_{\mathbf{r}_{ij}} f_i = -\sum_{j \in \mathcal{N}_i} \nabla_{\mathbf{r}_{ij}} f_{ij} = -\sum_{j \in \mathcal{N}_i} \mu_{ij} \mathbf{q}_{ij} \ .$$

Finally, we have the following proposition for reaching the balanced circular formation with collision avoidance:

Proposition 4.2: *Consider a system of n agents with dynamics (1) and applying the control input*

$$\begin{aligned} \omega_i &= \omega_o + u_i + \alpha_i \qquad (21) \\ &= \omega_o - \kappa_b \sum_{j \in \mathcal{N}_i} < \mathbf{v}_i, \mathbf{q}_{ij} > + \kappa_s \sum_{j \in \mathcal{N}_i} \mu_{ij} < \mathbf{v}_i^{\perp}, \mathbf{q}_{ij} > \end{aligned}$$

where $\kappa_b > 0$ and $\kappa_s > 0$. Given that $\mathbb{G}$ remains connected, the n-agent system asymptotically reaches the balanced formation, and collisions between the interconnected agents are avoided.

V. EXPERIMENTS

In this section we show the results of experimental tests for two important cases: (a) the complete-graph topology and (b) the ring topology. But first, let us describe the experimental testbed.

A. Experimental Testbed Components

The experimental testbed consists of many components that are interfaced together to create the total system. In the discussion that follows, we present the robots, software and infrastructure of the testbed.

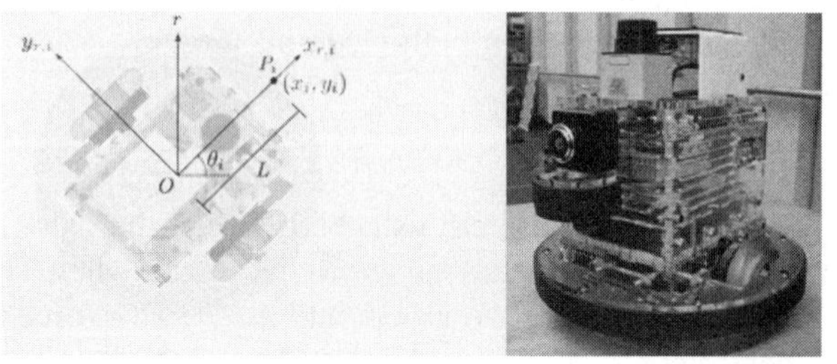

Fig. 5. *Scarab* is a small robot with a differential drive axil. LED markers are placed on top of each *Scarab* for tracking and ground-truth verification.

Robots: We use a series of small form-factor robots called *Scarab* [17]. The *Scarab* is a 20 x 13.5 x 22.2 cm^3 indoor ground platform with a mass of 8 kg. Each *Scarab* is equipped with a differential drive axle placed at the center of the length of the robot with a 21 cm wheel base (See Figure 5). Each *Scarab* is equipped with an onboard computer, power management system and wireless communication. Each robot is actuated by stepper motors that allows us to model it as a point robot with unicycle kinematics (15) for its velocity range. The linear velocity of each robot is bounded at $0.2\ m/s$. Each robot is able to rotate about its center of mass at speeds below $1.5\ rad/s$. Typical angular velocities resulting from the control law were below $0.5\ rad/s$.

Software: Every robot is running identical modularized software with well defined interfaces connecting modules via the *Player* robot architecture system [18], which consists of libraries that provide access to communication and interface functionality. The *Player* also provides a close collaboration with the three-dimensional physics-based simulation environment *Gazebo*. *Gazebo* provides the powerful ability to transition transparently from code running on simulated hardware to real hardware.

Infrastructure: In the experiments, visibility of the robot's set of neighbors is the main issue. Using omnidirectional cameras seems to be a natural solution. However, in order to reduce the on-board computation, a tracking system consisting of LED markers on the robots and eight overhead cameras is designed. This ground-truth verification system can locate and track the robots with position error of approximately 2 cm and an orientation error of $5°$. The overhead tracking system allows control algorithms to assume pose is known in a global reference frame. The process and measurement models fuse local odometry information and tracking information from the camera system. Each robot locally estimates its pose based on the globally available tracking system data and local motion, using an extended Kalman filter. We process global overhead tracking information but hide the global state of the system from each robot, providing only the current state of the robot as well as the positions of each robot's set of neighbors. In this way, we use the tracking system in lieu of an inter-robot sensor implementation.

In all the experiments the neighborhood relations, *i.e.,* the connectivity graphs, are fixed and undirected. Each robot computes the bearing angles with respect to its neighbors from equation (16), and applies the vision-based control input (17). The conclusions for each set of experiments are drawn from significant number of successful trials that supported the effectiveness of the designed controller. The results of the experiments are provided in the following subsections.

B. Complete-Graph Topology

First we applied the bearing-only control law (17) to a group of $n = 5$ robots without considering collision avoidance among the agents. In Figures 6 (a) through 6 (d) snapshots from the actual experiment are shown, and in Figures 6 (e)

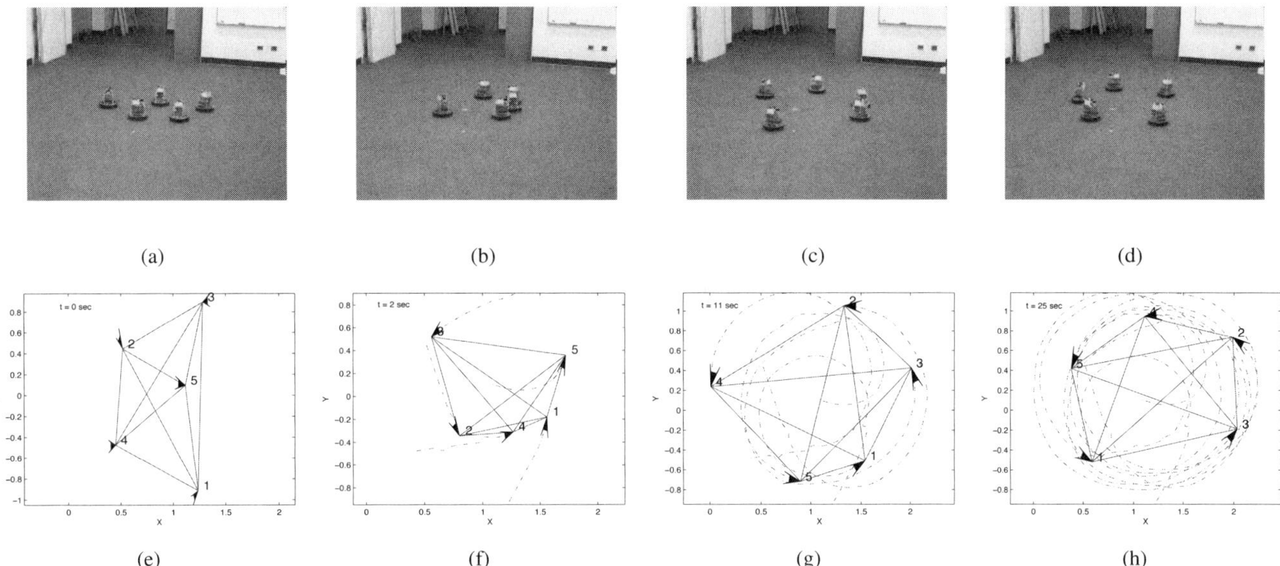

Fig. 6. Five *Scarabs* form a circular formation starting with a complete-graph topology. (a) At time $t = 0$ robots starts at random positions and orientations. (b) $t = 2$ sec. (c) $t = 11$ sec. (d) At $t = 25$ sec. the robots reach a stable balanced configuration around a circle with radius of 1m. Figures (e) through (h) show the actual trajectories of the robots and their connectivity graph at the times specified above. Figure 6(h) shows that the final configuration is a regular polygon.

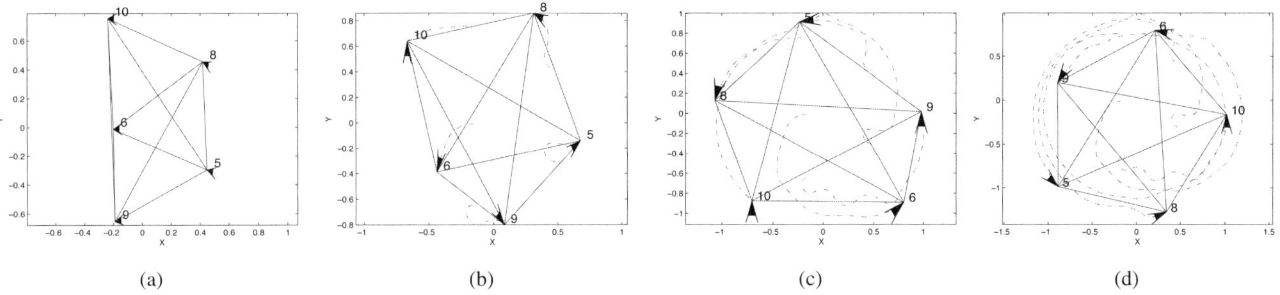

Fig. 8. Five *Scarabs* form a circular formation starting with a complete-graph topology while avoiding collisions. (a) $t = 0$ sec. (b) $t = 8$ sec. (c) $t = 20$ sec. (d) At $t = 36$ sec. the robots reach a stable balanced configuration around a circle with radius of 1m. Figures (a) through (d) show the actual trajectories of the robots and their connectivity graph at the times specified above.

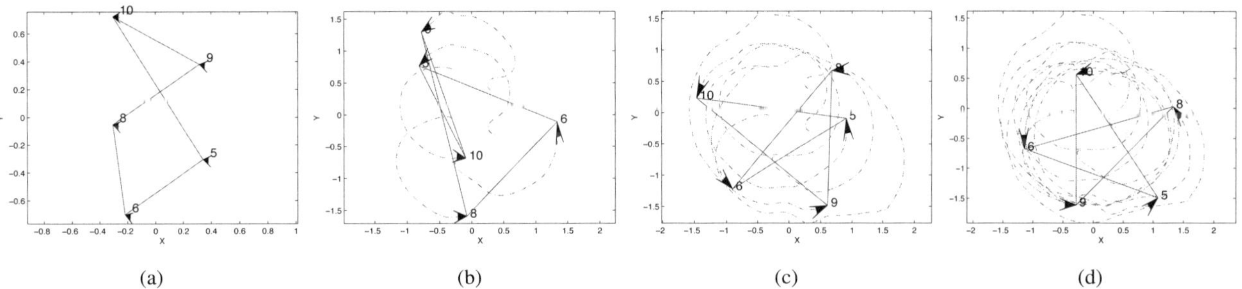

Fig. 9. Five *Scarabs* form a circular formation starting with a ring topology while avoiding collisions. (a) $t = 0$ sec. (b) $t = 16$ sec. (c) $t = 40$ sec. (d) At $t = 80$ sec. the robots reach a stable balanced configuration, which is the star polygon $\{5/3\}$, around a circle with radius of 1m. Figures (a) through (d) show the actual trajectories of the robots and their connectivity graph at the times specified above.

through 6 (h) the corresponding trajectories, generated from overhead tracking information, are demonstrated. Note that for the complete-graph topology the ordering of the robots in the final configuration is not unique, and it depends on the initial positions.

Since there was no collision avoidance implemented in the experiments of Figure 6 the robots could become undesirably close to one another as it can be seen in Fig. 6 (b). However, by applying control input (21) no collisions occur among the robots as they reach the equilibrium. The actual trajectories

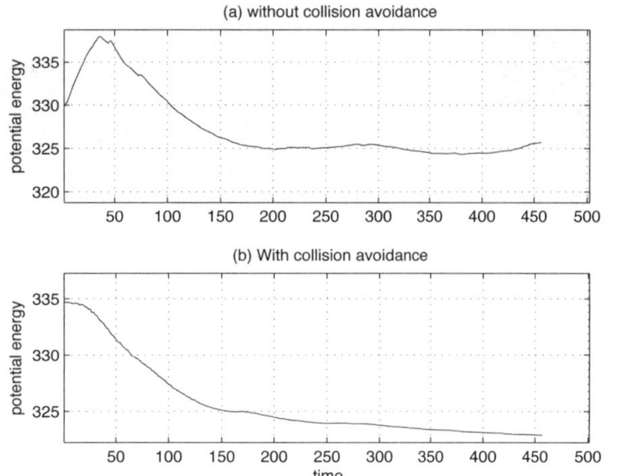

Fig. 7. Comparison of the values of the 5-agent system's potential energy while robots are applying (a) control input (17) and (b) control input (21) with collision avoidance.

of $n = 5$ robots for this scenario are shown in Figure 8. The comparison of the potential energies of the system with and without α_i term (20) are presented in Figure 7. The potential energy of the system is computed from $f = \sum_{i=1}^{n} f_i$ where f_i is given by (18). The peak in Fig. 7 (a) corresponds to the configuration observed in Fig. 6 (b) where robots become too close to each other. By using the control input (21) the potential energy of the 5-agent system monotonically decreases (see Fig. 7 (b)) and the system stabilizes on a state that the potential energy of the entire system is minimized.

C. Ring Topology

If every robot can only "sense" two other robots in the group, the topology of the connectivity graph will be a ring topology. Since the connectivity graph is assumed fixed, the agents need to be numbered during the experiments. For n even, the largest region of attraction is an $\{n/d\}$ polygon with $d = n/2$, which is not physically possible, because it requires that robots with even indices to stay on one side of a line segment and robots with odd indices to stay at the other side. For n odd, both simulations and experiment suggest that the largest region of attraction belong to star polygon $\{n/d\}$ with $d = (n \pm 1)/2$, therefore, there are only two possible ordering of the robots in the final circular formation. Figure 9 shows that in our experiment the robots are stabilized to the star polygon $\{5/3\}$.

Remark 5.1: When the communication graph is a fixed, directed graph with a ring topology, where agent i could only see agent $(i+1)/\text{mod}(n)$, then the n-agent system would behave like a team of robots in cyclic pursuit [9].

VI. Conclusions

We developed a control input for balanced circular formations of a group of ground robots that required only the measurements of the bearing angles with respect to the set of neighbors. Since the bearing angles could be simply measured using basic visual sensors on a robot, this control input could be considered a vision-based input. The results show how we can generate interesting *global* patterns using only *local* information, and without knowing a global reference frame. To improve the experimental results, we added collision avoidance capabilities to our control input for balanced formations.

In future we would like to implement the proposed control algorithm on robots with vision sensors. If the robot-mounted visual sensor for bearing measurements is a camera with a limited field of view, the underlying connectivity graph will be a *directed* graph. The study of circular formations with a directed graph is an ongoing work.

References

[1] A. Okubo, "Dynamical aspects of animal grouping: swarms, schools, flocks, and herds," *Advances in Biophysics*, vol. 22, pp. 1–94, 1986.
[2] D. J. Low, "Following the crowd," *Nature*, vol. 407, pp. 465–466, 2000.
[3] I. Couzin, J. Krause, R. James, G. Ruxton, and N. Franks, "Collective memory and spatial sorting in animal groups," *Journal of Theoretical Biology*, vol. 218(1), p. 111, 2002.
[4] H. Levine, E. Ben-Jacob, I. Cohen, and W.-J. Rappel, "Swarming patterns in microorganisms: Some new modeling results," in *IEEE Conference on Decision and Control*, December 2006, pp. 5073 – 5077.
[5] D. A. Paley, N. E. Leonard, R. Sepulchre, D. Grnbaum, and J. K. Parrish, "Oscillator models and collective motion," *IEEE Control Systems Magazine*, vol. 27, no. 4, pp. 89 – 105, August 2007.
[6] R. Sepulchre, D. Paley, and N. Leonard, "Stabilization of planar collective motion: All-to-all communication," *IEEE Transaction of Automatic Control*, vol. 52, no. 5, pp. 811 – 824, May 2007.
[7] N. Moshtagh and A. Jadbabaie, "Distributed geodesic control laws for flocking of nonholonomic agents," *IEEE Transaction on Automatic Control*, vol. 52, April 2007.
[8] J. Marshall, M. Broucke, and B. Francis, "Formations of vehicles in cyclic pursuit," *IEEE Transaction of Automatic Control*, November 2004.
[9] J. Marshall, T. Fung, M. Broucke, G. D'Eleuterio, and B. Francis, "Experiments in multirobot coordination," *Robotics and Autonomous Systems*, vol. 54, pp. 265–275, November 2006.
[10] M. Pavone and E. Frazzoli, "Decentralized policies for geometric pattern formation and path coverage," *ASME Journal on Dynamic Systems, Measurement, and Control*, 2007.
[11] J. Jeanne, N. Leonard, and D. Paley, "Colletive motion of ring-coupled planar particles," in *44th IEEE Conference on Decision and Control*, Seville, Spain, December 2005, pp. 3929 – 3934.
[12] E. Justh and P. Krishnaprasad, "Equilibria and steering laws for planar formations," *Systems and Control letters*, vol. 52, no. 1, pp. 25–38, May 2004.
[13] N. Moshtagh, A. Jadbabaie, and K. Daniilidis, "Vision-based control laws for distributed flocking of nonholonomic agents," in *IEEE International Conference of Robotics and Automation*, Orlando, Florida, May 2006, pp. 2769 – 2774.
[14] P. Ogren, E. Fiorelli, and N. Leonard, "Cooperative control of mobile sensing networks: Adaptive gradient climbing in a distributed environment," *IEEE Transaction of Automatic Control*, vol. 49(8), pp. 1292–1302, August 2004.
[15] H. Tanner, A. Jadbabaie, and G. Pappas, "Flocking in fixed and switching newtorks," *IEEE Transaction of Automatic Control*, vol. 52, pp. 863–868, 2007.
[16] E. Justh and P. Krishnaprasad, "Natural frames and interacting particles in three dimensions," in *IEEE Conference on Decision and Control*, December 2005, pp. 2841 – 2846.
[17] N. Michael, J. Fink, and V. Kumar, "Controlling a team of ground robots via an aerial robot," in *Proc. of the IEEE/RSJ Int. Conf. on Intelligent Robots and Systems*, San Diego, CA, Nov. 2007.
[18] B. Gerkey, R. Vaughan, and A. Howard, "The player/stage project: Tools for multi-robot and distributed sensor systems," in *Proc. of the Int. Conf. on Advanced Robotics*, Portugal, June 2003, pp. 317–323.

CPG-based Control of a Turtle-like Underwater Vehicle

Keehong Seo, Soon-Jo Chung, Jean-Jacques E. Slotine

Abstract—**We present a new bio-inspired control strategy for an autonomous underwater vehicle by constructing coupled nonlinear oscillators, similar to the animal central pattern generators (CPGs). Using contraction theory, we show that the network of oscillators globally converges to a specific pattern of oscillation. We experimentally validate the proposed control law using a turtle-like underwater vehicle, whose fin actuators successfully exhibit a pattern that resembles the motion of fore limbs of a swimming sea turtle. In order to further fulfill the potential of the CPG-based control, we propose to feed back the actuator states to the coupled oscillators, thereby achieving not only the synchronization of the oscillators, but also the synchronization of actual foil states. Such a closed-loop version of CPGs makes the controller more robust and practical in the presence of external disturbances.**

I. Introduction

Biologically inspired approaches to locomotion have been studied in robotics to develop robots like snake [1], fish [2], salamander [3], and so on. The flexibility and adaptability of the bio-inspired mechanism in dealing with the environment has been discussed in the literature, especially in the context of an alternative solution to traditional means of locomotion such as wheels and propellers. To control these biologically inspired types of robotic locomotion, a plausible approach is to mimic or get inspired from animal central pattern generators (CPGs), leading to modular designs.

A CPG is a neuronal network that exists in animal spinal cord to govern the locomotion [4], [5], [6]. CPGs are believed to relieve the burdens of the central nerve system in controlling the locomotion. In other words, animals walk or run even without paying much attention to the periodic movement of their legs. The CPG system has inspired many robotic researchers since it can reduce the control bandwidth required from the main controller to its actuators.

In engineering applications, the dominant approach is using governing oscillators to represent the neurons in the CPG network and the outputs of the oscillators are used to generate torque inputs or reference signals for servos. Some [7], [8] use feedback from sensors to adapt phase of the governing oscillators while others [9], [10], [11] use open-loop approach without the feedback from sensors to the oscillators, depending on the complexity of the application.

For many types of robots, the actuation for locomotion is essentially an oscillatory motion. It is also true for our testbed of interest, a turtle-like autonomous underwater vehicle (AUV) as introduced in [12]. Its fluidic maneuvers are controlled by the roll and pitch motion of its four fins that mimic the fore limbs of a sea turtle. The roll and pitch motions used for the operation of vehicle in [12] was basically harmonic oscillations.

Our new approach for biologically inspired control simplifies the conventional CPG-based control to establish a robust coordination of the actual fin motions. While our method is based on limit cycle oscillators, we induce the oscillation to emerge by feeding the fin states directly back to the fin actuators. Independent from our study, the emerging oscillation in the close-loop has been studied in a neuro-mechanical system as in [13]. The resulting oscillator exhibits a limit cycle behavior and the oscillation properties can be adjusted by modifying the system parameters. Once the oscillation of the fin motion is established, we can let the motions synchronize among the fins by using diffusive coupling of velocities from other fins. The rolling motion of fins synchronize themselves and the pitch motion in a fin synchronize with its rolling motion with a 90-degree phase difference.

One advantage of the proposed approach over the open-loop CPG method is that we enforce the synchronization of the actuators directly while the open-loop CPG approach enforces only the synchronization of the reference inputs to the actuators. In the open-loop approach, if some disturbance was applied to a fin, the reference input from CPG still produces sinusoidal waves regardless of the current state of the fin, thereby potentially inducing large errors at some moment, which could in turn result in unnecessarily large control effort for recovery. In contrast, if the fin itself is already a self-sustaining oscillator, it does not need the clock-like reference input. In case of the recovery from the disturbance, it can return to its limit cycle from where it was, which would require less control effort than tracking a time-specific reference trajectory. During the recovery phase, the synchronization with other fin is simultaneously achieved through the diffusive velocity couplings, as we prove theoretically and demonstrate experimentally. In essence, the proposed approach serves as a flexible means to recover from disturbances, which is an important characteristic of the robustness to rapidly changing environments.

The paper is organized as follows. Section II describes the

K. Seo was in the Nonlinear Systems Laboratory, Massachusetts Institute of Technology, Cambridge, MA 02139; currently in the Department of Aerospace Engineering, Iowa State University, Ames, IA 50011, Email: kseo@iastate.edu

S. -J. Chung is in the Aerospace Robotics Laboratory, the Department of Aerospace Engineering, Iowa State University, Ames, IA 50011, Email: sjchung@iastate.edu

J. -J. E. Slotine is in the Nonlinear Systems Laboratory, the Department of Mechanical Engineering, Massachusetts Institute of Technology, Cambridge, MA 02139, Email: jjs@mit.edu

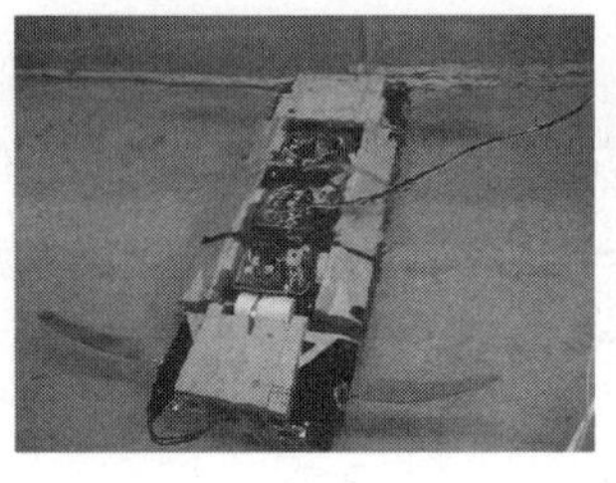

(a)

(b)

Fig. 1. "The biomimetic flapping foil autonomous underwater vehicle (BFFAUV) was conceived as a test platform and proof of concept for the use of flapping foils as the sole source of propulsion and maneuvering forces in an underwater vehicle," quoting [12]

turtle-like AUV as the testbed for the experimental validation. In section III, we introduce previous CPG-based control approaches. In section IV, the new CPG-based control strategy is proposed and its performance of synchronization is discussed. In section V, we present experimental results by implementing the proposed approach in the AUV.

II. The Bio-inspired Underwater Vehicle System

As the test platform of our approach for the biologically inspired control of locomotion, we use an autonomous underwater vehicle (AUV) propelled by flapping foils that resemble the fore fins of a sea turtle. As shown in Figure 1, it has four flapping foils with two degrees of freedom for each. Each fin has freedom in the roll and the pitch directions, actuated by two independent electric motors. The size of the vehicle is as large as 2m $\times$ 0.5m $\times$ 0.5m. The top operating speed is near 2 m/s and the flapping foils provide the whole propulsion as well as the control of the attitude and position. The detailed description of the turtle underwater vehicle can be found in [12].

III. Top-down CPG-based Control Models

In this section, we propose models of CPG-based control of the AUV, based on a top-down architecture. In the top-down CPG-based control architecture, there are two separate layers. One is the CPG layer composed of governing oscillators and the other is the mechanical layer composed of actuators and sensors. Sometimes the architecture is enhanced with sensory feedback from the mechanical layer to the CPG layer.

Similar models are also widely used in the bio-inspired robotics community. In our specific model, the CPG layer is based on the coupled Hopf oscillators. Notice that the artificial CPG model we use is rather a simplification of animal CPGs since we only capture properties essential to our purpose. The fins connected to motors and sensors comprise the mechanical layer.

Based on such an architecture, we propose two CPG-based controllers — one is the open-loop method without any feedback from the mechanical layer to the CPG layer while the other is a feedback approach where the coupling in the CPG layer is wholly composed of mechanical layer elements.

A. Top-down Open-loop approach

One can implement a CPG-based control law with a tracking control law that follows any oscillatory reference signal. In our top-down models, coupled Hopf oscillators form the CPG layer. A Hopf oscillator is a limit-cycle oscillator with circular symmetry on a two-dimensional plane [10], [14]. By feeding states of the oscillator to the servo system of each fin, we can coordinate the ensemble of fin motions and thus control the locomotion of the vehicle. An advantage of using coupled oscillators as the reference is, when we want to extend the system by connecting multiple modules, we have the authority for global synchronization or concurrent synchronization as discussed in [10], [14]. The model introduced in the following is a special case of [10] modified for the AUV.

Let us denote the state vector of Hopf oscillator associated with the roll control of the i-th fin as $\mathbf{x}_i$, and the one associated with the pitch control of the fin as $\mathbf{y}_i$. The state vectors $\mathbf{p}_i$ and $\mathbf{q}_i$ are the roll and the pitch state vectors of the i-th fin, respectively. Each state vector consists of angular velocity and position. A top-down CPG-based control law is proposed as

$$\begin{aligned}\{\dot{\mathbf{x}}\} &= \{\mathbf{f}_x\}(\{\mathbf{x}\}) - k_x \mathbf{L}_x \{\mathbf{x}\} &(1)\\ \{\dot{\mathbf{y}}\} &= \{\mathbf{f}_y\}(\{\mathbf{y}\}) - k_y \mathbf{L}_y \{\mathbf{y}\} \\ &\quad + k_{xy}\left(\left[\mathbf{R}\left(\frac{\pi}{2}\right)\right]\{\mathbf{x}\} - \{\mathbf{y}\}\right) &(2)\\ \{\dot{\mathbf{p}}\} &= \{\mathbf{f}_p\}(\{\mathbf{p}\}) + [\mathbf{B}_p]\{\mathbf{x}\} &(3)\\ \{\dot{\mathbf{q}}\} &= \{\mathbf{f}_q\}(\{\mathbf{q}\}) + [\mathbf{B}_q]\{\mathbf{y}\}, &(4)\end{aligned}$$

where [ ] denotes a block diagonal matrix of appropriate dimension, { } an aggregation of the state vectors in a column vector, and $\mathbf{R}(\phi)$ a planar rotational transformation of angle ϕ. $\mathbf{L}$ is the coupling matrix in each network of oscillators distinguished by its subscript, k is the scalar coupling strength for each network. The dynamics of the oscillators and the tracking controllers are given as

$$\begin{aligned}\mathbf{f}_x(\mathbf{x}_i) &= \mathbf{f}_H(\mathbf{x}_i; \rho_x, \lambda_x) &(5)\\ \mathbf{f}_y(\mathbf{y}_i) &= \mathbf{f}_H(\mathbf{y}_i; \rho_y, \lambda_y) &(6)\\ \mathbf{f}_p(\mathbf{p}_i) &= \mathbf{f}_{PD}(\mathbf{p}_i; B_p, D_p, K_p, P_p, I_p) &(7)\\ \mathbf{f}_q(\mathbf{q}_i) &= \mathbf{f}_{PD}(\mathbf{q}_i; B_q, D_q, K_q, P_q, I_q) &(8)\end{aligned}$$

with input matrices

$$\begin{aligned}\mathbf{B}_p &= \begin{pmatrix} \omega D_p/I_p & P_p/I_p \\ 0 & 0 \end{pmatrix} &(9)\\ \mathbf{B}_q &= \begin{pmatrix} \omega D_q/I_q & P_q/I_q \\ 0 & 0 \end{pmatrix}, &(10)\end{aligned}$$

where

$$\mathbf{f}_H((u_i, v_i); \rho, \lambda) = \begin{pmatrix} -\omega v_i - \lambda\left(\frac{u_i^2+v_i^2}{\rho^2} - 1\right) u_i \\ \omega u_i - \lambda\left(\frac{u_i^2+v_i^2}{\rho^2} - 1\right) v_i \end{pmatrix} \quad (11)$$

and $\mathbf{f}_{PD}((u_i, v_i); B, K, D, P, I)$ is a general PD control law; ω is the common frequency for roll and pitch; ρ the limit cycle radius; λ the convergence rate to the limit cycle; B

the damping coefficient; D the differential gain; K the spring constant; P the proportional gain; and I the moment of inertia.

Notice that (1) and (2) represent a network of coupled oscillators as a special case of [10] while (3) and (4) represent PD controlled mass-spring-damper systems. Hence, the first two consist the CPG layer and the latter two the mechanical layer.

Using partial contraction analysis as in [10], [14], [15], one can find a lower limit of $k > 0$ that ensures the exponential synchronization of the oscillators in the CPG layer. Once synchronized, the diffusively coupled terms (e.g., $q_1 - q_2$) vanish and thus each oscillator behaves as if it were uncoupled to exhibit its intrinsic limit-cycle behavior. The sinusoidal output v_i of $\mathbf{f}_H$ is used as a reference input for the position tracking system of the roll and the pitch controllers. One condition on $\mathbf{L}$ is that it should represent a *connected* network [14].

We can also set an arbitrary phase bias between oscillators in the CPG layer by adjusting $\mathbf{L}$. The synchronization with phase difference is proved in [10], [14]. For example, one can set $\mathbf{L}_x$ as follows for two-way ring couplings:

$$\mathbf{L}_x = \begin{pmatrix} 2\mathbf{I} & -\mathbf{R}(\phi_{12}) & -\mathbf{R}(\phi_{13}) & \mathbf{0} \\ -\mathbf{R}(\phi_{21}) & 2\mathbf{I} & \mathbf{0} & -\mathbf{R}(\phi_{24}) \\ -\mathbf{R}(\phi_{31}) & \mathbf{0} & 2\mathbf{I} & -\mathbf{R}(\phi_{34}) \\ \mathbf{0} & -\mathbf{R}(\phi_{42}) & -\mathbf{R}(\phi_{43}) & 2\mathbf{I} \end{pmatrix} \tag{12}$$

with $\phi_{ij} = -\phi_{ji}$. By setting $\phi_{ij} = 0$ for all i and j, one can establish the in-phase synchronization for roll motions. Setting $\phi_{ij} = 0$ for $i + j = 0 \bmod 2$ and $\phi_{ij} = \pi$ for $i + j = 1 \bmod 2$ yields the bound gait where the fins are synchronized ipsilaterally out of phase and contralaterally in phase. Exponential synchronization implies that the change of phase bias at any moment yields exponentially fast convergence to a new pattern.

Essentially, in the top-down open-loop approach, the coupled oscillators generate coordinated reference signals for the fin actuators. The amplitude and the frequency can be modulated by commanding a small number of parameters such as ρ and ω, thereby reducing the control space.

B. Top-down CPG with Feedback Coupling

One drawback of the open-loop CPG is that the synchronization occurs only in the CPG level and the synchronization of fin motions depends on the performance of the position tracking controller. One can easily consider a scenario where the fins are not ideally identical or the position tracking systems have slightly different performance among fins. As a result, the fins remain slightly unsynchronized while the reference signals are synchronized. To overcome the limit of the top-down open-loop CPG controller, the couplings in CPG are modified to accommodate the state of the fins as in

$$\{\dot{\mathbf{x}}\} = \{\mathbf{f}_x\}(\{\mathbf{x}\}) - k_x \mathbf{L}_x \{\mathbf{p}\} \tag{13}$$

$$\{\dot{\mathbf{y}}\} = \{\mathbf{f}_y\}(\{\mathbf{y}\}) - k_y \mathbf{L}_y \{\mathbf{q}\} + k_{xy}\left(\left[\mathbf{R}\left(\frac{\pi}{2}\right)\right]\{\mathbf{p}\} - \{\mathbf{q}\}\right) \tag{14}$$

$$\{\dot{\mathbf{p}}\} = \{\mathbf{f}_p\}(\{\mathbf{p}\}) + [\mathbf{B}_p]\{\mathbf{x}\} \tag{15}$$

$$\{\dot{\mathbf{q}}\} = \{\mathbf{f}_q\}(\{\mathbf{q}\}) + [\mathbf{B}_q]\{\mathbf{y}\}. \tag{16}$$

By applying differential coordinate transformations in (3) and (4), one can see that the position tracking system is semi-contracting in the absence of the input. Hence, the solutions forget the initial conditions asymptotically, and after some transient time they oscillate at the frequency of the input signals. The amplitude and the phase lag can be computed given the input frequency. By ignoring the transient behavior, we can reduce the preceding model in (13-16) as

$$\{\dot{\mathbf{x}}\} = \{\mathbf{f}_x\}(\{\mathbf{x}\}) - k_x \mathbf{L}_x \{A_p \mathbf{R}(\phi_{xp})\mathbf{x}\} \tag{17}$$

$$\{\dot{\mathbf{y}}\} = \{\mathbf{f}_y\}(\{\mathbf{y}\}) - k_y \mathbf{L}_y \{A_q \mathbf{R}(\phi_{yq})\mathbf{y}\} + k_{xy}\left(\left[\mathbf{R}\left(\frac{\pi}{2}\right)\right]\{A_p \mathbf{R}(\phi_{xp})\mathbf{x}\} - \{A_q \mathbf{R}(\phi_{yq})\mathbf{y}\}\right), \tag{18}$$

where $A_p = A_p(\omega)$ (or $A_q = A_q(\omega)$) is the amplitude as a function of the eigenfrequency ω, and $\phi_{xp} = \phi_{xp}(\omega)$ (or $\phi_{yq} = \phi_{yq}(\omega)$) is the phase lag from $\mathbf{x}$ to $\mathbf{p}$ (or $\mathbf{y}$ to $\mathbf{q}$, respectively), also a function of ω.

The performance of the proposed model depends on the dynamics of $\mathbf{p}$ and $\mathbf{q}$. If the dynamics is more complex than a mass-spring-damper as we assumed, the direct coupling e.g., between $\mathbf{x}$ and $\mathbf{p}$ could disturb the CPG layer. When ϕ_{xp} is larger than π, the system bifurcates to anti-synchronization, where the coupling acts like repulsion instead of attraction.

IV. Single Layer Architecture

While the top-down approach can serve as a good motion-planning method from its simplicity and modularity, the synchronization via coupling is limited only to the CPG oscillators. Actual synchronization of the fins are affected by how identical the dynamics of the fins are. To extend the synchronization of the coupled oscillators beyond the layer of the CPG oscillators to the actual fins, we propose a control law that induces self-sustained oscillation in the fin motions and also enforces direct synchronization of the oscillations.

In this section, we let oscillations emerge from the mechanical layers by feeding a nonlinear function of velocity back to the motor torque. The result is comparable to the neuro-mechanical oscillation discussed in [13] in that the oscillation is induced by coupling non-oscillatory elements. As a result, the CPG layer is eliminated from the top-down architecture to form a single-layer architecture. On top of the oscillating fins, coupling input for synchronization is applied within the mechanical layer. In summary, we can synchronize the oscillation of fins without using a reference oscillator.

One can model the servo-actuated fins by using Euler-Lagrange equations. The aim here is to apply nonlinear state feedback to construct a limit cycle oscillator. Consider the following second-order coupled roll-pitch actuator dynamics

$$\mathbf{M}(\mathbf{p}, \mathbf{q}) + \mathbf{C}(\mathbf{p}, \mathbf{q}, \dot{\mathbf{p}}, \dot{\mathbf{q}}) + \mathbf{K}(\mathbf{p}, \mathbf{q}) = (\tau_{roll}, \tau_{pitch})^T \tag{19}$$

In order to focus on verifying the feasibility of the proposed closed-loop CPG method, let us assume that the coupling between $\mathbf{p}$ and $\mathbf{q}$ is relatively small. Then, each decoupled dynamics can be represented by

$$I\ddot{x} + B(\dot{x})\dot{x} + Kx = \tau, \quad (20)$$

where I is the moment of inertia, B the nonlinear damping term as a function of $\dot{x}$, K the spring constant, and τ the input torque. x represents angular displacement, and $\dot{x}$ angular velocity.

The nonlinear damping of the fin in the fluid can be modelled as

$$B(\dot{x}) = \beta_0 + \beta_1|\dot{x}|, \quad (21)$$

where $\beta_0, \beta_1 > 0$. The term led by β_1 is justified by the experimental observation [12], [16], where the lift force of a single fin oscillating in the fluid is in phase with the angular velocity and its magnitude is proportional to the angular velocity squared.

A. Velocity feedback

Let us design a torque control law as

$$\tau = -Px + \gamma_0\dot{x} - \gamma_1|\dot{x}|\dot{x} + Is(t), \quad (22)$$

where $s(t)$ is a synchronizing input to be discussed further below. The closed-loop dynamics of (20) and (22) is

$$I\ddot{x} + (\beta_0 - \gamma_0 + \beta_1|\dot{x}| + \gamma_1|\dot{x}|)\dot{x} + (K+P)x = Is(t). \quad (23)$$

After dividing the equation by I and setting $\omega_0 = \sqrt{\frac{K+P}{I}}$ and $C(\dot{x}) = \frac{\beta_0 - \gamma_0 + \beta_1|\dot{x}| + \gamma_1|\dot{x}|}{I}$, we have

$$\ddot{x} + C(\dot{x})\dot{x} + \omega_0^2 x = s(t). \quad (24)$$

By choosing γ_0 and γ_1 to satisfy $\beta_0 - \gamma_0 < 0$ and $\beta_1 + \gamma_1 > 0$, the system shows limit cycle behavior. If we denote $(\gamma_0 - \beta_0)/I = \sigma_0$ and $(\beta_1 + \gamma_1)/I = \sigma_1$, then

$$\ddot{x} + (\sigma_1|\dot{x}| - \sigma_0)\dot{x} + \omega_0^2 x = s(t). \quad (25)$$

The resulting dynamics shows limit cycle property. Different feedback controllers yield different types of limit cycle oscillators and we may also use them to control the AUV.

We can deliberately select the values σ_1, σ_0, and ω_0 for the feedback controller to shape the limit cycle and its frequency. Depending on the scale of the nonlinearity in the damping term, we can compute its amplitude and frequency as follows.

B. Weak Nonlinearity

For a weakly nonlinear case, i.e., for $\sigma_1 \ll 1$ and $\sigma_0 \ll 1$, the phase portrait is close to a circle. We can apply singular perturbation theory for two-time scales (see [17], [18]) to estimate the amplitude and frequency of the oscillation when it is uncoupled ($s(t) = 0$).

The small parameters $\sigma_0, \sigma_1 \ll 1$ can be written using $\epsilon \ll 1$ and $0 < B_0, B_1 \sim \mathcal{O}(1)$ as in

$$\ddot{x} + \epsilon(B_1|\dot{x}| - B_0)\dot{x} + w_0^2 x = 0. \quad (26)$$

The amplitude of oscillation is an important parameter to generate proper swimming motion. We can assume a sinusoidal solution with amplitude r. The amplitude r is found to have "slow" dynamics as

$$r' = -\frac{1}{2}w_0^{-1}B_0 r + B_1 r|r|\frac{4}{3\pi}, \quad (27)$$

where r' is the time derivative of r with respect to the "slow" time. Its stable equilibrium is found at $r_\infty = \frac{3\pi}{8w_0}\frac{B_0}{B_1}$. The frequency of motion is not modulated during the swimming. However, we need to set the frequency at a certain range to ensure agility of the vehicle. It is dealt with the perturbation theory as well. The result is found as $\omega = \omega_0 + \mathcal{O}(\epsilon^2)$.

C. Strong Nonlinearity

In the strong nonlinearity, represented by $\sigma_0, \sigma_1 \gg 1$, the phase portrait looks distorted compared to that of weakly nonlinear oscillator. We replace ϵ in (26) with $\mu \gg 1$ to have

$$\ddot{x} + \mu(B_1|\dot{x}| - B_0)\dot{x} + \omega_0^2 x = 0. \quad (28)$$

It is often called a relaxation oscillator. The name follows from the fact that the cycle of oscillation composed of slow build-up and fast relaxation. After some work on the analysis on the phase portrait, one can find an approximate period and amplitude of the solution $x(t)$ as follows. The period is found as

$$T = \frac{2\mu}{\omega_0^2}\left(B_0 \ln \frac{B_0}{B_0 + \sqrt{B_0(B_0+\omega_0)}} + \sqrt{B_0(B_0+\omega_0)}\right) \quad (29)$$

and the amplitude of oscillation can also be approximated by considering its nullclines as $r_\infty = \frac{\mu B_0}{4\omega_0 B_1}$.

D. Bias

The center of oscillation has been at the origin through the discussion so far, which can be extended to a biased oscillation centered at $x = x_0$. To implement a bias x_0 in x, the feedback input must be modified as

$$\tau = -Px + \gamma_0\dot{x} - \gamma_1|\dot{x}|\dot{x} + Is(t) + (K+P)x_0 \quad (30)$$

and the closed loop dynamics becomes

$$\ddot{x} + C(\dot{x})\dot{x} + \omega_0^2(x - x_0) = s(t). \quad (31)$$

One can define a new variable such as $x' = x - x_0$ to apply the results above.

E. Integration with Underwater Vehicle Control System

For each cycle of the oscillation, the main CPU determines the oscillation parameters such as the amplitude, frequency, and bias by comparing the current attitude and the desired attitude. Such updates of the oscillation parameters are performed at a much slower rate than the sampling rate of the feedback controller that governs the oscillatory motion of fins. The oscillation parameters commanded by the main CPU need to be mapped to the parameters for the feedback controller. Although we derived all the equations regarding how the parameters of the feedback controller determines the oscillator parameters, there are some unknown constants that present the physical property of the system. It is also possible that there are some dynamics that are not accounted for in (20). Hence, for the successful implementation of the proposed CPG-based controller, we chose to determine the relation of controller parameters versus oscillation parameters through experimental tests and the curve fitting method. To determine the order of the curve, the equations derived above are helpful.

F. Coupling Input for Synchronization

The synchronizing input for i-th system $s_i(t)$ can be designed in various ways. If we assume the dynamics of the oscillators are identical, then we can use

$$s_i(t) = \kappa(y_j - y_i), \tag{32}$$

where $y_i = \dot{x}_i/\omega_0$. Partial contraction analysis, similar to that in [15], can show that they will synchronize asymptotically for strong enough coupling gain $\kappa > 0$. Furthermore, to force the oscillators to synchronize with a phase difference of ϕ, one can use

$$s_i(t) = \kappa(\sin(\phi)x_j + \cos(\phi)y_j - y_i). \tag{33}$$

Notice that the phase ϕ_i of the state (x_i, y_i) can be defined in terms of x_i and y_i as $\phi_i = \tan^{-1}\frac{y_i}{x_i}$.

As for the synchronization of the roll oscillator and the pitch oscillator, it is often desired that their amplitudes remain independent. In that case, one can simply modify the coupling by scaling with the estimate amplitudes r_i and r_j of the oscillations as in

$$s_i(t) = \kappa\left(\frac{r_i}{r_j}y_j - y_i\right). \tag{34}$$

To present the coupling input $s_i(t)$ for synchronization in a brief form, let us define $\mathbf{x}_i = (x_i, y_i)^{\mathrm{T}}$, $\mathbf{R}(\phi) \in \mathcal{SO}(2)$ to be a planar rotational transformation of angle ϕ, and $\mathbf{S} = (0\ 1)$. Then, after integrating the discussion above on $s_i(t)$, we can propose the coupling input for synchronization to the fin i as

$$s_i(t) = \kappa\mathbf{S}\left(\frac{r_i}{r_j}\mathbf{R}(\phi)\mathbf{x}_j - \mathbf{x}_i\right). \tag{35}$$

G. Synchronization Analysis for One-way Ring Network

Using partial contraction theory and its extended theorems in [15], one can analyze the stability of synchronization.

1) Special Case with $\phi = 0, \pi$: The analysis for asymptotical synchronization with phase difference of $\phi = 0$ or π is found in [15] for van der Pol oscillators. Below, we derive the same conclusion using the method of projected jacobian introduced in [14].

Let us start with the synchronization of roll oscillations with $\phi = 0$. The result can be easily extended for $\phi = \pi$. The angular position and the angular velocity of the roll motions are represented by a vector $\mathbf{x}_i$ for fin i. For $\phi = 0$, we have $\mathbf{R}(0) = \mathbf{I}$ and the dynamics of the coupled oscillators can be written as

$$\dot{\mathbf{x}}_i = \begin{pmatrix} 0 & \omega_0 \\ -\omega_0 & \sigma_0 - \sigma_1\omega_0|y_i| \end{pmatrix}\mathbf{x}_i + k\mathbf{K}\left(\mathbf{x}_j - \mathbf{x}_i\right), \tag{36}$$

where $k = \kappa\omega^{-1}$ and

$$\mathbf{K} = \begin{pmatrix} 0 & 0 \\ 0 & 1 \end{pmatrix} \tag{37}$$

with $j = i + 1$ (modulo 4) for the one-way ring network.

The subspace for synchronization is $\mathcal{M} = \{\mathbf{x}_1 = \mathbf{x}_2 = \mathbf{x}_3 = \mathbf{x}_4\}$, which is verified to be flow-invariant under (36). According to [14], if the jacobian of the projected dynamics on $\mathcal{M}^{\perp}$ has negative definite symmetric part, then the system is contracting toward $\mathcal{M}$, i.e., synchronizing.

The jacobian $\mathbf{J}_i$ of uncoupled dynamics of an oscillator is

$$\mathbf{J}_i = \begin{pmatrix} 0 & \omega_0 \\ -\omega_0 & \sigma_0 - 2\sigma_1\omega_0|y_i| \end{pmatrix} \tag{38}$$

and its symmetrical part $\mathbf{J}_{is}$ is

$$\mathbf{J}_{is} = \begin{pmatrix} 0 & 0 \\ 0 & \sigma_0 - 2\sigma_1\omega_0|y_i| \end{pmatrix}. \tag{39}$$

Define

$$\mathbf{L} = \begin{pmatrix} \mathbf{K} & -\mathbf{K} & \mathbf{0} & \mathbf{0} \\ \mathbf{0} & \mathbf{K} & -\mathbf{K} & \mathbf{0} \\ \mathbf{0} & \mathbf{0} & \mathbf{K} & -\mathbf{K} \\ -\mathbf{K} & \mathbf{0} & \mathbf{0} & \mathbf{K} \end{pmatrix} \tag{40}$$

and its symmetric part as $\mathbf{L}_s$. Now the jacobian of the coupled oscillator has its symmetric part as

$$\mathbf{J}_s = [\mathbf{J}_{is}] - k\mathbf{L}_s, \tag{41}$$

where $[\ ]$ denotes a block diagonal form. An orthogonal projection $\mathbf{V}$ to $\mathcal{M}^{\perp}$ can be found as [19]

$$\mathbf{V} = \frac{1}{2}\begin{pmatrix} \mathbf{I} & -\mathbf{I} & \mathbf{I} & -\mathbf{I} \\ \mathbf{0} & -\sqrt{2}\mathbf{I} & \mathbf{0} & \sqrt{2}\mathbf{I} \\ -\sqrt{2}\mathbf{I} & \mathbf{0} & \sqrt{2}\mathbf{I} & \mathbf{0} \end{pmatrix} \tag{42}$$

using the eigenvectors of $\mathbf{L}_s$.

From [14], if the projected jacobian $\mathbf{P}_s = \mathbf{V}\mathbf{J}_s\mathbf{V}^{\mathrm{T}}$ is uniformly negative definite, then $\mathbf{V}\{\mathbf{x}\}$ converges to $\mathbf{0}$, which is equivalent to $\{\mathbf{x}\}$ converging to $\mathcal{M}$.

Ruling out the zero columns and rows from $\mathbf{P}_s$ yields

$$\overline{\mathbf{P}}_s = \overline{\mathbf{V}[\mathbf{J}_{is}]\mathbf{V}^{\mathrm{T}}} - k\begin{pmatrix} 2 & 0 & 0 \\ 0 & 1 & 0 \\ 0 & 0 & 1 \end{pmatrix}, \tag{43}$$

where $\overline{(\ )}$ denotes the remaining part after removing the zero columns and rows. The eigenvalues of $\overline{\mathbf{V}\mathbf{L}_s\mathbf{V}^{\mathrm{T}}}$ is $(2, 1, 1)$ and $\overline{\mathbf{V}[\mathbf{J}_{is}]\mathbf{V}^{\mathrm{T}}}$ is upper-bounded by $\sup_{y_i}(\sigma_0 - 2\omega_0\sigma_1|y_i|) = \sigma_0$.

Hence, a sufficient condition for $\overline{\mathbf{P}}_s$ to be negative definite is $k > \sigma_0$. Since the removed columns and rows correspond to positions x_i, the negative definiteness of $\overline{\mathbf{P}}_s$ implies that $\mathbf{P}_s$ negative semi-definite (semi-contracting). By Barbalat's lemma, it is straightforward to show that the velocities $\dot{x}_i$ synchronize asymptotically from any initial condition. Once the oscillators synchronize their velocities, i.e., $\dot{x}_i - \dot{x}_j \to 0$, the coupling inputs s_i vanish, resulting in a stable limit cycle. Since $\delta\{\mathbf{x}\}^T\delta\{\mathbf{x}\}$ tends to a lower limit asymptotically, its higher-order Taylor expansion, similar to [15], indicates that the angular positions on the resulting limit cycle synchronize as well, i.e., $x_i - x_j \to 0$.

2) General Case with Arbitrary Phase Bias: The following is a general form in the sense that it accommodates an arbitrary phase bias ϕ in the coupling as well as scaling for amplitude difference as r_i/r_j.

$$\begin{aligned}\dot{\mathbf{x}}_i &= \begin{pmatrix} 0 & \omega_0 \\ -\omega_0 & \sigma_0 - \sigma_1\omega_0|y_i| \end{pmatrix}\mathbf{x}_i \\ &+ k\mathbf{K}\left(\frac{r_i}{r_j}\mathbf{R}(\phi)\mathbf{x}_j - \mathbf{x}_i\right), \end{aligned} \quad (44)$$

where $\mathbf{K}$ is defined in (37) and set $j = i+1$ with modulo 4 for a one-way ring network. We use shorthand notation $\mathbf{T}_{i-1} = \frac{r_i}{r_{i+1}}\mathbf{R}((i-1)\phi)$. Also, we assume that the phase bias ϕ is the same for each oscillator. For the same one-way ring topology represented by (40), we present the proof of the synchronization of $\mathbf{x}_i$ to $\mathcal{M} = \{\mathbf{T}_i\mathbf{x}_{i+1} = \mathbf{T}_{i-1}\mathbf{x}_i, \forall i \bmod 4\}$. Hence, the flow-invariant set $\mathcal{M}$ contains phase-shifted variables such as $\mathbf{x}_1 = \mathbf{T}_1\mathbf{x}_2 = \mathbf{T}_2\mathbf{x}_3 = \mathbf{T}_3\mathbf{x}_4$. Therefore, $\prod_i^n \mathbf{T}_{i-1} = \mathbf{I}$ is required as a constraint.

For simplicity, let us assume $r_j = r_i$. (For $r_j \neq r_i$, one can use the coordinate transformation introduced in [10]). If we define $\mathbf{z}_i = \mathbf{R}((i-1)\phi)\mathbf{x}_i = \mathbf{T}_{i-1}\mathbf{x}_i$, then

$$\mathbf{x}_{i+1} = \mathbf{R}(-\phi)\mathbf{x}_i \Leftrightarrow \mathbf{z}_{i+1} = \mathbf{z}_i. \quad (45)$$

The virtual system dynamics for $\delta\mathbf{z}_i$ can be obtained by left-multiplying (44) with $\mathbf{T}_{i-1}$:

$$\delta\dot{\mathbf{z}}_i = \left(\mathbf{T}_{i-1}\mathbf{J}_i\mathbf{T}_{i-1}^T\right)\delta\mathbf{z}_i + k\mathbf{T}_{i-1}\mathbf{K}\mathbf{T}_{i-1}^T(\mathbf{z}_{i+1} - \mathbf{z}_i) \quad (46)$$

Let us introduce the shorthand notation $\mathbf{K}_{Ti} = \mathbf{T}_{i-1}\mathbf{K}\mathbf{T}_{i-1}^T$. The jacobian $\mathbf{J}_z$ of the congregated system in the space of $\mathbf{z}$ is

$$\mathbf{J}_z = \left[\mathbf{T}_{i-1}\mathbf{J}_i\mathbf{T}_{i-1}^T\right] - k\,\mathbf{L}_z, \quad (47)$$

where [ ] is a notation for block diagonal matrix, $\mathbf{J}_i$ is defined from (38), and

$$\mathbf{L}_z = \begin{pmatrix} \mathbf{K}_{T1} & -\mathbf{K}_{T1} & 0 & 0 \\ 0 & \mathbf{K}_{T2} & -\mathbf{K}_{T2} & 0 \\ 0 & 0 & \mathbf{K}_{T3} & -\mathbf{K}_{T3} \\ -\mathbf{K}_{T4} & 0 & 0 & \mathbf{K}_{T4} \end{pmatrix}. \quad (48)$$

Notice that the eigenvalues of the symmetric part of $\mathbf{T}_{i-1}\mathbf{J}_i\mathbf{T}_{i-1}^T$ are equal to those of the symmetric part of $\mathbf{J}_i$. The eigenvalues of the symmetric part of $\mathbf{L}_z$ also agree with the Laplacian $\mathbf{L}$ in (40). Hence, the proof in the previous section still holds. As a result, we can conclude that the oscillators asymptotically converge to the flow-invariant manifold $\mathcal{M}$ of phase synchronization for any $k > \sigma_0$. This result holds generally for an arbitrarily large number of oscillators with phase shift ϕ.

If we assign the first oscillator to the fore-left fin, the second to the hind-left fin, the third to the fore-right fin, and the fourth to the hind-right fin, then using the one-way ring structure allows us to implement such gaits [20] as *walk*, *bound*, and *pronk* by setting $\phi = \frac{\pi}{2}, \pi$ and 0, respectively.

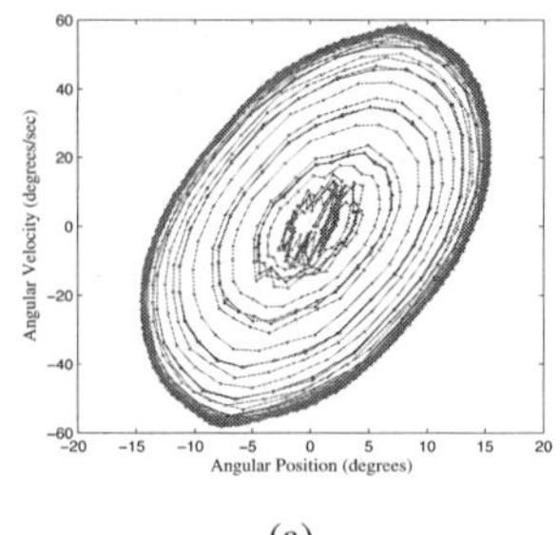

(a)

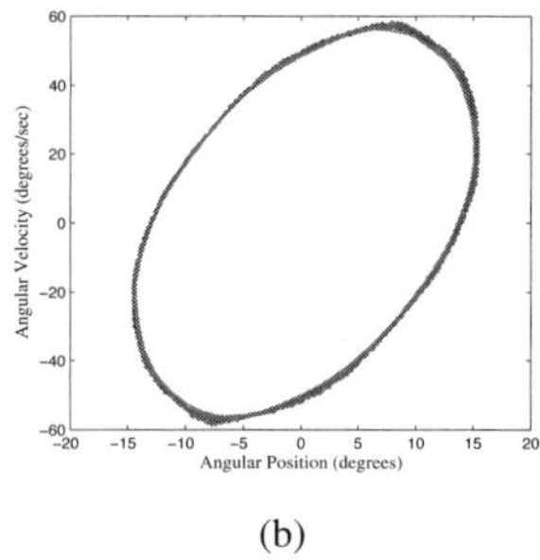

(b)

Fig. 2. The states of the four fins are plotted for angles versus filtered angular rate with respect to the roll axis. The circular trajectory and slow convergence to the limit cycle are the characteristics of weak nonlinearity. (a) The oscillation grows from the origin. (b) The trajectories after 25 seconds are plotted to illustrate limit cycle clearly.

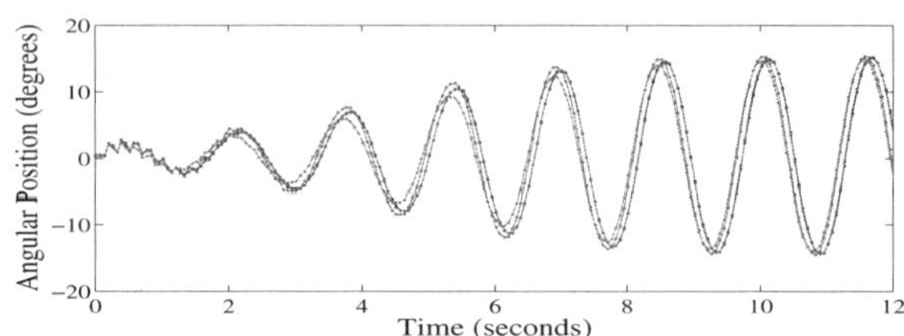

Fig. 3. Roll positions of the four fins are plotted when the synchronization is not applied. The phase differences persist.

H. Network of the oscillators for AUV Fins

The following is the principles that we put on the design of the CPG network for AUV fins.

1) The coupling from a roll oscillator to a pitch oscillator of the same fin is one-way and one-to-one with a 90-degree phase shift.
2) Roll oscillators are coupled with each other and the phase bias between oscillators can be arbitrarily chosen.
3) Pitch oscillators are not connected to each other.

Once the roll oscillators synchronize themselves, we can show that the pitch oscillators also synchronize with the associated roll oscillators in a leader-follower fashion. We refer the readers to [14], [15], [21] for the detailed proof of the leader-follower synchronization using contraction theory. Although the pitch and roll oscillators have different dynamics and different frequencies, after simplifying the model as phase oscillators, one can show that they synchronize to oscillate at a common frequency with some constant phase delay that depends on the coupling strength and the difference between the intrinsic frequencies [22].

V. Experiments using the Single-Layer Approach

We experimentally validated the feasibility of the proposed single-layer CPG-based control for synchronized fin motions with the turtle AUV. After observing that the oscillation actually occurs by using the velocity feedback and that the roll and pitch motions of a fin show stable limit-cycle behaviors, the coupling among the oscillators was activated. Figure 2(a) shows the phase portrait of the oscillation. The x-axis

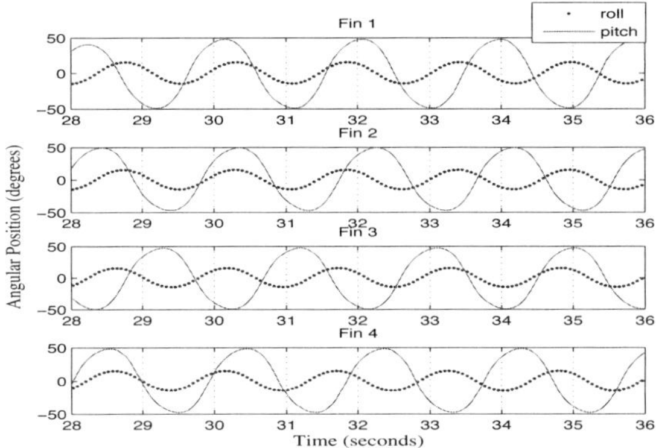

Fig. 4. Unsynchronized behaviors of roll and pitch motions of the four fins when the coupling for synchronization is not applied.

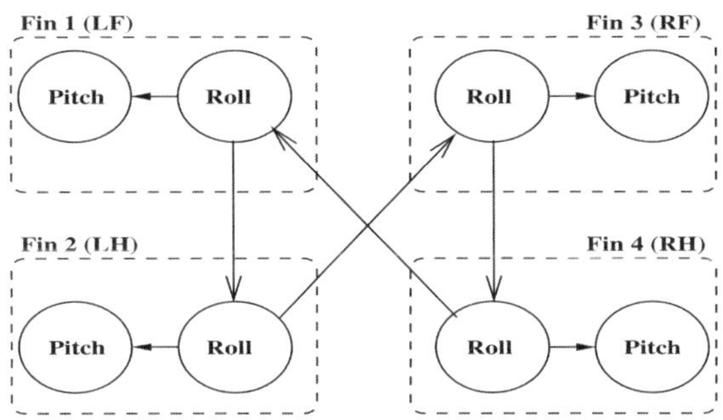

Fig. 5. One-way ring architecture was used to achieve the synchronization of roll motions among fins. Pitch motion synchronizes to its corresponding roll motion.

represents the angular position and the y-axis the angular velocity. The initial condition is near the origin, and the spiral trajectories grow outward to converge to the limit cycle. Figure 2(b) shows the trajectories after 25 seconds to clearly show the resulting limit cycle. The time series of the roll positions are plotted in Figure V, where one can see the motions are not fully synchronized without the couplings between them. The roll motions and pitch motions are plotted together without any couplings to indicate their unsynchronized behaviors (see Figure 4). Notice that their frequencies are also different.

To achieve the synchronization of the fin motions, we applied one-way diffusive ring couplings for the roll controllers of the turtle AUV. For the pitch controllers, we added one-way diffusive coupling with a 90-degree phase lag from the corresponding roll controllers. The coupling scheme is illustrated in Figure 5.

Figure 6 shows the synchronized roll and pitch motions of all four fins. We commanded the bound gait after 25 seconds by applying a phase bias of π in the one-way ring network of roll motions. In Figure 7(a), the pattern starts to shift from the pronk gait (4-fin in-phase synchronization) to the bound gait (the fore fins are out of phase from the hind fins). Since the transition occurred slowly, Figure 7(b) shows the correct bound gait about 30 seconds after the phase bias is changed to π.

Figure 8 demonstrates the property of robustness of the

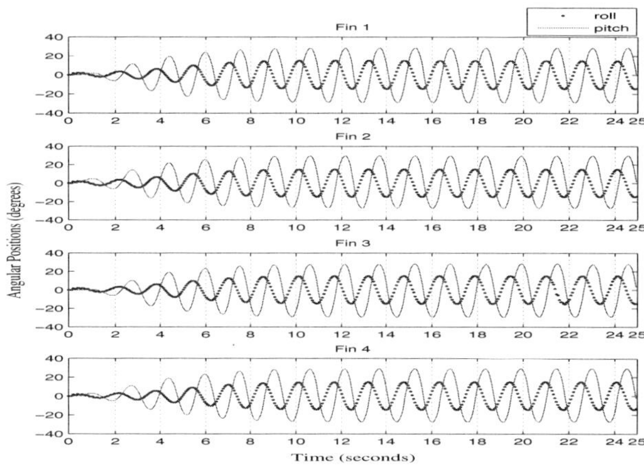

Fig. 6. Synchronized behaviors of roll and pitch motions of the four fins: all the roll motions are synchronized among themselves and the pitch motions are phase locked to the roll motions with 90-degree phase lag

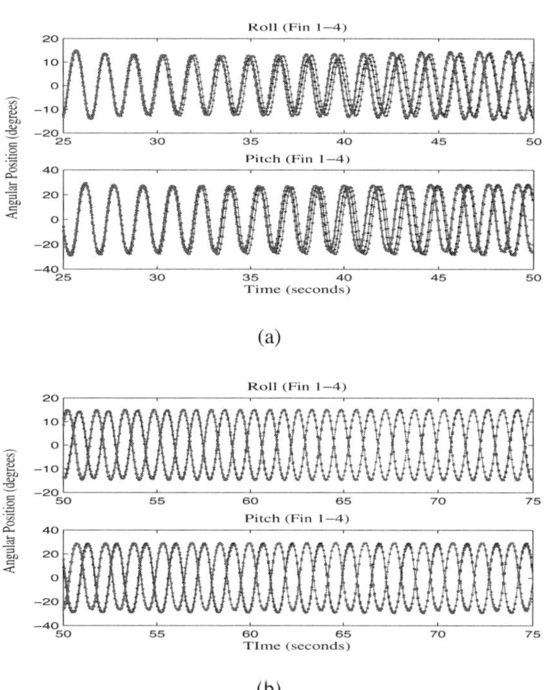

Fig. 7. Gait transition starts at t=25s. (a)The synchronization starts bifurcation. (b)The new pattern "bound" gait settles around t=55s.

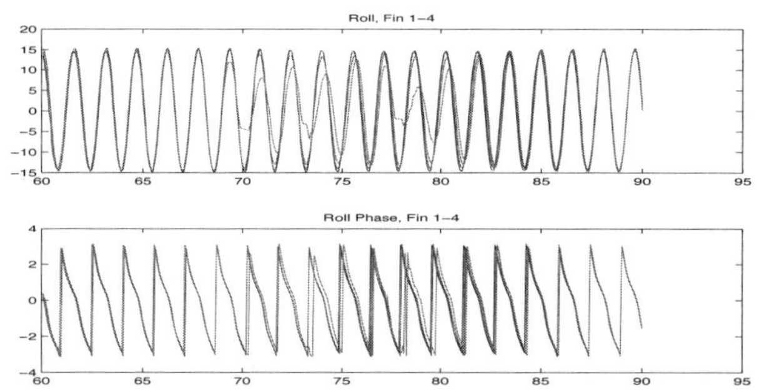

Fig. 8. A fin was disturbed and then recovered back to its synchronized rolling motion.

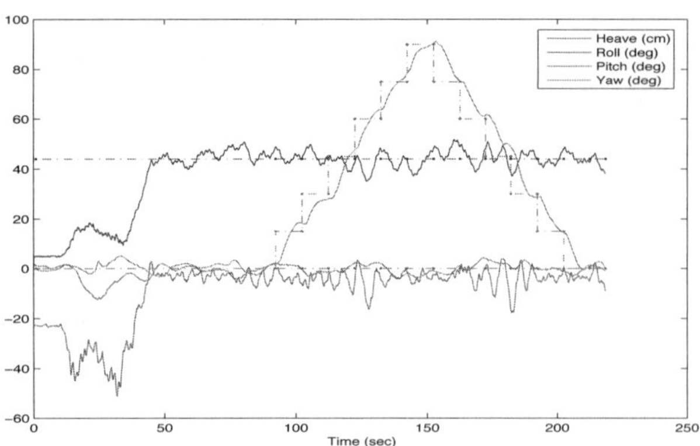

Fig. 9. Underwater mission to follow the reference yaw, pitch, roll angles and heave (depth) was performed using the proposed CPG-based controller.

proposed controller. The rolling motion was disturbed by human intervention and the motion recovers to its limit cycle while all the four fins recover to synchronization.

Finally, to demonstrate the feasibility of the proposed CPG-based controller for underwater missions, the attitude control of the vehicle was tested in a water tank in the MIT Tow Tank Lab. The vehicle is neutrally buoyant and the center of the gravity is located higher than the center of buoyancy, hence the vehicle cannot maintain its attitude without proper controller. Figure 9 shows results of underwater experimentation, where the turtle robot follows the reference heave, pitch, roll, and yaw angles by synchronized oscillatory motions of the foil fins.

VI. Conclusion

We first introduced several conventional top-down CPG-based control strategies for biologically inspired robot locomotion. In order to improve the synchronization performance of the actual fin states, we developed and experimentally validated the new CPG-based approach for a biomimetic AUV. The proposed method can be summarized as follows. By applying a nonlinear velocity feedback, we rendered each fin actuator to exhibit stable limit cycle dynamics. Further, the coupling inputs were added to synchronize multiple fins from any initial conditions. By adjusting the phase bias parameters in the coupling gains, we could also implement gait transitions. The proposed approach was experimentally validated using the turtle-like underwater vehicle. The proposed CPG-based method successfully controlled the attitude and altitude of the underwater vehicle by synchronizing the actuator foil fins in the presence of external disturbances.

Acknowledgment

The authors thank Prof. Michael Triantafyllou for letting them borrow the robotic turtle developed in his laboratory, as well as Stephen Licht for extensive help with the implementation and inspiring discussions on the experiments at the MIT Tow Tank. The first and second authors were partially supported by the Air Force Office of Scientific Research, and thank Prof. James Oliver at the Virtual Reality Application Center, Iowa State University.

References

[1] S. Hirose, *Biologically inspired robots : snake-like locomotors and manipulators*. Oxford University Press, Oxford, New York, 1993.

[2] M. Triantafyllou and G. Triantafyllou, "An efficient swimming machine," *Scientific American*, pp. 64–70, March 1995.

[3] A. Crespi and A. Ijspeert, "Amphibot II: an amphibious snake robot that crawls and swims using a central pattern generator," in *Proceedings of the 9th International Conference on Climbing and Walking Robots (CLAWAR 2006)*, Brussels, Belgium, September 2006, pp. 19–27.

[4] G. S. Stent, W. B. Kristan, Jr., W. O. Friesen, C. A. Ort, M. Poon, and R. L. Calabrese, "Neuronal generation of the leech swimming movement," *Science, New Series*, vol. 200, no. 4348, pp. 1348–1357, June 1978.

[5] N. Krouchev, J. F. Kalaska, and T. Drew, "Sequential activation of muscle synergies during locomotion in the intact cat as revealed by cluster analysis and direct decomposition," *Journal of Neurophysiology*, vol. 96, pp. 1991–2010, October 2006.

[6] I. A. Rybak, N. A. Shevtsova, M. Lafreniere-Roula, and D. A. McCrea, "Modelling spinal circuitry involved in locomotor pattern generation: insights from deletions during fictive locomotion," *Journal of Physiology*, vol. 577, no. 2, pp. 617–639, December 2006.

[7] G. Taga, "A model of the neuro-musculo-skeletal system for anticipatory adjustment of human locomotion during obstacle avoidance," *Biological Cybernetics*, vol. 78, no. 1, pp. 9–17, 1998.

[8] J. Morimoto, G. Endo, J. Nakanishi, S.-H. Hyon, G. Cheng, D. Bentivegna, and C. G. Atkeson, "Modulation of simple sinusoidal patterns by a coupled oscillator model for biped walking," in *Proceedings of the 2006 IEEE International Conference on Robotics and Automation*, May 2006, pp. 1579–1584.

[9] A. Ijspeert, A. Crespi, D. Ryczko, and J.-M. Cabelguen, "From swimming to walking with a salamander robot driven by a spinal cord model," *Science*, vol. 315, no. 5817, pp. 1416–1420, 2007.

[10] K. Seo and J.-J. E. Slotine, "Models for global synchronization in CPG-based locomotion," in *Proceedings of 2007 IEEE International Conference on Robotics and Automation*, Rome, Italy, April 2007, pp. 281–286.

[11] M. A. Lewis, F. Tenore, and R. Etienne-Cummings, "CPG design using inhibitory networks," in *Proceedings of the 2005 IEEE International Conference on Robotics and Automation*, Barcelona, Spain, April 2005, pp. 3682–3687.

[12] S. Licht, V. Polidoro, M. Flores, F. Hover, and M. Triantafyllou, "Design and projected performance of a flapping foil AUV," *IEEE Journal of Oceanic Engineering*, vol. 29, no. 3, July 2004.

[13] M. Sekerli and R. J. Butera, "Oscillations in a simple neuromechanical system: Underlying mechanisms," *Journal of Computational Neuroscience*, vol. 19, no. 2, pp. 181–197, October 2005.

[14] Q.-C. Pham and J.-J. E. Slotine, "Stable concurrent synchronization in dynamic system networks," *Neural Networks*, vol. 20, no. 1, pp. 62–77, 2007.

[15] W. Wang and J.-J. E. Slotine, "On partial contraction analysis for coupled nonlinear oscillators," *Biological Cybernetics*, vol. 92, no. 1, pp. 38–53, 2005.

[16] D. N. Beal and P. R. Bandyopadhyay, "A harmonic model of hydrodynamic forces produced by a flapping fin," *Experiments in Fluids*, vol. 43, no. 5, pp. 675–682, Nonvember 2007.

[17] S. Strogatz, *Nonlinear Dynamics and Chaos: With Applications to Physics, Biology, Chemistry, and Engineering (Studies in Nonlinearity)*. Perseus Books Group, 2001.

[18] H. K. Khalil, *Nonlinear Systems*, 3rd ed. Prentice Hall, 2002.

[19] S.-J. Chung and J.-J. E. Slotine, "Cooperative robot control and concurrent synchronization of lagrangian systems," *IEEE Transactions on Robotics*, vol. in review, 2008.

[20] M. Golubitsky, I. Stewart, P. Buono, and J. Collins, "Symmetry in locomotor central pattern generators and animal gaits," *Nature*, vol. 401, no. 19, pp. 693–695, October 1999.

[21] W. Lohmiller and J.-J. E. Slotine, "Contraction analysis for nonlinear systems," *Automatica*, vol. 34, no. 6, pp. 683–696, June 1998.

[22] A. Pikovsky, M. Rosenblum, and J. Kurths, *Synchronization: A Universal Concept in Nonlinear Sciences*. Cambridge University Press, 2001.

HyPE: Hybrid Particle-Element Approach for Recursive Bayesian Searching-and-Tracking

Benjamin Lavis and Tomonari Furukawa

ARC Centre of Excellence for Autonomous Systems
School of Mechanical and Manufacturing Engineering
The University of New South Wales
Sydney 2052, Australia
Phone: +61-2-9385-4125
Email: benjamin.lavis@student.unsw.edu.au

***Abstract*—This paper presents a hybrid particle-element approach, HyPE, suitable for recursive Bayesian searching-and-tracking (SAT). The hybrid concept, to synthesize two recursive Bayesian estimation (RBE) methods to represent and maintain the belief about all states in a dynamic system, is distinct from the concept behind "mixed approaches", such as Rao-Blackwellized particle filtering, which use different RBE methods for different states. HyPE eliminates the need for computationally expensive numerical integration in the prediction stage and allows space reconfiguration, via remeshing, at minimal computational cost. Numerical examples show the efficacy of the hybrid approach, and demonstrate its superior performance in SAT scenarios when compared with both the particle filter and the element-based method.**

I. Introduction

Recursive Bayesian estimation (RBE) of the state of a dynamic system, under uncertain observation and state transition processes, forms the basis for a variety of autonomous estimation and control problems, including mobile robot localization, environment mapping and exploration, target tracking and optimal searching [1]. RBE techniques recursively update and predict a probability density function (PDF) over the system's state with respect to time. Recursive Bayesian searching-and-tracking (SAT) refers to those RBE problems involving incorporation of sensor data, both when the target is observed (tracking) and when it is not (searching) [2]. SAT may be applied to any searching or tracking tasks (in the general sense of the terms), or multi-objective tasks such as those requiring a lost target to be found and then subsequently tracked.

Many of the fundamental concepts of search theory were first posed by B. O. Koopman and colleagues in the Antisubmarine Warfare Operations Research Group (ASWORG) during World War II [3]. Since the search problem is primarily concerned with the area to be searched, initial studies simplified the search problem to an area coverage problem. The introduction of the probability of detection along with advances in computational hardware led to more optimal allocation of search effort [4], [5], [6]. Later years saw the implementation of RBE for manned search and rescue and anti-submarine search operations [7]. More recently, techniques have been formulated for decentralized search using multiple vehicles [8] and optimal autonomous search using the grid-based method for RBE [9].

On the other hand, target tracking, which initially consisted of simple feedback motion tracking, has evolved with the development of a variety of RBE techniques such as the Kalman filter (KF) [10], the extended Kalman filter (EKF) [11], sequential Monte Carlo (SMC) methods [12] and sequential quasi-Monte Carlo (SQMC) methods [13], [14], and their variants. These techniques for tracking seek computational efficiency in representing the sharp and often near-Gaussian PDF of an observable target, with little thought about representing the boundary of the target search space, which is an important consideration for search missions. While the KF and EKF represent the target PDF with a mean and a covariance matrix, the SMC and SQMC methods represent it with a set of particles (the particle filter, PF), which move freely with a resampling technique such as sequential importance sampling [15], [16].

Recently, the unified SAT approach was introduced using the grid-based method (GBM), and subsequently the element-based method (EBM) [17]. However, the maintenance of a large search space, necessary to include all the possible states of the moving targets, yields an excessively large amount of computational effort. Thus SAT approaches involving reconfigurable search spaces have been developed, using the EBM [18] or PF [19]. The element-based space reconfiguration guarantees the inclusion of the full extent of the target's motion, but introduces significant computational overheads in doing so. Alternatively, PFs inherently reconfigure themselves, but the representation of the full target space, desirable for searching, is limited because only a finite set of discrete samples are considered.

This paper presents HyPE, a hybrid particle-element approach, suitable for recursive Bayesian SAT. The hybrid concept, to *synthesize* two RBE methods to represent and maintain the belief about all states in a dynamic system, is distinct from the concept behind "mixed approaches", such as Rao-Blackwellized particle filtering [20], which use different RBE methods for different states. HyPE performs a *static conversion* between an element-based representation, used during update and evaluation of the PDF, and a particle-based representation, used for Bayesian prediction, and thus eliminates the need for computationally expensive numerical integration. The static conversion process also allows reconfiguration, via remeshing of the target space, at minimal computational cost (the cost associated with remeshing only).

This paper is organized as follows. RBE and SAT are outlined in Sect. II along with a description of PFs and the EBM. Section III details the concept behind HyPE, and its implementation. Numerical examples are shown in Sect. IV and conclusions and future work are contained in the final section.

II. Recursive Bayesian Searching-and-Tracking

This section outlines the general form of RBE, and describes the formulation for the SAT class of problems. Various methods for representing PDFs, the belief metric generated with RBE, are also reviewed.

A. Recursive Bayesian Estimation

RBE seeks to estimate the state, $\mathbf{x}$, of a dynamical system by considering sensor data in light of all previously collected data. The belief about $\mathbf{x}_k$, the state of the system at a discrete time step k, is represented by $p(\mathbf{x}_k|\mathbf{z}_{1:k}, \mathbf{u}_{1:k-1})$, the posterior probability density over the state space $\mathcal{X}$. Here, $\mathbf{z}_{1:k} = \{\mathbf{z}1, \ldots, \mathbf{z}_k\}$ and $\mathbf{u}_{1:k-1} = \{\mathbf{u}1, \ldots, \mathbf{u}_{k-1}\}$ are, respectively, the sequence of all observations, including the current observation, and the sequence of all previous control actions.

The posterior is recursively *updated* using Bayes' Theorem,

$$p(\mathbf{x}_k|\mathbf{z}_{1:k}, \mathbf{u}_{1:k-1}) = \frac{p(\mathbf{z}_k|\mathbf{x}_k)p(\mathbf{x}_k|\mathbf{z}_{1:k-1}, \mathbf{u}_{1:k-1})}{\int p(\mathbf{z}_k|\mathbf{x}_k)p(\mathbf{x}_k|\mathbf{z}_{1:k-1}, \mathbf{u}_{1:k-1})d\mathbf{x}}, \tag{1}$$

where $p(\mathbf{z}_k|\mathbf{x}_k)$ is the likelihood of the observation, described by a sensor model, and $p(\mathbf{x}_k|\mathbf{z}_{1:k-1}, \mathbf{u}_{1:k-1})$ is the *predicted* probability density. Given a posterior, the prediction may be performed in light of a state transition probability, $p(\mathbf{x}_{k+1}|\mathbf{x}_k, \mathbf{u}_k)$, using the Chapman-Kolmogorov equation

$$p(\mathbf{x}_{k+1}|\mathbf{z}_{1:k}, \mathbf{u}_{1:k}) = \int p(\mathbf{x}_{k+1}|\mathbf{x}_k, \mathbf{u}_k)p(\mathbf{x}_k|\mathbf{z}_{1:k}, \mathbf{u}_{1:k-1})d\mathbf{x}_k. \tag{2}$$

Note that when $k = 1$, the prior belief, $p(\mathbf{x}_0)$, is used in place of the prediction in (1).

This general form of RBE is a basis for probabilistic state estimation, however the implementation of RBE requires specific observation likelihood and state transition probability, $p(\mathbf{z}_k|\mathbf{x}_k)$ and $p(\mathbf{x}_k|\mathbf{z}_{1:k-1}, \mathbf{u}_{1:k-1})$, but also a method for representing the updated and predicted PDFs, $p(\mathbf{x}_k|\mathbf{z}_{1:k}, \mathbf{u}_{1:k})$ and $p(\mathbf{x}_{k+1}|\mathbf{z}_{1:k-1}, \mathbf{u}_{1:k-1})$.

B. Searching-and-Tracking

1) SAT observation likelihood: In SAT problems the observation likelihood for a sensor platform s making observations about a target t, is denoted $p({}^s\tilde{\mathbf{z}}_k^t|\tilde{\mathbf{x}}_k^t, \tilde{\mathbf{x}}_k^s)$, as it depends on not only the state of the target, $\tilde{\mathbf{x}}_k^t \in \mathcal{X}^t$, but also the state of the sensor platform, $\tilde{\mathbf{x}}_k^s \in \mathcal{X}^t$. Note that the tilde is used to signify an instance ($\tilde{\cdot}$) of a variable ($\cdot$).

Furthermore, the SAT observation likelihood also depends on ${}^s d_k^t \in \{0, 1\}$, where ${}^s d_k^t = 1$ signifies a detection event and ${}^s d_k^t = 0$ signifies a non-detection event. The SAT observation likelihood is therefore given by

$$p({}^s\tilde{\mathbf{z}}_k^t|\tilde{\mathbf{x}}_k^t, \tilde{\mathbf{x}}_k^s) = \begin{cases} l_d(\tilde{\mathbf{x}}_k^t|\tilde{\mathbf{x}}_k^s, {}^s\tilde{\mathbf{z}}_k^t), & {}^s d_k^t = 1 \\ l_{nd}(\tilde{\mathbf{x}}_k^t|\tilde{\mathbf{x}}_k^s), & {}^s d_k^t = 0 \end{cases} \tag{3}$$

where $l_d(\mathbf{x}_k^t|\tilde{\mathbf{x}}_k^s, {}^s\tilde{\mathbf{z}}_k)$ is the detection, or tracking, likelihood function, and $l_{nd}(\mathbf{x}_k^t|\tilde{\mathbf{x}}_k^s)$ is the non-detection, or searching, likelihood function. Note that for a sensor with a field of view, or 'detection space', ${}^s\mathcal{X}_k^d \subset \mathcal{X}^t$, there is a possibility of missing a detection or falsely detecting a target, that is $\Pr({}^s d_k^t = 0|\exists\tilde{\mathbf{x}}_k^t \in {}^s\mathcal{X}_k^d, \tilde{\mathbf{x}}_k^s) > 0$ and $\Pr({}^s d_k^t = 1|\nexists\tilde{\mathbf{x}}_k^t \in {}^s\mathcal{X}_k^d, \tilde{\mathbf{x}}_k^s) > 0$. Therefore the detection and non-detection likelihoods are typically defined to represent the uncertainty in both the positional observation error, $e \neq 0$, and the truth of the detection. Example detection and non-detection likelihoods and shown in figure 1. One consequence of considering such observation likelihoods is that the resulting update and prediction PDFs are typically non-linear, non-Gaussian and potentially multimodal. For this reason RBE methods which can accommodate arbitrary PDFs are generally preferred for SAT.

2) SAT state transition probability: Often the search vehicle cannot know the control actions of the target vehicle, however many SAT can be considered to be "one-sided" problems. One-sided problems describe SAT tasks where the target cannot deliberately act to aid nor avoid detection. An example of a one-sided SAT problem would be a search and rescue mission involving a powerless vessel adrift at sea. In such cases the state transition probability depends only on the target state, and may be written as $p(\mathbf{x}_k^t|\mathbf{x}_{k-1}^t)$.

The SAT update and prediction equations are therefore given

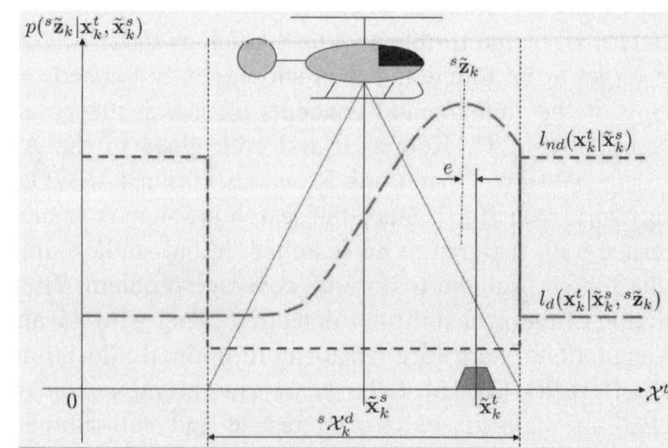

Fig. 1. SAT observation likelihoods.

by

$$p(\mathbf{x}_k^t|{}^s\tilde{\mathbf{z}}_{1:k}^t, \tilde{\mathbf{x}}_{1:k}^s) = \frac{p({}^s\tilde{\mathbf{z}}_k^t|\tilde{\mathbf{x}}_k^t, \tilde{\mathbf{x}}_k^s)p(\mathbf{x}_k^t|{}^s\tilde{\mathbf{z}}_{1:k-1}^t, \tilde{\mathbf{x}}_{1:k-1}^s)}{\int p({}^s\tilde{\mathbf{z}}_k^t|\tilde{\mathbf{x}}_k^t, \tilde{\mathbf{x}}_k^s)p(\mathbf{x}_k^t|{}^s\tilde{\mathbf{z}}_{1:k-1}^t, \tilde{\mathbf{x}}_{1:k-1}^s)d\mathbf{x}_k^t} \quad (4)$$

and

$$p(\mathbf{x}_{k+1}^t|{}^s\tilde{\mathbf{z}}_{1:k}^t, \tilde{\mathbf{x}}_{1:k}^s) = \int p(\mathbf{x}_{k+1}^t|\mathbf{x}_k^t)p(\mathbf{x}_k^t|{}^s\tilde{\mathbf{z}}_{1:k}^t, \tilde{\mathbf{x}}_{1:k}^s)d\mathbf{x}_k^t, \quad (5)$$

respectively.

Furthermore, autonomous SAT may be achieved by utilizing RBE for SAT in the selection of search vehicle control actions. For an objective function J and a finite planning horizon of n_k time steps, a sequence of control actions can be found by solving

$$\arg\max J(\mathbf{u}_{k:k+n_k-1}| \tilde{\mathbf{x}}_k^s, \{p(\mathbf{x}_{k+\kappa}^t|{}^s\tilde{\mathbf{z}}_{1:k}^t, \tilde{\mathbf{x}}_{1:k}^s), \forall\kappa \in \{1, \ldots, n_k\}\}) \quad (6)$$

where $p(\mathbf{x}_{k+\kappa}^t|{}^s\tilde{\mathbf{z}}_{1:k}^t, \tilde{\mathbf{x}}_{1:k}^s)$ can be recursively predicted using the Chapman-Kolmogorov equation,

$$p(\mathbf{x}_{k+\kappa}^t|{}^s\tilde{\mathbf{z}}_{1:k}^t, \tilde{\mathbf{x}}_{1:k}^s) = \int p(\mathbf{x}_{k+\kappa}^t|\mathbf{x}_{k+\kappa-1}^t)p(\mathbf{x}_{k+\kappa-1}^t|{}^s\tilde{\mathbf{z}}_{1:k}^t, \tilde{\mathbf{x}}_{1:k}^s)d\mathbf{x}_{k+\kappa-1}^t. \quad (7)$$

C. Particle and Element Based PDF Representations

1) Particle Filtering: Particle filtering approximates the posterior distribution with a finite set of particles, $\mathcal{P}_k^t = \{{}^t\mathbf{p}_k^1, {}^t\mathbf{p}_k^2, \ldots, {}^t\mathbf{p}_k^M\}$, where each particle, ${}^t\mathbf{p}_k^m$, $m \in \{1, \ldots M\}$, represents a hypothetical state of the target t. The update is performed by sampling the M particles in the filter with a probability proportional to an importance weighting, corresponding to the latest observation. As a result, the particles are distributed according to the current posterior, ${}^t\mathbf{p}_k^m \sim p(\mathbf{x}_k^t|{}^s\tilde{\mathbf{z}}_{1:k}^t, \tilde{\mathbf{x}}_{1:k}^s)$. The prediction stage is carried out by taking M samples from the state transition probability and applying a single instance of the target's motion to each particle, thus avoiding any costly numerical integration in the prediction stage.

2) Element Based Method: The element-based method continuously approximates the target space and PDF using irregularly shaped elements described by shape functions. Generally the target space is first defined by a number of nodes which are then connected so as to create elements. For two-dimensional search spaces the simplest such elements are linear triangular elements generated via Delaunay triangulation. However elements need not be limited by shape or linearity; triangular or quadrilateral elements and higher-order elements with more nodes are all possible, as shown in Figs. 2 and 3.

Let an approximate target space $\mathcal{X}^e$, consisting of n_e elements, be described by

$$\mathcal{X}^e \equiv \{\mathcal{X}_1^e, \ldots, \mathcal{X}_{n_e}^e\} \approx \mathcal{X}^t, \quad (8)$$

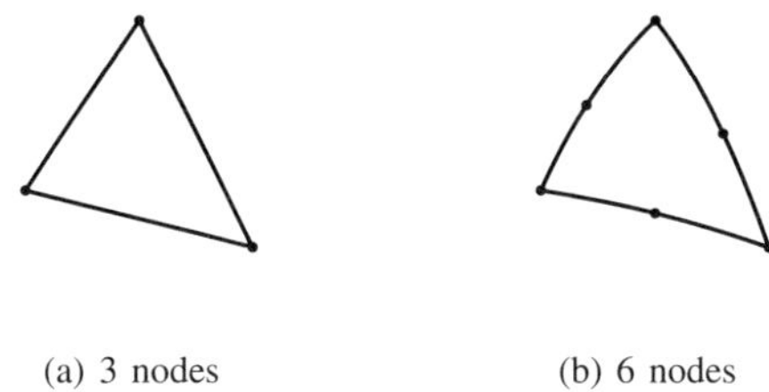

(a) 3 nodes (b) 6 nodes

Fig. 2. Triangular Element Types

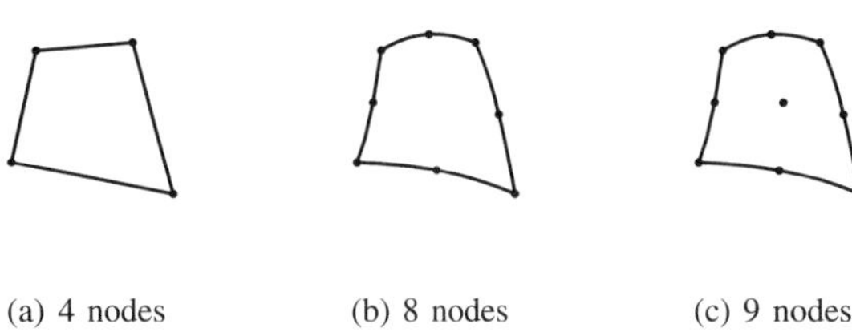

(a) 4 nodes (b) 8 nodes (c) 9 nodes

Fig. 3. Quadrilateral Element Types

where $\bigcup_{i=1}^{n_e} \mathcal{X}_i^e = \mathcal{X}^e$ and $\bigcap_{i=1}^{n_e} \mathcal{X}_i^e = \emptyset$. As such, any point in the target space, $\tilde{\mathbf{x}}^t \in \mathcal{X}^t$ may be located in one of the elements. For a point in the the ith element, $\tilde{\mathbf{x}}^t \in \mathcal{X}_i^e$, the point may be expressed in terms of the n_v nodes of the element. For nodes, $\check{\mathbf{x}}_{ij}^e = [\check{x}_{ij}^e, \check{y}_{ij}^e]^T$, $\forall j \in \{1, \ldots, n_v\}$, $\tilde{\mathbf{x}}^t$ may be expressed as

$$x^t = \varphi_x(\xi, \eta) \equiv \sum_{j=1}^{n_v} \check{x}_{ij}^e N_j(\xi, \eta)$$
$$y^t = \varphi_y(\xi, \eta) \equiv \sum_{j=1}^{n_v} \check{y}_{ij}^e N_j(\xi, \eta) \quad (9)$$

where $N_j(\xi, \eta)$ is the shape function, which must satisfy

$$0 < N_j(\xi, \eta) < 1$$
$$\sum_{j=1}^{n_v} N_j(\xi, \eta) = 1 \quad (10)$$

and $\xi \in \Xi = [\xi_{\min}, \xi_{\max})$ and $\eta \in H = [\eta_{\min}, \eta_{\max})$ are known as the natural coordinates.

The shape function and the ranges of the natural coordinates vary according to the type of element. In general the shape function takes the form

$$N_j(\xi, \eta) = \sum_{k=1}^{n_v} a_{jk} b_k(\xi, \eta) \quad (11)$$

where a_{jk} is a coefficient determined by the constraints (10) and $b_k(\xi, \eta)$ is the basis function of monomials in the natural coordinates. $b_k(\xi, \eta)$ may be determined using the binomial theorem:

$$
\begin{gathered}
b_1(\xi, \eta) = 1 \\
b_2(\xi, \eta) = \xi, b_3(\xi, \eta) = \eta \\
b_4(\xi, \eta) = \xi\eta \\
b_5(\xi, \eta) = \xi^2, b_6(\xi, \eta) = \eta^2 \\
b_7(\xi, \eta) = \xi^2\eta, b_8(\xi, \eta) = \xi\eta^2 \\
b_9(\xi, \eta) = \xi^2\eta^2 \\
\dots
\end{gathered} \tag{12}
$$

For triangular elements the ranges of the natural coordinates are $[\xi_{\min}, \xi_{\max}] = [0, 1]$ and $[\eta_{\min}, \eta_{\max}] = [0, 1 - \xi]$, and for quadrilateral elements the ranges of the natural coordinates are $[\xi_{\min}, \xi_{\max}] = [-1, 1]$ and $[\eta_{\min}, \eta_{\max}] = [-1, 1]$.

In order to perform RBE using the EBM, one must be able to both evaluate a function at a point in the state space, and integrate a function over the search space. Using the EBM the evaluation of a function at a point and the integration of a function may be carried out by considering the natural coordinates. For a point in the ith element, $\tilde{\mathbf{x}}^t \in \mathcal{X}_i^e$, the natural coordinates of the point in the element may be determined using

$$
\left[\tilde{\xi}, \tilde{\eta}\right]^T = \varphi^{-1}(\tilde{\mathbf{x}}^t) \tag{13}
$$

where φ^{-1} is the inverse of the set of functions $\varphi = \{\varphi_x, \varphi_y\}$. The function value at $\tilde{\mathbf{x}}^t$ is then given by

$$
f(\tilde{\mathbf{x}}^t) \approx f^e(\tilde{\mathbf{x}}^t) = \sum_{j=1}^{n_v} f(\check{\mathbf{x}}_{ij}^e) N_j(\tilde{\xi}, \tilde{\eta}). \tag{14}
$$

Integration is performed with respect to the natural coordinates according to the transformation

$$
d\mathbf{x}^t = \det \mathbf{J}(\xi, \eta) d\xi d\eta \tag{15}
$$

where $\mathbf{J}$ is the Jacobian matrix

$$
\mathbf{J}(\xi, \eta) = \begin{bmatrix} \frac{\partial x}{\partial \xi} & \frac{\partial y}{\partial \xi} \\ \frac{\partial x}{\partial \eta} & \frac{\partial y}{\partial \eta} \end{bmatrix}. \tag{16}
$$

Integration over the target space is given by

$$
I = \int_{\mathcal{X}^t} f(\mathbf{x}^t) d\mathbf{x}^t \approx \sum_{i=1}^{n_e} I_i^e \tag{17}
$$

where I_i^e, the integral over an element, is

$$
I_i^e = \int_{\mathcal{X}_i^e} f^e(\mathbf{x}^t) d\mathbf{x}^t = \int_{\Xi, H} f^e(\varphi(\xi, \eta)) \det \mathbf{J} d\xi d\eta. \tag{18}
$$

Note that this integral is only analytically derivable for triangular elements with three nodes. In general, Gauss integration may be used to numerically calculate the integral over each element.

III. HyPE: Hybrid Particle-Element Approach

HyPE seeks to imbue the RBE process with the strengths of each of its constituent methods. The key idea being the synthesis of the two methods, in order to utilize the most appropriate representation at different stages of the estimation process. Through processes of *static conversion* between particle and element representations, HyPE is able to switch to either representation, without the need to maintain the other.

Before HyPE is described in detail, certain terms must be defined. The set of n_n nodes forming the element-based representation of the search space is given by,

$$
\mathcal{N} = \{\mathbf{n}_1, ..., \mathbf{n}_{n_n}\} = \{\check{\mathbf{x}}_{ij}^e | \forall i \in \{1, .., n_e\}, \forall j \in 1, ..., n_v\}\}. \tag{19}
$$

The sets of posterior and prediction values, evaluated at each node in the mesh, represent element-based beliefs, and are given by

$$
\mathcal{B}_k = \{p(\mathbf{x}_k^t = \mathbf{n}_i | {}^s\tilde{\mathbf{z}}_{1:k}^t, \tilde{\mathbf{x}}_{1:k}^s) | \forall i \in \{1, \dots, n_n\}\} \tag{20}
$$

and

$$
\overline{\mathcal{B}}_k = \{p(\mathbf{x}_k^t = \mathbf{n}_i | {}^s\tilde{\mathbf{z}}_{1:k-1}^t, \tilde{\mathbf{x}}_{1:k-1}^s) | \forall i \in \{1, \dots, n_n\}\}, \tag{21}
$$

respectively. Also, the set of observation likelihood values, evaluated at the nodes, is given by,

$$
\mathcal{Z}_k = \{p({}^s\tilde{\mathbf{z}}_k^t = \mathbf{n}_i | \tilde{\mathbf{x}}_k^t, \tilde{\mathbf{x}}_k^s) | \forall i \in \{1, \dots, n_n\}\} \tag{22}
$$

The HyPE approach is described by Algorithm 1. The algorithm takes as input the prediction for the current state, the current observation likelihood, the state transition probability, the set of nodes and the number of particles. The update is performed using the element-based representation, whereas prediction is carried out using the particle-based representation. The algorithm returns the element-based representation of the prediction for the next time step.

Therefore line 1 of Algorithm 1 calls the element-based update function, `update`, to determine the posterior values

Algorithm 1: HyPE: Hybrid Particle-Element Approach

Input : $\overline{\mathcal{B}}_k$, $\mathcal{Z}_k$, $p(\mathbf{x}_{k+1}^t | \mathbf{x}_k^t)$, $\mathcal{N}$, M

Output: $\overline{\mathcal{B}}_{k+1}$

1 $\mathcal{B}_k$ = `update(`$\overline{\mathcal{B}}_k$, $\mathcal{Z}_k$`);`

2 $\mathcal{W}$ = `calculate_weights(`$\mathcal{B}_k$, $\mathcal{N}$`);`

3 $\mathcal{P}_k$ = `generate_particles(`$\mathcal{W}$, $\mathcal{N}$, M`);`

4 $\mathcal{P}_{k+1}$ = `particle_prediction(`$\mathcal{P}_k$, $p(\mathbf{x}_{k+1}^t | \mathbf{x}_k^t)$`);`

5 [Optional] $\mathcal{N}$ = `remesh(`$\mathcal{P}_{k+1}$`);`

6 $\overline{\mathcal{B}}_{k+1}$ = `extract_prediction(`$\mathcal{P}_{k+1}$,$\mathcal{N}$`);`

Function `update(`$\overline{\mathcal{B}}_k$, $\mathcal{Z}_k$`)`

Output: $\mathcal{B}_k$

forall *nodes*, i **do**

  $\mathcal{B}(i)_k = \alpha \overline{\mathcal{B}}(i)_k \mathcal{Z}(i)_k$;

end

at each of the nodes. The α which appears in the update stage is a normalization constant, used to ensure the PDF integrates to unity. Note that neglecting to normalize the PDF using α will not alter the performance of HyPE.

Line 2 of Algorithm 1 calls the function `calculate_weights` in order to determine a sampling weight for each of the nodes in the mesh. The set of sampling weights is denoted $\mathcal{W}$. The sampling weight for each node is its posterior density, given by the product of its posterior density value and its relative volume. Voronoi cells are used to calculate the relative volume associated with each node. Line 1 of `calculate_weights` computes $\mathcal{X}^v$ the Voronoi tessellation of $\mathcal{N}$, where $\mathcal{X}_i^v$ is the Voronoi cell corresponding to node i.

The Voronoi tessellation of a set of nodes randomly distributed in the plane is shown in Fig. 4. Each Voronoi cell $\mathcal{X}_i^v$ defines the space within which all points are closer to $\mathbf{n}_i$ than $\mathbf{n}_{j \neq i}$. Some Voronoi cells are unbounded, such as those shown unshaded in Fig. 4, and therefore have infinite volume. For that reason the relative volume, calculated in line 2, takes the intersection of the Voronoi cell and the search space $\mathcal{X}^t$. Note that if the nodes are equally spaced, such as in a regular mesh, then the relative volumes of all nodes are equal. In such cases, calculation of the relative volume, and also therefore the Voronoi tessellation, is unnecessary and can be neglected.

The M particles to be used for the prediction stage are then generated by calling `generate_particles` in line 3 of Algorithm 1. The particles are generated by sampling with replacement from the n_n nodes. The probability of a node i being selected is proportional to its weighting, $\mathcal{W}(i)$, resulting in the particle set being distributed according to $\mathcal{B}_k$. The standard particle filter prediction is then performed by calling `particle_prediction`, resulting in $\mathcal{P}_{k+1}$. All that remains to complete the HyPE approach is to extract the density of the predicted particle set node locations by calling `extract_prediction`. Existing density extraction techniques which extract particle densities at certain points may be called in line 6. An example of such a technique is *kernel density estimation*, where each particle represents the center of a "kernel", with a known density function. The mixture (sum) of all kernel densities at a point in the search space, gives the overall density at that point.

Function `calculate_weights` ($\mathcal{B}_k$, $\mathcal{N}$)

Output: $\mathcal{W}$
1 $\mathcal{X}^v (= \{\mathcal{X}_1^v, \ldots, \mathcal{X}_{n_n}^v\}) =$ `voronoi` $(\mathcal{N})$;
2 **forall** *nodes, i* **do**
3 $\quad v =$ `volume` $(\mathcal{X}_i^v \bigcap \mathcal{X}^t)$ / `volume` $(\mathcal{X}^t)$;
4 $\quad \mathcal{W}(i) = v\mathcal{B}(i)_k$;
5 **end**

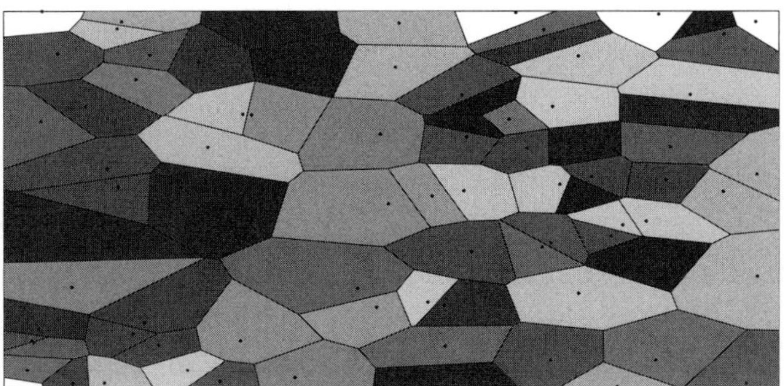

Fig. 4. 2D Voronoi Tessellation.

Function `generate_particles` ($\mathcal{W}$, $\mathcal{N}$, M)

Output: $\mathcal{P}_k$
$\mathcal{P}_k = \emptyset$;
for $m = 1$ *to* M **do**
$\quad$ draw i with probability $\propto \mathcal{W}(i)$;
$\quad$ add ${}^t\mathbf{p}_k^m = \mathbf{n}_i$ to $\mathcal{P}_k$;
end

Function `particle_prediction` ($\mathcal{P}_k$, $p(\mathbf{x}_{k+1}^t | \mathbf{x}_k^t)$)

Output: $\mathcal{P}_{k+1}$
$\mathcal{P}_{k+1} = \emptyset$;
for $m = 1$ *to* M **do**
$\quad$ sample ${}^t\mathbf{p}_{k+1}^m \sim p(\mathbf{x}_{k+1}^t | \mathbf{x}_k^t = {}^t\mathbf{p}_k^m)$;
$\quad$ add ${}^t\mathbf{p}_{k+1}^m$ to $\mathcal{P}_{k+1}$;
end

An alternative density extraction technique is described by the function `extract_prediction` (Voronoi Extraction). This technique may be considered as an element-based generalization of the grid-based technique in which a grid is superimposed on the search space and the number of particles which fall in each grid cell gives the density of the cell. In the Voronoi extraction approach the density is given by the number of particles which fall within a node's Voronoi cell, weighted by the cell's relative volume, (and normalized using a new α value if necessary). Again, if a regular mesh is employed the relative volume need not be calculated. Line 4 in `extract_prediction` makes use of the indicator function,

$$\delta_i({}^t\mathbf{p}_{k+1}^m - \mathbf{n}_i) = \begin{cases} 1, & {}^t\mathbf{p}_{k+1}^m \in \mathcal{X}_i^v \\ 0, & {}^t\mathbf{p}_{k+1}^m \notin \mathcal{X}_i^v. \end{cases} \tag{23}$$

Function `extract_prediction` ($\mathcal{P}_{k+1}$, $\mathcal{N}$)
(Voronoi Extraction)

Output: $\overline{\mathcal{B}}_{k+1}$
1 $\mathcal{X}^v (= \{\mathcal{X}_1^v, \ldots, \mathcal{X}_{n_n}^v\}) =$ `voronoi` $(\mathcal{N})$;
2 **forall** *nodes, i* **do**
3 $\quad v =$ `volume` $(\mathcal{X}_i^v \bigcap \mathcal{X}^t)$ / `volume` $(\mathcal{X}^t)$;
4 $\quad w = \sum_{m=1}^{M} \delta_i({}^t\mathbf{p}_{k+1}^m - \mathbf{n}_i)$;
5 $\quad \overline{\mathcal{B}}(i)_{k+1} = \alpha w / v$;
6 **end**

Both kernel density estimation and the Voronoi extraction technique can be used to extract the density at arbitrary points in the search space. For that reason the option of *remeshing* the search space has been included in line 5 of the HyPE algorithm, before the density extraction function is called. Doing so eliminates the need for extraneous interpolation at the new node locations, and enables the search space boundary and interior nodes to be reconfigured, in order to better capture the nature of the target state, at only the computational cost required to generate the new mesh.

Also, HyPE can perform the multistep prediction often required for the SAT control problem, simply by repetition of the particle prediction step, with density extraction as needed.

IV. Numerical Examples

Ten two-dimensional SAT scenarios were considered in order investigate the efficacy of HyPE and for comparison with the EBM and PF. In each scenario a single sensor platform searched for a single target. The sensor was assumed to have perfect detection capabilities (no false negatives or false positives), with a range of 30 meters, but observations of the target location were assumed to be noisy. The prior density used in all the scenarios was a mixture of two Gaussian distributions with means at $[250, 250]^T$ and $[450, 450]^T$ and covariances

$$\Sigma_1 = \begin{bmatrix} 150 & 0 \\ 0 & 100 \end{bmatrix}, \Sigma_2 = \begin{bmatrix} 90 & 0 \\ 0 & 90 \end{bmatrix}.$$

The element and particle based representation of the prior density are shown in Figs. 5(a)&(b), along with the initial position of the sensor platform and the targets from all ten scenarios. Figure 5(c) shows the velocity map which the targets follow. The state transition probability considered in the RBE used a Gaussian distributions of the velocity at each point, with means given by the velocity map. The control limits of the sensor platform and targets are given in Table I. A time step of $\Delta k = 1$ second was used, and 210 instances of the sensor control actions (speed and steering angle) were used to evaluate a single step lookahead control strategy. The maximum allowable iteration time to complete the the observation, update, prediction and control was set to Δk.

TABLE I
Vehicle Model Control Limits

	Sensor Platform	Target
Maximum Speed [knots]	100	50
Minimum Speed [knots]	10	0
Maximum Steer Angle [deg/s]	15	N/A

Two implementations of HyPE were evaluated, HyPE(a) and HyPE(b), both using regular meshes and the optional remeshing step, to resize the mesh based on the underlying particle distribution. The only difference in implementation between HyPE(a) and HyPE(b) was that HyPE(a) used fewer nodes in the mesh than HyPE(b). Despite running faster than HyPE(b), only one iteration of HyPE(b) (including control optimization) was performed per time step. The search space

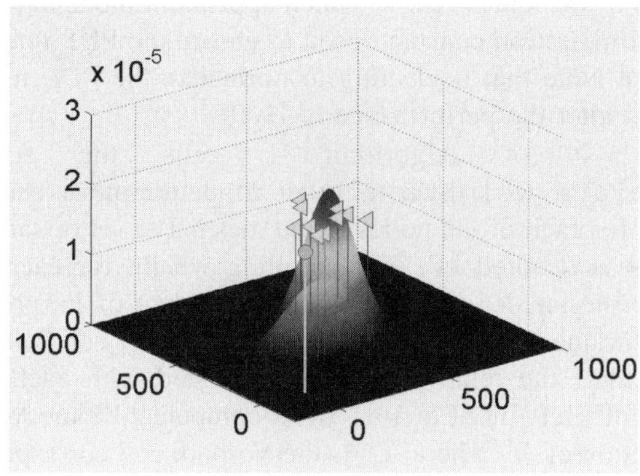

(a) Element-Based Prior

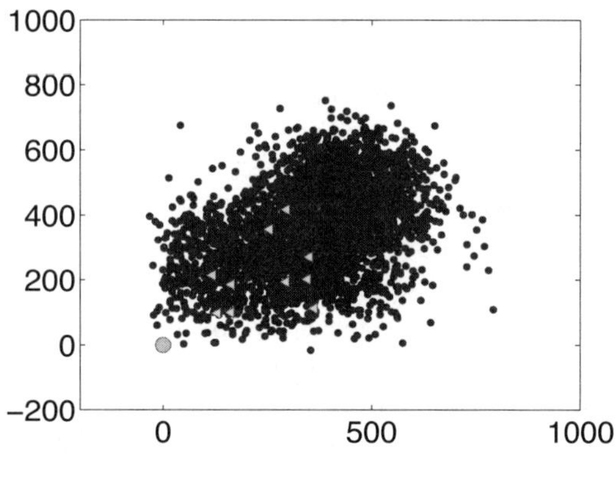

(b) Particle Filter Prior

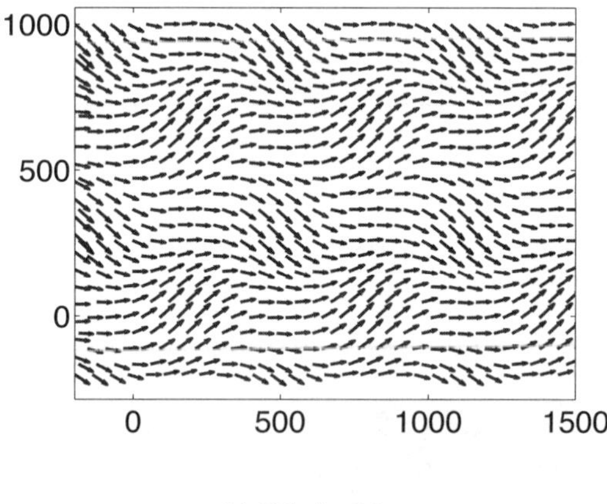

(c) Velocity Map

Fig. 5. Search and Tracking Scenarios

reconfiguration technique for the EBM was prohibitively slow for meeting the iteration time requirement, thus a static mesh was used for the EBM. A fixed sample size was used in the PF. The number of sample points used for each representation are shown in Fig. 6. It can be seen that the median number of sample points used for HyPE(a) was 30% of the number used for the EBM and 14% of the number used for the PF. For HyPE(b) the percentages were 76% compared with the EBM and 35% compared with the PF.

Figure 7 shows the evolution of position errors over time for a single scenario. It can be seen that all approaches enabled the sensor to find the target (indicated by the distance between sensor and target falling below the 30m range line). Furthermore, despite using fewer sample points, the fast remeshing ability of HyPE allowed higher resolution of sample points during tracking, resulting in smaller estimation errors. This is demonstrated in terms of the error between the mode of the posterior density and true target state (note that the mean of the particle positions was used calculate this error for the PF due to the ease its computation and because, for tracking, the discrepancy between the two was negligible). Furthermore, whilst the EBM and PF approaches had trouble maintaining sensor contact with the found target, both implementations of HyPE remained well within sensor contact subsequent to finding the target.

Figure 8 shows the performance of each approach during tracking (subsequent to the first detection of the target). The boxplot represents the distribution of the average sensor and posterior position errors taken over the tracking periods of each scenario. It can be seen that the median sensor position error for HyPE(a) was 33% of the median error for the EBM and 25% of the error for the PF. For HyPE(b) the percentages were 29% compared with the EBM and 22% compared with the PF. The median posterior position error for HyPE(a) was 39% of the median error for the EBM and 26% of the error for the PF. For HyPE(b) the percentages were 26% compared with the EBM and 16% compared with the PF.

In terms of searching performance, the time taken to make the first detection of the target was recorded for each scenario and each approach. The detection times for each scenario were then ranked to determine in which order the approaches found the target. Figure 9(a) shows that on three occasions HyPE(a) was quicker than both the EBM and PF in detecting the target. In the other seven scenarios HyPE(a) was second behind either the EBM (6 times) or the PF (once). Figure 9(b) shows that in half of the scenarios HyPE(b) was quicker than both the EBM and PF in detecting the target. In in the other five scenarios HyPE(b) was second behind either the EBM (4 times) or the PF (once).

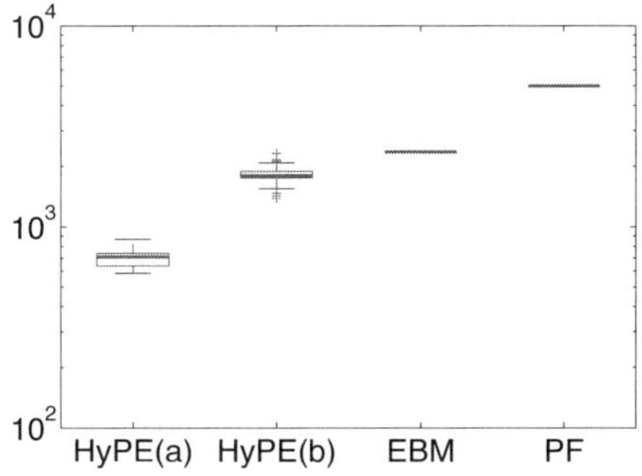

Fig. 6. Number of nodes used in each SAT implementation.

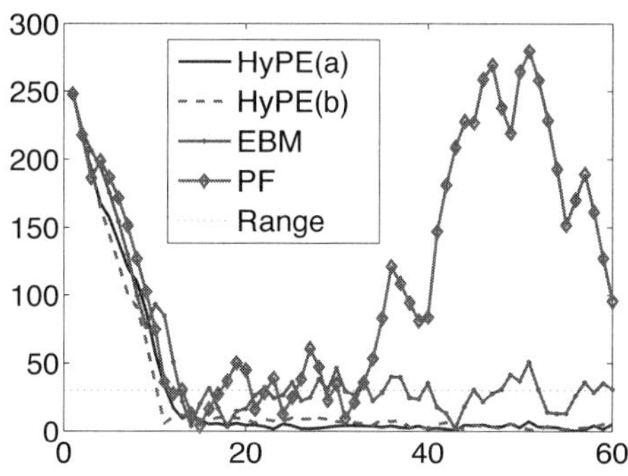

(a) 2D position error between sensor and target (m) vs. time step k.

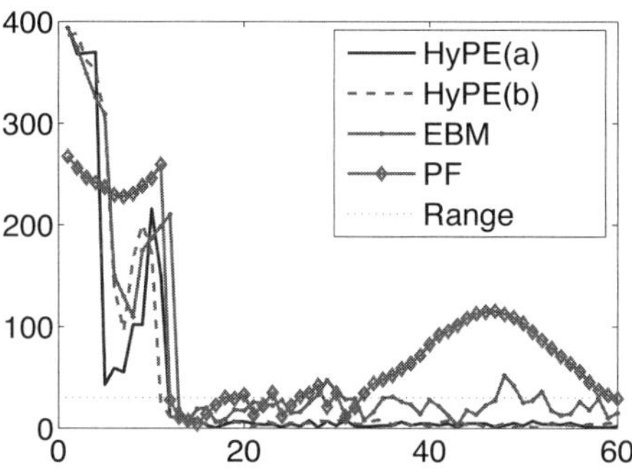

(b) 2D position error between the posterior mode (mean for PF) and target (m) vs. time step k.

Fig. 7. Example of the evolution of position errors over time.

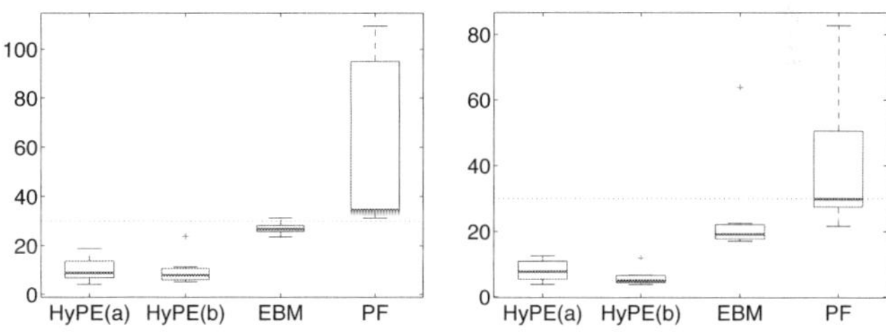

(a) Average 2D position error between sensor and target during tracking.

(b) Average 2D position error between posterior mode (mean for PF) and target during tracking.

Fig. 8. Average position errors during tracking.

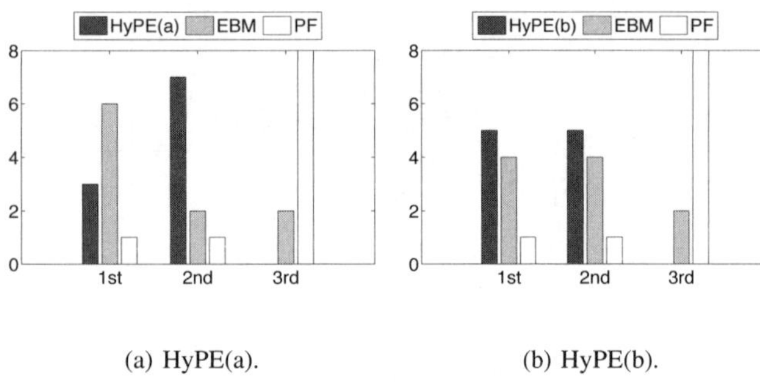

(a) HyPE(a). (b) HyPE(b).

Fig. 9. Search Performance: Number of scenarios ranked 1,2 or 3.

V. Conclusion

This paper presented HyPE, a hybrid particle-element approach, suitable for recursive Bayesian SAT. The hybrid concept, to *synthesize* two RBE methods to represent and maintain the belief about all states in a dynamic system, is distinct from the concept behind "mixed approaches", such as Rao-Blackwellized particle filtering, which use different RBE methods for different states. HyPE performs a *static conversion* between an element-based representation, used during update and evaluation of the PDF, and a particle-based representation, used for Bayesian prediction, and thus eliminates the need for computationally expensive numerical integration. The static conversion process also allows reconfiguration, via remeshing of the target space, at minimal computational cost (the cost associated with remeshing only). The efficacy of the proposed approach was shown through a number of simulated SAT scenarios. Furthermore, it was shown that the ability to quickly remesh the search space allowed HyPE to reduce estimation and position errors by over 39%, whist also using as little as 14% of the number of sample points used by existing methods for RBE.

It should be noted that the two static conversion processes necessarily introduce a degree of approximation error into the estimation which would not otherwise occur. Furthermore, the resolution of the mesh with respect to the range of motion in the target states must be considered, as within the GBM and EBM. If the mesh is too coarse, the predicted particles may never escape the influence of the nodes from which they are sampled.

As this paper represents a first proof of concept for the hybrid particle-element approach, investigation of some of the practical consequences of using the approach remains. For example, an investigation into how the time and memory requirements and transmissibility scale with the scope and dimension of the problem, has been left for future work. Also, the reduction in the number of sampling points required using HyPE may hold significance for the area of data communication in multi-robot cooperation and coordination missions. This will be a focus of future work.

References

[1] S. Thrun, W. Burgard, and D. Fox, *Probabilistic Robotics*. Cambridge, MA: MIT Press, 2005.

[2] T. Furukawa, F. Bourgault, B. Lavis, and H. F. Durrant-Whyte, "Recursive Bayesian search-and-tracking using coordinated UAVs for lost targets," in *Proc. IEEE Int. Conf. Robot. Autom.*, Orlando, Florida, May 2006, pp. 2521–2526.

[3] J. M. Dobbie, "A survey of search theory," *Operations Research*, vol. 16, no. 3, pp. 525–537, May 1968.

[4] L. D. Stone, "Search theory: a mathematical theory for finding lost objects," *Mathematics Magazine*, vol. 50, no. 5, pp. 248–256, Nov. 1977.

[5] ——, *Theory of Optimal Search*. Arlington, VA: Operations Research Society of America (ORSA) Books, 1989.

[6] ——, "What's happened in search theory since the 1975 Lanchester prize?" *Operations Research*, vol. 37, no. 3, pp. 501–506, May-Jun 1989.

[7] H. R. Richardson, L. D. Stone, W. R. Monach, and J. H. Discenza, "Early maritime applications of particle filtering," O. E. Drummond, Ed., vol. 5204, no. 1. SPIE, 2003, pp. 165–174.

[8] F. Bourgault, T. Furukawa, and H. F. Durrant-Whyte, "Coordinated decentralized search for a lost target in a Bayesian world," in *IEEE/RSJ Int. Conf. Intel. Robot. Sys.*, 2003, pp. 48–53.

[9] ——, "Process model, constraints and the coordinated search strategy," in *IEEE Int. Conf. Robot. Autom.*, vol. 5, 2004, pp. 5256–5261.

[10] D. Salmond, "Target tracking: introduction and Kalman tracking filters," in *Proc. IEE Target Tracking: Algorithms and Applications*, vol. Workshop, Oct. 2001, pp. 1/1–1/16 vol.2, (Ref. No. 2001/174).

[11] A. E. Nordsjo, "Target tracking based on Kalman-type filters combined with recursive estimation of model disturbances," in *Proc. IEEE Int. Radar Conf.*, May 2005, pp. 115–120.

[12] C. Hue, J.-P. Le Cadre, and P. Pérez, "Sequential Monte Carlo methods for multiple target tracking and data fusion," *IEEE Transactions on Signal Processing*, vol. 50, no. 2, pp. 309–325, Feb. 2002.

[13] S. Julier, J. Uhlmann, and H. F. Durrant-Whyte, "A new method for the nonlinear transformation of means and covariances in filters and estimators," *IEEE Transactions on Automatic Control*, vol. 45, no. 3, pp. 477–482, Mar. 2000.

[14] D. Guo and X. Wang, "Quasi-Monte Carlo filtering in nonlinear dynamic systems," *IEEE Transactions on Signal Processing*, vol. 54, no. 6, pp. 2087–2098, Jun. 2006.

[15] D. Siegmund, "Importance sampling in the Monte Carlo study of sequential tests," *The Annals of Statistics*, vol. 4, no. 4, pp. 673–684, Jul. 1976.

[16] A. Doucet, S. Godsill, and C. Andrieu, "On sequential Monte Carlo sampling methods for Bayesian filtering," *Statistics and Computing*, vol. 10, no. 3, pp. 197–208, Jul. 2000.

[17] T. Furukawa, H. F. Durrant-Whyte, and B. Lavis, "The element-based method - theory and its application to Bayesian search and tracking," in *Proc. IEEE/RSJ Int. Conf. Intelligent Robots and Systems*, San Diego, California, Oct. 2007, pp. 2807–2812.

[18] B. Lavis, T. Furukawa, and H. F. Durrant-Whyte, "Dynamic space reconfiguration for Bayesian search and tracking with moving targets," *Int. Journ. Auton. Robot.*, Jan. 2008, in print.

[19] B. Lavis and T. Furukawa, "Particle filters for estimation and control in search and rescue using heterogeneous UAVs," in *Proc. 4th Int. Conf. Comp. Intel., Robotics and Auton Sys.*, 28-30 November 2007, pp. 217–222.

[20] C. Stachniss, G. Grisetti, and W. Burgard, "Information gain-based exploration using rao-blackwellized particle filters," in *Robotics: Science and Systems*, 2005, pp. 65–72.

Abstractions and Algorithms for Cooperative Multiple Robot Planar Manipulation

Peng Cheng Jonathan Fink Vijay Kumar
GRASP Laboratory
University of Pennsylvania
Philadelphia, PA 19104 USA
{chpeng, jonfink, kumar}@grasp.upenn.edu

***Abstract*— In this paper, we will study abstractions and algorithms for planar manipulation systems using two cooperating robots under uncertainties. We propose a formal framework for developing abstractions, which are simpler models of the original systems that preserve properties of interest to facilitate the development of planning and control algorithms. Our abstractions are derived from robust motion primitives that correspond to control inputs leading to system trajectories which preserve the properties of interest under uncertainties. We then use the proposed framework to construct an abstraction and design planning and control algorithms for a multiple robot cooperative manipulation system. Finally, we present experimental results to validate our approach.**

I. Introduction

It is well known that conventional approaches to robotic manipulation, where deliberative planning is augmented by feedback controllers, are difficult to implement except in the simplest of cases. This is primarily because of non smooth dynamics engendered by frictional contacts and uncertainties in the parameters governing the contact dynamics. Experiments in robotic juggling [8], locomotion [11, 25], non prehensile manipulation [35], manipulation via caging (Fig. 2) [13], and part-feeding [29] have shown that feedback controllers, behaviors or designs, which are specially designed to preserve a specific set of properties (*e.g.*, convergence to sub manifolds or limit sets), are more robust to uncertainties than those that follow optimally-planned trajectories in the full state space. Indeed, this philosophy of designing components that each drive the system to a state that satisfies a specific property is used extensively in manufacturing operations, where designers carefully structure the environment to ensure that devices like bowl-feeders [14], conveyors [1], traps [5], and pick-and-place arms work in concert to accomplish the given task. Many paradigms in robotics such as caging [7], the one-joint-over-conveyor part positioning [1], and remote-center-of-compliance assembly [12] are also illustrative of this philosophy. While these examples are arguably special-purpose solutions, they illustrate a very important point. By designing planners/controllers that drive the system to submanifolds in the state space, one can derive *abstractions* of complex processes, *i.e.*, conceptual models that are much simpler than the complex real-world system, that lend themselves to the design of *algorithms* that can reason about these abstractions and the composition of these complex processes.

We use the simple example of multi-fingered or multi-robot manipulation in the plane via caging to illustrate the role of abstractions and algorithms (Fig. 2). The modeling of multi-fingered hands or multi-robot manipulation is complicated by the fact it involves multi-body dynamics with frictional contacts. Static indeterminacy and frictional impacts introduce additional difficulty making the design of provably-correct planners and controllers impractical. However, in many manipulation tasks the main goal is to position and orient an object to some destination with a specified tolerance. Since the main *property of interest* is the geometric property of containing or enclosing the manipulated object, one is motivated to derive geometric *abstractions* for the complex, multi-dimensional dynamics problem. This is the central idea in configuration-space abstractions used to derive *algorithms* for multi-robot manipulation: motion planning algorithms for caging [33, 32], control algorithms for object closure [21], and composition of controllers for multi-robot manipulation [13]. Each robot or finger is abstracted into a geometric model. And the planning/control problem is to determine how to move/control these geometric entities to enforce geometric closure.

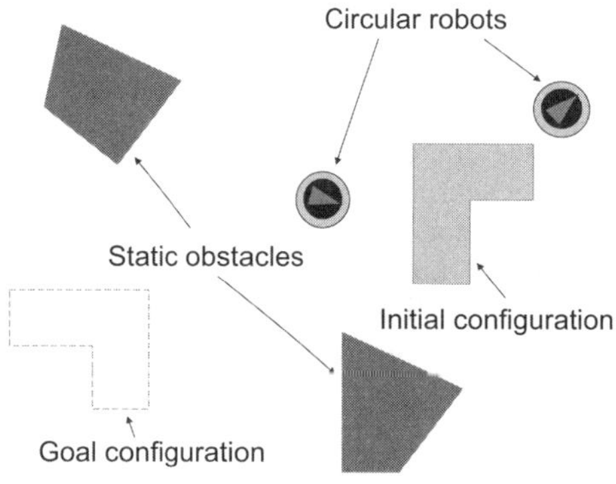

Fig. 1. A representative assembly problem.

In this paper, we will construct abstractions and design planning and control algorithms for multi-contact, planar manipulation tasks in which multiple nonholonomic mobile robots cooperate to manipulate a 2.5-dimensional object on an even, rough surface (see Fig. 1). The manipulation problem in such scenario is very challenging due to non-smooth dynamics and frictional contacts as well as uncertainties in sensing, actuation, and system parameters (*e.g.* friction coefficient and unknown support distribution). It has been studied in [19, 2, 4, 18, 33, 22]. Our work is very similar to [33, 13] in application (using circular robots to manipulate polygonal parts). However, geometric abstractions of caging are used in [33,

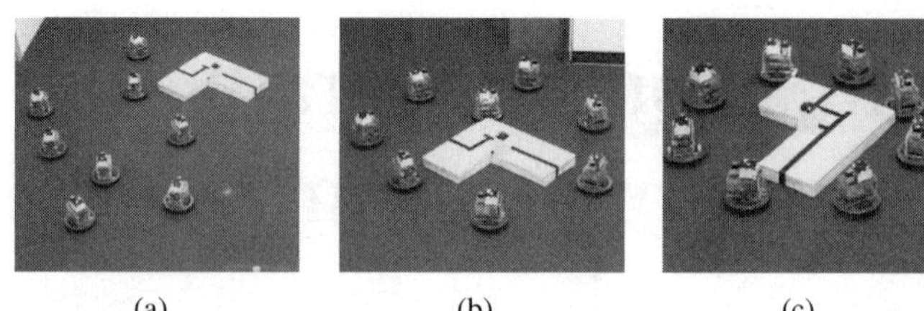

Fig. 2. Approaches to cooperative manipulation and multi-fingered grasping that rely on form or force closure [6, 23, 24, 26] are not as robust to uncertainties as *object closure*, in which the robots or fingers enclose or cage the object. Robots can approach (Fig. 2(a)), surround (Fig. 2(b)), cage (Fig. 2(c)) and manipulate or transport the object reliably using geometric abstractions associated with caging [13].

13], which require at least three robots and large operational space. Also, caging in [13] provides few guarantees on part orientation. In assembly tasks like the one in Fig. 1, it is hard to use caging to drive the part to a goal configuration within a specified tolerance in a constrained environment. The property preserved by our abstraction is neither enclosing nor caging, but to maintain contacts between two manipulation robots and the part. This idea is similar to stable pushing [18]. However, instead of preserving sticking contact between the part and a single pushing bar, we are using two robots to cooperatively manipulate the part by preserving contacts (either rolling or sliding) between both robots and part.

The remainder of the paper is organized as follows. Section II provides a formal framework of abstractions for the manipulation system. The multiple robot cooperative manipulation problem is described in Section III. Abstraction and algorithms for such system are provided in Section IV-B with a focus on abstraction. In Section V, we provide experimental results to validate the proposed approaches.

II. Abstractions of Manipulation Systems

We define a manipulation system by the tuple, $\mathcal{M} = \{f, X, U, \mathcal{P}, T\}$, where X denotes the state space, U the input space, $\mathcal{P}$ the space of (possibly time-varying) model parameters, T a finite time interval, and f the differential equation characterizing the flow of the system. To distinguish between the value of controls ($u \in U$), parameters ($p \in \mathcal{P}$) or state ($x \in X$) from the corresponding trajectories, we use the notation $\tilde{(\cdot)}$ to indicate histories or trajectories. Thus $\tilde{u}$ is the input history, while $\tilde{p}$ is the history of parameter variation and will be used to represent uncertainties. Given a control $\tilde{u} : T \rightarrow U$, a parameter history $\tilde{p} : T \rightarrow \mathcal{P}$, and an initial state $x_0 \in X$, the trajectory is given by $\tilde{x}(x_0, \tilde{u}, \tilde{p}, t) = x_0 + \int_0^t f(\tilde{x}(\eta), \tilde{u}(\eta), \tilde{p}(\eta))\, d\eta, t \in T$.

We use $\tilde{X}$ to denote the set of trajectories with all possible initial states, controls, and parameter histories. We now define the *property* of interest for the system that characterizes the successful execution of a task or subtask as a polymorphic characteristic function, $\Phi : \tilde{X} \rightarrow \{0, 1\}$, which determines whether or not a trajectory of model $\mathcal{M}$ satisfies the given property. It is polymorphic (in analogy to polymorphism in object-oriented programming [3]) because, as we will see, the property function can be used to characterize either the original model or its abstraction. We can also define a subset, a collection of trajectories, $S \subset \tilde{X}$, satisfying a given property: $S = \{\tilde{x} \in \tilde{X} \mid \Phi(\tilde{x}) = 1\}$. In particular, we will be interested in the trivial property, Φ_0, that is satisfied by all trajectories

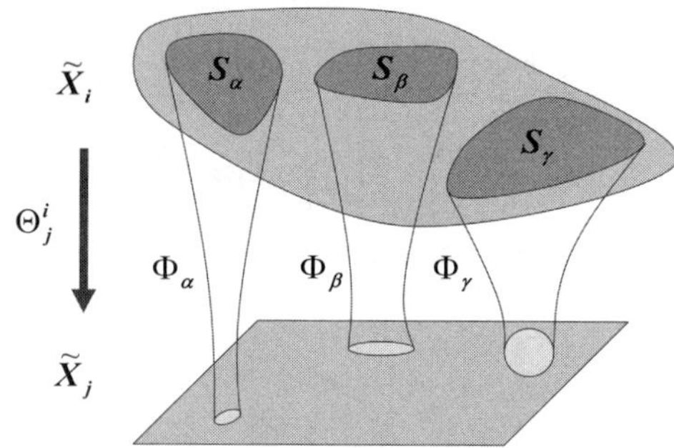

Fig. 3. The surjective map Θ^i_j maps $\tilde{X}_i$ to $\tilde{X}_j$. S_α, S_β, and S_γ are sets of trajectories preserving the properties Φ_α, Φ_β, and Φ_γ.

satisfying the system equations for the model $\mathcal{M}$. In this case, $S = \tilde{X}$. We now establish conditions under which a model $\mathcal{M}_j$ is an abstraction of $\mathcal{M}_i$ with respect to a property, Φ. We use subscript i and j to distinguish components from model $\mathcal{M}_i$ and $\mathcal{M}_j$. Thus, $x_i \in X_i$ is a state in model $\mathcal{M}_i$ and $x_j \in X_j$ is associated with $\mathcal{M}_j$. To keep matters simple, we assume the system is time-invariant and the system dynamics are characterized by a vector of constant parameters for both models and we will omit the dependence on p in the discussion in this subsection. We construct $\mathcal{M}_j$ so that the underlying state space X_j is an image of X_i under the surjective map Θ^i_j. This in turn induces a map in trajectory space as shown in Fig. 3. We say that $\mathcal{M}_j$ is a sufficient abstraction[1] of $\mathcal{M}_i$ if, for any trajectory $\tilde{x}_j(x_j, \tilde{u}_j, t)$ in $S_j \subset \tilde{X}_j$ satisfying the property Φ, there exist $\tilde{u}_i$ and $x_i \in X_i$ so that $\Theta^i_j(x_i) = x_j$ and

$$\Phi\left(\tilde{x}_j(\Theta^i_j(x_i), \tilde{u}_j, t)\right) = \Phi\left(\tilde{x}_i(x_i, \tilde{u}_i, t)\right) = 1. \quad (1)$$

A simple example of this sufficient condition is seen in fully-actuated, six degree-of-freedom robot arms. We frequently use kinematic abstractions ($\mathcal{M}_j$) and inverse-kinematics-based algorithms to plan tasks and trajectories for the tasks because we know that computed-torque-based nonlinear feedback controllers for the real dynamic system ($\mathcal{M}_i$) can be used to realize paths synthesized by simpler kinematic controllers. In other words, these two models satisfy the sufficient condition with respect to the trivial property Φ_0.

III. Cooperative Planar Manipulation

A. The task

We consider the representative problem, depicted in Fig. 1, in which multiple robots are able to manipulate the object into the desired goal. The robots are position-controlled without force or contact sensors. They are able to sense the relative position and orientation of the object and coordinate via communication before manipulation. Because of latency in the network and imperfect sensing, the control during the pushing motion must be open-loop. This paradigm is typical of assembly tasks in industry where robot tasks often involve sequences of subtasks each involving sensing before the subtask, computation, followed by execution. Our goal is to design

[1]One can also define necessary abstractions using a necessary condition in a similar way but since we will not use this concept, we will not discuss this further in this paper.

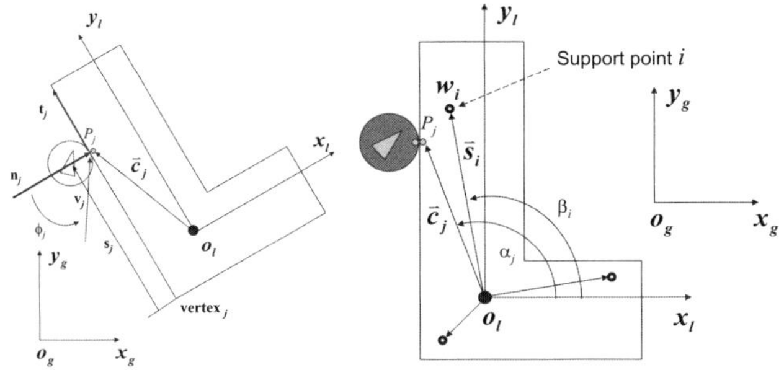

Fig. 4. Geometry and system with the inertial frame (left) and position vectors in the body-fixed frame (right). The object has three, unknown support points ($i = 1, 2, 3$) denoted by unfilled circles. The center of mass is denoted by $\mathbf{o}$, denoted by a filled circle. The contact point between the object and the robot is denoted by P_j with the inward-pointing normal $\mathbf{n}_j$.

controls for multiple robots to manipulate the part to a given configuration within specified tolerances.

Alternative approaches based on force closure require force sensors which lead to expensive and unreliable hardware. Instead we use Roomba-like nonholonomic robots that are position-controlled to follow desired trajectories.

Accordingly, we restrict ourselves to the quasi-static regime where the inertial forces are small compared to the contact forces applied by the robots. Further, we use circular robots to simplify the geometry and the algorithms required for planning and control. Because the application of more than two frictional contacts always results in static indeterminacy we only use two robots at any given time.

B. Modeling and notation

Consider the representative part shown in Fig. 4. We adopt the frictional, three-point support model from [22] but recognize that these support points can change as the part moves and their locations are unknown. The robot(s) exhibit frictional, point contact with the object. All coefficients of friction are unknown but lie within a known set. The part geometry, its inertial properties, and the location of the center of mass are known.

The weight of the part, $w = mg$, is supported by three, unknown support points S_i ($i = 1, 2, 3$) with coordinates (x_i, y_i). The position and orientation of the part is denoted by $q = (x, y, \theta)$ and its velocity in the inertial frame is $\dot{q} = (\dot{x}, \dot{y}, \dot{\theta})$. The body-fixed frame, $x_l - y_l$, has its origin at the center of mass o_l. The jth contact with the jth robot occurs at P_j whose position vector in the body-fixed frame is $\vec{c}_i$.

The robot velocity is $\mathbf{v}_{R,j}$ while the velocity of the point P_j on the part is $\mathbf{v}_{P,j}$. The relative velocity at P_j is given by components $(v_{n,j}, v_{t,j})$ denoting the separation velocity and sliding velocity respectively:

$$v_{n,j} = (\mathbf{v}_{P,j} - \mathbf{v}_{R,j}).\vec{n}_j, \; v_{t,j} = (\mathbf{v}_{P,j} - \mathbf{v}_{R,j}).\vec{t}_j.$$

The forces on the object include the normal forces w_i, the tangential frictional forces at support S_i $(f_{s,i,x}, f_{s,i,y})$, as well as the robot-object contact force at P_j, with components $(\lambda_{n,j}, \lambda_{t,j})$ along the inward-pointing normal $\vec{n}_j$ and tangent $\vec{t}_j$. μ_s is the coefficient of surface friction while μ_c is the coefficient of friction at the robot-object contact.

C. Uncertainties

Although we use the three-point support model to predict the force distribution, we allow the support points S_i to vary. They are chosen to lie within a specified set E_s with the constraint that the center of mass falls within the support triangle.

The friction coefficients between the part and the support and between the part and robots are unknown, but they are assumed to belong to a known, compact set E_f.

The errors in sensing the position and orientation of the object/part and the errors in controlling individual robots must be modeled. The errors in positioning and orienting are denoted by E_t and E_θ. E_d denotes the errors on the relative positions of the robots. In our case, since this is related to the sensing error, $E_d = 2E_t$. We use E_v to denote the error in relative velocity.

D. Quasi-static model for planar manipulation

The non negative normal force at S_i denoted by w_i are uniquely determined from the force equilibrium in the vertical (out-of-plane) direction and the coordinates of the support points:

$$\sum w_i = w, \; \sum w_i \vec{s}_i = 0. \tag{2}$$

The force-balance equations (forces and moments about o_l) in the plane are:

$$W_n(q)\lambda_n + W_t(q)\lambda_t = w_s \tag{3}$$

where $\lambda_n = [\lambda_{n,1}, \; \lambda_{n,2}]^T$ and $\lambda_t = [\lambda_{t,1}, \; \lambda_{t,2}]^T$ and $w_s = \sum_{i=1}^{3} w_{s,i}$ is the resultant support wrench. The wrench matrices W_n and W_t are given by:

$$W_n = \begin{bmatrix} \vec{n}_1 & \vec{n}_2 \\ \vec{c}_1 \times \vec{n}_1 & \vec{c}_2 \times \vec{n}_2 \end{bmatrix}, \; W_t = \begin{bmatrix} \vec{t}_1 & \vec{t}_2 \\ \vec{c}_1 \times \vec{t}_1 & \vec{c}_2 \times \vec{t}_2 \end{bmatrix}$$

We write the tangential sliding velocity as the difference of two non negative quantities:

$$v_{t,j} = v_{t,j}^{+} - v_{t,j}^{-}. \tag{4}$$

We can now write the following complementarity conditions [30]:

$$0 \le v_{n,j} \cdot n_j \perp \lambda_{n,j} \ge 0. \tag{5}$$

$$0 \le v_{t,j}^{+} \perp \lambda_{n,j}\mu_c + \lambda_{t,j} \ge 0 \tag{6}$$

$$0 \le v_{t,j}^{-} \perp \lambda_{n,j}\mu_c - \lambda_{t,j} \ge 0. \tag{7}$$

Note that Equations (2-7) provide a comprehensive description of the system independent of whether each contact is separating, rolling or sliding [34]. Although the uniqueness and existence properties for this set of equations has not been established for the general case, it is possible to show that under conditions of *positive-linear independence* [28], there is a unique solution. This is discussed again in the next section.

E. Practical considerations

In order to ensure the quasi-static assumption is satisfied, we must ensure that the kinetic energy of the object never exceeds the energy that can be dissipated due to friction in some small time interval. Specifically, we are concerned with errors in sensing E_t and E_θ and we want to make sure that the kinetic energy of the object does not cause it to translate more than E_t or rotate more than E_θ. Accordingly we require

$$\dot{x}^2 + \dot{y}^2 \ll E_t g \mu_s \qquad R\dot{\theta}^2 \ll E_\theta g \mu_s. \tag{8}$$

which in turn restricts the velocity of our robots. Second, we cannot require forces that exceed the maximum frictional force or traction between the robot and the support surface.

$$\sqrt{\lambda_{n,i}^2 + \lambda_{t,i}^2} < t_{\max} \tag{9}$$

The robot sensors and controllers, their dynamic properties and the properties of the object will impose further constraints. To ensure robustness to communication latencies and delays we assume that all robots coordinate their execution but do not exchange state information during the manipulation task. The robots used for experiments are approximately 8 Kg. We choose an L-shaped object for manipulation whose mass is around 2.5 Kg. The coefficients of friction are $\mu_s = 0.08 \pm 0.02$ and $\mu_c = 0.6 \pm 0.02$. In order to satisfy Eq. (8), robot speeds are restricted to approximately 10 ± 1 cm/sec with positioning errors of $E_t = 2$ cm and orienting errors of $E_\theta = 5°$. As we will see, only those motion primitives that result in a contact force less than $t_{\max} = 5$ N are allowed.

IV. Abstraction and Algorithms for Cooperative Planar Manipulation

In this section, we will focus on using the proposed abstraction framework in Section II to construct a simple kinematic sufficient abstraction for the original complex quasi-static model in Section III-B. Then, we will briefly describe the planning and tracking control algorithms using such abstraction.

A. Overview and purpose of abstraction

With the notation of Section II, the original complex quasi-static model in Section III-B is represented by $\mathcal{M}_i = \{f_i, X_i, U_i, \mathcal{P}_i, T_i\}$, where f_i is given in Equations (2-7), X_i includes all configurations of the robots and part, U_i includes all the inputs $\{v_i, \phi_i, n_i, c_i\}$ for robots, $\mathcal{P}_i$ includes all system parameters (part geometry, friction coefficients μ_s and μ_c, and three support points $\{x_i, y_i\}$), and T_i includes the time intervals of arbitrary length.

The objective of abstraction is to be able to predict the motion of the part via an efficient approximation of the reachable set. Since the original system has complicated motion behaviors under uncertainties in sensing, actuation, and system parameters, (*e.g.* unknown and changing three support points), it is impossible to compute a reachable set approximation for general inputs. Instead, we construct an abstraction model, which consists the motions of a finite set of *robust motion primitives* for which the following properties of interest are satisfied and preserved under uncertainties such that their reachable sets can be computed.

1) Two contacts between the robots and part, *i.e.*,.

$$\lambda_{n,j} > 0, j = 1, 2. \tag{10}$$

2) Straight line motion at speeds that imply quasi-static dynamics, (Eqs. 8 and 9).

In other words, for a trajectory $\tilde{x}_i : [0, t_f] \to X_i$, the property function Φ returns 1 if Eqs. 10, 8 and 9 are satisfied for any $t \in [0, t_f]$.

A robust motion primitive for the model $\mathcal{M}_i$ is a nominal control $\tilde{u}_i : [0, t_f] \to U_i$ with $\tilde{u}_i(0) = u_0$ and nominal $x_0 \in X_i$ such that with respect to nominal parameter history $\tilde{p}_i \in \mathcal{P}_i$ and $\tilde{p}_i(0) = p_0$

$$\Phi(\tilde{x}_i(x_0 + \delta x_i, \tilde{u}_i + \delta u_i, \tilde{p}_i + \delta\tilde{p}_i, t)) = 1 \tag{11}$$

for $t \in [0, t_f]$ and uncertainties δx_i in sensing, δu_i in actuation, and $\delta\tilde{p}_i$ in system parameters.

After constructing a finite set of robust motion primitives, the resulting sufficient abstraction model $\mathcal{M}_j = \{f_j, X_j, U_j, \mathcal{P}_j, T_j\}$ will be: f_j only include kinematic part of f_i. X_j is still the same as X_i under the identify map Θ_j^i. U_j is a discrete subset of U_i, each of which corresponds to a resulting motion of a robust motion primitive. T_j only includes a set of time intervals of specific length corresponding to each input in U_j.

Because the dynamics of $\mathcal{M}_j$ are invariant (or robust) to the three support points and friction coefficient, we are able to ignore the three point support parameters and friction coefficients in the system parameter set $\mathcal{P}_j$. This greatly reduces the uncertainty in the abstraction model.

While we will use the specific example and parameters described in Section III-E in our development, the same ideas are extensible to *any* planar object and to any set of position controlled robots.

B. Construction of robust motion primitives

The fundamental question in searching for robust motion primitives is to whether the properties of interest are preserved over continuous uncertainty sets. This is a very challenging verification problem for which the state-of-the-art does not include a solution for general systems [15]. Instead, motivated by recent work on sampling-based verification and falsification [10], we use the numerical Monte Carlo simulation method to construct robust motion primitives by checking whether the properties of interest are preserved with respect to a finite number of samples. This at least provides us with a computational tool for complex manipulation systems.

The construction process is carried out in the following steps:

1) Sample $\tilde{u}$ from U_i and x_0 from X_i.
2) Check whether $\Phi(\tilde{x}(x_0, \tilde{u}, \tilde{p}, 0)) = 1$.
   This step check whethers Φ is satisfied at the starting moment by solving the Mixed Complementarity Problem defined by Eqs. 2-7. Theoretically, there are no results on the uniqueness and existence of the solution for the quasi-static problem with multiple rigid bodies under two pushing contact. So, when there is no solution for a given manipulation, we simply say that Φ is not satisfied. When multiple solutions are observed, we say that Φ is preserved if it is satisfied in all solutions.
3) Check whether $\Phi(\tilde{x}(x_0 + \delta x, \tilde{u} + \delta u, \tilde{p} + \delta\tilde{p}, 0)) = 1$ for uncertainties δx, δu, and $\delta\tilde{p}$.
   This step checks whether the properties of interest are satisfied at the starting moment with respect to uncertainties in sensing, actuation, and system parameters. We consider the uncertainties in the three support points, robot velocity magnitude, and friction coefficients that are chosen from three bounded continuous sets. We first generate a finite set of three support point samples,

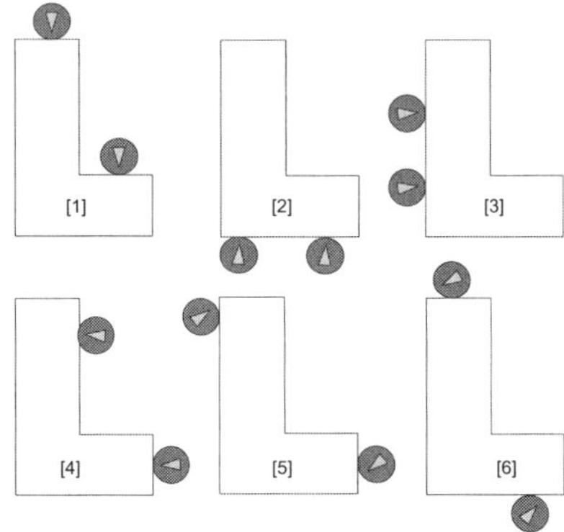

Fig. 5. Robust translational ([1]-[4]) and rotational ([5], [6]) motion primitives

velocity samples, and friction coefficient samples, and then use the procedure in (2) to check whether Φ is satisfied for all these samples. If true, the manipulation is identified as robust; otherwise, it is non robust.

4) Check whether $\Phi(\tilde{x}(x_0+\delta x, \tilde{u}+\delta u, \tilde{p}+\delta\tilde{p}, t)) = 1$ for uncertainties δx, δu, and $\delta\tilde{p}$ in at a finite set of discrete times $t \in [0, t_f]$.
This step checks whether Φ is preserved over the entire duration of a robust motion primitive. The time interval is a bounded continuous set.

Note that for a high dimensional input space, this search process is very computationally expensive. However, we only need such a computation once in the preprocessing. Alternatively, in this paper, we use a set of candidates for robust motion primitives from human intuition instead of random samples.

C. Constructed robust motion primitives

Robust motion primitives for an L-shape part are shown in Fig. 5 (see detailed parameters in Section V). We will now analyze the reachable sets of these primitives under uncertainties, which will help design of the tracking control along a given path.

Abstraction reduces the uncertainties due to the friction coefficient, intermittent contact, and unknown three support points. Uncertainties in sensing still exist and cause the part to vary from its nominal trajectory. However, the preserved properties in the abstraction enable us to predict the motions of the part under these uncertainties by estimating the bounds on their reachable set. In the following, we will compute bounds on these variations as a function of pushing distance d based on kinematic analysis of the abstraction model. These bounds will help to design planning and tracking control algorithms on top of the abstraction model.

1) Bounding the reachable set for the motion in Fig. 5[1-4]: The concept is illustrated with the example shown in Fig. 6. When the intermittent frictional contact modes switch from the left to the right, the maximal offset in y is generated by the following situation. We will assume that two robots push with the same nominal velocity v along the positive x direction with nominal separation distance d_s. Initially, the top and bottom robots respectively have rolling (**R**) and sliding (**S**) contacts (Fig. 6.1). When the coordination errors between two robots results in an error of E_d (Fig. 6.2), the contact modes switch to **S** (top) and **R** (bottom) (Fig. 6.3). Then, when the errors drive

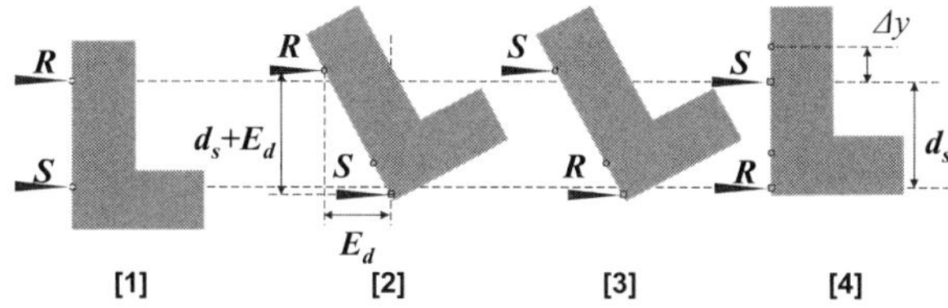

Fig. 6. The position offset in y caused by intermittent contact modes from the left to the right, in which **R** and **S** respectively stand for rolling and sliding.

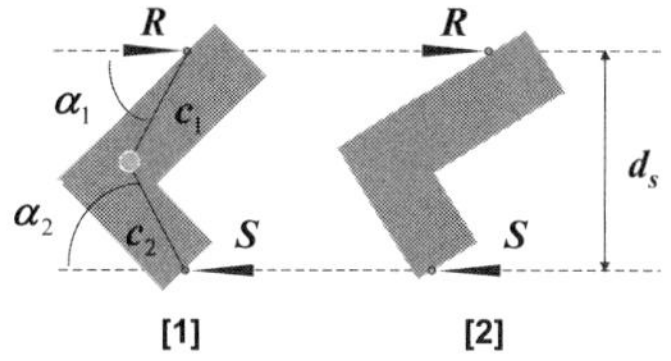

Fig. 7. Constant contact modes cause maximal changes of the configuration.

the robots back to their nominal separation, it can be observed that an offset in y has been generated (Fig. 6.4). Furthermore, when such switching pattern is executed multiple times, the offset in y will be accumulated.

More generally, if both contacts are not on the same edge, the top and bottom contact points are on edges respectively with orientation γ_1 and γ_2. The vector from the top contact point to the bottom has length l and angle χ. We have the following upper bounds over the whole pushing distance d,

$$y \in \begin{array}{l}[-\max(\Delta d_1 \sin\gamma_1, \Delta d_2 \sin\gamma_2)\lceil \frac{E_v d}{4E_d v}\rceil - E_t, \\ \max(\Delta d_1 \sin\gamma_1, \Delta d_2 \sin\gamma_2)\lceil \frac{E_v d}{4E_d v}\rceil + E_t].\end{array} \tag{12}$$

Similarly, the reachable sets of x and θ are respectively bounded by

$$x \in [-E_d + d, E_d + d], \tag{13}$$

$$\theta \in [-E_\theta - \tan^{-1}\frac{E_d}{d_s - E_d}, E_\theta + \tan^{-1}\frac{E_d}{d_s - E_d}]. \tag{14}$$

2) Bounding the reachable set for the motion in Fig. 5 [5-6]: The reachable set under this pushing is also bounded by analyzing the potential contact model switching patterns.

In Fig. 7, when the top and bottom contacts are respectively always rolling and sliding, x reaches its maximal value, which is less than $d + E_d$. Similarly, x is larger than $-d - E_d$ when the top and bottom contact modes are respectively sliding and rolling. Therefore, we have

$$x \in [-d - E_d, d + E_d], \tag{15}$$

$$\theta \in [-E_\theta, \tan^{-1}\frac{2d + d_h + E_d}{d_s - E_d} - \tan^{-1}\frac{d_h - E_d}{d_s + E_d} + E_\theta = \Delta\theta], \tag{16}$$

$$y \in \begin{array}{l}[-c_2(\sin(\alpha 2 + E_\theta) - \sin(\alpha 2 - \Delta\theta - E_\theta)) - E_t, \\ c_1(\sin(\alpha 1 + E_\theta) - \sin(\alpha 1 - \Delta\theta - E_\theta)) + E_t],\end{array} \tag{17}$$

in which d_h is the horizontal distance between two contact points at the starting moment of the manipulation in Fig. 7.1.

D. Planning and tracking control algorithms

For the L shape part in this paper, the four robust translation motion primitives and two robust rotational motion primitives in Section IV-C are sufficient to achieve small time local controllability of the part.

For a general polygonal shape part, we need three translational robust motion primitives and two rotational motion primitives to achieve the small time locally controllable property. Any two of them should not be collinear and the dot product of three direction vectors should be negative. Two rotational motion primitives should be in opposite directions.

Given a path with a given tracking precision E_p, we are able to iteratively track the path using these motion primitives. In each iteration, we compute the pushing distance d for a given robust motion primitive with respect to the required tracking precision. The reachable set of the part after pushing distance d should be bounded inside the E_p-neighborhood of the nominal trajectory. After the pushing, if the part is off the nominal path, robust motion primitives are used to push the part back to the nominal trajectory. In this way, we are able to track any given path with precision up to the sensing and actuation limit.

With this tracking control algorithm, we solve the challenging cooperative manipulation problem by first using a sampling based path planning algorithm, *e.g.* PRM [16] or RRT [17], to compute a collision-free path and then execute by path tracking that relies on robust motion primitives.

V. Experimental Results

We have collected experimental results to demonstrate both the validity of robust motion primitives for mobile robot manipulation as well as the effectiveness of these primitives applied to manipulation/assembly tasks. First, we demonstrate that robust motions (due to the definition in Section IV-B) are feasible for our system then we go on to validate the reachable sets for the primitives derived in Section IV-C. Because we have a conservative estimate of the set of states that can be reached by applying a motion primitive, we can construct a tracking control system that can use robust motion primitives to follow an arbitrary path. Finally, we demonstrate that a simple planning algorithm in conjunction with these techniques can be used to complete a cooperative manipulation task.

A. System parameters for the experimental platform

All experiments are conducted on a multi-robot testbed [20] utilizing a team of small differential drive robots (radius 0.15m) and an overhead tracking system for localization of both the manipulated object and the robots. The position and orientation sensing error due to the tracking system are respectively $E_t = 2$cm and $E_\theta = 5°$. Each robot is controlled using a feedback linearization scheme to tow along a desired trajectory but there is no feedback of relative state information. In other words, while each robot is controlling its own state to execute a straight line maintaining the abstraction in Section IV-A, the cooperative manipulation primitive is executed in an open loop fashion. This introduces additional error in their relative position control bounded by $E_d = 4$cm. Each robot is able to control its velocity within an error of $E_v = 1$ cm/s.

As mentioned earlier, the robots are manipulating an L-shape with a characteristic length of 1m, mass of 2.5Kg, and an approximate coefficient of friction with the floor of $\mu_s = 0.08$. Each robot has a mass of 8.6Kg and coefficient of friction with the L-shape of $\mu_c = 0.6$.

	$vert_1$	s_1	v_1	ϕ_1	$vert_2$	s_2	v_2	ϕ_2
+ X	5	0.9	0.1	0.0	5	0.1	0.1	0.0
+ Y	4	0.7	0.1	0.0	4	0.1	0.1	0.0
- X	1	0.1	0.1	0.0	3	0.3	0.1	0.0
- Y	0	0.1	0.1	0.0	2	0.3	0.1	0.0
+ θ	0	0.1	0.1	−0.7	4	0.1	0.1	−0.7
- θ	5	0.9	0.1	0.7	3	0.3	0.1	0.7

TABLE I

Details for our robust motion primitives, in which s_1 and s_2 are measured in m, ϕ_1 and ϕ_2 are in rad, and v_1 and v_2 are in m/s.

B. Non-robust motions

To highlight the advantage of using robust motion primitives, Fig. 8 depicts non-robust manipulations. Two-point contact is not maintained in these non-robust examples and the motion is unpredictable under the uncertainty inherent to the system. On the other hand, as we will now show, the result of robust motion primitives can be analytically bounded based on the assumptions of system uncertainty.

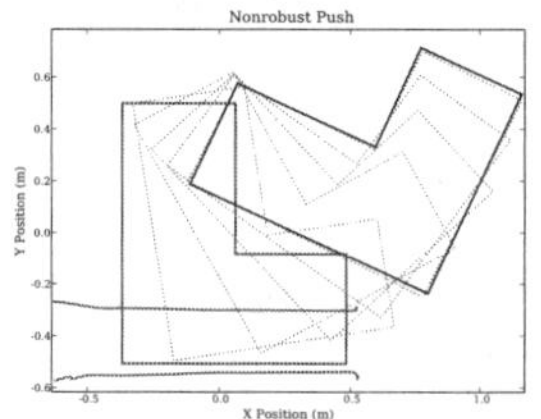

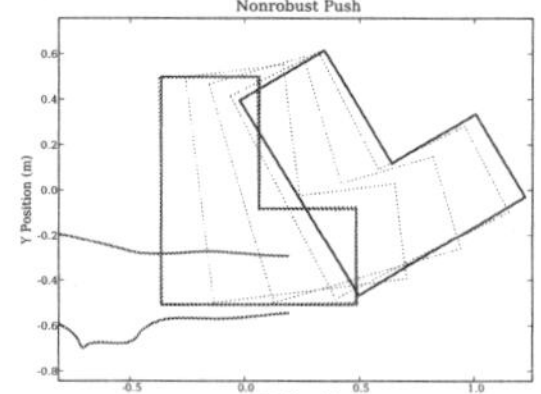

Fig. 8. Non-robust motion primitives.

C. Validation of reachable sets

To compute the bounds on the reachable sets due to motion primitives on the L-shape, we must evaluate the equations presented in Section IV-C. For translation, there is a nonlinear system of equations that must be solved numerically. Other than this step, the rest of the calculations are straight forward given the parameters of the primitive and the uncertainty of the system. Each motion primitive is parameterized as specified in Fig. 5 with the values in Table I. Additionally, the reachable set is a function of the pushing distance d.

We conducted several trials with different configurations of the two canonical manipulation primitives for translation and rotation to show that the resulting trajectories of the manipulated object always lie within the computed bounds. Fig. 9 depicts the part trajectories under different robust manipulation primitives to demonstrate that the final part positions lie within the analytically calculated bounds. Tables II and III provide details on the reachability sets for two sample motion primitives.

We have observed that in experiments, the motion primitives nearly always overshoot the upper bound along the direction of the desired motion. This is the result of assuming a kinematically controlled robot when there are, in fact, acceleration limits. However, it is a simple task to account for this with adequately enlarged reachability bounds or better low-level position control.

D. Validation of the tracking control

Since, for a given motion primitive and pushing distance d, we can calculate bounds the reachable set of the part, it

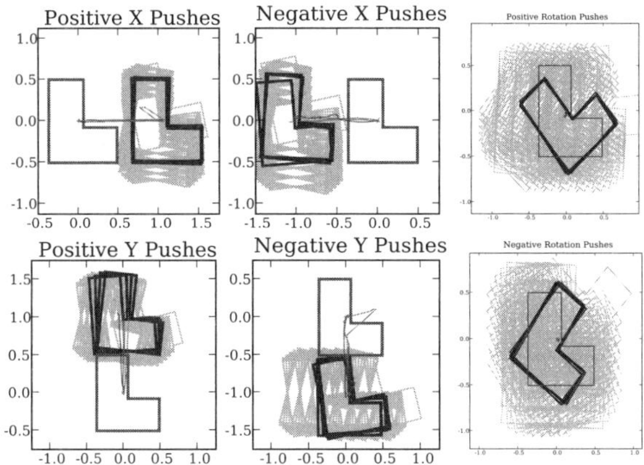

Fig. 9. Part trajectories respectively due to four configurations of the translation primitive ($\pm x$, $\pm y$) and two configurations of the rotation primitive ($\pm\theta$) overlaid on a sampling of the reachable set.

Positive X			
	x(m)	y(m)	θ (°)
Predicted			
Max	1.02	0.13	8.40
Min	0.98	-0.13	-8.40
Measured			
Avg	1.05	0.01	-0.11
Max	1.06	0.02	1.72
Min	1.04	-0.01	-1.32

Positive Y			
	x(m)	y(m)	θ (°)
Predicted			
Max	0.19	1.02	8.67
Min	-0.19	0.98	-8.67
Measured			
Avg	0.01	1.05	-0.11
Max	0.02	1.06	1.72
Min	-0.01	1.04	-1.32

TABLE II

BOUNDS ON ROBUST TRANSLATION PRIMITIVE OVER FOUR TRIALS

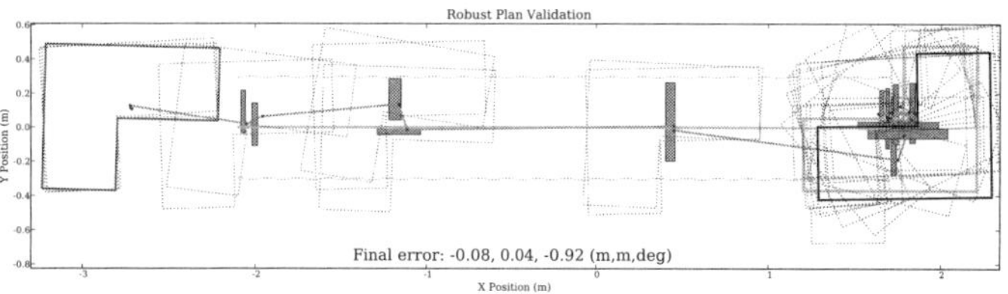

Fig. 10. Example of tracking control with robust motion primitives along a plan with small collision free zone. The bounding box for the reachability set of each translational push is shown.

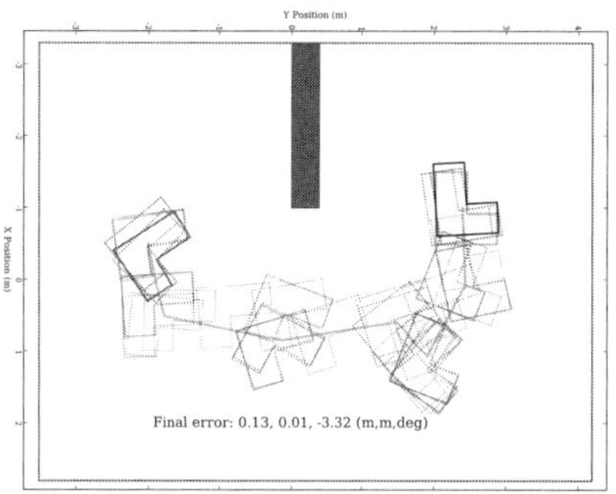

Fig. 11. Trajectory of L-Shape while being pushed along planned obstacle free path with robust motion primitives.

is possible to design a tracking control algorithm that can efficiently follow a path within an E_p-neighborhood. Such a tracking controller can be thought of as a hybrid system that switches between modes for (1) correcting orientation to align the part tangent to path, (2) providing correction perpendicular to the path, and (3) pushing the part along a segment of the path. Each mode must ensure that d is chosen such that the reachable set after pushing will lie within E_p of the path while attempting to minimize tracking error or maximize distance traveled along the path. Thus the tracking controller will employ larger magnitude pushes along paths with larger E_p-neighborhoods.

Transitions between robust primitives are currently handled by a simple circular trajectory that each robot can follow to reach the initial conditions necessary to begin the next desired robust motion primitive. A less conservative but more general approach for transitioning between primitives in a complex environment is a challenging problem that we will address in future work.

Positive θ			
	x(m)	y(m)	θ (°)
Predicted			
Max	0.56	0.21	42.47
Min	-0.58	-0.19	-4.58
Measured			
Avg	-0.02	-0.07	40.11
Max	-0.00	-0.06	43.14
Min	-0.03	-0.08	37.87

Negative θ			
	x(m)	y(m)	θ (°)
Predicted			
Max	0.27	0.49	4.58
Min	-0.31	-0.44	-47.50
Measured			
Avg	0.01	-0.01	-37.07
Max	0.02	0.00	-34.38
Min	-0.01	-0.03	-41.37

TABLE III

BOUNDS ON ROBUST ROTATION PRIMITIVE OVER SEVEN TRIALS

Fig. 10 depicts an example path with a relatively small E_p-neighborhood to show how the tracking controller must use a sequence of pushes and corrections to accurately follow the path.

E. Validation with a manipulation task

Finally, we solve the multiple robot manipulation problem in which two robots must cooperatively push a part from an initial to goal configuration though an environment with obstacles such as that shown in Fig. 1.

We use a sample-based algorithm, such as PRM or RRT, to generate a collision-free path from the initial configuration to the goal configuration which is then tracked with robust motion primitives. Snapshots of the experimental results are shown in Fig. 12, in which the solid piecewise-linear line connecting the initial and goal configurations is the collision free path from the path planning, the curves followed by the robots are the nominal controls to achieve robust motion primitives, and the wire-frame rectangular box represents a virtual obstacle. Fig. 11 shows the resultant trajectory of the L-shape during manipulation.

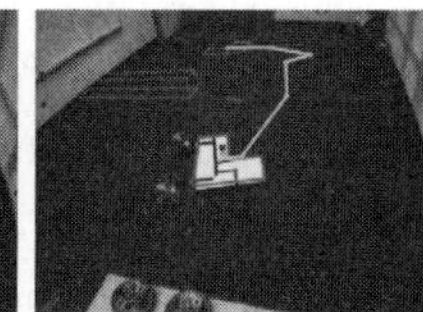

Fig. 12. Snapshots of cooperative manipulation of the L-shape part with two robots

VI. CONCLUSION

In this paper, we proposed a framework to develop abstractions for quasi-static manipulation tasks with uncertainty arising primarily from friction and unknown support points and from errors in control and sensing. The abstractions were used to design algorithms for planar manipulation with cooperating mobile robots and the proposed approach was successfully validated with extensive experimental results. The manipulation system enables two autonomous robots to cooperatively push a part to a given goal configuration with a precision given by the errors in sensing and control.

There are several directions for ongoing work. First, our planning algorithm is very simple and generates very conservative paths. Clearly a better planner will achieve paths in more constrained environments. Because our focus was mainly on abstractions and control, we used a relatively simple planner. However, we are working on refining our planner for more cluttered environments. We are studying robust motion primitives with contact between the part and environment which will lead to a better abstraction for planning in the constrained space. We are also considering extension of the manipulation planning algorithm in [27, 31] to incorporate constraints from the movable part during the manipulation. Second, we used communication-less motion primitives with straight line robot trajectories in this work. Clearly, if robots can communicate, more complex trajectories can be generated and better performance can be obtained. However, this leads to more complexity in the formulation. We are currently studying if it is possible to derive more powerful abstractions that will exploit these additional capabilities.

The main conclusion of this work is simple. If the right abstractions can be derived for manipulation systems, powerful algorithms can also be derived to solve manipulation problems with uncertainties. Indeed, if we look at the examples in this paper and those in [9], it should be clear that the *same* planning algorithms can be used to solve manipulation problems across multiple length scales.

REFERENCES

[1] S. Akella, W. H. Huang, K. M. Lynch, and M. T. Mason. Parts feeding on a conveyor with a one joint robot. *Algorithmica*, 26(3/4):313–344, March/April 2000.

[2] S. Akella and M. Mason. Posing polygonal objects in the plane by pushing. In *Proceedings IEEE International Conference on Robotics & Automation*, pages 2255–2262, 1992.

[3] D. J. Armstrong. The quarks of object-oriented development. *Communications of the ACM*, 49(2):123–128, 2006.

[4] J. D. Bernheisel and K. M. Lynch. Stable pushing of assemblies. In *Proceedings IEEE International Conference on Robotics & Automation*, 2005.

[5] R.-P. Berretty, K. Goldberg, M. Overmars, and A. van der Stappen. Computing fence designs for orienting parts. *Computational geometry: theory and applications*, 10(4):249–262, 1998.

[6] A. Bicchi and V. Kumar. Robotic grasping and manipulation. In N. Siciliano, Bicchi and Valigi, editors, *Ramsete: Articuated and Mobile Robots for Services and Technology*, volume 270 of *Lecture Notes in Control and Information Sciences*. Springer Verlag, 2001.

[7] S. Blind, C. McCullough, S. Akella, and J. Ponce. Manipulating parts with an array of pins: A method and a machine. *International Journal of Robotics Research*, 20(10):808–818, Dec. 2001.

[8] M. Buehler, D. E. Koditschek, and P. J. Kindlmann. Planning and control of robotic juggling and catching tasks. *International Journal of Robotics Research*, 13(2):101–118, 1994.

[9] P. Cheng, D. Cappelleri, B. Gavrea, and V. Kumar. Planning and control of meso-scale manipulation tasks with uncertainties. In *Robotics: Science and Systems*, 2007.

[10] P. Cheng and V. Kumar. Sampling-based falsification and verification of controllers for continuous dynamic systems. In S. Akella, N. Amato, W. Huang, and B. Misha, editors, *Workshop on Algorithmic Foundations of Robotics VII*, 2006.

[11] S. Collins, A. Ruina, R. Tedrake, and M. Wisse. Efficient bipedal robots based on passive dynamic walkers. *Science Magazine*, 307:1082–1085, 2005.

[12] M. R. Cutkosky. *Robotic Grasping and Fine Manipulation*. Kluwer Academic Publishers, Norwell, Massachusetts, 1985.

[13] J. Fink, N. Michael, and V. Kumar. Composition of vector fields for multi-robot manipulation via caging. In *Robotics: Science and Systems Conference*, 2007.

[14] O. C. Goemans, K. Goldberg, and A. F. v. d. Stappen. Blades for feeding 3d parts on vibratory tracks. *Assembly Automation Journal*, 26, 2006.

[15] T. A. Henzinger, P. W. Kopke, A. Puri, and P. Varaiya. Whats decidable about hybrid automata? *J. Comput. Syst. Sci.*, 57:94–124, 1998.

[16] L. E. Kavraki, P. Svestka, J.-C. Latombe, and M. H. Overmars. Probabilistic roadmaps for path planning in high-dimensional configuration spaces. *IEEE Transactions on Robotics & Automation*, 12(4):566–580, June 1996.

[17] S. M. LaValle and J. J. Kuffner. Rapidly-exploring random trees: Progress and prospects. In *Proceedings Workshop on the Algorithmic Foundations of Robotics*, 2000.

[18] K. M. Lynch and M. T. Mason. Stable pushing: Mechanics, controllability, and planning. *International Journal of Robotics Research*, 15(6):533–556, 1996.

[19] M. T. Mason. Mechanics and planning of manipulator pushing operations. *International Journal of Robotics Research*, 5(3):53–71, 1986.

[20] N. Michael, J. Fink, and V. Kumar. Experimental testbed for large multi-robot teams: Verification and validation. *IEEE Robotics and Automation Magazine*, Mar. 2008.

[21] G. A. S. Pereira, V. Kumar, and M. F. M. Campos. Decentralized algorithms for multirobot manipulation via caging. *International Journal of Robotics Research*, 2004.

[22] M. Peshkin and A. Sanderson. The motion of a pushed sliding workpiece. *International Journal of robotics and automation*, 4:569–598, 1988.

[23] J. Ponce and B. Faverjon. On computing three-finger force-closure grasps of polygonal objects. *IEEE Transactions on Robotics & Automation*, 11(6):868–881, 1995.

[24] J. Ponce, S. Sullivan, A. Sudsang, J.-D. Boissonnat, and J.-P. Merlet. On computing four-finger equilibrium and force-closure grasps of polyhedral objects. *International Journal of Robotics Research*, 16(1):11–35, Feb. 1997.

[25] M. Raibert. *Legged Robots That Balance*. The MIT Press, 1986.

[26] J. K. Salisbury and J. J. Craig. Articulated hands: Force control and kinematic issues. *International Journal of Robotics Research*, 1(1):4–17, Spring 1982.

[27] T. Siméon, J.-P. L. J. Cortés, and A. Sahbani. Manipulation planning with probabilistic roadmaps. *International Journal of Robotics Research*, 23(1-8), 2004.

[28] P. Song, J. Pang, and V. Kumar. A semi-implicit time-stepping model for frictional compliant contact problems. *International Journal for Numerical Methods in Engineering*, 60:2231–2261, 2004.

[29] P. Song, J. Trinkle, V. Kumar, and J.-S. Pang. Design of part feeding and assembly processes with dynamics. In *Proceedings IEEE International Conference on Robotics & Automation*, Apr. 2004.

[30] D. Stewart and J. Trinkle. An implicit time-stepping scheme for rigid body dynamics with inelastic collisions and coulomb friction. *International Journal of Numerical Methods in Engineering*, 39:2673–2691, 1996.

[31] M. Stilman and J. Kuffner. Planning among movable obstacles with artificial constraints. In *Proceedings Workshop on Algorithmic Foundations of Robotics*, 2006.

[32] A. Sudsang, J. Ponce, and N. Srinivasa. Grasping and in-hand manipulation: Geometry and algorithms. *Algorithmica*, 26:466–493, 2000.

[33] A. Sudsang, F. Rothganger, and J. Ponce. Motion planning for disc-shaped robots pushing a polygonal object in the plane. *IEEE Transactions on Robotics & Automation*, 18(4):550–562, 2002.

[34] J. Trinkle, S. Berard, and J. Pang. A time-stepping scheme for quasistatic multibody systems. In *Proceedings, ASME International Design Engineering Technical Conferences*, September 2005.

[35] N. Zumel and M. Erdmann. Nonprehensible two palm manipulation with non-equilbrium transitions between stable states. In *Proceedings IEEE International Conference on Robotics & Automation*, pages 3317–3323, Apr. 1996.

A Local Collision Avoidance Method for Non-strictly Convex Polyhedra

Fumio Kanehiro*, Florent Lamiraux†, Oussama Kanoun†, Eiichi Yoshida* and Jean-Paul Laumond†
IS/AIST-ST2I/CNRS Joint Japanese-French Robotics Laboratory(JRL)
*Intelligent Systems Research Institute, National Institute of Advanced Industrial Science and Technology(AIST)
Tsukuba Central 2, 1-1-1 Umezono, Tsukuba, Ibaraki 305-8568 Japan
Email: {f-kanehiro, e.yoshida}@aist.go.jp
†LAAS-CNRS, University of Toulouse
7 Avenue du Colonel Roche, 31077 Toulouse, France
Email: {florent, okanoun, jpl}@laas.fr

***Abstract*— This paper proposes a local collision avoidance method for *non-strictly convex* polyhedra with *continuous* velocities. The main contribution of the method is that non-strictly convex polyhedra can be used as geometric models of the robot and the environment without any approximation. The problem of the continuous interaction generation between polyhedra is reduced to the continuous constraints generation between polygonal faces and the continuity of those constraints are managed by the combinatorics based on Voronoi regions of a face. A collision-free motion is obtained by solving an optimization problem defined by an objective function which describes a task and linear inequality constraints which do geometrical constraints to avoid collisions. The proposed method is examined using example cases of simple objects and also applied to a humanoid robot HRP-2.**

I. Introduction

Detecting and avoiding collisions is fundamental for the development of robots that can be safely operated in human environments. This issue has given rise to contributions in the 1980's in the context of robotic manipulators. One of the most famous and most used contribution in this domain is certainly [1]. [1] proposes a method which plans a motion by minimizing an error between a desired velocity and the planned velocity under inequality constraints to avoid collision. The robot and the environment must consists of strictly convex objects. [2] extends this method to avoid local minima by modifying the task description in heavily cluttered environment. [3] proposes another approach for avoiding collision by following the distance gradient. The robot is approximated by a set of strictly convex objects (ellipsoids).

Recent developments of humanoid robots have made the issue of collision detection and avoidance very critical again. [4] proposes a path planning method which computes dynamically stable and collision-free trajectories. It prepares a set of statically stable postures in advance and finds a path by exploring it with RRT-connect[5]. And the path is transformed into a dynamically stable trajectory by applying a dynamics filter[6]. [7] proposes a fast method to ensure that there is no self-collision in a trajectory. Shapes of the robot are approximated by convex hulls and the minimum distances between them are tracked using V-Clip[8]. [9] proposes a local collision avoidance method using repulsion fields defined by the minimum distance. [10] combines three methods to detect self-collisions online, (1) table look-up to detect self-collisions between adjacent joints and (2) reduction of pairs of links to be checked using heuristics, and (3) collision check using approximated shapes by convex hulls. [11] approximates shapes by spheres and swept sphere lines and used the minimum distances between them to avoid self-collision. [12] computes a strictly convex bounding volume called STP-BV by patching spheres and toruses for each body of a humanoid robot and builds collision-free postures.

Fig. 1. "Pick up an object under the table" example

Many works have focused on approximating the robot by strictly convex objects but to our knowledge, very few has been done to deal with non-strictly convex objects without geometric approximation.

In this paper, we extend the method described in [1] to non-necessarily convex polyhedral objects, in such a way that the resulting velocity of the robot is continuous. The basic framework is same as in [1]. The robot velocity is computed by minimizing difference between the desired task velocity and the planned one under linear inequality constraints implied by pairs of objects close to each other. The main difference is the way of generating constraints. In the case of strictly convex objects, one constraint over the velocity of the closest points between two objects is enough. The minimum distance and the closest points between non-strictly convex objects can be computed using efficient algorithms and robust implementations[13, 14, 15, 16]. But performing collision avoidance by applying a linear inequality constraint over the velocity between the closest points might result in discontinuous velocities for the objects. Because the closest points between two strictly convex objects move continuously whereas the closest points between non-strictly convex objects do not. If one object is a robot body, discontinuous velocity cannot be applied. So in the case of polyhedra, several constraints are generated for each pair of objects unlike the method in [1]. Our method reduces the problem of the continuous interaction generation between polyhedra to the continuous constraints generation between faces and manages the continuity of those constraints using the combinatorics based on Voronoi regions of a face.

The paper is organized as follows. In section II, we recall Faverjon and Tournassoud's method for strictly convex objects. In section III, we extend the method to make it possible to accept polyhedra. In section IV, the extended method is examined using example cases of simple objects and a humanoid robot HRP-2[17]. In section V, we summarize and conclude the paper.

II. A LOCAL METHOD FOR STRICTLY CONVEX OBJECTS

A. Strictly convex objects

First, let us recall the following definition.

Definition: *Strictly convex object. Let $\mathcal{O}$ be a closed subset of $\mathbb{R}^3$ and int($\mathcal{O}$) be the interior of (greater open subset of) $\mathcal{O}$. $\mathcal{O}$ is strictly convex if and only if*

$$\forall A \in \mathcal{O}, \forall B \in \mathcal{O}, \forall \lambda \in \mathbb{R}, 0 < \lambda < 1, \ \ \lambda A + (1-\lambda) B \in \mathit{int}(\mathcal{O})$$

For instance, a convex polyhedron is not strictly convex. If two points on a facet are selected as A and B, the line segment linking them is not inside the interior of the polyhedron but on the facet.

B. Outline of the method

Let us recall the principle of Faverjon and Tournassoud's method [1]. Let $\mathcal{O}_1$ and $\mathcal{O}_2$ be two strictly convex objects. For ease of explanation, let's consider $\mathcal{O}_1$ as a movable object and $\mathcal{O}_2$ as a static one. Let $\boldsymbol{p}_1$ and $\boldsymbol{p}_2$ denote the closest points between $\mathcal{O}_1$ and $\mathcal{O}_2$ and d does the distance $\|\boldsymbol{p}_1 - \boldsymbol{p}_2\|$ between them (Fig. 2). Since $\mathcal{O}_1$ and $\mathcal{O}_2$ are strictly convex objects, $\boldsymbol{p}_1$ and $\boldsymbol{p}_2$ move continuously on $\mathcal{O}_1$ and $\mathcal{O}_2$ boundaries and d is continuously differentiable.

If d is smaller than a threshold called *influence distance* and denoted by d_i, the following constraint is defined for velocity of d:

$$\dot{d} \geq -\xi \frac{d - d_s}{d_i - d_s} \quad (1)$$

where ξ is a positive coefficient for adjusting convergence speed, $d_s (< d_i)$ is a positive value called *security distance.*

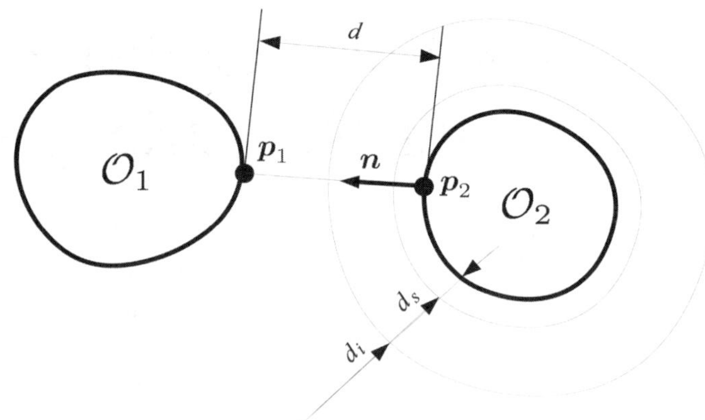

Fig. 2. Faverjon and Tournassoud's method

Inequality (1) is called *velocity damper* and it expresses that d must not decrease too fast when it is smaller than d_i. As the result, d never be smaller than d_s. The region where d is smaller than d_i is called *influence zone*. Note that $\dot{d}$ must satisfy Inequality (1) when $\boldsymbol{p}_1$ enters *influence zone*. If it doesn't, $\dot{d}$ is constrained discontinuously. Therefore, the value of ξ must be tuned according to applications.

$\dot{d}$ is computed by the following equation.

$$\dot{d} = (\dot{\boldsymbol{p}}_1 | \boldsymbol{n})$$

where $\boldsymbol{n}$ is the unit vector $(\boldsymbol{p}_1 - \boldsymbol{p}_2)/d$ and notation $(\boldsymbol{u}|\boldsymbol{v})$ refers to the inner product of vectors $\boldsymbol{u}$ and $\boldsymbol{v}$.

Let $\boldsymbol{q}$ denote the configuration and $\dot{\boldsymbol{q}}$ the velocity of the robot. The velocity of $\boldsymbol{p}_1$ can be expressed as:

$$\dot{\boldsymbol{p}}_1 = \boldsymbol{J}(\boldsymbol{q}, \boldsymbol{p}_1)\dot{\boldsymbol{q}}$$

where $\boldsymbol{J}(\boldsymbol{q}, \boldsymbol{p}_1)$ is a Jacobian matrix of $\mathcal{O}_1$ at $\boldsymbol{p}_1$. Inequality (1) thus becomes a linear inequality constraint over the robot velocity $\dot{\boldsymbol{q}}$:

$$(\dot{\boldsymbol{q}} | \boldsymbol{J}(\boldsymbol{q}, \boldsymbol{p}_1)^T \boldsymbol{n}) \geq -\xi \frac{d - d_s}{d_i - d_s}$$

A task is described by the control of a measure of the problem, a vector $\boldsymbol{\tau}(\boldsymbol{q})$ in a task space. The task is achieved by finding $\boldsymbol{q}$ which satisfies $\boldsymbol{\tau}(\boldsymbol{q}) = \boldsymbol{0}$. Given a desired task velocity $\underline{\dot{\boldsymbol{\tau}}}$, the robot velocity to achieve the task while avoiding collision is computed by solving the following optimization problem over $\dot{\boldsymbol{q}}$:

$$\begin{aligned} \min_{\dot{\boldsymbol{q}}} \quad & \|\boldsymbol{J}_\tau(\boldsymbol{q})\dot{\boldsymbol{q}} - \underline{\dot{\boldsymbol{\tau}}}\|^2 \\ \text{subject to} \quad & (\dot{\boldsymbol{q}} | \boldsymbol{J}(\boldsymbol{q}, \boldsymbol{p}_1)^T \boldsymbol{n}) \geq -\xi \frac{d - d_s}{d_i - d_s}. \end{aligned} \quad (2)$$

where $\boldsymbol{J}_\tau(\boldsymbol{q})$ is a Jacobian matrix of $\boldsymbol{\tau}(\boldsymbol{q})$ at $\boldsymbol{q}$.

Of course, considering each body of the robot and several obstacles yields several inequality constraints.

C. Lower bound of the distance between $\mathcal{O}_1$ and $\mathcal{O}_2$

If we denote by $d(t)$ the minimum distance between $\mathcal{O}_1$ and $\mathcal{O}_2$ along time, $d(t)$ is continuously differentiable. If a condition on the derivative $\dot{d}$ of d:

$$\forall t > 0, \dot{d}(t) \geq -\xi \frac{d(t) - d_s}{d_i - d_s}$$

and the initial condition $d(0) \geq d_s$ are satisfied, then the following condition is derived.

$$\forall t > 0,\ d(t) \geq d_s + (d(0) - d_s)e^{-\frac{\xi}{d_i - d_s}t} > d_s \qquad (3)$$

This proves that the distance between objects constrained by *velocity damper* never be smaller than d_s.

III. A Local Method for Non-strictly convex polyhedra

A. A discontinuous case of non-strictly convex polyhedra

If we apply Faverjon and Tournassoud's method to non-strictly convex polyhedra, the robot velocity changes discontinuously since the closest points between two objects move discontinuously. Fig. 3 and Fig. 4 show snapshots and results of an example case. In this example, a rectangle(0.2×0.8[m]) moves above an horizontal floor. The task of the rectangle is to move its center $\boldsymbol{p}_c(\boldsymbol{q})$ from $(0.0, 0.7)^T$[m] to $\boldsymbol{p}_g = (0.0, -1.0)^T$[m]. Parameters of *velocity damper*, d_i, d_s and ξ are set to 0.4[m], 0.2[m] and 0.5[m/s] respectively. $\dot{\underline{\tau}}$ is given by:

$$\dot{\underline{\tau}} = \delta\tau_{max} \frac{\boldsymbol{p}_g - \boldsymbol{p}_c(\boldsymbol{q})}{\|\boldsymbol{p}_g - \boldsymbol{p}_c(\boldsymbol{q})\|}$$

where $\delta\tau_{max}$ is set to 0.2.

Top left picture of Fig. 4 shows the minimum distance between the rectangle and the floor. Top right picture displays the vertical position of the rectangle. Bottom left and right pictures show the linear velocity along Y axis and the angular velocity respectively. One of vertices of the bottom edge of the rectangle, C_1 enters into *influence zone* at $t = 0.25$. But the velocity of C_1 is not affected since its velocity satisfies Inequality (1). From $t = 0.5$, the object velocity is affected by *velocity damper*. The center must move downward with the constant speed to achieve the task but the admissible speed of C_1 is limited. As a result, the object rotates. Around $t = 2.0$, the bottom edge of the rectangle becomes almost parallel to the ground, the closest points start to oscillate between C_1 and C_2. The object rotates clockwise to achieve the task when C_1 is constrained and does counterclockwise when C_2 is constrained. As a result, the minimum distance becomes smaller than d_s and the rectangle eventually collides with the floor at $t = 3.0$.

The collision-freeness is not assured anymore when the robot bodies and obstacles are not strictly convex as shown. The oscillation of the closest points causes discontinuous changes of $\boldsymbol{p}_1$ and $\boldsymbol{n}$ in Problem (2) and it leads to discontinuity in the solution of Problem (2) as we can see in Fig. 4.

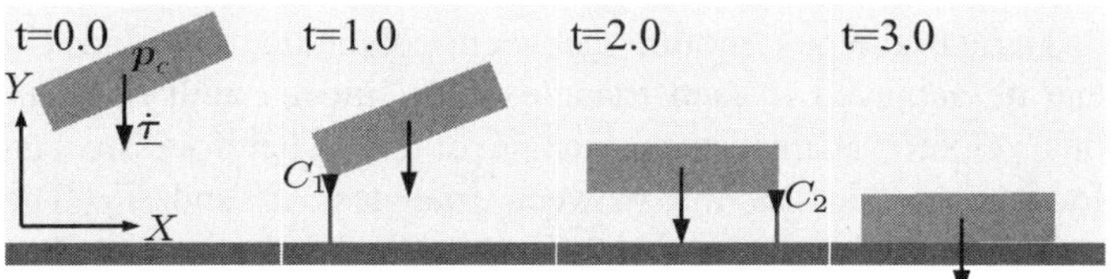

Fig. 3. Example of a discontinuous constraint

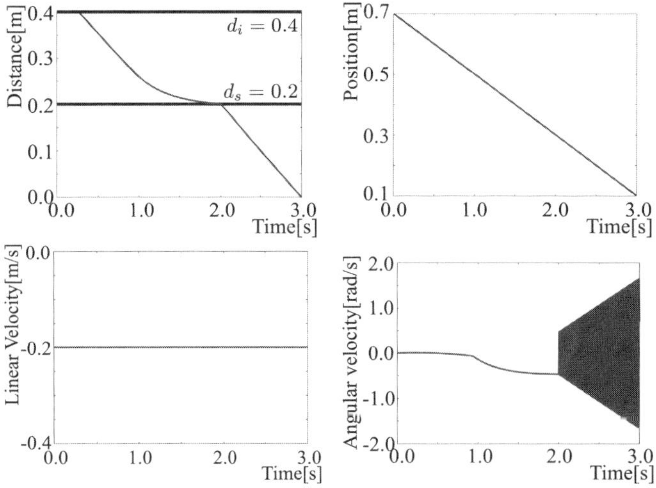

Fig. 4. Results of the example

B. Decomposition of interaction between polyhedra

The goal of this section is to define the inequality constraints in such a way that $\dot{\boldsymbol{q}}$ remains continuous. To cope with this issue, we propose to keep track of several pairs of points that move continuously on the facets of the polyhedra composing the obstacles and robot.

Discontinuity of constraints happens in the following cases.

1) a new constraint appears suddenly
2) a constraint disappears suddenly

The closest points jump discontinuously if these two cases happen at the same time.

In order to prevent these cases and generate a collision-free motion with continuous velocities, pairs of points must be selected by complying with the following rules.

1) the closest points between a robot body and an obstacle must be constrained to guarantee that the robot never collides.
2) the potential closest points must have been constrained before they become the closest points.
3) the closest points must continue to be constrained even if they are not closest anymore.

Let us decompose the interaction between polyhedra into a set of interactions between faces. Polygonal faces are assumed to be decomposed into triangles[1].

[1]In the following, *features* of a triangle are the triangular (open)face, the three edges and the three vertices.

The continuous motion between polyhedra, $\mathcal{O}_1$ and $\mathcal{O}_2$ can be achieved if each triangle of $\mathcal{O}_1$ moves with continuous velocity against each triangle of $\mathcal{O}_2$. Therefore we can focus on an interaction between triangles, $\mathcal{T}_1$ and $\mathcal{T}_2$. The continuous motion between $\mathcal{T}_i$ and $\mathcal{T}_j$ can be achieved if each edge of $\mathcal{T}_1$ moves with continuous velocity against $\mathcal{T}_2$ and each edge of $\mathcal{T}_2$ moves with continuous velocity against $\mathcal{T}_1$. Finally the interaction between polyhedra is decomposed into interactions between an edge and a triangle as shown in Fig. 5. It means that the continuous interaction between polyhedra can be achieved if we can find a method to realize a continuous interaction between an edge and a triangle. In the same way, the continuous interaction between a robot and the environment which consists of several polyhedra respectively can be achieved if each polyhedron of the robot moves in continuous way against each polyhedron of the environment.

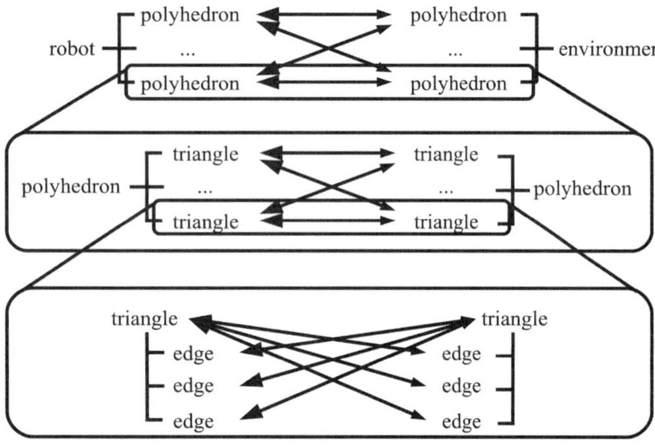

Fig. 5. Decomposition of interaction

C. Constraint generation using Voronoi regions

Next, let us find pairs of points to be constrained to realize the continuous interaction between an edge and a triangle. The pairs of points to be constrained depend on the Voronoi regions in which the edge lies. The Voronoi region is defined as follows.

Definition: *Voronoi region $\mathcal{VR}(X)$ for feature X. A Voronoi region associated with a feature X of a triangle is a set of points that are closer to X than any other feature.*

The Voronoi plane is also defined as follows.

Definition: *Voronoi plane $\mathcal{VP}(X, Y)$ between neighboring features X and Y. $\mathcal{VP}(X, Y)$ is the plane containing $\mathcal{VR}(X) \cap \mathcal{VR}(Y)$.*

Since a triangle consists of a face $\mathcal{F}$, three edges $\mathcal{E}_i(i = 1, 2, 3)$ and three vertices $\mathcal{V}_i(i = 1, 2, 3)$, 3D space around the triangle is separated into 7 Voronoi regions.

Case1 : The edge is in $\mathcal{VR}(\mathcal{F})$

The closest point jumps from one of end points of the edge to the other when the edge and the triangle are almost parallel. Therefore, two pairs, $(\mathcal{V}_1, \mathcal{V}_1')$ and $(\mathcal{V}_2, \mathcal{V}_2')$ are constrained, where $\mathcal{V}_1$ and $\mathcal{V}_2$ are end points of the edge. (a, b) denotes a pair of points, a and b and $\mathcal{V}_i'(i = 1, 2)$ denotes a projection of $\mathcal{V}_i$ onto $\mathcal{F}$ along its normal vector. Any point on the edge can be the closest point when the edge and the triangle are parallel. But we don't need any additional pair since both end points are already constrained and they are also the closest points.

Case2 : The edge is in $\mathcal{VR}(\mathcal{E})$

The closest point jumps from one of end points of the edge to the other when the edge and $\mathcal{E}$ are almost parallel. So two pairs, $(\mathcal{V}_1, \mathcal{V}_1')$ and $(\mathcal{V}_2, \mathcal{V}_2')$ are constrained, where $\mathcal{V}_i'(i = 1, 2)$ is a projected point of $\mathcal{V}_i$ onto $\mathcal{E}$. The closest points between the edge and $\mathcal{E}$ coincides with one of two pairs in some cases, but doesn't in other cases. Therefore, one more pair for the closest points is added. Three pairs are created as a consequence.

Case3 : The edge is in $\mathcal{VR}(\mathcal{V})$

The closest point moves continuously on the edge. So the pair of the closest points is constrained.

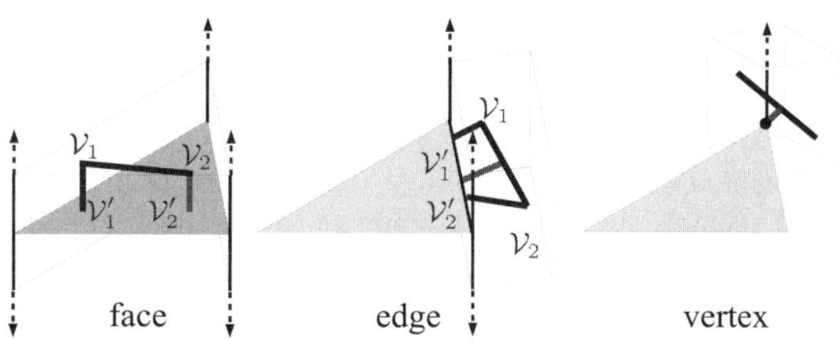

Fig. 6. Constraints generated between an edge and a triangle

So far, we considered cases the edge lies in one of the Voronoi regions of the triangle. But in most cases, the edge may move to another Voronoi region and lie in several Voronoi regions at the same time. Therefore, the edge is decomposed into several line segments again by clipping it with Voronoi planes.

Before this decomposition, we need to confirm that continuity of constraints is kept when an edge goes into another Voronoi region. When the edge moves between $\mathcal{VR}(\mathcal{F})$ and $\mathcal{VR}(\mathcal{E})$, continuity of constraints are maintained since end points of decomposed line segments are on $\mathcal{VP}(\mathcal{F}, \mathcal{E})$ and they produce the same constraints. However, in other cases, constraints may appear or disappear suddenly. An example is shown in Fig. 7. The edge is moving from $\mathcal{VR}(\mathcal{V})$ to $\mathcal{VR}(\mathcal{E})$. When an end point of the edge touches $\mathcal{VP}(\mathcal{V}, \mathcal{E})$ and a new constraint appears suddenly. In the reverse case, the constraint disappears suddenly.

This discontinuity can be solved by adding two more constraints on both end points of the edge when it is in $\mathcal{VR}(\mathcal{V})$.

Finally, the algorithm to pick up pairs of points to be constrained is described as in Algorithm 1-4.

Algorithm 1 decomposes interaction between polyhedra, $\mathcal{O}_1$ and $\mathcal{O}_2$ into interactions between triangles. Function DISTANCE_BOUND($\mathcal{O}_1$, $\mathcal{O}_2$, d_i) filters out such pairs of triangles that distances between triangles are bigger than d_i. This function is very important to improve efficiency and it can

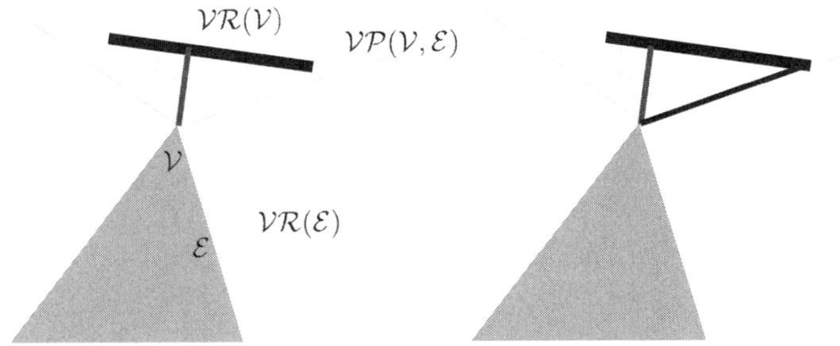

Fig. 7. Appearance/disappearance of a constraint

Algorithm 1 PAIRS_POLYHEDRA($\mathcal{O}_1$, $\mathcal{O}_2$)

$pairs \leftarrow \emptyset$
$triangle_pairs \leftarrow$DISTANCE_BOUND($\mathcal{O}_1$, $\mathcal{O}_2$, d_i)
for all $(\mathcal{T}_1, \mathcal{T}_2) \in triangle\_pairs$ **do**
  $pairs \leftarrow pairs \cup$ PAIRS_TRIANGLES($\mathcal{T}_1$, $\mathcal{T}_2$)
end for
return $pairs$

be implemented easily using techniques for collision detection such as OBB-Tree.

Algorithm 2 PAIRS_TRIANGLES($\mathcal{T}_1$, $\mathcal{T}_2$)

$pairs \leftarrow \emptyset$
for all $\mathcal{E}_2 \in$EDGES($\mathcal{T}_2$) **do**
  $pairs \leftarrow pairs \cup$ PAIRS_EDGE_TRIANGLE($\mathcal{E}_2$, $\mathcal{T}_1$)
end for
for all $\mathcal{E}_1 \in$EDGES($\mathcal{T}_1$) **do**
  $pairs \leftarrow pairs \cup$ PAIRS_EDGE_TRIANGLE($\mathcal{E}_1$, $\mathcal{T}_2$)
end for
return $pairs$

Algorithm 2 decomposes interaction between triangles, $\mathcal{T}_1$ and $\mathcal{T}_2$ into interactions between an edge and a triangle. Function EDGE($\mathcal{T}$) returns the set of edges that compose $\mathcal{T}$.

Algorithm 3 decomposes interaction between an edge and a triangle into interactions between a line segment and a triangle. A function VORONOI_CLIP($\mathcal{E}$, $\mathcal{VR}(f)$) clips a part of $\mathcal{E}$ which is in $\mathcal{VR}(f)$.

Algorithm 4 generates pairs of points to be constrained. Function VERTICES($\mathcal{S}$) returns the set of end points of $\mathcal{S}$, a function PARALLEL($\mathcal{S}_1$, $\mathcal{S}_2$) checks two segments are parallel or not, and a function CLOSEST_PAIR(A, B) computes the closest points between geometric elements A and B.

After $pairs$ are computed, *velocity damper* is inserted between each pair of points if the distance between those points is smaller than d_i.

In this procedure, constraints are generated on all end points of line segments. Since end points are shared by several line segments, duplicated pairs are generated. A duplicated pair is also generated when the closet point coincides with one of the end points. Therefore, we need to check duplication before adding a new pair.

Algorithm 3 PAIRS_EDGE_TRIANGLE($\mathcal{E}$, $\mathcal{T}$)

$pairs \leftarrow \emptyset$
for all $f \in \{\mathcal{F}, \mathcal{E}_1, \mathcal{E}_2, \mathcal{E}_3, \mathcal{V}_1, \mathcal{V}_2, \mathcal{V}_3\}$ **do**
  $\mathcal{S} \leftarrow$VORONOI_CLIP($\mathcal{E}$, $\mathcal{VR}(f)$)
  if $\mathcal{S}$ **then**
    $pairs \leftarrow pairs \cup$ PAIRS_SEGMENT_FEATURE($\mathcal{S}$, f)
  end if
end for
return $pairs$

Algorithm 4 PAIRS_SEGMENT_FEATURE($\mathcal{S}$, f)

$pairs \leftarrow \emptyset$
for all $\mathcal{V} \in$VERTICES($\mathcal{S}$) **do**
  $pairs \leftarrow pairs \cup$ {CLOSEST_PAIR($\mathcal{V}$, f)}
end for
if $f \in \{\mathcal{V}_1, \mathcal{V}_2, \mathcal{V}_3\}$ **then**
  $pairs \leftarrow pairs \cup$ {CLOSEST_PAIR($\mathcal{S}$, f)}
else if $f \in \{\mathcal{E}_1, \mathcal{E}_2, \mathcal{E}_3\}$ **then**
  if not PARALLEL($\mathcal{S}$, f) **then**
    $pairs \leftarrow pairs \cup$ {CLOSEST_PAIR($\mathcal{S}$, f)}
  end if
end if
return $pairs$

D. Sources of discontinuous robot velocities

Solving Problem 2 is equivalent to finding the closest point between the desired task velocity $\dot{\underline{\tau}}$ and the space of admissible task velocities $\mathcal{S}_{\dot{\tau}}$(light blue region in Fig.8). The robot velocities which satisfy all the linear inequality constraints(yellow regions in Fig.8) exist in the convex sub-space(orange region in Fig.8). It is projected into $\mathcal{S}_{\dot{\tau}}$ by $\boldsymbol{J}_\tau$. And when the task is defined in the lower dimensional space than the robot velocity space, a point in the task velocity space corresponds to the subspace in the robot velocity space $\mathcal{S}_{\dot{q}}$(red region in Fig.8).

There are three kinds of sources of discontinuous robot velocities.

1) If a constraint changes discontinuously, the shape of $\mathcal{S}_{\dot{\tau}}$ also does. As the result, the robot velocity might change discontinuously. This source can be removed using the constraint generation method described in this section.
2) Even all the constraints move continuously in the robot velocity space, discontinuous robot velocities might be generated when the task is defined in the lower dimensional space than the robot velocity space. This situation is similar with solving inverse kinematics at singular postures. SR-Inverse[18] is proposed to prevent the robot velocity from going to infinity. The same thing can be realized by modifying the objective function of Problem 2 as follows:

$$\|\boldsymbol{J}_\tau(\boldsymbol{q})\dot{\boldsymbol{q}} - \dot{\underline{\tau}}\|^2 + \lambda\|\dot{\boldsymbol{q}}\|^2 \qquad (4)$$

where λ is a positive coefficient for adjusting strength of the penalty for big velocities. The added second term put a damping effect to the robot velocity and remove discontinuities.

3) Every point in $\mathcal{S}_{\dot{q}}$ is the optimal solution of Problem 2. A point in $\mathcal{S}_{\dot{q}}$ might be chosen discontinuously by the optimization algorithm. The modified objective function changes $\mathcal{S}_{\dot{q}}$ into a point and this discontinuity is also removed.

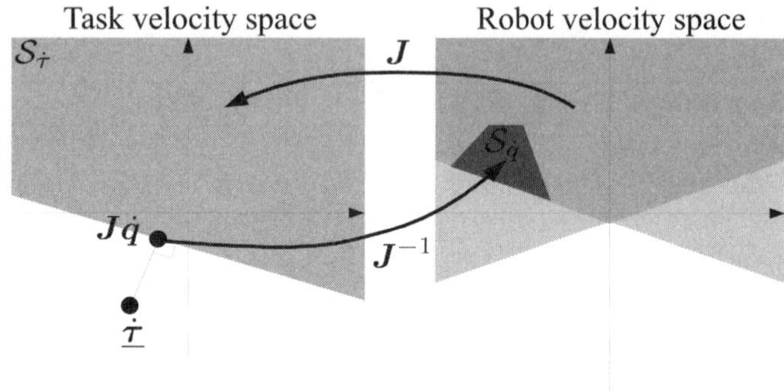

Fig. 8. Mappings between velocity spaces

IV. Examples

A. Collision Avoidance of a Single Object

The proposed method is applied to the example shown in Section III-A to check that a collision avoidance motion with continuous velocity can be generated. Fig. 9 shows snapshots of the generated motion and Fig. 10 shows results corresponding to Fig. 4. The minimum distance converges to d_s and there is no collision. And linear velocity along Y axis and angular velocity changes in continuous way. In this case, a constraint is generated when C_1 enters into *influence zone* and one more constraint is added when C_2 does.

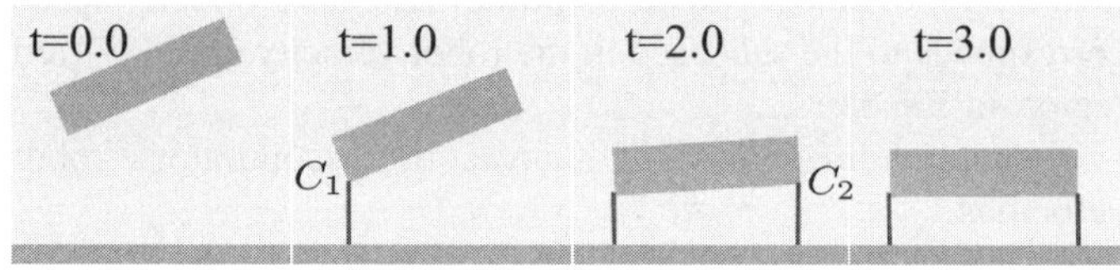

Fig. 9. Snapshots of interaction between a rectangle and the ground

Fig. 11 shows another example which includes a concave shape and a shape with a hole. A "L" shape object which consists of 12 triangles passes through a torus which consists of 512 triangles. It is impossible for existing methods to plan a collision free motion between these kinds of shapes without approximations.

Top left of Fig. 12 shows the minimum distance. The distance between non-strictly convex objects is continuous and piecewise smooth. Since the closest points are constrained at any time, the condition in Eq.(3) holds and the minimum distance between objects never become smaller than d_s. Top right of Fig. 12 shows the number of pairs of triangles which are found by DISTANCE_BOUND(). The total number of

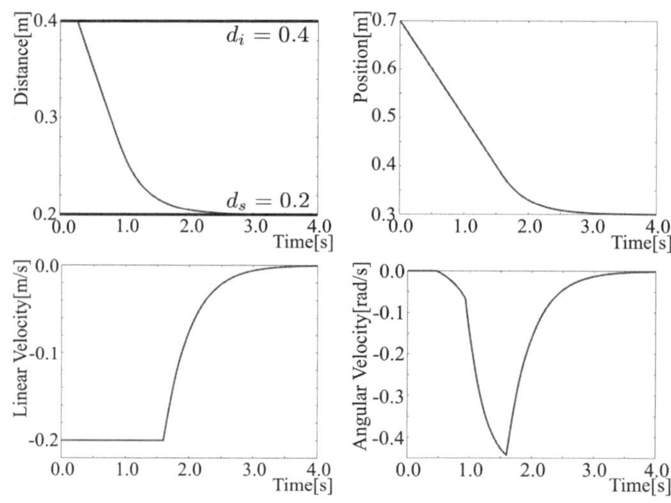

Fig. 10. Improved results of an example #1

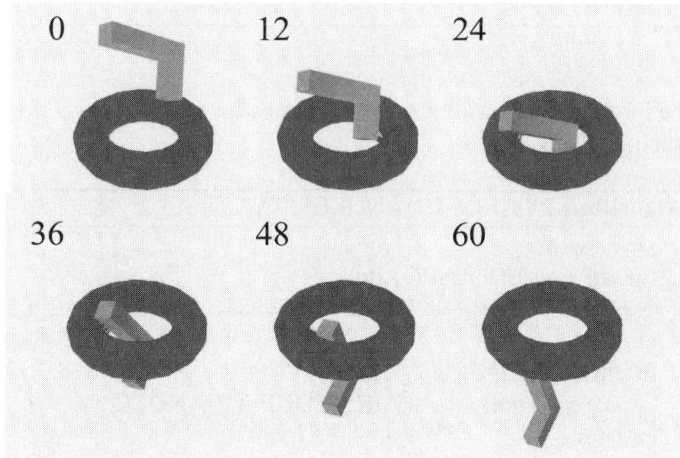

Fig. 11. Snapshots of an interaction between a concave shape and a torus

pairs of triangles is $12 \times 512 = 6144$ whereas it is less than 100. Bottom left of Fig. 12 shows the number of constraints. It changes along time since they are activated only if the distance between points is smaller than d_i. In this example, 117 constraints are generated at a maximum. Bottom right of Fig. 12 shows the computational time. It is measured on a PC equipped with Intel Core 2 Duo 2.13[GHz]. It is almost proportional to the number of pairs of triangles.

B. Collision Avoidance of a Humanoid Robot

The proposed method is also applied to a humanoid robot HRP-2. The task of the robot is to move its left hand to the specified position using its whole body. In addition to constraints for collision avoidance, three kinds of constraints are added. (1) A relative transformation of feet is kept while reaching since the robot stands on both legs. (2) A horizontal position of the center of mass is kept to keep static balance(We assume the center of mass is above the support polygon at the initial configuration.). (3) Joint angles are kept in their movable ranges. As a result, this collision-free reaching task

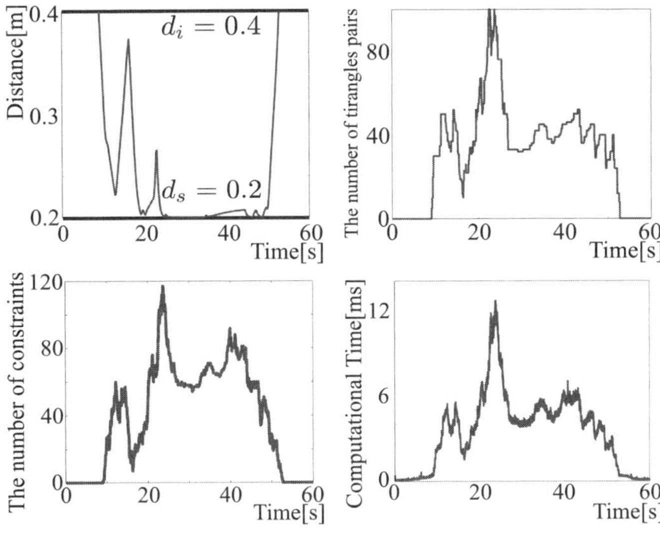

Fig. 12. Results of an example #2

can be achieved by solving the following QP problem:

$$\begin{aligned}
\text{minimize} \quad & \|\boldsymbol{J}_{hand}\dot{\boldsymbol{q}} - \underline{\dot{\boldsymbol{p}}}\|^2 + \lambda\|\dot{\boldsymbol{q}}\|^2 && \text{(5a)}\\
\text{subject to} \quad & (\dot{\boldsymbol{q}}|\boldsymbol{J}_{dist_j}^T\boldsymbol{n}) \geq -\xi\frac{d_j - d_s}{d_i - d_s}, \\
& \qquad \text{for } j \in \{1, \cdots, n_c\}, && \text{(5b)}\\
& \boldsymbol{J}_{feet}\dot{\boldsymbol{q}} = \boldsymbol{0}, && \text{(5c)}\\
& \boldsymbol{J}_{com}\dot{\boldsymbol{q}} = \boldsymbol{0}, && \text{(5d)}\\
& vmax_j(q_j) \geq \dot{q}_j \geq vmin_j(q_j), \\
& \qquad \text{for } j \in \{1, ..., n_{dof}\}. && \text{(5e)}
\end{aligned}$$

$\boldsymbol{J}_{hand}, \boldsymbol{J}_{com}, \boldsymbol{J}_{feet}$ and $\boldsymbol{J}_{dist_j}$ are Jacobian matrices for the hand position(3DOF), for the horizontal position of the center of mass(2DOF), for the relative transformation between feet(6DOF) and for the distance between jth pair of points(1DOF) respectively. Inequality (5b) defines geometric constraints to avoid collision where n_c is the number of pairs of points to be constrained. In this example, d_i, d_s and ξ are set to 0.05[m], 0.03[m] and 0.5[m/s] respectively. Equality (5c) defines a kinematic constraint to keep the relative transformation between feet and Equality (5d) does a dynamic one to keep the center of mass on a vertical line. Inequality (5e) defines kinematic constraints for joint limits, where n_{dof} is the number of joints. The joint velocity is limited when the joint angle comes close to its limit. The limit is also computed by *velocity damper*:

$$vmax_j(q_j) = \begin{cases} \xi\dfrac{(q_j^+ - q_j) - q_s}{q_i - q_s} & \text{if } q_j^+ - q_j \leq q_i, \\ v_j^+ & \text{otherwise} \end{cases} \quad (6)$$

$$vmin_j(q_j) = \begin{cases} -\xi\dfrac{(q_j - q_j^-) - q_s}{q_i - q_s} & \text{if } q_j - q_j^- \leq q_i, \\ v_j^- & \text{otherwise} \end{cases} \quad (7)$$

where q_j^+ and q_j^- are physical upper bound and lower bound of joint angle and v_j^+ and v_j^- are those of joint velocity of jth joint respectively. In this example, q_i, q_s and ξ are set to 0.2[rad], 0.02[rad] and 0.3[rad/s] respectively.

Fig. 13 shows snapshots of the generated motion. The frame 1 shows an initial configuration. A trapezoid in front of the robot is an obstacle and a small box close to the robot foot indicates the target position of the hand. At the frame 2, the left upper arm comes close to the obstacle, and constraints for collision avoidance become active. The left shoulder avoids the obstacle from the frame 2 to 4 and the head directs upward to avoid collision in the frame 5. We can see that the whole body is fully used to avoid collision and achieve the task simultaneously. If we don't include Inequality (5b), the left shoulder collides with the table as shown in the frame 3'.

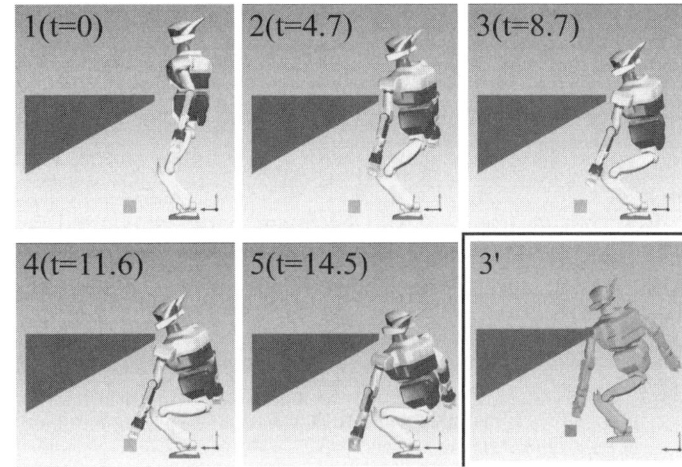

Fig. 13. Snapshots of "pick up an object under the table" motion

Fig. 14 shows the number of constraints(left) and the computational time(right). The average computational time for one step is about 100[ms] on the same PC with previous examples. We can get this motion(the total duration is 14[s]) in 28[s] when we select 50[ms] as the time step. The computational time is not so long in this case since shape of the obstacle is very simple. It is expected that it becomes longer drastically if the number of obstacles and complexity of shapes increase. If the number of pairs of triangles is too many to get a solution within reasonable time, we can reduce the number of pairs by using thin *influence zone* or simplified shapes. Even in the latter case, simplified shapes can be non-strictly convex.

V. Conclusion

In this paper, we proposed a local method for collision avoidance between non-strictly convex polyhedra with continuous velocities. The continuity is achieved by decomposing the interaction between polyhedra into a set of interactions between line segments clipped by Voronoi regions and triangles and constraining several pairs of points on those geometrical elements. These pairs of points can be used to define constraints in other collision avoidance methods like [9].

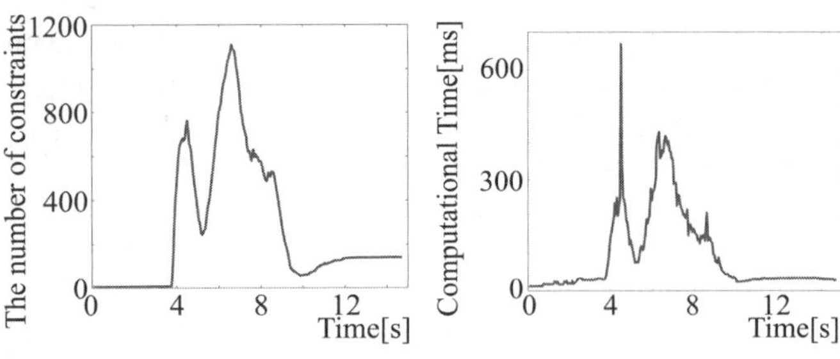

Fig. 14. Results of a picking up motion example

In case of a humanoid robot, dynamic stability of the robot is a very critical issue. But it is not guaranteed by our method since ZMP [19] is not constrained directly. A generated motion can be stable at least if it is executed with sufficiently small speed since the center of mass is constrained above its support polygon. In order to get a fast and dynamically stable motion, we are trying to use the motion as an initial path of an optimization method.

Acknowledgment

Researchers from LAAS-CNRS are partly supported by the project ANR RNTL PerfRV2. We gratefully acknowledge AEM Design, developer of a powerful QP solver, C-FSQP[20].

References

[1] B. Faverjon and P. Tournassoud, "A Local Based Approach for Path Planning of Manipulators With a High Number of Degrees of Freedom," in *Proc. of IEEE International Conference on Robotics and Automation*, 1987, pp. 1152–1159.

[2] C.Helguera and S.Zeghloul, "A Local-based Method for Manipulators Path Planning in Heavy Cluttered Environments," in *Proc. of International Conference on Robotics & Automation*, 2000, pp. 3467–3472.

[3] M. Schlemmer and G. Gruebel, "A Distance Function and its Gradient for Manipulator On-Line Obstacle Detection and Avoidance," in *Proc. of International Conference on Advanced Robotics*, 1997, pp. 427–432.

[4] J. Kuffner, K. Nishiwaki, S. Kagami, M. Inaba, and H. Inoue, "Motion Planning for Humanoid Robots Under Obstacle and Dynamic Balance Constraints," in *Proc. of International Conference on Robotics & Automation*, 2001, pp. 692–698.

[5] J. James J. Kuffner and S. M. LaValle, "RRT-Connect: An Efficient Approach to Single-Query Path Planning," in *In Proc. IEEE International Conference on Robotics & Automation*, 2000, pp. 995–1001.

[6] S.Kagami, F.Kanehiro, Y.Tamiya, M.Inaba, and H.Inoue, "AutoBalancer: An Online Dynamic Balance Compensation Scheme for Humanoid Robots," in *Proc. of the Fourth International Workshop on the Algorithmic Foundations on Robotics(WAFR'00)*, 2000.

[7] J. Kuffner, K. Nishiwaki, S. Kagami, Y. Kuniyoshi, M. Inaba, and H. Inoue, "Self-Collision Detection and Prevention for Humanoid Robots," in *Proc. of International Conference on Robotics and Automation*, 2002, pp. 2265–2270.

[8] M.Mirtich, "VClip: Fast and robust polyhedral collision detection," *ACM Transactions on Graphics*, vol. 17, no. 3, pp. 177–208, 1998.

[9] L. Sentis and O. Khatib, "A Whole-Body Control Framework for Humanoids Operating in Human Environments," in *Proc. of International Conference on Robotics and Automation*, 2006, pp. 2641–2648.

[10] K. Okada and M. Inaba, "A Hybrid Approach to Practical Self Collision Detection System of Humanoid Robot," in *Proc. of International Conference on Intelligent Robots and Systems*, 2006, pp. 3952–3957.

[11] H. Sugiura, M. Gienger, H. Janssen, and C. Goerick, "Real-Time Collision Avoidance with Whole Body Motion Control for Humanoid Robots," in *Proc. of International Conference on Intelligent Robots and Systems*, 2007, pp. 2053–2058.

[12] A. Escande, S. Miossec, and A. Kheddar, "Continuous gradient proximity distance for humanoids free-collision optimized-postures," in *Proc. of International Conference on Humanoid Robots*, 2007.

[13] E. G. Gilbert, D. W. Johnson, and S. S. Keerthi, "A Fast Procedure for Computing the Distance Between Complex Objects in Three-Dimensional Space," *IEEE Journal of Robotics and Automation*, vol. 4, no. 2, pp. 193–203, 1988.

[14] M. C. Lin and J. F. Canny, "A Fast Algorithm for Incremental Distance Calculation," in *Proceedings of the 1991 IEEE International Conference on Robotics and Automation*, 1991, pp. 1008–1014.

[15] E. Larsen, S. Gottschalk, M. C. Lin, and D. Manocha, "Fast Proximity Queries with Swept Sphere Volumes," in *Proc. of International Conference on Robotics and Automation*, 2000, pp. 3719–3726.

[16] S. A. Ehmann and M. C. Lin, "Accelerated Proximity Queries Between Convex Polyhedra By Multi-Level Voronoi Marching," in *Proc. of International Conference on Intelligent Robots and Systems*, 2000, pp. 2101–2106.

[17] K. Kaneko, F. Kanehiro, S. Kajita, H. Hirukawa, T. Kawasaki, M. Hirata, K. Akachi, and T. Isozumi, "Humanoid Robot HRP-2," in *Proc. of IEEE International Conference on Robotics and Automation*, 2004, pp. 1083–1090.

[18] Y.Nakamura and H.Hanafusa, "Inverse kinematic solutions with singularity robustness for robot manipulator control," *ASME, Transactions, Journal of Dynamic Systems, Measurement, and Control*, vol. 108, pp. 163–171, 1986.

[19] M.Vukobratović and B.Borovac, "Zero-moment point –thirty five years of its life," *International Journal of Humanoid Robotics*, vol. 1, no. 1, pp. 157–173, 2004.

[20] C. Lawrence, J. L. Zhou, and A. L. Tits, *User's Guide for CFSQP Version 2.5: A C Code for Solving (Large Scale) Constrained Nonlinear (Minimax) Optimization Problems, Generating Iterates Satisfying All Inequality Constraints*.

BiSpace Planning: Concurrent Multi-Space Exploration

Rosen Diankov and Nathan Ratliff
Robotics Institute
Carnegie Mellon University
Pittsburgh, PA
{rdiankov, ndr}@cs.cmu.edu

Dave Ferguson and Siddhartha Srinivasa
Intel Research Pittsburgh
4720 Forbes Ave
Pittsburgh, PA
{dave.ferguson, siddhartha.srinivasa}@intel.com

James Kuffner
Robotics Institute
Carnegie Mellon University
Pittsburgh, PA
kuffner@cs.cmu.edu

***Abstract*— We present a planning algorithm called *BiSpace* that produces fast plans to complex high-dimensional problems by simultaneously exploring multiple spaces. We specifically focus on finding robust solutions to manipulation and grasp planning problems by using BiSpace's special characteristics to explore the work and configuration spaces of the environment and robot. Furthermore, we present a number of techniques for constructing informed heuristics to intelligently search through these high-dimensional spaces. In general, the *BiSpace* planner is applicable to any problem involving multiple search spaces.**

I. Introduction

One of the long-term goals of robotics is to develop a general purpose physical agent that can co-exist with, and provide assistance to, human beings. Substantial progress has been made toward creating the *physical* components of such an agent, resulting in a wide variety of humanoid robots that possess amazing potential for dexterity and finesse. But progress toward controlling such agents in real-time in unstructured, inhabited environments has been slower.

Planning algorithms represent the state-of-the-art in control strategies for these agents. However, in many real-world scenarios the action possibilities for the agent can become very high dimensional and contorted, rendering many algorithms ineffective. Moreover, the environments in which these agents need to operate are often not known a priori and are often dynamic. A successful planning algorithm, therefore, must perform quickly so that the resulting solution can be executed before the environment changes substantially.

In response to these planning challenges, a number of sampling-based search algorithms, such as the Rapidly-exploring Random Trees (RRT) family of algorithms [1], have been developed which demonstrate encouraging empirical performance on high-dimensional planning problems. RRTs in particular are easy to implement and have shown fast convergence to feasible solutions on a wide variety of motion planning scenarios [1, 2, 3, 4, 5].

In this work, we focus on the problem of mobile robotic manipulation. Specifically, the robot must be able to maneuver to an object and grasp it without a human specifying a priori where and how the object should be picked up. In these manipulation scenarios, the robot must find a feasible motion trajectory from its initial configuration to a grasp-achieving configuration. While most sampling-based motion planning algorithms assume the goal is a point within the configuration space of the robot, the goal in manipulation problems can consist of a continuous subset of the configuration space — any configuration of the robot resulting in a feasible grasp can be considered a goal configuration.

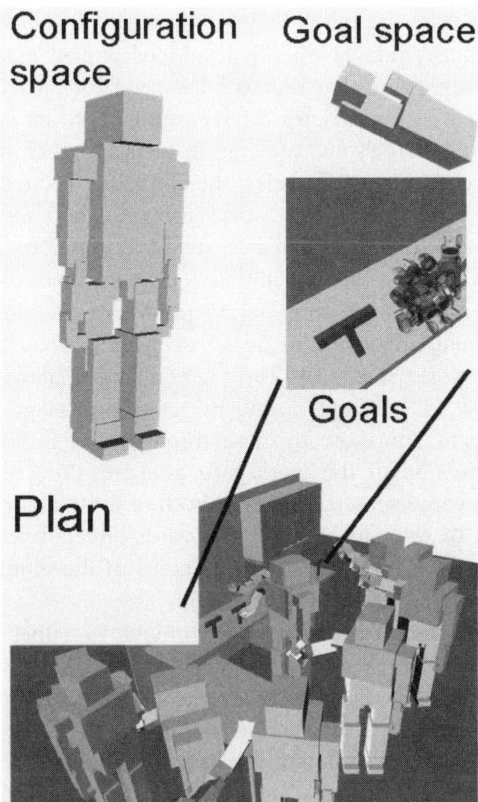

Fig. 1. Humanoid robot grasps an object by autonomously searching through the space of possible grasps and torso movements.

For example, consider a grasping problem where the robotic manipulator has to pick up a cup from the sink to put in a cupboard. The manipulator first has to move its end-effector close to the cup and then it has to achieve the correct contacts with the cup, constraints that are inherently defined in the workspace. In general, there is a fundamental mismatch between the space used to describe the goals of the manipulation planning problem and the configuration space used to search for a solution.

Methods exist to overcome this mismatch, such as using inverse kinematics (IK) to transform the workspace constraints into configuration space goals [6, 7]. However, as the dimensionality of the problem increases the effective dimensionality of the goal set increases as well, often causing these methods to become prohibitively slow. Further, it has been shown that such approaches can fail to converge to a valid goal and are usually limited to finding only one of the potentially infinite number of goals [8]. When planning through obstacle-laden environments, having only one goal configuration is problematic, as this configuration is not guaranteed to be reachable from the initial configuration of the robot.

Instead, Bertram et al. [8] developed a nice extension to the RRT algorithm that removes the need for an IK solution and instead accommodates a goal specified in the workspace of the manipulator's end effector. In their approach they use a workspace goal metric to select the configuration in the search tree that is closest to the workspace goal and then extend out from this configuration in a random direction. This goal extension occurs randomly throughout the growth of the tree.

A more recent approach by Vande Weghe et. al. [9] uses the Jacobian Transpose to do an even more focused search toward a workspace goal. Their approach operates similar to Bertram et al.'s except during the extension stage they use the Jacobian Transpose to move through configuration space in the direction of the workspace goal, resulting in a more efficient overall search. However, both of these algorithms are restricted to growing a single, forwards-directed search tree, and therefore do not capture the benefits of the more efficient bi-directional RRT approaches [10].

Thus, current approaches are limited to either approximating the desired goal configurations and concentrating the entire search in configuration space, or to using an accurate workspace goal representation and growing a single goal-biased search tree. Alternatively, in the following section we present an approach that allows us to search in multiple spaces simultaneously, with a forwards-directed search tree grown through one space, known as the *configuration space*, and a backwards-directed tree grown in another space, known as the *goal space*. This algorithm, known as *BiSpace*, relaxes the need for an explicit mapping from one space to another and in our present application is able to significantly reduce the amount of searching that occurs in the full configuration space.

After describing the algorithm in depth we provide comparative results involving a collection of complex robotic mechanisms and present an experimental results involving a physical 11 degree of freedom manipulator arm.

II. The Bi-space Algorithm

The core idea of the BiSpace algorithm is to grow two different search trees at the same time. One tree explores the full configuration space starting from the initial configuration

Algorithm 1: BiSpace(q_{init}, b_{goals})

```
/* ρ ∈ [0,1] - uniform random variable */
forward ← false
Init(T_f, q_init); Init(T_b, b_goals)
for iter = 1 to maxIter do
    if forward then
        for fiter = 1 to J do
            q ← Extend(T_f)
            if ρ < FollowProbability(q) then
                b_follow ← NearestNeighbor(T_f, q)
                {success, q'} ← FollowPath(q, b_follow)
                if success then
                    return success
        end
    else
        for biter = 1 to K do Extend(T_b)
    forward ← not forward
end
return failure
```

and guarantees feasible, executable, and collision-free trajectories, while the other tree explores the backspace starting from the set of goal configurations and acts as an adaptive, well informed heuristic. The BiSpace algorithm proceeds by extending RRTs in both spaces. Once certain conditions are met, the forward tree attempts to *follow* the goal space tree path to the goal (Figure 2). The algorithm combines elements of both bidirectional RRTs and the RRT-JT algorithm [9].

For clarity, we denote a configuration with q and a goal space configuration with b. We assume that there exists a mapping $F(\cdot)$ from the configuration space to the goal space such that $F(q)$ maps to exactly one goal space configuration. Using this notation, given a goal space distance metric $\delta_b(F(q), b)$, the goal of planning is to find a path to a configuration q such that $\delta_b(F(q), b_{goals}) < \epsilon_{\text{goal}}$.

The flow of the BiSpace algorithm is summarized by Algorithm 1. The **forward** variable is used to keep track of which tree to grow. If **forward** is **true**, then the configuration space tree is extended **J** times, using the standard RRT extension algorithm Extend [1]. Alternatively, if **forward** is **false**, then the goal space tree is extended **K** times. After each iteration, the value of **forward** is flipped so that the opposite tree is extended during the subsequent iteration. After a new node q is added to the configuration space tree, a *follow* step is performed from q with probability FollowProbability(q). An example of such a distribution is described in detail in Section IV-C. If a *follow* step is performed, then q is extended toward b_{follow} and its parents.

The differences between BiSpace and BiRRTs become clear in the *follow* step. In the BiRRT case, *following* consists of connecting the two trees along the straight line joining q and b_{follow}; this is possible since b_{follow} is also in the configuration space. Because each branch of the both the forward and backward trees in the BiRRT algorithm represent a valid

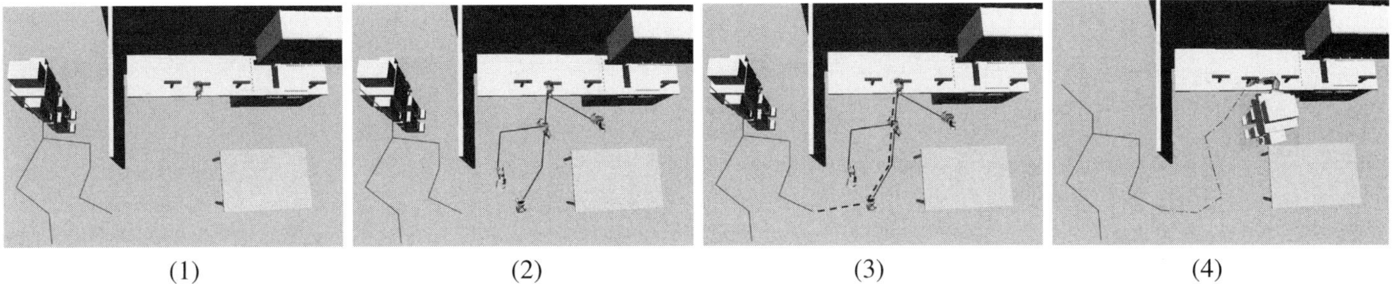

(1) (2) (3) (4)

Fig. 2. BiSpace Planning: A full configuration space tree is grown out from the robot's initial configuration (1). Simultaneously, a goal back tree is randomly grown out from a set of goal space nodes (2). When a new node is created, the configuration tree can choose to *follow* a goal space path leading to the goal (3). Following can directly lead to the goal (4); if it does not, then repeat starting at (1).

collision free path in the configuration space, connecting the two trees immediately implies a path can be found from the start configuration to the goal configuration. However, that is not true with the BiSpace algorithm. Since the goal space is different from the configuration space, the path suggested by the goal space tree must be validated in the configuration space.

Each unique path from a node in the goal space tree to a goal can be used by the forward tree as a heuristic to informatively bias extension toward the goal. Starting from b_{follow}, such a path can be extracted by recursively following its parents. The forward tree can use the goal space path generated by b_{follow} as a bias to greedily *follow* it. If the forward tree succeeds in reaching the goal, a solution is returned (Figure 2). Otherwise, the search continues as before.

Algorithm 2: $\{success, q\} \leftarrow$ FOLLOWPATH(q, b)

```
/* ρ ∈ [0,1] - uniform random variable */
success ← false
for iter = 1 to maxFollowIter do
    best ← null
    bestdist ← γ_inflation * δ_b(F(q), b)
    for i = 1 to N do
        q' ← SAMPLENEIGHBORHOOD(q)
        if δ_b(F(q'), b) < bestdist then
            bestdist ← δ_b(F(q'), b)
            best ← q'
    end
    if best is null then
        if b.parent is null then
            break
        b ← b.parent
    else
        q ← T_f.add(q, best)
end
success ← δ_b(F(q), b.root) < ε_goal
/* Optional IK test */
if not success and (q' ← IKSOLUTION(q, T_b.goals)) then
    {success, q} ← BiRRT(T_f, q, q')
```

Path following is an integral part of the BiSpace algorithm. It generates a very powerful bias as to where the configuration tree should grow by using the nodes in the goal space tree. Each goal space node has already validated a subset of the conditions necessary for the configuration tree to follow it. Although the FOLLOWPATH function is general, we present a simple, but effective, implementation of a stochastic gradient approach for it (Algorithm 2). The forward tree slowly makes progress by randomly sampling configurations that get close to the target goal space node b. Whenever the forward tree stops making progress, it checks if b has any parents. If it does, b is set to its parent and the loop repeats. If there are no more parents, the goal space distance from q to the final parent $b.root$ is checked: if this distance is within the goal threshold, the function returns success; otherwise it returns false.

FOLLOWPATH can require a lot of samples if SAMPLENEIGHBORHOOD uniformly samples the neighborhood of q. This is especially a problem for the high-dimensional configuration spaces used in manipulation planning. Instead, we sample each of the dimensions one at a time while leaving the rest fixed. This type of coordinate descent method has been shown to perform better than regular uniform sampling in optimization and machine learning algorithms [11]. Furthermore, SAMPLENEIGHBORHOOD can incorporate the Jacobian transpose idea from [9] to further bias samples in the correct direction. Because it is not always beneficial to be greedy due to many local minima, we introduce $\gamma_{inflation}$ to relax the distance metric we are minimizing[1].

As an optional addition to the FOLLOWPATH algorithm, we propose using IK solutions, if they exist, to speed up planning. After the forward tree terminates at a configuration q, an IK solution can be checked for a subset of the DOFs of the configuration space. If there exists a solution, we can run a bidirectional RRT using the subset of DOFs used for IK to find a path from q to the new goal configuration. For example, if a 7 DOF arm is mounted on a mobile platform, its full configuration space becomes 10 dimensional, however, the arm's IK equations will still remain 7 dimensional. In this case, IKSOLUTION($q, goals$) will use the arm's position from the last configuration q and check the standard arm IK. Having such a check greatly reduces planning times and is

[1]This has a similar effect to inflating the goal heuristic in A*. We use $\gamma_{inflation} = 1.4$ for all results.

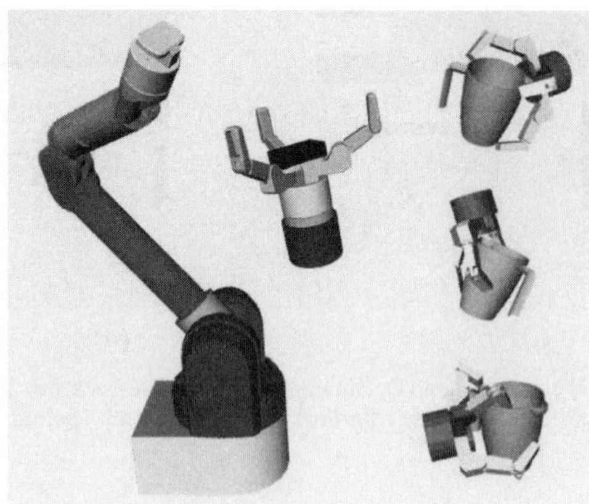

Fig. 3. The hand and arm area treated separately in the grasp planning framework. Only the hand is allowed contact with the environment. The arm is only used for planning.

not prohibitively expensive if the IK equations are in closed form. While some algorithms ignore IK solutions, BiSpace can naturally use inverse kinematics to its advantage. Empirical results suggest that BiSpace can experience a 40% decrease in planning time when exploiting available IK solutions. However, it should be noted that having inverse kinematics equations for a robot does not guarantee feasible solutions can be efficiently found.

III. Applications to Manipulation and Grasp Planning

In grasp planning, the task is to plan for an arbitrarily complex robot to move to pick up any object in the environment. Ideally, the planner decides how to best grasp the object and how to manipulate the robot to achieve the desired grasp. Recently, [6] proposed a method to solve this problem when the robot is stationary and there exist IK equations that provide an efficient mapping from workspace to configuration space. They use a two-tiered approach: they first find the workspace positions of any feasible grasps by sampling from a precomputed table and testing in the real environment, then seed a BiRRT planner with the IK solution of each of those workspace positions. Applying their method to a mobile robot poses several challenges because it relies on producing fast configuration space goals from workspace positions of the end-effector of the robot. As we show, this two-tiered approach is much slower than using BiSpace. Nevertheless, our systems level approach to solving manipulation and grasp planning for mobile robots is inspired from their method (Figure 4).

We divide each robot into two semantic pieces: the *hand* and the *arm* (Figure 3). Only the hand can make contact with the target object. Following [6], grasp tables can be precomputed for every hand-object pair. These grasp tables are computed with the detached *hand* approaching the object from all possible directions with all possible preshapes; usually the final tables are on the order of 300-800 good grasps per object (Figure 1). In the real environment, each grasp in the table is tested against the object and environment for collisions. Note that collisions are only checked with the detached *hand*

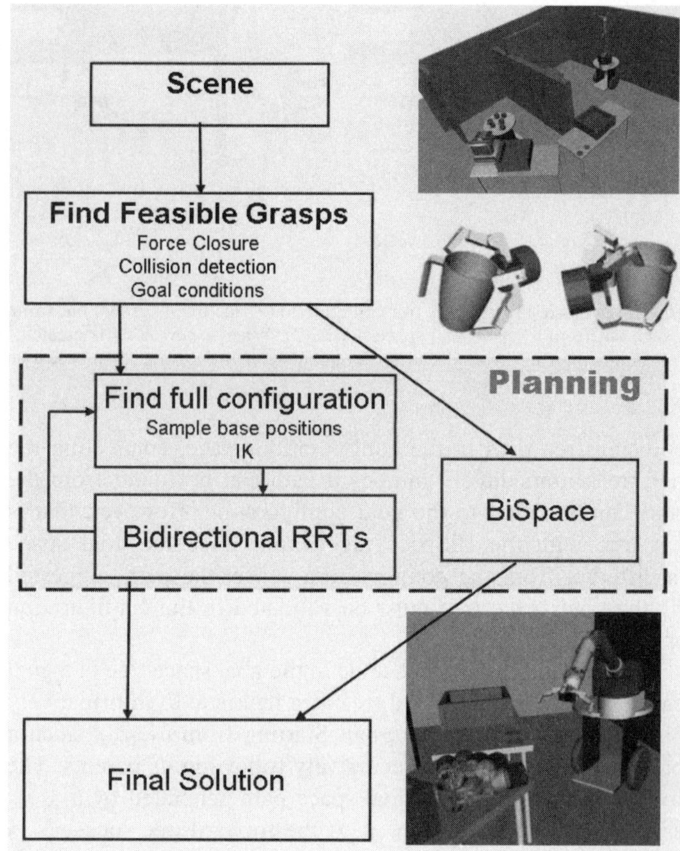

Fig. 4. A framework for grasp planning. BiSpace allows exploration of the space without locking down on a particular IK solution.

since the full configuration of the robot is unknown. Once a collision-free grasp is found, we treat the final hand pose as the goal. The goal of the BiSpace planner is now to move the robot so that the hand attached to it achieves the desired grasp. For simplicity we assume that the planner only moves the *arm* and the mobile base it is mounted on.

Since BiSpace is a randomized algorithm, in general it cannot detect in a finite amount of time that a given collision-free grasp is impossible to reach. Therefore, seeding BiSpace with only one grasp at a time is dangerous as the planner might never find a solution. Instead, it is favorable to seed the BiSpace planner from the beginning with as many feasible grasps as possible using the precomputed grasp tables, increasing the likelihood that at least one of the grasps can be reached. Since the Extend operation is not affected by the number of trees being grown, incorporating multiple goals in the goal space does not affect efficiency [12].

IV. Manipulation Planning Heuristics

We introduce several heuristics to the BiSpace planner in order to improve planning efficiency. Each of these heuristics assumes that every robot is composed of a mobile base and at least one arm whose end-effector is used to make contact with the target objects. Note that no assumptions are made about the kinematics of any of the robot's parts, and both humanoid and wheeled robots are applicable in this framework (as we demonstrate in Section V). Furthermore, each heuristic can be automatically derived for any robot platform, making them ideal for general use.

A. Base Reachability

Many researchers have shown that using some form of goal biasing by modifying the configuration space sampling distribution greatly reduces planning times. This section tackles the problem of intelligently biasing configuration space samples when the only goals given are the final grasps.

Given a target grasp g, we would like to quickly compute a distribution $P_g(p,\theta)$ over the 2D placement (p,θ) of the base of the robot for which g will be successful.

We first perform a kinematics workspace analysis for the arm similar to [13]. Figure 5 shows the hand reachability volume generated for the HRP2 humanoid robot. This was computed by randomly sampling a 6D end-effector position around the space of the humanoid's shoulder and querying for an IK solution. It can similarly be computed by randomly sampling arm configurations and storing their resultant 6D end-effector. We store all the valid 6D end-effector positions in $\mathcal{X} = \{(x_q, x_t)\}$ where x_q and x_t are the rotation and translation associated with the transformation of the end-effector. Clearly, to succeed in the planning for a specific grasp, the robot should move its body so that its reachability volume coincides with the particular grasp.

Each grasp g represents an affine transformation where g_t is the translation and g_q is the rotation. Our goal is to find similar grasps to g in $\mathcal{X}$ and perform simple counting to extract a probability of existence of an IK solution.

We begin by defining a rotation on the plane as $R_n(\theta)$ where n is the normal vector to the plane the robot rotates on. The group of all rotations on the plane is denoted by

$$\mathcal{Q}_n = \{R_n(\theta) \mid \theta \in \mathbb{S}^1\} \tag{1}$$

We can now define an equivalence class of rotations that differ only by a rotation about the plane as

$$q_q\, \mathcal{Q}_n = \{r * g_q \mid r \in \mathcal{Q}_n\}. \tag{2}$$

where $*$ denotes the action of applying one rotation after another. We then define the equivalence class of all similar grasps up to a rotation on the plane as

$$\mathcal{X}_g = \{(g_q, (g_q * x_q^{-1})(x_t)) \mid x \in \mathcal{X},\ x_q \in g_q\, \mathcal{Q}_n\} \tag{3}$$

where $(g_q * x_q^{-1})(x_t)$ transforms the position of the end effector from the frame of the robot (in the reachability map) to the frame of the grasp.

We compute equivalence classes for the entire set $\mathcal{X}$ by sampling x_i, storing its equivalence class $\mathcal{X}_i$ and continuing sampling from $\mathcal{X} - \mathcal{X}_i$. Doing this greatly reduces the number of grasps from over $100,000$ in $\mathcal{X}$ to about 100 equivalence classes. In practice, we accept grasps if they are within a threshold of the rotation x_q.

We now compute the inverse reachability volume $\mathcal{D}_g(p,\theta)$ for each equivalence class g. Note that each end-effector

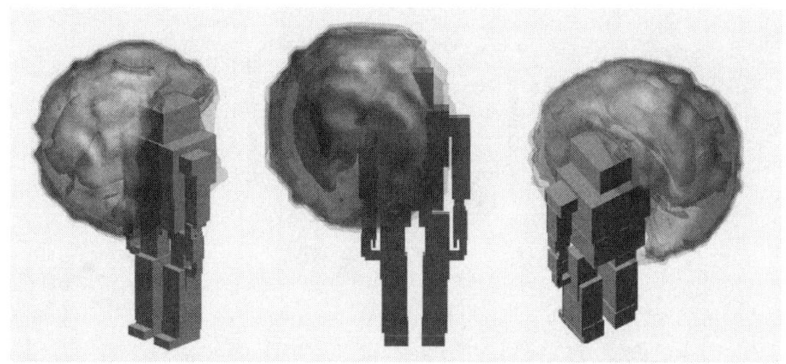

Fig. 5. 6D hand reachability from a given base placement projected in 3D. Shown are three different views of the reachability volume. Dark opaque areas contain more reachable end-effector positions.

position x_t where $x \in \mathcal{X}_g$ has been aligned to the frame of the grasp g. p and θ are still in world coordinates and need to be converted to the grasp coordinate system induced by g. This is achieved by

$$\mathcal{D}_g(p,\theta) = \{\ \|x_t - (R_n(\theta) * g_q)^{-1}(g_t - p)\| < \epsilon \mid x_t \in X_g\} \tag{4}$$

Finally the inverse reachability map converts $\mathcal{D}_g(p,\theta)$ into a probability distribution as

$$P_g(p,\theta) = \exp\left\{-\omega\left(\frac{\sum_{d\in\mathcal{D}_g(p,\theta)} d}{|\mathcal{D}_g(p,\theta)|}\right)^2\right\} \tag{5}$$

Views of the map for various equivalence classes g in a sample scene for the HRP2 humanoid are shown in Figure 6.

To test the effectiveness of the inverse reachability map, we randomly sampled base positions using $P_g(p,\theta)$ to see if they would contain feasible IK solutions. Empirical results showed that $P_g(p,\theta)$ was able to generate a feasible base placement 2.5 times faster than uniform sampling around the grasp.

B. Workspace Exploration

As humans, we employ different navigation strategies based on our distance to a goal object. When a person is far away from an object of interest, they care primarily about moving their body in a direction that will get them close to the object. When they are close, they usually plant their feet and use their arms to make contact with the object. We can achieve the same behavior in BiSpace by modifying the configuration space distance metric such that

$$\delta(q) = |\omega(q) q_{arm}| + |q_{base}| \tag{6}$$

where q_{arm} is the degrees of freedom associated with the arm. When the robot base is far away from the goal, the weight ω should be small so that the robot takes bigger steps on average. This suggests a simple monotonic function for ω:

$$\omega(q) \propto \exp\left\{-\frac{\min_i |goal_i - BasePosition(q)|^2}{2\sigma^2}\right\} \tag{7}$$

where σ is proportional to the length of the arm. Figure 7 demonstrates the behavior of BiSpace when using the modified

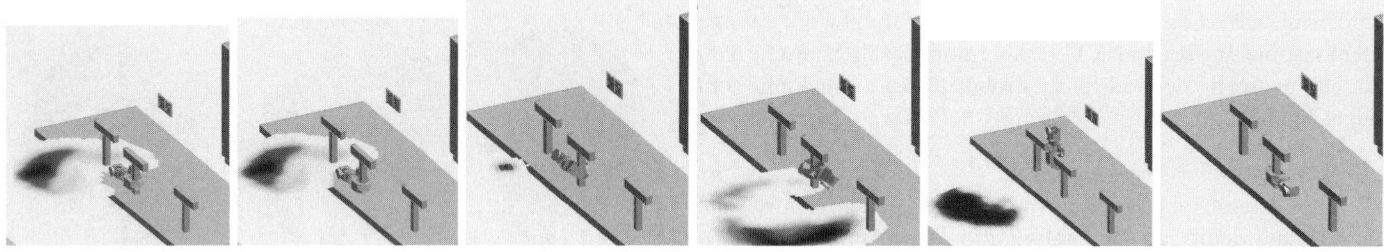

Fig. 6. Base placements derived from the goal space goals. As described in the text, these placements can be efficiently generated using the pre-computed hand reachability volume.

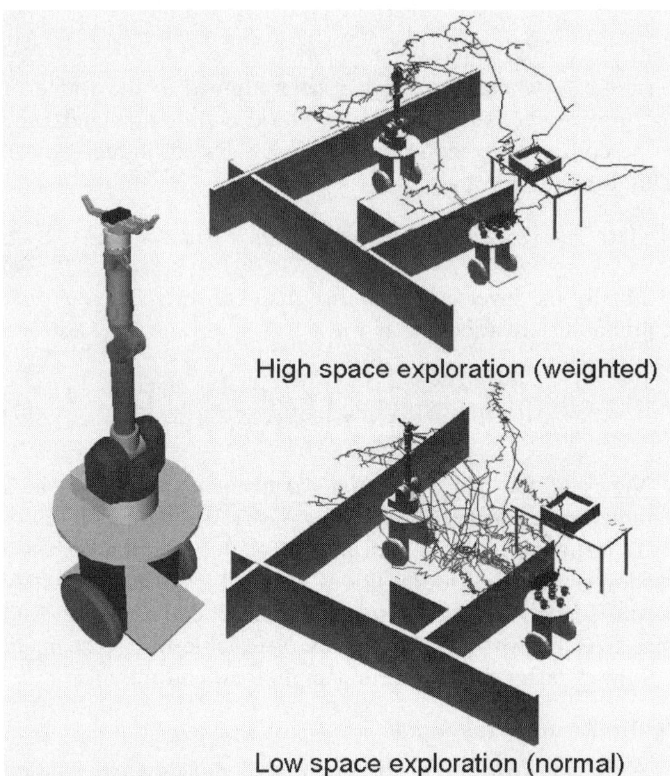

Fig. 7. Comparison of how the distance metric can affect the exploration of the arm. The top image shows the search trees (red/black) generated when the distance metric and follow probability is weighted according to Equation 6. The bottom image shows the trees when the distance metric stays uniform across the space; note how it repeatedly explores areas. The goal space trees are colored in blue.

distance metric, and empirical results show that planning times reduce by 20% when this metric is used.

C. Follow Probability

The farther the robot is away from the goal, the less chance it will have of reaching it through FOLLOWPATH. The reason is because FOLLOWPATH itself is not exploration-centric like RRTs; it is meant for greedily approaching the goal when the body and hand of the robot are relatively unobstructed by complex environment obstacles. We propose two metrics to compute the follow probability: the hand reachability volume (Figure 5) or the distance falloff $\omega(q)$ (Equation 7). Both metrics monotonically decrease as the robot gets farther from the goal. The hand reachability is more informed since it is a 6D table reflecting the real arm kinematics while $\omega(q)$ is much easier to compute and often very effective (Figure 7). The correct follow probability can have a dramatic effect on planning times, sometimes reducing it by 60-70%.

V. RESULTS

For all planners, simulations, and real-robot experiments, we used an open-source planning test-bed called OpenRAVE [14]. There are few existing algorithms that work in high-dimensional spaces and cope with goals that are not specified explicitly in the configuration space, making direct comparison of BiSpace a little challenging. We chose to compare BiSpace with RRT-JT [9] and the two-tiered BiRRT approach described in Section III. Whenever the robot is mobile in a test scene, it adds 3 degrees of freedom to its configuration since its base can translate and rotate freely on the floor. Because randomized algorithms are known to have a long convergence tail, we terminate the search after 10-20 seconds and restart. This termination strategy produces much faster average times for all algorithms. Note however that every termination counts against the final planning time for that particular algorithm. Termination times were uniquely set for each algorithm in order to give it the fastest possible average time. Each algorithm is run on each scene 16-30 times, and the average planning time is recorded in Table I. Other parameters like RRT step size and goal thresholds were kept the same for all algorithms. To demonstrate the generality of the proposed algorithms, we produced results using both the HRP2 humanoid and a WAM arm loaded on a segway (Figure 8).

Since BiRRTs operate only in the full configuration space it would be unfair if they were seeded with the final solutions without any penalties. In order to make comparison fair, we randomly sample full configuration solutions for a given target grasp until a collision-free, feasible configuration is generated. The recorded time is added to the final planning time. The sampling takes somewhere from 2-9 seconds for HRP2 and less than 1 second for the WAM on segway[2].

[2]The large difference in sampling times is because the reachability area for the WAM is much larger

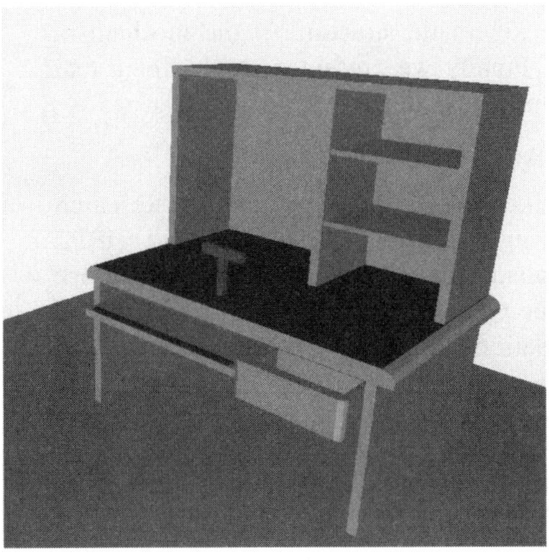

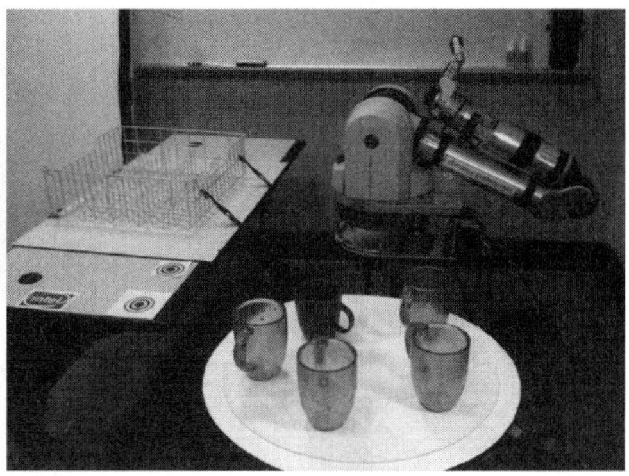

Fig. 8. Scenes used to compare BiSpace, RRT-JT, and BiRRTs.

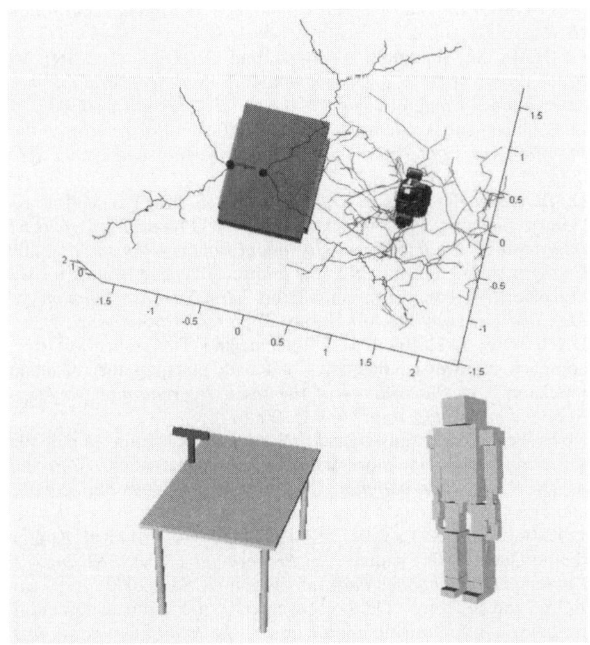

Fig. 9. Hard scene for BiSpace. The forward space tree (red) does not explore the space since it is falsely led over the table by the goal space tree (blue).

	BiSpace	RRT-JT	BiRRTs
HRP2 table (11 DOF, easy)	**33**	53	68
HRP2 table (11 DOF, harder)	**45**	528	78
HRP2 random (11 DOF)	**37**	170	40
WAM/segway (10 DOF)	**17.25**	25.2	22.93
WAM (7 DOF)	0.44	11	**0.37**

TABLE I

AVERAGE PLANNING TIME IN SECONDS FOR EACH SCENE.

A. HRP2

When planning for the HRP2 robot, we make the assumption that its base can freely travel on the floor and the legs do not need to move. Once BiSpace has planned a global trajectory, later footstep planners can add the necessary leg movements and dynamics to make the HRP2 move. In order to allow for leg space, an invisible cylinder is super-imposed over the lower body. Thus the planning space for HRP2 is reduced to 11 degrees-of-freedom: 3 for the base, 1 for the waist, and 7 for the arm. As can be seen from Figure 5, most of the hand reachability lies shoulder height to the side of the robot. This makes it hard for the robot to manipulate objects in front of it at waist height, which is why all the planners require significant planning time.

One of the hardest scenes for BiSpace is when the target object is on a table and HRP2 has to circle the table to get to it (Figure 9). Here, the goal space tree produces many false paths directly over the table, which the HRP2 cannot follow to the end. This process goes on until the rest of the configuration space tree finally explores the space on the other side of the table. This limitation is characteristic of bi-directional RRTs also and provides a good example of why exploration is always a crucial ingredient in sampling-based planners.

B. WAM

We tested two main scenes for the WAM: a living room scenario where the WAM is mobile, and a scenario where the WAM has to put cups in a dishwasher. The WAM arm has 7 degrees of freedom and very high reachability making planning very fast. BiSpace compares relatively well with BiRRTs, however it is a little slower due to the extra overhead in the FOLLOWPATH function. We also tested the entire grasp planning framework using the BiSpace planner on a real WAM arm setup (Figure 8). The WAM arm runs in real-time and can compensate immediately for changes in the environment.

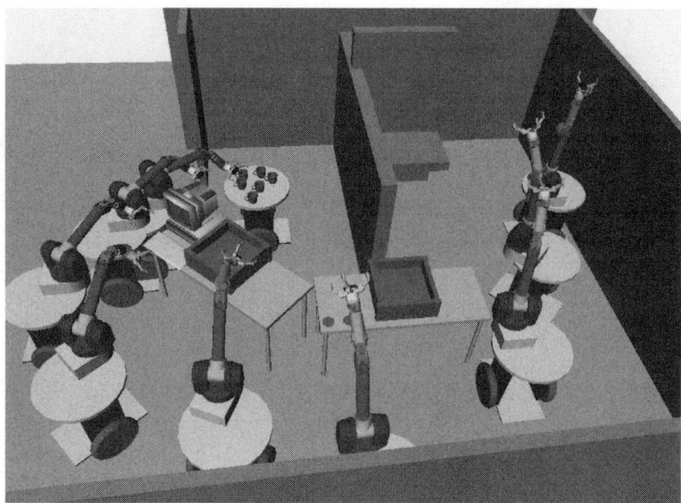

Fig. 10. Manipulation and Grasp Planning solutions for one of the test scenes.

Fig. 11. The real WAM arm grasping mugs.

VI. Conclusions

We presented the BiSpace algorithm for efficiently producing solutions to complex path planning problems involving a goal space that is different from the configuration space. We used this algorithm to plan for several mobile robots to perform manipulation tasks. One key feature of the grasp framework we employed is that it makes very few assumptions about how the robot should move to manipulate its target object, which makes it ideal for autonomous robot scenarios. Furthermore, we showed several heuristics that exploit various information about the kinematic structure of the mechanism to speed up planning. Finally we presented results for a real WAM arm loading cups into a dishwasher rack.

VII. Acknowledgements

The authors would like to thank the Digital Human Research Center for the allowing the use of the humanoid HRP2 model in these experiments. This material is based upon work supported in part by the Quality of Life Technology Research Center as part of the National Science Foundation under EEC-0540865.

References

[1] S. LaValle and J. Kuffner, "Randomized kinodynamic planning," *International Journal of Robotics Research*, vol. 20, no. 5, pp. 378–400, 2001.

[2] J. Kuffner, K. Nishiwaki, S. Kagami, M. Inaba, and H. Inoue, "Motion planning for humanoid robots," in *Proceedings of the International Symposium on Robotics Research (ISRR)*, 2003.

[3] J. Kim and J. Ostrowski, "Motion planning of aerial robots using Rapidly-exploring Random Trees with dynamic constraints," in *Proceedings of the IEEE International Conference on Robotics and Automation (ICRA)*, 2003.

[4] G. Oriolo, M. Vendittelli, L. Freda, and G. Troso, "The SRT Method: Randomized strategies for exploration," in *Proceedings of the IEEE International Conference on Robotics and Automation (ICRA)*, 2004.

[5] D. Ferguson and A. Stentz, "Anytime RRTs," in *Proceedings of the IEEE International Conference on Intelligent Robots and Systems (IROS)*, 2006.

[6] D. Berenson, R. Diankov, K. Nishiwaki, S. Kagami, and J. Kuffner, "Grasp planning in complex scenes," in *Proceedings of IEEE-RAS International Conference on Humanoid Robots (Humanoids)*, 2007.

[7] C. Klein and C. Huang, "Review on pseudoinverse control for use with kinematically redundant manipulators," *IEEE Transactions on Systems, Man and Cybernetics*, vol. 13, no. 3, pp. 245–250, 1983.

[8] D. Bertram, J. Kuffner, R. Dillmann, and T. Asfour, "An integrated approach to inverse kinematics and path planning for redundant manipulators," in *Proceedings of the IEEE International Conference on Robotics and Automation (ICRA)*, 2006.

[9] M. V. Weghe, D. Ferguson, and S. Srinivasa, "Randomized path planning for redundant manipulators without inverse kinematics," in *Proceedings of IEEE-RAS International Conference on Humanoid Robots (Humanoids)*, 2007.

[10] J. Kuffner and S. LaValle, "RRT-Connect: An Efficient Approach to Single-Query Path Planning," in *Proceedings of the IEEE International Conference on Robotics and Automation (ICRA)*, 2000.

[11] Z. Luo and P. Tseng, "On the convergence of coordinate descent method for convex differentiable minimization," *Journal of Optimization Theory and Applications*, vol. 72, no. 1, pp. 7–35, 1992.

[12] K. Okada, T. Ogura, A. Haneda, J. Fujimoto, F. Gravot, and M. Inaba, "Humanoid motion generation system on hrp2-jsk for daily life environment," in *International Conference on Machantronics and Automation (ICMA)*, 2004.

[13] F. Zacharias, C. Borst, and G. Hirzinger, "Capturing robot workspace structure: Representing robot capabilities," in *Proceedings of the IEEE International Conference on Intelligent Robots and Systems (IROS)*, 2007.

[14] R. Diankov and J. Kuffner, "Openrave: A planning architecture for autonomous robotics," Robotics Institute, Carnegie Mellon University, Pittsburgh, PA, Tech. Rep. CMU-RI-TR-08-34, July 2008.

Structural Improvement Filtering Strategy for PRM

Roger Pearce[†], Marco Morales[‡], Nancy M. Amato[†]
[†]Parasol Laboratory, Department of Computer Science
Texas A&M University, College Station, Texas, 77843-3112, USA
[‡]Department of Digital Systems, Instituto Tecnológico Autónomo de México
Rio Hondo 1, Tizapán, México D.F., 01080, México
{rpearce,amato}@cs.tamu.edu
marco.morales@itam.mx

Abstract—**Sampling based motion planning methods have been highly successful in solving many high degree of freedom motion planning problems arising in diverse application domains such as traditional robotics, computer-aided design, and computational biology and chemistry. Recent work in metrics for sampling based planners provide tools to analyze the model building process at three levels of detail: sample level, region level, and global level. These tools are useful for comparing the evolution of sampling methods, and have shown promise to improve the process altogether [15], [17], [24].**

Here, we introduce a filtering strategy for the Probabilistic Roadmap Methods (PRM) with the aim to improve roadmap construction performance by selecting only the samples that are likely to produce *roadmap structure improvement*. By measuring a new sample's maximum potential structural improvement with respect to the current roadmap, we can choose to only accept samples that have an adequate potential for improvement. We show how this approach can improve the standard PRM framework in a variety of motion planning situations using popular sampling techniques.

I. Introduction

The general *motion planning* problem consists of finding a valid path for an object from a start configuration to a goal configuration. In the traditional application of robotics, a valid path is defined as a collision-free path. Many of the techniques originally designed for robotics have been extended to other applications such as the study of protein folding in Biology and Chemistry [3], [5], [21]–[23], virtual prototyping in manufacturing and mechanical design [4], [8], and the simulation of characters for animation and games [13], [14].

Unfortunately, exact or complete motion planners are intractable for most practical problems because the complexity grows exponentially with the problem's dimensionality [20]. This led researchers to explore sampling based methods and create incomplete approximate solutions. One popular sampling-based method is the Probabilistic Roadmap Method (PRM) [11] which randomly builds a *roadmap* representation of the planning space. Many heuristics have been added to the PRM framework [1], [6], [9], with the overall goal to increase the distribution of samples in regions of the space that model highly constrained robot motions. As a result, there are many planners to choose from and it is not always clear how to choose among them.

Dynamically adapting the planning strategy to features discovered in each problem [7], [10], [16], [18], [24] has been successful in addressing the shortcomings of a particular sampling strategy. However these methods need metrics that gather relevant information about the planning process in order to make effective decisions to adapt the planning strategy.

Another area of research focuses on creating minimal roadmaps. VisPRM [19] is a strategy that achieves minimal roadmaps by accepting only the necessary samples; samples are accepted only if they improve the structure of the roadmap, and this criterion is tested in brute-force manner.

In our previous work, we have developed online metrics [15], [17] to monitor the sampling process to estimate the performance of the model building process at three levels of detail: at the sample level, how "good" and efficient are the samples created? At a small region level, how well are the small locally-homogeneous regions in the large high-dimensional space being explored? At a global level, how well is the "global view" of the high-dimensional space being modeled? Unfortunately, these questions cannot be directly answered because there is not a perfect solution to compare with. These questions are answered indirectly by measuring the relative performance change over small time intervals for each of the three levels of detail. The overall goal of this work is to use the online performance metrics in creating adaptive sampling algorithms to solve more complex problems.

In this paper we develop a metric and filtering strategy to estimate the potential *roadmap structural improvement* of a new sample. This estimate of improvement can be used with a threshold to bias the sampling process towards new samples that have a higher potential for structural improvement. We show that this addition to the PRM framework can increase the speed of roadmap convergence but may come at the price of roadmap quality.

The remainder of this paper is organized as follows: Section II introduces previous work in filtering strategies for PRM; Section III covers how we define and calculate *Roadmap*

This research supported in part by NSF Grants EIA-0103742, ACR-0081510, ACR-0113971, CCR-0113974, ACI-0326350, CRI-0551685, by the DOE, IBM, and Intel. This research used resources of the National Energy Research Scientific Computing Center, which is supported by the Office of Science of the U.S. Department of Energy under Contract No. DE-AC02-05CH11231.

Pearce supported in part by a Dept. of Education Graduate Fellowship (GAANN).

Structure Improvement; Section IV describes the experimental analysis of this method, and compares the speed of convergence and roadmap quality of the different strategies; Section V discusses our final conclusions.

II. PRM FILTERING STRATEGIES

The basic PRM framework [11] builds a C-Space *roadmap* model in two main steps: node generation and node connection. Node generation consists of randomly sampling configurations, testing them, and keeping the valid ones as roadmap vertices. For the valid samples kept in the roadmap, a neighborhood of potential edge candidate samples in the roadmap is gathered using some simple heuristic (e.g. k-closest neighbors). Edges between the sample and its neighborhood are tested with a local planner, and the valid transitions are kept as roadmap edges. The resulting roadmap can be queried as many times as needed.

To improve the performance of roadmap construction, many filtering techniques have been proposed. The general aim of this filtering is to reduce sampling and local planning in areas that are easily mapped (oversampling) while biasing the exploration to the difficult regions of C-Space. At the node level, sometimes called node generation, filtering is done to bias sampling to difficult regions of the C-Space; GAUSS-PRM [6], BRIDGE-TEST [9], and OBPRM [1] are examples of node-level filtering. Similarly, edge pairs can be filtered in different ways from k-closest or neighborhood-radius to more advanced methods like VISPRM [12]. Table I briefly summarizes the PRM filtering techniques mentioned in this work.

TABLE I

ROADMAP-BASED PLANNERS STUDIED

Method	Node-level Filtering	Edge-level Filtering
BASIC-PRM [11]	only basic CD filter	k-closest neighborhood
GAUSS-PRM [6]	after CD check	k-closest neighborhood
BRIDGE-TEST [9]	after CD check	k-closest neighborhood
OBPRM [1]	only basic CD filter	k-closest neighborhood
VISPRM [12]	after all edge checks	all nodes not in current connected component
STRUCTURAL IMPROVEMENT FILTERING	based on roadmap structure, before edge checks	based on roadmap structure, before edge checks

Most of the filtering strategies shown in table I filter based on information gathered during collision detection tests. For example, in GAUSS-PRM and BRIDGE-TEST many samples may be discarded based on the local region of C-Space explored during collision detection tests (whether in open free-space, or deep inside obstacle-space); the only samples that remain are on the boundaries of C-Obstacles. In VISPRM many samples (along with its fully computed edges) are discarded if the sample and its edges fail to either merge two connected components or to create a new one.

In this work we are proposing a new filtering strategy based only on the current roadmap and an estimate of the structural improvement. In this way filtering is done before expensive edge local planning.

III. ROADMAP STRUCTURE IMPROVEMENT

In our previous work, we monitored the evolution of the PRM process and classify each new sample. These classifications have been used to compare the quality of samples generated by different sampling strategies. As a quick summary, each new sample X is classified as follows: (See [15], [17] for more details.)

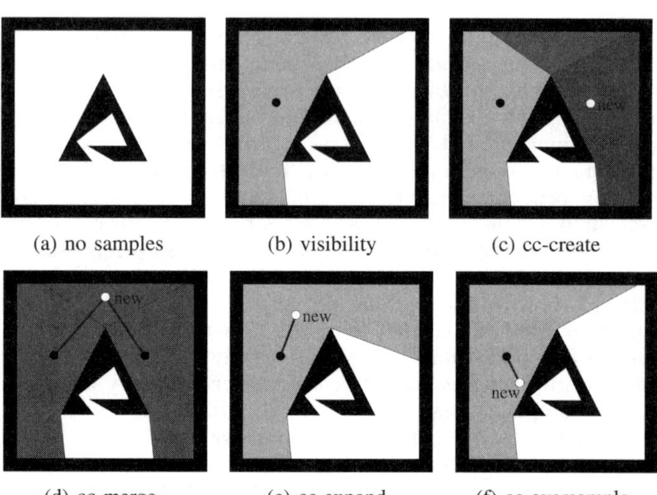

(a) no samples (b) visibility (c) cc-create

(d) cc-merge (e) cc-expand (f) cc-oversample

Fig. 1. Classification of new nodes when modeling the C-Space of a point robot moving in the plane shown in (a). (b) The first sample in the model with its visibility region. (c) A new sample lying outside the visibility region of any other sample creates another component with its own visibility region. (d) A new sample lying in the overlap of the visibility region of two components allows to merge them. (e) A new sample lying inside the visibility region of one component expanding its visibility: cc-expand. (f) A new sample lying inside the visibility region of one component without changing its visibility: cc-oversample.

- cc-create — A new component CC with X as its only node is created as seen in Figure 1(c). The coverage of the roadmap increases by the coverage of X and the connectivity and topology improve due to the new component.
- cc-merge — X merges two or more existing components in the roadmap as seen in Figure 1(d). The coverage, connectivity, and topology improve due to the new pathways found.
- cc-expand — X expands an existing component in the roadmap as seen in Figure 1(e). The coverage and topology improve due to the new pathways found.
- cc-oversample — X fails to expand the coverage of a component in the roadmap as seen in Figure 1(f). The coverage and connectivity remain constant.

VISPRM [19] is an aggressive strategy to eliminate *cc-expand* and *cc-oversample* nodes. It bypasses the neighborhood information for a brute-force all-pairs method to classify the sample type. In this paper we will make comparisons to this work as a case where only *cc-create* or *cc-merge* nodes are accepted.

The remainder of this section discusses how we define structural improvement, estimate the potential structural improvement of a sample, and design acceptance polices for the PRM framework.

A. Defining Structural Improvement

In this work we filter the sampling process and bias the samples towards areas that improve the structure of the roadmap. We define structural improvement as:

- The addition of pathways previously nonexistent: *cc-merge*
- Finding shorter pathways between existing nodes: some *cc-expand*

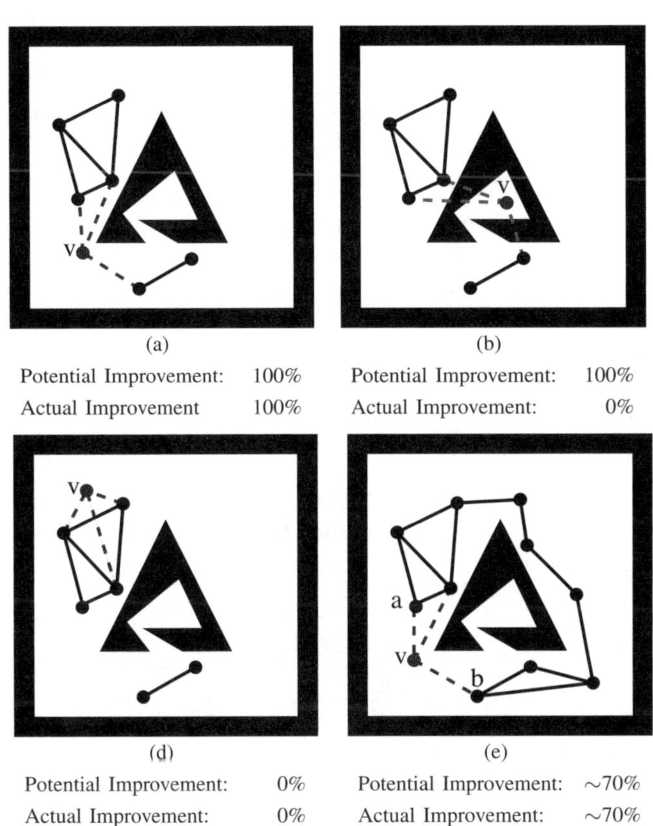

Fig. 2. Cases of Potential Structural Improvement vs. Actual Structural Improvement. Solid edges represent existing roadmap pathways, dashed edges represent potential neighbors of the new sample v.

When the addition of a new sample and its corresponding neighbor edges reduces the distance between any two neighbors there is structural improvement. For each new sample we estimate its maximum potential to create such a structural improvement before the edge pairs are actually checked to improve the quality of roadmap nodes and reduce costs.

Estimating the potential before edge pairs are tested is what differentiates this work from previous work [15], [17] or VISPRM [12]. Previously, expensive local planners for the edge pairs were needed to make reliable classifications or to calculate the improvement of a new sample.

B. Estimating Sample Potential Improvement

Following the PRM framework of sample selection X and neighborhood identification $\{N_1, ..., N_k\}$, we use the existing roadmap model to evaluate the potential structural improvement of a sample X.

Figure 3 illustrates how the potential improvement of a new sample X is computed. First, the existing pairwise pathways between all neighbors $\{N_1, ..., N_k\}$ are evaluated. If it is found that some neighbors belong to different connected components (CC), then X is a potential *cc-merge* node and its potential structural improvement is set to 100%. Each existing graph pathway between N_i and N_j must be evaluated, and a *Single Source Shortest Path* (SSSP) algorithm can be used to obtain the pathway $P_{i,j}$. The weight of this pathway can then be compared to the new potential pathway through X: $P'_{i,j} = N_i \rightarrow X \rightarrow N_j$. The potential improvement of X is the maximum percentage improvement of all $P'_{i,j}$ over the existing $P_{i,j}$. Algorithm 1 details how the maximum potential improvement is calculated.

Algorithm 1 Calculation of potential *structural improvement*

Input: The new sample X, the existing roadmap R, and the sample's neighborhood $\{N_1, ..., N_k\}$

Output: max_imp – Maximum potential improvement, as a percentage

$max\_imp = 0$
if $\{N_1, ..., N_k\}$ are not in the same CC of R **then**
  $max\_imp = 100\%$
  RETURN
end if
for every N_i in $\{N_1, ..., N_k\}$ **do**
  Find SSSP(R, N_i, $\{N_{i+1}, ..., N_k\}$)
  for every N_j in $\{N_{i+1}, ..., N_k\}$ **do**
    $P_{i,j}$ = SSSP from N_i to N_j through R
    $P'_{i,j}$ = distance from $N_i \rightarrow X \rightarrow N_j$
    $improvement$ = % improvement of $P'_{i,j}$ over $P_{i,j}$
    max_imp = max(max_imp, $improvement$)
  end for
end for

C. Sample Acceptance Policy

Based on a new sample's potential improvement we can make informed decisions about the fate of the sample. By creating sample improvement thresholds we can effectively filter the samples and accept the samples that yield a desired potential improvement.

At one extreme, an improvement threshold of 0% will accept any sample which offers any amount of improvement. This does not accept all samples, because cases arise where all pathways through the sample offer longer pathways than currently exist in the roadmap.

On the other extreme, an improvement threshold of 100% will only accept samples which are potential *cc-merge* nodes. This aggressive threshold comes with quality considerations discussed in Section IV-B.

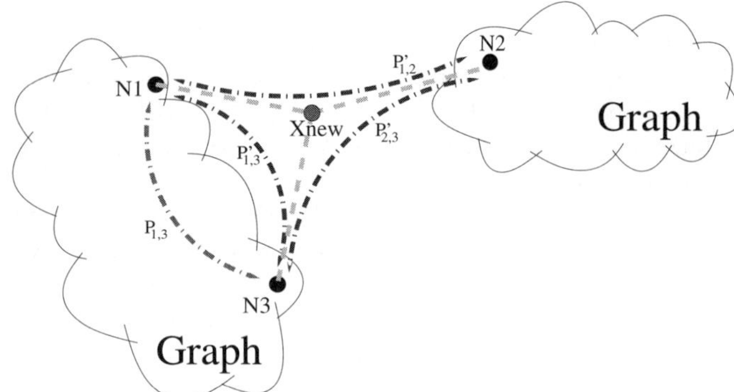

Fig. 3. Illustrates calculating the potential *structural improvement* of sample X. $N1$, $N2$, $N3$ are neighbors of X. $P_{i,j}$ represents existing roadmap paths between N_i and N_j. $P'_{i,j}$ represents potential new pathways between N_i and N_j through X.

It is important to note that delaying the *sample acceptance policy* for an initial time period is necessary. During the initial phase of roadmap building the model is in a "Quick Learning" stage [15], [17]. Here, estimations and classifications of samples are not accurate due to the primitive knowledge in the model. It is not until the majority of the planning space is covered (but not necessarily connected) that the sample estimations and classifications become accurate. In this work the initial window is set to 20 samples; we plan to automate this in future work.

IV. EXPERIMENTAL ANALYSIS

In this study we show the effect of both *Structural Improvement Filtering* in PRM and VISPRM as two different *sample acceptance policies*. Our experimental setup, described in detail below, shows the effect of aggressive policies that accept few samples and how it leads to cheaper roadmap construction at the cost of roadmap quality.

A. Experimental Setup

1) Motion Planning Environments: Throughout this paper we study instances of the motion planning problem with different valid and invalid densities, and with a mixture of open spaces and narrow passages. The instances discussed are:

- The *rigid-walls* problem, Fig. 4 (left), has a 6-DOF rigid-body box robot that should pass through the small openings (slightly larger than the robot) in the walls that divide the environment into five chambers from one side to the other side. Three of the chambers are cluttered with small cube-shaped obstacles. This problem has a C-Space that is similar to its workspace, with four narrow passages and open and cluttered spaces in between.
- The *rigid-maze* problem, Fig. 4 (center), has a 6-DOF rigid-body robot that should pass through a series of tunnels with some dead-ends from the top to the bottom. This problem is interesting because its C-Space resembles the workspace with two clear free areas, the tunnels form a long and narrow passage with dead ends, and the obstacle occupies the majority of the planning space.
- The *rigid-hook* problem, Fig. 4 (right), has a 6-DOF rigid-body hook robot that should pass through the narrow openings in the two walls that divide the environment into three chambers from one side to the other side of the environment. This is a difficult problem that requires simultaneous translational and rotational motions.

2) Node Generation Strategies: We study the *Node Generation* strategies described in Table II.

TABLE II
ROADMAP-BASED PLANNERS STUDIED

Planner	Sampling Strategy
BASIC-PRM [11]	Uniform, keeping all valid configurations
OBPRM [1]	Generate invalid samples and push them away to get valid samples around obstacles
GAUSS-PRM [6]	Uniform sampling, find valid samples within distance d from invalid samples. Valid samples have a Gaussian distribution around obstacles
BRIDGE-TEST [9]	Uniform sampling, find pairs of invalid samples separated a distance d and keep valid samples between pairs.

3) Neighborhood Selection and Local Planner Strategies: We use simple heuristics for neighborhood selection and local planning. We implemented these strategies as described in [24]: every new sample in the roadmap attempts to connect to the 10-closest nodes already in the roadmap by using the straight-line and rotate-at-s local planners [2].

4) Sample Acceptance Policies: Throughout the experiments in this paper, we will make comparisons between four *sample acceptance policies* (Table III) in the PRM framework.

TABLE III
STYLES OF PRM STUDIED

PRM Style	Description
Pure [11]	The original PRM strategy, this will be our standard or baseline comparison.
Imp=50%	This will accept any sample with at least 50% potential improvement.
Imp=100%	This will accept any sample with 100% potential improvement.
VISPRM [12]	Visibility PRM is another style of PRM which aims to create a minimal roadmap.

5) Methods of Comparison: We compare the different acceptance policies in two ways. In a table we average the statistics gathered over 20 random iterations of each set of parameters. Table IV describes the statistics we gathered and averaged for comparison.

We also show the effects of filtering on a single run by plotting the evolution of the diameter of the largest connected component. For these plots we fix the random seed across the different parameter tests to ensure that the different *acceptance policies* are working with the same stream of random samples.

B. Quality Compromises

The decision to discard samples with low potential for improvement comes at the cost of reduced roadmap quality. To

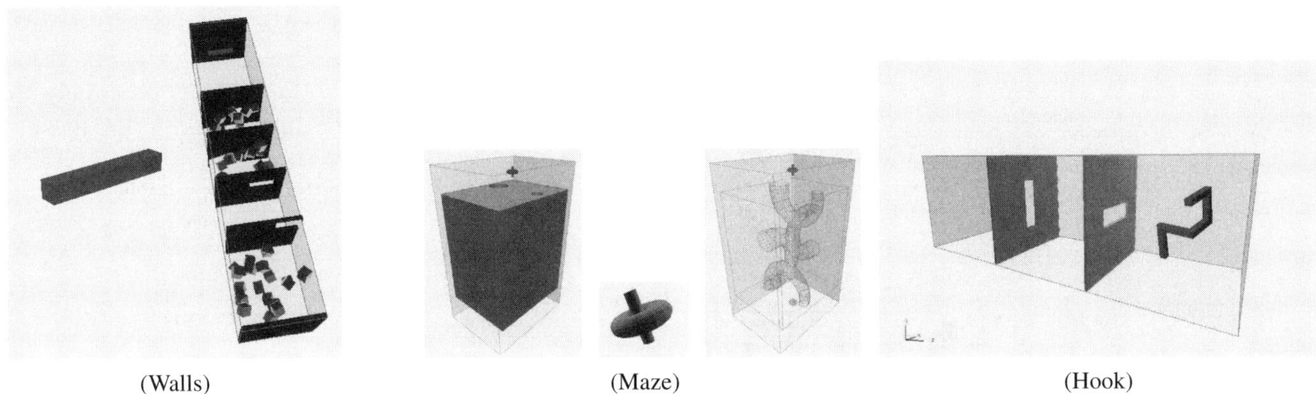

(Walls) (Maze) (Hook)

Fig. 4. (*Walls*) 6-DOFcubic robot, four short passages; (*Maze*) solid view, 6-DOF robot, and wire view; (*Hook*) 6-DOFhook robot, two medium passages.

TABLE V

AVERAGE STATISTICS GATHERED WHILE SOLVING THE *Maze* ENVIRONMENT

Sampler	Measure	Sample Acceptance Policy: Pure	Imp = 50%	Imp = 100%	VISPRM [12]
BASIC-PRM [11]	% Samples Accepted	100% †	2.92%	1.60%	0.28% †
	CD-Calls	1,224,784 †	96,522	57,642	1,010,809 †
	Largest CC Dia	328.2 †	327.2	338.5	434.3 †
	Time Struct Imp (sec)	none †	40.32	10.60	none †
	Total Time (sec)	1,207.3 †	154.9	38.1	239.9 †
BRIDGE-TEST [9]	% Samples Accepted	100% †	54.79%	49.87%	19.30% †
	CD-Calls	873,011 †	622,088	622,447	1,100,653
	Largest CC Dia	280.4 †	287.8	245.7	509.2
	Time Struct Imp (sec)	none †	0.19	0.18	none
	Time (sec)	256.3 †	180.2	179.8	317.2
GAUSS-PRM [6]	% Samples Accepted	100% †	12.50%	7.64%	1.66% †
	CD-Calls	308,871 †	93,168	82,232	328,388
	Largest CC Dia	321.7 †	325.5	326.5	436.7
	Time Struct Imp (sec)	none †	2.86	2.08	none
	Time (sec)	99.1 †	46.5	40.1	103.9
OBPRM [1]	% Samples Accepted	100% †	82.9%	9.63%	4.41% †
	CD-Calls	720,818 †	619,470	456,050	957,076
	Largest CC Dia	332.9 †	346.8	409.6	917.3
	Time Struct Imp (sec)	none †	10.23	2.69	none
	Time (sec)	219.6 †	214.7	138.5	265.8

† — represents implementations as defined in [1], [6], [9], [11], [12].

TABLE IV

STATISTICS AVERAGED FOR COMPARISON

Statistic	Description
% Samples Accepted	The ratio of samples accepted by the policy to the total generated.
CD-Calls	The total number of Collision Detections preformed.
Largest CC Dia	The diameter of the largest connected component which represents the vast majority of the connectable roadmap. The distance is measured in the number of resolution ticks.
Time Struct Imp (sec)	The time spent calculating the *estimated structural improvement* described in this paper.
Time (sec)	The total time spent in PRM: Sampling, Local Planning, and Struct Imp. Additional time generating statistics (e.g. Largest CC Dia) is not included.

compare the quality between the different *sample acceptance policies* in PRM, we examine how the largest connected component's diameter evolves over the number of valid-samples evaluated (as shown in figure 5). The *Pure* PRM sample acceptance policy selects every attempt; this serves as our quality baseline. *Pure* PRM provides the best roadmap quality here because the filtering process inevitably removes nodes and edges that would have refined existing paths in the roadmap.

Measuring the largest component's diameter is not an exact measure of roadmap quality, but can be used as an approximation [24]. The diameter is the longest shortest-path in the roadmap, and the quality we consider refers to how efficiently the robot can move between the two extreme points in the roadmap.

In Tables V, VI, and VII, we can evaluate the average Largest CC Diameter. Here we see that the sample acceptance policies *Imp=50%* and *Imp=100%* lead to roadmaps where the diameter is only slightly larger than that of *Pure*, while VISPRM's more aggressive sample acceptance policy significantly reduces the roadmap quality compared with other policies; this can be seen in the significantly larger component diameters.

TABLE VI

AVERAGE STATISTICS GATHERED WHILE SOLVING THE *Hook* ENVIRONMENT

Sampler	Measure	Sample Acceptance Policy Pure	Imp = 50%	Imp = 100%	VISPRM [12]
BASIC-PRM [11]	% Samples Accepted	n/a* †	n/a*	4.72%	0.47% †
	CD-Calls	n/a* †	n/a*	200,635	4,552,916 †
	Largest CC Dia	n/a* †	n/a*	189.0	301.6 †
	Time Struct Imp (sec)	none †	n/a*	846.1	none †
	Total Time (sec)	n/a* †	n/a*	1,937.4	333.4 †
BRIDGE-TEST [9]	% Samples Accepted	100% †	81.8%	71.6%	11.5% †
	CD-Calls	2,438,478 †	2,406,267	1,859,074	2,633,966
	Largest CC Dia	145.9 †	136.4	145.5	249.5
	Time Struct Imp (sec)	none †	1.15	0.84	none
	Time (sec)	343.0 †	341.7	269.3	346.5
GAUSS-PRM [6]	% Samples Accepted	100% †	63.9%	43.6%	5.32% †
	CD-Calls	341,562 †	130,773	121,713	1,018,312
	Largest CC Dia	171.8 †	182.4	181.1	364.4
	Time Struct Imp (sec)	none †	14.27	11.43	none
	Time (sec)	317.5 †	208.9	196.4	211.4
OBPRM [1]	% Samples Accepted	100% †	99.1%	83.5%	17.6% †
	CD-Calls	122,466 †	109,750	107,483	330,648
	Largest CC Dia	177.5 †	185.4	180.2	259.4
	Time Struct Imp (sec)	none †	0.65	0.92	none
	Time (sec)	9.66 †	9.39	9.21	20.95

† — represents implementations as defined in [1], [6], [9], [11], [12].
n/a* — the planner was unable to consistently solve the problem.

TABLE VII

AVERAGE STATISTICS GATHERED WHILE SOLVING THE *Walls* ENVIRONMENT

Sampler	Measure	Sample Acceptance Policy Pure	Imp = 50%	Imp = 100%	VISPRM [12]
BASIC-PRM [11]	% Samples Accepted	100% †	34.4%	7.76%	4.03% †
	CD-Calls	813,877 †	83,573	48,669	599,239 †
	Largest CC Dia	476.3 †	527.7	518.2	936.3 †
	Time Struct Imp (sec)	none †	8.53	5.60	none †
	Total Time (sec)	207.6 †	27.59	12.97	40.58 †
BRIDGE-TEST [9]	% Samples Accepted	100% †	89.6%	44.4%	13.7% †
	CD-Calls	395,514 †	324,548	303,785	599,438
	Largest CC Dia	535.7 †	582.2	610.7	1,002.7
	Time Struct Imp (sec)	none †	2.58	3.52	none
	Time (sec)	33.19 †	29.74	27.39	42.23
GAUSS-PRM [6]	% Samples Accepted	100% †	88.5%	19.6%	10.8% †
	CD-Calls	358,524 †	202,494	99,039	401,057
	Largest CC Dia	486.4 †	516.5	578.7	1,034.9
	Time Struct Imp (sec)	none †	8.49	7.90	none
	Time (sec)	51.12 †	44.56	21.01	28.27
OBPRM [1]	% Samples Accepted	100% †	95.2%	38.1%	11.4% †
	CD-Calls	399,078 †	181,127	111,142	572,487
	Largest CC Dia	540.2 †	590.0	649.4	1,121.7
	Time Struct Imp (sec)	none †	3.92	3.96	none
	Time (sec)	69.83 †	25.05	14.79	40.07

† — represents implementations as defined in [1], [6], [9], [11], [12].

C. Speed of Improvement

To compare the *speed of improvement* between different *sample acceptance policies* in PRM, we examine how the largest connected component's diameter evolves over the number of Collision Detections (shown in Figure 6). Both a rise and a fall in the component's diameter signify an improvement: a rise corresponds to an addition of connected C-Space, and a fall corresponds to improvements in shorter pathways. We also show when the point at which the query for the environment is solved, although this is a poor metric for comparisons [15], [17], [24]. Additionally, tables of average statistics gathered when solving the queries are shown for: *Maze* in Table V, *Hook* in Table VI, and *Walls* in Table VII.

All sample acceptance policies for a given planner provide some level of improvement over *Pure*; the amount of improvement depends on the type of planner. A combination of node-level filtering and then *structural improvement* filtering can also be advantageous. For the *Maze* environment, BASIC-PRM with *Imp=100%* preformed best with GAUSS-PRM and *Imp=100%* at a close second. In the *Hook* environment,

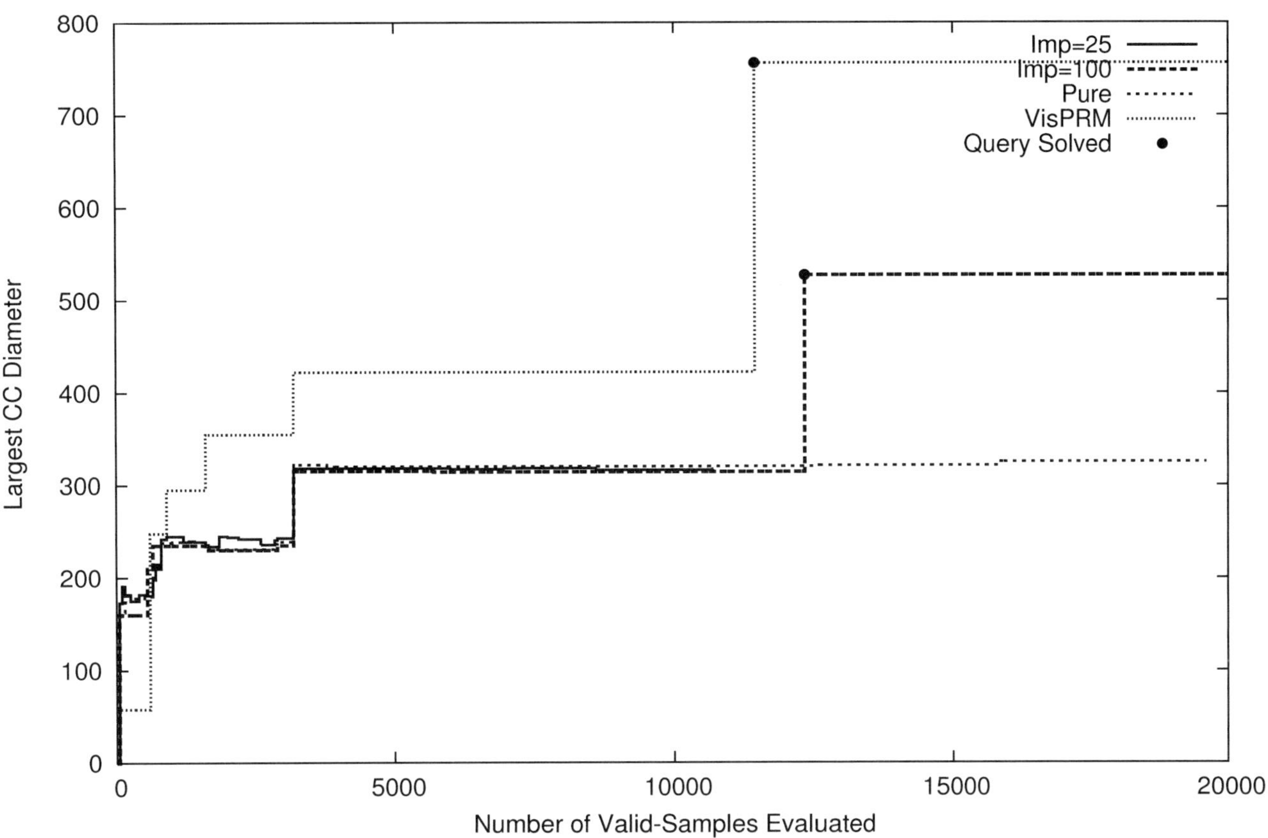

Fig. 5. Maximum roadmap diameter over number of attempts in the *Maze* environment with BASIC-PRM.

OBPRM with *Imp=100%* outperformed the other samplers by a large margin. In the *Walls* environment, BASIC-PRM and OBPRM with *Imp=100%* preformed well.

All of the tests shown here show the power of intelligent sample acceptance policies. In most cases, an intelligent policy combined with node-level filtering can lead to significant improvements. The trade-off in roadmap quality can be significant for an aggressive acceptance policy. Fortunately *Structural Improvement Filtering* allows the policy to be tuned.

V. CONCLUSIONS

In this paper we introduced an addition to the PRM framework that effectively filters samples and edges thereby increasing the percentage of samples which improve the roadmap structure. We have created a metric that can estimate the maximum potential roadmap structural improvement for each new sample and we have used this metric to improve PRM roadmap construction.

We have shown how varying the *sample acceptance policy* can affect both the speed of roadmap improvement and the overall quality of the roadmap. In particular, we have shown that, for all samplers tested, good samples are effectively identified without a drastic sacrifice in roadmap quality. The largest benefit was observed when using uniform samples produced by BASIC-PRM, making it very competitive with the other strategies.

In the future, we plan to explore adaptive ways to control the *sample acceptance policy*. In addition to filtering, we are exploring strategies to search the roadmap and locate areas of potential improvement to guide the sampling process. We also plan to estimate improvement of other criteria, such as path clearance, when filtering samples and edges.

REFERENCES

[1] N. M. Amato, O. B. Bayazit, L. K. Dale, C. V. Jones, and D. Vallejo. OBPRM: An obstacle-based PRM for 3D workspaces. In *Robotics: The Algorithmic Perspective*, pages 155–168, Natick, MA, 1998. A.K. Peters. Proc. Third Workshop on Algorithmic Foundations of Robotics (WAFR), Houston, TX, 1998.

[2] N. M. Amato, O. B. Bayazit, L. K. Dale, C. V. Jones, and D. Vallejo. Choosing good distance metrics and local planners for probabilistic roadmap methods. *IEEE Trans. Robot. Automat.*, 16(4):442–447, August 2000.

[3] N. M. Amato and G. Song. Using motion planning to study protein folding pathways. *J. Comput. Biol.*, 9(2):149–168, 2002. Special issue of Int. Conf. Comput. Molecular Biology (RECOMB) 2001.

[4] O. B. Bayazit, G. Song, and N. M. Amato. Enhancing randomized motion planners: Exploring with haptic hints. In *Proc. IEEE Int. Conf. Robot. Autom. (ICRA)*, pages 529–536, 2000.

[5] O. B. Bayazit, G. Song, and N. M. Amato. Ligand binding with OBPRM and haptic user input: Enhancing automatic motion planning with virtual touch. In *Proc. IEEE Int. Conf. Robot. Autom. (ICRA)*, pages 954–959, 2001. This work was also presented as a poster at *RECOMB 2001*.

[6] V. Boor, M. H. Overmars, and A. F. van der Stappen. The Gaussian sampling strategy for probabilistic roadmap planners. In *Proc. IEEE Int. Conf. Robot. Autom. (ICRA)*, volume 2, pages 1018–1023, May 1999.

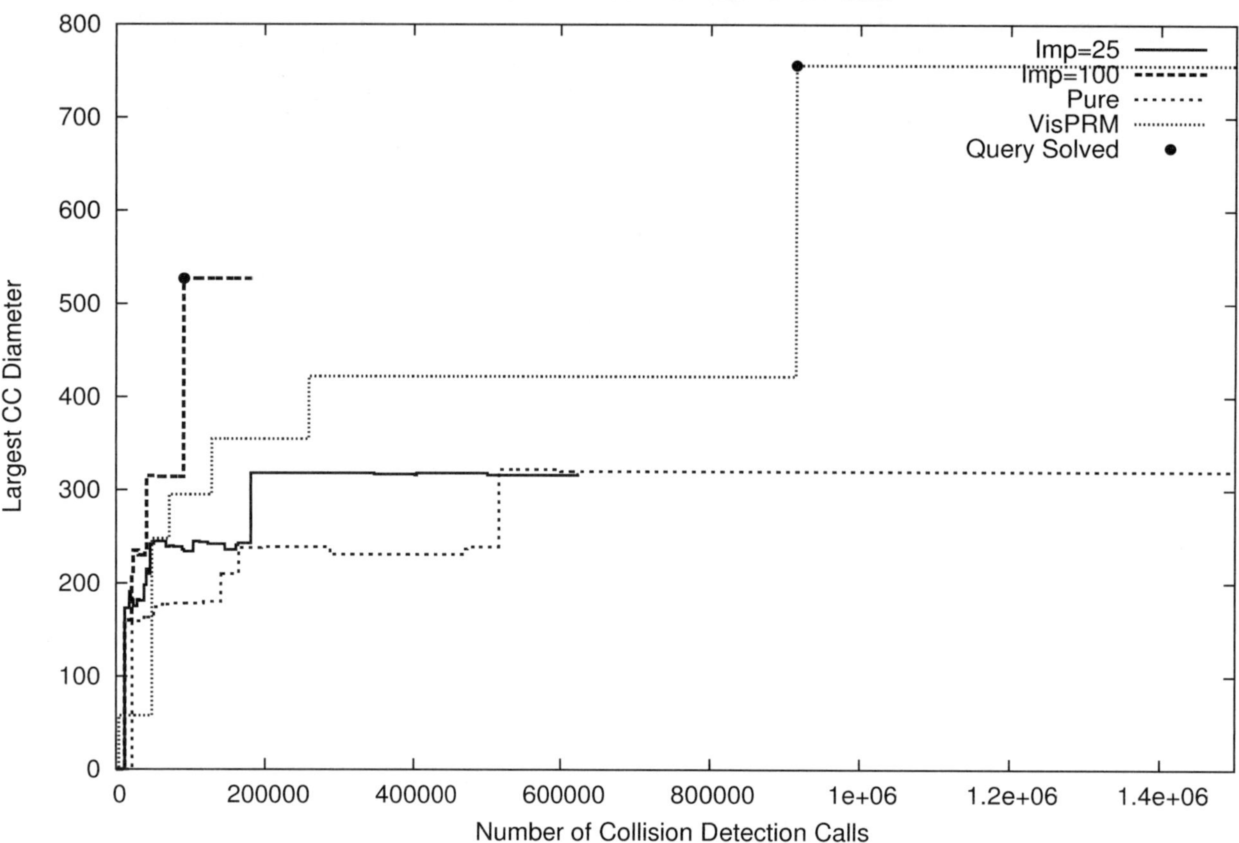

Fig. 6. Maximum roadmap diameter over number of Collision Detections in the *Maze* environment with BASIC-PRM.

[7] Brendan Burns and Oliver Brock. Sampling-based motion planning using predictive models. In *Proc. IEEE Int. Conf. Robot. Autom. (ICRA)*, 2005.

[8] H. Chang and T. Y. Li. Assembly maintainability study with motion planning. In *Proc. IEEE Int. Conf. Robot. Autom. (ICRA)*, pages 1012–1019, 1995.

[9] D. Hsu, T. Jiang, J.H. Reif, and Z. Sun. Bridge test for sampling narrow passages with probabilistic roadmap planners. In *Proc. IEEE Int. Conf. Robot. Autom. (ICRA)*, pages 4420–4426, 2003.

[10] D. Hsu, G. Sánchez-Ante, and Z. Sun. Hybrid PRM sampling with a cost-sensitive adaptive strategy. In *Proc. IEEE Int. Conf. Robot. Autom. (ICRA)*, pages 3885–3891, 2005.

[11] L. E. Kavraki, P. Svestka, J. C. Latombe, and M. H. Overmars. Probabilistic roadmaps for path planning in high-dimensional configuration spaces. *IEEE Trans. Robot. Automat.*, 12(4):566–580, August 1996.

[12] J.-P. Laumond and T. Siméon. Notes on visibility roadmaps and path planning. In *Proc. Int. Workshop on Algorithmic Foundations of Robotics (WAFR)*, 2000.

[13] Jyh-Ming Lien, O. B. Bayazit, R.-T. Sowell, S. Rodriguez, and N. M. Amato. Shepherding behaviors. In *Proc. IEEE Int. Conf. Robot. Autom. (ICRA)*, pages 4159–4164, April 2004.

[14] Jyh-Ming Lien, Samuel Rodriguez, Jean-Philippe Malric, and Nancy M. Amato. Shepherding behaviors with multiple shepherds. In *Proc. IEEE Int. Conf. Robot. Autom. (ICRA)*, pages 3413–3418, April 2005.

[15] Marco Morales, Roger Pearce, and Nancy M. Amato. Analysis of the evolution of C-Space models built through incremental exploration. In *Proc. IEEE Int. Conf. Robot. Autom. (ICRA)*, pages 1029–1034, April 2007.

[16] Marco Morales, Lydia Tapia, Roger Pearce, Samuel Rodriguez, and Nancy M. Amato. A machine learning approach for feature-sensitive motion planning. In *Proc. Int. Workshop on Algorithmic Foundations of Robotics (WAFR)*, pages 361–376, Utrecht/Zeist, The Netherlands, July 2004.

[17] Marco A. Morales A., Roger Pearce, and Nancy M. Amato. Metrics for analyzing the evolution of C-Space models. In *Proc. IEEE Int. Conf. Robot. Autom. (ICRA)*, pages 1268–1273, May 2006.

[18] Marco A. Morales A., Lydia Tapia, Roger Pearce, Samuel Rodriguez, and Nancy M. Amato. C-space subdivision and integration in feature-sensitive motion planning. In *Proc. IEEE Int. Conf. Robot. Autom. (ICRA)*, pages 3114–3119, April 2005.

[19] C. Nissoux, T. Simeon, and J.-P. Laumond. Visibility based probabilistic roadmaps. In *Proc. IEEE Int. Conf. Intel. Rob. Syst. (IROS)*, pages 1316–1321, 1999.

[20] J. H. Reif. Complexity of the mover's problem and generalizations. In *Proc. IEEE Symp. Foundations of Computer Science (FOCS)*, pages 421–427, San Juan, Puerto Rico, October 1979.

[21] A.P. Singh, J.C. Latombe, and D.L. Brutlag. A motion planning approach to flexible ligand binding. In *7th Int. Conf. on Intelligent Systems for Molecular Biology (ISMB)*, pages 252–261, 1999.

[22] X. Tang, B. Kirkpatrick, S. Thomas, G. Song, and N. M. Amato. Using motion planning to study RNA folding kinetics. In *Proc. Int. Conf. Comput. Molecular Biology (RECOMB)*, pages 252–261, 2004.

[23] Shawna Thomas, Xinyu Tang, Lydia Tapia, and Nancy M. Amato. Simulating protein motions with rigidity analysis. In *Proc. Int. Conf. Comput. Molecular Biology (RECOMB)*, pages 394–409, 2006.

[24] Dawen Xie, Marco Morales, Roger Pearce, Shawna Thomas, Jyh-Ming Lien, and Nancy M. Amato. Incremental map generation (IMG). In *Proc. Int. Workshop on Algorithmic Foundations of Robotics (WAFR)*, July 2006.

Model Based Vehicle Tracking for Autonomous Driving in Urban Environments

Anna Petrovskaya and Sebastian Thrun
Computer Science Department
Stanford University
Stanford, California 94305, USA
{ anya, thrun }@cs.stanford.edu

***Abstract*— Situational awareness is crucial for autonomous driving in urban environments. This paper describes moving vehicle tracking module that we developed for our autonomous driving robot Junior. The robot won second place in the Urban Grand Challenge, an autonomous driving race organized by the U.S. Government in 2007. The tracking module provides reliable tracking of moving vehicles from a high-speed moving platform using laser range finders. Our approach models both dynamic and geometric properties of the tracked vehicles and estimates them using a single Bayes filter per vehicle. We also show how to build efficient 2D representations out of 3D range data and how to detect poorly visible black vehicles. Experimental validation includes the most challenging conditions presented at the UGC as well as other urban settings.**

I. Introduction

Autonomously driving cars have been a long-lasting dream of robotics researchers and enthusiasts. Self-driving cars promise to bring a number of benefits to society, including prevention of road accidents, optimal fuel usage, comfort and convenience. In recent years the Defense Advanced Research Projects Agency (DARPA) has taken a lead on encouraging research in this area. DARPA has organized a series of competitions for autonomous vehicles. In 2005, autonomous vehicles were able to complete a 131 mile course in the desert. In 2007 competition, the Urban Grand Challenge, the robots were presented with an even more difficult task: autonomous safe navigation in urban environments. In this competition the robots had to drive safely with respect to other robots, human-driven vehicles and the environment. They also had to obey the rules of the road as described in the California rulebook. One of the most significant changes from the previous competition is that for urban driving, robots need to have situational awareness of both static and dynamic parts of the environment. Our robot won the second prize at the 2007 competition. In this paper we describe the approach we developed for tracking of moving vehicles.

Vehicle tracking has been studied for several decades. A number of approaches focused on the use of vision exclusively [1, 2, 3]. Whereas others utilized laser range finders sometimes in combination with vision [4, 5, 6, 7]. Typically these approaches perform data segmentation and data association prior to performing a filter update. Usually only position and velocity of each vehicle are tracked. The vehicle tracking literature almost universally relies on variants of Kalman filters, although particle filters and hybrid approaches have been widely used in other tracking applications [8, 9, 10].

For our application we are concerned with laser based vehicle tracking from our autonomous robotic platform Junior,

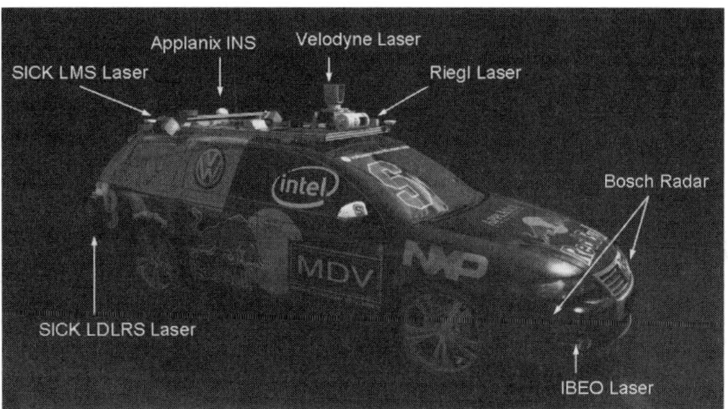

Fig. 1. Junior, our entry in the DARPA Urban Challenge. Junior is equipped with five different laser measurement systems, a multi-radar assembly, and a multi-signal inertial navigation system, as shown in this figure.

to which we will also refer as the ego-vehicle (Fig. 1). In contrast to prior art we propose a model based approach which encompasses both geometric and dynamic properties of the tracked vehicle in a single Bayes filter. The approach naturally handles data segmentation and association, so that these pre-processing steps are not required. To properly model the dependence between geometric and dynamic vehicle properties, we introduce *anchor point coordinates*. Further, we introduce an abstract sensor representation we call the *virtual scan*, that allows for efficient computation and can be used for a wide variety of laser sensors. We present techniques for building virtual scans from 3D range data and show how to detect poorly visible black vehicles in laser scans. Our approach runs in real time with an average update rate of 40Hz, which is 4 times faster than the common sensor frame rate of 10Hz. The results show that our approach is reliable and efficient even in challenging traffic situations presented at the Urban Grand Challenge.

II. Representation

A. Probabilistic model and notation

Our goal for this work is to track multiple vehicles in an urban environment. Our ego-vehicle has been outfitted with the Applanix navigation system that can provide pose localization with 1m error as well as produce a locally consistent pose estimates based on an inertial measurement unit (IMU). Hence we will leave ego-vehicle localization outside the scope of the paper. Instead we will assume that a reasonably precise pose of the ego-vehicle is always available.

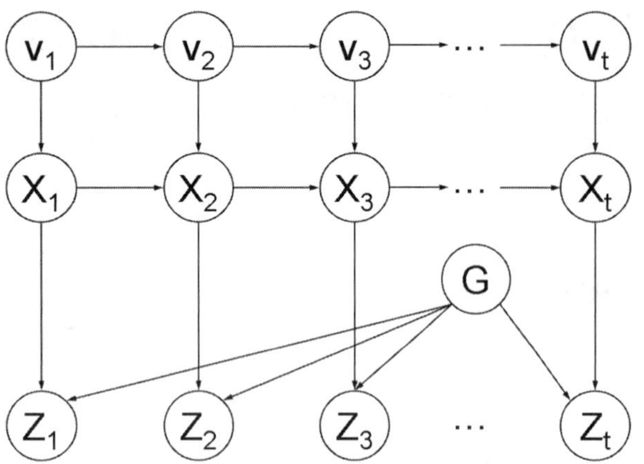

Fig. 2. Dynamic Bayesian network model of the tracked vehicle pose X_t, forward velocity v_t, geometry G, and measurements Z_t.

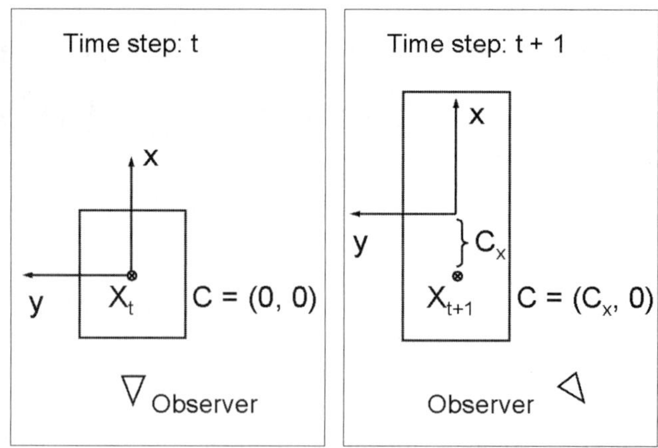

Fig. 3. As we move to observe a different side of a stationary car, our belief of its shape changes and so does the position of the car's center point. To compensate for the effect, we introduce local anchor point coordinates $C = (C_x, C_y)$ so that we can keep the anchor point X_t stationary in the world coordinates.

From the theoretical standpoint, multiple vehicle tracking entails a single joint probability distribution over the state parameters of all of the vehicles. Unfortunately, such a representation is not practical because it quickly becomes intractable as the number of vehicles grows. Also, since the number of vehicles is unknown and variable, it is in fact challenging to model the problem in this way. We note that dependencies between vehicles are strong when vehicles are close together, but become extremely weak as the distance between vehicles increases. Hence it is wasteful to model dependencies between vehicles that are far from each other. Instead, following the common practice in vehicle tracking, we will represent each vehicle with a separate Bayesian filter, and represent dependencies between vehicles via a set of local spatial constraints. Specifically we will assume that no two vehicles overlap, that there is a free space of at least 1m around each vehicle and that all vehicles of interest are located on or near the road. This representation is efficient because its complexity grows linearly with the number of vehicles. It also easily accommodates a variable number of tracked vehicles.

For each vehicle we estimate its 2D position and orientation $X_t = (x_t, y_t, \theta_t)$ at time t, its forward velocity v_t and its geometry G (further defined in Sect. II-B). Also at each time step we obtain a new measurement Z_t. See Fig. 2 for a dynamic Bayes network representation of the resulting probabilistic model. The dependencies between the parameters involved are modeled via probabilistic laws discussed in detail in Sects. II-C and II-E. For now we briefly note that the velocity evolves over time according to

$$p(v_t|v_{t-1}).$$

The vehicle moves based on the evolved velocity according to a dynamics model:

$$p(X_t|X_{t-1}, v_t).$$

The measurements are governed by a measurement model:

$$p(Z_t|X_t, G).$$

For convenience we will write $X^t = (X_1, X_2, ..., X_t)$ for the vehicle's trajectory up to time t. Similarly, v^t and Z^t will denote all velocities and all measurements up to time t.

B. Vehicle geometry

The exact geometric shape of a vehicle can be complex and difficult to model precisely. For simplicity we approximate it by a rectangular shape of width W and length L. The 2D representation is sufficient because the height of the vehicles is not important for driving applications.

For vehicle tracking it is common to track the position of a vehicle's center within the state variable X_t. However, there is an interesting dependence between our belief about the vehicle's shape and position (Fig. 3). As we observe the object from a different vantage point, we change not only our belief of its shape, but also our belief of the position of its center point. Allowing X_t to denote the center point can lead to the undesired effect of obtaining a non-zero velocity for a stationary vehicle, simply because we refine our knowledge of its shape.

To overcome this problem, we view X_t as the pose of an *anchor point* who's position with respect to the vehicle's center can change over time. Initially we set the anchor point to be the center of what we believe to be the car shape and thus its coordinates in the vehicle's *local* coordinate system are $C = (0, 0)$. We assume that the vehicle's local coordinate system is tied to its center with the x-axis pointing directly forward. As we revise our knowledge of the vehicle's shape, the local coordinates of the anchor point will also need to be revised accordingly to $C = (C_x, C_y)$. Thus the complete set of geometric parameters is $G = (W, L, C_x, C_y)$.

C. Vehicle dynamics model

Given a vehicle's velocity v_{t-1} at time step $t-1$, the velocity evolves via addition of random bounded noise based on maximum allowed acceleration a_{max} and the time delay Δt between time steps $t-1$ and t. Specifically, we sample Δv uniformly from $[-a_{max}\Delta t,\ a_{max}\Delta t]$.

The pose evolves via linear motion - a motion law that is often utilized when exact dynamics of the object are unknown. The motion consists of perturbing orientation by $\Delta\theta_1$, then moving forward according to the current velocity by $v_t\Delta t$, and making a final adjustment to orientation by $\Delta\theta_2$. Again we sample $\Delta\theta_1$ and $\Delta\theta_2$ uniformly from $[-d\theta_{max}\Delta t,\ d\theta_{max}\Delta t]$ for a maximum allowed orientation change $d\theta_{max}$.

D. Sensor data representation

In this paper we focus on laser range finders for sensing the environment. Recently these sensors have evolved to be more suitable for driving applications. For example IBEO Alasca sensors allow for easy ground filtering by collecting four parallel horizontal scan lines and marking which of the readings are likely to come from the ground. Velodyne HDL-64E sensors do not provide ground filtering, however they take a 3D scan of the environment at high frame rates (10Hz) thereby producing 1,000,000 readings per second. Given such rich data, the challenge has become to process the readings in real time. Vehicle tracking at 10 - 20Hz is desirable for driving decision making.

A number of factors make the use of raw sensor data inefficient. As the sensor rotates to collect the data, each new reading is made from a new vantage point due to ego-motion. Ignoring this effect leads to significant sensor noise. Taking this effect into account makes it difficult to quickly access data that pertains to a specific region of space. Much of the data comes from surfaces uninteresting for the purpose of vehicle tracking, e.g. ground readings, curbs and tree tops. Finally, the raw 3D data wastes a lot of resources as vehicle tracking is a 2D application where the cars are restricted to move on the ground surface. Therefore it is desirable to pre-process the data to produce a representation tailored for vehicle tracking.

To expedite computations, we construct a grid in polar coordinates - a *virtual scan* - which subdivides 360° around a chosen origin point into angular grids (Fig. 4). In each angular grid we record the range to the closest obstacle. Hence each angular grid contains information about free, occupied, and occluded space. We will often refer to the cone of an angular grid from the origin until the recorded range as a *ray* due to its similarity to a laser ray.

Virtual scans simplify data access by providing a single point of origin for the entire data set, which allows constant time look-up for any given point in space. As we mentioned earlier it is important to compute correct world coordinates for the raw sensor readings. However, once the correct positions of obstacle points have been computed, adjusting the origin of each ray to be at the common origin for the virtual scan produces an acceptable approximation. Constructed in this manner a virtual scan provides a compact representation of the space around the ego-vehicle classified into free, occupied and occluded. The classification helps us properly reason about what parts of an object should be visible as we describe in Sect. II-E.

For the purpose of vehicle tracking it is crucial to determine what changes take place in the environment over time. With virtual scans these changes can be easily computed in spite of the fact that ego-motion can cause two consecutive virtual scans to have different origins. The changes are computed by checking which obstacles in the old scan are cleared by rays in the new scan and vice versa. This computation takes time linear in the size of the virtual scan and only needs to be carried out once per frame. Fig. 4(d) shows results of a virtual scan differencing operation with red points denoting new obstacles, green points denoting obstacles that disappeared,

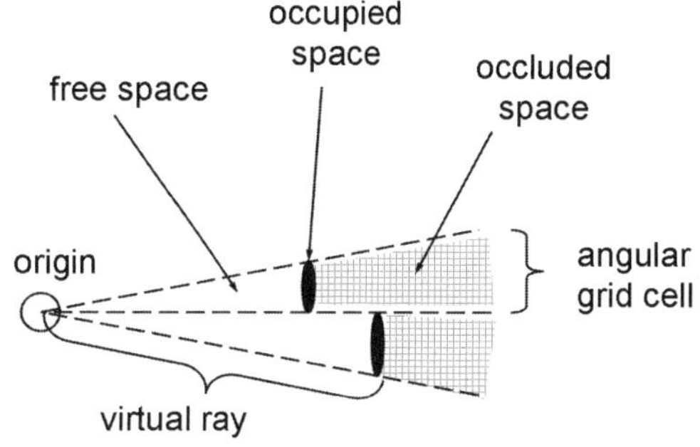

(a) schematic of a virtual scan

(b) actual scene

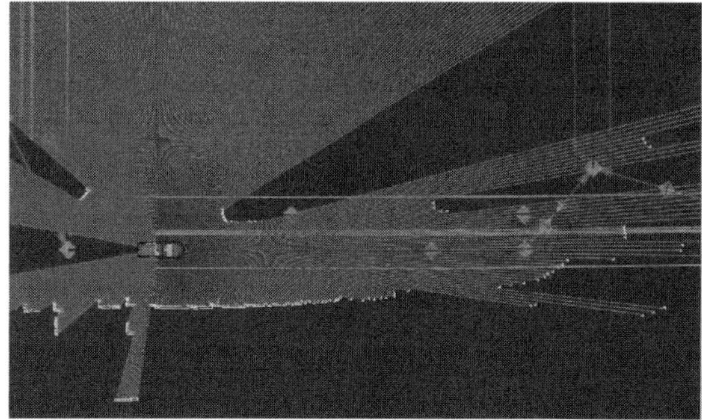

(c) virtual scan

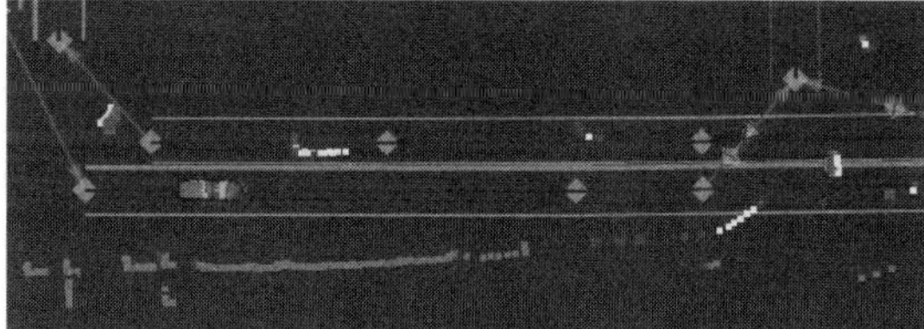

(d) scan differencing

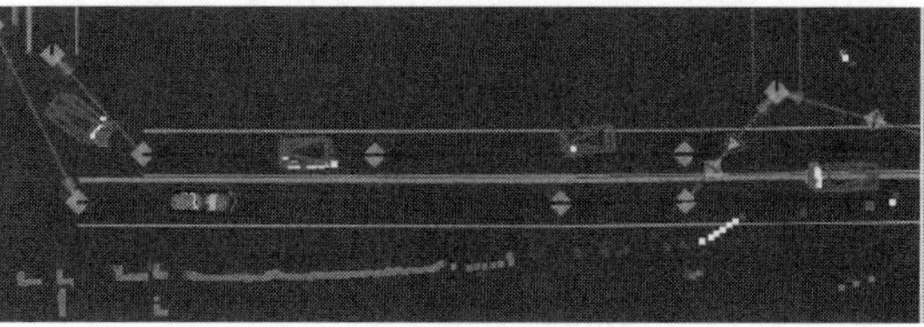

(e) tracking results

Fig. 4. Virtual scan construction. In (c) green line segments represent virtual rays. In (d) red points are new obstacles, green points are obstacles that disappeared, and white points are obstacles that remained unchanged. In (e) the purple boxes denote the tracked vehicles. (Best viewed in color.)

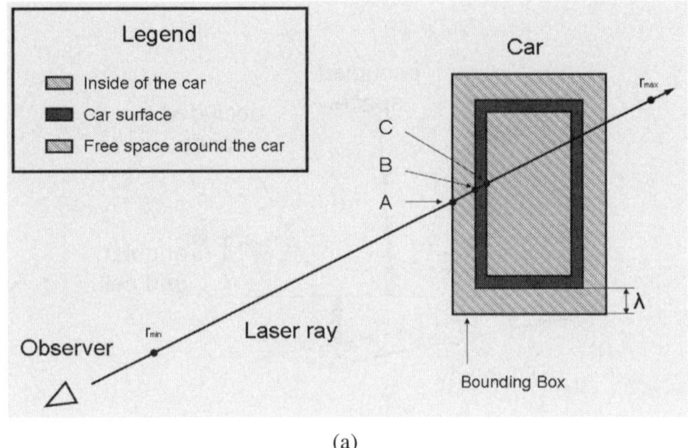

(a)

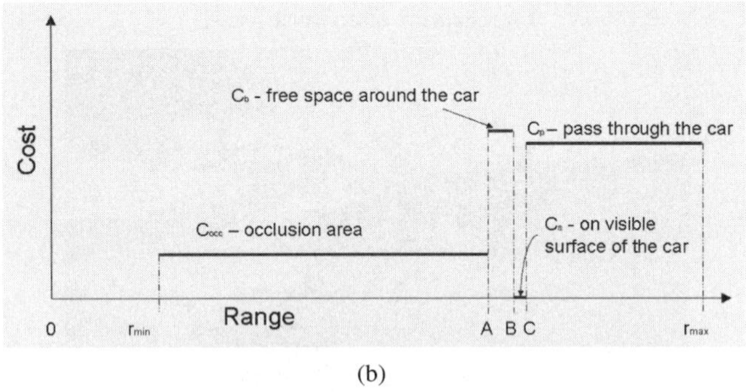

(b)

Fig. 5. Measurement likelihood computations. (a) shows the geometric regions involved in the likelihood computations. (b) shows the costs assignment for a single ray.

and white points denoting obstacles that remained in place or appeared in previously occluded areas.

E. Measurement model

Given a vehicle's pose X, geometry G and a virtual scan Z we compute the measurement likelihood $p(Z|G,X)$ as follows. We position a rectangular shape representing the vehicle according to X and G. Then we build a bounding box to include all points within a predefined distance λ[1] around the vehicle (see Fig. 5). Assuming that there is an actual vehicle in this configuration, we would expect the points within the rectangle to be occupied or occluded, and points in its vicinity to be free or occluded, because vehicles are spatially separated from other objects in the environment.

Following the common practice for modeling laser range finders, we consider measurements obtained along each ray independent of each other. Thus if we have a total of N rays in the virtual scan Z, the measurement likelihood factors as follows:

$$p(Z|G,X) = \prod_{i=1}^{N} p(z_i|G,X).$$

We model each ray's likelihood as a zero-mean Gaussian of variance σ_i computed with respect to a cost c_i selected based on the relationship between the ray and the vehicle (η_i is a normalization constant):

$$P(z_i|G,X) = \eta_i \; \exp\{ \; -\frac{c_i^2}{\sigma_i^2} \; \}.$$

The costs and variances are set to constants that depend on the region in which the reading falls into (see Fig. 5 for illustration). c_{occ}, σ_{occ} are the settings for range readings that fall short of the bounding box and thus represent situations when another object is occluding the vehicle. c_b and σ_b are the settings for range readings that fall short of the vehicle but inside of the bounding box. c_s and σ_s are the settings for readings on the vehicle's visible surface (that we assume to be of non-zero depth). c_p, σ_p are used for rays that extend beyond the vehicle's surface.

The domain for each range reading is between minimum range r_{min} and maximum range r_{max} of the sensor. Since the costs we select are piece-wise constant, it is easy to integrate the unnormalized likelihoods to obtain the normalization constants η_i. Note that for the rays that do not target the vehicle or the bounding box, the above logic automatically yields uniform distributions as these rays never hit the bounding box.

Note that the above measurement model naturally handles partially occluded objects including objects that are "split up" by occlusion into several point clusters. In contrast these cases are often challenging for approaches that utilize separate data segmentation and correspondence methods.

III. Inference

Most vehicle tracking methods described in the literature apply separate methods for data segmentation and correspondence matching before fitting model parameters via extended Kalman filter (EKF). In contrast we use a single Bayesian filter to fit model parameters from the start. This is possible because our model includes both geometric and dynamic parameters of the vehicles and because we rely on efficient methods for parameter fitting. We chose the particle filter method for Bayesian estimation because it is more suitable for multi-modal distributions than EKF. Unlike the multiple hypothesis tracking (MHT) method commonly used in the literature, the computational complexity for our method grows linearly with the number of vehicles in the environment, because vehicle dynamics dictates that vehicles can only be matched to data points in their immediate vicinity. The downside of course is that in our case two targets can in principle merge into one. In practice we have found that it happens rarely and only in situations where one of the targets is lost due to complete occlusion. In these situations target merging is acceptable for our application.

We have a total of eight parameters to estimate for each vehicle: $X = (x, y, \theta)$, v, $G = (W, L, C_x, C_y)$. Computational complexity grows exponentially with the number of parameters for particle filters. Thus to keep computational complexity low, we turn to Rao-Blackwellized particle filters (RBPFs) first introduced in [11]. We estimate X and v by samples and keep Gaussian estimates for G within each particle. Below we give a brief derivation of the required update equations.

A. Derivation of update equations

At each time step t we produce an estimate of a Bayesian belief about the tracked vehicle's trajectory, velocity and

[1] We used the setting of $\lambda = 1m$ in our implementation.

geometry based on a set of measurements:

$$Bel_t = p(X^t, v^t, G|Z^t).$$

The derivation provided below is similar to the one used in [12]. We split up the belief into two conditional factors:

$$Bel_t = p(X^t, v^t|Z^t) \;\; p(G|X^t, v^t, Z^t).$$

The first factor encodes the vehicle's motion posterior:

$$R_t = p(X^t, v^t|Z^t).$$

The second factor encodes the vehicle's geometry posterior, conditioned on its motion:

$$S_t = p(G|X^t, v^t, Z^t).$$

The factor R_t is approximated using a set of particles; the factor S_t is approximated using a Gaussian distribution (one Gaussian per particle). We denote a particle by $q_m^t = (X^{t,[m]}, v^{t,[m]}, S_t^{[m]})$ and a collection of particles at time t by $Q_t = \{q_m^t\}_m$. We compute Q_t recursively from Q_{t-1}. Suppose that at time step t, particles in Q_{t-1} are distributed according to R_{t-1}. We compute an intermediate set of particles $\bar{Q}_t$ by sampling a guess of the vehicle's pose and velocity at time t from the dynamics model (described in detail in Sect. II-C). Thus, particles in $\bar{Q}_t$ are distributed according to the vehicle motion prediction distribution:

$$\bar{R}_t = p(X^t, v^t|Z^{t-1}).$$

To ensure that particles in Q_t are distributed according to R_t (asymptotically), we generate Q_t by sampling from $\bar{Q}_t$ with replacement in proportion to importance weights given by $w_t = R_t/\bar{R}_t$. Before we can compute the weights, we need to derive the update equations for the geometry posterior.

We use a Gaussian approximation for the geometry posterior, S_t. Thus we keep track of the mean μ_t and the co-variance matrix Σ_t of the approximating Gaussian in each particle: $q_m^t = (X^{t,[m]}, v^{t,[m]}, \mu_t^{[m]}, \Sigma_t^{[m]})$. We have:

$$\begin{aligned} S_t &= p(G|X^t, v^t, Z^t) \\ &\propto p(Z_t|G, X^t, v^t, Z^{t-1}) \;\; p(G|X^t, v^t, Z^{t-1}) \\ &= p(Z_t|G, X_t) \;\; p(G|X^{t-1}, v^{t-1}, Z^{t-1}). \end{aligned} \tag{1}$$

The first step above follows from Bayes' rule; the second step follows from the conditional independence assumptions of our model (Fig. 2). The expression (1) is a product of the measurement likelihood and the geometry prior S_{t-1}. To obtain a Gaussian approximation for S_t we linearize the measurement likelihood as will be explained in Sect. III-C. Once the linearization is performed, the mean and the co-variance matrix for S_t can be computed in closed form, because S_{t-1} is already approximated by a Gaussian (represented by a Rao-Blackwellized particle from the previous time step).

Now we are ready to compute the importance weights. Briefly, following the derivation in [12], it is straightforward to show that the importance weights w_t should be:

$$w_t = R_t/\bar{R}_t = \frac{p(X^t, v^t|Z^t)}{p(X^t, v^t|Z^{t-1})} = I\!E_{S_{t-1}}[\; p(Z_t|G, X_t) \;].$$

In words, the importance weights are the expected value (with respect to the vehicle geometry prior) of the measurement likelihood. Using Gaussian approximations of S_{t-1} and $p(Z_t|G, X_t)$, this expectation can be expressed as an integral over a product of two Gaussians, and can thus be carried out in closed form.

B. Motion inference

As we mentioned in Sect. II-A, a vehicle's motion is governed by two probabilistic laws: $p(v_t|v_{t-1})$ and $p(X_t|X_{t-1}, v_t)$. These laws are related to the motion prediction distribution as follows:

$$\begin{aligned} \bar{R}_t &= p(X^t, v^t|Z^{t-1}) \\ &= p(X_t, v_t|X^{t-1}, v^{t-1}, Z^{t-1}) \; p(X^{t-1}, v^{t-1}|Z^{t-1}) \\ &= p(X_t|X^{t-1}, v^t, Z^{t-1}) \; p(v_t|X^{t-1}, v^{t-1}, Z^{t-1}) \; R_{t-1} \\ &= p(X_t|X_{t-1}, v_t) \; p(v_t|v_{t-1}) \; R_{t-1}. \end{aligned}$$

The first and second steps above are simple conditional factorizations; the third step follows from the conditional independence assumptions of our model (Fig. 2).

Note that since only the latest vehicle pose and velocity are used in the update equations, we do not need to actually store entire trajectories in each particle. Thus the memory storage requirements per particle do not grow with t.

C. Shape inference

In order to maintain the vehicle's geometry posterior in a Gaussian form, we need to linearize the measurement likelihood $p(Z_t|G, X_t)$ with respect to G. Clearly the measurement likelihood does not lend itself to differentiation in closed form. Thus we turn to Laplace's method to obtain a suitable Gaussian approximation. The method involves fitting a Gaussian at the global maximum of a function. Since the global maximum is not readily available, we search for it via local optimization starting at the current best estimate of geometry parameters. Due to construction of our measurement model (Sect. II-E) the search is inexpensive as we only need to recompute the costs for the rays directly affected by a local change in G.

The dependence between our belief of the vehicle's shape and position (discussed in Sect. II-B) manifests itself in a dependence between the local anchor point coordinates C and the vehicle's width and length. The vehicle's corner closest to the vantage point is a very prominent feature that impacts how the sides of the vehicle match the data. When revising the belief of the vehicle's width and length, we keep the closest corner in place. Thus a change in the width or the length leads to a change in the global coordinates of the vehicle's center point, for which we compensate with an adjustment in C to keep the anchor point in place. This way a change in geometry does not create phantom motion of the vehicle.

D. Initializing new tracks

Before vehicle tracking can begin, we need to initialize new vehicle tracks. Detection of new vehicles is the most expensive part of vehicle tracking. However a number of optimizations can be made to achieve detection in real time, including spatial constraints and sensor data analysis. Detailed description of our vehicle detection algorithm is given in [13]. Here we provide a brief summary.

To detect new vehicles, we search the area within sensor range of our ego-vehicle to find good matches using the measurement model described in Sect. II-E. A total of three (3) frames are required to acquire a new tracking target. The

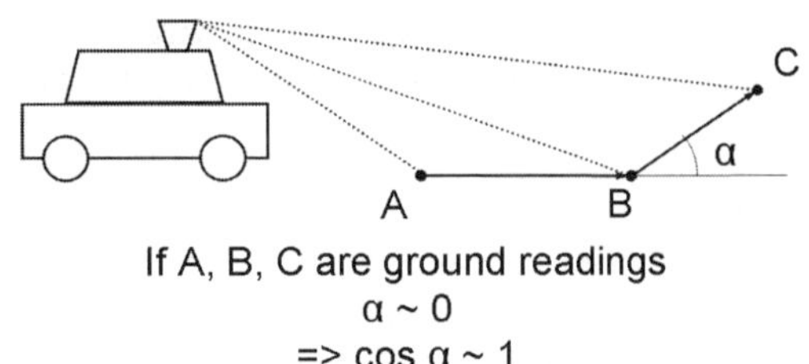

Fig. 6. We determine ground readings by comparing angles between consecutive readings.

first two frames are required to detect motion of an object. The third frame is required to check that the motion is consistent over time and follows vehicle dynamics laws described in Sect. II-C.

E. Discontinuing tracks

Under certain conditions it is desirable to discontinue tracking of a target. We discontinue tracks if the target vehicle gets out of sensor range or moves too far away from the road (a digital street map was available for our application). Additionally we implemented logic that merged hypothesis of two particle filters if the tracked targets were too close together. However, it turned out that this condition occurs only very rarely.

We also discontinue tracks if the unnormalized weights have been low for several turns. Low unnormalized weights signal that the sensor data is insufficient to track the target, or that our estimate is too far away from the actual vehicle. This logic keeps the resource cost of tracking occluded objects low, yet it still allows for a tracked vehicle to survive bad data or complete occlusion for several turns. Since new track acquisition only takes three frames, it does not make sense to continue tracking objects that are occluded for significantly longer periods of time.

IV. Implementation and Results

A. Building virtual scans from 3D range data

As we explained in Sect. II-D, vehicle tracking is a 2D problem, for which efficient 2D virtual scans are sufficient. These virtual scans are easy to build for 2D range sensors with ground filtering, such as IBEO. However for 3D sensors, such as Velodyne, it is a less trivial task. These sensors provide immense 3D data sets of the surroundings, making computational efficiency a high priority when processing the data. However, in our experience, the hard work pays off and the resulting virtual scans carry more information than for 2D sensors.

Given a 3D data set, which of the data points should be considered obstacles? From the perspective of driving applications we are interested in the slice of space directly above the ground and about 2m high, as this is the space that a vehicle would actually have to drive through. Objects elevated more than 2m above ground - e.g. tree tops or overpasses - are not obstacles. The ground itself is not an obstacle (assuming the terrain is drivable). Moreover, for tracking applications low obstacles such as curbs should be excluded from virtual scans, because otherwise they can prevent us from seeing more important obstacles beyond them. The remaining objects in

(a) actual scene

(b) Velodyne data after classification

(c) generated virtual scan

Fig. 7. In (b) Velodyne data is colored by type: orange - ground, yellow - low obstacle, red - medium obstacle, green - high obstacle. Note the white van parked at a distance in (a) and (c).

the 2m slice of space are obstacles for a vehicle, even if these objects are not directly touching the ground.

In order to classify the data into the different types of objects described above we first build a 3D grid in spherical coordinates. Similarly to a virtual scan, it has a single point of origin and stores actual world coordinates of the sensor readings. Just as in the 2D case, this grid is an approximation of the sensor data set, because the actual laser readings in a scan have varying points of origin. In order to downsample and reject outliers, for each spherical grid cell we compute the median range of the readings falling within it. This gives us a single obstacle point per grid cell. For each spherical grid cell we will refer to the cone from the grid origin to the obstacle point as a virtual ray.

The first classification step is to determine ground points. For this purpose we select a single slice of vertical angles from the spherical grid (i.e. rays that all have the same bearing

(a) actual appearance of the vehicle

(b) the vehicle gives very few laser returns

(c) generated virtual scan after black object detection

(d) successful tracking of the black vehicle

Fig. 8. Detecting black vehicles in 3D range scans. White points represent raw Velodyne data. In (c) green lines represent the generated virtual scan. In (d) the purple box denotes the estimated pose of the tracked vehicle.

angle). We cycle through the rays in the slice from the lowest vertical angle to the highest. For three consecutive readings A, B, and C, the slope between AB and BC should be near zero if all three points lie on the ground (see Fig. 6 for illustration). If we normalize AB and BC, their dot product should be close to 1. Hence a simple thresholding of the dot product allows us to classify ground readings and to obtain estimates of local ground elevation. Thus one useful piece of information we can obtain from 3D sensors is an estimate of ground elevation.

Using the elevation estimates we can classify the remaining non-ground readings into low, medium and high obstacles, out of which we are only interested in the medium ones (see Fig. 7). It turns out that there can be medium height obstacles that are still worth filtering out: birds, insects and occasional readings from cat-eye reflectors. These obstacles are easy to filter, because the BC vector tends to be very long (greater than 1m), which is not the case for normal vertical obstacles such as buildings and cars. After identifying the interesting obstacles we simply project them on the 2D horizontal plane to obtain a virtual scan.

Laser range finders are widely known to have difficulty seeing black objects. Since these objects absorb light, the sensor never gets a return. Clearly it is desirable to "see" black obstacles for driving applications. Other sensors could be used, but they all have their own drawbacks. Here we present a method for detecting black objects in 3D laser data. Figure 8 shows the returns obtained from a black car. The only readings obtained are from the license plate and wheels of the vehicle, all of which get filtered out as low obstacles. Instead of looking at the little data that is present, we can detect the presence of a black obstacle by looking at the data that is absent. If no readings are obtained along a range of vertical angles in a specific direction, we can conclude that the space must be occupied by a black obstacle. Otherwise the rays would have hit some obstacle or the ground. To provide a conservative estimate of the range to the black obstacle we place it at the last reading obtained in the vertical angles just before the absent readings. We note that this method works well as long as the sensor is good at seeing the ground. For the Velodyne sensor the range within which the ground returns are reliable is about 25 - 30m, beyond this range the black obstacle detection logic does not work.

B. Tracking results

The most challenging traffic situation at the Urban Grand Challenge was presented on course A during the qualifying event (Fig. 9(a) and Fig. 9(b)) . The test consisted of dense human driven traffic in both directions on a course with an outline resembling the Greek letter θ. The robots had to merge repeatedly into the dense traffic. The merge was performed using a left turn, so that the robots had to cross one lane of traffic each time. In these conditions accurate estimates of positions and velocities of the cars are very useful for determining a gap in traffic large enough to perform the merge safely. Cars passed in close proximity to each other and to stationary obstacles (e.g. signs and guard rails) providing plenty of opportunity for false associations. Partial and complete occlusions happened frequently due to the traffic density. Moreover these occlusions often happened near merge points which complicated decision making.

During extensive testing the performance of our vehicle tracking module has been very reliable and efficient (see Fig. 4 and Fig. 9). It proved capable of handling complex traffic situations such as the one presented on course A. The computation time of our approach averages out at 25ms per frame, which is faster than real time for most modern laser range finders.

We also gathered empirical results of the tracking module performance on data sets from several urban environments: course A of the UGC, Stanford campus and a port town in Alameda, CA. For each frame of data we counted how many vehicles a human is able to identify in the laser range data. The vehicles had to be within 50m of the ego-vehicle, on or near the road, and moving with a speed of at least 5mph. We summarize the tracker's performance in Fig. 10. Note that the maximum theoretically possible true positive rate is lower

Datasets	Total Frames	Total Vehicles	Correctly Detected	Falsely Detected	Max TP (%)	TP (%)	FP (%)
Area A	1,577	5,911	5,676	205	97.8	96.02	3.35
Stanford	2,140	3,581	3,530	150	99.22	98.58	4.02
Alameda	1,531	901	879	0	98.22	97.56	0
Overall	5,248	10,393	10,085	355	98.33	97.04	3.3

Fig. 10. Tracker performance on data sets from three urban environments. Max TP is the theoretically maximum possible true positive percent for each data set. TP and FP are the actual true positive and false positive rates attained by the algorithm.

(a) test conditions on course A at the UGC

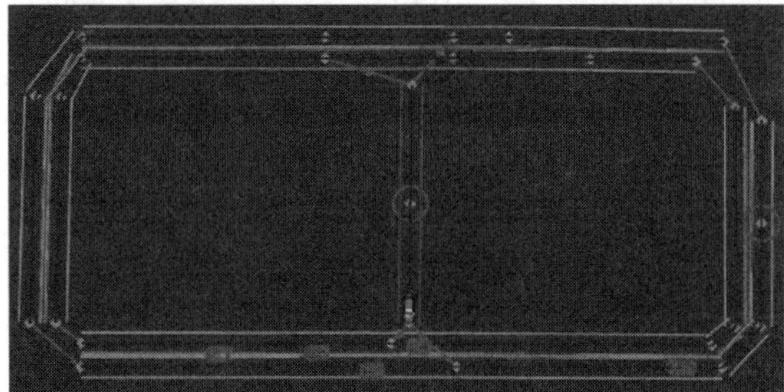

(b) Junior at intersection on course A

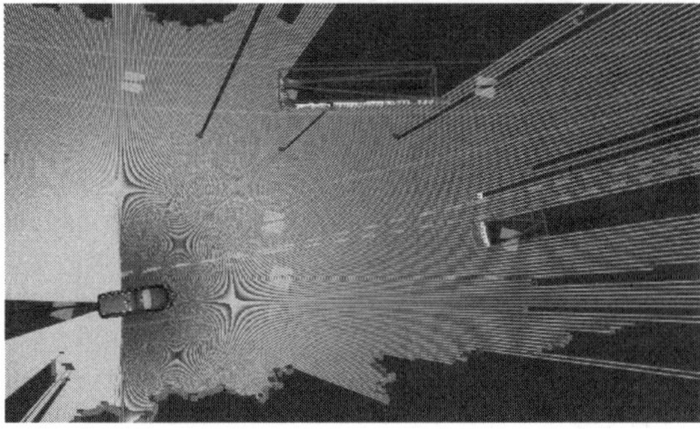

(c) vehicle size estimation on Stanford campus

Fig. 9. Test conditions and results of tracking. Purple boxes represent tracked vehicles. In (c) yellow lines represent the virtual scan.

than 100% because three frames are required to detect a new vehicle. On all three data sets the tracker performed very close to the theoretical bound. Overall the true positive rate was 97% compared to the theoretical maximum of 98%.

Several videos of vehicle detection and tracking using the techniques presented in this paper are available at the website

http://cs.stanford.edu/∼anya/uc.html

V. Conclusions

We have presented the vehicle detection and tracking module developed for Stanford's autonomous driving robot Junior. Tracking is performed from a high-speed moving platform and relies on laser range finders for sensing. Our approach models both dynamic and geometric properties of the tracked vehicles and estimates them with a single Bayes filter per vehicle. In contrast to prior art, the common data segmentation and association steps are carried out as part of the filter itself. The approach has proved reliable, efficient and capable of handling challenging traffic situations, such as the ones presented at the Urban Grand Challenge.

Clearly there is ample room for future work. The presented approach does not model pedestrians, bicyclists, or motorcyclists, which is a prerequisite for driving in populated areas. Another promising direction for future work is fusion of different sensors, including laser, radar and vision.

Acknowledgment

This research has been conducted for the Stanford Racing Team and would have been impossible without the whole team's efforts to build the hardware and software that makes up the team's robot Junior. The authors thank all team members for their hard work. The Stanford Racing Team is indebted to DARPA for creating the Urban Challenge, and for its financial support under the Track A Program. Further, Stanford University thanks its various sponsors. Special thanks also to NASA Ames for permission to use their air field.

References

[1] T. Zielke, M. M. Brauckmann, and W. von Seelen. Intensity and edge based symmetry detection applied to car following. In *ECCV, Berlin, Germany*, 1992.

[2] E. Dickmanns. Vehicles capable of dynamic vision. In *IJCAI, Nagoya, Japan*, 1997.

[3] F. Dellaert and C. Thorpe. Robust car tracking using kalman filtering and bayesian templates. In *Conference on Intelligent Transportation Systems*, 1997.

[4] L. Zhao and C. Thorpe. Qualitative and quantitative car tracking from a range image sequence. In *Computer Vision and Pattern Recognition*, 1998.

[5] D. Streller, K. Furstenberg, and K. Dietmayer. Vehicle and object models for robust tracking in traffic scenes using laser range images. In *Intelligent Transportation Systems*, 2002.

[6] C. Wang, C. Thorpe, S. Thrun, M. Hebert, and H. Durrant-Whyte. Simultaneous localization, mapping and moving object tracking. *The International Journal of Robotics Research, Sep 2007; vol. 26: pp. 889-916*, 2007.

[7] S. Wender and K. Dietmayer. 3d vehicle detection using a laser scanner and a video camera. In *6th European Congress on ITS in Europe, Aalborg, Denmark*, 2007.

[8] D. Schulz, W. Burgard, D. Fox, and A. Cremers. Tracking multiple moving targets with a mobile robot using particle filters and statistical data association. In *ICRA*, 2001.

[9] S. S. Blackman. Multiple hypothesis tracking for multiple target tracking. *IEEE AE Systems Magazine*, 2004.

[10] S. Särkkä, A. Vehtari, and J. Lampinen. Rao-blackwellized particle filter for multiple target tracking. *Inf. Fusion*, 8(1), 2007.

[11] A. Doucet, N. de Freitas, K. Murphy, and S. Russell. Rao-blackwellised filtering for dynamic bayesian networks. In *UAI, San Francisco, CA*, 2000.

[12] M. Montemerlo. *FastSLAM: A Factored Solution to the Simultaneous Localization and Mapping Problem with Unknown Data Association.* PhD thesis, Robotics Institute, Carnegie Mellon University, 2003.

[13] A. Petrovskaya and S. Thrun. Efficient techniques for dynamic vehicle detection. In *ISER, Athens, Greece*, 2008.

Improving Localization Robustness in Monocular SLAM Using a High-Speed Camera

Peter Gemeiner
Automation and Control Institute
Vienna University of Technology
gemeiner@acin.tuwien.ac.at

Andrew J. Davison
Department of Computing
Imperial College London
ajd@doc.ic.ac.uk

Markus Vincze
Automation and Control Institute
Vienna University of Technology
vincze@acin.tuwien.ac.at

Abstract— **In the robotics community localization and mapping of an unknown environment is a well-studied problem. To solve this problem in real-time using visual input, a standard monocular Simultaneous Localization and Mapping (SLAM) algorithm can be used. This algorithm is very stable when smooth motion is expected, but in case of erratic or sudden movements, the camera pose typically gets lost. To improve robustness in Monocular SLAM (MonoSLAM) we propose to use a camera with faster readout speed to obtain a frame rate of 200Hz. We further present an extended MonoSLAM motion model, which can handle movements with significant jitter. In this work the improved localization and mapping have been evaluated against ground truth, which is reconstructed from off-line vision. To explain the benefits of using a high frame rate vision input in MonoSLAM framework, we performed repeatable experiments with a high-speed camera mounted onto a robotic arm. Due to the dense visual information MonoSLAM can faster shrink localization and mapping uncertainties and can operate under fast, erratic, or sudden movements. The extended motion model can provide additional robustness against significant handheld jitter when throwing or shaking the camera.**

I. Introduction

For mobile robotics it is essential to continuously localize and estimate 3D positions of new landmarks in an unknown environment. The *localization* and *mapping* can be addressed with an incremental probabilistic approach, which is known as SLAM (for an overview please refer to [1]).

As input for SLAM different kinds of sensors (e.g. laser [2], sonars [3]) can be used. One of the most interesting (cost, weight, etc.) and challenging sensors is a single perspective-projective camera. When observing the environment with a camera, the depth information of new landmarks can not be directly acquired. To recover this depth information the camera has to move, and observe these landmarks from different viewpoints.

Davison et al. introduced the first real-time Monocular SLAM (MonoSLAM) (recently summarized in [4]) algorithm. The camera motion estimation and incremental map building (from new landmarks) are computed within a standard Extended Kalman Filter (EKF) SLAM framework. An alternative SLAM framework is typically based on FastSLAM-type particle filter algorithms.

One of the underlying assumptions in MonoSLAM is that the camera is expected to move smoothly. This classical EKF SLAM framework is prone to fail as soon as sudden or erratic camera movements occur, and can not reliably recover when the pose is lost. The smooth motion assumption attracted recently attention, and several authors proposed solutions to this problem.

Williams et al. presented in [5] an additional relocalization algorithm, which can operate parallel to MonoSLAM, and increases robustness against camera shakes and occlusions. A recent improvement of this algorithm is using randomized lists classifier and RANSAC to determine the pose robustly [6].

To handle erratic camera motion and occlusions, Pupilli and Calway [7] presented a visual SLAM framework based on a FastSLAM-type particle filter. This work tackles mainly the problem of camera robust localization, and Chekhlov et al. [8] extended this SLAM framework to operate over a large range of views using a SIFT-like spatial gradient descriptor.

Another visual SLAM framework based on a FastSLAM-type particle filter introduced by Eade and Drummond [9] can incorporate hundreds of features in real-time. However, the filter needs to be adapted for closing loops over large trajectories.

The monocular SLAM algorithms discussed above [5, 6, 7, 8, 9] are using a standard 30Hz camera.

Interesting high-speed applications already enabled to e.g. measure the motion of a waving flag or a flying ball [10]. Komuro and Ishikawa proposed in [11] a method for three-dimensional object tracking from noisy images. The noise typically produced by high-speed cameras can be tackled with a proper noise models.

Another way to improve the robustness of tracking and estimating features is to increase the sampling rate of the camera. This has been beautifully demonstrated by Ishii et al. [12]. They introduced a 1ms visual feedback system using massively parallel processing. Since image acquisition time is countered with equally fast image processing (both operate at 1kHz), very fast motion of a bouncing ball can be tracked. However, objects need to be bright in front of dark background.

Our idea is to show that monocular SLAM can operate at a frame rate of 200Hz and that this improves the robustness of localization considerably. The intention is to show that with a higher frame rate a performance similar to the recent improvements discussed above [5, 6, 7, 8, 9] can be achieved, while additionally all these measures could again be used to even further improve performance.

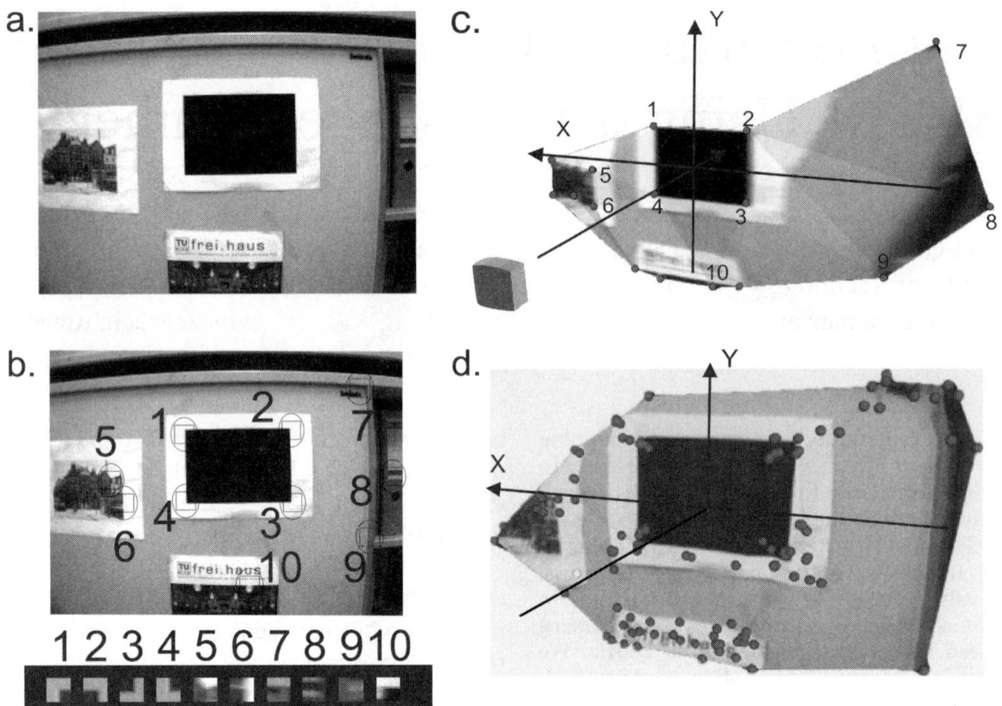

Fig. 1. The only sensor input in MonoSLAM are images from a single camera (a). As the camera moves, new distinctive features are detected (b). The output of MonoSLAM is the camera pose and a sparse three-dimensional map of these distinctive features (c). An alternative to MonoSLAM sparse mapping is a more dense structure reconstruction using nonlinear optimization techniques (d).

From the variety of state-of-the-art monocular SLAM frameworks mentioned before [4, 7, 8, 9] we are using in this work MonoSLAM algorithm, because we think that it is the most advanced.

The contribution of this paper has three parts:

- using a high-speed camera for monocular SLAM for the first time,
- more robust MonoSLAM localization using an extended motion model and
- experiments exploiting the repeatable motion of a robotic arm compared with accurate ground truth from off-line vision and an experiment to throw the camera to indicate robustness.

In this paper the standard MonoSLAM algorithm and the ground truth calculations are briefly introduced in Section II. Details and discussions related to the proposed more robust motion model are introduced in Section III. Section IV presents experimental results, performing well-controlled motions with the camera mounted on the robotic arm and handheld sequences. Section V closes with a discussion and an outlook to the future work.

II. Localization and Mapping Algorithms

Before explaining the proposed more robust localization and benefits of using a high-speed camera, a short outline of the top-down Bayesian MonoSLAM system (see Fig. 1) is given in two parts. Firstly, we summarize MonoSLAM from the system perspective (input, initial assumptions and output). Secondly, a short introduction to additional existing modules used in this work is presented. This section concludes with a brief description of offline computer vision algorithms used in this work for the ground truth acquisition.

A. MonoSLAM and Additional Modules

The only sensor input in MonoSLAM used in this work are images from a single perspective-projective camera as displayed in Fig. 1-a. When the camera moves new distinctive features can be detected (e.g. corners with numbers 5-10 in Fig. 1-b and Fig. 1-c). The output of MonoSLAM is the camera poses and a sparse map of recovered features, as depicted in Fig. 1-c. Due to the real-time demand *mapping* in MonoSLAM is not playing a crucial role, and should rather support *localization*.

In MonoSLAM, the following conditions are assumed:

- a well-calibrated camera,
- rigid scene with textured objects (not moving),
- constant lighting conditions,
- one initial object with known geometry (e.g. features with numbers 1-4 in Fig. 1-b and Fig. 1-c) and
- camera is expected to move smoothly.

Three additional modules have been used in this work to improve the performance of MonoSLAM.

- To permit initialization of features of all depths, Montiel et al. introduced an *inverse-depth* [13] parameterization.

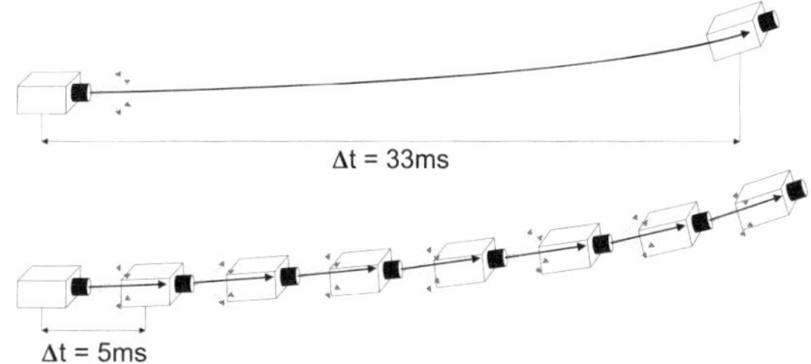

Fig. 2. MonoSLAM can handle smooth camera movements using the standard camera (difference of timestamps - Δt is equal 33ms), and the high-speed camera (Δt is equal 5ms) is not needed.

This parameterization can cope with features, which are very distant from the camera.

- Instead of detecting Shi and Tomasi [14] features, the *Fast* feature detector [15] is applied. The best *Fast* features are then found using the Shi and Tomasi [14] cornerness measure.
- Joint Compatibility Branch and Bound (JCBB) [16] is employed to search for the largest number of jointly compatible pairings.

B. Ground-truth Computation

To precisely evaluate experiments with MonoSLAM motion models and the high-speed camera, we need accurate *localization* and *mapping* ground truth. For the computation of accurate ground truth we used two offline, iterative and *computationally intensive* algorithms from the field of computer vision. The first is a camera pose algorithm presented by Schweighofer and Pinz [17]. This algorithm needs a object with known geometry (black rectangle in Fig. 1-b), which is visible in every frame. The second is a structure reconstruction algorithm based on nonlinear optimization, summarized by Triggs et al. in [18]. An example of the scene reconstruction computed with the combination of these two algorithms is depicted in Fig. 1-d. These computer vision techniques provide more dense and more accurate results then MonoSLAM reconstruction (see Fig. 1-c). The detailed explanation of these algorithms is out of scope of this paper, but in the experimental section an accuracy example is provided.

III. Improving Localization Robustness

To explain the details related to the more robust *localization*, this section comprises four parts. Firstly, the benefits of the *localization* using the high-speed camera are presented. Secondly, the proposed more robust MonoSLAM motion model is explained. Thirdly, a discussion of motion parameters is included. Fourthly, a problem of skipping vision information when using an asynchronous high-speed camera is reviewed.

A. Dense Visual Information

The standard camera (difference of timestamps - Δt is equal 33ms) provides rather sparse vision input, but when a smooth movement is expected (see Fig. 2), it is sufficient for MonoSLAM to robustly *localize*. If an erratic camera motion is performed, EKF based MonoSLAM very likely lose the pose. Observing the scene features with faster readout speed (e.g. Δt is equal 5ms) allows to update the camera state more frequently, and this can prevent MonoSLAM to lose the pose as depicted schematically in Fig. 3.

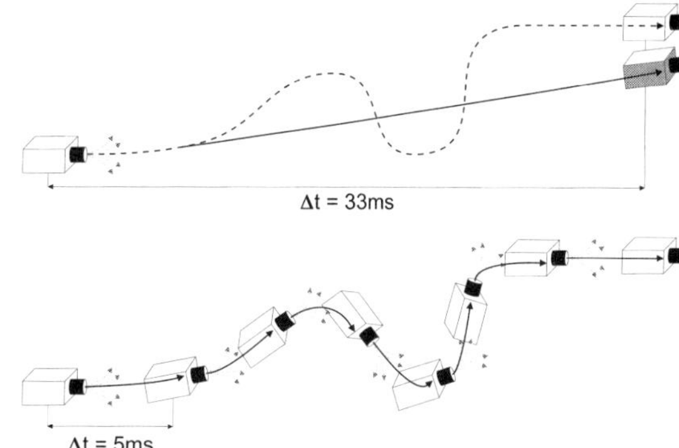

Fig. 3. If erratic camera movements occur, MonoSLAM using the standard camera (difference of timestamps - Δt is equal 33ms) loses the pose. An example is displayed in the upper row, where the camera pose (solid line) drifted away from the true pose (dashed line). A high-speed camera provides more dense visual information, which helps to handle this erratic motion, as depicted in the bottom row.

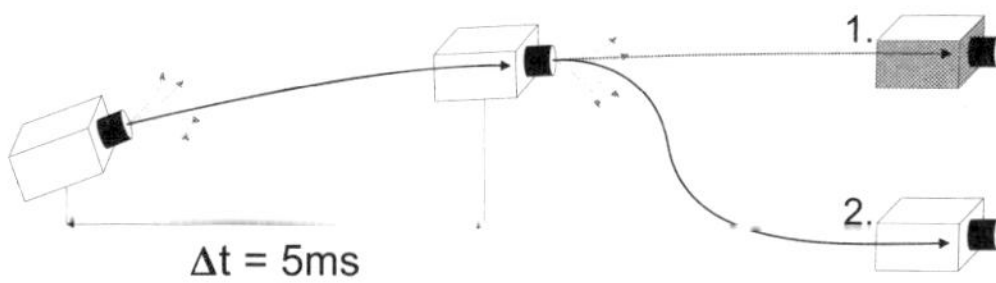

Fig. 4. If erratic or jitter movements occur and dense vision information are available, the second order motion model (2) is assumed to be more robust than the first order model (1).

B. Higher Order Motion Model

In MonoSLAM, it is assumed that the camera linear and angular velocities may change in every frame, but they are expected to be *constant* in average. In other words, the camera movements are approximated using the so-called *constant linear and angular velocity* motion model. This model assumes that in each time step the unknown linear ($\vec{a}^W$) and the unknown angular ($\vec{\alpha}^R$) accelerations cause impulses of linear

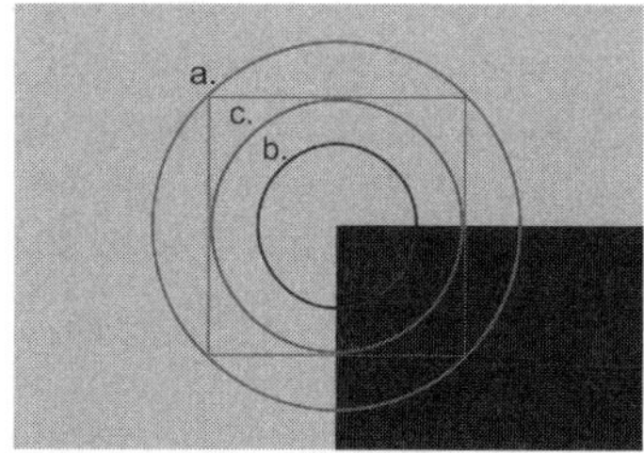

Fig. 5. To update the camera pose in each time step, known scene features' descriptors (square) are matched with predicted features' positions (a). *Localization* using the camera with fast readout speed (e.g. Δt is equal 5ms) causes these predicted features' positions to shrink obviously (b). To handle more erratic movements the motion noise uncertainty P_n should contain larger values, which cause the features' searching regions to enlarge (c).

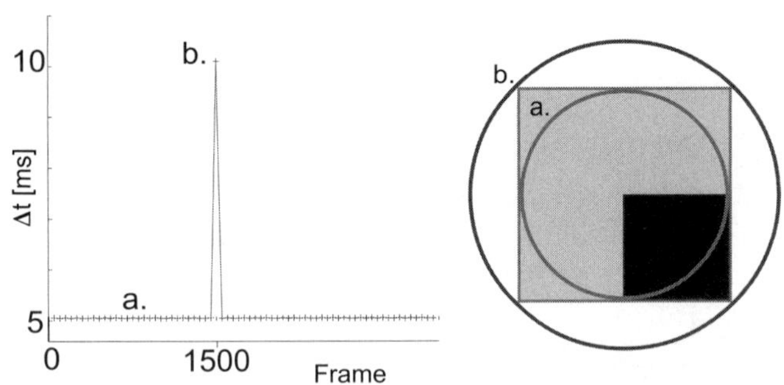

Fig. 6. The asynchronous high-speed camera can not guarantee constant time between consecutive frames. We expect that the readout speed unpredictably decreases for some frames (left image). This Δt diminution causes the features' searching ellipses to grow (right image).

($\vec{V}^W$) and angular ($\vec{\Omega}^W$) velocities. The noise vector $\vec{n}$ in MonoSLAM equals:

$$\vec{n} = \begin{pmatrix} \vec{V}^W \\ \vec{\Omega}^R \end{pmatrix} = \begin{pmatrix} \vec{a}^W \Delta t \\ \vec{\alpha}^R \Delta t \end{pmatrix}, \tag{1}$$

and the unknown accelerations are assumed to be zero mean Gaussian processes.

However, the high-speed camera can readout images several times faster (Δt is equal e.g. 5ms) than a standard camera (Δt is 33ms). Due to this high frame rate we assume that the velocities ($\vec{v}^W$ and $\vec{\omega}^R$) can *change* in average, and the accelerations ($\vec{a}^W$ and $\vec{\alpha}^R$) are expected to be *constant* in average.

The *constant linear and angular velocity* motion model can be extended to a second order model. The second order model includes linear ($\vec{a}^W$) and angular ($\vec{\alpha}^R$) accelerations in the camera state vector $\vec{x_v}$. We expect that this model can handle more erratic and jitter motion as depicted schematically in Fig. 4.

The camera vector state using the *constant linear and angular velocity* model has 13 states and comprises:

$$\vec{x}_v = \begin{pmatrix} \vec{r}^W & \vec{q}^{WR} & \vec{v}^W & \vec{\omega}^R \end{pmatrix}, \tag{2}$$

where the metric 3D position vector $\vec{r}^W$ and linear velocity $\vec{v}^W$ are estimated relative to a fixed world frame W. Quaternion $\vec{q}^{WR}$ represents the orientation between the robot frame R carried by the camera and the world frame. The angular velocity $\vec{\omega}^R$ is estimated in the robot frame R.

The extended camera state vector $\vec{x_v}$ in the *constant linear and angular acceleration* model has 6 new states and includes:

$$\vec{x}_v = \begin{pmatrix} \vec{r}^W & \vec{q}^{WR} & \vec{v}^W & \vec{\omega}^R & \vec{a}^W & \vec{\alpha}^R \end{pmatrix}, \tag{3}$$

where the zero order ($\vec{r}^W$ and $\vec{q}^{WR}$) and the first order ($\vec{v}^W$ and $\vec{\omega}^R$) states are inherited from the first order motion model. The new states comprises the linear ($\vec{a}^W$) and angular ($\vec{\alpha}^R$) accelerations.

We assume that in each time step when using the *constant linear and angular acceleration* model, an unknown linear jitter $\vec{j}^W$ and unknown angular jitter $\vec{\eta}^R$ cause an impulse of

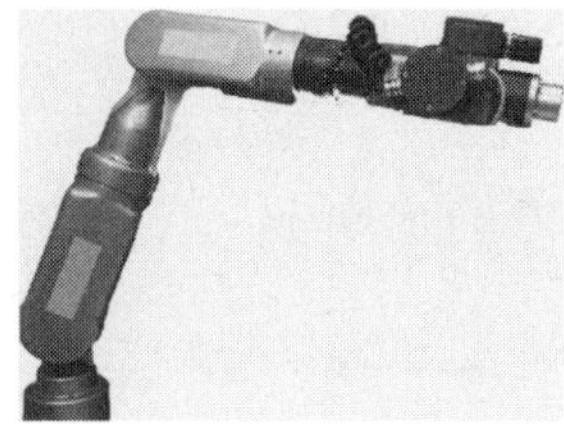

Fig. 7. The setup for the well-controlled experiments consists of the gigabit ethernet camera mounted onto a 7DOF robotic arm.

linear ($\vec{A}^W$) and angular ($\vec{\Psi}^R$) accelerations. These accelerations are again zero mean Gaussian processes, and the noise vector is equal:

$$\vec{n} = \begin{pmatrix} \vec{A}^W \\ \vec{\Psi}^R \end{pmatrix} = \begin{pmatrix} \vec{j}^W \Delta t \\ \vec{\eta}^R \Delta t \end{pmatrix}. \tag{4}$$

The camera state update is computed as follows:

$$\begin{aligned} \vec{f}_v &= \begin{pmatrix} \vec{r}^W_{new} & \vec{q}^{WR}_{new} & \vec{v}^W_{new} & \vec{\omega}^R_{new} & \vec{a}^W_{new} & \vec{\alpha}^R_{new} \end{pmatrix}^T \\ &= \begin{pmatrix} \vec{r}^W + \vec{v}^W \Delta t + \frac{1}{2}(\vec{a}^W + \vec{A}^W)\Delta t^2 \\ \vec{q}^{WR} \otimes \vec{q}(\vec{\omega}^R \Delta t + \frac{1}{2}(\vec{\alpha}^R + \vec{\Psi}^R)\Delta t^2) \\ \vec{v}^W + (\vec{a}^W + \vec{A}^W)\Delta t \\ \vec{\omega}^R + (\vec{\alpha}^R + \vec{\Psi}^R)\Delta t \\ \vec{a}^W + \vec{A}^W \\ \vec{\alpha}^R + \vec{\Psi}^R \end{pmatrix}, \end{aligned} \tag{5}$$

where the quaternion product $\otimes$ is defined in [19]. The notation:

$$\vec{q}(\vec{\omega}^R \Delta t + \frac{1}{2}(\vec{\alpha}^R + \vec{\Psi}^R)\Delta t^2)$$

represents the quaternion trivially defined by the addition of two angle-axis rotation vectors:

$$\vec{\omega}^R \Delta t \text{ and } \frac{1}{2}(\vec{\alpha}^R + \vec{\Psi}^R)\Delta t^2.$$

The EKF process noise covariance Q_v is equal to:

$$Q_v = \frac{\partial \vec{f}_v}{\partial \vec{n}} P_n \frac{\partial \vec{f}_v}{\partial \vec{n}}^T, \tag{6}$$

where $\frac{\partial \vec{f}_v}{\partial \vec{n}}$ is computed as follows:

$$\frac{\partial \vec{f}_v}{\partial \vec{n}} = \begin{pmatrix} \frac{\partial \vec{r}}{\partial \vec{A}} & \frac{\partial \vec{r}}{\partial \vec{\Psi}} \\ \frac{\partial \vec{q}}{\partial \vec{A}} & \frac{\partial \vec{q}}{\partial \vec{\Psi}} \\ \frac{\partial \vec{v}}{\partial \vec{A}} & \frac{\partial \vec{v}}{\partial \vec{\Psi}} \\ \frac{\partial \vec{\omega}}{\partial \vec{A}} & \frac{\partial \vec{\omega}}{\partial \vec{\Psi}} \\ \frac{\partial \vec{a}}{\partial \vec{A}} & \frac{\partial \vec{a}}{\partial \vec{\Psi}} \\ \frac{\partial \vec{\alpha}}{\partial \vec{A}} & \frac{\partial \vec{\alpha}}{\partial \vec{\Psi}} \end{pmatrix} = \begin{pmatrix} \frac{1}{2} I \Delta t^2 & 0 \\ 0 & \frac{\partial \vec{q}}{\partial \vec{\Psi}} \\ I \Delta t & 0 \\ 0 & I \Delta t \\ I & 0 \\ 0 & I \end{pmatrix}. \tag{7}$$

The implementation of the *constant linear and angular acceleration* motion model requires nontrivial Jacobians. Similarly as in [4], the complex differentiation of these Jacobians is tractable, but is out of the scope of this paper.

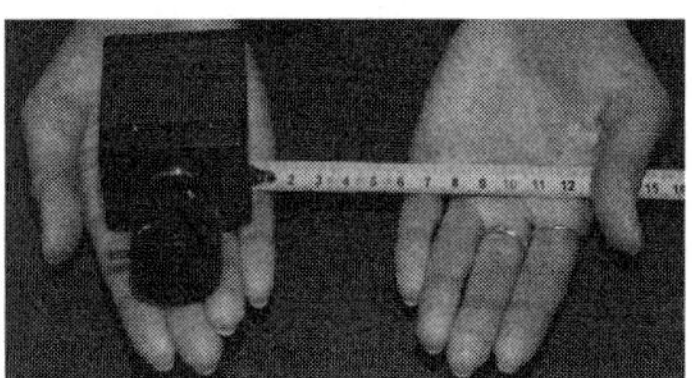

Fig. 8. The gigabit ethernet camera has been thrown and caught for a distance approximately equal 0.1m.

C. Motion Model Parameters

Choosing EKF process noise Q_v is an important part of the filter deployment. The only part of Q_v, which can be parameterized is the covariance matrix P_n of the noise vector $\vec{n}$. This matrix is diagonal, as required when the linear and angular components are uncorrelated.

In MonoSLAM, using the *constant linear and angular velocity* motion model, the covariance noise matrix P_n is equal:

$$P_n = \begin{pmatrix} SD_a^2\ \Delta t^2 & 0 \\ 0 & SD_\alpha^2\ \Delta t^2 \end{pmatrix}, \tag{8}$$

where the standard deviation parameters (SD_a and SD_α) define the smoothness of motion we expect.

If the *constant linear and angular acceleration* motion model is applied, the covariance noise matrix P_n is equal:

$$P_n = \begin{pmatrix} SD_j^2\ \Delta t^2 & 0 \\ 0 & SD_\eta^2\ \Delta t^2 \end{pmatrix}, \tag{9}$$

and standard deviation parameters (SD_j and SD_η) again define the kind of motion we assume.

If the camera with fast readout speed (e.g. Δt is 5ms) is used and the noise covariance matrix P_n contains the same values as when using the standard camera (Δt is 33ms), the motion model expects very smooth motion. To increase the *localization* robustness using the high-speed camera, setting P_n parameters to larger values in both motion models is sufficient. An example is displayed in Fig. 5.

D. Skipping Vision Information

The standard camera (Δt is 33ms) is typically providing a fairly stable vision input, and missing frames are unlikely to occur. However, the asynchronous high-speed camera (e.g. Δt is 5ms) is not so reliable. A skipped or missing frame causes the Δt to increase, and this will effect on enlarged searching regions, as depicted in Fig. 6.

To handle erratic motion using the high-speed camera, the values in the noise covariance matrix P_n (for both motion models) should be adjusted to large values. However, it is suitable to keep a balance between the motion model robustness and the real-time demand. The larger P_n values require to search for feature matches in larger regions, and that is a more time consuming operation.

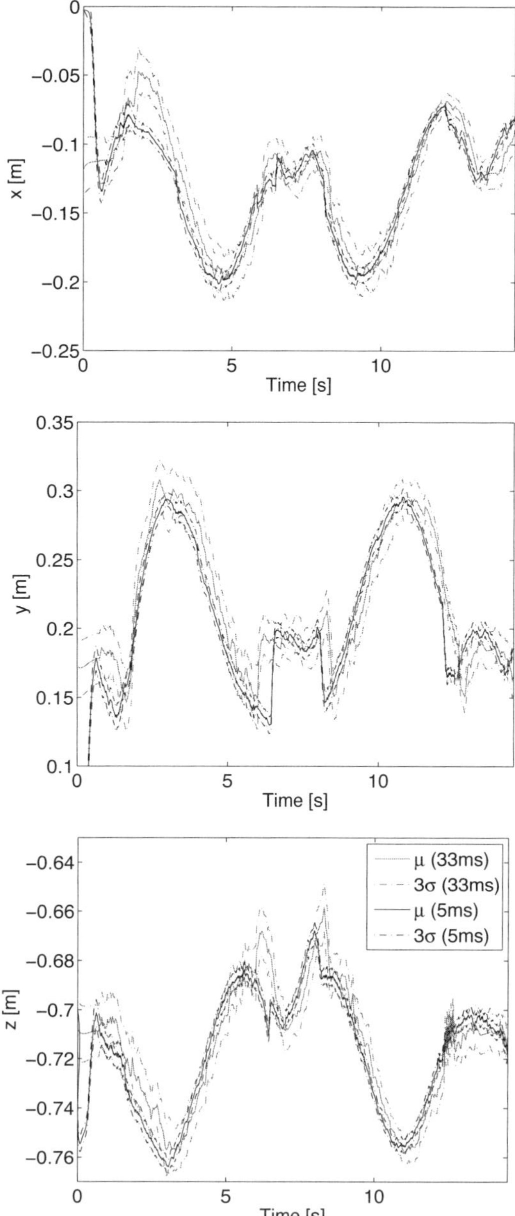

Fig. 10. The *localization* uncertainties using two different vision readout speeds (Δt): 33ms and 5ms. The camera performed the same movements repeatedly as depicted in Fig. 9, but only the camera world frame positions of the first 15s are displayed here.

IV. Experimental Results

To present the improved robustness in MonoSLAM in practice, the performed experiments are explained in three parts. Firstly, the setup and the performed precise robotic arm and high acceleration handheld movements are briefly introduced. Secondly, the accuracy of ground truth is summarized. Thirdly, the *localization* and *mapping* evaluations are compared with the calculated ground truth.

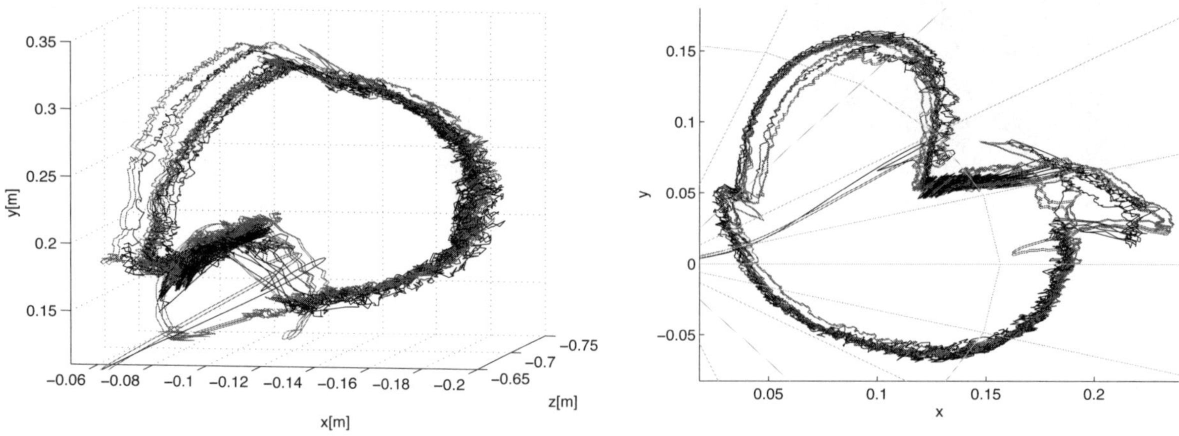

Fig. 9. Three loops of the well-controlled movement, using the high-speed camera (Δt equals 5ms) carried by the robotic arm. Left are the three-dimensional camera world frame positions and right are the camera rotations. The rotations in MonoSLAM are represented as normed quaternions, which can be visualized in a unit sphere. The red lines are the estimated camera poses using the *constant linear and angular velocity* motion model, and the green lines are the computed ground truth. The black lines are the estimated camera poses using the *constant linear and angular acceleration* motion model.

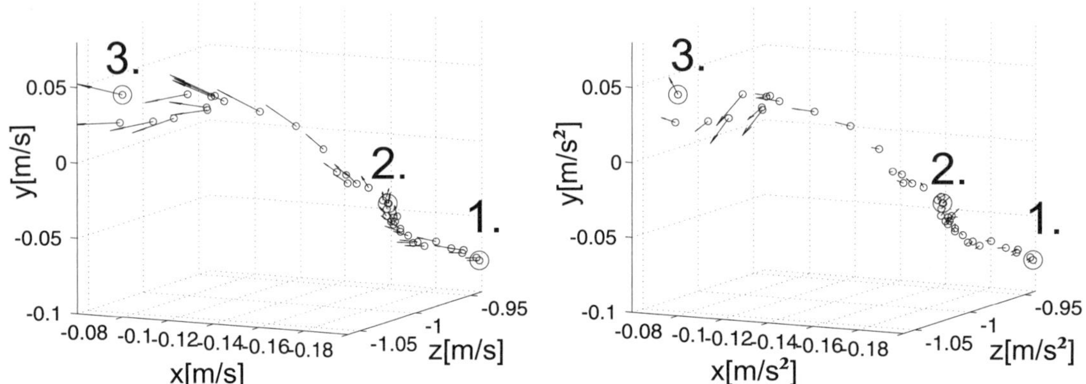

Fig. 11. The distance between the initial (1) and final (3) camera positions is equal 0.175m. The high-speed camera has been thrown (2) and caught (3) over a distance of approximately 0.1m. In MonoSLAM, the *constant linear and angular acceleration* motion model has been used. The velocity (left) and acceleration vectors (right) are scaled.

A. Experimental Setup

The setup for the well-controlled experiments consists of: a commercial Amtec[1] 7DOF robotic arm (see Fig. 7), a robotics software package[2] and the high-speed camera. The arm has been used to perform an accurate, pre-defined and smooth movement, which was repeated with different camera readout speeds.

For the two types of handheld experiments the high-speed camera has been needed. Firstly, the camera has been shortly thrown and caught (see Fig. 8). Secondly, a more erratic movement has been performed to compare the motion models.

In this work, we have been using a very fast GE680C Prosilica[3] gigabit ethernet camera (see Fig. 7 and Fig. 8), which - thanks to the 1/3" CCD sensor - offers a very good image quality.

This gigabit ethernet camera can provide VGA resolution vision output at 200Hz. In MonoSLAM, 320x240px images are more suitable, because VGA resolution would require more computationally intensive vision processing. This camera can do the binning on the chip, and this allows to address smaller images directly, faster, and at *requested* frame rates.

[1]www.amtec-robotics.com

[2]www.amrose.dk

[3]www.prosilica.com

	True			**Reconstructed**		
	X[m]	Y[m]	Z[m]	X[m]	Y[m]	Z[m]
1	0.105	0.07425	0.0	0.10497	0.07428	-0.00010
2	-0.105	0.07425	0.0	-0.10499	0.07429	-0.00013
3	0.105	-0.07425	0.0	0.10502	-0.07428	0.00008
4	-0.105	-0.07425	0.0	-0.10502	-0.07426	-0.00005

TABLE I

THREE-DIMENSIONAL FEATURES' TRUE VS. RECONSTRUCTED POSITIONS. THE ROOT MEAN SQUARE ERROR (RMSE) OF THE RECONSTRUCTED POSITIONS IS SMALLER THAN 0.1MM.

B. Ground Truth Accuracy

To present the ground truth accuracy, an example of a comparison of four reconstructed features' positions and their known values is given in Tab. I. This shows that the estimation using the procedure described in Section II-B is accurate to

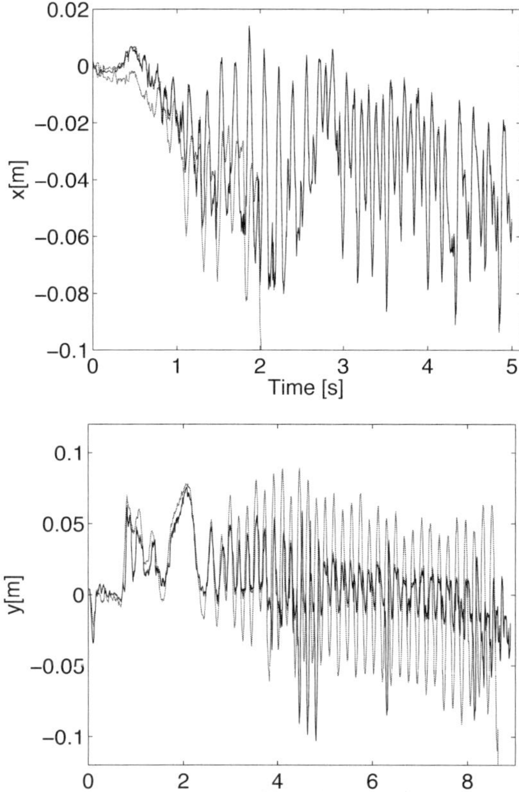

Fig. 12. In the upper row a handheld jitter motion has been performed in world x direction, and the camera world frame x positions are depicted. In the bottom row a jitter movement in world y direction is displayed as camera y positions. The *constant linear and angular velocity* model (red lines) lost the camera pose during both motions (time equals 2s in upper row and short after 8s in bottom row). The *constant linear and angular acceleration* model (black lines) proved to be more robust to these movements. The green lines are the computed ground truth.

	Constant Velocity Motion Model	
Δt[ms]	Translation [m/s^2]	Rotation [rad/s^2]
33	4	6
5	40	60
	Constant Acceleration Motion Model	
Δt[ms]	Translation [m/s^3]	Rotation [rad/s^3]
33	40	60
5	4000	6000

TABLE II

STANDARD DEVIATION PARAMETERS FOR BOTH MOTION MODELS.

within 0.1 percent of the true values, which refers to an estimate of better than 0.1mm.

C. Localization Evaluations

Prior to the evaluation of MonoSLAM *localization* results against the ground truth we explain the used criterions, and present the motion models' parameters.

To evaluate the improved robustness of MonoSLAM using the high-speed camera, these experiments are introduced:

- repeated, well-controlled and precise robotic arm movements have been executed,
- the camera has been thrown and caught and

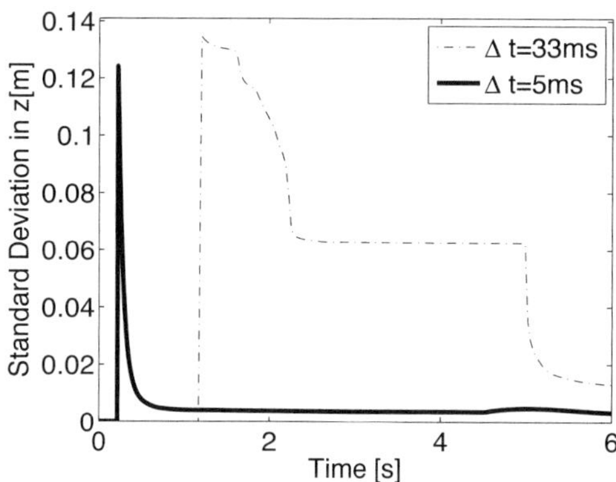

Fig. 13. In MonoSLAM, the uncertainty of a feature location is decreasing as presented in [1]. If the fast readout speed is used, then the feature three-dimensional position uncertainty is converging several times faster.

- handheld jitter motions in different world frame directions have been performed.

The criterions to compare the true and estimated camera poses are:

- the root mean square error (RMSE) to calculate the deviations of *positions* and
- the RMSE to compute the discrepancies of *rotations*.

For the comparison purposes the camera rotation representation is transformed from quaternions to Roll-Pitch-Yaw (RPY) radians.

An overview of both MonoSLAM motion models' parameters is given in Tab. II. These parameters have been adjusted according to the sections III-C and III-D.

1) Robotic Arm Movements: When performing smooth robotic arm movements using 5ms camera readout speed (Δt), both motion models were providing robust results as depicted in Fig. 9. The positions' RMSE was 0.227m and the rotations' RMSE was 0.2922rad, when using the *constant linear and angular velocity* motion model. However, the *constant linear and angular acceleration* motion model provided less accurate results: the positions' RMSE was 1.461m and the rotations' RMSE was 1.9595rad.

An important *localization* advantage when using the fast vision readout speed is that the uncertainties of the camera poses are obviously smaller as depicted in Fig. 10.

2) Thrown Camera: The setup for camera throwing and catching is displayed in Fig. 8, where the distance between two hands was approximately 0.1m. This experiment has been performed repeatedly, and the result is that MonoSLAM can operate in real-time capturing both the free as well as high accelerations in the throwing and catching motions.

As an example, Fig. 11 displays the individual track points and for each point the estimated motion using the *constant linear and angular acceleration* model. The throwing of the camera does not contain significant jitter, so the accuracy differences between the two motion models are similar to the robotic arm movements. Again the initialization used is a black rectangular object (see 1-b) in the first field of view.

3) High Acceleration Handheld Movements: Finally we want to demonstrate the behavior for fast and shaky motion. Fig. 12 depicts the motion following rapid up and down movements of the camera in hand. Maximum frequency of the up/down motion is about 5Hz. The graphs show the MonoSLAM reconstruction using the two motion models. As the Figure indicates, the *constant linear and angular acceleration* model (Δt equals 5ms) successfully tracks such a motion while the *constant linear and angular velocity* model first shows considerably high inaccuracy and then looses track.

D. Mapping Evaluations

MonoSLAM using dense visual information can faster converge features' location uncertainties as depicted in Fig. 13.

The known MonoSLAM problem is that the processing time associated with the EKF update is $O(n^2)$, where n is the number of features in the map. Due to the real-time demand all experiments in this paper have been performed in a small scene with approximately 20 features. MonoSLAM has been able to operate in such an environment in real-time using the high frame rate vision input (Δt equals 5ms).

V. Conclusion

The main contribution of this paper is the introduction of a high-speed camera to monocular SLAM. In addition we proposed to use a combination of this fast readout speed with a second order motion model to allow MonoSLAM to operate under motions with significant jitter. Such a very fast MonoSLAM algorithm running at a frame rate of 200Hz was investigated. We performed repeated and well-controlled experiments with a robot arm and various handheld movements. The comparison of the camera poses is made against accurate ground truth from off-line vision. As expected the constant velocity model is better suited for rather smooth motions. In handheld motions with an up and down motion at a frequency of about 5Hz, the acceleration model is superior and successfully tracks throughout the sequence. Additionally, the high frame rate operation of MonoSLAM makes it possible to throw the camera from one hand to the other.

By showing that high frame rate is able to considerably improve the localization robustness of monocular SLAM, new applications become possible. For example, in a future project it will be evaluated to consumer robotics and wearable computing applications.

A. Future Work

The goal of this work is the high-speed *localization* and *mapping* in a larger indoor environment (e.g. office scene). Recently Clemente et al. presented in [20] an interesting and relevant approach, which can build local maps in near real-time. The most exciting about this approach is that it can keep the computational time of the filter bounded.

Our future work includes improving the mapping using an approach based on local coordinate frames, as this can reduce the current computational complexity $O(n^2)$ to $O(n)$ [20].

Acknowledgments

This research was supported by the European Union project XPERO IST-29427 and EPSRC Grant GR/T24685/01.

Authors are very grateful to Paul Smith and other members of Oxford's Active Vision Laboratory for software collaboration.

References

[1] H. F. Durrant-White and T. Bailey, "Simultaneous localization and mapping: Part I," *IEEE Robotics and Automation Magazine*, vol. 13, no. 2, pp. 99–110, June 2006.

[2] J. Weingarten and R. Siegwart, "EKF-based 3D SLAM for structured environment reconstruction," in *IEEE International Conference on Intelligent Robots and Systems*, Edmonton, Canada, 2005, pp. 3834–3839.

[3] G. Zunino and H. Christensen, "Simultaneous localization and mapping in domestic environments," in *Multisensor Fusion and Integration for Intelligent Systems*, Baden-Baden, Germany, 2001, pp. 67–72.

[4] A. J. Davison, I. Reid, N. Molton, and O. Stasse, "MonoSLAM: Real-time single camera SLAM," *IEEE Transactions on Pattern Analysis and Machine Intelligence*, vol. 29, no. 6, pp. 1052–1067, June 2007.

[5] B. Williams, P. Smith, and I. Reid, "Automatic relocalisation for a single-camera simultaneous localisation and mapping system," in *IEEE International Conference on Robotics and Automation*, vol. 1, April 2007, pp. 2784–2790.

[6] B. Williams, G. Klein, and I. Reid, "Real-time SLAM relocalisation," in *IEEE International Conference on Computer Vision*, 2007.

[7] M. Pupilli and A. Calway, "Real-time visual SLAM with resilience to erratic motion," in *IEEE International Conference on Computer Vision and Pattern Recognition*, vol. 1, June 2006, pp. 1244–1249.

[8] D. Chekhlov, M. Pupilli, W. Mayol, and A. Calway, "Robust real-time visual SLAM using scale prediction and exemplar based feature description," in *IEEE International Conference on Computer Vision and Pattern Recognition*, June 2007, pp. 1–7.

[9] E. Eade and T. Drummond, "Scalable monocular SLAM," in *IEEE International Conference on Pattern Recognition*, vol. 1, 2006, pp. 469–476.

[10] Y. Watanabe, T. Komuro, and M. Ishikawa, "955-fps real-time shape measurement of a moving/deforming object using high-speed vision for numerous-point analysis," in *IEEE International Conference on Robotics and Automation*, April 2007, pp. 3192–3197.

[11] T. Komuro and M. Ishikawa, "A moment-based 3D object tracking algorithm for high-speed vision," in *IEEE International Conference on Robotics and Automation*, April 2007, pp. 58–63.

[12] I. Ishii, Y. Nakabo, and M. Ishikawa, "Target tracking algorithm for 1ms visual feedback system using massively parallel processing," in *IEEE International Conference on Robotics and Automation*, April 1996, pp. 2309–2314.

[13] J. M. M. Montiel, J. Civera, and A. J. Davison, "Unified inverse depth parametrization for monocular SLAM," in *Proceedings of Robotics: Science and Systems*, Philadelphia, USA, August 2006.

[14] J. Shi and C. Tomasi, "Good features to track," in *IEEE International Conference on Computer Vision and Pattern Recognition*, 1994, pp. 593–600.

[15] E. Rosten and T. Drummond, "Machine learning for high-speed corner detection," in *European Conference on Computer Vision*, vol. 1, May 2006, pp. 430–443.

[16] J. Neira and J. D. Tardós, "Data association in stochastic mapping using the joint compatibility test," *IEEE Transactions on Robotics and Automation*, vol. 17, no. 6, pp. 890–897, December 2001.

[17] G. Schweighofer and A. Pinz, "Robust pose estimation from a planar target," *IEEE Transactions on Pattern Analysis and Maching Intelligence*, vol. 28, no. 12, pp. 2024–2030, 2006.

[18] B. Triggs, P. McLauchlan, R. Hartley, and A. Fitzgibbon, "Bundle adjustment a modern synthesis," 1999.

[19] A. Ude, "Filtering in a unit quaternion space for model-based object tracking," *Robotics and Autonomous Systems*, vol. 28, no. 2-3, pp. 163–172, 1999.

[20] L. Clemente, A. J. Davison, I. Reid, J. Neira, and J. D. Tardós, "Mapping large loops with a single hand-held camera," in *Robotics Science and Systems*, 2007.

Simplex-Tree Based Kinematics of Foldable Objects as Multi-body Systems Involving Loops

Li Han and Lee Rudolph
Department of Mathematics and Computer Science
Clark University, Worcester, MA 01610
Email: lhan@clarku.edu, lrudolph@black.clarku.edu

Abstract— **Many practical multi-body systems involve loops. Studying the kinematics of such systems has been challenging, partly because of the requirement of maintaining loop closure constraints, which have conventionally been formulated as highly nonlinear equations in joint parameters. Recently, novel parameters defined by trees of triangles have been introduced for a broad class of linkage systems involving loops (*e.g.*, spatial loops with spherical joints and planar loops with revolute joints); these parameters greatly simplify kinematics related computations and endow system configuration spaces with highly tractable piecewise convex geometries. In this paper, we describe a more general approach for multi-body systems, with loops, that allow *construction trees of simplices*. We illustrate the applicability and efficiency of our simplex-tree based approach to kinematics by a study of foldable objects. We present two sets of new parameters for single-vertex rigid fold kinematics; like the parameters in the triangle-tree prototype, each has a geometrically meaningful and computationally tractable constraint formulation, and each endows the configuration space with a nice geometry.**

I. Introduction

Many practical multi-body systems involve one or more loops—physical (in parallel platforms, ring-type molecules, ...) or virtual (in inverse kinematics of serial manipulators or molecular chains,...). The kinematics of loop systems is complicated by the so-called loop closure constraint, *i.e.*, the need to maintain the closed chain structure (see *e.g.* books [1, 2, 3, 4] and references therein). There are many other practical issues, like joint limits (and other system limits) and collision avoidance, but here we focus on loop closure to the exclusion of all other constraints. In general, progress in any subset of system related issues contributes to progress in overall system knowledge; and the loop closure constraint is a recognized stumbling block in the study of multi-body loop systems.

The main difficulty with loop systems is the generally complex constraint formulation with respect to system parameters. To date, the most widely used linkage parameters are joint parameters, such as joint angles for rotational joints and linear displacements for prismatic joints. Conventionally, loop closure constraints have been formulated as equality constraints (of highly non-linear functions) over joint parameters. This formulation shows that for generic linkages the set of closure configurations is a smooth submanifold of the ambient joint parameter space, and in many cases its topology is partly or completely known (see, *e.g.*, [5, 6] and [7, 8], which treat planar and spatial linkages with spherical-type joints). Smooth manifolds are characterized by the existence of local coordinates, but although some manifolds are equipped with global coordinates (like the joint angle parameters, in the cases of a serial chain without closure constraints or [7, 8] a planar loop satisfying the technical "3 long links" condition), typically a manifold has neither global coordinates nor any standard atlas of local coordinate charts. Thus, calculations that depend on coordinates are often difficult to perform.

Some recent kinematic work uses new parameters that are not conventional joint parameters. One series of papers [9, 10, 11, 12, 13] presented novel formulations and techniques, including distance geometry, linear programming and flag manifolds, to solve inverse kinematics, identify trilaterable 6-DOF parallel and serial manipulators, and parametrize configurations of flag manipulators. Loosely speaking, a system is "trilaterable" as addressed there if it can be decomposed into tetrahedra in such a way that all unknown edge-lengths of the tetrahedra can be systematically computed from known edge-lengths using distance constraints (triangle inequalities and Cayley–Menger constraints; see below). In those papers, the trilaterable systems to be solved are already given in trilaterable form: the kinematic structure explicitly includes all distance parameters needed to determine system configurations (*e.g.*, lengths of the legs between base and platform of a parallel manipulator). We believe those papers were the first to recognize and utilize trilaterability of the systems they discuss.

In papers [14, 15, 16], we presented a different set of parameters for a class of linkages including planar loops with revolute joints and spatial loops with spherical joints. Our parameters are diagonal lengths (inter-joint distances) and triangle orientation parameters (discrete signs in the plane, dihedral angles in space). In essence, to define the parameters one joint of each loop is used as an anchor; diagonals are drawn from it to all non-adjacent joints, partitioning the loop into an open chain of triangles. The diagonal lengths and triangle orientation parameters are precisely enough to determine the shapes and relative configurations of these anchored triangles, which in turn determine the loop configurations. We proved that the defined parameters are indeed coordinates on the set of closure configurations. Further, we observed that the resulting atlas of local coordinate charts endows that space with a nice geometric structure we called "practical convexity", and remarked that our approach generalizes to any linkage system that can be decomposed into a tree of triangles.

Patent pending.

II. Novel Idea: Construction Trees of Simplices

Here we will present a general simplex-tree based parametrization approach for multi-body systems allowing construction trees of simplices; it includes the triangle-based approach of [14, 15] as a special case. Due to space limits, we write this paper somewhat intuitively and informally.

Before describing our new parametrization, we make a few comments. First, following the approach in [14, 15], we will focus on multi-body systems' deformations, that is, configurations with rigid motions factored out. The set of all deformations, called the deformation space, is mathematically the quotient space of the configuration space modulo the group of rigid motions that respect system constraints: $DSpace = CSpace/RM$.

Second, we note that many multi-body systems can be studied as multi-point systems, *e.g.*, by reducing each rigid member in the system to at most 4 general points in the member. Further, there may exist distance constraints among the points, which can be modeled as links between points: each pair of points subject to distance constraints can be modeled by a link that joins them, of fixed or variable length depending on the nature of the constraints. By such means, we can use linkage concepts, terms, and notations to study multi-object and multi-point systems under distance constraints.

Last, we recall some basic definitions concerning simplices, and facts about their geometry. In this paper we need only simplices of dimension at most 3: a 0-simplex is a point; a 1-simplex is a line segment; a 2-simplex is a triangle; a 3-simplex is a tetrahedron. A 0-dimensional face of a simplex is called a vertex, and a 1-dimensional face is called an edge. For $k \geq 2$, the edge lengths (*i.e.*, inter-vertex distances) of a k-simplex are subject to non-trivial constraints; the most familiar and simplest are the "triangle inequality" constraints for $k = 2$, and, in the Euclidean case, non-negativity of its Cayley–Menger determinant [17] (see Eq. (2) below) for all k. If all edge lengths of a simplex σ are fixed, then the shape of σ is essentially fixed. In the language of deformations, all points of $DSpace(\sigma)$ are isolated; *e.g.*, in Euclidean space, $DSpace(\sigma)$ contains just 1 or 2 points, the latter case happening only for deformations of an n-simplex in $\mathbb{R}^n$ (its 2 deformations are distinguished by their orientation).

Definition. Our new approach to kinematics of linkage systems involving loops (and other multi-point systems under distance constraints) is based on representing the system under study by a construction tree of simplices. We say a tree of simplices is a construction tree of a given linkage system if the simplices satisfy the following three conditions. (1) Each link in the linkage system is an edge of at least one simplex in the tree. (2) The set of points of the multi-point linkage system equals the set of all vertices of all simplices in the tree. (3) The deformations of the linkage system can be constructed from the shapes of the simplices and relative configurations of simplices adjacent in the tree.

Simplex Placement Procedure and Deformation Construction. We can use a tree traversal process to construct any deformation of a multi-point system with a construction tree $\mathcal{T} = (V(T), E(T))$ of simplices, given the following necessary and sufficient data about the deformation: (i) for each node $\sigma \in V(\mathcal{T})$, the shape of the simplex σ, and (ii) for each edge $\{\sigma, \tau\} \in E(\mathcal{T})$, the relative configuration of the simplices σ and τ in the ambient space $\mathbb{R}^n$. Indeed, placing a simplex in an ambient space is equivalent to determining the coordinates of its vertices. Now, given the data (i) and (ii), we construct the corresponding deformation of the multi-point system recursively as follows. (I) Place any simplex $\sigma \in V(\mathcal{T})$ anywhere in its ambient space (in case $\dim(\sigma)$ equals the ambient dimension, an orientation parameter specifies one of its two orientations). (II) If for some edge $\{\rho, \tau\} \in E(\mathcal{T})$, the simplex ρ has already been placed in space but the simplex τ has not yet been placed, then use the data (ii) to place τ. When this simplex placement procedure terminates, the deformation has been constructed.

New Parameters. Two types of parameters hold the deformation data (i) and (ii): shape parameters (associated to nodes of $\mathcal{T}$) and orientation parameters (associated to edges of $\mathcal{T}$).

The shape of a simplex is determined by its edge lengths. In any given simplex in $V(\mathcal{T})$, some edges may be links (some with fixed lengths and others with variable lengths); we call a construction tree simplex edge a diagonal of the linkage if it is not a link. Our shape parameters for a linkage comprise the lengths of all variable edges and diagonals from the tree.

Two adjacent simplices in a construction tree share a sub-simplex (*e.g.*, two triangles with a common edge). Thus there is no relative translation between two adjacent simplices, only at most a relative reorientation (about the common sub-simplex)—in essence, an element of that subgroup of the orthogonal group of the ambient Euclidean space which acts as the identity on a linear subspace having the dimension of the common sub-simplex. Our orientation parameters for a linkage comprise relative reorientation data for adjacent simplices.

In summary, our simplex-based parameters for deformations comprise (a) lengths of diagonals and links of variable length (to give shapes of simplices), and (b) orientation parameters (to give relative configurations of adjacent simplices).

Results. The aforementioned simplex placement procedure for deformation determination indicates how to use simplex-based parameters to completely determine system deformations.

If we define the forward kinematics (FK) of a multi-point system as the determination of system point positions from given parameter values, the procedure is an algorithm that solves the FK problem. Conversely, the inverse kinematics (IK) problem of determining valid simplex-based parameters that satisfy loop closure constraints is equivalent to solving the system deformation space in those parameters.

This approach is very general and applies in many ambient geometries. In this paper, limited to simplices in Euclidean space or the 2-sphere, our main results are as follows.

Theorem 1: Consider a multi-point (or multi-body) system that allows a construction tree $\mathcal{T}$ of simplices. Then:

(A) The deformations of the system are described by simplex-based parameters.

(B) The FK problem for the system is solved by a simplex placement procedure, with shapes and relative orientations of simplices directly determined by simplex-based parameters.

(C) The IK problem for the system is solved by giving an explicit description of its deformation space (*DSpace*) in terms of simplex-based parameters. More precisely, *DSpace* is essentially the product of *DStretch* and *DFlip*, where (1) *DStretch* comprises shape parameters satisfying explicit, simply evaluated constraints (triangle or Cayley–Menger determinant inequalities, and range inequalities) required for successful simplex formation, and is a convex body, while (2) *DFlip* comprises relative orientation parameters, and is independent of loop closure constraints.

Proof: If (a) the system under investigation is a planar or spatial linkage with (respectively) revolute or spherical joints, and no links of variable length, (b) $\mathcal{T}$ has no node of valence ≥ 3 (*i.e.*, is a subdivided interval), and (c) each simplex in $V(\mathcal{T})$ is a triangle in $\mathbb{R}^2$ or $\mathbb{R}^3$, then we gave detailed proofs of (A), (B), and (C2) (using the relevant versions of our simplex placement procedure) in papers [14] and [15]; the proofs in the general case are entirely analogous.

Essentially the same is true of (C1), with an important technical difference. In our earlier papers, *DStretch* is a *convex polytope*, because only triangle inequalities (involving link lengths and shape parameters) are involved, and triangle inequalities are linear. Here, in case $V(\mathcal{T})$ includes one or more tetrahedra, Cayley–Menger inequalities (in link lengths and shape parameters) are involved, and these are non-linear—in fact, for tetrahedra they are of total degree at most 3 in the squares of the diagonal lengths (and at most quadratic in the square of any one diagonal length). They are, however, still convex; the proof is an exercise in low-dimensional real algebraic geometry (relying heavily on elementary properties of cubics and quadratics). Thus *DStretch* need not be a polytope (though it will be as long as all nodes are triangles, even if some are spherical triangles) but it is always a piecewise-smooth semi-algebraic convex body. ■

As stated, in papers [14, 15] we made extensive use of the relevant special cases of Theorem 1. Here, just as in those papers, the general theorem shows that the solution of loop closure constraints can be much more efficient in simplex-based parameters than in conventional joint parameters. This gain in efficiency is due to both the convexity of *DStretch* and the independence of loop closure from orientation parameters.

In the remainder of this paper, we illustrate the applicability and efficiency of simplex-based parameters by studying foldable objects, especially single-vertex rigid folds, for which we present two sets of new parameters, each with geometrically-meaningful constraint formulations.

III. Foldable Objects and Prior Work

In our daily life, we encounter such foldable objects as paper bags, umbrellas, and space-station antennas. Normal use of foldable objects involves folding and unfolding, but not cutting up; thus many complex constraints are involved in maintaining system structures and allowing system deformations.

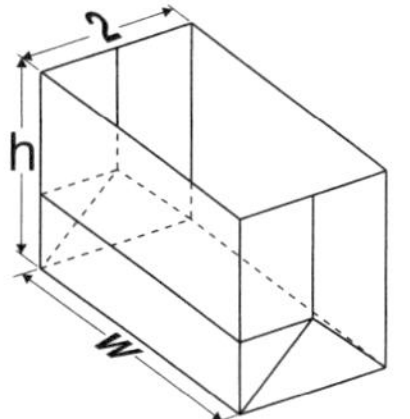

(a) Standard paper bag.

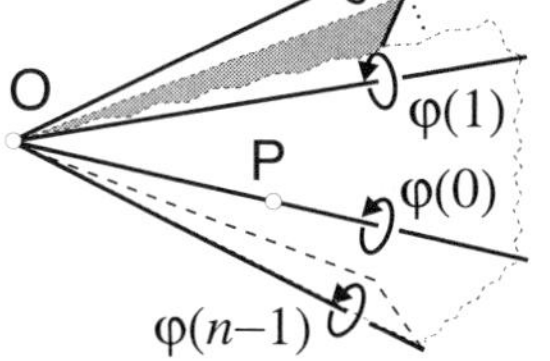

(b) Single-vertex origami.

Fig. 1. Foldable objects.

Many foldable objects have crease patterns associated with them; see Fig. 1(a) for the creases of a standard paper bag. Each crease line allows relative rotations between two panels sharing the line; it can be viewed as a revolute joint (hinge). Assuming that the panels are rigid, the relative configuration of adjacent panels can be parametrized by an angle that measures their relative rotation about the shared crease line.

Vertices of the crease pattern occur at intersections of crease lines with other crease lines or with the boundary of the object (or both); *e.g.*, the standard bag crease pattern (Fig. 1(a)) has 8 interior vertices and 6 boundary vertices. An interior vertex O on which are incident n crease lines is adjacent to n panels, which surround it in a circuit, or closed chain; if no tearing is allowed this chain has to maintain the closure constraint. In terms of the angle parameters, closure can be understood intuitively as follows: pick any point on any one crease line, and subject that point to a sequence of rotations about each of the crease lines; then the point's final position should coincide with its original position. Mathematically, this becomes (as in [18]) the equation $R(v_0, \phi_0) \cdots R(v_{n-1}, \phi_{n-1})P = P$, where P denotes a point on a crease line, v_i denotes the directional vector for crease line i, and $\phi(i)$ denotes the rotation angle about crease line i (see Fig. 1(b)). As this equation must be satisfied for all P, it is equivalent to require that the ordered product of the rotation matrices be the identity matrix,

$$R(v_0, \phi_0) \cdots R(v_{n-1}, \phi_{n-1}) = \mathrm{I}. \tag{1}$$

Each such equation imposes 3 non-linear constraints on n angular parameters. This means that, for a foldable object with only one non-boundary vertex (*i.e.*, a cycle of n panels surrounding a single common vertex), the space of deformations will generally be of dimension $n-3$. The angular formulation above amounts to studying this deformation space as the subset of the ambient angle space (an n-dimensional torus) defined by the constraint (1). Just as for a loop of linear links, the highly non-linear nature of this constraint on the angular parameters makes it technically difficult to understand and compute the structure of *DSpace* in these traditional coordinates. Foldable objects with more than one non-boundary vertex have yet more complicated descriptions in these parameters. Notwithstanding these difficulties, progress has been made. One interesting result [19] is that *DSpace* of a paper bag, creased as in Fig. 1(a) and with rigid panels, has isolated points corresponding to the folded and completely unfolded states: in this model, a flattened shopping bag cannot be opened.

Kinematic related issues of foldable objects have been studied in various communities. In robotics, in addition to origami folding [18], sheet-metal and carton box folding (*e.g.* [20, 21, 22, 23]) have also been studied, mainly from the manipulation planning point of view, sometimes with no loop closure constraints for the folded objects. Folding, especially origami folding, has been studied in and outside of scientific communities, and commands a rich literature. The combinatorics and geometry communities have interesting kinematic results on origami folding (see *e.g.* [24, 25, 26, 27] and references therein), especially on rigid flat origamis, which we use below to give a clear and representative example of our approach. Rigid origamis are modeled to have rigid and planar panels, and flat origami folds are those that can be pressed in a book without (in theory) introducing new creases. Among many interesting results on single-vertex rigid flat origamis, we will use the following properties later (see, *e.g.*, [24, 25] for proofs). Assume the cone angles about the vertex add up to 2π (as in [27], by cone angle we mean an angle between adjacent crease lines.) Then (a) the number of crease lines, and thus the number of cone angles, for a rigid flat origami vertex is even, and (b) the sum of even-indexed cone angles (thus also the sum of odd-indexed cone angles) is π.

We also refer readers to a very important and closely related paper [27], where the authors model a single-vertex origami as a spherical polygonal linkage and further prove that all single-vertex origami shapes are reachable from one to another via simple, non-crossing motions. They also consider general conical paper, where the total sum of the cone angles centered at the vertex is not 2π, and obtain similarly remarkable results for conical paper with certain properties. Their approach relies on natural extensions to the sphere of planar Euclidean rigidity results regarding the existence and combinatorial characterization of expansive motions, which have been used in the recent breakthrough on collision-free convexifying (straightening) Euclidean polygonal loops (open chains) [28, 29]. One of our methods for single vertex origamis also models such a system as a spherical loop. But we use spherical triangles and corresponding parameters to study these loops, and explicitly parametrize their *DSpaces*, a very different approach from that in [27]. Further comparing results, paper [27] proves that the valid subset of *DSpace* (*i.e.*, *DFree*, to use a naming scheme parallel to the well-known *CFree* for the valid subset of *CSpace*) is connected, and provides an efficient collision-free path planner; our approach can solve the complete *DSpace* structure, but does not apply immediately to *DFree*.

IV. Simplex-Tree Based Kinematics for Folds

Most of this section is devoted to the kinematics of a single-vertex fold, that is, a foldable object with multiple rigid and planar panels incident on one vertex O, each panel having two crease lines incident on it at that vertex (see Fig. 1(b)). The rigid panels and crease lines incident on O define a generalized cone with apex at O, so O is called the cone apex. Crease lines correspond to edges of the cone, and panels can rotate about crease lines without tearing, so cone deformations are subject to the loop closure constraint if panels and crease lines are to stay intact. Rather than describe *DSpace* of this cone in terms of angles subject to a constraint of the form (1), we modify our ideas from [14] and use a construction tree of simplices; we present two approaches, one based on Euclidean tetrahedra, the other on spherical triangles. Both approaches reveal close structural similarities between the kinematics of single-vertex folds and the kinematics of Euclidean planar loops.

A. Approach One: Trees of Euclidean Tetrahedra

System Modeling. Pick a non-apex point $P(i)$ on the ith crease line (in cyclic order, with i taken modulo n as needed). The line segment $\overline{P(i)P(i+1)}$ is contained in the ith panel. The panels are rigid and rotation around a crease line is a rigid motion, so the length l_i of this segment is constant over all deformations; thus taken together all these segments form a (spatial) loop of n rigid links. Moreover, the length of $\overline{OP(i)}$ (on the ith crease line) is also constant, for the same reason. Thus each triangle $Tri(O, P(i), P(i+1))$ maintains its congruence class throughout all deformations of the generalized cone, so from the point of view of deformation space the original generalized cone may be replaced with a polyhedral cone comprising the n triangular panels $Tri(O, P(i), P(i+1))$, $i = 0, \ldots, n-1$. We now find a construction tree of simplices for this cone, and use simplex-based parameters to study its *DSpace*.

Tetrahedra Trees and Parameters. One method is to use any construction tree of triangles for the loop $(P(0), P(1), \ldots, P(n-1))$. For example, a 6-bar loop with vertices $P(0), \ldots, P(5)$ is decomposed by the three diagonals $(P(1), P(3))$, $(P(3), P(5))$, and $(P(5), P(1))$ into a construction tree of 4 triangles: $Tri(1,3,5)$, $Tri(1,2,3)$, $Tri(3,4,5)$, and $Tri(5,0,1)$, with $Tri(1,3,5)$ adjacent to all other three. (We abbreviate $Tri(P(1), P(2), P(3))$ by $Tri(1,2,3)$ and so on.) Now, each of these triangles $Tri(P(i), P(j), P(k))$ determines a tetrahedron $Tet(O, P(i), P(j), P(k))$, whose three edges (additional to those of the triangle) have already been noted to be of fixed length. These tetrahedra fit together, in a tree combinatorially identical to that chosen to construct the loop, so as to construct the polyhedral cone (see Fig. 2.) The corresponding tetrahedra-based parameters for *DSpace* of the cone are closely related, but not identical, to the triangle-based

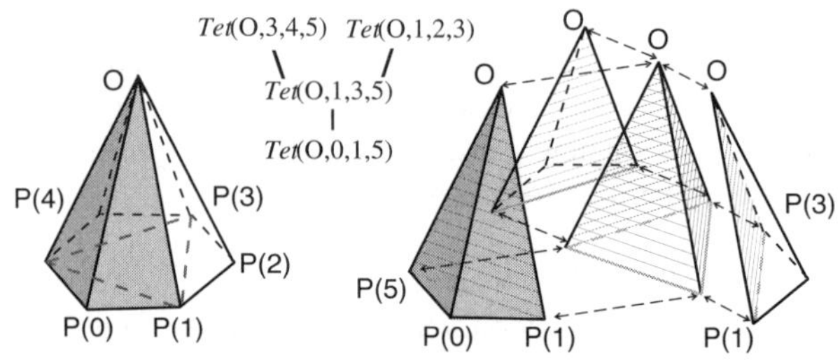

Fig. 2. Decomposition of a polyhedral cone into a tree of tetrahedra.

parameters for *DSpace* of the loop. Namely, we use (a) the squared diagonal lengths $D[\mathcal{T}] = [D[\mathcal{T}](1), \ldots, D[\mathcal{T}](n-3)]$, and (b) the tetrahedron orientation signs $s[\mathcal{T}] = [s[\mathcal{T}](1), \ldots, s[\mathcal{T}](n-2)]$. Here $\mathcal{T}$ denotes a construction tree of tetrahedra for the polyhedral cone, $D[\mathcal{T}](i) = d^2[\mathcal{T}](i)$ is the squared length of the ith diagonal, and $s[\mathcal{T}](j)$ is the orientation sign of the jth tetrahedron comprised in $\mathcal{T}$. (The number $n-3$ of squared diagonal lengths for the tree of tetrahedra equals the number $n-3$ of diagonal lengths for the tree of triangles, and similarly for the number $n-2$ of orientation signs.)

Constraints and *DSpace* Structures in Tetrahedral Parameters. In this formulation, the loop closure constraint on the panels of our polyhedral cone becomes that set of constraints on the diagonal lengths which is necessary and sufficient for the tetrahedra comprised in $\mathcal{T}$ to exist. These constraints can be phrased in terms of the Cayley–Menger determinant. As is well known, given real numbers $D(i,j)$ $(0 \le i,j \le 3)$ with $D(i,j) = D(j,i) \ge 0$, $D(i,i) = 0$, points $P(i) \in \mathbb{R}^3$ $(0 \le i \le 3)$ with $D(i,j) = \|P(j) - P(i)\|^2$ exist if and only if

$$\begin{vmatrix} 0 & 1 & 1 & 1 & 1 \\ 1 & 0 & D(0,1) & D(0,2) & D(0,3) \\ 1 & D(1,0) & 0 & D(1,2) & D(2,3) \\ 1 & D(2,0) & D(2,1) & 0 & D(2,3) \\ 1 & D(3,0) & D(3,1) & D(3,2) & 0 \end{vmatrix} \ge 0. \quad (2)$$

Equality holds if and only if the points $P(i)$ are coplanar. Each of the $n-2$ tetrahedra $\sigma \in V(\mathcal{T})$ gives us one such inequality constraint. The set of all $n-2$ inequalities defines the set of feasible squared diagonal lengths for the n-sided polyhedral cone, which we call *DStretch*.

The tetrahedron orientation signs are likewise defined along the lines set out in [14] for triangle orientation signs for a loop with a construction tree of triangles. If we use $+$ and $-$ for the two orientations of non-degenerate tetrahedra, and label singular tetrahedra with both $+$ and $-$, we obtain (roughly) an identification of $DSpace[\mathcal{T}]$ of the polyhedral cone parametrized by the tree $\mathcal{T}$ of tetrahedra with $DStretch[\mathcal{T}] \times DFlip$, where $DFlip = \{+,-\}^{n-2}$. (As in [14], eliminating the "roughness" requires analysis of "super-singular" deformations.) Such a *DSpace* parametrization is very similar to that for a planar loop with n links given in [14]. An important difference is that, whereas [14] proves that *DStretch* of a planar loop with n revolute joints and fixed link lengths is a convex polyhedron of dimension $n-3$, in the present case *DStretch* is still convex but (for $n \ge 5$) it is no longer a polyhedron.

The geometry of its various curvilinear faces is revealed by separate analyses of the constraint (2) in the cases where the corresponding tetrahedron includes exactly 1, 2, or 3 diagonals: a single diagonal's squared length is constrained to lie in an interval; a pair of diagonals' squared lengths are constrained to lie inside or on a certain ellipse; and a triple of diagonals' squared lengths lie inside or on the boundary of a certain semi-algebraic (cubic) convex body.

B. Approach Two: Trees of Spherical Triangles

For the tetrahedra above, the non-apex points $P(i)$ on the crease lines, and thus their distances to the cone apex O and the link lengths, can be chosen quite arbitrarily, though astute choices might afford extra convenience in computation.

System Modeling. For the second approach, we demand that all $P(i)$'s be at the same distance from the cone apex O. Thus the set of $P(i)$'s is the intersection of the crease lines with a sphere centered at O. As noted earlier, the distance between O and any point on a crease line is constant across all deformations, so this sphere is also constant; further, since the cone panels are assumed to be rigid and planar, each of them intersects the sphere in a spherical line segment (*i.e.*, an arc of a great circle through the corresponding points $P(i)$ and $P(i+1)$), and the spherical length of that segment is also constant. For simplicity, instead of spherical length we use central angle to measure spherical line segments (the spherical length is the radius of the sphere multiplied by the radian measure of the central angle). Following standard usage from spherical geometry, we also call a spherical line segment minor if its central angle is in $[0, \pi]$, and major if not.

In this language, what we have observed is that the intersection of each cone panel intersects the sphere in a spherical line segment of constant central angle. Therefore the intersection of the sphere with the entire system of rigid and planar panels incident on the cone apex O is a spherical n-gon loop (n being the number of panels) with links of fixed central angles; and the kinematics of the single-vertex fold is equivalent, in the sense of having identical *DSpace*s, to that of this spherical loop. The converse, that every spherical loop corresponds in this way to a single-vertex fold, is obvious.

There is a subtlety which we have no room to discuss properly here, but which must be mentioned. It is entirely standard to define the spherical distance between two points of a sphere to be the length of a minor spherical line segment with those endpoints, and with that definition the sphere becomes a metric space in the usual way—in particular, the triangle inequality holds; but there is no compelling reason to require the edges of a spherical polygonal n-gon to be minor segments. Nonetheless, in the remainder of this paper we restrict our attention to spherical polygonal n-gon loops in which each edge is minor; equivalently, in the language of foldable objects, to single-vertex folds in which the angle of each panel at the cone apex is at most π.

Spherical geometry shares many theorems with Euclidean plane geometry, but not all; so our triangle approach to spherical n-gon loop kinematics is very similar, but not identical, to our triangle approach in [14] for Euclidean planar n-gon loops. Below, we emphasize the points of difference.

Spherical Triangle Trees and Parameters.

The construction trees of triangles in [14] have no branching because the triangles are anchored (share a common vertex); however, as noted in the proof of Theorem 1, the results and arguments in [14] extend to arbitrary construction trees of triangles, Euclidean or spherical. In particular, for an n-gon loop each construction tree has $n-2$ nodes and $n-3$ edges (loop diagonals); Fig. 3 shows this for a spherical 6-gon with central angles $(50, 97, 151, 82, 35, 59)$ (in degrees).

Following our general simplex-based approach, we define spherical-triangle-based parameters for the deformations of a spherical loop, and thus that of a polyhedral cone allowing rotation about the edges. Namely, we use (a) the spherical diagonal lengths $\beta[\mathcal{T}](1), \ldots, \beta[\mathcal{T}](n-3)$, and (b) the spherical triangle orientation signs $s[\mathcal{T}](1), \ldots, s[\mathcal{T}](n-2)$, where $\mathcal{T}$ is a construction tree of spherical triangles for the spherical loop, $\beta[\mathcal{T}](i)$ is the central angle of diagonal i, and $s[\mathcal{T}](i)$ is the orientation sign ($+$, $-$, or 0) of the ith spherical triangle in $\mathcal{T}$. One way to define this sign is as the orientation sign of the tetrahedron formed by the triangle vertices (in their given order) preceded by the center of the sphere; as usual 0 means that the tetrahedron is degenerate (equivalently, that the triangle is contained in a great circle). Reasoning like that in [14] shows that the orientation sign values $s[\mathcal{T}](i)$ are essentially uncoupled from the central angle parameters $\beta[\mathcal{T}](j)$; reversing the sign of a non-degenerate triangle corresponds to reversing its orientation (compare Figs. 3(a) and 4(c)).

Constraints and *DSpace* Structures in Spherical Triangle Parameters. To generalize the method given in [14], we need to find the necessary and sufficient conditions on three angles less than or equal to π such that there exist a spherical triangle with those central angles. This is easy to do in light of the observation (true in both spherical and Euclidean geometry) that the length of one edge of a triangle is at most the sum of the lengths of the other two edges. One issue here is that the spherical distance between any two points is restricted to $[0, \pi]$. Given two distances β_1, β_2, each between 0 and π, their usual sum $\beta_1 + \beta_2$ will have a direct range of $[0, 2\pi]$. However, if we define their *spherical sum* by

$$\beta_1 \oplus \beta_2 - \min(\beta_1 + \beta_2, 2\pi - \beta_1 + \beta_2) \tag{3}$$

then we find that their spherical sum is again between 0 and π; and we see that given a spherical triangle with vertices $P(i)$, $P(j)$, $P(k)$ at spherical distances $\beta(i,j)$, $\beta(j,k)$ and $\beta(k,i)$ in $[0, \pi]$, the triangle inequalities become

$$\beta(i,j) \leq \beta(j,k) \oplus \beta(k,i) \tag{4}$$

$$\beta(j,k) \leq \beta(k,i) \oplus \beta(i,j) \tag{5}$$

$$\beta(k,i) \leq \beta(i,j) \oplus \beta(j,k) \tag{6}$$

Moreover, in the cube $[0, \pi]^3$, these inequalities (which are all linear in the coordinates) cut out a convex polyhedron.

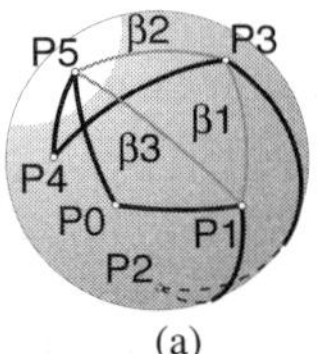

(a)

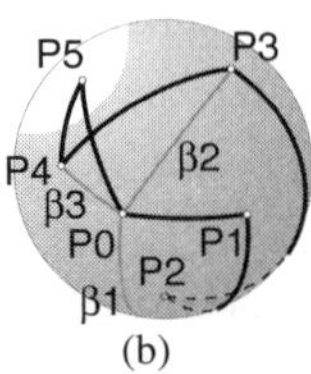

(b)

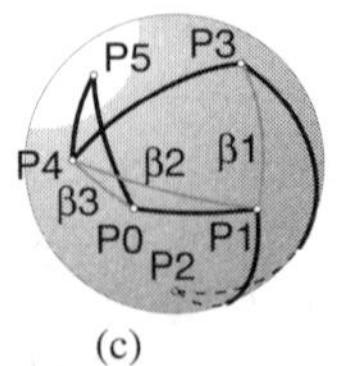

(c)

Fig. 3. One deformation of a spherical 6-gon loop with 3 sets of shape parameters derived from 3 different trees of 2-simplices.

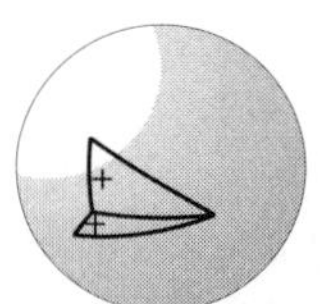
(a) Like.

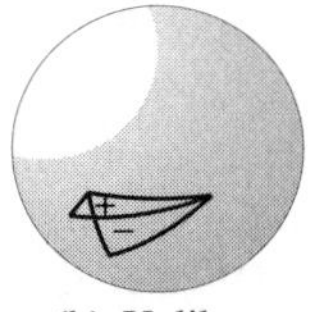
(b) Unlike.

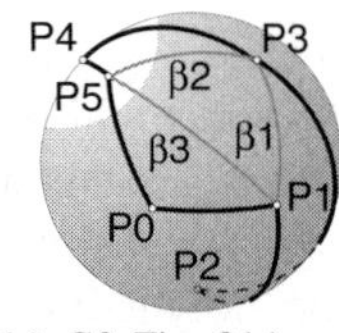

(c) Cf. Fig. 3(a)

Fig. 4. (a), (b) Two embeddings, with different relative orientations, of one pair of spherical triangles. (c) A deformation of a 6-gon loop, differing in the orientation of exactly one triangle from that shown in Fig. 3(a).

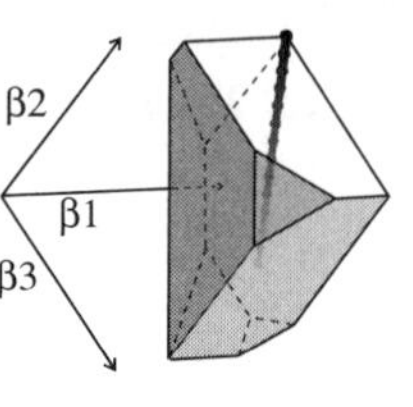

(a)

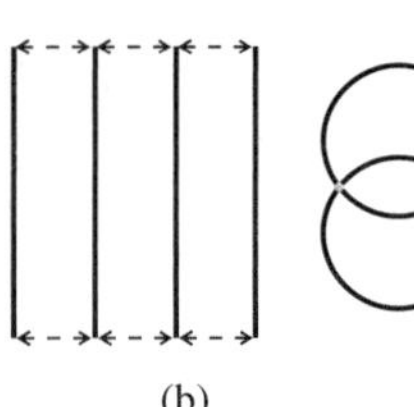
(b)

Fig. 5. (a) *DStretch* for a flat origami vertex with six creases and cone angles (45; 30; 65; 50; 70; 100) degrees, and a linear path of shape parameter values joining a gray interior point $(60, 60, 80)$ to a black corner $(95, 120, 145)$. (b) *DSpace* for a flat origami vertex with four creases; it is not a manifold.

Now, returning to spherical loops, we note that the loop closure constraint is the set of triangle inequality constraints on the diagonal lengths required for successful formation of all spherical triangles in the construction tree $\mathcal{T}$. As in [14] this implies that *DStretch*, the set of feasible diagonal central angles, is a convex polytope with a natural stratification. (Fig. 5(a) illustrates the situation for a spherical 6-gon loop with central angles of $(45, 30, 65, 50, 70, 100)$ degrees.)

Again as in [14], this construction endows *DSpace* with a stratification: roughly, *DSpace* is constructed by gluing together 2^{n-2} copies of *DStretch* (one for each assignment of orientation signs to triangles in $\mathcal{T}$) along boundary strata corresponding to singular deformations (with one or more singular triangles). One essential fact used here is that reversing the orientation of a non-singular triangle, if possible, requires passing through a singular deformation of the triangle.

As a simple example of *DSpace* structures, consider a flat origami vertex with 4 panels. Let its cone angles be $\beta(0,1)$, $\beta(1,2)$, $\beta(2,3) = \pi - \beta(0,1)$, and $\beta(3,0) = \pi - \beta(1,2)$ (here we use properties (a) and (b) of flat origami vertices as quoted in section III). As the only shape parameter, we use the central angle between $P(0)$ and $P(2)$. It is easy to verify that *DStretch* is the interval with endpoints $|\beta(0,1) - \beta(1,2)|$ and $\beta(0,1) \oplus \beta(1,2)$. Further, at these endpoints the two triangles of the tree, one with vertices $P(0)$, $P(1)$, $P(2)$ and the other with vertices $P(2)$, $P(3)$, $P(0)$, are simultaneously singular. (Trying some concrete values for $\beta(0,1)$ and $\beta(1,2)$ may help understand this fact.) Thus when *DSpace* is constructed from 4 copies of the interval *DStretch*, at each end of the interval all four copies are glued together; the resulting *DSpace* is singular (*i.e.*, not a manifold), and looks like the union of 2 circles intersecting at 2 points, as shown in Fig. 5(b). We see that all flat origami vertices of 4 panels (with generic cone angles) have topologically identical *DSpaces*.

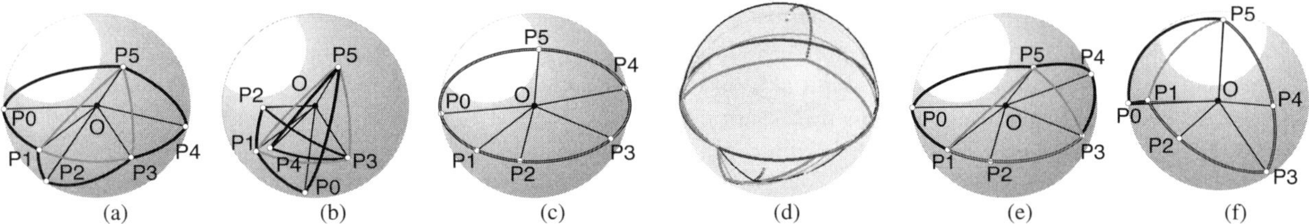

Fig. 6. Deformations and motion of the flat origami vertex with *DStretch* illustrated in Fig. 5(a); diagonal angles (in degrees) and orientation signs are as indicated. (a) Query deformation 1 $((60, 60, 80), (-, -, -, +))$; (b) query deformation 2 $((60, 60, 80), (+, +, +, +))$; (c) a flat deformation $((95, 120, 145), (0, 0, 0, +))$; (d) a linear path between query deformation 1 and the flat deformation (also shown in the copy in *DSpace* of Fig. 5(a) that has orientation sign $(-, -, -, +)$); (e) singular deformation 1 $((95, 60, 80), (0, -, -, +))$; (f) singular deformation 2 $((95, 120, 80), (0, 0, -, +))$.

C. Results and Discussions

The tetrahedron and spherical-triangle approaches exhibit similarity in *DSpace* structures for single-vertex fold kinematics, as summarized in the following theorem.

***Theorem** 2:* For a single-vertex fold with n rigid panels, *DStretch* is an $(n-3)$-dimensional convex body, *DFlip* is the discrete set $\{+, -\}^{n-2}$, and *DSpace* consists of 2^{n-2} copies of *DStretch* properly identified along their boundaries. ■

Both approaches have significant advantages over approaches that use joint angle parameters, thanks both to the explicit constraints in our new parameters and the practical convexity of *DSpace* in these parameters. As an illuminating example for the efficiency of our simplex-based approach, consider the generation of loop deformations. In our new parameters, this can be done in two steps.

First, we solve the shape parameters. Both Cayley–Menger determinant and spherical triangle inequality constraints can be efficiently solved, *e.g.*, with convex programming (or, for triangle inequalities, linear programming). We have also developed our own solving methods that take advantage of the highly structured constraints; in our computation study using Matlab, they beat the general methods by orders of magnitude.

Second, we pair the shape parameter values with the relative orientation sign parameters. For a generic single-vertex fold with n rigid panels, each set of diagonal length parameter values of a non-singular deformation pairs with 2^{n-2} distinct sets relative orientation signs, generating 2^{n-2} distinct (though related) system deformations, all with the same simplex shapes but differing simplex orientations. On a laptop computer, using linear programming in Matlab, a set of valid shape parameters for a generic single-vertex 1000-panel fold can be generated in about 1 millisecond. Paired with the points of *DFlip*, this yields 2^{998} different deformations.

Knowledge and nice geometry of system *DSpace*s have significant impact on many kinematics related issues such as system design and path planning. For example, it is foreseeable for certain design tasks to favor systems in which having all (or important) system configurations fall into a single connected component of *DSpace* (or of *DFree*, when collision avoidance is taken into consideration). Explicit and efficient parametrization of *DSpace*s provides invaluable tools to designers for the evaluation and improvements of design schemes. As another example of the broad implications of our approach and results, the piecewise convexity of *DSpace* in simplex-based parameters greatly simplifies path planning for single-vertex folds. That is, two query deformations in the closure of one *DSpace* stratum (roughly, the points of *DSpace* corresponding to one copy of *DStretch*) can be joined by a straight-line path because *DStretch* is convex. Two deformations merely in the same component of *DSpace* can be joined by a piecewise-linear path, once we determine critical singular deformations through which to move successively between strata; again, the simple nature of the constraints on our shape parameters leads to efficient computation of singular deformations.

Figs. 6(a) and 6(b) show two query deformations of a flat origami vertex with 6 panels, again having central angles of $(45, 30, 65, 50, 70, 100)$ degrees: both deformations have diagonal length values of $(60, 60, 80)$ degrees, and have opposite orientations for all triangles except the one with vertices $P(1)$, $P(3)$, $P(5)$. Each query deformation can be linearly connected to a flat deformation, like that in Fig. 6(c) with diagonal lengths of $(95, 120, 145)$ degrees. Fig. 6(d) illustrates a linear path from the deformation in Fig. 6(a) to the flat deformation. While a linear path between two points in a convex set is shortest in the convex metric and mathematically optimal, it generally involves simultaneous motion (folding) of all crease lines. Since in practice many folding motions are most easily done by folding only one or a few creases at once, for folding problems Manhattan paths are likely to be more practical than linear paths. The query deformation in Fig. 6(a) can reach the flat deformation by using a 3-segment Manhattan path through two singular deformations as in Figs. 6(e) and 6(f).

For single-vertex fold systems, the spherical triangle approach will generally be more efficient than the tetrahedron approach, since triangle constraints (inequalities (4)–(6)) are generally simpler than determinant constraints (inequality (2)). We introduced both approaches since each has its own advantages in various situations; *e.g.*, the tetrahedron approach generalizes directly to multi-vertex folds (like that in Fig. 7)

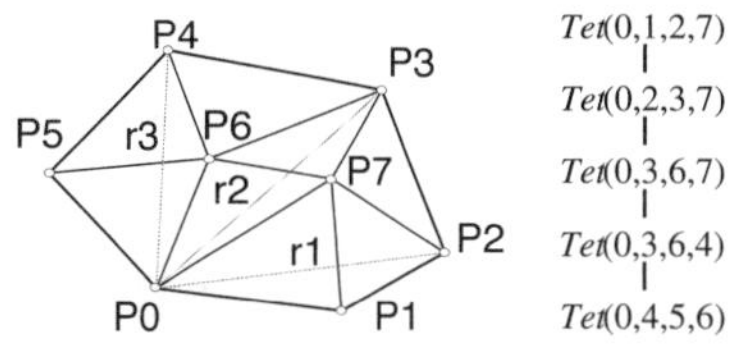

Fig. 7. A multi-vertex fold with a construction tree of simplices.

that allow construction trees of tetrahedra. Note that a multi-vertex fold contains multiple loops of panels; Fig. 7 is just one of the simplest examples of a multi-loop, multi-point system allowing a construction tree of simplices. Our approach and results, described in section II and illustrated in this section for single-vertex folds, apply directly to all such systems.

V. Summary

In this paper, we have described our novel simplex-tree based approach for the kinematics of multi-body systems involving loops, and illustrated it with an example system of single-vertex folds. For systems allowing construction trees of simplices, we efficiently use simplex-based shape and orientation parameters for system deformations, formulate loop constraints as constraints on shape parameters needed for successful formation of the simplices (*e.g.*, the triangle inequality and Cayley–Menger determinant constraints for triangles and tetrahedra), solve the set of valid length parameters, and explicitly construct the *DSpace* structures essentially as the product of the shape parameter set and the orientation parameter set (which carries no loop constraints). For systems involving loops, knowledge of *DSpace* and *CSpace* structures with explicit parametrizations is invaluable for many kinematics related issues, *e.g.*, motion planning, system design, analysis, simulation, and control; until recently (*cf.* [7, 8, 14, 15]) these structures and parametrizations, and techniques for working with them, have largely remained elusive. In this paper we have greatly extended our work in [14, 15].

Part of our ongoing research is to identify and study systems allowing simplicial construction trees. For a multi-point linkage system (again, with links modeling distance constraints), the conditions on the existence of a construction tree can be intuitively understood as follows: its constraints as reflected in the existing links shall allow the addition of virtual links that decompose the system into a collection of simplices (or more broadly, a collection of simplices with other rigid system components, as dictated by existing system constraints, attached to the simplices) with edges consisting of all existing and virtual links, and with free relative configurations allowed among adjacent simplices. Lengths of virtual links and existing links with non-trivial ranges are captured by shape parameters, relative configurations by orientation parameters. We are currently working on further properties and algorithms useful to identify these systems and their parameters. We emphasize that, while the representative systems solved in simplex-tree based approaches so far (including Euclidean polygonal loops in our earlier papers [14, 15] and the systems in this paper) use construction trees with equidimensional simplices as nodes, our simplex-based approach allows non-equidimensional simplices as nodes of a single tree. In future papers we will describe practical robotic systems that can be usefully studied using construction trees of non-equidimensional simplices.

Acknowledgement

We thank the reviewers and area chairs for help and advice. We were supported in part by NSF Grant IIS-0713335.

References

[1] J. J. Craig, *Introduction to Robotics: Mechanics and Control, 2nd Edition*. Reading, MA: Addison-Wesley Publishing Company, 1989.
[2] R. M. Murray, Z. Li, and S. S. Sastry, *A Mathematical Introduction to Robotic Manipulation*. Boca Raton, FL: CRC Press, 1994.
[3] J.-P. Merlet, *Parallel Robots*. Springer, 2000.
[4] M. Mason, *Mechanics of Robotic Manipulation*. The MIT Press, 2001.
[5] M. Kapovich and J. Millson, "On the moduli spaces of polygons in the euclidean plane," *J. Diff. Geom.*, vol. 42, pp. 133–164, 1995.
[6] ——, "On the moduli space of a spherical polygonal linkage," *Canadian Mathematical Bulletin*, vol. 42, pp. 307–320, 1999.
[7] R. Milgram and J. Trinkle, "The geometry of configuration spaces for closed chains in two and three dimensions," *Homology Homotopy and Applications*, 2002.
[8] J. Trinkle and R. Milgram, "Complete path planning for closed kinematic chains with spherical joints," *Int. J. Robot. Res.*, vol. 21, no. 9, pp. 773–789, 2002.
[9] J. Porta, L. Ros, and F. Thomas, "Inverse kinematics by distance matrix completion," *Proc. of 12th International Workshop on Computational Kinematics*, 2005.
[10] ——, "On the trilaterable six-degree-of-freedom parallel and serial manipulators," in *Proc. IEEE Int. Conf. Robot. Autom. (ICRA)*, 2005.
[11] ——, "Multiple-loop position analysis via iterated linear programming," *Robotics: Science and Systems*, 2006.
[12] C. Torras, F. Thomas, and M. Alberich, "Stratifying the singularity loci of a class of parallel manipulators," *IEEE Trans. Robot.*, pp. 23–32, 2006.
[13] M. Alberich, F. Thomas, and C. Torras, "Flagged parallel manipulators," *IEEE Trans. Robot.*, pp. 1013–1023, 2007.
[14] L. Han, L. Rudolph, J. Blumenthal, and I. Valodzin, "Convexly stratified deformation spaces and efficient path planning for planar closed chains with revolute joints," in *Proc. Seventh International Workshop on Algorithmic Foundation of Robotics*, 2006.
[15] L. Han and L. Rudolph, "Inverse kinematics for a serial chain with joints under distance constraints," in *Robotics: Science and Systems (RSS)*, 2006.
[16] ——, "A unified geometric approach to inverse kinematics of a spatial chain with spherical joints," 2006, accepted to ICRA 2007.
[17] L. M. Blumenthal, *Theory and Applications of Distance Geometry, 2nd edition*. American Mathematical Society, 1970.
[18] D. Balkcom and M. Mason, "Introducing robotic origami folding," in *Proc. IEEE Int. Conf. Robot. Autom. (ICRA)*, 2004.
[19] D. J. Balkcom, E. Demaine, and M. Demaine, "Folding paper bags," in *Annual Workshop on Computational Geometry*, 2004.
[20] S. Gupta, D. Bourne, K. K. Kim, and S. S. Krishnan, "Automated process planning for robotic sheet metal bending operations," *Journal of Manufacturing Systems*, 1998.
[21] L. Lu and S. Akella, "Folding cartons with fixtures: A motion planning approach," *IEEE Trans. Robot. Automat.*, pp. 346–356, 2000.
[22] G. Song and N. M. Amato, "A motion planning approach to folding: From paper craft to protein folding," in *Proc. IEEE Int. Conf. Robot. Autom. (ICRA)*, 2001, pp. 948–953.
[23] J. Liu and J. Dai, "An approach to carton-folding trajectory planning using dual robotic fingers," *Robotics and Autonomous Systems*, vol. 42, pp. 47—63, 2003.
[24] T. C. Hull, "On the mathematics of flat origamis," *Congressus Numerantium*, vol. 100, pp. 215–224, 1994.
[25] ——, "The combinatorics of flat folds: a survey," in *Proc. 3rd International Meeting of Origami Science, Math, and Education*, 2001.
[26] E. D. Demaine and M. L. Demaine, "Recent results in computational origami," in *Proc. 3rd International Meeting of Origami Science, Math, and Education*, 2001.
[27] I. Streinu and W. Whiteley, "Single-vertex origami and spherical expansive motions," in *Proc. Japan Conf. on Discrete and Comp. Geometry Tokai University, Tokyo, Oct. 8-11 2004. Lecture Notes in Computer Science 3742, pp. 161-173, Springer 2005.*, 2004.
[28] R. Connelly, E. D. Demaine, and G. Rote, "Straightening polygonal arcs and convexifying polygonal circles," *Discrete and Computational Geometry*, 2003.
[29] I. Streinu, "A combinatorial approach to planar non-colliding robot arm motion planning," in *Proc. IEEE Symp. Foundations of Computer Science (FOCS)*, 2000, pp. 443–453.

fMRI-Compatible Robotic Interfaces with Fluidic Actuation

Ningbo Yu [1], Christoph Hollnagel [1], Armin Blickenstorfer [2], Spyros Kollias [2], Robert Riener [1]

[1] Sensory-Motor Systems Lab, ETH and University Zurich, Switzerland
[2] Institute of Neuroradiology, University Hospital Zurich, Switzerland

Abstract— **Actuation is a major challenge in the development of robotic systems intended to work in functional Magnetic Resonance Imaging (fMRI) procedures, due to the high magnetic fields and limited space in the scanner. Fluidic actuators can be made fMRI-compatible and are, thus, promising solutions. In this work we developed two robotic interface devices, one with hydraulic and another with pneumatic actuation, to control one degree-of-freedom translational movements of a user that performs fMRI tasks. Due to the fMRI-compatibility restrictions, special materials were used for the endeffector which works in the scanner bore, and active components such as the control valves and pressure sensors, had to be placed far away from the endeffector with long transmission lines in between. Therefore, the two fMRI-compatible setups differed from conventional fluidic actuation systems and brought control difficulties. Both systems have been proved to be fMRI-compatible and yield no image artifacts in a 3T scanner. Passive as well as active subject movements were realized by classical position and impedance controllers. With the hydraulic system we achieved smoother movements, higher position control accuracy and improved robustness against force disturbances than with the pneumatic system. In contrast, the pneumatic system was back-drivable, showed faster dynamics with relatively low pressure, and allowed force control. Furthermore, it is easier to maintain and does not cause hygienic problems after leakages. In general, pneumatic actuation is favorable for fast or force-controlled applications, whereas hydraulic actuation is recommended for applications that require high position accuracy, or slow and smooth movements.**

I. Introduction

Robotic systems and devices that are compatible with magnetic resonance imaging (MRI) technique find wide range of applications in academic and industrial fields [1, 2]. Functional MRI (fMRI) is an advanced research and clinical tool in neuroscience. An fMRI-compatible robot could perform well controlled and reproducible sensorimotor tasks, while the subject's motor interactions with the robot are recorded by fMRI procedures and translated into brain images (Fig. 1). Therefore, fMRI-compatible robots can be applied with fMRI procedures to map brain functions [3, 4], investigate human motor control [5, 6], monitor rehabilitation induced cortical reorganization in neurological patients [7], etc. Such kind of fMRI-robotic systems could provide insights into the cortical reorganization mechanism after damage to the nervous system, offer a better understanding of therapy-induced recovery, and help to derive more efficient rehabilitation strategies.

To construct fMRI-compatible devices is rather challenging. First, the device must not disturb the scanner magnetic fields and ensure image quality. Second, the device should function properly in the scanner room. Third, the device is compact to fit into the scanner bore with the diameter of 55~70 cm [1, 3].

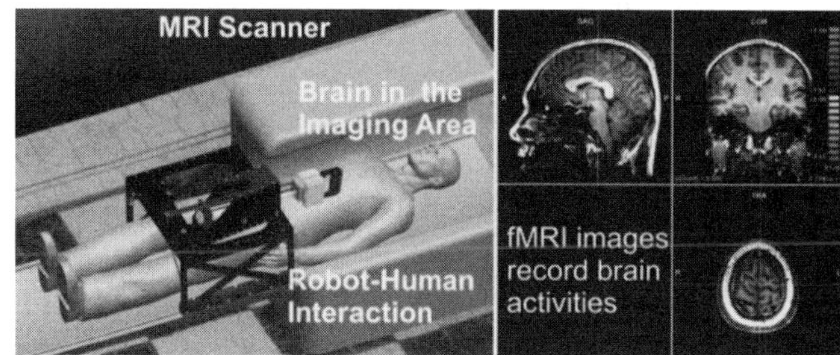

Fig. 1 fMRI-compatible robot working with fMRI procedures

The strong magnetic fields limit the choice of materials, sensors and actuators to be used in the MRI environment. Stiff polymer materials are a good alternative of magnetic metals for applications in the scanner environment. Sensors and actuators using strong electrical currents should also be avoided. Electrical components may be brought into the scanner environment if their electrical signals are of low frequency and low amplitude, and if the components are placed at a certain distance from the scanner and/or they are shielded [5, 6, 8]. Sensors with optical principles have been employed to measure position [6, 9], force / torque [6, 10, 11].

Typical fMRI-compatible actuation technologies are based on hydraulic or pneumatic principles, special electromagnetic principles, shape memory alloys, contractile polymers, piezoelectric actuation, materials with magnetostriction properties, electro-rheological fluids (ERFs) , or bowden cables [1, 2, 12, 13]. Among these working principles, fluidic actuations are promising solutions for fMRI-compatible robots that are intended to perform defined functional movement tasks, because 1) the fluids are magnetically inert in nature and the moving endeffector can be made fMRI-compatible, 2) the power can be generated distantly from the endeffector and sent to the endeffector inside the scanner via transmission hoses, 3) the actuators can provide large movement ranges and large forces, 4) the force-to-mass ratio is high, and 5) the transmission can be made flexible so that they can be placed adaptively to the work environment [2, 12].

In literature, many efforts have been made for the application of pneumatic actuation technologies to fMRI-compatible robotic systems [14] and devices [4, 15, 16]. Hydrostatic actuation was applied in master-slave setups to interact with human[9] or to position a forceps for surgery [17]. Reported problems were leakages, resulting in pollution, performance degeneration, and entrance of air bubbles. Furthermore, image deterioration occurred due to the high magnetic susceptibility of materials used for the systems [17, 18].

Traditional hydraulic or pneumatic actuation techniques cannot be directly transferred to fMRI-compatible applications. The fluid power generators, i.e., hydraulic or pneumatic compressors, consist of ferromagnetic materials. They must be placed outside of the scanner room for safety reason. Control valves are normally actuated by magnetically driven solenoids. Furthermore, valves and pressure sensors also contain ferromagnetic materials. Thus, they must be positioned far away from the scanner and the endeffector to avoid electromagnetic interferences causing malfunction and/or image artifacts. Therefore, long hoses have to be used to transmit the fluid power from the compressor to the control valves and then to the endeffector.

This arrangement results in several challenges for both construction and control. First, the endeffector must be made of fMRI-compatible materials so that it can work close to or inside the scanner bore. This can result in friction and stiffness problems at the fluidic cylinder, which is required to transfer fluidic pressure into force and motion. Second, valves and pressure sensors are distant from the endeffector, causing delay and measurement inaccuracies. Third, long hoses result in high inertia and compliance. Fourth, the system will interact with the user, so that the working pressure must be limited to ensure safety. Reduced pressure may also increase the compliance of the system. Finally, position and force sensors used inside the MRI scanner must be made MRI-compatible, which may reduce their signal quality. The mechatronic setup, including sensor, actuator and controller must be able to cope with these challenges and work in an, accurate, stable and robust way.

In this work two comparable robotic interface devices with hydraulic and pneumatic actuation respectively, were developed and implemented to control a translational one degree of freedom movement for fMRI studies. The interface devices are equipped with fMRI-compatible position and force sensors. Position and impedance/admittance controllers were realized to achieve active as well as passive subject movements, which are both required to investigate different fMRI-relevant motion tasks. The two systems were evaluated and compared with respect to control performance. Furthermore, both manipulandum systems were examined for MRI-compatibility in a 3 Tesla MRI scanner.

II. Design and Realization of The fMRI-Compatible Robotic Systems

A. Design Considerations and System Structure

According to the prospective clinical applications, the design considerations are defined as:

- fMRI-compatible;
- Haptic interaction force and position measurement;
- Movement range: 0…20 cm;
- Actuator velocity range: -15…15 cm/s;
- Actuator force range: -100…100 N;
- Subject passive movement: guide the subject's hand to follow a designed position curve;
- Subject active movement: simulate a virtual spring so that the subject can push or pull against the system.

The fluidic system structure was taken as in Fig. 2.

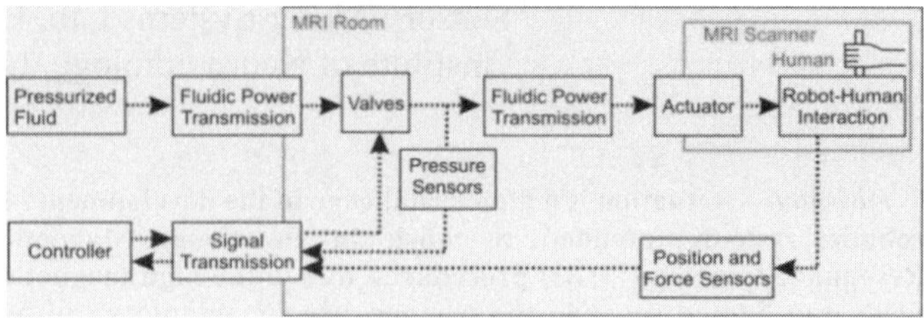

Fig. 2 Concept of fluidic-actuated robots to work with fMRI

B. Material Consideration and Measurement Principles

The materials put inside or close to the MRI scanner must have low magnetic susceptibility and low electric conductivity. Therefore, PET and PVC plastic were taken as the main construction material for frames and mechanical adapters. Nevertheless, metals have to be used for some parts required to be stiff, such as cylinders that will work under high pressure and force. Both cylinders were specially manufactured, with aluminum being the housing material. The piston of the pneumatic cylinder is made of PET, while that of the hydraulic cylinder is made of bronze to sustain the higher forces due to the significantly higher pressures. Both aluminum and bronze have low magnetic susceptibilities.

Table 1 Physical properties of several materials

Material	Magnetic susceptibility	Electrical conductivity [m/Ωmm^2, or 10^6s/m]
Bronze (CuSn8*)	-0.879×10^{-6}	7.5
Zinc	-15.7×10^{-6}	16.6
Brass	-8.63×10^{-6}	17
Aluminum	20.7×10^{-6}	36
Copper	-9.63×10^{-6}	57
Drinking water	-9.05×10^{-6}	$0.05\ldots5\times10^{-2}$
Nickel	600	14.4
Iron	200 000	9.9

*: values provided by the supply company.

Both manipulandum systems are equipped with one force and two position sensors. The force sensor consists of three optical fibers, one with emitting laser light and two with receiving laser light [19]. When a pull or push force is applied to the handbar, the emitting fiber is slightly displaced, thus, changing the light intensities in the two receiving fibers. The measured force is a function of the ratio of light intensities I_1 and I_2. Laser signals I_1 and I_2 are sent out via glass fibers, converted to voltage signal by the processing circuit, and then read into the control computer. An optical encoder measures the handbar position, and a potentiometer works as a redundant position sensor for safety consideration.

C. Fluidic Actuators

The oil used in hydraulic actuation is Orcon Hyd 32, which is accepted as a lubricant with incidental food contact. Hence, it is appropriate for biomedical applications.

The supply oil pressure from the compressor is 25bar. A directional valve regulates oil flow and, thus, controls the movement of the actuation cylinder. Two pressure sensors were mounted on the valve manifold. Oil is nearly incompressible and the actuation system is not back-drivable, i.e., the piston cannot be easily moved when the directional valve is closed.

For pneumatic actuation, the supply air pressure is 4bar. Both flow control and pressure control can be implemented. Pressure control is considered superior to flow control to overcome limitations of compressibility, friction and external disturbances [12]. In our application the manipulandum interacts with human subjects and the interaction force varies within a large range, so that we preferred pressure control. For each cylinder chamber, one valve regulates the pressure with the feedback from a pressure sensor.

The hydraulic and pneumatic transmission hoses between the control valves and the cylinders are 6m and 5m long, respectively. The valves were located at the corner of the scanner room, far from the scanner isocenter. The scanner magnetic field decreases rather quickly with increasing distance from the scanner bore and comes to be only 0.2 mT at the valve location [20] (For comparison, the magnetic field of the earth is about 0.06 mT). This special setup, different from conventional fluidic actuators, was taken to fulfill the fMRI-compatibility requirements.

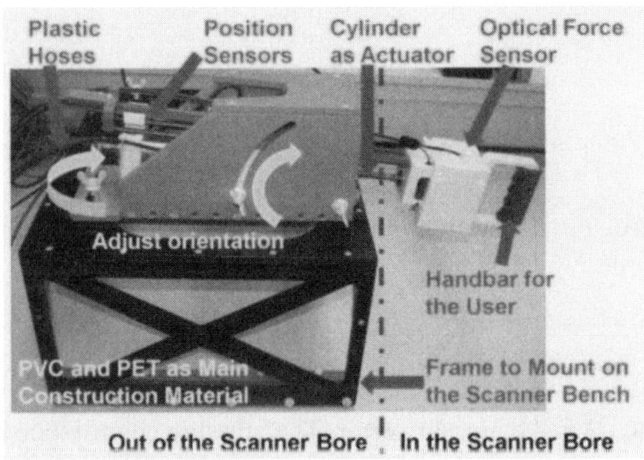

Fig. 3 Composition of the fMRI-compatible haptic interface

D. fMRI-Compatibility Test

The fMRI-Compatibility of the two robotic systems must be examined by fMRI experiments. The working position of robots has already been shown by *Fig. 1* and *Fig. 3*. The test consists of three parts:

1. No force or torque is induced by the magnetic fields and the robotic system when it is placed at the working position.
2. The robotic system functions properly as designed when it is placed at the working position. All components work properly, and the whole system can perform the desired movements.
3. The robotic system does not deteriorate fMRI image quality. A mineral water phantom is to be scanned in each of the following experimental conditions: 1) No device, in which the robotic system is not placed into the MRI room; 2) silent device, in which the robotic interface is at the working position but not in operation; 3) functioning device, in which the robotic system is at the working position and in operation.

Two methods are taken to evaluate whether artifacts have been introduced into the fMRI images [19]. The signal-to-noise ratio (SNR) in dB [21]

$$20\log_{10}\left(\frac{0.66\times\text{mean signal}}{\text{Average of noise region standard deviations}}\right)$$

quantitatively estimates whether additional noise has been introduced into fMRI procedures by the robot. We define the SNR variation threshold to be *5%*. A second method is image subtraction, which qualitatively checks whether image shift or deformation has occurred.

III. Controller Design

A. Hydraulic

Hydraulic oil compressibility is characterized by the bulk modulus K. Changes of pressures P_1 and P_2 in the cylinder chambers can be written as

$$\dot{P}_k=\frac{K}{V_k}(-\dot{V}_k+q_k),k=1,2 \qquad (1)$$

Here $V_1=V_{10}+xA_1$ and $V_2=V_{20}+(L-x)A_2$ are the total fluid volumes on two sides of the cylinder, L is the stroke of the cylinder, x is the position of the piston, V_{10} and V_{20} are the dead volumes, A_1 and A_2 are the cross sections of cylinder chambers, q_1 and q_2 are oil flows that are dependent on the chamber oil pressure, supply oil pressure or reservoir oil pressure, and also on the control signal u [22].

From equations (1), the piston velocity can be derived as:

$$\dot{x}=\frac{1}{A_1}q_1-x\frac{\dot{P}_1}{K}-\frac{V_{10}}{A_1}\frac{\dot{P}_1}{K}-\frac{1}{A_1}\dot{V}_{10}$$

or

$$\dot{x}=-\frac{1}{A_2}q_2+(L-x)\frac{\dot{P}_2}{K}+\frac{V_{20}}{A_2}\frac{\dot{P}_2}{K}+\frac{1}{A_2}\dot{V}_{20}$$

First, we consider the steady situation. Pressure changes and dead volume variations are ignored. In this case, $\dot{P}_1$, $\dot{P}_2$ and $\dot{V}_{10}$, $\dot{V}_{20}$ are all equal to zero. Thus, the velocity of the piston is fully determined by the oil flows q_1 and q_2:

$$\dot{x}=\frac{1}{A_1}q_1=-\frac{1}{A_2}q_2 \qquad (2)$$

When the piston moves at a constant speed, the pressures P_1 and P_2 are both constants, too. Thus, the oil flows $q_1$and q_2 only depend on the proportional valve. The control voltage to the proportional valve regulates the piston velocity, and this can be modeled as a lookup table. To deal with model uncertainties, external disturbances, and compliance from the hydraulic system, a velocity controller was designed which consists of a compliance compensation term and a proportional term.

In our hydraulic system, compliance comes from pressure variations $\dot{P}_1$, $\dot{P}_2$, long hose volumes V_{10}, V_{20} and their variations $\dot{V}_{10}$, $\dot{V}_{20}$. It can significantly affect the system performance. The long hoses are the main source of high compliance. Additionally, it can be observed by visual inspection that the hose volumes also change as the inside pressures change, but this change cannot be well detected. We design the compliance compensation component as:

$$\dot{x}_c = -\frac{c}{2}\left[-x\frac{\dot{P}_1}{K} - \frac{V_{10}}{A_1}\frac{\dot{P}_1}{K} + (L-x)\frac{\dot{P}_2}{K} + \frac{V_{20}}{A_2}\frac{\dot{P}_2}{K}\right] \quad (3)$$

Here, $c \in [0,1]$ determines to which extent the velocity is compensated. The model errors, external disturbances as well as uncompensated compliance components, were handled by the proportional control term. The proportional term was determined by experiments. The user force F_h affects pressures P_1 and P_2, and causes a shift in the voltage-velocity lookup table, which gets corrected by the velocity controller.

A PD position controller was designed to work in cascade with the velocity controller to guide the user's hand and track the given position trajectory (Fig. 4). It is not possible to realize impedance control on the hydraulic system, because it is not backdrivable due to the incompressibility of oil. However, the virtual spring for user active movements can be simulated by the following admittance control law (Fig. 4):

$$\dot{x} = \frac{1}{K_v}\left[F_h - K_x(x - x_0)\right] \quad (4)$$

Since the manipulandum moves in a low speed range, we can set K_v to be small such that the viscous term $K_v\dot{x}$ is relatively insignificant in the admittance relationship:

$$F_h - K_x(x - x_0) = K_v\dot{x} \approx 0,$$

and the hydraulic system behaves like a virtual spring with stiffness K_x. Here K_v was experimentally defined to be 2 N/(cm/s), and K_x is 3…30 N/cm. If K_x was set to be very small to simulate a soft spring, the term $K_x(x - x_0)$ goes close to $K_v\dot{x}$, and the viscous effect becomes obvious. With these parameters the system remained stable.

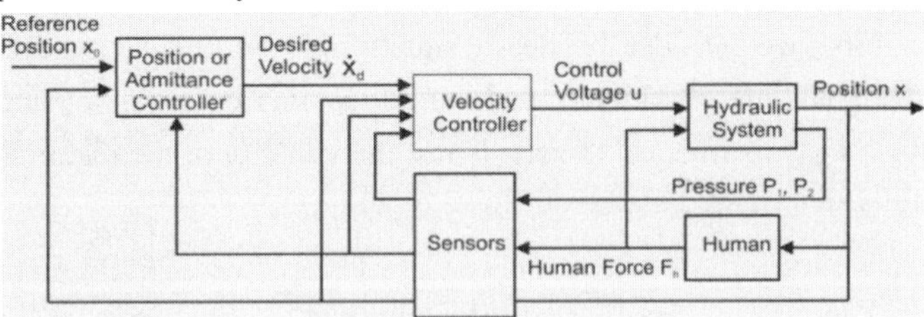

Fig. 4 Hydraulic system Control: position control can be achieved by a PD position controller in cascade with a velocity controller, and the virtual spring can be achieved by setting the virtual admittance as Eq. (4)

Table 2 Parameters for the hydraulic and pneumatic systems

Properties	Hydraulic	Pneumatic
Power generation		
Supply pressure *Ps*	15/25 bar*	4 bar*
Exhaust pressure *Pe*	0.4 bar*	0*
Fluid media		
Bulk modulus *K*	1.25×10^4 bar	Pressure *P*
Density *ρ*	856 kg/m^3	$P/Pe\times1.2$ kg/m^3
Kinetic viscosity *υ*	3.1×10^{-5} m^2/s	1.5×10^{-5} m^2/s
Double acting cylinder		
Cross section A_1	2.54 cm^2	9.62 cm^2
Cross section A_2	1.41 cm^2	7.85 cm^2
Stroke *L*	0.24 m	0.25 m
Work pressure limit	25 bar	6 bar
Transmission Hose		
Length L_t	6 m	5 m
Cross section A_t	0.317 cm^2	0.283 cm^2
Dynamics		
Force range	-194…356 N	-314…384 N
Velocity range	-0.2…0.3 m/s	-1.51…1.67 m/s

*Pressure value relative to environmental pressure 1.013bar

B. Pneumatic

Since the pressure sensor measures the cylinder pressure relative to the environmental pressure, we also use relative pressure. The force by the pneumatic cylinder is

$$F_c = P_1A_1 - P_2A_2.$$

Here, we regulate the pressures P_1 and P_2 in two cylinder chambers by two independent valves, and thus regulate the force produced by the cylinder.

Given the desired force F_d, the desired pressures P_{1d} and P_{2d} are calculated in the following way:

$$\begin{cases} P_{1d} = \frac{1}{A_1}(F_0 + \max(F_d, 0)) \\ P_{2d} = \frac{1}{A_2}(F_0 + \max(-F_d, 0)) \end{cases} \quad (5)$$

Here, F_0 is the minimum chamber force. A first order controller was designed for pressure control:

$$u_{1,2} = \frac{2}{\frac{1}{2\pi\times f}s + 1}e_P \quad (6)$$

And e_P is the pressure error. The pressure control loop is the inner-most loop of the pneumatic system for both position and impedance control. Then, we close the force control loop for force and impedance control, and close the position loop for position control.

A PD position controller with friction compensation worked in cascade with the force-pressure regulator to obtain user passive movement.

$$F_d = k_v(\dot{x}_0 - \dot{x}) + k_x(x_0 - x) - F_h - F_f \quad (7)$$

Due to manufacture and material properties, the friction force F_f depends not only on velocity, but also on position. The friction was modeled as the summation of velocity related and position related friction forces, and then compensated by force-pressure control. The user force was measured by the optical force sensor and got corrected afterwards.

Both admittance control and impedance control can be implemented on the pneumatic system [23, 24] to simulate the spring. Admittance control requires a good position/velocity controller that is robust against force disturbances, as the velocity controller in our hydraulic system. Here the position controller depends on the nested force-pressure regulator and suffers from the long distance between the valves, pressure sensors and the cylinder. Thus, admittance control is not the optimal option. Besides, pneumatic systems are natural impedances due to the compressibility of air, and impedance control can be realized directly by pressure regulation.

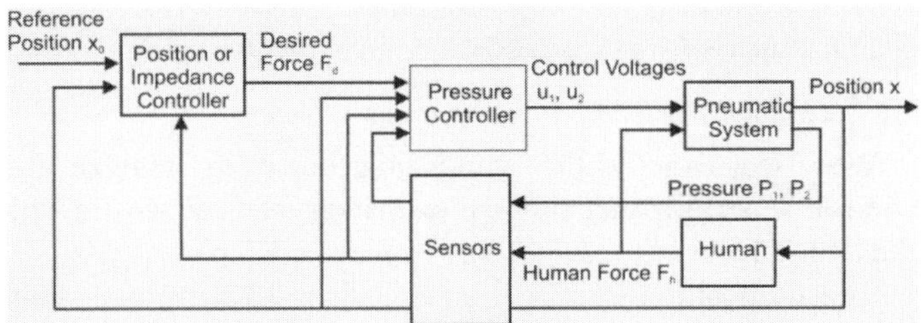

Fig. 5 Pneumatic position control can be achieved by a PD position controller in cascade with a pressure controller, and the virtual spring can be achieved by setting the virtual impedance as Eq. (8)

The impedance control law is quite straightforward

$$F_d = -K_x(x - x_0). \tag{8}$$

It calculates the desired force from the measured position and the specified stiffness, and then feed this signal to force-pressure regulation to achieve the desired force.

C. Control Software and Data Acquisition

The controllers were implemented in MATLAB Simulink and then compiled to a computer that runs an xPC target and communicates with the system by a data acquisition card (AD614, HUMUSOFT). The sampling frequency was 1 kHz.

IV. Results and Discussion

A. fMRI-Compatibility Evaluation

Both robotic systems were tested in a 3.0 T MRI system (Philips Medical Systems, Eindhoven, The Netherlands), with a phantom imaging object [19]. Their working position is not in the imaging area.

1. No force or torque was observed when the hydraulic or pneumatic system was placed at their working position.
2. The force and positions sensors worked properly. Both robotic systems functioned properly when they were placed at the working position, as will be shown in the next subsections.
3. In each of the three experimental conditions, the phantom body was imaged as 31 slices. 20 images were acquired for every slice of the phantom. Since slide 31 was closest to the robotic systems, the worst image deterioration would happen to this slice. Therefore, we checked slice 31 for SNR and images subtraction.

   It has been shown that high SNR values were obtained in all fMRI experiments (Table 3). After introduction of the hydraulic or pneumatic devices into the scanner environment, variations of SNR were all below 5%. The decrease of SNR values were minor and could be attributed to statistical errors.

   At slice 31, images from the fMRI experiments on hydraulic as well as pneumatic robotic systems were shown in Fig. 6, together with the subtraction results by the corresponding control image. No significant differences were observed.

The experimental results have verified that both robotic systems are fMRI-compatible.

Table 3 SNR in (dB) for slice 31: Mean (Standard Deviation)

Condition	Hydraulic*		Pneumatic*	
No Device	45.7(0.4)	N/A	47.7(0.4)	N/A
Silent Device	45.6(0.3)	-0.22%	47.5(0.5)	-0.42%
Functioning Device	45.5(0.5)	-0.44%	47.4(0.4)	-0.63%

*: Two experiments were not done at the same day

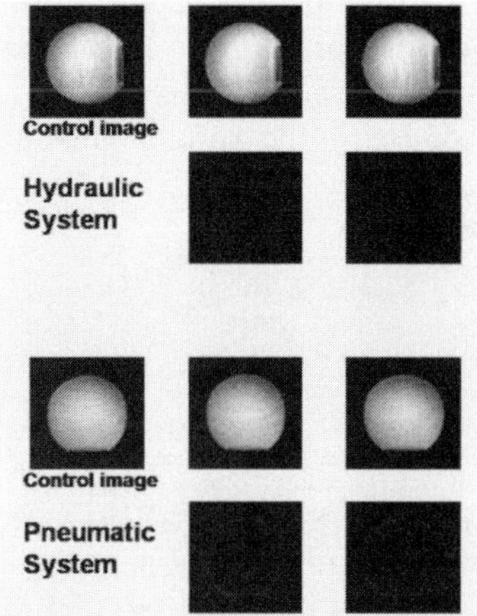

Fig. 6 Phantom images from the fMRI-compatibility test of the hydraulic (upper part) and pneumatic (lower part) robotic systems. For each part, the left image was obtained in the 'No device' condition and taken as the control image. The other two columns of images were obtained for the 'Silent device' and 'Functioning device' conditions, and their difference with the control image by direct subtraction.

B. Control Performance: Hydraulic System

To analyze the influence of working pressure on the dynamic performance, we tested the hydraulic system at two supply pressures of 15 bar and 25 bar, respectively. Here, 15 bar is the minimal working pressure for the hydraulic system to fulfill the defined velocity requirement, while 25 bar is the limit pressure for the hydraulic system to work safely.

The position control performance was first examined for step responses (Fig. 7). The reference step curve jumped twice from 5 cm to 15 cm and back, and then jumped twice from 5cm to 10 cm and back. When the hydraulic system worked at 15 bar, the steady position error was smaller than 0.06 cm, overshoot was smaller than 0.02 cm, and rise time was about 3.14 s. When the hydraulic system worked at 25 bar, the steady position error was still smaller than 0.06 cm, but the overshoot went up to 0.27 cm and the rise time decreased to 0.86 s.

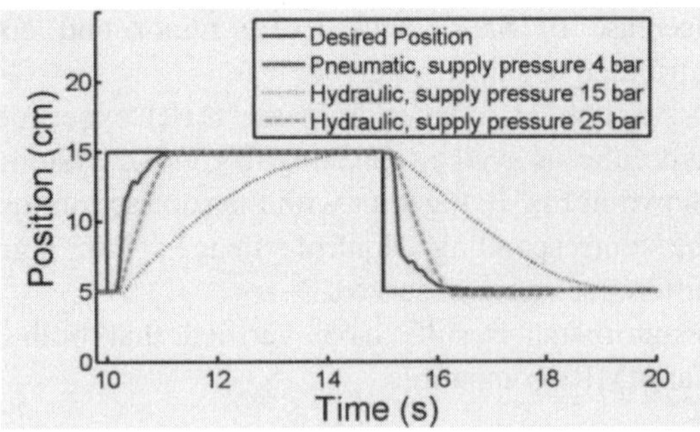

Fig. 7 Position control: step responses of the robotic systems

We then checked the position controlled hydraulic system for dynamic tracking performance. A so-called "chirp" signal from MATLAB Simulink was taken as the reference trajectory (*Fig. 8*). The signal was of sinusoidal shape, fixed amplitude of 10 cm, and offset 12 cm. The frequency of this signal linearly increased from 0 to 1 Hz as time went from 0 to 100 s. The actual position curve was recorded and compared with the reference "chirp" signal for bandwidth information (Fig. 8). The position bandwidth for the given signal was 0.48 Hz when the hydraulic system worked at 15 bar, and went up dramatically to 0.65 Hz for the working pressure of 25 bar.

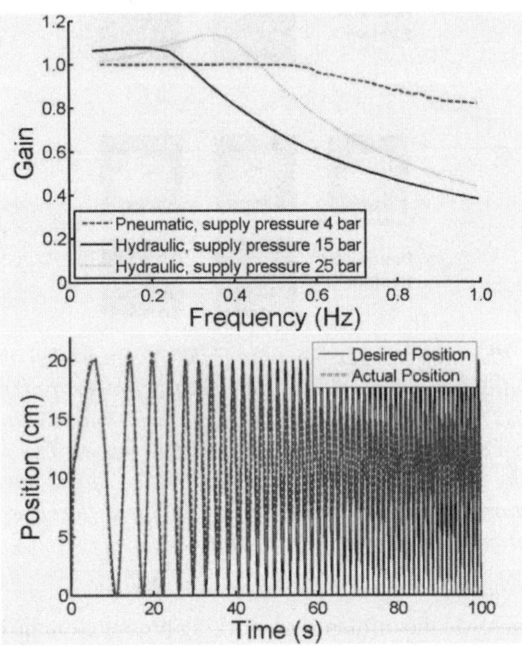

Fig. 8 Position control bandwidth of the robotic systems

User active movements were achieved by the simulated virtual spring. Fig. 9 shows an example spring of stiffness 5 N/cm when the hydraulic system worked at 15 bar and 25 bar of supply pressure. In the ideal case, the computed virtual force should equal to the user force. That is, $F_h = F_v = K_x(x - x_0)$. It can be seen from the plot that the virtual force curve coincided quite well with the user force curve at 25 bar working pressure, and was slightly postponed at 15 bar working pressure. When the spring constant is small to simulate a soft spring or the device moves fast, the neglected viscous term becomes significant and blurs the spring feeling. This resulted from the admittance control law we used.

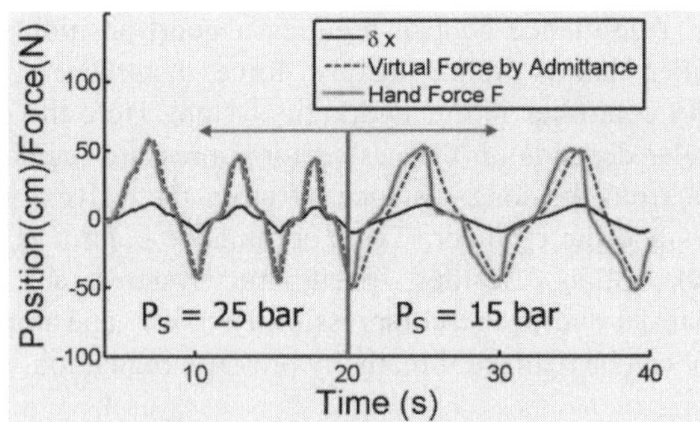

Fig. 9 Virtual spring simulation: an example

C. Control Performance: Pneumatic System

We used exactly the same procedures to analyze the controlled performance of the pneumatic system as we did with the hydraulic system. According to the step responses (Fig. 7), the steady position error was smaller than 0.25 cm, overshoot smaller than 0.01 cm, and the rise time was about 0.86 s. The position bandwidth for the given "chirp" signal was around 0.9 Hz, higher than the bandwidth of the hydraulic system working at 15 bar or 25 bar.

The simulated spring was achieved by controlling the cylinder force F_c to closely follow the desired virtual force by the impedance control law $-K_x(x - x_0)$. The spring effect was more natural and obvious than that of the hydraulic system.

D. Comparison of the Two fMRI-Compatible Systems

We summarize the characteristics of hydraulic and pneumatic actuation in Table 4.

Table 4 Comparison of hydraulic and pneumatic actuation for fMRI-compatible applications

Aspects	Hydraulic Actuation	Pneumatic Actuation
fMRI-Compatibility and Related Challenges		
Fluid media	Oil and air are both magnetically inert	
Cylinder	MRI-compatible materials such as Bronze, aluminum, plastic, etc.	
Hose length	≥ 5m – Active components (e.g., valves) are far from the scanner for fMRI-compatibility – This increases compliance of the system – Pressure sensors are far away from actuator, causing inaccuracies and time delay	
Fluid Power		
Power generation	Compressor	Compressor
Flow	Laminar	Laminar & turbulent
Working pressure	≥15bar*	≤6bar*
Force	Large	Medium
Working Mode		
Component	Directional valve	Pressure regulation valve
Control target	Flow control, regulate velocity and position	Pressure control, regulate force
Position control	High accuracy Low bandwidth	Medium accuracy Medium bandwidth
Velocity range	Small	Big
Friction or force disturbances	Robust	Sensitive
Back drivability	Not backdrivable	backdrivable
Others		
Leakage	Rare	Not a problem
Complexity&Cost	High	Medium
Maintenance	Medium	Simple

*: Relative to environment air pressure.

The design requirements have been fulfilled by both the hydraulic system and the pneumatic system with different working pressures. With the hydraulic system, we were able to achieve smoother movements, higher position control accuracy and improved robustness against force disturbances than with the pneumatic system. In contrast, the pneumatic system is backdrivable and shows better and faster force control performance. Furthermore, it is easier to maintain and has no serious consequences by leakages. In general, pneumatic actuation is more favorable for fast or force-controlled fMRI-compatible applications, whereas hydraulic actuation can be recommended for applications that require higher position accuracy and slow and smooth movements.

V. Conclusion

We conclude that both hydraulic and pneumatic actuation systems can be developed for fMRI-compatible applications. The fMRI-compatibility requirements can be fulfilled by special selection of materials and a nonconventional fluidic setup. The resulted limiting factors, such as long distance between cylinders and valves/pressure sensors, long transmission hoses as well as the usage of second quality fMRI-compatible components, increased control difficulties. Nevertheless, satisfactory control performances have been achieved by classical control strategies. Performances of the hydraulic and pneumatic actuation systems differ from each other due to the different physical properties of oil and air. The user has to decide, which system better fits the requirements of a specific application.

In future, stability and robustness of the system during robot-human interaction still need further study.

Acknowledgment

The authors would like to thank our colleagues Dr. Tobias Nef, Mr. Joachim von Zitzewitz, Mr. Severin Eisner and Mr. Andreas Brunschweiler for their support for this work. The authors would also thank Prof. Dr. Peter Bösiger, Dr. Roger Lüchinger and the MR center of ETH and University Zurich for providing the MRI scanner facility.

This work was supported in part by the Swiss National Science Foundation NCCR on Neural Plasticity and Repair, project P8 Rehabilitation Technology Matrix as well as ETH Research Grant TH-34 06-3 MR-robotics

References

[1] N. V. Tsekos, A. Khanicheh, E. Christoforou, and C. Mavroidis, "Magnetic Resonance-Compatible Robotic and Mechatronics Systems for Image-Guided Interventions and Rehabilitation: A Review Study," *Annu Rev Biomed Eng,* vol. 9, pp. 351-387, 2007.

[2] N. Yu and R. Riener, "Review on MR-compatible robotic systems," in *IEEE/RAS-EMBS International Conference on Biomedical Robotics and Biomechatronics* Pisa, Italy, 2006.

[3] G. S. Harrington, C. T. Wright, and J. H. Downs, 3rd, "A new vibrotactile stimulator for functional MRI," *Hum Brain Mapp,* vol. 10, pp. 140-5, Jul 2000.

[4] S. M. Golaszewskir, C. M. Siedentopf, E. Baldauf, F. Koppelstaetter, W. Eisner, J. Unterrainer, G. M. Guendisch, F. M. Mottaghy, and S. R. Felber, "Functional Magnetic Resonance Imaging of the Human Sensorimotor Cortex Using a Novel Vibrotactile Stimulator," in *Neuroscience Letters*. vol. 17, 2002, pp. 421-430.

[5] R. Gassert, L. Dovat, G. Ganesh, E. Burdet, H. Imamizu, T. Milner, and H. Bleuler, "Multi-joint arm movements to investigate motor control with fMRI," *Conf Proc IEEE Eng Med Biol Soc,* vol. 5, pp. 4488-91, 2005.

[6] A. Khanicheh, A. Muto, C. Triantafyllou, B. Weinberg, L. Astrakas, A. Tzika, and C. Mavroidis, "fMRI-compatible rehabilitation hand device," *J Neuroengineering Rehabil,* vol. 3, p. 24, 2006.

[7] R. Riener, T. Nef, and G. Colombo, "Robot-aided neurorehabilitation of the upper extremities," *Med Biol Eng Comput,* vol. 43, pp. 2-10, Jan 2005.

[8] M. Flueckiger, M. Bullo, D. Chapuis, R. A. G. R. Gassert, and Y. A. P. Y. Perriard, "fMRI compatible haptic interface actuated with traveling wave ultrasonic motor," in *Industry Applications Conference, 2005. Fourtieth IAS Annual Meeting. Conference Record of the 2005*, 2005, pp. 2075-2082 Vol. 3.

[9] R. Gassert, R. Moser, E. Burdet, and H. Bleuler, "MRI/fMRI-compatible robotic system with force feedback for interaction with human motion," *IEEE-ASME Transactions on Mechatronics,* vol. 11, pp. 216-224, Apr 2006.

[10] R. Riener, T. Villgrattner, R. Kleiser, R. Nef, and S. Kollias, "fMRI-Compatible Electromagnetic Haptic Interface," in *Proc. IEEE/EMBS Annual International Conference*, 2005.

[11] D. Chapuis, R. Gassert, L. Sache, E. Burdet, and H. Bleuler, "Design of a Simple MRI/fMRI Compatible Force/Torque Sensor," in *Proc. IEEE International Conference on Robotics and Intelligent Systems*, 2004.

[12] J. M. Hollerbach, I. W. Hunter, and J. Ballantyne, "A comparative analysis of actuator technologies for robotics," in *The robotics review 2*: MIT Press, 1992, pp. 299-342.

[13] R. Gassert, A. Yamamoto, D. Chapuis, L. Dovat, H. Bleuler, and E. Burdet, "Actuation methods for applications in MR environments," *Concepts in Magnetic Resonance Part B-Magnetic Resonance Engineering,* vol. 29B, pp. 191-209, Oct 2006.

[14] J. Diedrichsen, H. Yasmin, R. Tushar, and S. Reza, "Neural Correlates of Reach Errors," in *Journal of Neuroscience*. vol. 25, 2005, pp. 9919 -9931.

[15] R. W. Briggs, I. Dy-Liacco, M. P. Malcolm, H. Lee, K. K. Peck, K. S. Gopinath, N. C. Himes, D. A. Soltysik, P. Browne, and R. Tran-Son-Tay, "A pneumatic vibrotactile stimulation device for fMRI," in *Magnetic Resonance in Medicine*. vol. 51, 2004, pp. 640-643.

[16] A. C. Zappe, T. Maucher, K. Meier, and C. Scheiber, "Evaluation of a pneumatically driven tactile stimulator device for vision substitution during fMRI studies," in *Magnetic Resonance in Medicine*. vol. 51, 2004, pp. 828-834.

[17] D. Kim, E. Kobayashi, T. Dohi, and I. Sakuma, "A new, compact MR-compatible surgical manipulator for minimally invasive liver surgery," *Medical Image Computing and Computer-Assisted Intervention-Miccai 2002, Pt 1,* vol. 2488, pp. 99-106, 2002.

[18] R. Wang, T. Foniok, J. I. Wamsteeker, M. Qiao, B. Tomanek, R. A. Vivanco, and U. I. Tuor, "Transient blood pressure changes affect the functional magnetic resonance imaging detection of cerebral activation," *Neuroimage,* vol. 31, pp. 1-11, May 15 2006.

[19] N. Yu, W. Murr, A. Blickenstorfer, S. Kollias, and R. Riener, "An fMRI compatible haptic interface with pneumatic actuation " in *International Conference on Rehabilitation Robotics* Nordwijk, The Netherlands, 2007.

[20] Philips, "Technical Description: Intera 1.5T Release 2.5 series, Achieva 1.5T / 3.0T / XR Release 2.5, Panorama HFO Release 2.5 series," *Magnetic Resonance 4522 132 68821,* May 2007.

[21] E. Burdet, R. Gassert, G. Gowrishankar, D. Chapuis, and H. Bleuler, "fMRI Compatible Haptic Interfaces to Investigate Human Motor Control," in *Experimental Robotics IX*, 2006, pp. 25-34.

[22] G. A. Sohl and J. E. Bobrow, "Experiments and simulations on the nonlinear control of a hydraulic servosystem," *Control Systems Technology, IEEE Transactions on,* vol. 7, pp. 238-247, 1999.

[23] R. Richardson, M. Brown, B. Bhakta, and M. C. Levesley, "Design and control of a three degree of freedom pneumatic physiotherapy robot," *Robotica,* vol. 21, pp. 589-604, 2003.

[24] Z. Yong and E. J. Barth, "Impedance Control of a Pneumatic Actuator for Contact Tasks," in *Robotics and Automation, 2005. ICRA 2005. Proceedings of the 2005 IEEE International Conference on*, 2005, pp. 987-992.

Proofs and Experiments in Scalable, Near-Optimal Search by Multiple Robots

Geoffrey Hollinger and Sanjiv Singh
Robotics Institute
Carnegie Mellon University, Pittsburgh, PA 15213
Email: gholling@ri.cmu.edu, ssingh@ri.cmu.edu

Abstract— **In this paper, we examine the problem of locating a non-adversarial target using multiple robotic searchers. This problem is relevant to many applications in robotics including emergency response and aerial surveillance. Assuming a known environment, this problem becomes one of choosing searcher paths that are most likely to intersect with the path taken by the target. We refer to this as the Multi-robot Efficient Search Path Planning (MESPP) problem. Such path planning problems are NP-hard, and optimal solutions typically scale exponentially in the number of searchers. We present a finite-horizon path enumeration algorithm for solving the MESPP problem that utilizes sequential allocation to achieve linear scalability in the number of searchers. We show that solving the MESPP problem requires the maximization of a nondecreasing, submodular objective function, which directly leads to theoretical guarantees on paths generated by sequential allocation. We also demonstrate how our algorithm can run online to incorporate noisy measurements of the target's position during search. We verify the performance of our algorithm both in simulation and in experiments with a novel radio sensor capable of providing range through walls. Our results show that our linearly scalable MESPP algorithm generates searcher paths competitive with those generated by exponential algorithms.**

I. Introduction

The problem of gaining line-of-sight to a non-adversarial target is one that is relevant to many applications in robotics. Emergency response teams may need to locate a lost first responder or survivor in a disaster scenario. Military operations in urban environments may seek to locate a friendly or neutral target during rendezvous or evacuation. The growing availability of emergency response robots and mechanized infantry necessitates the development of algorithms that can solve these problems autonomously. In such scenarios, noisy information about the target's position may be available. In indoor environments, non-line-of-sight sensors like ultra-wideband ranging radios can provide range to a target through walls. A similar scenario arises during outdoor operations if an air vehicle provides surveillance while ground vehicles search for a target. The major application that has motivated our work is that of using a team of autonomous searchers to locate a lost first responder in a disaster scenario while receiving noisy measurements to his or her location [1].

In the scenarios mentioned above, the target is non-adversarial (i.e. not actively avoiding the searchers). In these cases, we can model the target's motion and plan searcher paths that maximize the probability of finding the target over a time interval. We refer to this problem as the Multi-robot Efficient Search Path Planning (MESPP) problem because robots must plan paths that *efficiently* find a non-adversarial target. This is in contrast with methods that seek to guarantee capture of an adversarial target. The MESPP problem is related to the Multi-robot Informative Path Planning (MIPP) problem in which robots must plan paths that best observe the environment [2]. These path planning problems maximize an objective (or reward) function that relates to how much information is gained by the robots' paths. Planning paths for multiple robots is an NP-hard problem in general because the number of searchable paths grows exponentially with the number of robots [3]. This motivates the development of approximation algorithms in these domains.

Solving the MESPP problem optimally requires planning in the joint space of searcher paths, which grows exponentially in the number of searchers. This is an example of *explicit coordination* during which the searchers explicitly plan for their teammates. Alternatively, if each searcher plans individually and shares information about its path for other searchers to consider, the search space no longer grows. We refer to this scenario as *implicit coordination*. Our algorithm uses sequential allocation, which is an instance of implicit coordination. During sequential allocation, searchers plan in an iterative fashion sharing their paths after planning and modifying their objective functions based on other searchers' shared paths. This algorithm scales linearly in the number of searchers. We present an algorithm using finite-horizon planning and sequential allocation that generates paths competitive with those generated by explicit coordination while using far fewer computational resources. Our algorithm also operates online allowing for the incorporation of measurements.

In this paper, we provide both theoretical and empirical results demonstrating the performance of finite-horizon planning with sequential allocation in the MESPP domain. We prove that MESPP requires the optimization of a nondecreasing, submodular objective function. An objective function is submodular if it follows an intuitive property of diminishing returns. The more areas (nodes) in the environment that the searchers have visited, the less incremental reward is gained. An objective function is nondecreasing if observing additional nodes can only increase the expected reward. The nondecreasing submodularity of the MESPP objective function directly leads to bounds on the performance of sequential allocation. We complement these theoretical results with empirical results both in simulation and using data from ultra-wideband ranging radio sensors.

The rest of this paper is organized as follows. We first put MESPP into context by discussing prior work in pursuit-evasion and coordinated search (Section II). We then formally define the MESPP problem and show that it requires the optimization of a nondecreasing, submodular objective function (Section III). We present a linearly scalable algorithm for solving MESPP using sequential allocation and finite-horizon path enumeration, and we derive theoretical bounds on the performance of this algorithm (Section IV). We further demonstrate the performance of our algorithm both in simulation and using data from novel ranging sensors (Section V). Finally, we summarize our results and discuss avenues for future work in efficient search and other domains (Section VI).

II. Related Work

Much research in coordinated search makes the adversarial assumption on the target's motion (i.e., assumes that the target is actively avoiding capture). This strong assumption on the target's motion necessitates algorithms that guarantee capture for worst-case target behavior. Algorithms for solving this problem both on graphs [4] and in polygonal environments [5] have been proposed. These methods do not directly estimate the target's position and consequently cannot properly handle uncertainty. They provide no mechanism for incorporating measurements or motion models during search. Thus, these methods provide worst-case solutions even if there is a high likelihood of a best-case scenario.

To better model uncertainty and target motion, an alternative is to use a purely probabilistic formulation of coordinated search problems. Partially Observable Markov Decision Processes (POMDPs) can be used to model both non-adversarial and adversarial coordinated search problems. Solving the POMDP formulation requires planning in the joint space of searcher paths. Roy et al. demonstrated how belief compression can be used to directly solve the POMDP formulation of the adversarial coordinated search problem [6]. Smith developed Heuristic Search Value Iteration (HSVI), a near-optimal POMDP solver capable of solving POMDPs with thousands of states [7]. While these solvers are capable of near-optimally solving small instances of MESPP, the exponential increase in the size of the joint planning space yields several million states for large instances of the MESPP problem. These instances are well outside the scope of even state-of-the-art POMDP solvers.

The algorithms described above plan in the joint space of searcher paths to provide explicit coordination between the searchers. This space grows exponentially with increasing searchers. One popular approach to mitigating the computational demands of joint planning is to hold synthetic auctions between agents. Auction-based methods are designed to inject joint planning into the problem where it is most needed. In other words, robots hold auctions to explicitly coordinate when it is most beneficial. Kalra proposed an auction-based method for sharing plans in domains requiring tight coordination [8]. In her method, robots auction a limited number of joint-space plans for consideration by other robots. Gerkey et al. also developed a parallel stochastic hill-climbing technique that dynamically forms teams, which is closely related to auction-based methods [9]. These techniques reduce the complexity of planning for problems such as MESPP. However, they rely on the overhead of auctions and/or team formation, which can consume considerable communication bandwidth and computation.

The poor scalability of joint planning methods has spurred research in heuristic methods for solving coordinated search problems. Sarmiento et al. proposed a heuristic method for finding stationary targets with multiple robots [10]. Their work is purely heuristic and does not derive optimality bounds on performance. Our proofs using submodularity complement prior work in heuristic search by providing theoretical guarantees.

In this paper, we show that the MESPP problem requires the optimization of a submodular objective function, and this key insight provides optimality bounds on sequential allocation, an algorithm linearly scalable in the number of searchers. Submodularity has been utilized in related domains to provide theoretical guarantees on sequential allocation. Guestrin et al. used submodular set functions to develop algorithms for sensor placement problems in Gaussian Processes [11] and in more general domains [12]. They also extended their algorithms to robust observation selection against sensor failure [13]. These applications deal primarily with placing sensors to monitor information in an environment (e.g. monitoring algae blooms in lakes and temperature in a building). These algorithms do not incorporate moving nodes (searchers) and thus are not suitable for ESPP. Singh et al. developed algorithms for solving the Multi-robot Informative Path Planning (MIPP) problem, which does allow for moving nodes [2]. We extend their theoretical bounds to the MESPP problem. To the best of our knowledge, our work is the first approximation algorithm with theoretical guarantees in the MESPP domain.

III. Problem Setup

This section formally defines the problem of locating a mobile, non adversarial target with multiple searchers (the MESPP problem). It also shows that the MESPP problem optimizes a submodular objective function.

To formulate the MESPP problem, we need to describe the environment in which the searchers and target are located. We first divide the environment into convex cells. The convexity of the cells guarantee that a searcher in a given cell will have line-of-sight to a target in the same cell. The searcher's goal is now to move into the same cell as the target. Gaining line-of-sight is relevant to most sensors that a mobile robot would carry including cameras and laser rangefinders. Our method for discretization takes advantage of the inherent characteristics of indoor environments. To discretize an indoor map by hand, label convex hallways and rooms as cells and arbitrarily collapse overlapping sections. Alternatively, a suitable discretization can be found automatically using a convex region finding algorithm (such as Quine-McClusky [14]). Taking into account the cell adjacency in a discretized map yields an undirected graph that the searchers can traverse. For the rest of this paper we assume that the graph has N convex cells. Figure 1 shows three example floorplans used our experiments. We use the two large floorplans for simulated trials and the smaller floorplan for experiments with ranging radios. The museum floorplan is particularly challenging because it contains many cycles by which the target can avoid line-of-sight contact with the searchers.

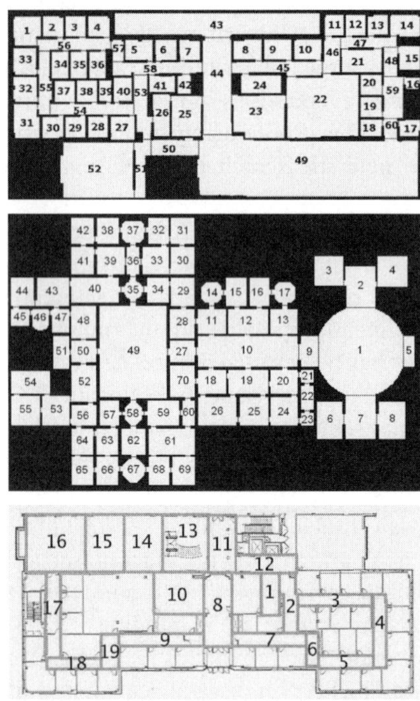

Fig. 1. Example floorplans of environments used for efficient search trials. The larger maps (top and middle) were used for simulated testing. The smaller map (bottom) was used for "hybrid" trials in which simulated searchers found a Pioneer robot with measurements from an experiment.

Let $G(N, E)$ be the undirected environment graph with vertices N and edges E. At any time t, a searcher exists on vertex $s(t) \in N$. The searcher's movement is deterministically controlled, and it may travel to vertex $s(t+1)$ if there exists an edge between $s(t)$ and $s(t+1)$. A target also exists on this graph on vertex $e(t) \in N$. The target moves probabilistically between vertexes. The searcher receives reward by moving onto the same vertex as the target, $s(t) = e(t)$, and no reward is gained after this occurs. The reward is discounted by γ^t so that the searcher receives more reward for finding the target at a lower t. This discount factor corresponds to the probability that the search will end at a given time. For instance, the target may leave the search area or expire. This necessitates locating the target in a short time.

The target's movement model is known to the searcher, and it is independent of the searcher's position on the graph. In this paper, we assume that the target's motion model is Markovian (i.e., it depends solely on its current cell). This assumption allows for a rich space of motion models including those followed by randomly moving and stationary targets. The searcher knows its own position and it has knowledge of the target's position at a time t in the form a belief distribution over all vertices, $b_N(t)$. Since $b_N(t)$ can be an arbitrary distribution, this formulation allows multi-modal estimates of the target's position. Call the problem so far the Efficient Search Path Planning (ESPP) problem. To extend to MESPP, place more than one searcher on vertices $s_k(t) \in N$. The searchers now gain reward if any of them are on the same vertex as the target. Incorporating additional searchers forces both the state and action space to grow exponentially. Both ESPP and MESPP can be formulated as a Partially Observable Markov Decision Process (POMDP) with deterministic actions, but we do not give the full formulation here due to space constraints.

This formulation shows that the MESPP requires the optimization of the objective function in Equation 1. Searchers choose a feasible set of paths that maximize the expected probability of intersecting the target's path at the earliest possible time (before reward is heavily discounted). To see why, expand the graph G into a time augmented graph G'. Each node N' now represents a cell in the environment at a discrete time point. The reward function can now be seen as the expected intersection between the searchers' paths and the target's path. Let $A \subset N'$ be a feasible set of searcher paths in G', Ψ be the space of all possible target paths, $P(Y)$ be the probability of the target taking path Y, and $F_Y(A)$ be the discounted reward received by path A if the target chooses path Y. Theorem 1 shows that the resulting objective function is nondecreasing and submodular.

Theorem 1: Equation 1 is a nondecreasing, submodular set function. This is the objective function optimized by the MESPP.

$$F(A) = \sum_{Y \in \Psi} P(Y) F_Y(A) \tag{1}$$

The proof of Theorem 1 along with a formal definition of submodularity is given in the Appendix. This result is used in the next section to prove optimality bounds on sequential allocation in this domain. Since the Markov assumption is made on the target's motion model, calculating the expectation in Equation 1 can be done using diffusion matrices rather than enumerating paths. This greatly simplifies the computation of $F(A)$ over the space of searcher paths. This is explained in more detail below.

IV. Algorithm Description

In this section, we describe our algorithm for non-adversarial coordinated search with multiple robots utilizing sequential allocation and finite-horizon planning. We show theoretical bounds by taking advantage of the nondecreasing submodularity of the MESPP objective function. This somewhat surprising result shows that sequential allocation, an algorithm linearly scalable in the number of searchers, generates near-optimal paths in the MESPP domain.

A. Sequential Allocation

As noted above, the joint space of searcher paths grows exponentially in the number of searchers. Explicitly coordinating by planning in this space quickly leads to intractably large planning spaces with multiple searchers. As an alternative, we propose that searchers choose their paths sequentially. Algorithm 1 gives pseudocode for our sequential allocation algorithm in the MESPP domain. The algorithm maintains a list of nodes $V \subset N'$ that have been visited by the searchers. Note that N' is the time-augmented version of the nodes N in the environment, which allows for revisiting areas in the environment more than once. The searchers choose paths A_k that maximize the objective function $F(V \cup A_k)$ and then add the nodes they have visited to V. Effectively, subsequent searchers treat the paths of previous searchers as "given", and they are not allowed to change them. Sharing nodes to update V is an instance of implicit coordination as described above. Since the search space does not grow with the number of searchers, the complexity of sequential allocation is linear in the number of searchers. It is important to note that while path planning occurs sequentially, the execution of paths is simultaneous.

Algorithm 1 Sequential allocation MESPP algorithm

```
Input: Multi-agent efficient search problem
% V ⊂ N' is the set of nodes visited by searchers
V ← ∅
for all searchers k do
    % A_k ⊂ N' is a feasible path for searcher k
    % Finding this arg max solves the ESPP for searcher k
    A_k ← arg max_{A_k} F(V ∪ A_k)
    V ← V ∪ A_k
end for
Return A_k for all searchers k
```

Algorithm 1 requires maximizing the objective function for the ESPP problem as a subroutine. Any algorithm for solving the ESPP problem can be inserted here. However, if the ESPP solver is bounded, the nondecreasing submodularity of the objective function leads to bounds on the performance of sequential allocation. Theorem 2 from previous work shows that sequential allocation leads to theoretical guarantees in the informative path planning domain.

Theorem 2: From Singh et al. [2]: Let κ be the approximation guarantee for the single path instance of the informative path planning problem for any nondecreasing, submodular function. Then sequential allocation achieves an approximation guarantee of $(1 + \kappa)$ for the multi-robot informative path planning problem.

The findings in Theorem 1 can be leveraged to extend these results to MESPP. Corollary 1 states that if an ESPP solver has an approximation guarantee of κ, then sequential allocation on the MESPP will yield an approximation guarantee of $(1+\kappa)$.

Corollary 1: If a solver achieves an approximation guarantee of κ for the ESPP problem, sequential allocation yields an approximation guarantee of $(1+\kappa)$ for the Multi-robot ESPP (MESPP) problem.

Proof: The proof of Corollary 1 is immediate from Theorem 2 and Theorem 1. Theorem 2 states that sequential allocation achieves this bound for any single-agent path planning problem optimizing a nondecreasing, submodular function. Theorem 1 shows that the ESPP problem requires the optimization of such an objective function. ■

Here an approximation guarantee κ states that if the ESPP solver returns a path $A \subset N'$, then $F(A) \geq \frac{1}{\kappa}F(A^{OPT})$, where A^{OPT} is the set of nodes visited by an optimal path. Clearly, $\kappa \geq 1$ since $F(A)$ cannot be greater than the optimal reward. The case where $\kappa = 1$ corresponds to solving the ESPP problem optimally. In this case, sequential allocation can achieve no worse than half ($\kappa = 2$) the optimal reward. This theoretical result allows single-agent performance bounds to be extended to the multi-agent case using a linearly scalable algorithm, albeit with a loss in approximation quality. The next section presents a bounded algorithm for solving the ESPP problem using finite-horizon path enumeration.

B. Finite-Horizon Planning

In large environments, even the single-agent ESPP may be intractable to solve optimally (or even near-optimally) due to the computational overhead of considering many infinite-horizon paths. In these cases, one option is for the searchers to plan a finite number of cells ahead and choose the best path to that horizon. Any time while traversing this path, the searcher can plan again utilizing new information on a new horizon. This leads to an online solution to MESPP, and it allows for the incorporation of measurements of the target's position as they become available. Algorithm 2 gives pseudocode for solving the ESPP problem using finite-horizon path enumeration. Because the finite-horizon method relies on path enumeration to solve ESPP, it scales exponentially with the search depth: $O(b^d)$, where b is the maximum branching factor of the search graph, and d is the search depth in cells.

Algorithm 2 Finite-horizon path enumeration for ESPP

```
Input: Single-agent efficient search problem
for All paths A to horizon d do
    Calculate F(A)
end for
Return A ← arg max_A F(A)
```

Lemma 1 derives optimality bounds for finite-horizon path enumeration, and the result extends to the multi-robot case with sequential allocation as in Theorem 3.

Lemma 1: Finite-horizon path enumeration on the ESPP problem achieves a lower bound of:

$$F(A^{FH}) \geq F(A^{OPT}) - \epsilon, \quad (2)$$

where A^{FH} is the path returned by finite-horizon path enumeration, A^{OPT} is the optimal feasible path, and $\epsilon = R\gamma^{d+1}$.

Proof: Finite-horizon path enumeration achieves the optimal reward inside the horizon depth for ESPP because it checks all paths. The maximum reward that could be gained outside the horizon depth is given by $\epsilon = R\gamma^{d+1}$, where R is the reward received for locating the target, γ is the discount factor, and d is the search depth. The bound is immediate. ■

Theorem 3: Finite-horizon path enumeration with sequential allocation on the K-robot MESPP problem achieves a lower bound of:

$$F(A_1^{FH}, \ldots, A_K^{FH}) \geq \frac{F(A_1^{OPT}, \ldots, A_K^{OPT}) - \epsilon}{2} \quad (3)$$

Proof:

The maximum reward outside the horizon remains the same as in Lemma 1 (i.e., ϵ is unchanged). Since the single robot case achieves the optimal reward within the finite-horizon (i.e., $\kappa = 1$), sequential allocation yields an approximation guarantee as in Corollary 1 as $\kappa + 1 = 2$. ∎

Intuitively, as search depth increases, the bound tightens. Additionally, decreasing the discount factor tightens the bound. This is because lesser discount factors more heavily weight reward gained earlier, which is more likely to be within the finite-horizon. It is important to note that the quality of the bound is independent of the number of searchers. This is a worst-case bound for arbitrary starting distributions and motion models. In practice, searchers can run finite-horizon path enumeration repeatedly on a receding horizon, which leads to performance that far exceeds this lower bound.

C. Measurement Incorporation

If non-line-of-sight measurements of the target's location are available during ESPP, searchers can utilize them to assist in search. The online capabilities of finite-horizon search allow for measurements to be easily incorporated. We use a Bayesian update method to calculate a new belief of the target's position given a measurement.

Denote the probability that the target is in cell i at time t as p_t^i. To incorporate measurements, we recursively estimate the posterior distribution $p_t^i = P(e_t = i|z_1 \ldots z_t)$ for all i, where $z_1 \ldots z_t$ are the measurements received thus far.[1] We are also given a known motion model, which provides $P(e_t = i|e_{t-1} = j)$ for all cells i and j. Once we calculate the posterior, we can renormalize and define a new belief distribution on the target's state. We use a method for modifying the target's belief distribution that calculates the likelihood of a measurement using a finely discretized grid. Using standard Bayesian recursion [15], the posterior can be rewritten as in Equation 4.

$$p_t^i = \eta P(z_t|e_t = i) \sum_j P(e_t = i|e_{t-1} = j) p_{t-1}^j, \quad (4)$$

where η is a normalizing constant.

Assuming a known motion model, this reduces the problem of calculating the posterior to that of calculating a likelihood term $P(z_t|e_t = i)$. Since each cell is represented as a continuous set of points in the map plane, this calculation is difficult. To reduce the complexity of the problem, further discretize each cell into small subcells and calculate a likelihood at the center of each subcell. We denote the M^i subcells of cell i as m^{ij} for all $j \in \{1, \ldots, M^i\}$. The calculation of $P(z_t|c_t = i)$ is now one of calculating a likelihood at many points and then taking the sum of these likelihoods.

For range measurements, we calculate $P(z_t|e_t = i)$ by determining the expected range value for the center of each subcell. Let q^{ij} be the Euclidean distance from the ranging sensor to subcell m^{ij}, and let r_t be the received range measurement with assumed Gaussian noise variance σ^2. The likelihood is then calculated as in Equation 5.

$$P(z_t|e_t = i) = \sum_{j=1}^{M^i} N(r_t; q^{ij}, \sigma^2) \quad (5)$$

After receiving a new measurement, searchers can use sequential allocation and finite-horizon path enumeration to replan using the new belief distribution on the target's location. Combining sequential allocation, finite-horizon path enumeration, and Bayesian measurement updating yields a scalable and online algorithm for solving the MESPP problem. Since searchers replan after receiving measurments, online measurement incorporation heuristically improves path quality but does not affect theoretical guarantees.

[1]Note: we denote the target's location $e(t)$ as e_t for this section to simplify notation.

V. Experimental Results

A. Simulated Results

To test our MESPP algorithm, we ran simulated trials on a multi-agent coordinated search simulation in C++ on a 3.2 GHz Pentium 4 processor. Our simulation allows for multiple searchers and both stationary and moving targets. We assumed that the average speed of the target is 1 m/s, and that it moves holonomically between cell boundaries. The searchers also move with a maximum speed of 1 m/s, which would be a reasonable speed for state-of-the-art autonomous vehicles. The searchers start in the same location for all trials, and the location of the target is initialized at random on the map. We ran simulated experiments in the museum (150 $m \times 100$ m) and office (100 $m \times 50$ m) environments shown in Figure 1. In all tests, the performance metric is the average reward received over many trials. For a given trial, reward received is calculated as $R(t_c) = R\gamma^{t_c}$, where R is the reward for locating the target, γ is the discount factor, and t_c is the time at which the target was found. We arbitrarily set $R = 1$ and $\gamma = 0.95$ for all experimental trials.

Our results in Figure 2 compare finite-horizon path enumeration (horizon depth five) to the infinite horizon POMDP solution for a single searcher. With a single searcher, the POMDP formulation of ESPP is still solvable using Heuristic Search Value Iteration (HSVI2) [7]. Our results show that, for the single searcher case, finite-horizon path enumeration yields average rewards competitive with those generated by the HSVI POMDP solution. We attempted running HSVI with two searchers, but these trials were unsuccessful because the exponentially increased state-action space would not fit in memory. This demonstrates the poor scalability of the POMDP formulation of the MESPP problem.

Since solving the POMDP formulation is intractable for multiple searchers, we introduce for comparison a finite-horizon explicit coordination algorithm that is identical to Algorithm 2 except that it enumerates paths for all searchers in the joint space. This scales $O(b^{dK})$, where b is the branching factor, d is the search depth, and K is the number of searchers. Figure 3 gives a comparison of reward received by sequential allocation and this explicit coordination algorithm. Since explicit coordination grows intractable at large lookahead depths, a depth of two was used for comparison. Figure 3 also shows a lower bound for sequential allocation calculated from Corollary 1 using the explicit coordination results. This bound is the lowest reward that sequential allocation could achieve if explicit coordination yielded the optimal reward. On both maps, sequential allocation greatly outperforms its lower bound.

These simulated experiments demonstrate that implicit coordination with sequential allocation yields results nearly equivalent to those achieved through explicit coordination. In sharp contrast with explicit coordination's exponential scalability, sequential allocation is linearly scalable in the number of searchers. Figure 4 demonstrates the scalability of sequential allocation by showing reward received with up to five searchers in the museum and office.

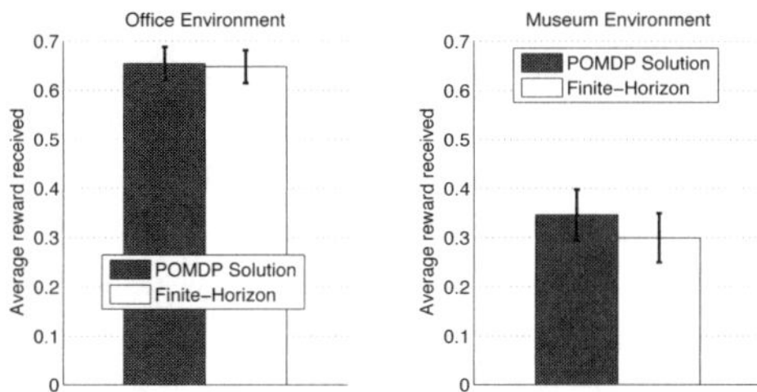

Fig. 2. Comparison of finite-horizon path enumeration (lookahead depth five) versus the POMDP solution for a single searcher in two complex simulated environments. Error bars are one standard error of the mean (SEM), and averages are over 200 trials. Target and searchers move at a maximum speed of 1 m/s.

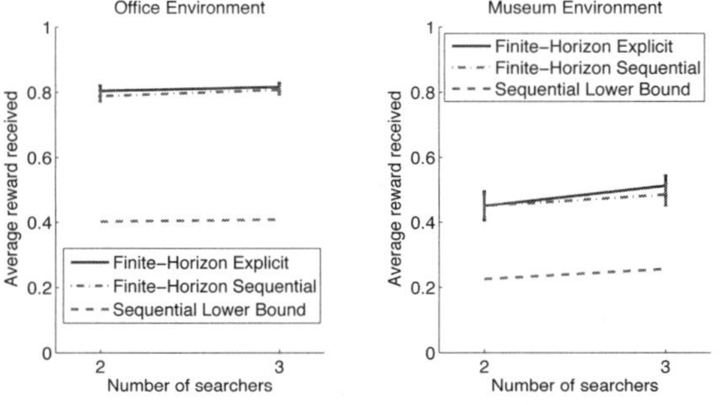

Fig. 3. Comparison of sequential allocation versus explicit coordination in two complex simulated environments. Finite-horizon path enumeration with lookahead depth two was used for both methods. Error bars are one standard error of the mean (SEM), and averages are over 200 trials. Target and searchers move at a maximum speed of 1 m/s. Sequential allocation greatly outperforms its lower bound.

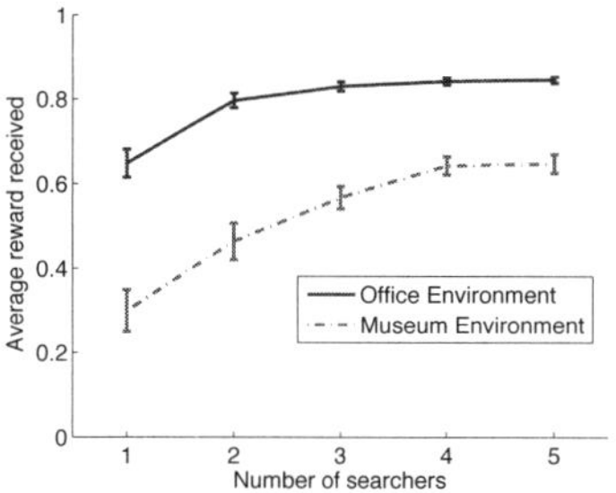

Fig. 4. Multiple searcher scalability trials for finite-horizon path enumeration (lookahead five) and sequential allocation in simulated environments. Error bars are one standard error of the mean (SEM), and averages are over 200 trials. Target and searchers move at a maximum speed of 1 m/s.

B. Ranging Radio Measurements

One major application of MESPP is that of finding lost first responders in disaster scenarios. To better model this scenario, we set up an urban response test environment using a Pioneer robot and five Multispectral ranging radio nodes [16]. These sensors use time-of-flight of ultra-wideband signals to provide inter-node ranging measurements through walls. They have an effective range of approximately 30 m indoors and provide accuracy approximately within $1 - 2$ m. In our experiments, the Pioneer robot acted as a lost first responder and was teleoperated around the environment carrying a ranging radio node. Four stationary nodes were placed in surveyed locations around the environment to provide range to the Pioneer. The Pioneer also carried a SICK laser rangefinder, and its location was found using laser AMCL-SLAM methods from the Carmen software package [15]. The Pioneer's laser localization was used for ground truth but was not used to assist in search. The Pioneer's maximum speed was set to 0.3 m/s, the maximum that provided consistent laser localization. Figure 5 shows a photograph of the Pioneer robot as well as the office environment used for testing.

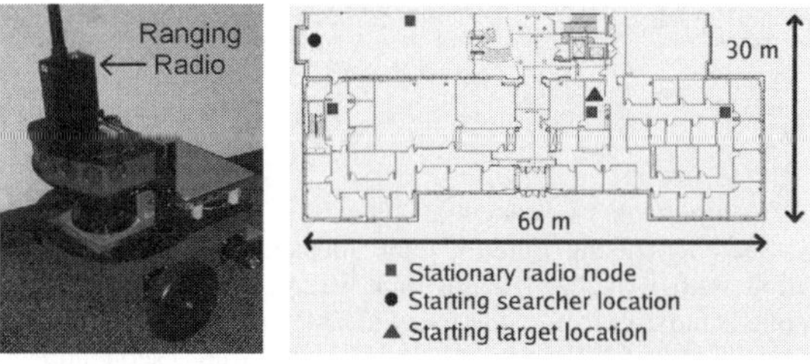

Fig. 5. Photograph of Multispectral ultra-wideband ranging radio mounted on Pioneer robot (left) and floorplan of testing environment (right). The robot was teleoperated around the environment to act as the moving target.

After gathering data from the ultra-wideband ranging sensors, simulated searchers were added to the environment. These searchers have access to the ranging measurements from the stationary nodes in the environment, which allows them to utilize real range data from the experiment to find the target in the simulated world. The searchers were given a maximum speed of 0.3 m/s to match that of the Pioneer target. Figure 6 shows the results for a single searcher in these "hybrid" trials. As in the purely simulated trials, the finite-horizon path enumeration method provides nearly equivalent reward as the POMDP solution. Figure 7 shows results with two searchers using sequential allocation. As above, sequential allocation is competitive with explicit coordination.

Since the experiments are run in playback, we can vary the number of sensors used by turning off some sensors' data streams. Figure 8 shows average rewards using an increasing number of ranging radio nodes. The zero node case corresponds to search without measurements. The results show that adding more searchers leads to decreasing capture times. Increasing the number of measurement beacons also leads to decreasing capture times. These results suggest that if a small number of searchers are available, this can be compensated with more measurement beacons, and vice versa.

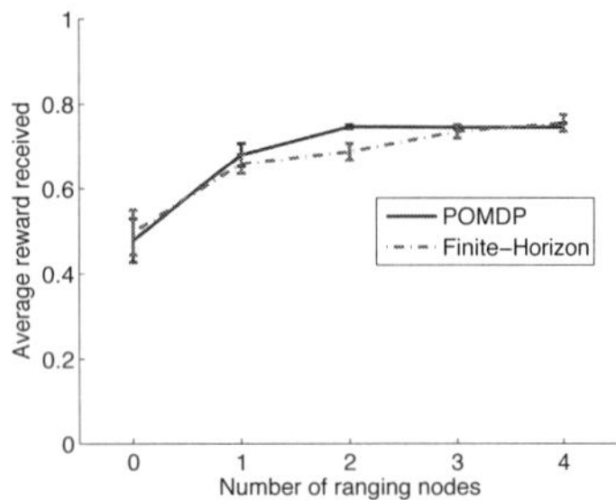

Fig. 6. Comparison of finite-horizon path enumeration (lookahead depth five) versus the POMDP solution for a single searcher using ultra-wideband ranging radio measurements from experimental trials. Error bars are one standard error of the mean (SEM), and the searcher moves at a maximum speed of $0.3\ m/s$.

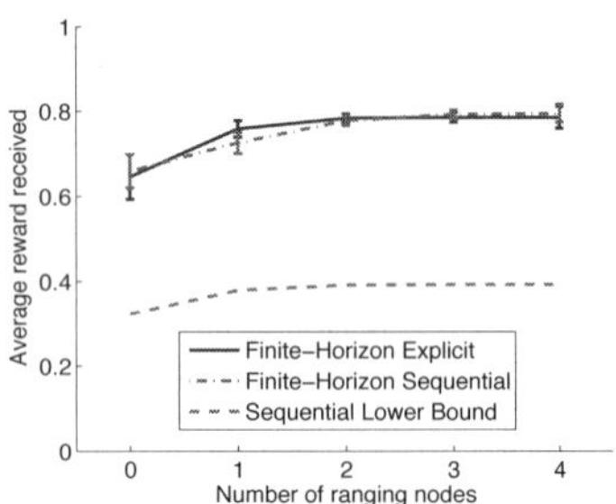

Fig. 7. Comparison of sequential allocation versus explicit coordination with two searchers using ranging radio measurements. Finite-horizon path enumeration with lookahead depth four was used for both methods. Error bars are one standard error of the mean (SEM). Target and searchers move at a maximum speed of $0.3\ m/s$. As in simulated trials, sequential allocation greatly outperforms its lower bound

VI. Conclusion and Future Work

This paper has presented a scalable algorithm for solving the Multi-robot Efficient Search Path Planning (MESPP) problem of locating a non-adversarial target using multiple robotics searchers. We have defined the MESPP problem and shown how it can be modeled using a Partially Observable Markov Decision Process (POMDP). We have also shown that current POMDP solvers are incapable of handling large instances of MESPP. Our proposed algorithm uses sequential allocation

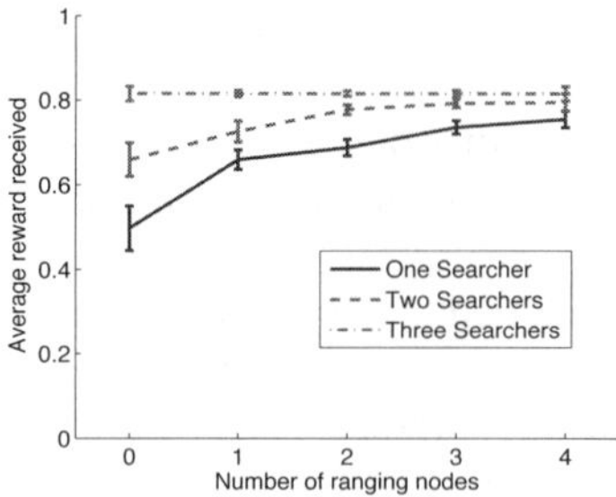

Fig. 8. Multiple searcher scalability trials for finite-horizon path enumeration (lookahead five) and sequential allocation with ranging radio measurements. Error bars are one standard error of the mean (SEM). Target and searchers move at a maximum speed of $0.3\ m/s$. The zero node case corresponds to the absence of measurements.

and finite-horizon path enumeration to remain computationally tractable for multiple searchers in large environments. We have given a rigorous theoretical analysis that shows the near-optimality of sequential allocation in this domain by exploiting the nondecreasing submodularity of the MESPP objective function. Sequential allocation is an instance of *implicit coordination* during which multiple robots share information rather than planning in the joint path space. Implicit coordination is linearly scalable, and it remains tractable in large problem instances when exponential methods using explicit coordination are far beyond computational limits. Our simulated and experimental results using ultra-wideband ranging radios show the performance of our algorithm in complex environments.

One extension is to apply our algorithm to the case where actions of searchers are no longer fully deterministic. For instance, a searcher may have a fifty percent chance of failing to move because of rubble blocking the way. The POMDP formulation of MESPP can easily express this scenario. The solution to this POMDP would no longer be a deterministic searcher path, but it would instead be a distribution over paths. Even though submodular set analysis does not directly apply, the resulting objective function on distributions over paths may still show qualities related to submodularity leading to theoretical guarantees for sequential allocation.

Throughout this paper, we have made the assumption that the target's motion model is known to the searchers. If instead a set of candidate motion models are known, an algorithm like SATURATE [13] can be used to generate a multi-searcher policy that performs well against any of the models. This extension would allow for both stationary and moving target models to be considered simultaneously.

Finally, we have assumed that the target's motion is non-adversarial. If the target may be actively avoiding the searchers, the properties of submodularity no longer hold. In such scenarios, an algorithm with both average-case and worst-case guarantees would be desired. To the best of our knowledge, the development of such a search algorithm is still an open problem. However, methods utilizing implicit coordination show great promise in providing near-optimal solutions even in such challenging multi-robot domains.

Acknowledgment

The authors are grateful to Andreas Krause, Athanasios Kehagias, and Joseph Djugash for their insightful comments. This work is funded by NSF Grant No. IIS-0426945.

References

[1] V. Kumar, D. Rus, and S. Singh, "Robot and sensor networks for first responders," *Pervasive Computing*, pp. 24–33, 2004.

[2] A. Singh, A. Krause, C. Guestrin, W. Kaiser, and M. Batalin, "Efficient planning of informative paths for multiple robots," in *Proc. Int'l Joint Conf. on Artificial Intelligence*, 2007.

[3] A. Blum, S. Chawla, D. Karger, T. Lane, A. Meyerson, and M. Minkoff, "Approximation algorithms for orienteering and discounted-reward tsp," *SIAM Journal on Computing*, vol. 37, no. 2, pp. 653–670, 2007.

[4] T. Parsons, "Pursuit-evasion in a graph," in *Theory and Applications of Graphs*, Y. Alavi and D. Lick, Eds. Springer, 1976, pp. 426–441.

[5] L. Guibas, J. Latombe, S. LaValle, D. Lin, and R. Motwani, "Visibility-based pursuit-evasion in a polygonal environment," *Int'l Journal of Comp. Geometry and Applications*, vol. 9, no. 5, pp. 471–494, 1999.

[6] N. Roy, G. Gordon, and S. Thrun, "Finding approximate pomdp solutions through belief compression," vol. 23, pp. 1–40, 2005.

[7] T. Smith, "Probabilistic planning for robotic exploration," Ph.D. dissertation, Carnegie Mellon University, 2007.

[8] N. Kalra, "A market-based framework for tightly-coupled planned coordination in multirobot teams," Ph.D. dissertation, Carnegie Mellon University, 2006.

[9] B. Gerkey, S. Thrun, and G. Gordon, "Parallel stochastic hill-climbing with small teams," in *Proc. 3rd Int'l NRL Workshop on Multi-Robot Systems*, 2005.

[10] A. Sarmiento, R. Murrieta-Cid, and S. Hutchinson, "A multi-robot strategy for rapidly searching a polygonal environment," in *Proc. 9th Ibero-American Conference on Artificial Intelligence*, 2004.

[11] C. Guestrin, A. Krause, and A. Singh, "Near-optimal sensor placements in gaussian processes," in *Proc. Int'l Conf. on Machine Learning*, 2005.

[12] A. Krause and C. Guestrin, "Near-optimal observation selection using submodular functions," in *Proc. 22nd Conf. on AI*, 2007.

[13] A. Krause, B. McMahan, C. Guestrin, and A. Gupta, "Selecting observations against adversarial objectives," in *Proc. Neural Information Processing Systems*, 2007.

[14] J. S. Singh and M. D. Wagh, "Robot path planning using intersecting convex shapes: Analysis and simulation," *IEEE Journal of Robotics and Automation*, vol. RA-3, no. 2, April 1987.

[15] S. Thrun, W. Burgard, and D. Fox, *Probabilistic Robotics*. Cambridge, MA: MIT Press, 2005.

[16] "Multispectral, company website: http://www.multispectral.com/," 2008.

Appendix: Proof of Theorem 1

Proof: Consider a time-augmented version of graph G, $G'(N',E')$. Level 1 of G' represents the vertexes of G at time 1, and directed edges from level 1 connect to reachable vertexes at time 2. Extend this graph down to a max time τ. The graph now contains vertices $N' = N \times T$ where $T = \{1,\ldots,\tau\}$. This result extends to the infinite case by making τ arbitrarily large. $\mathbf{P}(N')$ denotes the powerset of N', i.e. the set of (time stamped) node subsets. A function $F : \mathbf{P}(N') \to \Re_0^+$ is called *nondecreasing* iff

$$A \subseteq B \Rightarrow F(A) \leq F(B).$$

It is called *submodular* iff

$$A \subseteq B \Rightarrow F(A \cup C) - F(A) \geq F(B \cup C) - F(B).$$

(In the above $A, B \in \mathbf{P}(N')$ and $C = \{(m,t)\} \subseteq N'$ – i.e. C is a singleton.)

In the following, for a given $Y \subseteq N'$ and any $A \subseteq N'$ we define $t_A = \min\{t : (m,t) \in A \cap Y\}$, $F_Y(A) = \gamma^{t_A}$, with the understanding that $\gamma \in (0,1)$, $\min \emptyset = \infty$, and $\gamma^{\infty} = 0$.

We now show that for every $Y \subseteq N'$ the function $F_Y(A)$ is nondecreasing and submodular. Take an arbitrary Y and fix it for the proof. Take any $A, B \subseteq N'$ and any $C = \{(m_0,t_0)\} \subseteq N'$. We have:
$t_A = \min\{t : (m,t) \in A \cap Y\}$,
$t_B = \min\{t : (m,t) \in B \cap Y\}$,
$t_C = \min\{t : (m,t) \in C \cap Y\}$.
$A \subseteq B \Rightarrow \{t : (m,t) \in A \cap Y\} \subseteq \{t : (m,t) \in B \cap Y\} \Rightarrow$
$t_A \geq t_B \Rightarrow F_Y(A) = \gamma^{t_A} \leq \gamma^{t_B} = F_Y(B)$.

Hence $F_Y(\cdot)$ is nondecreasing.

Now, regarding submodularity, note that, since C is a singleton, we have two cases: either $C \cap Y \neq \emptyset$ and so $t_c = t_0 < \infty$; or $C \cap Y = \emptyset$ and so $t_c = \infty$. We examine the two cases separately.

Case I, $t_C < \infty$. In this case we have three subcases.

1) $t_B \leq t_A \leq t_C$. Then $F_Y(B \cup C) = \gamma^{t_B}$, $F_Y(B) = \gamma^{t_B}$, $F_Y(A \cup C) = \gamma^{t_A}$, $F_Y(A) = \gamma^{t_A}$ and $F_Y(A \cup C) - F_Y(A) = \gamma^{t_A} - \gamma^{t_A} = 0 = \gamma^{t_B} - \gamma^{t_B} = F_Y(B \cup C) - F_Y(B)$.
2) $t_B \leq t_C \leq t_A$. Then $F_Y(B \cup C) = \gamma^{t_B}$, $F_Y(B) = \gamma^{t_B}$, $F_Y(A \cup C) = \gamma^{t_C}$, $F_Y(A) = \gamma^{t_A}$ and $F_Y(A \cup C) - F_Y(A) = \gamma^{t_C} - \gamma^{t_A} > 0 = \gamma^{t_B} - \gamma^{t_B} = F_Y(B \cup C) - F_Y(B)$.
3) $t_C \leq t_B \leq t_A$. Then $F_Y(B \cup C) = \gamma^{t_C}$, $F_Y(B) = \gamma^{t_B}$, $F_Y(A \cup C) = \gamma^{t_C}$, $F_Y(A) = \gamma^{t_A}$ and $F_Y(A \cup C) - F_Y(A) = \gamma^{t_C} - \gamma^{t_A} \geq \gamma^{t_C} - \gamma^{t_B} = F_Y(B \cup C) - F_Y(B)$, since $t_A \geq t_B \Rightarrow \gamma^{t_A} \leq \gamma^{t_B} \Rightarrow -\gamma^{t_A} \geq -\gamma^{t_B}$.

Case II, $t_C = \infty$. Then we have a single subcase: $t_B \leq t_A \leq t_C$ from which follows $F_Y(A \cup C) - F_Y(A) = 0 = F_Y(B \cup C) - F_Y(B)$ as already seen.

In every case the submodularity inequality holds.

The one searcher reward function $F(A)$ (where A is the searcher's path) is defined by

$$F(A) = \sum_{Y \in \Psi} P(Y) F_Y(A),$$

where the summation is over all possible target paths Ψ, and $P(Y)$ is the probability of the target taking path Y. The K-searcher reward function $F(A_1, ..., A_K)$ (where A_k is the path of the k-th searcher) is defined by

$$F(A_1 \cup \ldots \cup A_K) = \sum_{Y \in \Psi} P(Y) F_Y(A_1 \cup \ldots \cup A_K).$$

We now show that for any $K = 1, 2, ...$ and every $Y \subseteq N'$ the function $F(A_1 \cup \ldots \cup A_K)$ is nondecreasing and submodular. Nondecreasing submodularity is closed under nonnegative linear combinations (and hence expectations). Here $F(A_1 \cup \ldots \cup A_K)$ is the expected value of $F_Y(A)$ where $A = A_1 \cup \ldots \cup A_K$, $F_Y(\cdot)$ is a nondecreasing, submodular function, and the expectation is taken over all possible target paths.

■

Planning Long Dynamically-Feasible Maneuvers for Autonomous Vehicles

Maxim Likhachev
School of Computer Science
University of Pennsylvania
Philadelphia, PA
maximl@seas.upenn.edu

Dave Ferguson
Intel Research Pittsburgh
4720 Forbes Ave
Pittsburgh, PA
dave.ferguson@intel.com

Abstract— **In this paper, we present an algorithm for generating complex dynamically-feasible maneuvers for autonomous vehicles traveling at high speeds over large distances. Our approach is based on performing anytime incremental search on a multi-resolution, dynamically-feasible lattice state space. The resulting planner provides real-time performance and guarantees on and control of the suboptimality of its solution. We provide theoretical properties and experimental results from an implementation on an autonomous passenger vehicle that competed in, and won, the Urban Challenge competition.**

I. INTRODUCTION

Autonomous vehicles navigating through cluttered, unstructured environments or parking in parking lots often need to perform complex maneuvers and reason over large distances. Furthermore, this reasoning usually needs to be performed very quickly so that the resulting maneuvers can be executed in a timely manner, particularly if the environment is inhabited, dynamic, or dangerous. In particular, our current focus is planning for autonomous urban driving including both off-road scenarios and large unstructured parking lots such as the ones in front of malls and large stores (on the order of 200×200 meters). Maneuvering at human driving speeds ($\backsim$ 15 mph) through such areas requires very efficient planning, especially if they contain static obstacles or other moving vehicles.

Roboticists have concentrated on the problem of mobile robot navigation for several decades, providing a large body of research. Early approaches concentrated on *local* planning, where very short term reasoning is performed to generate the next action for the vehicle. These include potential field-based techniques, where obstacles exert repulsive forces on the vehicle while the goal exerts an attractive force [1], and the curvature velocity [2] and dynamic window [3] approaches, where planning is performed in control space to generate dynamically-feasible actions. One major limitation of these purely local approaches was their capacity to get the vehicle stuck in local minima en route to the goal (for instance, cul-de-sacs). Further, these approaches are unable to perform complex multi-stage maneuvers, such as three-point turns, as these maneuvers are not within the set of local actions considered by the planner.

To reduce the susceptibility to local minima of these approaches, algorithms were developed that incorporated *global* as well as local information [4, 5, 6, 7]. Typically, these approaches generate a set of candidate simple local actions and evaluate each based on both their local traversability cost and the desirability of their endpoints based on a global value function (e.g. the expected distance to the goal based on known obstacle information). Although these approaches perform better with respect to local minima, their simple local planning can still cause the vehicle to get stuck or take highly suboptimal paths. Subsequent approaches have focused on improving this local planning by using more sophisticated local action sets that better follow the global value function [8, 9], and by generating sequences of actions to perform more complex local maneuvers [10, 11, 12]. The most complex of these approaches are able to perform very precise local maneuvering but are limited by the mismatch between their powerful local planning and their approximate global planning, resulting once more in a susceptibility to local minima.

Recognizing this mismatch, other researchers have concentrated on improving the quality of global planning, so that a global path can be easily tracked by the vehicle [13, 14, 15, 16, 17]. However, the computational expense of generating complex global plans over large distances has remained very challenging, and these approaches are restricted to either small distances, fairly simple environments, or highly suboptimal solutions.

In this paper, we present an efficient, global planning approach that attempts to overcome these challenges. First, we employ a multi-resolution lattice search space to reduce the complexity of the global search while still providing extremely high-quality solutions. Second, we use an efficient anytime, incremental search to quickly generate bounded suboptimal solutions, then improve these solutions while deliberation time allows and repair them when new information is received. The resulting approach is able to plan complex, dynamically-feasible maneuvers over hundreds of meters and improve and repair them in real-time for vehicles traveling at high ($\backsim$ 15 mph) speeds.

We first describe the key ideas and components of our approach, then provide key theoretical properties and results from both simulation and the Urban Challenge competition.

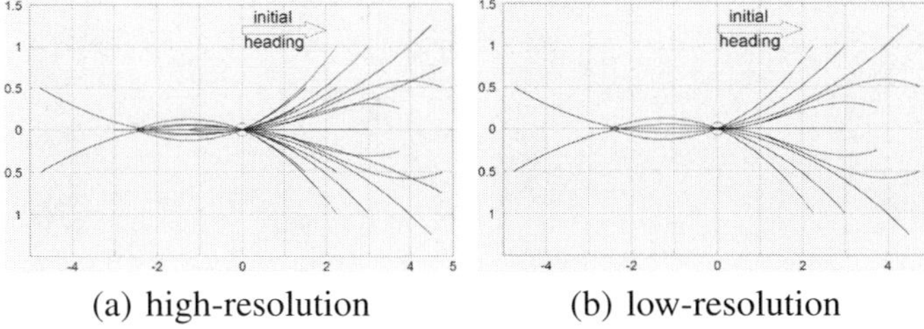

(a) high-resolution (b) low-resolution

Fig. 1. High- and low-resolution action spaces.

II. Multi-resolution Lattice State-space

A state lattice [18] is a discretization of the configuration space into a set of states, representing configurations, and connections between these states, where every connection represents a feasible path. As such, lattices provide a method for motion planning problems to be formulated as graph searches. However, in contrast to many graph-based representations (such as 4-connected or 8-connected grids), the feasibility requirement of lattice connections guarantees that any solutions found using a lattice will also be feasible. This makes them very well suited to planning for non-holonomic and highly-constrained robotic systems, such as passenger vehicles.

State-space. The two key considerations in constructing a lattice are the discretization (or sampling) strategy used for representing the states in the lattice, and the action space (or control set) used for the inter-state connections. For our application we employ a four dimensional (x, y, θ, v) state representation, where (x, y) represent the position of the center of the vehicle in the world, θ represents the orientation of the vehicle, and v represents its translational velocity. The (x, y, θ) coordinates are important for computing the validity of the poses of the vehicle in the world and making sure that no path in the lattice requires an instantaneous change in the orientation of the vehicle. For the velocity v we use two possible values: maximum forwards velocity and maximum reverse velocity. We take velocity into account because the time involved in switching between forward and backward directions is substantial so reasoning about this cost is important for generating fast, smooth paths[1].

Action Space. The action space for each state in the lattice is intended to be dense enough that every possible feasible path through the lattice can be constructed by combining sequences of these actions. However, because this action space represents the branching factor of the subsequent graph search, in practice it must be carefully constructed to provide flexibility in path selection while maintaining computational tractability.

The offline construction of our action space is based on work by Pivtoraiko and Kelly [18] that attempts to create near-minimal spanning action spaces. Given a state s, we compute the action space by first calculating a subset of states within a distance d of s that are reachable via some feasible action. To generate the feasible actions we use a trajectory generation algorithm originally developed by Howard and Kelly [9]. This algorithm employs an accurate vehicle model to produce feasible, directly-executable actions and an optimization technique to minimize the endpoint error of these actions with respect to a desired endpoint state. We use this approach to 'snap' the actions to the lattice so that the endpoint of each action lands on a lattice state. Next, we look at this set of actions and calculate whether any single action can be approximately recomposed out of a combination of other, shorter actions. If so, these longer actions are discarded from our set. This provides us with a compact set of actions that approximate the full reachable space. However, in contrast to the approach in [18], we maintain multiple straight segments of varying lengths to improve the speed of the subsequent search, as we will discuss in Section III-A. Figure 1(a) illustrates the action space for a single state (oriented to the right) in our lattice.

Multi-resolution Lattice. Even with a compact action space, planning long complex maneuvers over lattices can be expensive in terms of both computation and memory. An important observation, however, is that usually, there exists a wide spectrum of smooth, dynamically-feasible paths between the vehicle and goal configurations and it is waste of time and memory to explore all of them. On the other hand, all of these paths start and end at the exact same configurations, and the challenge is in finding a path that satisfies the current vehicle configuration and the specific goal configuration precisely.

This motivated us to take a novel, multi-resolution approach, where we use a high-resolution action space in the vicinity of the robot and the goal, and a low-resolution action space elsewhere. We call the resulting combination a multi-resolution lattice. With this approach, we can harness most of the benefit of the high-resolution representation without paying anything near the full computational cost. The trick is making sure that the high-resolution and low-resolution lattices connect together smoothly.

Our multi-resolution approach maintains the same dimensionality (x, y, θ, v) for both resolutions, but the action space for the low-resolution lattice is a strict subset of the action space for the high-resolution lattice. Figure 1(a) shows the action space used in the high-resolution lattice and Figure 1(b) shows the action space used in the low-resolution lattice[2]. Using this method ensures that the low-resolution lattice is utilized fully and that paths in the multi-resolution lattice are guaranteed to be feasible, which is a strong advantage over existing combined local and global approaches for navigation.

Theorem 1: Every path in a lattice that uses only a low-resolution action space is also a valid path in our multi-resolution lattice. Further, every path in the multi-resolution lattice is a valid path in a lattice that uses only the high-resolution action space.

Proof. The proof of the first claim follows trivially from the fact that any action in the low-resolution lattice is a valid

[1]We do not reason about curvature (the orientation of wheels) because we found this to be less critical for the speeds we are interested in traveling at, as discussed in the results section.

[2]In practice, choosing the appropriate set can be achieved with a basic check: if the (x, y) location of a state is not within some distance d of the vehicle or goal, its action set is the low-resolution set.

action in both the low-resolution and high-resolution lattices, and therefore is a valid action in the multi-resolution lattice. A similar argument applies for the second claim. ■

Enforcing the low-resolution action space to be a subset of the high-resolution action space decreases the *branching factor* of the graph constructed by the search, which is certainly important, but it does not necessarily decrease the *size* of the graph. However, it is also possible to decrease the size of the graph as follows. Suppose A_h is an action space used in the high-resolution space, and A_l is an action space used in the low-resolution space. Thus, $A_l \subset A_h$. Then, we can construct A_l by picking only the actions from A_h that end at states with a coarser discretization than the end states of actions in A_h. For example, we can choose for A_l only those actions whose end states have θ equal to one of 16 possible angles, while actions in A_h can connect states with 32 possible values of θ. (This is precisely what we used in our system.) Mathematically, the construction of the action space A_l can be expressed as follows in terms of a high-resolution discretization Q_h and a lower-resolution discretization Q_l of variables x, y, θ, v: an action a connecting states $s_1 = (x_1, y_1, \theta_1, v_1)$ and $s_2 = (x_2, y_2, \theta_2, v_2)$ belongs to A_l if and only if $a \in A_h$ and $(x_2, y_2, \theta_2, v_2) \in Q_l$.

Restricting Q_l to a coarser discretization for (x, y) or θ corresponds to using a discretization that adapts based on the vehicle and goal configurations. This technique can also be used to explicitly constrain the behavior of the vehicle in the different areas. For instance, restricting Q_l to contain only positive v-values prevents the vehicle from moving backward when far from the initial and goal configurations. This general approach allows for an arbitrarily-reduced state and action space in the low-resolution portion of the lattice, and can also be trivially extended to more than two levels of resolution if desired.

III. Anytime, Incremental Search

Given a search space (in our case, in the form of a multi-resolution lattice) and a cost function associated with each action, we need an efficient method for searching through this space for a solution path. A* search is perhaps one of the most popular methods for doing this [19]. It utilises a heuristic to focus the search towards the most promising areas of the search space. While highly efficient, A* aims to find an optimal path which may not be feasible given time constraints and the size of environments autonomous vehicles need to operate in. To cope with very limited deliberation time, anytime variants of A* search have been developed [20, 15]. These algorithms generate an initial, possibly highly-suboptimal solution very quickly and then concentrate on improving this solution while deliberation time allows. Furthermore, these anytime algorithms are able to provide bounds on the suboptimality of the solution at any point of time during the search.

A* and its anytime variants work best when the search space, and thus environment, is mostly known a priori. In robotic path planning this is rarely the case, and the robot typically receives updated environmental information through onboard and/or offboard sensors during execution. To cope with imperfect initial information and dynamic environments, efficient incremental variants of A* search have been developed that update previous solutions based on new information (e.g. from sensors) [21, 22, 23]. These algorithms repair existing solutions for a fraction of the computation required to generate such solutions from scratch.

When faced with limited deliberation time *and* imperfectly-known or dynamic environments, it is extremely useful to have a search algorithm that is both anytime and incremental. The Anytime Dynamic A* algorithm developed by Likhachev et al. is a version of A* search that combines these two properties into a single approach and has been shown to be very effective for a range of robotic planning tasks [16]. We employ this algorithm for planning and re-planning paths in our multi-resolution lattice.

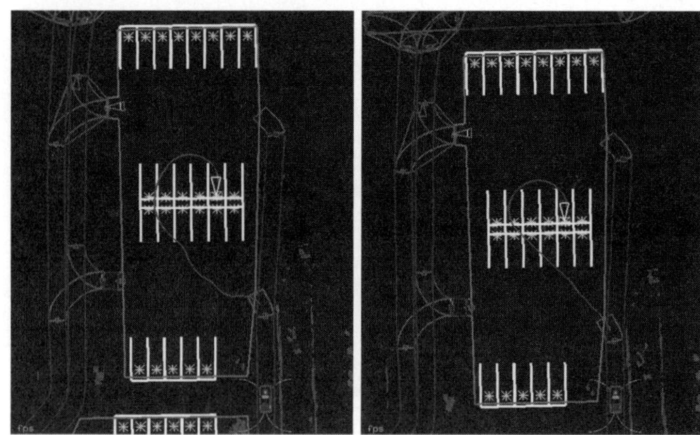
Fig. 2. Pre-planning a path into a parking spot and improving this path in an anytime fashion.

A. Anytime Dynamic A*

Anytime Dynamic A* (AD*) exploits a property of A* that can result in much faster generation of solutions, namely that if consistent heuristics are used and multiplied by an inflation factor $\epsilon > 1$, then A* can often generate a solution much faster than if no inflation factor is used [24], and the cost of the solution generated by A* will be at most ϵ times the cost of an optimal solution [25]. AD* operates by performing a series of these inflated A* searches with decreasing inflation factors, where each search reuses information from previous searches. By doing so, it is able to provide suboptimality bounds on all solutions generated and allows for control of these bounds, since the user can decide how much the inflation factor is decreased between searches. To cope with updated information, AD* also borrows ideas from the D* and D* Lite algorithms [21, 22] and only propagates updated information through the affected and relevant (given the current search) portions of the search space.

To enable efficient anytime planning and replanning as the vehicle moves, we use AD* to search backwards from the goal configuration towards the current configuration of the vehicle. The heuristic used thus needs to estimate the cost of a shortest path from the vehicle configuration (rather than goal) to each state in question.

The effectiveness of Anytime Dynamic A* is highly dependent on its use of an informed heuristic to focus its search. An accurate heuristic can reduce the time and memory required to generate a solution by orders of magnitude, while a poor heuristic can diminish the benefits of the algorithm. It is thus important to devote careful consideration to the heuristic used for a given search space. Further, because we are inflating heuristic values, it is useful to have long actions that can skip over several nodes and reduce the number of states in the search. It is for this reason we add several straight line actions of varying length in both the forwards and backwards directions to our action set (see Section II).

B. Informative Heuristics

The purpose of a heuristic is to improve the efficiency of the search by guiding it in promising directions. A common approach for constructing a heuristic is to use the results from a simplified search problem (e.g. from a lower-dimensional search problem where some of the original constraints have been relaxed). In selecting appropriate heuristics, it is important to analyze the original search problem and determine the key factors contributing to its complexity. In robotic path planning these are typically the complexity inherited from the constraints of the mechanism and the complexity inherited from the nature of the environment.

To cope with the complexity inherited from the mechanism constraints, a very useful general heuristic is the cost of an optimal solution through the search space assuming a completely empty environment. This can be computed offline and stored as a heuristic lookup table, and several efficiencies can be used to reduce the required memory for this table [17]. This is a very well informed heuristic for operating in sparse environments and is guaranteed to be an optimistic (or admissible) approximation of the actual path cost.

To cope with the complexity inherited from the nature of the environment, it is not practical to pre compute heuristic values for all possible environment configurations, as there are an effectively infinite number of possibilities for any reasonably-sized environment. However, in this case it is beneficial to solve online a simplified search problem given the actual environment and use the result of this search as a heuristic to guide the original, complex search. In particular, we solve a 2D ((x, y)) version of the problem by running a single Dijkstra's search starting at the cell that corresponds to the center of the current vehicle position. The search computes the costs of shortest paths to all other cells in the environment[3].

AD* requires the heuristics to be admissible and consistent. This holds if $h(s_{start}) = 0$ and for every pair of states s, s' such that s' is an end state of a single action executed at state s, $h(s) + c(s, s') \geq h(s')$, where $h(s)$ is a heuristic of state s, s_{start} is a state that corresponds to the vehicle configuration and $c(s, s')$ is the cost of the action that connects s to s'. The cost $c(s, s')$ of the action is typically computed as the length of the action times the average of the costs of the cells covered by the vehicle when moving from state s to state s'. The heuristic based on the 2D search, however, may overestimate these costs since it estimates the cost of moving the center of the vehicle only. To resolve this, the cost of each cell in the 2D grid used for computing the 2D heuristic is set to the average of cells covered by the largest circle than can be inscribed into the vehicle perimeter. The cost of each transition $c(s, s')$ is then computed as the length of the transition times the maximum of two quantities: (a) the average value of the costs of the cells covered by the vehicle when moving from state s to state s' (as before), and (b) the maximum of the 2D grid cell costs, used to compute heuristics, traversed through by the center of the vehicle when moving from s to s'. Intuitively, this cost function penalizes slightly more when vehicle traverses high-cost areas (e.g., obstacles) residing right under the center of the vehicle. In addition, the heuristics are scaled down by a factor of 1.08 to compensate for the suboptimality of optimal paths in 8-connected grids. It can be then shown that our 2D heuristic function is admissible and consistent with respect to this cost function.

[3]However, even though it is very fast, we still restrict this search to only compute the states that are no more than twice as far (in terms of path cost) from the vehicle cell as the goal cell.

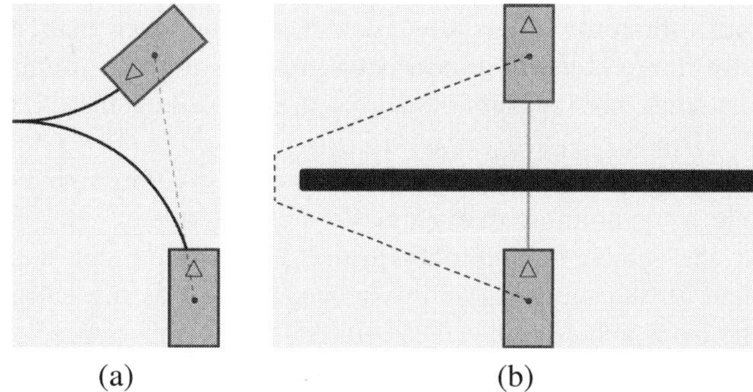

(a) (b)

Fig. 3. Mechanism-constrained (solid) and environment-constrained (dashed) heuristic paths. In each case, the initial and desired vehicle poses are shown as blue and red rectangles, respectively (with the interior triangles specifying the headings). (a) The mechanism-constrained heuristic is perfectly informed when no obstacles are present in the environment. (b) The environment-constrained 2D heuristic can provide significant benefit when obstacles exist. Here, an obstacle (shown in black) resides over the direct path to the desired pose.

Each of these heuristic generation approaches, mechanism-relative and environment-relative, have strong and complementary benefits (see Figure 3). Rather than selecting one, it is possible to combine the two. We do this by constructing a new heuristic that, for each state s, returns the value $h(s) = \max(h_{fsh}(s), h_{2D}(s))$, where $h_{fsh}(s)$ is the heuristic value of state s according to the mechanism-constrained heuristic (freespace heuristic), and $h_{2D}(s)$ is the value according to the environment-constrained heuristic (2D heuristic). As shown in the experimental results, this combined heuristic function can be an order of magnitude more effective than either of the component heuristic functions. Since both $h_{fsh}(s)$ and

$h_{2D}(s)$ are admissible and consistent, the combined heuristic is also admissible and consistent [26]. This property implies the bounds on the suboptimality of the paths returned by AD* [16]:

Theorem 2: The cost of a path returned by Anytime Dynamic A* is no more than ϵ times the cost of a least-cost path from the vehicle configuration to the goal configuration using actions in the multi-resolution lattice, where ϵ is the current value by which Anytime Dynamic A* inflates heuristics.

IV. Optimizations

Typically, one of the most computationally expensive parts of planning for vehicles is computing the cost of actions, as this involves convolving the geometric footprint of the vehicle for a given action with a map from perception. In our application, we used a 0.25m resolution 2D perception map and the (x, y) dimensions of our vehicle were $5.5m \times 2.25m$. Thus, even a short $1m$ action requires collision checking roughly 300 cells. Further, the specific cells need to be calculated based on the action and the initial pose of the vehicle.

To reduce the processing required for this convolution, we performed two optimization steps. First, for each action a we pre-computed the cells covered by the vehicle when executing this action. During online planning, these cells are quickly extracted and translated to the appropriate position when needed. Second, we generated two configuration space maps to be used by the planner to avoid performing convolutions. The first of these maps expanded all obstacles in the perception map by the inner radius of the robot; this map corresponded to an optimistic approximation of the actual configuration space. Given a specific action a, if any of the cells through which the center of the robot executing action a passes are obstacles in this inner map, then a is guaranteed to collide with an obstacle. The second map expanded all obstacles in the perception map by the outer radius of the robot and therefore corresponded to a pessimistic approximation of the configuration space. If all of the cells through which the center of the vehicle passes when executing action a are obstacle-free in this map, then a is guaranteed to be collision-free. Only those actions that do not produce a conclusive result from these simple tests need to be convolved with the perception map. Typically, this is a severely reduced percentage, thus saving considerable computation. To create these auxiliary maps efficiently, we performed a single distance transform on the perception map and then thresholded the distances using the corresponding radii of the robot for each map.

V. Experimental Results

We have implemented our approach on an autonomous passenger vehicle (lower-left image in Figure 5) where it has been used to drive over 3000 kilometers in urban environments, including competing in the DARPA Urban Challenge. The multi-resolution lattice planner was used for planning through parking lots and into parking spots, as well as for geometric road following in off-road areas, and in error recovery scenarios. During these scenarios, the vehicle traveled speeds of

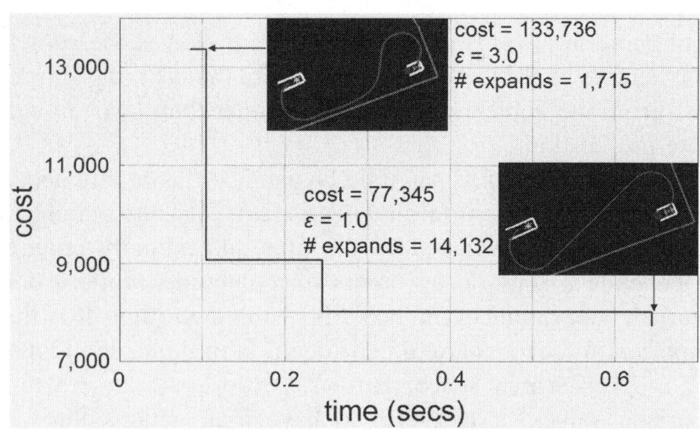

(a) anytime behavior

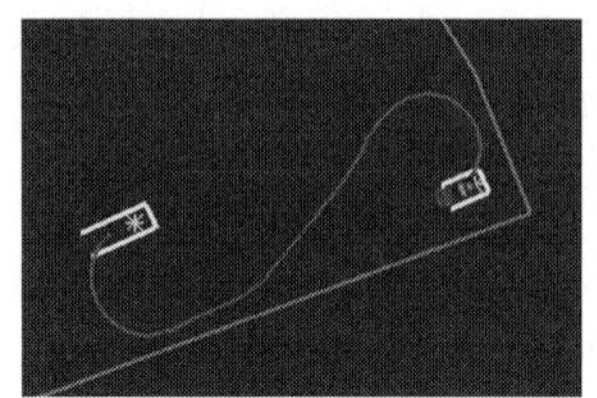

lattice	states expanded	time (secs)
high-res	2,933	0.19
multi-res	1,228	0.06

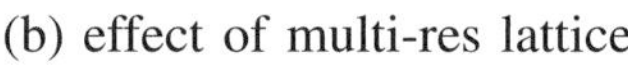

(b) effect of multi-res lattice

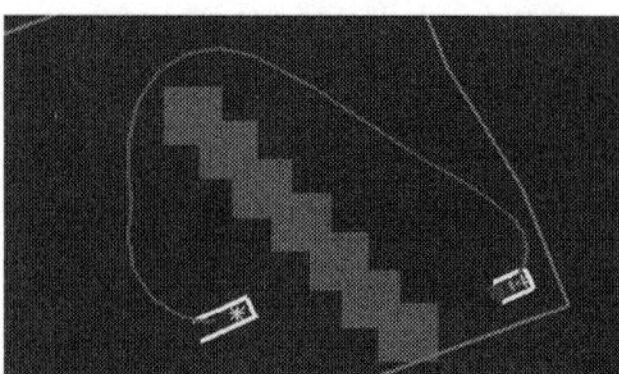

heuristic	states expanded	time (secs)
h	2,019	0.06
h_{2D}	26,108	1.30
h_{fsh}	124,794	3.49

(c) effect of heuristic

Fig. 4. An example highlighting our approach's anytime behavior and the benefits of the multi-resolution lattice and the combined heuristic function.

up to 15 miles per hour while performing complex maneuvers and avoiding static and dynamic obstacles.

In all cases, the multi-resolution lattice planner searches backwards out from the goal pose (or set of goal poses) and generates a path consisting of a sequence of feasible high-fidelity maneuvers that are collision-free with respect to the static obstacles observed in the environment. This path is also biased away using cost function from undesirable areas such as curbs and locations in the vicinity of dynamic obstacles.

When new information concerning the environment is received (for instance, a new static or dynamic obstacle is observed), the planner is able to incrementally repair its existing solution to account for the new information. This repair process is expedited by performing the search in a backwards direction, as in such a scenario updated information in the vicinity of the vehicle affects a smaller portion of the search space and so less repair is required. The lattice plan is typically updated once per second, however in trivial or very difficult scenarios this time may vary.

As mentioned earlier, the lattice used in this application does not explicitly represent curvature. Theoretically, this means that the paths produced over this lattice are guaranteed feasible only if we allow the vehicle to stop at each lattice state and re-orient its steering wheel. However, in practice we reduce (by a small fraction) the maximum curvature used

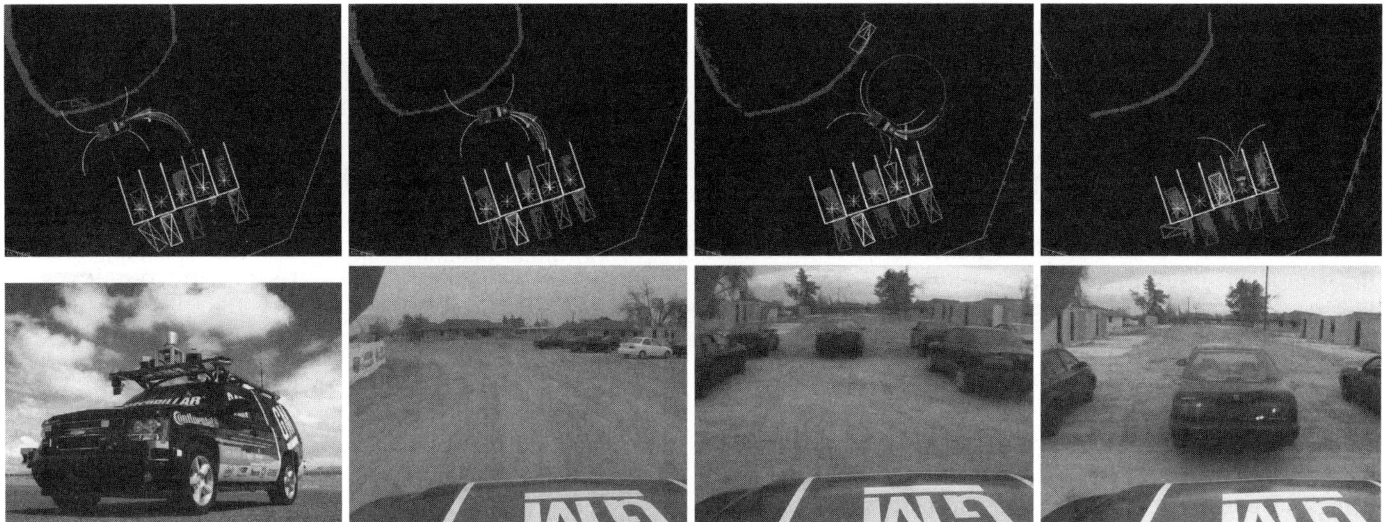

Fig. 5. Replanning when new information is received

in generating connections between states and we reduce the maximum speed at which we execute higher-curvature sections of lattice paths (from 5 m/s down to 2 m/s) so that this curvature discontinuity is not a critical issue. We also use a lookahead during execution to slow down and stop when switching velocity directions[4]. As a result, we don't need to stop during execution unless the path contains velocity sign changes.

The lattice path is tracked using a local planner that employs the same trajectory generation algorithm used to provide the action space for the lattice. Although a simple, single-trajectory tracker would suffice given the feasibility of the lattice plan, multiple trajectories are produced to account for dynamic obstacles and new observations that could require immediate reaction (the local planner runs at 10 Hz).

To ensure that a high-quality path is available for the vehicle as soon as it enters a parking lot, the lattice planner begins planning for the desired goal pose while the vehicle is still approaching the lot. By planning a path from the entry point of the parking lot in advance, the vehicle can seamlessly transition into the lot without needing to stop, even for very large and complex lots. Further, the anytime property of the search enables the solution to be improved during the pre-planning stage and, depending on how much time is available for pre-planning, the resulting path for the vehicle can converge to a (provably) optimal solution.

As well as providing smooth navigation amongst partially-known static objects, the efficiency of the multi-resolution lattice planner makes it possible to intelligently interact with several dynamic obstacles in the environment. In our application, we were able to not only avoid such obstacles but through updating regions of high cost as the obstacles moved, we could stay well clear of them unless necessary and also exhibit intelligent yielding behavior in unstructured areas (e.g. keeping to the right when approaching oncoming vehicles).

[4]A maximum lookahead of $2m$ is required given our vehicle's maximum deceleration and the top speed used for following lattice paths, but we use a slightly higher lookahead for smooth deceleration.

The multi-resolution lattice planner was also used for performing complex maneuvers in error recovery scenarios during on-road driving, such as when a lane or intersection is partially blocked with vehicles or obstacles, or a road is fully blocked and a u-turn is required. It was also used when there was some uncertainty as to where the road was; in these scenarios it uses the geometric perceptual information to bias the vehicle towards the center of the road (when there are perceivable curbs or berms).

We have included here a number of examples from the Urban Challenge and our testing to illustrate key characteristics of the approach.

a) Pre-planning: Figure 2 illustrates the pre-planning used by the lattice planner, as well as its anytime performance. The left image shows our vehicle approaching a parking lot (parking lot boundary shown in green, road lanes shown in blue), with its intended parking spot indicated by the white triangle. While the vehicle is still outside the parking lot it begins planning a path from one of the parking lot entries to the desired spot (path shown in red). Although the initial path shown in this left image is feasible, it is not ideal as it involves more turning than necessary. The right image shows how this path is improved over time as the vehicle approaches. This path is optimal with respect to our cost function and is generated well before the vehicle enters the parking lot.

b) Anytime Planning: Figure 4 is intended to provide insights into the benefits provided by each of the main components of our approach. Figure 4(a) illustrates the anytime behavior of the approach when planning between two parking spots. We have included a plot of the cost of the solution produced by Anytime D* as a function of computation time. Here, the initial suboptimality bound ϵ was set to 3. The upper image shows the first path Anytime D* finds. This path was found in less than 100 msecs (and after $1,715$ state expansions). The cost of the path was $133,736$. Given additional deliberation time, Anytime D* improves upon this solution, and after 650 msecs, the search converges to an optimal solution. This solution is significantly shorter than the initial path (as seen in the bottom image) and has a cost of 77,345.

c) Multi-resolution Planning: Figure 4(b) shows the benefits of using our multi-resolution lattice approach on the same simple example. The top row in the table represents a uniformly high-resolution lattice, while the bottom row represents our multi-resolution lattice (in both cases, $\epsilon = 2$). Planning with the multi-resolution lattice is more than three times faster. Note that the improvement in states expanded is less than a factor of three. This is because using a multi-resolution lattice decreases not only the number of states expanded but also the time spent expanding each state, since the number of possible actions from each state is decreased.

d) Combining Mechanism-relative and Environment-relative Heuristics: Figure 4(c) demonstrates the benefits of using our combined heuristic function on a simple example. The first row in the table represents our combined heuristic function. It combines the 2D environment-constrained heuristic (2nd row) and freespace mechanism-constrained heuristic (3rd row). Using this combination is over 21 times faster than using the 2D heuristic alone and over 58 times faster than using the freespace heuristic alone.

e) Replanning: Figure 5 illustrates the replanning capability of the lattice planner. These images were taken from a parking task performed during the National Qualification Event. The top-left image shows the initial path planned for the vehicle to enter the parking spot indicated by the white triangle. Several of the other spots were occupied by other vehicles (shown as rectangles of varying colors), with detected obstacles shown as red areas. The trajectories generated to follow the path are shown emanating from our vehicle (the selected trajectory is shown in blue). As the vehicle gets closer to its intended spot, it observes more of the vehicle parked in the right-most parking spot (top, second image from left). At this point, it realizes its current path is infeasible and replans a new path that has the vehicle perform a loop and pull in smoothly. This path was favored in terms of time over stopping and backing up to re-position. The three right-most photographs on the bottom row were taken by an onboard camera during the run.

f) Long-range Planning: As with other teams participating in the Urban Challenge, our vehicle underwent extensive testing before and during the competition. During the competition, the planner was able to continuously plan and replan without having the vehicle ever stop to wait for a plan. The scenarios we used for testing before the competition were numerous and included expansive obstacle-laden parking lots as well as narrow, highly-constrained parking lots. An example of the former is shown in Figure 6(a-b). This parking lot is $200m$ by $200m$. Initially, it is unknown and as the robot traverses the lot, it discovers a series of obstacles (shown as white dots in the image on the right). The robot has to replan in real-time to account for these obstacles. The time for replanning in this scenario varied from a few milliseconds for small re-planning adjustments to the path to a few seconds for finding drastically different trajectories, such as the one shown in Figure 6(b).

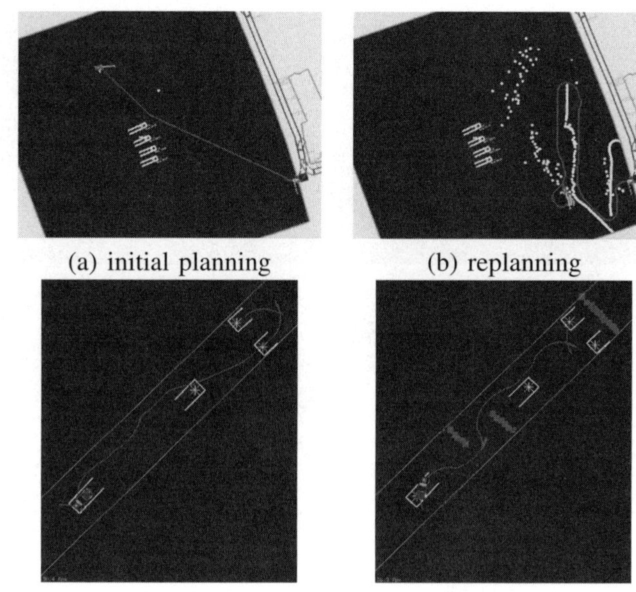

Fig. 6. Planning and replanning in large (a,b) and highly-constrained (c,d) environments

g) Complex Maneuvering: An example of a testing scenario involving a highly-constrained parking lot is shown in Figure 6(c-d). The trajectory planned involves the robot making an initial narrow U-turn and then making another one immediately before pulling into the final parking spot. While executing the trajectory, the robot discovers a series of obstacles and has to re-plan as shown in Figure 6(d). The new trajectory now requires the robot to backup a number of times. Moreover, it requires the robot to enter the desired spot in reverse since the discovered obstacles prohibit the robot from pulling in.

h) Coping with Dynamic Obstacles: Figure 7 shows the lattice planner being used to plan amongst several other moving vehicles in simulation. In these images, the current goal is shown as the white triangle and the inferred short-term trajectories of the other vehicles are included as fading polygons.

i) Coping with Static Obstacles: Figure 8 provides an example testing scenario for our physical vehicle. The left image shows the layout of the parking lot, the static obstacles (initially unknown to the vehicle), and the parking spots to be visited in order (1 through 5). The vehicle entered the lot through the left entrance between spots 3 and 4. The other images show snapshots from an onboard camera during the vehicle's traverse through this difficult environment.

VI. Conclusions

We have presented a general approach for complex planning involving large, high-dimensional search spaces. Our approach employs a novel multi-resolution action and state space that significantly reduces complexity while providing a seamless interface between the resolutions, as well as guarantees of solution feasibility. The approach also relies

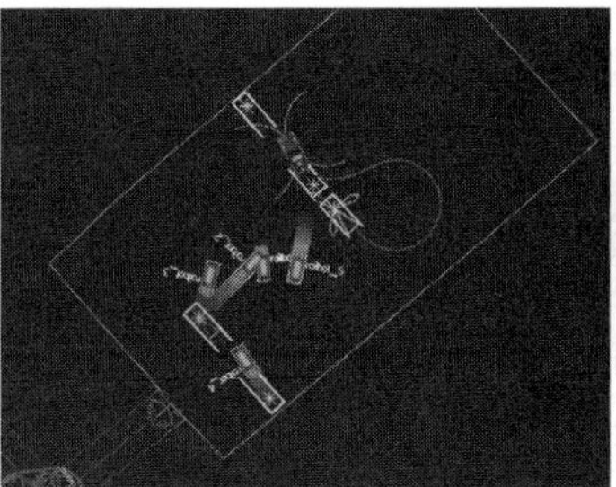
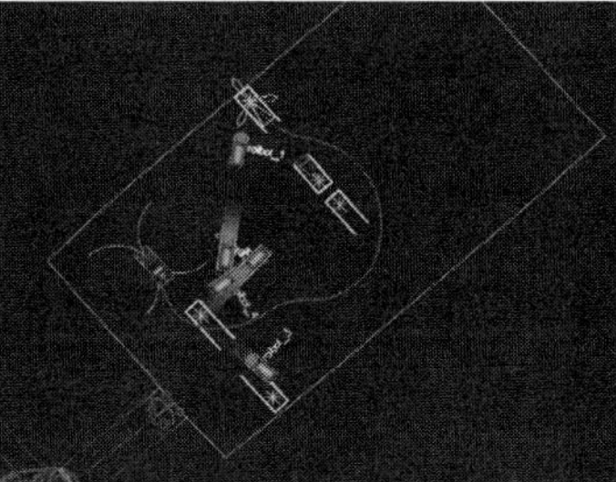
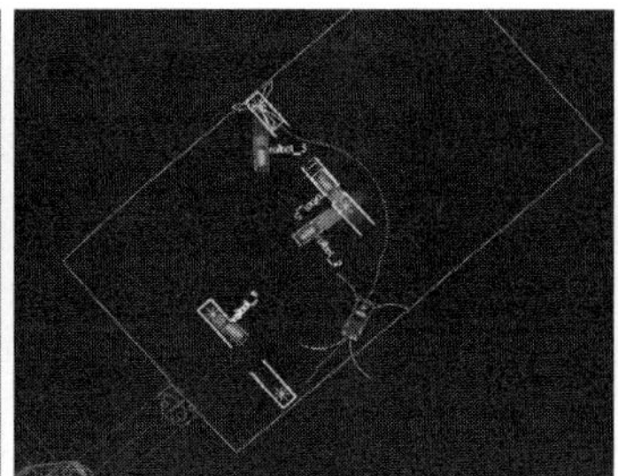
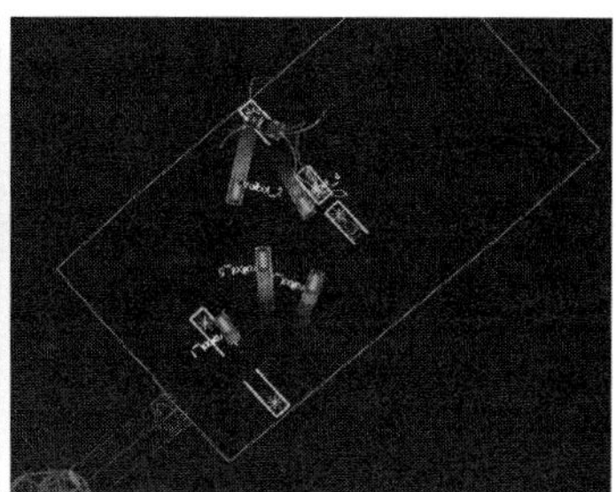

Fig. 7. Planning amongst moving obstacles

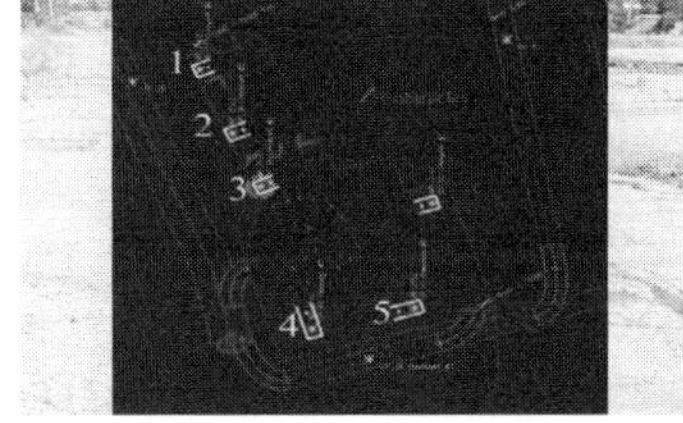

Fig. 8. Planning in complex obstacle environments

on an anytime, incremental search algorithm for generating solutions in partially-known or dynamic environments when deliberation time is limited. This search exploits a low-dimensional environment-dependent heuristic coupled with a full-dimensional freespace heuristic for efficient focusing, a powerful technique applicable to any high-dimensional planning problem. The resulting approach provides global, feasible solutions to challenging navigation tasks, and all the core techniques presented are applicable to a wide range of complex planning problems.

REFERENCES

[1] O. Khatib, "Real-time obstacle avoidance for manipulators and mobile robots," *International Journal of Robotics Research*, vol. 5, no. 1, pp. 90–98, 1986.

[2] R. Simmons, "The curvature velocity method for local obstacle avoidance," in *Proceedings of the IEEE International Conference on Robotics and Automation (ICRA)*, 1996.

[3] D. Fox, W. Burgard, and S. Thrun, "The dynamic window approach to collision avoidance." *IEEE Robotics and Automation*, vol. 4, no. 1, 1997.

[4] S. Thrun *et al.*, "Map learning and high-speed navigation in RHINO," in *AI-based Mobile Robots: Case Studies of Successful Robot Systems*, D. Kortenkamp, R. Bonasso, and R. Murphy, Eds. MIT Press, 1998.

[5] O. Brock and O. Khatib, "High-speed navigation using the global dynamic window approach," in *Proceedings of the IEEE International Conference on Robotics and Automation (ICRA)*, 1999.

[6] A. Kelly, "An intelligent predictive control approach to the high speed cross country autonomous navigation problem," Ph.D. dissertation, Carnegie Mellon University, 1995.

[7] R. Philippsen and R. Siegwart, "Smooth and efficient obstacle avoidance for a tour guide robot," in *Proceedings of the IEEE International Conference on Robotics and Automation (ICRA)*, 2003.

[8] S. Thrun *et al.*, "Stanley: The robot that won the DARPA Grand Challenge," *Journal of Field Robotics*, vol. 23, no. 9, pp. 661–692, August 2006.

[9] T. Howard and A. Kelly, "Optimal rough terrain trajectory generation for wheeled mobile robots," *International Journal of Robotics Research*, vol. 26, no. 2, pp. 141–166, 2007.

[10] C. Stachniss and W. Burgard, "An integrated approach to goal-directed obstacle avoidance under dynamic constraints for dynamic environments," in *Proceedings of the IEEE International Conference on Intelligent Robots and Systems (IROS)*, 2002.

[11] C. Urmson *et al.*, "A robust approach to high-speed navigation for unrehearsed desert terrain," *Journal of Field Robotics*, vol. 23, no. 8, pp. 467–508, August 2006.

[12] D. Braid, A. Broggi, and G. Schmiedel, "The TerraMax autonomous vehicle," *Journal of Field Robotics*, vol. 23, no. 9, pp. 693–708, August 2006.

[13] S. LaValle and J. Kuffner, "Rapidly-exploring Random Trees: Progress and prospects," *Algorithmic and Computational Robotics: New Directions*, pp. 293–308, 2001.

[14] G. Song and N. Amato, "Randomized motion planning for car-like robots with C-PRM," in *Proceedings of the IEEE International Conference on Intelligent Robots and Systems (IROS)*, 2001.

[15] M. Likhachev, G. Gordon, and S. Thrun, "ARA*: Anytime A* with provable bounds on sub-optimality," in *Advances in Neural Information Processing Systems*. MIT Press, 2003.

[16] M. Likhachev, D. Ferguson, G. Gordon, A. Stentz, and S. Thrun, "Anytime Dynamic A*: An Anytime, Replanning Algorithm," in *Proceedings of the International Conference on Automated Planning and Scheduling (ICAPS)*, 2005.

[17] R. Knepper and A. Kelly, "High performance state lattice planning using heuristic look-up tables," in *Proceedings of the IEEE International Conference on Intelligent Robots and Systems (IROS)*, 2006.

[18] M. Pivtoraiko and A. Kelly, "Generating near minimal spanning control sets for constrained motion planning in discrete state spaces," in *Proceedings of the IEEE International Conference on Intelligent Robots and Systems (IROS)*, 2005.

[19] N. Nilsson, *Principles of Artificial Intelligence*. Tioga Publishing Company, 1980.

[20] R. Zhou and E. Hansen, "Multiple sequence alignment using A*," in *Proceedings of the National Conference on Artificial Intelligence (AAAI)*, 2002, Student abstract.

[21] A. Stentz, "The Focussed D* Algorithm for Real-Time Replanning," in *Proceedings of the International Joint Conference on Artificial Intelligence (IJCAI)*, 1995.

[22] S. Koenig and M. Likhachev, "Improved fast replanning for robot navigation in unknown terrain," in *Proceedings of the IEEE International Conference on Robotics and Automation (ICRA)*, 2002.

[23] M. Barbehenn and S. Hutchinson, "Efficient search and hierarchical motion planning by dynamically maintaining single-source shortest path trees," *IEEE Transactions on Robotics and Automation*, vol. 11, no. 2, pp. 198–214, 1995.

[24] J. G. Gaschnig, "Performance measurement and analysis of certain search algorithms," Ph.D. dissertation, Carnegie Mellon University, 1979.

[25] H. W. Davis, A. Bramanti-Gregor, and J. Wang, "The advantages of using depth and breadth components in heuristic search," in *Methodologies for Intelligent Systems, 3*, Z. W. Ras and L. Saitta, Eds. New York: North-Holland, 1988, pp. 19–28.

[26] J. Pearl, *Heuristics: Intelligent Search Strategies for Computer Problem Solving*. Addison-Wesley, 1984.

Friction-Induced Velocity Fields for Point Parts Sliding on a Rigid Oscillated Plate

Thomas H. Vose, Paul Umbanhowar, and Kevin M. Lynch

Abstract— **We show that small-amplitude periodic motion of a rigid plate causes point parts on the plate to move as if they are in a position-dependent velocity field. Further, we prove that every periodic plate motion maps to a unique velocity field. By allowing a plate to oscillate with six-degrees-of-freedom, we can create a large family of programmable velocity fields. We examine in detail sinusoidal plate motions that generate fields with either isolated sinks or squeeze lines. These fields can be exploited to perform tasks such as sensorless part orientation.**

I. INTRODUCTION

When parts are in contact with a rigid oscillating surface, frictional forces induce the parts to move in a predictable manner. A single rigid plate is therefore a simple and appealing platform on which to perform a variety of parts manipulation tasks such as transporting, orienting, positioning, sorting, mating, etc. Because the motion of the plate is programmable, it is possible to perform these tasks on parts of various shapes and sizes without the need to reconfigure hardware for each new task or part geometry.

In this paper we propose a general model of part motion for point parts on a six-degree-of-freedom (DoF) oscillating plate. The key discovery is that small-amplitude periodic plate motions map to position-dependent velocity fields on the plate surface, which we refer to as *asymptotic velocity fields*. For many plate motions and coefficients of friction, part motion is well described by the asymptotic velocity field. Although we do not yet know how to characterize the set of all asymptotic velocity fields obtainable with a six-DoF oscillating plate, we do know that it includes fields with nonzero divergence (i.e., fields with sinks and sources). The video accompanying this paper shows parts on our six-DoF prototype device undergoing motion in such fields.

The rest of the paper is laid out as follows: in Section II we discuss related work that focuses on programmable force fields and vibratory surfaces; in Section III we present a full dynamic model of the part-plate system; in Section IV we present a simplified dynamic model which is the basis for our theorem asserting the existence of asymptotic velocity fields; in Sections V and VI we explain how to estimate asymptotic velocity fields and offer examples for sinusoidal plate motions; and in Section VII we conclude with remarks on future work.

T. H. Vose, P. Umbanhowar, and K. M. Lynch are with the Department of Mechanical Engineering, Northwestern University, Evanston, IL 60208, USA (e-mail: {t-vose, umbanhowar, kmlynch}@northwestern.edu)

This work was supported by NSF grants IIS-0308224 and CMMI-0700537 and Thomas Vose's NSF fellowship.

II. BACKGROUND

For some tasks, such as positioning and orienting parts, planar force fields with nonzero divergence (e.g., squeeze fields and sink fields) can be designed to interact with the part so that the task can be completed without the use of sensors. Significant theoretical work has gone into developing algorithms that exploit programmable force fields for sensorless positioning and orienting of planar parts (e.g., [1], [2], [3], [4]). Most devices designed to generate programmable force fields do so using a planar array of actuators, such as MEMS elements [5], [6], [7], rolling wheels [8], air jets [9], or vibrating plates [10]. Although these systems can create a wide range of fields, the fields are necessarily discrete, whereas most of the theoretical work assumes continuity. To approximate a continuous field, the array must contain a large number of actuators, often making fabrication and control difficult. Our current prototype vibratory device, which consists of just a single rigid plate and six voice coil actuators that allow for six-DoF motion of the plate, is comparatively simple yet powerful enough to create continuous fields with nonzero divergence.

Our work is a natural extension of the Universal Planar Manipulator (UPM) designed by Canny and Reznik [11]. The UPM consists of a single rigid horizontal plate that moves with three degrees-of-freedom (two translational and one rotational). Systems that restrict the plate motion to the horizontal plane, such as the UPM, can only generate frictional force fields with zero divergence [12]. Therefore, position sensing is required to orient and position parts. What allows our system to create fields with nonzero divergence is the plate's ability to move with all six degrees of freedom. In particular, the plate's ability to rotate out of the horizontal plane while simultaneously translating in the horizontal plane allows us to generate fields with sinks and sources.

The relationship between a periodic plate motion and the resultant frictional force acting on a planar part is generally quite complicated. However, when the plate motion is purely translational, point parts at all locations experience the same forces. This results in a single feed rate that is independent of position. In the simplest one-DoF case, the plate is horizontal and translates longitudinally. Reznik and Canny examined a particular type of bang-bang motion for this case in [13] and [14]. Okabe et al. looked at one-DoF plate motion in which the plate is also angled with respect to the horizontal [15]. Two-DoF systems, allowing translation both longitudinally and normal to the surface, are examined in [16] and [10]. In [17], Lynch and Umbanhowar derived optimal plate motions that maximize part speed on one- and two-DoF translating rigid plates. The three-DoF UPM is capable of generating certain position-dependent force fields [12] including localized ones [18].

In [19] we first demonstrated that fields with nonzero divergence can be created with a six-DoF rigid plate that oscillates periodically. This paper extends that work in three ways: it formally defines asymptotic velocity, it proves that every periodic plate motion maps to a unique asymptotic velocity field, and it introduces new plate motions that simplify the way some of the fields in [19] can be generated.

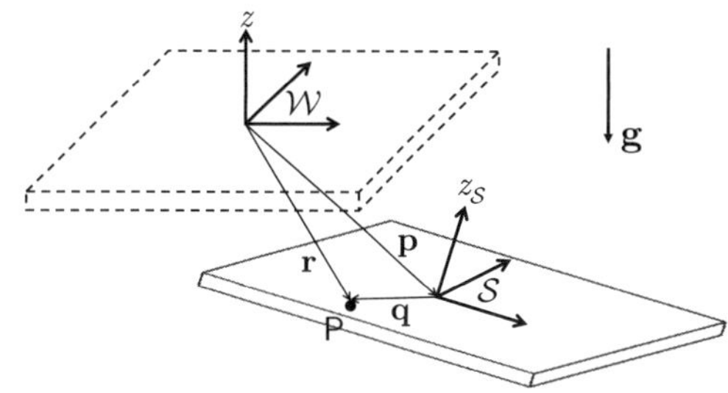

Fig. 1. An extremely exaggerated picture of the plate displaced from the fixed $\mathcal{W}$ frame by the vector $\mathbf{p}$. The position of the part P is given by $\mathbf{r}$ in the $\mathcal{W}$ frame and by $\mathbf{q}$ in $\mathcal{S}$ frame.

III. System Model

A. Plate Kinematics

Consider a rigid plate undergoing small-amplitude vibration. We define three coordinate systems: a fixed inertial frame $\mathcal{W}$, a local frame $\mathcal{S}$ attached to the origin of the plate, and an inertial frame $\mathcal{S}'$ instantaneously aligned with $\mathcal{S}$ (Figure 1). The z-axis of $\mathcal{W}$ is in the direction opposite the gravity vector, which is represented as $\mathbf{g} = [0, 0, -g]^T$ in the $\mathcal{W}$ frame. The $z_{\mathcal{S}}$-axis of $\mathcal{S}$ is perpendicular to the plate surface.

We choose to describe the kinematics of the plate in the $\mathcal{W}$ frame. The configuration of the plate is given by

$$\begin{bmatrix} \mathbf{R} & \mathbf{p} \\ 0 & 1 \end{bmatrix} \in SE(3),$$

where $\mathbf{R} \in SO(3)$. Both $\mathbf{R}$ and $\mathbf{p}$ are periodic C^1 functions of time with period T. In the *home position*, $\mathbf{p} = \mathbf{0}$ and $\mathbf{R} = \mathbf{I}$, where $\mathbf{I}$ is the identity matrix. The linear velocity of the origin of the plate is $\dot{\mathbf{p}}$ and the angular velocity of the plate is $\boldsymbol{\omega}$. The linear acceleration of the origin of the plate is $\ddot{\mathbf{p}} = [\ddot{p}_x, \ddot{p}_y, \ddot{p}_z]^T$ and the angular acceleration of the plate is $\boldsymbol{\alpha} = [\alpha_x, \alpha_y, \alpha_z]^T$. In general, we choose to specify the plate's motion in terms of $\ddot{\mathbf{p}}$ and $\boldsymbol{\alpha}$.

B. Part Kinematics

Let P be a point part with mass m in contact with the plate. As illustrated in Figure 1, let $\mathbf{q} = [x_{\mathcal{S}}, y_{\mathcal{S}}, 0]^T$ be a vector in $\mathcal{S}$ to P, and $\mathbf{r} = [x, y, z]^T$ be a vector in $\mathcal{W}$ to P such that

$$\mathbf{r} = \mathbf{p} + \mathbf{R}\mathbf{q}. \tag{1}$$

Let P* be the point on the plate directly underneath P. The position of P* is given by the vector $\mathbf{r}^*$ in the $\mathcal{W}$ frame. The velocity and acceleration of P* in the $\mathcal{W}$ frame are given by

$$\dot{\mathbf{r}}^* = \dot{\mathbf{p}} + \boldsymbol{\omega} \times \mathbf{R}\mathbf{q} \tag{2}$$
$$\ddot{\mathbf{r}}^* = \ddot{\mathbf{p}} + \boldsymbol{\omega} \times \boldsymbol{\omega} \times \mathbf{R}\mathbf{q} + \boldsymbol{\alpha} \times \mathbf{R}\mathbf{q}. \tag{3}$$

The velocity and acceleration of P in the $\mathcal{W}$ frame are given by

$$\begin{aligned} \dot{\mathbf{r}} &= \dot{\mathbf{p}} + \boldsymbol{\omega} \times \mathbf{R}\mathbf{q} + \mathbf{R}\dot{\mathbf{q}} \\ &= \dot{\mathbf{r}}^* + \mathbf{R}\dot{\mathbf{q}} \end{aligned} \tag{4}$$
$$\begin{aligned} \ddot{\mathbf{r}} &= \ddot{\mathbf{p}} + \boldsymbol{\omega} \times \boldsymbol{\omega} \times \mathbf{R}\mathbf{q} + \boldsymbol{\alpha} \times \mathbf{R}\mathbf{q} + 2\boldsymbol{\omega} \times \mathbf{R}\dot{\mathbf{q}} + \mathbf{R}\ddot{\mathbf{q}} \\ &= \ddot{\mathbf{r}}^* + 2\boldsymbol{\omega} \times \mathbf{R}\dot{\mathbf{q}} + \mathbf{R}\ddot{\mathbf{q}}. \end{aligned} \tag{5}$$

C. Part Dynamics

Three forces act on the part: gravity, friction, and the normal force from the plate (Figure 2). Applying Newton's second law in the $\mathcal{S}'$ frame gives

$$\mathbf{f}_{\mathrm{N}_{\mathcal{S}'}} + \mathbf{f}_{\mathrm{F}_{\mathcal{S}'}} + \mathbf{f}_{\mathrm{G}_{\mathcal{S}'}} = m\mathbf{R}^T\ddot{\mathbf{r}} \tag{6}$$
$$= m\mathbf{R}^T\left(\ddot{\mathbf{r}}^* + 2\boldsymbol{\omega} \times \mathbf{R}\dot{\mathbf{q}}\right) + m\ddot{\mathbf{q}}, \tag{7}$$

where $\mathbf{f}_{\mathrm{N}_{\mathcal{S}'}} = [0, 0, N]^T$, $\mathbf{f}_{\mathrm{F}_{\mathcal{S}'}} = [F_{x\mathcal{S}}, F_{y\mathcal{S}}, 0]^T$, and $\mathbf{f}_{\mathrm{G}_{\mathcal{S}'}} = m\mathbf{R}^T\mathbf{g}$ are the normal, frictional, and gravitational forces on the part in the $\mathcal{S}'$ frame. Solving (6) for $\ddot{\mathbf{r}}$ yields an expression for the part's acceleration in the $\mathcal{W}$ frame:

$$\ddot{\mathbf{r}} = \frac{1}{m}\mathbf{R}\left(\mathbf{f}_{\mathrm{N}_{\mathcal{S}'}} + \mathbf{f}_{\mathrm{F}_{\mathcal{S}'}}\right) + \mathbf{g}. \tag{8}$$

Explicit expressions for $\mathbf{f}_{\mathrm{N}_{\mathcal{S}'}}$ and $\mathbf{f}_{\mathrm{F}_{\mathcal{S}'}}$ are derived in the following two sections.

The state vector $\mathbf{x}_{\mathcal{W}} = [\mathbf{r}, \dot{\mathbf{r}}]^T$ can be computed by integrating (8). The state vector $\mathbf{x}_{\mathcal{S}} = [\mathbf{q}, \dot{\mathbf{q}}]^T$ can be computed from $\mathbf{x}_{\mathcal{W}}$ noting that (1) and (4) imply

$$\mathbf{q} = \mathbf{R}^T(\mathbf{r} - \mathbf{p}) \tag{9}$$
$$\dot{\mathbf{q}} = \mathbf{R}^T(\dot{\mathbf{r}} - \dot{\mathbf{r}}^*). \tag{10}$$

Our analysis is restricted to situations in which the part always remains in contact with the plate. Contact is maintained as long as the magnitude of the normal force is positive. Additionally, contact implies the acceleration of the part perpendicular to the plate surface is zero at all times in the $\mathcal{S}$ frame. Mathematically, we express this as

$$\mathbf{z}^T\ddot{\mathbf{q}} = 0, \tag{11}$$

where $\mathbf{z} \triangleq [0, 0, 1]^T$.

D. Normal Force

As noted previously, the normal force has the form $\mathbf{f}_{\mathrm{N}_{\mathcal{S}'}} = [0, 0, N]^T$ in the $\mathcal{S}'$ frame. Pre-multiplying (7) by $\mathbf{z}^T$ and noting (11) yields the following expression for the magnitude of the normal force, N:

$$N = m\mathbf{z}^T\left[\mathbf{R}^T(\ddot{\mathbf{r}}^* + 2\boldsymbol{\omega} \times \mathbf{R}\dot{\mathbf{q}} - \mathbf{g})\right]. \tag{12}$$

We define the *effective gravity* as

$$g_{\mathrm{eff}} = \mathbf{z}^T\left[\mathbf{R}^T(\ddot{\mathbf{r}}^* + 2\boldsymbol{\omega} \times \mathbf{R}\dot{\mathbf{q}} - \mathbf{g})\right], \tag{13}$$

so that $N = mg_{\mathrm{eff}}$.

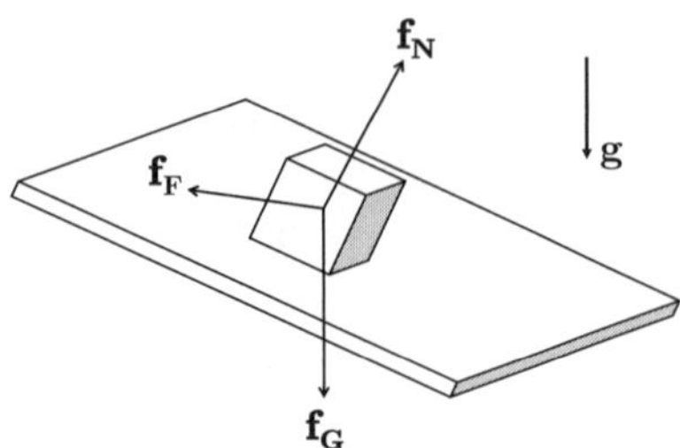

Fig. 2. The three forces that act on the part are due to gravity, friction, and the normal force from the plate. The gravitational force $\mathbf{f}_G$ always acts in the negative z-direction of the $\mathcal{W}$ frame, the frictional force $\mathbf{f}_F$ always acts tangent to the plate surface, and the normal force $\mathbf{f}_N$ always acts perpendicular to the plate surface.

E. Frictional Force

We assume Coulomb friction in our model. Since frictional forces can only act in the x-y plane of the $\mathcal{S}'$ frame we define the matrix $\mathbf{S}_{xy}$ that projects vectors in $\mathbb{R}^3$ onto the x-y plane.

The frictional force acting on a part located at $\mathbf{r}$ depends on the state of the system. There are three cases, which we summarize mathematically as:

$$\mathbf{f}_{\mathrm{F}_{\mathcal{S}'}} = \begin{cases} -\mu_k N \dfrac{\dot{\mathbf{q}}}{\|\dot{\mathbf{q}}\|}, & \|\dot{\mathbf{q}}\| > 0; \\ \mu_s N \dfrac{\mathbf{S}_{xy}\mathbf{R}^T\ddot{\mathbf{r}}^*}{\|\mathbf{S}_{xy}\mathbf{R}^T\ddot{\mathbf{r}}^*\|}, & \|\dot{\mathbf{q}}\| = 0 \\ & \|\mathbf{S}_{xy}\mathbf{R}^T\ddot{\mathbf{r}}^*\| > \mu_s g_{\mathrm{eff}}; \\ m\mathbf{S}_{xy}\mathbf{R}^T\ddot{\mathbf{r}}^*, & \|\dot{\mathbf{q}}\| = 0 \\ & \|\mathbf{S}_{xy}\mathbf{R}^T\ddot{\mathbf{r}}^*\| \le \mu_s g_{\mathrm{eff}}; \end{cases} \quad (14)$$

where μ_k and μ_s are the respective kinetic and static coefficients of friction between the part and the plate.

IV. Mapping Periodic Plate Motions to Position-Dependent Part Velocities

It is usually difficult to gain conceptual insight into the relationship between plate motion and part motion using the full dynamic model presented in the previous section. However, by running numerical simulations of the system, we observed that there is a unique average velocity $\mathbf{v}_{\mathrm{a}}(\mathbf{r})$ such that a point part at $\mathbf{r}$, moving with any other average velocity (when averaged over one cycle), tends toward $\mathbf{v}_{\mathrm{a}}(\mathbf{r})$. We call $\mathbf{v}_{\mathrm{a}}(\mathbf{r})$ the *asymptotic velocity at* $\mathbf{r}$. Thus, a part's motion on the plate is given approximately by the position-dependent *asymptotic velocity field*, where the quality of the approximation depends on the rate of convergence to the asymptotic velocity at each location.

The simplified system model presented in this section allows us to justify the position-dependent asymptotic velocity observed in the simulations and leads to further insight about part motion induced by small-amplitude periodic plate motions.

A. Simplified System Model

To simplify the system model, let us assume that the part is sliding at all times. Let us also operate in a regime where the period, linear displacement, and angular displacement of the plate are small enough so that we may assume $\mathbf{p} \approx \mathbf{0}$ and $\mathbf{R} \approx \mathbf{I}$. It follows that the part's position vector in the $\mathcal{W}$ frame can be approximated as $\mathbf{r} \approx \mathbf{q} = [x, y, 0]^T$, and that the gravitational, frictional, and normal forces acting on the part can be considered aligned with the $\mathcal{W}$ axes. In other words, the configuration of the plate is assumed to correspond to the home position at all times.

With the assumptions above, the approximate acceleration of the part in the horizontal plane, denoted by $\mathbf{a}$, is obtained by simplifying the x and y components of (8):

$$\mathbf{a} = \begin{bmatrix} \ddot{x} \\ \ddot{y} \\ 0 \end{bmatrix} = S_{xy}\ddot{\mathbf{r}} \approx -\mu_k g_{\mathrm{eff}} \frac{\dot{\mathbf{q}}}{\|\dot{\mathbf{q}}\|}. \quad (15)$$

The effective gravity g_{eff} and the relative velocity vector $\dot{\mathbf{q}}$ that respectively dictate the magnitude and direction of $\mathbf{a}$ can be approximated by simplifying (13) and (10) and further assuming that Coriolis and centripetal accelerations are insignificant:

$$g_{\mathrm{eff}} \approx \mathbf{z}^T (\ddot{\mathbf{p}} + \boldsymbol{\alpha} \times \mathbf{r} - \mathbf{g}) \quad (16)$$

$$\dot{\mathbf{q}} \approx \mathbf{S}_{xy} (\dot{\mathbf{r}} - \dot{\mathbf{r}}^*). \quad (17)$$

In summary, the simplified system model assumes the part is always sliding, the configuration of the plate always corresponds to the home position, and Coriolis and centripetal accelerations are negligible.

B. Asymptotic Velocity of Sliding Parts

Let $\mathbf{v}$ and $\mathbf{v}^*$ be the respective velocities of the part and the plate at a location $\mathbf{r}$ projected onto the x-y velocity plane in the $\mathcal{W}$ frame. We refer to this plane of velocities as $\mathcal{V}_{xy}$.

Let $\mathbf{r}_{xy} = [x, y]^T$. At a given $\mathbf{r}_{xy}$, $\mathbf{v}^*$ sweeps out a closed trajectory in $\mathcal{V}_{xy}$ for all periodic plate motions. Let $\mathcal{CH}^*$ denote the convex hull of this trajectory in the $\mathcal{V}_{xy}$ plane.

Even if the part has a nonzero velocity, we assume its displacement is negligible during a cycle of plate motion. Let us therefore denote by Σ a system consisting of

1) a plate undergoing periodic motion;
2) a part with a fixed value of $\mathbf{r}_{xy}$, but with a velocity that may be nonzero, and an acceleration given by (15) of the simplified dynamic model.

Thc following theorem asserts the existence of a unique asymptotic velocity for this type of system.

Theorem 1: For a system Σ, the part asymptotically converges from any initial velocity to a unique stable limit cycle of period T on or inside $\mathcal{CH}^*$.

Sketch of Proof: We first show that the part's velocity converges asymptotically to $\mathcal{CH}^*$ in the $\mathcal{V}_{xy}$ plane. From (15) and (17), the part accelerates such that $\mathbf{v}$ moves in the direction of $\mathbf{v}^*$ at each instant in the $\mathcal{V}_{xy}$ plane. Recalling that $\mathbf{v}^*$ sweeps out a closed trajectory, it follows that if $\mathbf{v}$ is outside of $\mathcal{CH}^*$ it must always move closer to $\mathcal{CH}^*$, and if $\mathbf{v}$ is contained in $\mathcal{CH}^*$ it can never escape.

Now we show that the part's velocity converges to a unique stable limit cycle in $\mathcal{CH}^*$. Let P_1 and P_2 be two point parts located at $\mathbf{r}_{xy}$ with identical coefficients of kinetic friction, μ_k. Let the velocities of P_1 and P_2 in the $\mathcal{V}_{xy}$ plane be $\mathbf{v}_1$

and $\mathbf{v}_2$. Let $\Delta v = \|\mathbf{v}_1 - \mathbf{v}_2\|$ be the distance between $\mathbf{v}_1$ and $\mathbf{v}_2$ in the $\mathcal{V}_{xy}$ plane (Figure 3).

From (16) the value of g_{eff} is the same for both P_1 and P_2. Thus, (15) implies that at all times P_1 and P_2 accelerate with equal magnitude in the direction of $\mathbf{v}^*$ in the $\mathcal{V}_{xy}$ plane. It immediately follows that $\frac{d}{dt}(\Delta v) \leq 0$. The nondecreasing case $\left(\frac{d}{dt}(\Delta v) = 0\right)$ corresponds to when $\mathbf{v}_1$, $\mathbf{v}_2$, and $\mathbf{v}^*$ are collinear such that $\mathbf{v}^*$ is not between $\mathbf{v}_1$ and $\mathbf{v}_2$. However, the periodic motion of the plate ensures that there will always be a duration of time $\tau > 0$ during the cycle when $\mathbf{v}_1$, $\mathbf{v}_2$, and $\mathbf{v}^*$ are not collinear, or $\mathbf{v}_1$, $\mathbf{v}_2$, and $\mathbf{v}^*$ are collinear such that $\mathbf{v}^*$ is between $\mathbf{v}_1$ and $\mathbf{v}_2$. In either case, $\frac{d}{dt}(\Delta v) < 0$ during this time, ensuring a contractive mapping over the course of a cycle. Thus, all parts located at $\mathbf{r}_{xy}$ must converge to a unique stable limit cycle on or inside $\mathcal{CH}^*$ that we call the *asymptotic trajectory at* $\mathbf{r}_{xy}$. The time-averaged velocity of the points on the asymptotic trajectory at $\mathbf{r}_{xy}$ is the unique asymptotic velocity $\mathbf{v}_\text{a}(\mathbf{r}_{xy})$.

Finally, we show that the period of the asymptotic trajectory must be the same as that of the plate. Consider an arbitrary periodic plate motion with period T. Assume that an asymptotic trajectory at $\mathbf{r}$ on or inside $\mathcal{CH}^*$ exists for this plate motion with period T' such that $T \neq T'$.

For the case in which $T > T'$ there must exist at least one point on the part's asymptotic trajectory that corresponds to at least two different points on the plate's trajectory in the $\mathcal{V}_{xy}$ plane. From (15) and (17), this implies that the part must accelerate in two different directions simultaneously, which is not physically possible.

For the case in which $T < T'$ there must exist at least two points on the part's asymptotic trajectory that correspond to a single point on the plate's trajectory. Suppose that two parts, P_1 and P_2, have initial velocities in the $\mathcal{V}_{xy}$ plane corresponding to $\mathbf{v}_1$ and $\mathbf{v}_2$ at the instant when the plate has velocity $\mathbf{v}^*$. By definition, the velocities of P_1 and P_2 are once again $\mathbf{v}_1$ and $\mathbf{v}_2$ every nT' seconds, where $n = 1, 2, 3, \ldots$. However, because n can be chosen such that $nT' > T$, this is a contradiction since the distance between $\mathbf{v}_1$ and $\mathbf{v}_2$ on or inside $\mathcal{CH}^*$ must decrease during any cycle of plate motion (i.e., any duration of T seconds). ■

Figure 4 shows simulation results that illustrate the ideas presented in Theorem 1. The plots are for the location $\mathbf{r}_{xy} = (0.06, 0)^T$ m on a plate undergoing the motion described in Figure 5(k). In Figure 4(a), two parts with different initial velocities are shown converging to the same asymptotic trajectory in $\mathcal{CH}^*$. This is also highlighted in Figure 4(b), which shows Δv decreasing over time, and in Figure 4(c), which shows the individual x and y velocities of the two parts converging in time. The markers in Figure 4(a) and (c) are plotted every half cycle to show that the asymptotic trajectory and the plate's trajectory have the same period.

C. Computing Asymptotic Velocity

Theorem 1 allows us to formally define the asymptotic velocity at $\mathbf{r}_{xy}$ as

$$\mathbf{v}_\text{a}(\mathbf{r}_{xy}) = \frac{1}{T}\int_t^{t+T} \mathbf{v}'(t)dt, \tag{18}$$

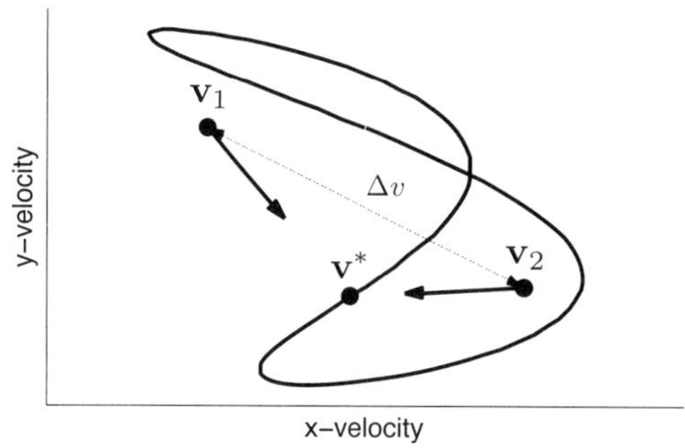

Fig. 3. The $\mathcal{V}_{xy}$ plane at an arbitrary location $\mathbf{r}$, in which the velocity of the plate $\mathbf{v}^*$ sweeps out a closed trajectory. From (15), the tangent to the trajectory of any point part (e.g., $\mathbf{v}_1$) passes through $\mathbf{v}^*$ at each instant.

where $\mathbf{v}'(t)$ is the unique limit cycle from Theorem 1. For some simple plate motions, the asymptotic velocity field as a function of $\mathbf{r}_{xy}$ can be determined analytically from (18) (see e.g., [19], [17], [20]). Otherwise, it can be determined numerically by computing the asymptotic velocity at a discrete set of points on the plate as follows:

1) Set the part's initial velocity to zero.
2) Simulate the part dynamics (without updating the position) for one cycle of plate motion.
3) Subtract the part's velocity at the end of the cycle from its velocity at the beginning of the cycle. If the magnitude of the difference is not within a predefined tolerance ϵ, repeat step 2, but use the part's velocity at the end of the current cycle as its initial velocity for the next cycle.
4) Average the part's velocity over the cycle.

V. Estimating Asymptotic Velocity

Although the asymptotic velocity can always be computed numerically, in this section we explain how to estimate it using the notion of *transient acceleration*. The advantage of this is that the transient acceleration is defined such that it is independent of the part's velocity and therefore does not require simulation to compute.

For some plate motions a qualitative estimate of the transient acceleration can be obtained by simple inspection. This is useful for gaining intuition about the properties of the fields generated by certain classes of plate motions, as discussed in Section VI. It also leads to insight about the inverse problem: finding a plate motion that approximately generates a desired field.

A. Transient Acceleration

Let a point part located at $\mathbf{r}_{xy}$ have an initial velocity $\mathbf{v} = \mathbf{0}$ in the $\mathcal{V}_{xy}$ plane. Unless the part happens to begin exactly on the asymptotic trajectory, there is a transient period during the first few cycles of plate motion in which the part converges to the asymptotic trajectory at $\mathbf{r}_{xy}$. During the transient period the average acceleration of the part is nonzero in order to bring the average cycle velocity closer to the asymptotic velocity. Thus, to a good approximation, the asymptotic velocity at $\mathbf{r}_{xy}$ is proportional to the average transient acceleration over a small

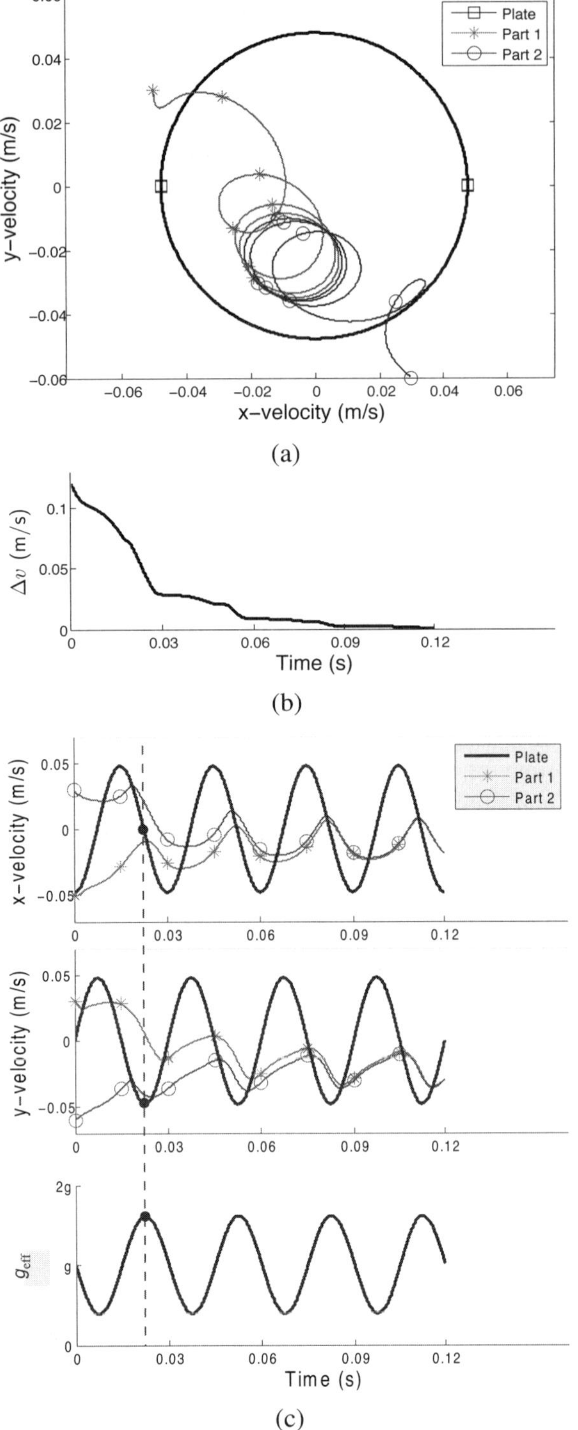

Fig. 4. A detailed look at two parts located at $(0.06, 0)$ m with different initial velocities on a plate undergoing the motion specified in Figure 5(k). The period of the plate's motion is $T = 0.03$ s, and the markers in (a) and (c) are plotted every half cycle (0.015 s). In (a), the trajectory of the two parts in the $\mathcal{V}_{xy}$ plane is shown for four cycles of plate motion. Both parts rapidly approach a nearly circular asymptotic trajectory centered around $(-0.01, -0.025)$ m/s that is fully contained the convex hull of the plate's trajectory. The asymptotic trajectory and the plate's trajectory each have two markers per cycle indicating that they have the same period. The distance between the parts' velocities in the $\mathcal{V}_{xy}$ plane rapidly approaches zero without ever increasing, as shown in (b). The x and y-components of the velocity are plotted individually vs. time in the top two graphs of (c); the effective gravity is plotted in the bottom graph of (c). The dashed vertical line in (c) denotes the time at which g_{eff} is a maximum.

number of cycles at $\mathbf{r}_{xy}$ arising from the initial condition $\mathbf{v} = \mathbf{0}$.

We use (15) to estimate the part's acceleration $\mathbf{a}$ during the transient. To estimate g_{eff} we use (16), rewritten below explicitly in terms of acceleration components:

$$g_{\text{eff}} \approx \mathbf{z}^T (\ddot{\mathbf{p}} + \boldsymbol{\alpha} \times \mathbf{r} - \mathbf{g}) = \ddot{p}_z + \alpha_x y - \alpha_y x + g. \quad (19)$$

To estimate $\dot{\mathbf{q}}$ we assume that during the transient period the magnitude of the plate's velocity $\mathbf{v}$ is much greater than the part's velocity $\mathbf{v}^*$. Thus, (17) reduces to

$$\dot{\mathbf{q}} \approx \mathbf{S}_{xy}(-\dot{\mathbf{r}}^*) = \mathbf{S}_{xy}(-\dot{\mathbf{p}} - \boldsymbol{\omega} \times \mathbf{r}) = \begin{bmatrix} -\dot{p}_x + \omega_z y \\ -\dot{p}_y - \omega_z x \\ 0 \end{bmatrix}. \quad (20)$$

We make the important observation that in our model of the transient acceleration neither g_{eff} nor $\dot{\mathbf{q}}$ depends on the motion of the part: from (19) the *magnitude* of $\mathbf{a}$ during the transient is a function of $\ddot{p}_z$, α_x, and α_y; from (20) the *direction* of $\mathbf{a}$ during the transient is a function of $\dot{p}_x$, $\dot{p}_y$, and ω_z. We refer to $\ddot{p}_z$, α_x, and α_y as the *out-of-plane acceleration components* of the plate's motion. We refer to $\dot{p}_x$, $\dot{p}_y$, and ω_z as the *in-plane velocity components*. In-plane velocity components can be obtained by integrating the *in-plane acceleration components* $\ddot{p}_x$, $\ddot{p}_y$, and α_z of the plate's motion.

Intuition about the qualitative properties of the field can often be gained by assuming that the part's dynamics are dominated by the portion of the cycle when g_{eff} (and thus the transient acceleration of the part) is largest. This is discussed below for sinusoidal plate motions.

B. Sinusoidal Motion Primitives

We now focus on the class of plate motions whose linear and angular acceleration components are sinusoidal with the same frequency, f:

$$\begin{aligned} \ddot{p}_x &= A_x \sin(2\pi f t + \phi_x) & \alpha_x &= A_\theta \sin(2\pi f t + \phi_\theta) \\ \ddot{p}_y &= A_y \sin(2\pi f t + \phi_y) & \alpha_y &= A_\varphi \sin(2\pi f t + \phi_\varphi) \\ \ddot{p}_z &= A_z \sin(2\pi f t + \phi_z) & \alpha_z &= A_\psi \sin(2\pi f t + \phi_\psi). \end{aligned}$$

This 11-dimensional space of plate motions is parameterized by the six amplitudes and five phases (one phase is chosen to be zero without loss of generality). We refer to plate motions of this form as *sinusoidal motion primitives*.

To qualitatively estimate the transient accelerations associated with sinusoidal motion primitives we assume that the frictional force during the instant in the cycle when g_{eff} is a maximum dominates the overall part dynamics. Thus, the average magnitude of $\mathbf{a}$ during the transient period is roughly proportional to the maximum value of g_{eff}. Similarly, the average direction of $\mathbf{a}$ during the transient period roughly corresponds to the direction of $\dot{\mathbf{q}}$ at the instant when g_{eff} is a maximum.

As an example, consider the point $\mathbf{r}_{xy} = (0.06, 0)^T$ m on a plate undergoing the motion given in Figure 5(k) with period $T = 0.03$ s. The circular trajectory of the plate's velocity $\mathbf{v}^*$ is depicted in Figure 4(a). In Figure 4(c) we see that g_{eff} is a maximum at $t = \frac{3}{4}T$. By examining the x and y velocities of the plate at this time we expect the asymptotic velocity to have a negative y-component and no x-component. This is very nearly true—the asymptotic trajectory to which the parts are converging in Figure 4(a) has an average y-velocity that is negative and an average x-velocity that is very close to zero, although slightly negative.

VI. Asymptotic Velocity Fields Arising from Sinusoidal Motion Primitives

In this section we look at two classes of sinusoidal motion primitives. For these classes, we use estimates of transient accelerations to predict the qualitative behavior of the asymptotic velocity fields. Numerically calculated versions of these fields, as well as several other fields, are shown in Figure 5. Below each field is the sinusoidal motion primitive from which the field is generated and the approximate form of the asymptotic velocity. Six seconds of simulated motion of a point part starting from rest is overlaid on each field. The simulation and the fields are based on the full dynamic system model given in Section III (using the simplified system model in Section IV to generate the fields gives visually indistinguishable results). After a brief transient period, agreement between the simulated motion and the asymptotic velocity field is very good in all cases.

A. Combining In-Plane Translation with Out-of-Plane Rotation: Nodal Line Fields

The class of *nodal line* sinusoidal motion primitives combine the in-plane acceleration components $\ddot{p}_x$ and $\ddot{p}_y$ with the out-of-plane acceleration components α_x and α_y as follows:

$$\begin{aligned} \ddot{p}_x &= A_x \sin(2\pi f t) \qquad & \alpha_x &= A_\theta \sin(2\pi f t + \phi) \\ \ddot{p}_y &= A_y \sin(2\pi f t) \qquad & \alpha_y &= A_\varphi \sin(2\pi f t + \phi). \end{aligned}$$

From (19) and (20), the effective gravity and relative velocity vector can be written as

$$g_{\text{eff}} \approx A_\theta \sin(2\pi f t + \phi) y - A_\varphi \sin(2\pi f t + \phi) x + g$$

$$\dot{\mathbf{q}} \approx \begin{bmatrix} -\dot{p}_x \\ -\dot{p}_y \\ 0 \end{bmatrix} = \begin{bmatrix} \frac{A_x}{2\pi f} \cos(2\pi f t) \\ \frac{A_y}{2\pi f} \cos(2\pi f t) \\ 0 \end{bmatrix}.$$

We note that g_{eff} is position-dependent; its maximum value increases with distance from the line through the origin in the direction of (A_θ, A_φ). We refer to this line as a nodal line. We also note that $\dot{\mathbf{q}}$ is position-independent and points in the direction of the vector (A_x, A_y). It follows that during the transient period a part will accelerate in a direction corresponding to $\pm(A_x, A_y)$ with a magnitude that scales with its distance from the nodal line.

Let us examine the special case where $A_y = A_\theta = 0$, implying that $\dot{\mathbf{q}}$ is always aligned with the x-axis and that the maximum value of g_{eff} increases with distance from the y-axis. Thus, we expect the magnitude of the transient acceleration to increase with distance from the y-axis. Further, whenever g_{eff} is a maximum on one side of the y-axis it is a minimum on the other side. This introduces an asymmetry that causes the direction of the part's transient acceleration to differ on opposite sides of the y-axis. Depending on the phase ϕ the part will accelerate toward or away from the nodal line.

For example, if $\phi = \frac{3}{2}\pi$, g_{eff} and $\dot{\mathbf{q}}$ are out of phase with each other for positions satisfying $x < 0$ (i.e., for parts with negative x-positions, the plate moves in the positive x-direction during the instant in the cycle when g_{eff} is a maximum). On the other hand, the plate moves in the negative x-direction during the instant in the cycle when g_{eff} is a maximum for parts satisfying $x > 0$. It follows that during the transient period parts with $x < 0$ tend to get accelerated in the positive x-direction whereas parts with $x > 0$ tend to get accelerated in the negative x-direction. As illustrated in Figure 5(c), the asymptotic velocity field for this case corresponds to a squeeze field converging on the y-axis. We refer to this as a `LineSink` field.

In general, the class of nodal line sinusoidal motion primitives create a nodal line of zero velocity in the direction of the rotation axis (i.e., the direction of the vector (A_θ, A_φ)). The value of ϕ determines whether the nodal line is attractive or repulsive. As illustrated in Figure 5(e)-(g), A_x, A_y, A_θ, A_φ, and ϕ can be chosen to create fields such as `SkewSink`, `SkewSource`, and `Shear`.

B. Combining In-Plane Translation with Out-of-Plane Rotation: Nodal Fields

The class of *nodal* sinusoidal motion primitives combine the in-plane acceleration components $\ddot{p}_x$ and $\ddot{p}_y$ with the out-of-plane acceleration components α_x and α_y as follows:

$$\begin{aligned} \ddot{p}_x &= A_x \sin(2\pi f t) & \alpha_x &= A_\theta \sin(2\pi f t + \phi) \\ \ddot{p}_y &= A_y \sin(2\pi f t + \pi/2) & \alpha_y &= A_\varphi \sin(2\pi f t + \pi/2 + \phi). \end{aligned}$$

From (19) and (20), the effective gravity and relative velocity vector can be written as

$$g_{\text{eff}} \approx A_\theta \sin(2\pi f t + \phi) y - A_\varphi \sin(2\pi f t + \pi/2 + \phi) x + g$$

$$\dot{\mathbf{q}} \approx \begin{bmatrix} -\dot{p}_x \\ -\dot{p}_y \\ 0 \end{bmatrix} = \begin{bmatrix} \frac{A_x}{2\pi f} \cos(2\pi f t) \\ -\frac{A_y}{2\pi f} \sin(2\pi f t) \\ 0 \end{bmatrix}.$$

The relative velocity $\dot{\mathbf{q}}$ is position-independent and rotates at a constant rate with a constant magnitude. This implies that the part can potentially accelerate in any direction at any location during the transient period. However, we can rule out many possibilities by examining g_{eff}. In particular, we expect the transient acceleration to be an odd function of position because when $g_{\text{eff}}(x, y, t)$ is a maximum $g_{\text{eff}}(-x, -y, t)$ is a minimum. Further, the magnitude of the transient acceleration should increase with distance from the origin because the maximum value of g_{eff} increases in this manner.

In general, nodal sinusoidal motion primitives create fields with a node of zero velocity at the origin of the plate. The value of ϕ determines whether the node is attractive or repulsive as well as whether the field is oriented clockwise or counterclockwise. The values of A_x, A_y, A_θ, and A_φ determine the strength, orientation, and eccentricity of the

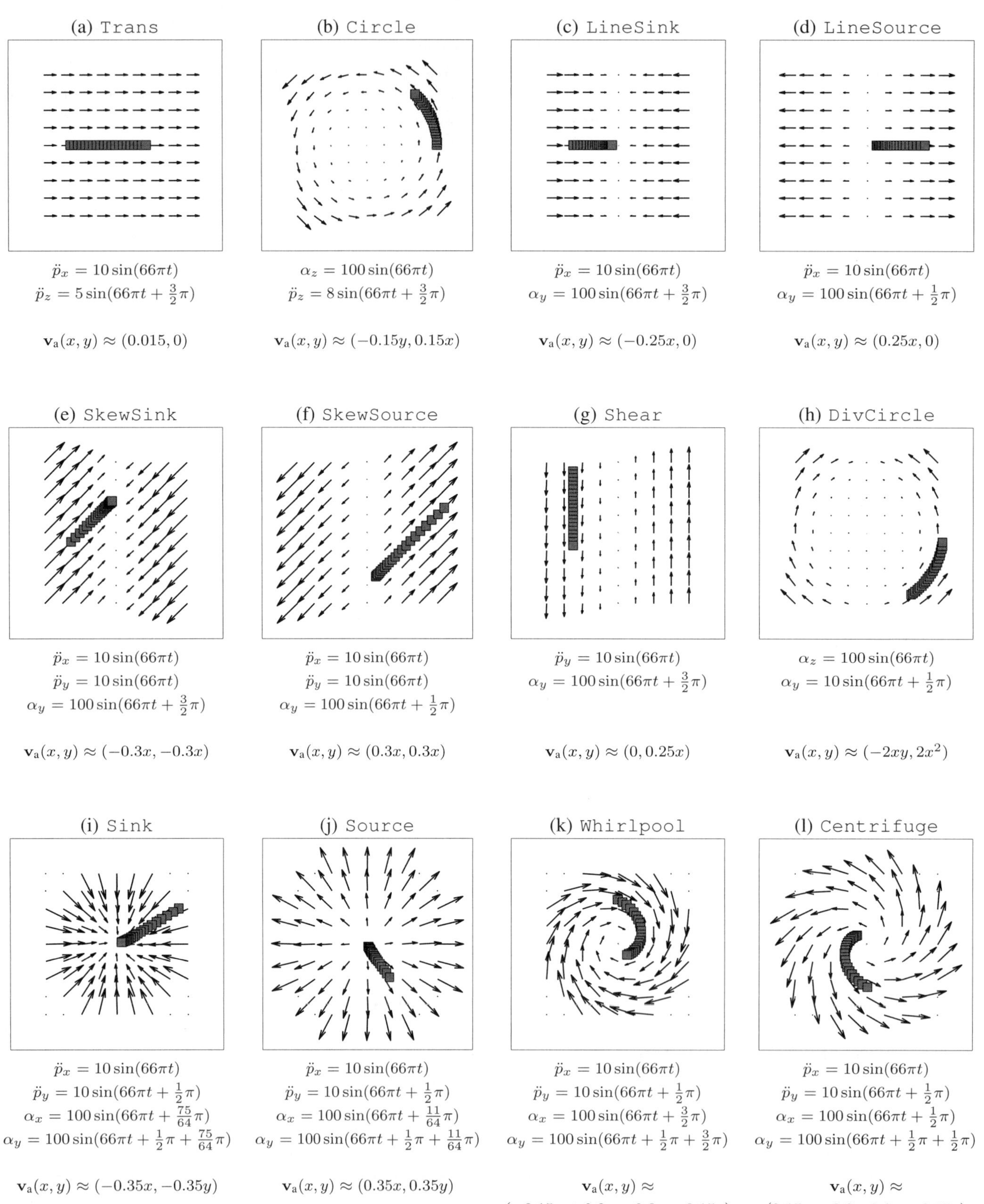

Fig. 5. Numerically calculated asymptotic velocity fields based on the full dynamic model corresponding to sinusoidal motion primitives (when the fields are calculated with the simplified dynamic model the results are visually indistinguishable). The fields are calculated for a point part with $\mu_k = \mu_s = 0.3$. Arrows are drawn in 2 cm increments. The arrows are missing in the corners of (i)–(l) because the part lost contact with the plate at those locations before reaching an asymptotic velocity. The sinusoidal motion primitive for each field is listed below it; linear and angular accelerations are in m/s^2 and rad/s^2 respectively. All acceleration components have a frequency of 33 Hz ($T = 0.03$ s). Below each sinusoidal motion primitive is an approximate form of the asymptotic velocity field given in units of m/s. Overlaid on each asymptotic velocity field is a six second (200 cycle) simulation of a point part starting from rest incorporating the full system dynamics. The position of the part is plotted every 0.3 seconds (every 10 cycles).

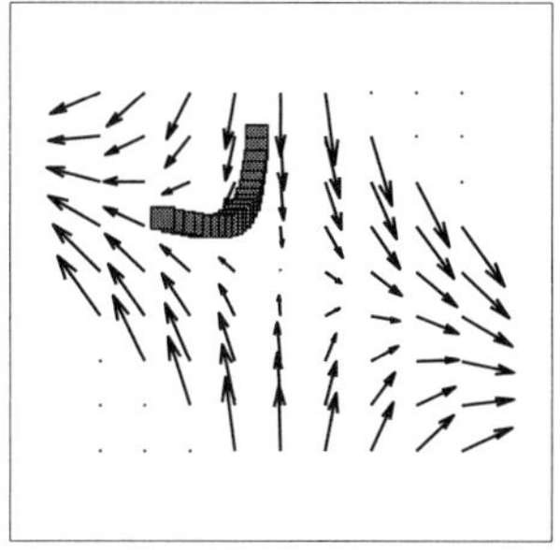

$$\ddot{p}_x = 10\sin(66\pi t)$$
$$\ddot{p}_y = 10\sin(66\pi t + \tfrac{1}{2}\pi)$$
$$\alpha_x = 100\sin(66\pi t + \tfrac{75}{64}\pi)$$
$$\alpha_y = 100\sin(66\pi t + \tfrac{27}{64}\pi)$$

$$\mathbf{v}_{\mathrm{a}}(x, y) \approx (0.25x, -0.3x - 45y)$$

$$\ddot{p}_x = 10\sin(66\pi t)$$
$$\ddot{p}_y = 10\sin(66\pi t + \tfrac{1}{2}\pi)$$
$$\ddot{p}_z = 2\sin(66\pi t)$$
$$\alpha_x = 100\sin(66\pi t + \tfrac{75}{64}\pi)$$
$$\alpha_y = 50\sin(66\pi t + \tfrac{107}{64}\pi)$$

$$\mathbf{v}_{\mathrm{a}}(x, y) \approx (-0.2x + 0.005, -0.4y + 0.01)$$

Fig. 6. Two asymptotic velocity fields with broken symmetries generated from sinusoidal motion primitives. The saddle on the left has nonorthogonal axes. The sink on the right biases motion in the y-direction and its node is shifted to the right and above the origin of the plate. Details about how the fields and simulated part motions were generated are given in the caption of Figure 5.

field. As illustrated in Figure 5(i)–(l), nodal fields include `Sink`, `Source`, `Whirlpool`, and `Centrifuge`.

C. The Set of Obtainable Fields for Sinusoidal Plate Motion

Based on the expressions for $\mathbf{v}_a$ given in Figure 5, we hypothesize that the class of sinusoidal motion primitives generate quadratic asymptotic velocity fields of the form

$$\mathbf{v}_{\mathrm{a}}(\mathbf{r}_{xy}) = \mathbf{r}_{xy}^T \mathbf{A} \mathbf{r}_{xy} + \mathbf{B} \mathbf{r}_{xy} + \mathbf{c}, \tag{21}$$

where $\mathbf{A} \in \mathbb{R}^{2\times 2\times 2}$, $\mathbf{B} \in \mathbb{R}^{2\times 2}$, and $\mathbf{c} \in \mathbb{R}^2$. The six combined elements of $\mathbf{B}$ and $\mathbf{c}$ can all be chosen independently, but there are constraints on $\mathbf{A}$ that are not yet fully understood.

This set of fields is a small subset of all possible asymptotic velocity fields obtainable with a six-DoF oscillating plate. Nonetheless, it is an interesting subset because it includes common fields with nonzero divergence. Figure 6 shows two asymptotic velocity fields that can be described by (21) that exhibit less symmetry than those in Figure 5.

VII. Conclusions

We have presented a model that predicts a relationship between small-amplitude periodic motions of a rigid plate and position-dependent velocity fields for point parts in contact with the plate. The model predicts the existence of velocity fields with nonzero divergence; the accompanying videos of our six-DoF prototype device qualitatively verify that these fields can indeed be generated in practice. Numerical simulations in this paper support the theoretical validity of the model.

There are four broad areas that we see as fruitful for future work. The first is to gain more insight about the scope of obtainable fields for more general periodic plate motions. The second is to develop a comprehensive understanding of how to map a desired field to its associated periodic plate motion(s). The third is to better understand how the coefficient of friction and the motion of the plate affect the rate of convergence to the asymptotic velocity. The fourth is to extend the work presented in this paper to handle parts with planar extent. These results will allow us to address a variety of applications including planar parts sorting, feeding, and assembly.

References

[1] K.-F. Böhringer, B. R. Donald, L. E. Kavraki, and F. Lamiraux, "A single universal force field can uniquely orient non-symmetric parts," in *International Symposium on Robotics Research*. Springer, 1999, pp. 395–402.

[2] M. Coutinho and P. Will, "A general theory for positioning and orienting 2d polygonal or curved parts using intelligent motion surfaces," in *IEEE International Conference on Robotics and Automation*, 1998, pp. 856–862.

[3] F. Lamiraux and L. Kavraki, "Positioning symmetric and non-symmetric parts using radial and constant force fields," in *Workshop on the Algorithmic Foundations of Robotics*, 2000.

[4] A. Sudsang, "Sensorless sorting of two parts in the plane using programmable force fields," in *IEEE International Conference on Intelligent Robots and Systems*, 2002, pp. 1784–1789.

[5] K.-F. Böhringer, B. Donald, and N. MacDonald, "Sensorless manipulation using massively parallel microfabricated actuator arrays," in *IEEE International Conference on Robotics and Automation*, 1994.

[6] K.-F. Böhringer, K. Goldberg, M. Cohn, R. Howe, and A. Pisano, "Parallel microassembly with electrostatic force fields," in *IEEE International Conference on Robotics and Automation*, 1998.

[7] S. Konoshi and H. Fujita, "A conveyance system using air flow based on the concept of distributed micro motion systems," *Journal of Microelectromechanical Systems*, vol. 3, no. 2, pp. 54–58, 1994.

[8] J. Luntz, W. Messner, and H. Choset, "Velocity field design for parcel manipulation on the modular distributed manipulator system," in *IEEE International Conference on Robotics and Automation*, 1999.

[9] J. Luntz and H. Moon, "Distributed manipulation with passive air flow," in *IEEE/RSJ International Conference on Intelligent Robots and Systems*, 2001.

[10] P. U. Frei, M. Wiesendanger, R. Büchi, and L. Ruf, "Simultaneous planar transport of multiple objects on individual trajectories using friction forces," in *Distributed Manipulation*, K. F. Böhringer and H. Choset, Eds. Kluwer Academic Publishers, 2000, pp. 49–64.

[11] D. Reznik, E. Moshkovich, and J. Canny, "Building a universal planar manipulator," in *Workshop on Distributed Manipulation at the International Conference on Robotics and Automation*, 1999.

[12] D. Reznik and J. Canny, "A flat rigid plate is a universal planar manipulator," in *IEEE International Conference on Robotics and Automation*, 1998, pp. 1471–1477.

[13] ——, "The Coulomb pump: a novel parts feeding method using a horizontally-vibrating surface," in *IEEE International Conference on Robotics and Automation*, 1998, pp. 869–874.

[14] D. Reznik, J. Canny, and K. Goldberg, "Analysis of part motion on a longitudinally vibrating plate," in *IEEE/RSJ International Conference on Intelligent Robots and Systems*, 1997, pp. 421–427.

[15] S. Okabe, Y. Kamiya, K. Tsujikado, and Y. Yokoyama, "Vibratory feeding by nonsinusoidal vibration—optimum wave form," *ASME Journal of Vibration, Acoustics, Stress, and Reliability in Design*, vol. 107, pp. 188–195, Apr. 1985.

[16] W. A. Morcos, "On the design of oscillating conveyers—case of simultaneous normal and longitudinal oscillations," *ASME Journal of Engineering for Industry*, vol. 92, no. 1, pp. 53–61, 1970.

[17] P. Umbanhowar and K. Lynch, "Optimal vibratory stick-slip transport," *IEEE Transactions on Automation Science and Engineering*, vol. 5, no. 3, 2008.

[18] D. Reznik and J. Canny, "C'mon part, do the local motion!" in *IEEE International Conference on Robotics and Automation*, 2001, pp. 2235–2242.

[19] T. Vose, P. Umbanhowar, and K. Lynch, "Vibration-induced frictional force fields on a rigid plate," in *IEEE International Conference on Robotics and Automation*, 2007.

[20] ——, "Friction-induced lines of attraction and repulsion for parts sliding on an oscillated plate," *IEEE Transactions on Automation Science and Engineering*, to appear.

Metastable Walking on Stochastically Rough Terrain

Katie Byl and Russ Tedrake

Abstract— **Simplified models of limit-cycle walking on flat terrain have provided important insights into the nature of legged locomotion. Real walking robots (and humans), however, do not exhibit true limit cycle dynamics because terrain, even in a carefully designed laboratory setting, is inevitably non-flat. Walking systems on stochastically rough terrain may not satisfy strict conditions for limit-cycle stability but can still demonstrate impressively long-living periods of continuous walking. Here, we examine the dynamics of rimless-wheel and compass-gait walking on randomly generated rough terrain and employ tools from stochastic processes to describe the 'stochastic stability' of these gaits. This analysis generalizes our understanding of walking stability and may provide statistical tools for experimental limit cycle analysis on real walking systems.**

I. Introduction

The science of legged locomotion is plagued with complexity. Many of the fundamental results for legged robots have come from detailed analytical and computational investigations of simplified models (e.g., [3, 5, 7, 10, 11]). These analyses reveal the limit cycle nature of ideal walking systems and employ Poincaré map analysis to assess the stability of these limit cycles. However, the very simplifications which have made these models tractable for analysis can limit their utility.

Experimental analyses of real machines based on these simple models [4] have revealed that real machines differ from these idealized dynamics in a number of important ways. Certainly the dynamics of impact and contact with the ground are more subtle than what is captured by the idealized models. But perhaps more fundamental is the inevitable stochasticity in the real system. More than just measurement noise, robots that walk are inherently prone to the stochastic influences of their environment by interacting with terrain which varies at each footstep. Even in a carefully designed laboratory setting, and especially for passive and minimally-actuated walking machines, the effects of this stochasticity can have a major effect on the long-term system dynamics. In practice, it is very difficult (and technically incorrect) to apply deterministic limit cycle stability analyses to our experimental walking machines - the real machines do not have true limit cycle dynamics.

In this paper, we extend the analysis of simplified walking models toward real machines by adding stochasticity into the our model. Although we have considered a number of sources of uncertainty, we will focus here on a compact and demonstrative model - where the geometry of the ground is drawn from a random distribution. Even with mild deviations in terrain from a nominal slope angle, the resulting trajectories of the machine are different on every step and for many noise distributions (e.g., Gaussian) the robot is guaranteed to eventually fall down (with probability one as $t \to \infty$). However, one can still meaningfully quantify stochastic stability, in terms of expected time to failure, and maximization of this metric in turn provides a parameter for optimization in the design of control for a walking robot on moderately rough, unmodeled terrain.

II. Background

Many stochastic dynamic systems exhibit behaviors which are impressively long-living, but which are also guaranteed to exit these behaviors ("fail") with probability one given enough time. Such systems cannot be classified as "stable", but it is also misleading and incomplete to classify them as "unstable". Physicists have long used the term *metastable* to capture this interesting phenomenon and have developed a number of tools for quantifying this behavior [8, 9, 12, 15]. Many other branches of science and engineering have also borrowed the terminology to describe dynamic systems in a wide variety of fields. Familiar metastable systems include crystalline structures (e.g. diamonds), flip-flop circuits, radioactive elements, oscillatory wave patterns in the brain, and ferromagnetic materials, such as spin glass or magnetic tape film (which explains why a taped recording sitting in storage still inevitably fades over time).

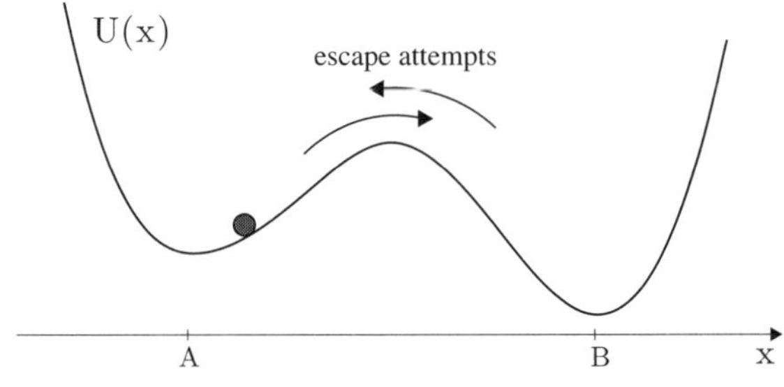

Fig. 1. Cartoon of a particle subject to Brownian motion in a potential $U(x)$ with two metastable states, A and B.

The canonical example of metastability is a particle in a potential well subject to Brownian motion, as cartooned in Figure 1. These systems have local attractors which tend to keep the dynamics within a particular neighborhood in state space. In the limit as such systems become deterministic (no noise), these local attractors are fixed points, and the system is truly stable whenever the dynamics begin with an initial condition somewhere inside the basin of attraction of the fixed point. In contrast, stochasticity constantly pushes the dynamics about within this neighborhood, and for some systems and noise types, this turns a stable system into a metastable one. Occasionally but repeatedly, such systems will deviate particularly far from a metastable attractor in state space (making "escape attempts"), and eventually, they will successfully exit (by which we mean entering a region where a different attractor is now a far more dominating influence).

III. Metastable Limit Cycle Analysis

The dynamics of walking systems are continuous, but they are punctuated by discrete impact events when a foot comes into contact with the ground. These impacts provide a natural time-discretization of a gait onto a Poincaré map. Therefore, we will consider walking systems governed by the discrete, closed-loop return-map dynamics:

$$\mathbf{x}[n+1] = \mathbf{f}(\mathbf{x}[n], \gamma[n]), \tag{1}$$

where $\mathbf{x}[n]$ denotes the state of the robot at step n and $\gamma[n]$ represents the slope of the ground, which is a random variable drawn independently from a distribution P_γ at each n. This model for stochastically rough terrain dramatically simplifies our presentation in this paper, but it also restricts our analysis to strictly forward walking[1]. These state evolution equations represent a discrete-time, continuous-state Markov process (or infinite Markov chain). For computational purposes, we will also discretize the state space into a finite set of states, x_i. Defining the state distribution vector, $\mathbf{p}[n]$, as

$$p_i[n] = \Pr(\mathbf{X}[n] = x_i), \tag{2}$$

we can describe the state distribution (master) equation in the matrix form:

$$\mathbf{p}[n+1] = \mathbf{p}[n]\mathbf{T}, \quad T_{ij} = \Pr(\mathbf{X}[n+1] = x_j \mid \mathbf{X}[n] = x_i). \tag{3}$$

$\mathbf{T}$ is the (stochastic) state-transition matrix; each row must sum to one. The n-step dynamics are revealed by the Chapman-Kolmogorov equation,

$$\mathbf{p}[n] = \mathbf{p}[0]\mathbf{T}^n.$$

We obtain the transition matrix numerically by integrating the governing differential equation forward from each mesh point, using barycentric interpolation [13] to represent the transition probabilities.

For walking, we will designate one special state, x_1, as an absorbing state representing all configurations in which the robot has fallen down. Transitions to this state can come *from* many regions of the state space, but there are no transitions *away* from this state. Assuming that it is possible to get to this absorbing state (possibly in multiple steps) from any state, then this absorbing Markov chain will have a unique stationary distribution, with the entire probability mass in the absorbing state.

The dynamics of convergence to the absorbing state can be investigated using an eigenmode analysis [1]. Without loss of generality, let us order the eigenvalues, λ_i, in order of decreasing magnitude, and label the corresponding (left) eigenvectors, $\mathbf{v}_i$, and characteristic times, $\tau_i = \frac{-1}{\log(\lambda_i)}$. The transition matrix from an absorbing Markov chain will have $\lambda_1 = 1$, with $\mathbf{v}_1$ representing the stationary distribution on the absorbing state. The magnitude of the remaining eigenvalues $(0 \le |\lambda_i| < 1, \forall i > 1)$ describe the transient dynamics and convergence rate (or mixing time) to this stationary distribution. Transient analysis on the walking models we investigate here will reveal a general phenomenon: λ_2 is very close to 1, and $\tau_2 \gg \tau_3$. This is characteristic of metastability: initial conditions (in eigenmodes 3 and higher) are forgotten quickly, and $\mathbf{v}_2$ describes the long-living (metastable) neighborhood of the dynamics. In metastable systems, it is useful to define the *metastable distribution*, ϕ, as the stationary distribution conditioned on having not entered the absorbing state:

$$\phi_i = \lim_{n\to\infty} \Pr(\mathbf{X}[n] = x_i \mid \mathbf{X}[n] \ne x_1).$$

This is easily computed by zeroing the first element of $\mathbf{v}_2$ and normalizing the vector to sum to one.

Individual trajectories in the metastable basin are characterized by random fluctuations around the attractor, with occasional "exits", in which the system enters a region dominated by a different attractor. For walking systems this is equivalent to noisy, random fluctuations around the nominal limit cycle, with occasional transitions to the absorbing (fallen) state. The existence of successful escape attempts suggests a natural quantification of the relative stability of metastable attractors in terms of first-passage times. The *mean* first-passage time (MFPT) to the fallen absorbing state describes the time we should expect our robot to walk before falling down.

Let us define the mean first-passage time vector, $\mathbf{m}$, where m_i is the expected time to transition from the state x_i into the absorbing state. Fortunately, the mean first-passage time is particularly easy to compute, as it obeys the relation:

$$m_i = \begin{cases} 0 & i = 1 \\ 1 + \sum_{j>1} T_{ij} m_j & \text{otherwise} \end{cases}$$

(the expected first-passage time must be one more than the expected first-passage time after a single transition into a non-absorbing state). In matrix form, this yields the one-shot calculation:

$$\mathbf{m} = \begin{bmatrix} 0 \\ (\mathbf{I} - \hat{\mathbf{T}})^{-1}\mathbf{1} \end{bmatrix}, \tag{4}$$

where $\hat{\mathbf{T}}$ is $\mathbf{T}$ with the first row and first column removed. $\mathbf{m}$ quantifies the relative stability of each point in state space. One interesting characteristic of metastable systems is that the mean first-passage time around an attractor tends be very flat; most system trajectories rapidly converge to the same metastable distribution (forgetting initial conditions) before escaping to the absorbing state. Therefore, it is also meaningful to define a system mean first-passage time, M, by computing the expected first-passage time over the entire metastable distribution,

$$M = \sum_i m_i \phi_i. \tag{5}$$

[1]Including backward steps is straightforward, but requires the model to include spatio-temporal correlations in the slope angle

When $\tau_2 \gg \tau_3$, we have $M \approx \tau_2$, and when $\lambda_2 \approx 1$, we have

$$M \approx \tau_2 = \frac{-1}{log(\lambda_2)} \approx \frac{1}{1-\lambda_2}.$$

IV. Numerical Modeling Results

This section uses two simple, classic walking models to demonstrate use of the methodology presented in Section III and to illustrate some of the important characteristics typical for metastable walking systems more generally. The two systems presented here are the rimless wheel and the passive compass gait walker, each of which is illustrated in Figure 2.

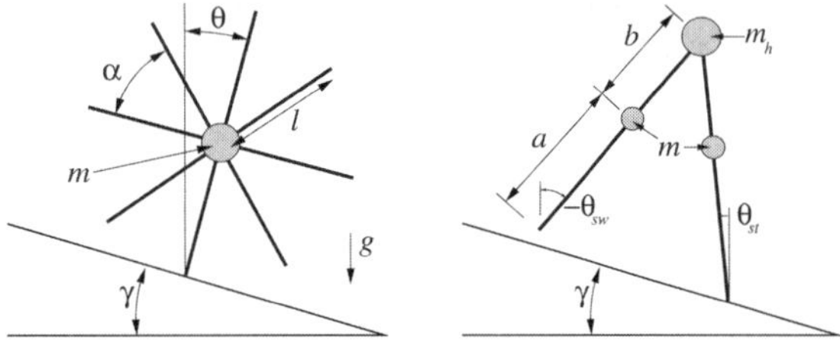

Fig. 2. The Rimless Wheel (left) and Compass Gait Walker (right) models.

A. Rimless Wheel

The rimless wheel (RW) model consists of a set of N massless, equally-spaced spokes about a point mass. Kinetic energy is added as it rolls downhill and is lost at each impulsive impact with the ground. For the right combination of constant slope and initial conditions, a particular RW will converge to a steady limit cycle behavior, rolling forever and approaching a particular velocity at any (Poincaré) "snapshot" in its motion (e.g., when the mass is vertically above a leg at $\theta = 0$ in Fig. 2). The motions of the rimless wheel on a constant slope have been studied in depth [3, 16].

In this section, we will examine the dynamics of the RW when the slope varies stochastically at each new impact. To do this, we discretize the continuous set of velocities, using a set of 250 values of ω, from 0.01 to 2.5 (rad/s). We also include an additional absorbing failure state, which is defined here to include all cases where the wheel did not have sufficient velocity roll past its apex on a particular step. Our wheel model has $N = 8$ spokes ($\alpha = \frac{\pi}{4}$). At each ground collision, we assume that the slope between ground contact points of the previous and new stance leg is drawn from an approximately[2] Gaussian distribution with a mean of $\bar{\gamma} = 8°$. For clarity, we will study only wheels which begin at $\theta = 0$ with some initial, downhill velocity, ω_o, and we consider a wheel to have failed on a particular step if it does to reach an apex in travel: $\theta = 0$ with $\omega > 0$. (Clockwise rotations go downhill, as depicted Fig. 2, and have positive values of ω.) Note that the dynamic evolution of angular velocity over time does not depend on the choice of a particular magnitude of the point mass, and we will use spokes of unit length, $l = 1$ meter, throughout.

[2]To avoid simulating pathological cases, the distribution is always truncated to remain within $\pm 10°$, or roughly 6σ, of the mean.

On a constant slope of $\gamma = 8°$, any wheel which starts with $\omega_o > 0$ has a deterministic evolution over time and is guaranteed to converge to a fixed point of $\omega = 1.2097$ (rad/s). The return map defining the step-to-step transitions from ω_n to ω_{n+1} is given as:

$$\omega_{n+1} = \sqrt{\cos^2\alpha\left(\omega_n^2 + \frac{2g}{L}(1 - cos\beta_1)\right) - \frac{2g}{L}(1 - cos\beta_2)}$$

where $\beta_1 = \frac{\alpha}{2} + \gamma$ and $\beta_2 = \frac{\alpha}{2} - \gamma$, with $\gamma > 0$ as the downhill slope. A plot of this return function is shown in Figure 3.

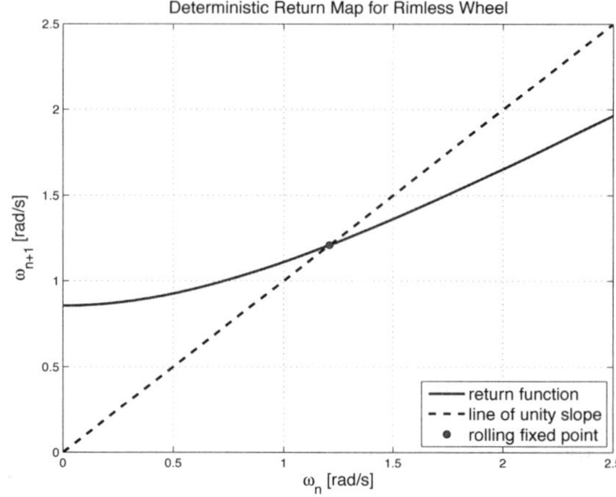

Fig. 3. Return map and fixed point for an 8-spoke rimless wheel on constant, downhill slope of 8°. Here, ω_n, is defined as angular velocity when the support spoke is exactly vertical.

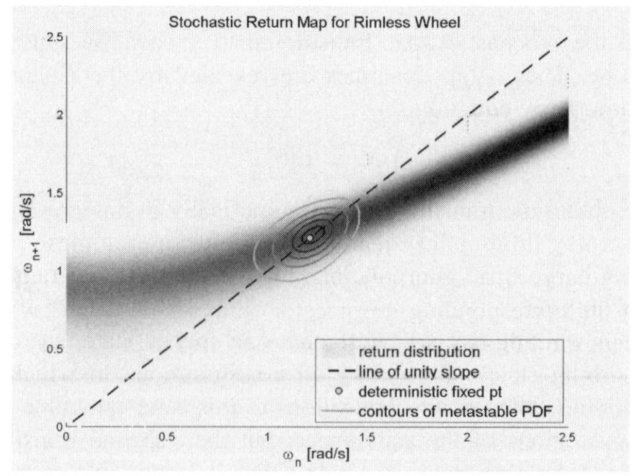

Fig. 4. Return distribution and metastable "neighborhood" for an 8-spoke rimless wheel on downhill terrain with a mean step-to-step slope of 8 degrees and $\sigma = 1.5°$. There is now a probability density function describing the transition from ω_n to ω_{n+1}.

When the slope between successive ground contacts is drawn from a stochastic distribution, the function given in Figure 3 is now replaced by a probabilistic description of the transitions, as illustrated in Figure 4. Given the current state is some particular ω_n, there is a corresponding probability density function (PDF) to describe what the next state, ω_{n+1}, will be. Figure 5 shows this set of PDF's clearly; it is a 3D plot of the same probabilistic return map shown from overhead in Figure 4. For our discretized model, each height value in Figure 5 is proportional to element T_{ij} of the transition matrix, where i is the state we are coming from (ω_n, on the x-axis) and j is the state we are going to (ω_{n+1}, on the y-axis); Figures 4 and 5 provide a graphical representation of the transition matrix describing this metastable dynamic system.

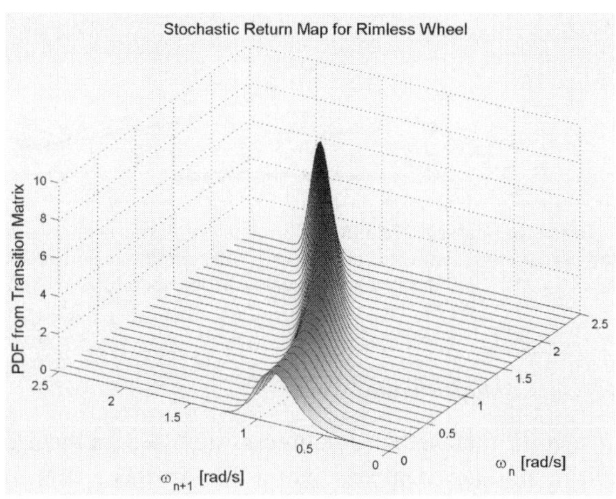

Fig. 5. 3D view of the return distribution for the stochastic rimless wheel system. This is a smoothed rendering of the step-to-step transition matrix, T, with the probability density functions for some particular states (ω_n) overlaid as lines for greater clarity.

To generate the discrete transition matrix, we calculate $\omega_{n+1} = f(\omega_n, \gamma)$ for each of a discrete set of 601 possible γ values, in the range of ± 10 degrees from the mean. Each new state is then represented in the mesh using barycentric weighting interpolation [13], which (we note) inherently adds a small level of additional (unintended) noise to the modeled dynamic. In Figures 4 and 5, the noise has a standard deviation of $\sigma = 1.5°$. Using MATLAB to take the 3 largest eigenvalues of the transpose of the transition matrix for this case, we find that the largest eigenvalue, λ_1, is within 10^{-14} of being exactly unity, which is within the mathematical accuracy expected. This eigenvalue corresponds to the absorbing failure state, and the corresponding eigenvector sums to 1, with all values except the failure state having essentially zero weight[3] in this vector (since all wheels will eventually be at this state, as $t \to \infty$). All other eigenvectors sum to zero (within numerical limits), since they must die away as $t \to \infty$. The second-largest eigenvalue is $\lambda_2 = 0.999998446$. Using the methods presented in Section III, this corresponds to a system-wide MFPT of about $1/0.000001554 = 643,600$ steps. Each initial condition has a particular MFPT, $m(\omega)$, which is obtained from Eq. 4 and plotted in Figure 6. Note that the value of the mean first-passage time is nearly flat throughout a large portion of state space. This is characteristic for metastable systems, which justifies the notion of a "system-wide" MFPT, $M \approx 1/(1 - \lambda_2)$, quantifying the overall stochastic stability of a particular dynamic system. For this particular case, there are no regions in state space (except the failure state) with MFPT significantly lower than the system-wide value, which is *not* typical more generally; the passive compass gait walker in Section IV-B is highly sensitive to initial conditions, for instance, although it too has regions of state space which share a nearly uniform MFPT value.

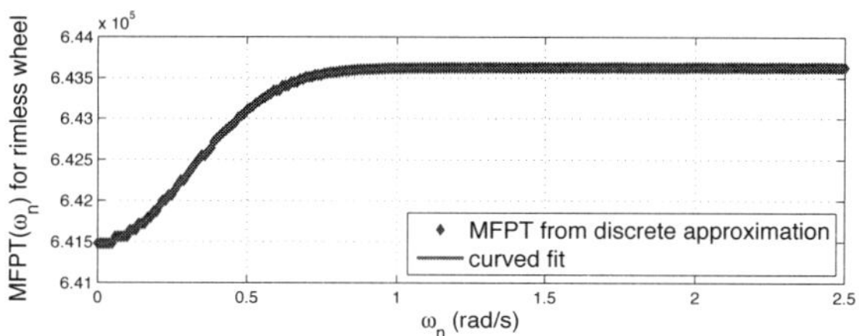

Fig. 6. Mean first-passage time as a function of the initial condition, ω_o. Data are for a rimless wheel on stochastic terrain with mean slope of 8 deg and $\sigma = 1.5°$. Points show the approximation obtained through eigen-analysis of the discretized system, and a smoothed line is overlaid. Note that MFPT is largely constant over a large portion of state space.

The eigenvector associated with λ_2 yields the PDF of the metastable dynamic process – the relative probability of being in any particular location in state space, given initial conditions have been forgotten and the walker has not yet failed. Figure 7 shows the resulting probability distribution functions for the rimless wheel for each of several levels of noise. Pictorially, each system-wide PDF for a metastable system is analogous to the fixed point for a stable, deterministic system. In the deterministic case, the probability of being exactly at the fixed point approaches unity as $t \to \infty$.

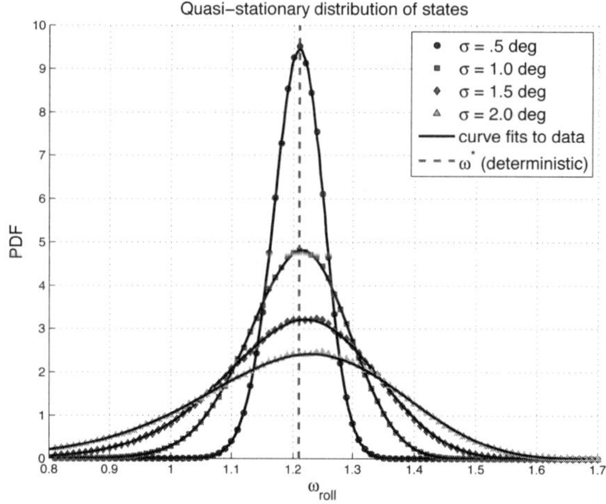

Fig. 7. Quasi-stationary probability density functions for the stochastic rimless wheel for each of several values of terrain noise, σ. Each distribution is estimated by renormalizing the eigenvector associated with the second-largest eigenvalue of the transpose of the transition matrix. Note that meshing inherently adds noise to the dynamic system; smoothed lines are drawn on top of the raw data (shown as points) from the scaled eigenvectors.

The third-largest eigenvalue of transition matrix, λ_3, quantifies the characteristic time scale in which initial conditions are forgotten, as the dynamics evolve toward the metastable

[3]All states except the failure state had a magnitude less than 10^{-10}, numerically.

distribution (or toward failure). For the case presented here ($\sigma = 1.5°$), $\lambda_3 \approx 0.50009$, which means almost half of the PDF of the initial condition composed of this eigenvector is lost ("forgotten") with each, successive step; an even larger fraction-per-step is lost for all remaining eigenvectors (with even smaller values of λ). Within a few steps, initial conditions for any wheel beginning in our range of analysis $0 < \omega_o \leq 2.5$ have therefore predominantly evolved into the metastable PDF (or have failed). If we multiply the metastable PDF, $\phi(\omega)$, by the transition matrix, we obtain a joint probability, $Pr(\omega_n, \omega_{n+1})$, of having just transitioned from ω_n to ω_{n+1}, given the wheel has not failed by step $n+1$. This is shown both as a 3D plot in Figure 8 and as a set of overlaid contour lines in Figure 4.

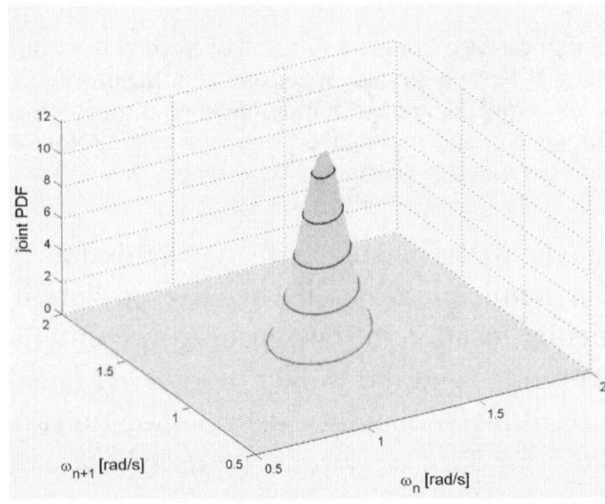

Fig. 8. 3D view of the metastable "neighborhood" of state-to-state transitions, (ω_n, ω_{n+1}). If a rimless wheel starts from some arbitrary initial condition and has not fallen after several steps, this contour map represents the joint probability density function of being in state ω_n now and transitioning to ω_{n+1}. The contour lines drawn are identical to those overlaid in Figure 4. They correspond to the neighborhood of likely (ω_n, ω_{n+1}) pairings, analogous to the unique fixed point of the deterministic case.

This particular system has a beautiful simplicity which allows us to extract some additional insight from the conditional probability in Figure 8. Because of the definition of ω_n as being the velocity when the mass is at its apex in a given step, the value of $\omega_{n+1} = 0$ represents the boundary to the absorbing failure state in this example. If we visualize the contours of the conditional probability as they extend toward $\omega_{n+1} = 0$ in Figure 4, we see that most failures do not occur because we transition from a very slow state (ω_n close to zero) to failure but are more typically due to sudden transitions from more dominant states in the metastable distribution to failure.

Finally, when this methodology is used to analyze the rimless wheel for each of a variety of noise levels (σ), the dependence of system-wide MFPT on σ goes as shown in Figure 9. For very low levels of noise, MATLAB does not find a meaningful solution (due to numerical limits). As the level of noise increases, the MFPT decreases smoothly but precipitously. (Note that the y-axis is plotted on a logarhithmic scale.) The stochastic stability of each particular system can be quantified and compared by calculating this estimate of MFPT which comes from λ_2 of the transition matrix.

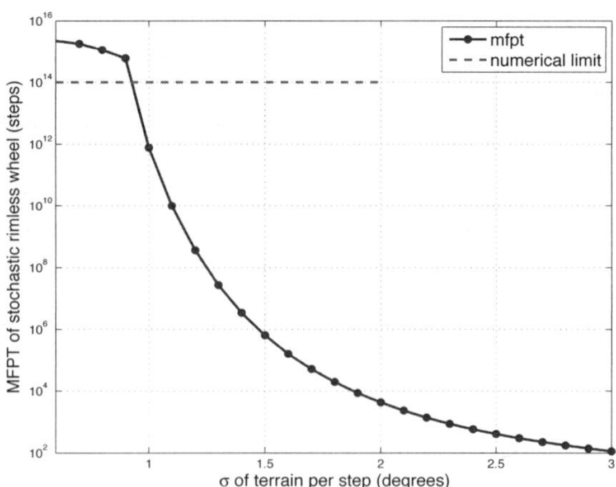

Fig. 9. Mean first-passage time (MFPT) for the rimless wheel, as a function of terrain variation, σ. Estimates above 10^{14} correspond to eigenvalues on the order of $1 - 10^{-14}$ and are beyond the calculation capabilities of MATLAB.

B. Passive Compass Gait Walker

The second metastable dynamic system we analyze in this paper is a passive compass gait (CG) walker. This system consists of two, massless legs with concentrated masses at the intersection of the legs ("the hip") and partway along each leg, and it has been studied in detail by several authors, e.g., [5, 7, 14]. Referring to Figure 2, the parameters used for our metastable passive walker are $m = 5$, $m_h = 1.5$, $a = .7$, and $b = .3$. Given an appropriate combination of initial conditions, physical parameters and constant terrain slope, this ideal model will walk downhill forever.

When each step-to-step terrain slope is instead selected from a stochastic distribution (near-Gaussian, as in Section IV-A), evolution of the dynamics becomes stochastic, too, and we can analyze the stochastic stability by creating a step-to-step transition matrix, as described in detail for the rimless wheel. The resulting system-wide MFPT as a function of terrain noise, $M(\sigma)$, is shown in Figure 10. Note that it is similar in shape to the dependence shown in Figure 9.

To analyze this system, our discretized mesh is defined using the state immediately after each leg-ground collision. The state of the walker is defined completely by the two leg angles and their velocities. On a constant slope, these four states are reduced to three states, since a particular combination of slope and inter-leg angle will exactly define the orientation of both the stance and swing leg during impact. Although the slope is varying (rather than constant) on stochastic terrain, we still use only three states to define our mesh. To do so, we simulate the deterministic dynamics (including impacts) a short distance forward or backward in time to find the robot state at the Poincaré section where the slope of the line connecting the "feet" of the legs is equivalent to our desired, nominal slope. Because the dynamics between collisions are entirely deterministic, these two states are mathematically equivalent for the stochastic analysis. If such a state does not exist for a particular collision (which occurs only very rarely),

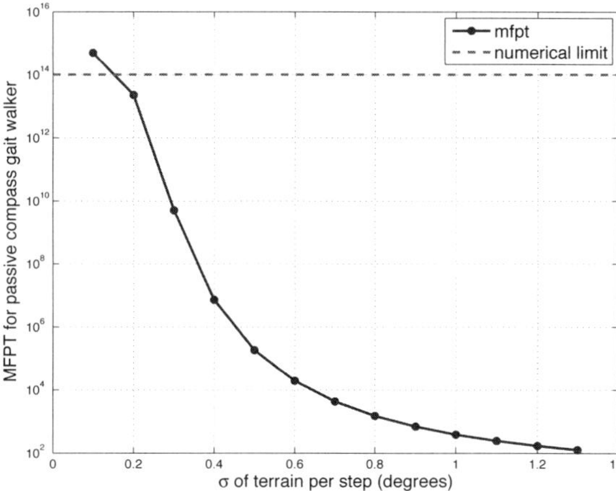

Fig. 10. Mean first-passage time as a function of terrain variation. Results for analysis of a compass gait walker using a discretized (meshed) approximation of the transitions. Average slope is 4 degrees, with the standard deviation in slope shown on the x-axis. Noise is a truncated Gaussian distribution, limited to between 0 and 8 degrees for all cases.

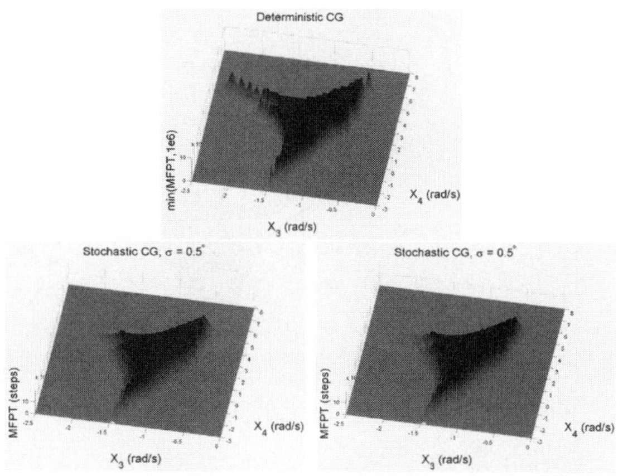

Fig. 11. Basin of attraction (top) for deterministic CG walker and map of MFPT for low-noise ($\sigma = 0.5°$, lower left) and high-noise ($\sigma = 1.0°$, lower right) examples. To aide in visual comparison, all 3 plots use the same mesh. The "near-constant MFPT basin" for each stochastic system is essentially a low-pass filtered version of the deterministic basin of attraction, and its shape does not change significantly, even when the magnitude of the MFPT itself varies greatly (e.g., 180,000 steps [left] vs 390 [right]). This region represents a boundary on the volume in state space from which a walker is likely to pulled into the metastable distribution.

we treat this as a member of the absorbing failure state. This approximation allows us to reduce the dimensionality from 4 states to 3, which improves numerical accuracy significantly. Specifically, it has allowed us to mesh finely enough to capture near-infinite MFPT for low-noise systems, while using four states did not. The three states we use in meshing are: (1) absolute angular velocity of the stance leg, X_3, (2) relative velocity of the swing leg, X_4, and (3) the inter-leg angle, α.

Figure 11 shows a slice of the basin of attraction for this compass gait on a constant slope (top), along with regions in state space with nearly-constant MFPT (bottom two) for two different magnitude of noise (σ) in terrain. Each slice is taken at the same inter-leg angle, $\alpha \approx 25.2°$. In the deterministic case, the basin of attraction defines the set of all states with infinite first-passage time: all walkers beginning with an initial condition in this set will converge toward the fixed point with probability 1. For stochastic systems which result in metastable dynamics, there is an analogous region which defines initial conditions having MFPT very close to the system wide value, M. Interestingly, the deterministic and stochastic basin shapes are quite similar here; we expect this may often be the case for systems such as this with discrete jumps in state space.

The image at the top of Figure 12 shows the deterministic basin of attraction for this CG walker more clearly. This plot was generated by sampling carefully over the state space and simulating the dynamics. The plot at the top of Figure 11 intentionally uses the same mesh discretization used for the stochastic system, to provide a better head-to-head comparison of the change in shape due to the addition of terrain noise (as opposed to the noise of the discretization itself). The second image in Figure 12 shows the deterministic basin of attraction for a different set of physical parameters ($m = m_h$; $a = b = .5$) on the same, constant slope of $4°$. This basin looks qualitatively more delicate and the resulting performance of this walker on stochastic terrain is in fact much worse (e.g., MFPT of about 20 steps when $\sigma = 0.5°$, where we find $M = 180,000$ for the other walker).

Just as in the case of the rimless wheel, the fixed point (for our deterministic compass gait system) is now replaced (in the stochastic case) by a probability density function, defining the likelihood of being in any particular state (conditioned on not having fallen) as $t \to \infty$. Figure 13 shows 2D contour plot sections of the PDF obtained from the eigen-analysis of the stochastic compass gait. The outermost contour defines a boundary containing 0.999 of the probability distribution in state space. The distribution spreads over more of state space as the level of noise increases, in a manner analogous to the widening of the distribution with noise seen in Figure 7.

Finally, we note that the relationship in state space between the PDF of the metastable dynamics, shown in Figure 13, and the region of nearly-uniform mean first-passage time, M, shown at the bottom of Figure 11, hints at where successful "escape attempts" are most likely to occur over time. Figure 14 overlays these two regions across a different dimensional slice of the 3D space for the $\sigma = .5°$ and $\sigma = 1.0°$ cases. As the tails of the metastable PDF (shown in yellow) approach the boundary of the uniform-MFPT basin (shown in blue), there is a higher probability of failing on any given step during the metastable process, resulting in turn in a less stochastically stable system (i.e., with a lower system-wide value of M).

V. Discussion

This section briefly discusses the use of the stochastic methods presented toward designing controllers for walking systems and also provides a few further observations on the properties of metastable systems which result in multiple attractors (e.g., period-n gaits).

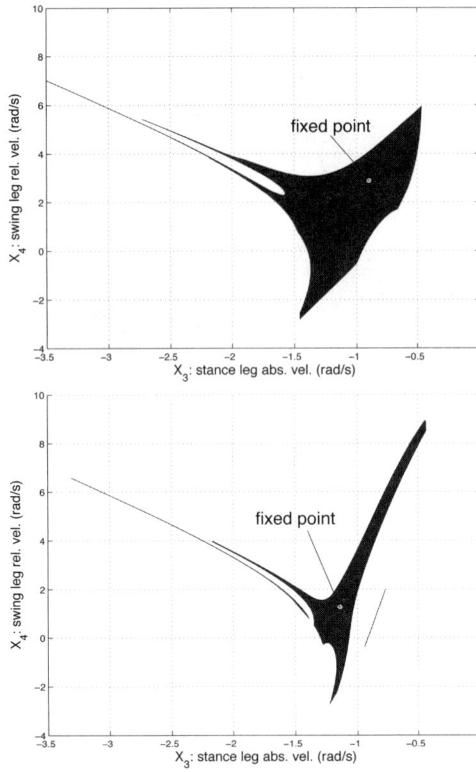

Fig. 12. Basins of attraction (blue region) and fixed point for two compass gait walkers, each on a constant slope of 4°. Walker with basin at top is more stable and uses the parameters defined for the stochastic system described throughout Section IV-B; for the other walker, $m = m_h$ and $a = b = .5$. MFPT is infinite inside the shaded region and is small (1-4 steps) outside of it. This image shows only a slice of the 3D basin, taken at the inter-leg angle of the fixed point for each respective walker. The fixed point is at $X_3 = -.89$ (rad/s), $X_4 = 2.89$ (rad/s), $\alpha = 25.2^{\circ}$ for the first walker, and it is at $X_3 = -1.14$ (rad/s), $X_4 = 1.26$ (rad/s), $\alpha = 33.4^{\circ}$ for the lower one. The deterministic basin of attraction for the second walker is narrower in shape, and this walker is significantly less stable on stochastic terrain.

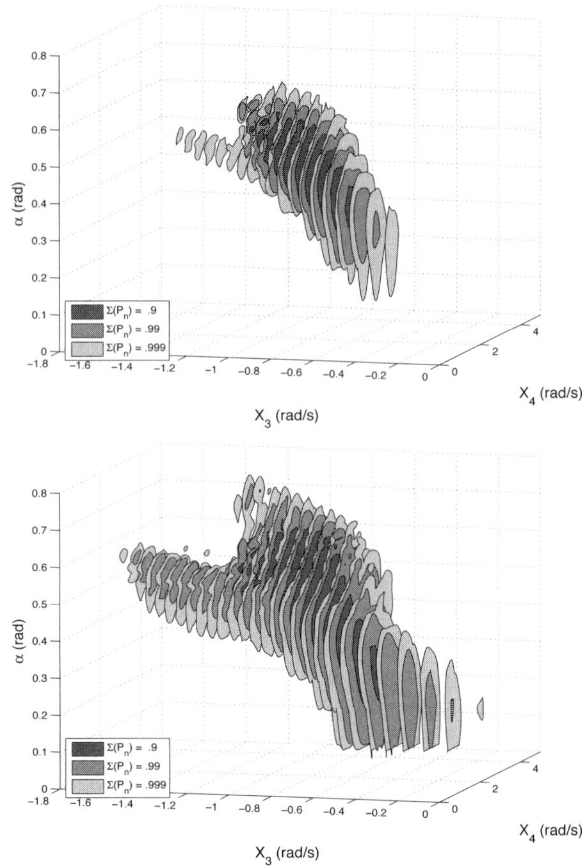

Fig. 13. On stochastic terrain, there is no fixed point for the compass gait walker. Instead, there are metastable "neighborhoods" of state space which are visited most often. As time goes to infinity, if a walker has not fallen, it will most likely be in a this region. The contours shown here are analogous to the PDF magnitude contours in Figure 7; they are drawn to enclose regions capturing 90%, 99%, and 99.9% of walkers at any snapshot during metastable walking. Top picture corresponds to $\sigma = 0.5^{\circ}$. Larger noise ($\sigma = 1.0^{\circ}$, bottom) results in larger excursions in state space, as expected.

A. Impacts on Control Design

One of the primary goals of a controller is to enhance the dynamic stability of a system. For walking systems, we propose throughout this paper that this should be defined as increasing the *stochastic stability*. We would like time-to-failure to be long, and we would like a system to converge toward the metastable distribution from a large set of initial conditions. The tools provided here can be used in optimizing controllers with either or both of these two aims in mind.

As an example, consider an active compass gait walker, with a torque source at the hip but with the ankles still unactuated at the ground contact. Putting this walker on a *repeating* terrain, as depicted in Figure 15, allows us to mesh across the entire state space of possible post-collision poses. By designing a low-level PD controller to regulate inter-leg angle, we can discretize the action space on a single once-per-step policy decision. The optimal high-level policy (to select desired inter-leg angle) for the system can now be solved via value iteration. Preliminary results for such a control methodology allow this underactuated compass gait model to walk continuously over impressively rough terrain [2].

B. Multiple stable limit cycles

Metastable dynamic systems sometimes have an inherent periodicity. We expect this may be the case on a slightly steeper slope, for instance, where compass gait models experience period-doubling bifurcations [7]. Another case where periodicity arises is for wrapping terrain, such as the terrain for the controlled walker in Figure 15. Wrapping is a realistic model for many in-laboratory walking robots, as they are often confined to walk on a boom – repeatedly covering the same terrain again and again. In our simulation of a hip-actuated CG walker on wrapping terrain, we observe that a repeating, n-step cycle results in multiple eigenvalues, λ_2 through λ_{n+1}, all with magnitude just under unity. They are complex eigenvalues, as are the corresponding eigenvectors. The top left image in Figure 15 shows such a set of eigenvalues, all lying just within the unit circle. The next-smallest set of eigenvalues are all significantly smaller in this example. The complex eigenvalues and eigenvectors mathematically capture an inherent period-

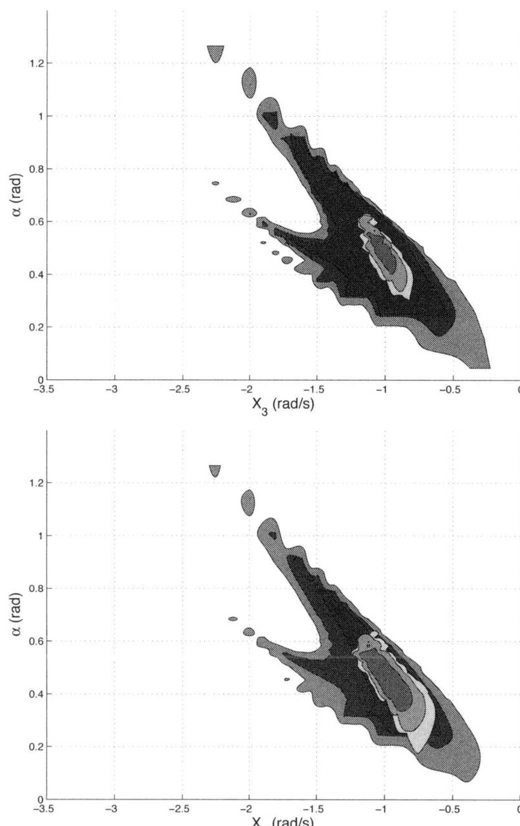

Fig. 14. Metastable system: Contours of the stochastic "basin of attraction" are shown where MFPT is $0.5M$, $0.9M$ and $0.99M$ (blue) versus contours where the integral of the PDF accounts for .9, .99, and .999 of the total metastable distribution (yellow). The metastable dynamics tend to keep the system well inside the "yellow" neighborhood. As the tails of this region extend out of the blue region, the system dynamics become less stochastically stable (lower M). The axis into the page represents the angle of the swing leg relative velocity, X_4, and a slice is taken at $X_4 = 2.33rad/s$. Terrain variation for the top plot is $\sigma = 0.5$ degrees (with $M \approx 180,000$ steps). For the noisier system at bottom ($\sigma = 1.0$ degrees), M is only 20 steps or so.

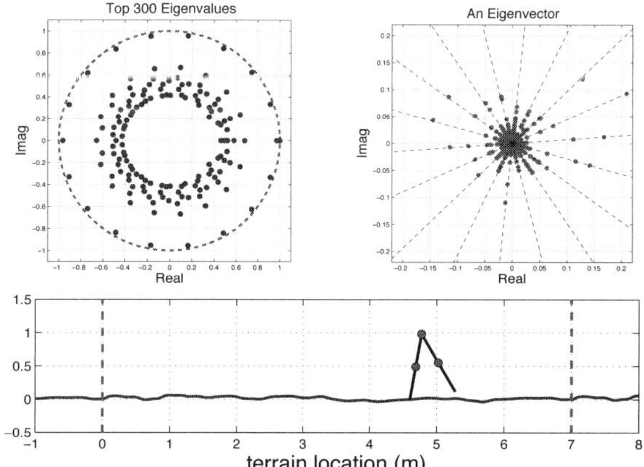

Fig. 15. Controlled compass gait walker, with torque at the hip. To solve for an optimal policy using value iteration, the terrain wraps every 7 meters. The optimization maximizes the MFPT from any given state. An eigenanalysis reveals a complex set of eigenvalues (top), spaced evenly about (but strictly inside of) the unit circle. Corresponding eigenvectors are also complex.

icity, in which the probability density function changes over time in a cyclical manner.

VI. Conclusions

The goal of this paper has been to motivate the use of stochastic analysis in studying and (ultimately) enhancing the stability of walking systems. Robots that walk are inherently more prone to the stochastic influences of their environment than traditional (e.g., factory) robots. Locomotory systems capable of interacting with the real world must deal with significant uncertainty and must perform well with limited energy budgets and despite limited control authority.

The stochastic dynamics of walking on rough terrain fit nicely into the well-developed study of metastability. The simplified models studied here elucidate the essential picture of a metastable limit cycle dynamics which makes occasional escape attempts to the fallen-down state. Metrics for stochastic stability, such as the mean first-passage time, may be potent metrics for quantifying both the relative stability across state-space and the overall system stability for real walking systems.

References

[1] Katie Byl and Russ Tedrake. Stability of passive dynamic walking on uneven terrain. In Art Kuo, editor, *Proceedings of Dynamic Walking 2006*, May 2006.

[2] Katie Byl and Russ Tedrake. Approximate optimal control of the compass gait on rough terrain. In *Proceedings IEEE International Conference on Robotics and Automation (ICRA)*, 2008.

[3] Michael J. Coleman, Anindya Chatterjee, and Andy Ruina. Motions of a rimless spoked wheel: a simple 3d system with impacts. *Dynamics and Stability of Systems*, 12(3):139–160, 1997.

[4] Steven H. Collins, Andy Ruina, Russ Tedrake, and Martijn Wisse. Efficient bipedal robots based on passive-dynamic walkers. *Science*, 307:1082–1085, February 18 2005.

[5] M Garcia, A Chatterjee, A Ruina, and M Coleman. The simplest walking model: Stability, complexity, and scaling. *Journal of Biomechanical Engineering – Transactions of the ASME*, 120(2):281–288, Apr 1998.

[6] C.W. Gardiner. *Handbook of Stochastic Methods for Physics, Chemistry and the Natural Sciences*. Springer-Verlag, third edition, 2004.

[7] Ambarish Goswami, Benoit Thuilot, and Bernard Espiau. Compass-like biped robot part I : Stability and bifurcation of passive gaits. Technical Report RR-2996, INRIA, October 1996.

[8] Peter Hanggi, Peter Talkner, and Michal Borkovec. Reaction-rate theory: fifty years after kramers. *Reviewsof Modern Physics*, 62(2):251–342, Apr 1990.

[9] N.G. Van Kampen. *Stochastic Processes in Physics and Chemistry*. Elsevier, third edition, 2007.

[10] Daniel E. Koditschek and Martin Buehler. Analysis of a simplified hopping robot. *International Journal of Robotics Research*, 10(6):587–605, Dec 1991.

[11] Tad McGeer. Passive dynamic walking. *International Journal of Robotics Research*, 9(2):62–82, April 1990.

[12] Reinhard Muller, Peter Talkner, and Peter Reimann. Rates and mean first passage times. *Physica A*, 247(247):338–356, Jun 1997.

[13] Remi Munos and Andrew Moore. Barycentric interpolators for continuous space and time reinforcement learning. In M. S. Kearns, S. A. Solla, and D. A. Cohn, editors, *Advances in Neural Information Processing Systems*, volume 11, pages 1024–1030. NIPS, MIT Press, 1998.

[14] Mark W. Spong and Gagandeep Bhatia. Further results on control of the compass gait biped. In *Proceedings of the IEEE International Conference on Intelligent Robots and Systems (IROS)*, pages 1933–1938, 2003.

[15] P. Talkner, P. Hangii, E. Freidkin, and D. Trautmann. Discrete dynamics and metastability: Mean first passage times and escape rates. *J. of Stat. Phys.*, 48(1/2):231–254, 1987.

[16] Russell L Tedrake. *Applied Optimal Control for Dynamically Stable Legged Locomotion*. PhD thesis, Massachusetts Institute of Technology, 2004.

Target Enumeration via Integration Over Planar Sensor Networks

Yuliy Baryshnikov
Mathematical and Algorithmic Sciences
Bell Laboratories
Murray Hill NJ, USA

Robert Ghrist
Department of Mathematics and
Coordinated Sciences Laboratory
University of Illinois
Urbana IL, USA

Abstract— **We solve the problem of counting the total number of observable targets (*e.g.*, persons, vehicles, etc.) in a region based on local counts performed by sensors which measure only the number of targets nearby and neither their identities nor any positional information. This theory is robust and accommodates ad hoc sensor networks and mobile robot sensors alike.**

I. Introduction: Target Counting

The prospect of small-scale sensor devices comes with the promise of sensor networks which can survey a region with dense coverage [5]. With this promise, however, comes many challenges, including power consumption, heat dissipation, and communication complexity. One strategy for remedying the situation is to focus on *minimal* sensing, engineering the individual sensors to be as simple as possible to accomplish the task and yet consume a minimum of resources.

We consider how to solve a simple data aggregation problem — counting unidentified targets — with a network of local minimal sensors. Specifically, we show that one can solve enumeration problems with sensors that can count nearby targets but cannot determine target identities, cannot estimate target range or bearing, and cannot record a time when a (moving) target came into view. Because the local sensors we envisage cannot discriminate targets, it is not obvious how to merge redundant counting by neighboring nodes.

It may seem surprising that a redundant array of simplistic sensors can solve the global enumeration problem. More surprising still is the fact that there are very few requirements on the sensors' detection specifications. We do not require that target visibility is purely a function of distance (cf. the typical use of the *unit disc* assumption in coverage problems). There are no hidden assumptions about convexity of the targets' detection zones, nor that the sensors or targets are uniform: some targets may have more 'impact' than others.

The reason for this combination of extreme robustness and simplicity of the sensor capabilities is the nature of our solution methods. We use a topological invariant — the *Euler characteristic* — molded into an integration theory.

A. Related work

There are few similar approaches to problems in target estimation or tracking, the literature on which almost always assumes the ability to identify different targets (along with other high-level functions, including distance estimation, bearing estimation, and sensor localization). For example, the large-scale wireless system implemented in [10] assumes an aggregation phase based on strict spatial separation of targets. Jung and Sukhatme [11] implement a multi-target robotic tracking system where the targets are labeled with colored lights. The survey paper of Guibas [8] pointing to the broader literature on geometric range-searching assumes the ability to aggregate target identities and concerns itself with computational complexity issues. The paper by Li *et al.* [13] on multi-target tracking via sensor networks notes that, *"target classification is arguably the most challenging signal processing task in the context of sensor networks."*

We are aware of two notable exceptions. One significant solution to a target enumeration problem is found in the work of Fang, Zhao, and Guibas [6], which gives a distributed algorithm for target enumeration without any target-identification capabilities on the part of the sensors. Their work assumes that all target supports are round balls in $\mathbb{R}^2$; that each sensor reads a $\mathbb{R}$-valued signal proportional to the inverse square of distance-to-target; and that target impacts are additive. Their algorithm counts the number of local maxima in the sensor signal field and therefore gives an accurate count so long as the target supports overlap minimally or not at all. Our work is complementary to this in that the theory we introduce is designed to handle very complex target support overlaps.

The other example of target counting without identification or localization arises in work of Singh *et al.*, who consider a network of sensors which return a value in $\{0, 1\}$ depending on target proximity [18]. Their technique involves using time-series data in the case of moving targets/sensors, since a target count in the stationary case is too difficult, even if all target supports are convex, round, fixed, etc.

We have ignored for the moment many of the important technical issues associated with network implementation of our methods. Much of the work in aggregation of data by a network concerns network protocols for signal processing [13], managing constraints on bandwidth and energy [4], and dealing with errors or node failures [20]. This introductory paper does not treat these important issues. We also assume noise-free sensor readings. This paper assumes an idealized setting to develop and highlight the formal tools.

B. Outline

The main results of this note consist of: (1) a theorem on target enumeration for continuum 'sensor fields' based on a topological integration theory; (2) bounds on the integrals for planar domains with holes; (3) a refinement theorem applicable to network discretizations; (4) a duality theorem for planar networks which provides a fast, distributed algorithm for ad hoc networks; (5) methods for computing expected target counts in the case of incomplete information; and (6) an outline of applications to time-dependent systems such as mobile robot sensing modalities. This note tersely summarizes ideas from [1], [2] in the restricted context of a planar network.

Our results follow from the classical and elegant theory of integration with respect to Euler characteristic [19], [16]. After surveying a simplified version of these methods in §II, we prove the fundamental enumeration theorem in §III. To solve the problem of sparse network discretization, we provide bounds in §V on integrals over planar domains with a hole. This yields a simple refinement theorem in §VI, and extensions in §VII. We prove a duality result for planar networks in §VIII that leads to fast numerical implementation, outlined in §XI.

II. Topological Integration

We present a simple, self-contained introduction to a topological integration theory. For simplicity, we work in the simplicial category. Let X denote a simplicial complex: a topological space built from a collection of closed simplices glued together along faces (see, *e.g.*, [9] for elementary definitions).

Definition 1: The EULER CHARACTERISTIC of a compact simplicial complex X has two equivalent definitions:

1) **combinatorial:**

$$\chi(X) = \sum_{k=0}^{\infty}(-1)^k \#\{k\text{-simplices in } X\}. \tag{1}$$

2) **homological:**

$$\chi(X) = \sum_{k=0}^{\infty}(-1)^k \dim(H_k(X)). \tag{2}$$

Here, $H_k(X)$ denotes the kth (simplicial) homology of X (in R coefficients), a vector space that measures the number of 'holes' in X that a k-dimensional subcomplex can detect [9]. As homology depends only on the homotopy type of X, the Euler characteristic χ is a topological invariant of a space, independent of how it is triangulated into a simplicial complex.

Example 2: The following examples are illustrative:

1) *Euler characteristic is a generalization of cardinality: for a discrete set X, $\chi(X) = |X|$.*
2) *If X is a compact contractible set — if it can be deformed continuously within itself to a single point — then $\chi(X) = 1$.*
3) *For a finite graph Γ, the Euler characteristic is $\chi(\Gamma) = \#V(\Gamma) - \#E(\Gamma)$.*
4) *For $X \subset \mathbb{R}^2$ a connected set with N holes, $\chi(X) = 1 - N$.*

The Euler characteristic satisfies an inclusion-exclusion principle (a consequence of the Mayer-Vietoris sequence on homology [9]): for A and B compact subcomplexes of X,

$$\chi(A \cup B) = \chi(A) + \chi(B) - \chi(A \cap B). \tag{3}$$

This equation evokes the definition of a measure and allows one to interpret χ as a generalized signed measure (generalized, as it is only finitely additive). As many authors have observed [15], [16], [19], this measure behaves as any conventional measure when restricted to the appropriate classes of integrands and domains. In the setting of $\mathbb{Z}$-valued functions over simplicial complexes, this measure theory is completely tame.

Definition 3: Let X denote a simplicial complex and $CF(X)$ the abelian group of functions from X to $\mathbb{Z}$ with generators $\mathbb{1}_\sigma$, where σ is a closed simplex of X. Given such a $\phi = \sum_\alpha c_\alpha \mathbb{1}_{\sigma_\alpha}$ in $CF(X)$, the INTEGRAL of ϕ with respect to Euler characteristic is defined to be

$$\int_X \phi \, d\chi := \sum_\alpha c_\alpha. \tag{4}$$

This integral is well-defined.

Lemma 4 ([19], [16]): The integral $\int_X \phi \, d\chi$ depends only on the function ϕ and not on its decomposition. Specifically, if $\phi = \sum_\alpha c_\alpha \mathbb{1}_{U_\alpha}$, where U_α is a subcomplex of X, then

$$\int_X \phi \, d\chi = \sum_\alpha c_\alpha \chi(U_\alpha). \tag{5}$$

Proof: Given two subcomplexes A and B, the relation $\mathbb{1}_{A\cup B} = \mathbb{1}_A + \mathbb{1}_B - \mathbb{1}_{A\cap B}$ is mirrored by Equation (3). It follows that $\int_X \mathbb{1}_{U_\alpha} d\chi = \chi(U_\alpha)$. By definition, $\int_X \cdot d\chi$ is a homomorphism from $CF(X)$ to $\mathbb{Z}$; the lemma follows. ■

Remark 5: It is by no means necessary to restrict to simplicial complexes. For a large class of topological spaces without an explicit cell structure, χ is well-defined using Eqn. (2) with singular (or, better still, Borel-Moore) homology. Likewise, the class of integrable functions $CF(X)$ above generalizes to the sheaf of CONSTRUCTIBLE functions on X [16]. This level of generality is not required for this paper.

III. Enumeration via Integration

We turn now to target-counting problems. The following mathematical formulation leads naturally to the integration theory of the previous section.

Consider a setting where the sensors are parameterized by a (reasonably nice) topological space X. One imagines a 'continuum field' setting in which a counting sensor resides at every point of X. It is helpful to keep in mind two cases: (1) $X = \mathbb{R}^2$ and is 'filled' with sensors; (2) X is a simplicial complex, where the counting sensors at the vertices of X 'pass' counting data to all other simplices of X. Assume a finite set of stationary targets are present and detectable by the sensor field. We do not specify detection ranges, etc., in terms of geometric constraints, but rather in terms of sets. For each target α, define its TARGET SUPPORT, $U_\alpha \subset X$, to be the subset of those sensors to which the target is 'visible' (rather, sensed: the actual sensor modality is irrelevant). The sensor field on X returns a counting function $h : X \to \mathbb{N}$, where

$$h(x) = \#\{\alpha : x \in U_\alpha\}.$$

Assuming knowledge of the target supports' topology, one has a simple means of enumerating the targets without localization or identification.

Theorem 6: If each target has a compact contractible support $\mathcal{U}_\alpha \subset X$, then the integral of the impact function $h(x) = \#\{\alpha : x \in U_\alpha\}$ with respect to $d\chi$ is the target count:

$$\#\alpha = \int_X h\, d\chi. \tag{6}$$

Proof: By definition, $h = \sum_\alpha \mathbb{1}_{U_\alpha}$. As U_α is compact and contractible, $\chi(U_\alpha) = 1$ and $\int_X h\, d\chi = \sum_\alpha 1 = \#\alpha$. ∎

The remarkable aspect of this result is that there are no constraints on the target supports other than the topological: each target has support with $\chi = 1$. In particular, targets can have different 'impact' on the sensor field, and there is no need for convexity or fixed-radius assumptions.

Theorem 6 does, however, assume that there is a well-defined counting function h over all of X — *e.g.*, given by a sensor at every point in the space X. In a less idealized setting, one has a finite number of nodes which, under the best circumstances, triangulates a region of the plane. The integrand 'counting function' h is known only on the vertex set of this triangulation $\mathcal{T}$. The target supports $U_\alpha \subset \mathbb{R}^2$ need not be well-placed with respect to $\mathcal{T}$ at all.

We resolve this discretization problem by extending the integration theory to $\mathbb{R}$-valued integrands. By taking the usual step-function upper semi-continuous approximation to a limit, one can define $\int h\, d\chi$ for real-valued functions $h : X \to [0, \infty)$ which are reasonably behaved (*e.g.*, which have a finite number of critical points). This extension of the theory is not without complications (*e.g.*, the integration operator is no longer linear), but it allows one to import perspectives from numerical analysis. In particular, given a sampling of an integrand h over a discrete set, the integral of the piecewise-linear (PL) interpolation of h, denoted h_{PL}, should be a good approximation if the sampling is of sufficient fidelity. This holds for integration with respect to $d\chi$.

Theorem 7: Fix a collection $\{U_\alpha\}$ of compact target supports in $\mathbb{R}^n$ in general position. For a triangulation $\mathcal{T}$ of $\mathbb{R}^n$, let h_{PL} denote the piecewise-linear extension of the restriction of $h = \sum_\alpha \mathbb{1}_{U_\alpha}$ to the vertices of $\mathcal{T}$. Then, for $\mathcal{T}$ sufficiently fine and regular,

$$\int_{\mathcal{T}} h_{PL}\, d\chi = \int_{\mathbb{R}^n} h\, d\chi = \#\alpha. \tag{7}$$

Proof: See §VI. ∎

IV. Computation

Theorems 6 and 7 are useless without effective means of computing integrals with respect to $d\chi$. Fortunately, there are several means of doing so.

Theorem 8: Given a compactly supported impact function $h : X \to \mathbb{N}$, the integral of h with respect to $d\chi$ may be computed as:

$$\int_X h\, d\chi = \sum_{s=0}^{\infty} \chi(\{h > s\}) \tag{8}$$

$$= \int_{s=0}^{\infty} \chi(\{h \geq s\})\, ds \tag{9}$$

$$= \sum_{p \in \mathcal{C}(h)} (-1)^{n-\mu(p)} h(p) \tag{10}$$

Eqns. (8) and (9) apply to $\mathbb{N}$-valued and $[0, \infty)$-valued impact functions respectively, and the notation $\{h > s\}$ represents the set $h^{-1}((s, \infty))$. Eqn. (10) applies to a $[0, \infty)$-valued Morse function on an n-dimensional manifold, where $\mathcal{C}(h)$ is the set of critical points of h, and $\mu(p)$ is its Morse index of $p \in \mathcal{C}(h)$ [14].

Proof: Eqns. (8) and (9) are elementary and follow directly from the definitions. For Eqn. (10), one has h Morse. Thus, the Euler characteristic of upper excursion sets is piecewise-constant, changing only at critical values. For $p \in \mathcal{C}(h)$, $s = h(p)$, and $\epsilon \ll 1$, elementary Morse theory [14] says that $\{h \geq s + \epsilon\}$ differs from $\{h \geq s - \epsilon\}$ by the addition of a product of discs $D^{\mu(p)} \times D^{n-\mu(p)}$ glued along $D^{\mu(p)} \times \partial D^{n-\mu(p)}$. The change in Euler characteristic resulting from this handle addition is $(-1)^{n-\mu(p)}$. This, applied to Eqn. (8) yields Eqn. (10). ∎

This theorem means, roughly speaking, that one can trade between Euler characteristic counts, integrals with respect to Lebesgue measure, and Morse theory at will.

Fig. 1(a) gives an example of a collection of target supports $\{U_\alpha\}$ with height function, which is sampled on a uniform hexagonal grid in (b). The upper excursion sets of h are easily computed and the integral with respect to Euler characteristic is thus:

$$\#\alpha = \int h\, d\chi = \overbrace{1}^{s=2} + \overbrace{3}^{s=1} + \overbrace{0}^{s=0} = 4. \tag{11}$$

Smoothing h to a function $\tilde{h}$ with nondegenerate critical points yields three maxima and three saddles, with minima (at height zero), see Fig. 1(c). Formula (10) implies

$$\int h\, d\chi = \overbrace{(3+2+2)}^{\mu=2} - \overbrace{(1+1+1)}^{\mu=1} + \overbrace{0}^{\mu=0} = 4. \tag{12}$$

Taking the PL extension h_{PL} yields upper excursion sets as illustrated in Fig. 1(d): Eqn. (9) yields a computation similar to that of (11).

V. Holes in the Network

It is common in sensor networks to encounter 'holes' within the network, through incomplete coverage or node failures. In this case, one wants to estimate the number of targets relative to the missing information. This translates to the following relative problem: if one knows $h : X \to \mathbb{N}$ only on some subset $A \subset X$, how well can one estimate $\int_X h\, d\chi$ from the restriction $h|A$? We give bounds for the planar case.

Denote by $\mathcal{V} = \{V_\beta\}$ the collection of nonempty connected components of intersections of all target supports U_α with $\overline{D}$. Since we work in $\mathbb{R}^2$, each V_β is a compact contractible set which intersects ∂D. By Theorem 6, $\int_{\overline{D}} h\,d\chi$ equals the number of components $|\mathcal{V}|$. There are at least $\max_{\partial D} h$ such pieces; hence

$$\int_{\overline{D}} \hat{h}\,d\chi \leq \int_{\overline{D}} h\,d\chi.$$

Consider $\min_{\partial D} h$ and remove from the collection $\mathcal{V}$ this number of elements, including all such V_β equal to $\overline{D}$ (which is possible since we remove $\min_{\partial D} h$ such elements). Each remaining $V_\beta \in \mathcal{V}$ is not equal to $\overline{D}$ and thus intersects ∂D in a set with strictly positive Euler characteristic. Thus,

$$\int_{\overline{D}} h\,d\chi \leq \int_{\overline{D}} \check{h}\,d\chi.$$

■

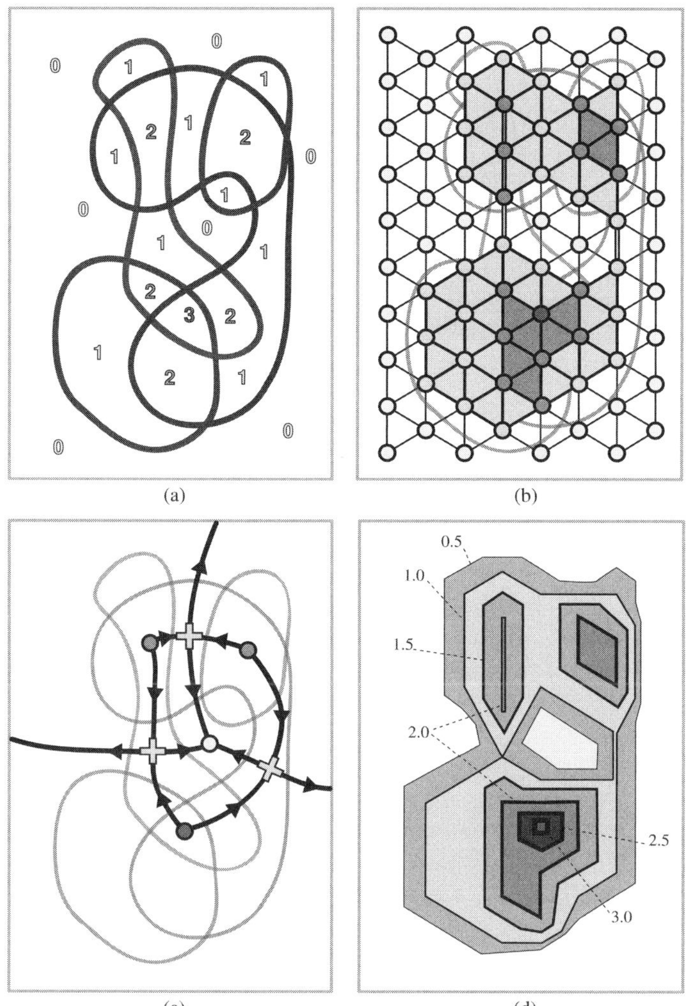

Fig. 1. The height function of a collection of target supports (a) is sampled on a regular triangulation (b). This can be smoothed to a Morse function with maxima/saddles/minima (c) or extended over the triangulation via PL interpolation (d).

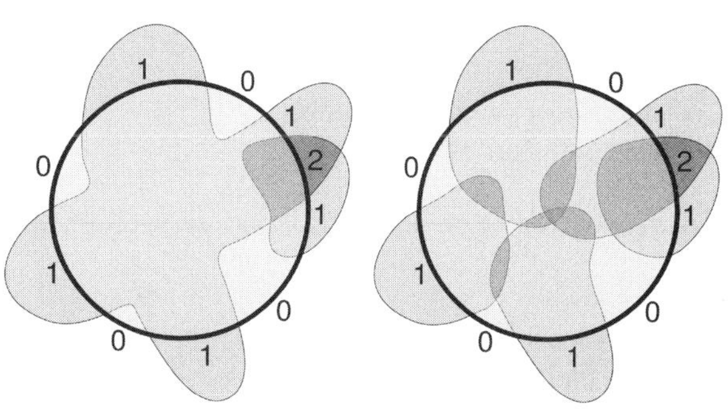

Fig. 2. An example for which the upper and lower bounds of Eqn. (13) are sharp.

Theorem 9: Assume $h : \mathbb{R}^2 \to \mathbb{N}$ is the sum of indicator functions over a collection of compact contractible sets in $\mathbb{R}^2$, none of which is contained entirely within D, a fixed open contractible disc. Then

$$\int_{\mathbb{R}^2} \hat{h}\,d\chi \leq \int_{\mathbb{R}^2} h\,d\chi \leq \int_{\mathbb{R}^2} \check{h}\,d\chi, \tag{13}$$

where

$$\hat{h}(y) = \begin{cases} \max_{\partial D} h & : y \in \overline{D} \\ h & : \text{else} \end{cases}$$

$$\check{h}(y) = \begin{cases} \min_{\partial D} h & : y \in D \\ h & : \text{else} \end{cases}$$

Proof: Via additivity of χ over domains, Eqn. (13) follows from the corresponding inequalities over the compact domain $\overline{D}$. Explicitly, if $\overline{h} = h$ on $\mathbb{R}^2 - \overline{D}$, then

$$\int_{\mathbb{R}^2} \overline{h}\,d\chi = \int_{\mathbb{R}^2 - D} h\,d\chi - \int_{\partial D} h\,d\chi + \int_{\overline{D}} \overline{h}\,d\chi.$$

Example 10: Consider the example illustrated in Fig. 2. The upper and lower estimates for the number of targets are 2 and 4 respectively. For this example, the estimates are sharp in that one can have collections of target supports over compact contractible sets which agree with h outside of $\overline{D}$ and realize the bounds.

Remark 11: The lower bound $\hat{h}$ can fail in several ways. For example, a target support can intersect D in multiple components, causing $\hat{h}$ to not have a decomposition as a sum of characteristic functions over contractible sets (but rather with annuli). One can even find examples for which each target support intersects $\overline{D}$ in a contractible set but for which $\int \hat{h}\,d\chi$ is negative. The fact that the lower bound $\int \hat{h}$ can be so defective follows from the difficulty associated with annuli in the plane — these are 'large sets of measure zero' in $d\chi$.

VI. The Refinement Theorem

The bounds of §V allow one to conclude when a hole is 'inessential' and no ambiguity about the integral exists.

Corollary 12: Under the hypotheses of Theorem 9, the upper and lower bounds are equal when there is a unique connected local maximum of h on ∂D.

Proof: In the case where h is constant on ∂D, $\check{h} = \hat{h}$ and the result is trivial. Otherwise, both $\check{h}$ and $\hat{h}$ have connected (and thus contractible) upper excursion sets. Applying Eqn. (8) yields

$$\begin{aligned}\int_{\mathbb{R}^2} \check{h}\, d\chi - \int_{\mathbb{R}^2} \hat{h}\, d\chi &= \sum_s \chi(\{\check{h} \geq s\}) - \chi(\{\hat{h} \geq s\}) \\ &= \sum_s 1 - 1 = 0.\end{aligned}$$

■

This permits an easy proof of Theorem 7 that, in accordance with one's intuition about integration, refinement of the network leads to convergence of the integrals.

Proof: (of Theorem 7) Since integration is local, we may compare $\int h\, d\chi$ and $\int h_{PL}\, d\chi$ over a single closed 2-simplex of the triangulation $\mathcal{T}$: if these are always equal, then the theorem follows. One observes that for the U_α in general position and $\mathcal{T}$ sufficiently fine and regular, the unique local maxima of h and h_{PL} on the boundary of any 2-simplex of $\mathcal{T}$ are equal and both level sets are connected. Corollary 12 completes the proof. ■

It is easy to extend this proof to $\mathbb{R}^n$ by proving the appropriate extension of Corollary 12.

VII. Harmonic Extension and Expected Target Counts

We continue the results of the previous section, considering the case of a planar domain with a contractible hole on which the integrand is unknown. As shown, upper and lower bounds are realized by extending the integrand across the hole via minimal and maximal values on the boundary of the hole. Inspired by the result that the PL-extension of a discretely sampled integrand yields correct integrals with respect to Euler characteristic, we consider extensions over holes via *continuous* functions.

The following result says that there is a principled interpolant between the upper and lower extensions. Roughly speaking, an extension to a harmonic function (discrete or continuous, solved over the hole with Dirichlet boundary conditions) provides an approximate integrand whose integral lies between the bounds given by upper and lower convex extensions. There is nothing magical about harmonic functions: any form of weighted averaging will lead to an extension which respects the bounds. A specific criterion follows.

Theorem 13: Given $h : \mathbb{R}^2 - D \to \mathbb{N}$ satisfying the assumptions of Theorem 9, let $\overline{h}$ be any extension of h which has no strict local maxima or minima on D. Then

$$\int_{\mathbb{R}^2} \hat{h}\, d\chi \leq \int_{\mathbb{R}^2} \overline{h}\, d\chi \leq \int_{\mathbb{R}^2} \check{h}\, d\chi, \tag{14}$$

Proof: Consider an open neighborhood of $\overline{D}$ in $\mathbb{R}^2$ and modify $\overline{h}$ so that it preserves critical values, is Morse, and falls off to zero quickly outside of D. This perturbed function, denoted $\tilde{h}$, has isolated maxima on ∂D, isolated saddles in the interior of D (since there are no local extrema in D by hypothesis) and no other critical points outside of D. Since $\tilde{h}$ is a small perturbation of $\overline{h}$, the integral of $\tilde{h}$ with respect to $d\chi$ is equal to $\int_{\overline{D}} \overline{h} d\chi$. Via the Morse-theoretic formula of Eqn. (10),

$$\int \tilde{h}\, d\chi = \sum_{p \in \mathcal{C}(\tilde{h})} (-1)^{2-\mu(p)} \tilde{h}(p).$$

The integral thus equals the sum of h over the maxima on ∂D minus the sum of $\overline{h}$ over the saddle points in the interior of D, since saddles have Morse index $\mu = 1$.

Denote by $\{p_i\}_1^M$ the maxima of $\tilde{h}$, ordered by their (increasing) $\tilde{h}$ values. Denote by $\{q_i\}_1^N$ the saddles of $\tilde{h}$, ordered by their (increasing) $\tilde{h}$ values. By the Poincaré index theorem,

$$1 = \chi(\overline{D}) = \#\text{maxima}(\tilde{h}) - \#\text{saddles}(\tilde{h}),$$

hence, $N = M - 1$. Note that, since there are no local minima, $\tilde{h}(q_i) < \tilde{h}(p_i)$ for all $i = 1 \ldots M-1$. Thus,

$$\begin{aligned}\int_{\overline{D}} \overline{h}\, d\chi &= \int_{\overline{D}} \tilde{h}\, d\chi \\ &= \tilde{h}(p_M) + \sum_{i=1}^{M-1} \tilde{h}(p_i) - \tilde{h}(q_i) \\ &\geq \tilde{h}(p_M) = \max_{\partial D} h = \int_{\overline{D}} \hat{h}\, d\chi.\end{aligned}$$

For the other bound,

$$\begin{aligned}\int_{\overline{D}} \overline{h}\, d\chi &= \tilde{h}(p_M) + \sum_{i=1}^{M-1} \tilde{h}(p_i) - \tilde{h}(q_i) \\ &\leq \sum_{i=1}^{M} \tilde{h}(p_i) = \int_{\overline{D}} \check{h}\, d\chi.\end{aligned}$$

■

A harmonic or harmonic-like function $\tilde{h}$ will often lead to an integral with non-integer value. Such an integral is best interpreted as an *expected* target count.

Example 14: Consider a hole D and a function h which is known only on ∂D and which has two maxima with value 1 and two minima with value 0. Without knowing more about the possible size and shape of the target supports which make up h, it is not clear whether this is more likely to come from one target support (which crosses the hole) or from two separate target supports. Computing a harmonic extension of this h over the interior of D yields a function $\tilde{h}$ with one saddle-type critical point in D. The value of the saddle is c and satisfies $0 < c < 1$, depending on the geometry of h on ∂D. This yields $\int \tilde{h}\, d\chi = 2 - c$, reflecting the uncertainty of either one or two targets. In the perfectly symmetric case of Fig. 3[left], $c = \frac{1}{2}$ and the expected target count is, naturally, $\frac{3}{2}$. In Fig. 3[right], the harmonic extension has $c < \frac{1}{2}$, meaning that it is more likely that there are two target supports.

In the network setting, holes often arise due to node failure or lack of sufficient node density. In these scenarios, one may reasonably employ any weighted local averaging scheme across dead nodes to recover a function which will respect the bounds of Theorem 9. Different weighting schemes may

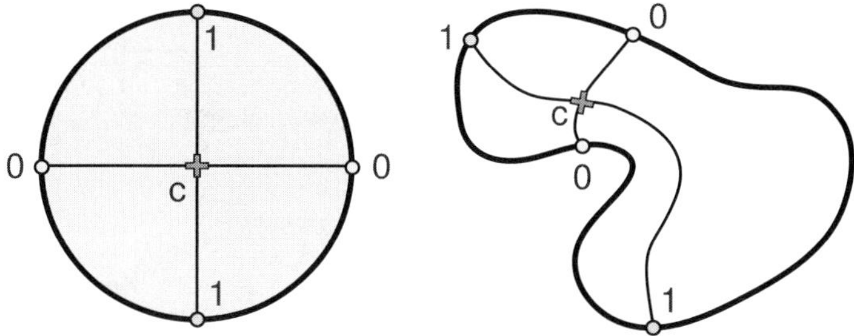

Fig. 3. An integrand with a hole has two minima at height 0 and two maxima at height +1. Filling in by a harmonic function $\tilde{h}$ has an interior saddle at height $0 < c < 1$, depending on the geometry of h on ∂D: [left] $c = \frac{1}{2}$; [right] $c < \frac{1}{2}$.

be more appropriate for different systems. For example, node readings can be assigned a "confidence" measure, which, when used as a weighting for the averaging over the dead zone, returns an expected value of the integral which reflects the fidelity of the data.

VIII. Holes via Duality

We augment Theorem 8 with a specialized formula for the plane which aids greatly with implementation. The strategy of this formula is to exploit the duality between holes and connected components of the complement. This duality has a formal expression in terms of algebraic topology.

Theorem 15: For $h : \mathbb{R}^2 \to \mathbb{N}$

$$\int_{\mathbb{R}^2} h \, d\chi = \sum_{s=0}^{\infty} \left(\beta_0\{h > s\} - \beta_0\{h \leq s\} + 1 \right), \qquad (15)$$

where β_0 is the number of connected components of the set.

Proof: Let A be a compact nonempty subset of $\mathbb{R}^2$. Since $A \subset \mathbb{R}^2$, $H_s(A) = 0$ for all $s \geq 2$. Thus, via Eqn. 2, it suffices to compute

$$\chi(A) = \dim H_0(A) - \dim H_1(A).$$

Note: $\dim H_0$ equals the number of connected components of A. The quantity $\dim H_1(A)$, the number of holes in A, is, by Alexander duality [9], equal to $\dim H_0(\mathbb{R}^2 - A) - 1$, the number of (bounded) connected components of the complement. The proof is completed by Eqn. (8), substituting in $A = \{h > s\}$ and $\mathbb{R}^2 - A = \{h \leq s\}$. ∎

Example 16: The duality formula (15) applied to the integrand of Fig. 1 yields

$$\int_{\mathbb{R}^2} h \, d\chi = \overbrace{1 - 2 + 1}^{s=0} + \overbrace{3 - 1 + 1}^{s=1} + \overbrace{1 - 1 + 1}^{s=2} = 4.$$

The formula in Theorem 15 is extremely applicable. We note that the determination of the number of connected components of the upper and lower excursion sets is a simple clustering problem, computable in logspace with respect to the number of network nodes.

IX. Ad hoc networks

We note that the strategy of converting the sampling of the true impact function h over $\mathcal{N}$ to a PL interpolation $\overline{h}$ does not necessarily require knowing the coordinates of the nodes. Indeed, the evaluation of $\int_Y \cdot \, d\chi$ is conspicuous in its freedom from coordinate geometry: it is a topological integral. If one is given a triangulation, the extension of the counting function h on vertices over the domain is automatic. However, if no geometry associated to $\mathcal{N}$ is known, it may not be possible to determine a canonical extension h_{PL} over the domain. Such a situation is not uncommon in sensor networks based on *ad hoc* wireless communications, an increasingly common protocol for distributed sensor networks and robotics.

Assume that one is given a network in the form of an abstract graph $\mathcal{G} = (\mathcal{N}, \mathcal{E})$. By "abstract" we mean that the projection of the 1-d cell complex $\mathcal{G}$ to the workspace is unknown. Edges should possess some proximity data. For example, one could assume that $\mathcal{G}$ is a UNIT DISC GRAPH, in which edges exist between nodes if and only if they are within unit distance in the workspace. A more realistic model is the QUASI unit disc graph, in which edges definitely exist below a certain distance, definitely do not exist above a certain distance, and may exist (say, according to some probability distribution) for nodes within a critical interval of distance. At any rate, the duality results of §VIII allow us to compute integrals based on ad hoc networks.

Corollary 17: Assume an integrand $h : \mathbb{R}^2 \to \mathbb{N}$, and let $\mathcal{G}$ be a network graph with nodes $\mathcal{N} \subset \mathbb{R}^2$, where the only thing known is the restriction of h to $\mathcal{N}$ (in particular, the coordinates of $\mathcal{N}$ in $\mathbb{R}^2$ are unknown). If the network $\mathcal{G}$ correctly samples the connectivity of the upper and lower excursion sets of h, then Eqn. (15) returns the exact number of targets.

An example appears in Fig. 4. Note that in this example, the topology of the excursion sets of h are *not* sampled correctly: sparsity leads to holes in the network. Nevertheless, since the connectivity of the upper and lower excursion sets is sampled faithfully, the integral is correct. Although the example drawn is a unit disc graph, this is by no means necessary for the result.

X. Mobile agents

The setting of this work has assumed stationary targets with fixed target supports, being sensed by a fixed network of stationary counting sensors. It is desirable to violate both assumptions, especially in the robotics context. We indicate how the results of this note are applicable to both settings in a sequence of remarks.

Remark 18: Consider the following scenario: a collection of fixed target supports $\{U_\alpha\}$ lie in the plane. One or more mobile robots R_i can maneuver in the plane along chosen paths $x_i(t)$, returning sensed counting functions $h_i(t) = \#\{\alpha : x_i(t) \in U_\alpha\}$. How should the paths x_i be chosen so as to effectively determine the correct target count? If target supports are extremely convoluted, no guarantees are possible: therefore, assume that some additional structure is known (*e.g.*,

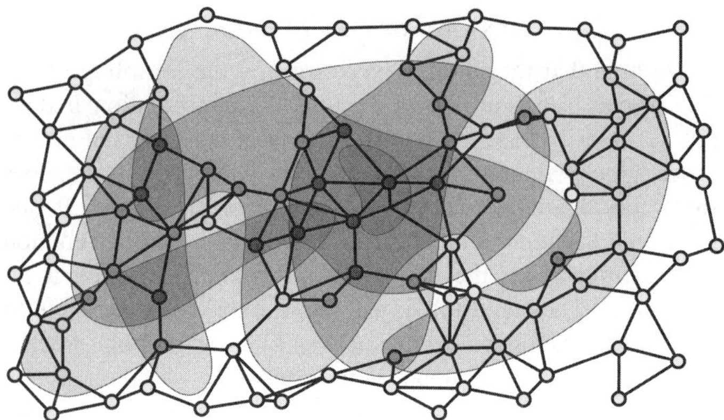

Fig. 4. A sparse sampling over an *ad hoc* network retains enough connectivity data to evaluate the integral exactly.

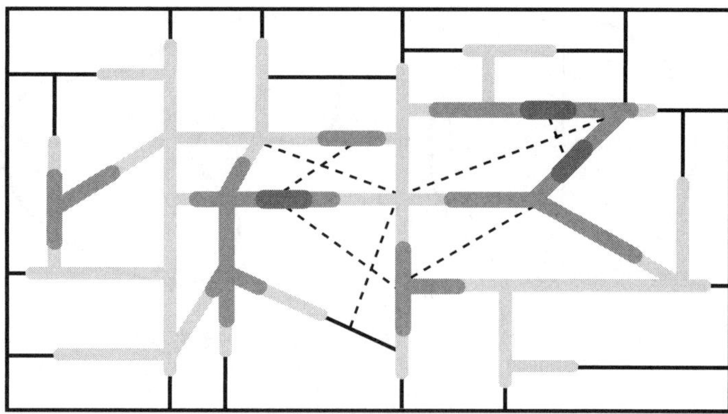

Fig. 5. Mobile agents determine target counts over a graph Γ. Holes with multiple maxima require further refinement (dashed lines).

an injectivity radius) giving a lower bound on how "thin" the target supports may be.

Assume that the robots initially explore the planar domain along a rectilinear graph Γ that tiles the domain into rectangles. If desired, one can make these rectangles have either width or height in order to guarantee that all the U_α intersect Γ. Consider the sensor function $h : \Gamma \to \mathbb{N}$. The integral $\int_\Gamma h\, d\chi$ is likely to give the wrong answer, even (especially!) for a dense Γ. Two means of getting a decent approximation are (1) use the duality formula of Eqn. (15); or (2) perform a harmonic extension over the holes of Γ as per §VII.

However, neither is guaranteed to give a good *a priori* approximation to the target count. How can one tell if Γ should be filled in more? The simplest criterion follows from Corollary 12. Consider a basic cycle $\Gamma' \subset \Gamma$ in the tiling induced by Γ. If there is a single connected local maximum on Γ', then (assuming that no small U_α lies entirely within the hole) the harmonic extension over Γ' gives an accurate contribution to the integral.

If, on the other hand, there are multiple maximal sets on Γ', then one must refine Γ into smaller cycles for which the criterion holds. The obvious approach is to guide the mobile sensors so as to try and connect disjoint maxima and/or disjoint minima. Fig. 5 gives the sense of the technique. We leave for future work detailing a complete algorithm and its analysis: the crucial observation is that Corollary 12 provides a stopping criterion.

Remark 19: One can imagine a much more complicated scenario. Consider the case where the target supports also vary (continuously) as a function of time: $U_\alpha(t) \subset \mathbb{R}^2$. However, the supports are unknown to the robots R_i, which can measure only a sampled count $h_i(t)$.

The problem is clearly unsolvable if there is a single, slow robot: such a sensor may never detect any (evasive) targets at all. On the other hand, if one assumes a dense network of sensors, the problem is trivial: at any fixed time, take a triangulation of the domain based on the robot positions, and compute the integral of the sensor function as per Theorem 7.

Where the problem is critically difficult is when the swarm of sensors is not dense enough to cover the plane, but does form a connected network with holes. These holes will change temporally, emerging, bifurcating, disappearing: all the while, mobile targets can slip in and out.

In this dynamic setting, the work in §V-VII suggests a natural strategy of computing an expected value of the integral as a function of time and keeping a running average of these approximants. More sophisticated tracking of targets within holes can be accomplished by examining localized temporal discontinuities of these integrals. This is the subject of a separate report.

Our discussion of mobile agents is necessarily brief: there are many more results possible about counting mobile targets without the need of clocks at all [1]. We leave these and implementation issues for a more detailed future treatment.

XI. Numerical issues

Space constraints forbid a comprehensive treatment of the topic of numerical integration with respect to Euler characteristic, a topic which seems to have been explored only in [12], and here from an integral-geometry perspective: there is much to be done. We present a few significant remarks, and leave the details for an archival work.

Remark 20: Implementation. We have implemented the integration formula of §VIII, Eqn. (15), for ad hoc planar networks based on a random unit disc graph: see Fig. 6. The code (written in Java and publicly available at [`hidden for review`]) allows the user to specify target support by drawing with the mouse. By using the obvious clustering algorithm, the code returns the quantity specified in Eqn. (15) in negligible time ($\sim$ 1s for a network of $\sim 100,000$ nodes).

Remark 21: Numerical errors. Of course, the guarantee that Eqn. (15) computes the correct value of the integral depends on having sampled the connectivity of the upper and lower excursion sets correctly. No a priori knowledge of this can be assured without knowing more about the network or the target support. Unfortunately, the duality formula computes a $\mathbb{Z}$-valued sum, any error in the computation is quantized. From the point of view of numerical errors, it is preferable to work with the 'expected' $\mathbb{R}$-valued integrals as in Eqn. (10) and §VII.

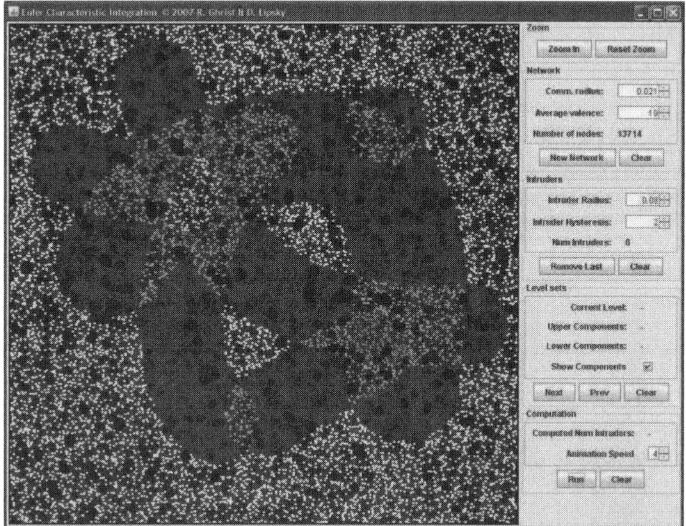

Fig. 6. Screenshot of a Java applet implementation [written by D. Lipsky], *cf.* Remark 20.

Remark 22: Distributed computation. Since our methods are based on an integration theory, the enumeration of targets detailed in this paper is a local computation. For A, B compact,

$$\int_{A\cup B} h\,d\chi = \int_A h\,d\chi + \int_B h\,d\chi - \int_{A\cap B} h\,d\chi.$$

Thus, enumeration can be performed in a distributed manner easily. This is particularly easy when the network is a lattice, as one can employ standard distributed protocols for localization and merging of target counts.

XII. Concluding remarks

The core message of this paper is that thinking of target enumeration in terms of a *topological integration theory* is much better than a raw combinatorial approach. One can import intuition, techniques, and perspectives from numerical analysis, algebraic topology, differential topology, and combinatorics at will.

This short paper has left many natural questions unanswered. We give a brief list of questions and remarks, to be expanded on in future papers.

Remark 23: Do these results extend to higher dimensions? Yes, and to reasonable topological spaces as well. The results on bounds for holes and the duality formula, unfortunately, do not generalize, being dependent on planar topology.

Remark 24: What about noise? This integration theory is robust to dead sensors: an empty node creates a 'hole' which the techniques of this paper resolve. However, as this integration theory counts the number and heights of critical points, it is very sensitive to integer-valued noise. A smoothing filter is required to preprocess noisy data in order to obtain accurate results.

Remark 25: How do you know if you've sampled the domain finely enough? As in the case of trying to approximate the Riemann integral of an unknown function from a finite point sample, one does not know without more data.

Remark 26: What about sensors which do not count but rather measure $[0,\infty)$-valued intensity? This integration theory is not immediately applicable, since the operator $\int \cdot\, d\chi$ is not linear on continuous integrands. However, one can obtain lower bounds using methods akin to Lusternik-Schnirelmann category [3].

We hope the reader finds that the increase in formalism for this integration theory more than pays for itself in terms of potential applications.

Acknowledgements

This work is funded by DARPA # HR0011-05-1-0008.

References

[1] Y. Baryshnikov and R. Ghrist, "Target enumeration via Euler characteristic integrals I: sensor fields," to appear, *SIAM J. Appl. Math.*

[2] Y. Baryshnikov and R. Ghrist, "Target enumeration via Euler characteristic integrals II: sensor networks," in preparation.

[3] Y. Baryshnikov and R. Ghrist, "Unimodal category and the topological decomposition of distributions," in preparation.

[4] A. Boulis, S. Ganeriwal, and M. Srivastava, "Aggregation in sensor networks: an energy - accuracy tradeoff," *J. Ad-hoc Networks*, 1, 2003, 317–331.

[5] D. Estrin, D. Culler, K. Pister, and G. Sukhatme, "Connecting the Physical World with Pervasive Networks," *IEEE Pervasive Computing* 1:1, 2002, 59–69.

[6] Q. Feng, F. Zhao, And L. Guibas, "Lightweight Sensing and Communication Protocols for Target Enumeration and Aggregation" in proceedings *MobiHoc*, 2003.

[7] H. Groemer, "Minkowski addition and mixed volumes," *Geom. Dedicata* 6, 1977, 141–163.

[8] L. Guibas. "Sensing, Tracking and Reasoning with Relations," *IEEE Signal Processing Magazine*, 19(2), Mar 2002, .

[9] A. Hatcher, *Algebraic Topology*, Cambridge University Press, 2002.

[10] T. He, P. Vicaire, T. Yan, L. Luo, L. Gu, G. Zhou, R. Stoleru, Q. Cao, J. Stankovic, T. Abdelzaher, "Achieving Real-Time Target Tracking Using Wireless Sensor Networks," in proceedings of *IEEE Real Time Technology and Applications Symposium*, 2006, 37–48.

[11] B. Jung and G. Sukhatme, "A Region-Based Approach for Cooperative Multi-Target Tracking in a Structured Environment," in proceedings of *IEEE/RSJ Conference on Intelligent Robots and Systems*, 2002.

[12] D. Klain, K. Rybnikov, K. Daniels, B. Jones, C. Neacsu, "Estimation of Euler Characteristic from Point Data," preprint 2006.

[13] D. Li, K. Wong, Y. Hu, and A. Sayeed, "Detection, classification, and tracking of targets," *IEEE Signal Processing Magazine*, 19(2), 2002, 17–30.

[14] J. Milnor, *Morse Theory*, Princeton University Press, 1963.

[15] G.-C. Rota, "On the combinatorics of the Euler characteristic," Studies in Pure Mathematics, Academic Press, London, 1971, 221–233.

[16] P. Schapira, "Operations on constructible functions," *J. Pure Appl. Algebra* 72, 1991, 83–93.

[17] P. Schapira, "Tomography of constructible functions," in proceedings of *11th Intl. Symp. on Applied Algebra, Algebraic Algorithms and Error-Correcting Codes*, 1995, 427-435.

[18] J. Singh, U. Madhow, R. Kumar, S. Suri, and R. Cagley, "Tracking multiple targets using binary proximity sensors." In *Proceedings of the 6th international Conference on Information Processing in Sensor Networks*, 2007, 529-538.

[19] O. Viro, "Some integral calculus based on Euler characteristic," Lecture Notes in Math., vol. 1346, Springer-Verlag, 1988, 127–138.

[20] J. Zhao, R. Govindan and D. Estrin, "Computing Aggregates for Monitoring Wireless Sensor Networks," in proceedings of *IEEE Intl. Workshop on Sensor Network Protocols and Applications (SNPA)*, 2003.

Using Recognition to Guide a Robot's Attention

Alexander Thomas*, Vittorio Ferrari†, Bastian Leibe†, Tinne Tuytelaars* and Luc Van Gool†*
*Katholieke Universiteit Leuven, Belgium
Email: alexander.thomas/tinne.tuytelaars@esat.kuleuven.be
†ETH Zürich, Switzerland
Email: ferrari/leibe/vangool@vision.ee.ethz.ch

Abstract— **In the transition from industrial to service robotics, robots will have to deal with increasingly unpredictable and variable environments. We present a system that is able to recognize objects of a certain class in an image and to identify their parts for potential interactions. This is demonstrated for object instances that have never been observed during training, and under partial occlusion and against cluttered backgrounds. Our approach builds on the Implicit Shape Model of Leibe and Schiele, and extends it to couple recognition to the provision of meta-data useful for a task. Meta-data can for example consist of part labels or depth estimates. We present experimental results on wheelchairs and cars.**

I. Introduction

People are very strong at scene understanding. They quickly create a holistic interpretation of their environment. In comparison, a robot's interpretation comes piecemeal. A major difference lies in the human ability to recognize objects as instances of specific classes, and to feed such information back into lower layers of perception, thereby closing a *cognitive loop* (see Fig. 1). Such loops seem vital to 'make sense' of the world in the aforementioned, holistic way [14]. The brain brings all levels, from basic perception up to cognition, into unison. A similar endeavour in robotics would imply less emphasis on strictly quantitative – often 3D – modeling of the environment, and more on a qualitative analysis.

Indeed, it seems fair to say that nowadays robotics still has a certain preoccupation with gathering explicit 3D information (typically in the form of range maps) about the environment. Not only is this often a rather tedious affair, but many surface types defy 3D scanning altogether (e.g. dark, specular, or transparent surfaces may pose problems, depending on the scanner). Taking navigation as a case in point, it is known from human strategies that the image-based *recognition* of landmarks plays a far more important role than distance-based localisation with respect to some world coordinate system. The first such implementations for robot navigation have already been published [4, 3, 19]. This paper argues that modern visual object class recognition can provide useful cognitive feedback for many tasks in robotics[1].

The first examples of cognitive feedback in vision have already been implemented [9, 7]. However, so far they only coupled recognition and crude 3D scene information (the position of the groundplane). Here we set out to demonstrate the wider applicability of cognitive feedback, by inferring 'meta-data' such as material characteristics, the location and extent of object parts, or even 3D object shape, based on object class recognition. Given a set of annotated training images of a particular object class, we transfer these annotations to new images containing previously unseen object instances of the same class.

Fig. 1. Humans can very quickly analyze a scene from a single image. Recognizing subparts of an object helps to recognize the object as a whole, but recognizing the object in turn helps to gather more detailed information about its subparts. Knowledge about these parts can then be used to guide actions. For instance, in the context of a car wash, a decomposition of the car in its subparts can be used to apply optimized washing methods to the different parts.

There are a couple of recent approaches partially offering such inference for 3D shape from single images. Hoiem et al. [8] estimate the coarse geometric properties of a scene by learning appearance-based models of surfaces at various orientations. The method focuses purely on geometry estimation, without incorporating an object recognition process. It relies solely on the statistics of small image patches. In [20], Sudderth et al. combine recognition with coarse 3D reconstruction in a single image, by learning depth distributions for a specific type of scene from a set of stereo training images. In the same vein, Saxena et al. [18] are able to reconstruct coarse depth maps from a single image of an entire scene by means of a Markov Random Field. Han and Zhu [5] obtain quite detailed 3D models from a single image through graph representations, but their method is limited to specific classes. Hassner and

[1]See also interview with Rodney Brooks in Charlie Rose 2004/12/21: http://www.youtube.com/watch?v=oEstOd8xyeQ, starting from 35:00

Basri [6] infer 3D shape of an object in a single image from known 3D shapes of other members of the object's class. Their method is specific to 3D meta-data though, and their analysis is not integrated with the detection and recognition of the objects, as is ours. The object is assumed to be recognized and segmented beforehand. Rothganger et al. [15] are able to both recognize 3D objects and infer pose and detailed 3D data from a single image, but the method only works for specific object instances, not classes.

In this work, object related parameters and meta-data are inferred from a single image, given prior knowledge about these data for other members of the same object class. This annotation is intensely linked to the process of object recognition and segmentation. The variations within the class are taken account of, and the observed object can be quite different from any individual training example for its class. We collect pieces of annotation from different training images and merge them into a novel annotation mask that matches the underlying image data. Take the car wash scenario of Fig. 1 as an example. Our technique allows to identify the positions of the windshields, car body, wheels, license plate, headlights etc. This allows the parameters of the car wash line to better adapt to the specific car. Similarly, for the wheelchairs in Fig. 5, knowing where the handles are to be expected yields strong indications for a service robot how to get hold of the wheelchair.

The paper is organized as follows. First, we recapitulate the Implicit Shape Model of Leibe and Schiele [10] for simultaneous object recognition and segmentation (section II). Then follows the main contribution of this paper, as we explain how we transfer meta-data from training images to a previously unseen image (section III). We demonstrate the viability of our approach by transferring both object parts for wheelchairs and cars, as well as depth information for cars (section IV). Section V concludes the paper.

II. Object Class Detection with an Implicit Shape Model

In this section we briefly summarize the *Implicit Shape Model* (ISM) approach proposed by Leibe & Schiele [10], which we use as the object class detection technique underlying our approach (see also Fig. 2).

Given a training set containing images of several instances of a certain category (e.g. sideviews of cars) as well as their segmentations, the ISM approach builds a model that generalizes over within-class variability and scale. The modeling stage constructs a codebook of local appearances, i.e. of local structures that appear repeatedly on the training instances. Codebook entries are obtained by clustering image features sampled at interest point locations. Instead of searching for exact correspondences between a novel test image and model views, the ISM approach maps sampled image features onto this codebook representation. We refer to the features in an image that are mapped onto a codebook entry as *occurrences* of that entry. The spatial intra-class variability is captured by modeling spatial occurrence distributions for each codebook entry. Those distributions are estimated by recording all locations where a codebook entry matches to the training images, relative to the annotated object centers. Together with each occurrence, the approach stores a local segmentation mask, which is later used to infer top-down segmentations.

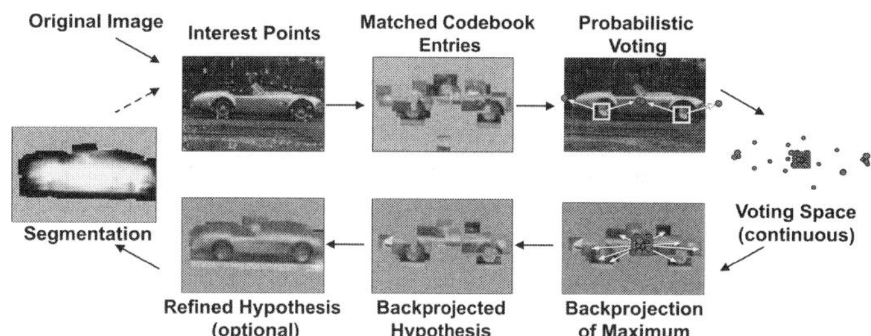

Fig. 2. The recognition procedure of the ISM system.

A. ISM Recognition.

The ISM recognition procedure is formulated as a probabilistic extension of the Hough transform [10]. Let e be a sampled image patch observed at location ℓ. The probability that it matches to codebook entry c_i can be expressed as $p(c_i|e)$. Each matched codebook entry then casts votes for instances of the object category o_n at different locations and scales $\lambda = (\lambda_x, \lambda_y, \lambda_s)$ according to its spatial occurrence distribution $P(o_n, \lambda|c_i, \ell)$. Thus, the votes are weighted by $P(o_n, \lambda|c_i, \ell)p(c_i|e)$, and the total contribution of a patch to an object hypothesis (o_n, λ) is expressed by the following marginalization:

$$p(o_n, \lambda|e, \ell) = \sum_i P(o_n, \lambda|c_i, \ell)p(c_i|e) \tag{1}$$

The votes are collected in a continuous 3D voting space (translation and scale). Maxima are found using Mean Shift Mode Estimation with a scale-adaptive uniform kernel [11]. Each local maximum in this voting space yields an hypothesis that an object instance appears in the image at a certain location and scale.

B. Top-Down Segmentation.

For each hypothesis, the ISM approach then computes a probabilistic top-down segmentation in order to determine the hypothesis' support in the image. This is achieved by backprojecting the contributing votes and using the stored local segmentation masks to infer the per-pixel probabilities that the pixel $\boldsymbol{p}$ is *figure* or *ground* given the hypothesis at location λ [10]. More precisely, the probability for a pixel $\boldsymbol{p}$ to be *figure* is computed as a weighted average over the segmentation masks of the occurrences of the codebook entries to which all features containing $\boldsymbol{p}$ are matched. The weights correspond to the patches' respective contributions to the hypothesis at location $\boldsymbol{x}$.

$$\begin{aligned} & p(\boldsymbol{p} = figure|o_n, \lambda) \\ & = \sum_{\boldsymbol{p} \in e} \sum_i p(\boldsymbol{p} = figure|e, c_i, o_n, \lambda) p(e, c_i|o_n, \lambda) \\ & = \sum_{\boldsymbol{p} \in e} \sum_i p(\boldsymbol{p} = figure|c_i, o_n, \lambda) \frac{p(o_n, \lambda|c_i) p(c_i|e) p(e)}{p(o_n, \lambda)} \end{aligned} \quad (2)$$

We underline here that a separate local segmentation mask is kept for every occurrence of each codebook entry. Different occurrences of the same codebook entry in a test image will thus contribute different segmentations, based on their relative location with respect to the hypothesized object center.

In early versions of their work [10], Leibe and Schiele included an optional processing step, which refines the hypothesis by a guided search for additional matches (Fig. 2). This improves the quality of the segmentations, but at a high computational cost. Uniform sampling was used, which became untractable once scale-invariance was introduced into the system. We therefore implemented a more efficient refinement algorithm as explained in Section III-C.

C. MDL Verification.

In a last processing stage, the computed segmentations are exploited to refine the object detection scores, by taking only *figure* pixels into account. Besides, this last stage also disambiguates overlapping hypotheses. This is done by a hypothesis verification stage based on Minimum Description Length (MDL), which searches for the combination of hypotheses that together best explain the image. This step precludes, for instance, that the same local structure, e.g. a wheel-like structure, is assigned to multiple detections, e.g. multiple cars. For details, we again refer to [10, 11].

III. Transferring Meta-data

The power of the ISM approach lies in its ability to recognize novel object instances as approximate jigsaw puzzles built out of pieces from different training instances. In this paper, we follow the same spirit to achieve the new functionality of transferring meta-data to new test images.

Example meta-data is provided as annotations to the training images. Notice how segmentation masks can be considered as a special case of meta-data. Hence, we transfer meta-data with a mechanism inspired by that used above to segment objects in test images. The training meta-data annotations are attached to the occurrences of codebook entries, and transferred to a test image along with each matched feature that contributed to the final hypothesis (Fig. 3). This strategy allows us to generate novel annotations tailored to the new test image, while explicitly accommodating for the intra-class variability.

Unlike segmentations, which are always binary, meta-data annotations can be either binary (e.g. for delineating a particular object part or material type), discrete (e.g. for identifying *all* object parts), real-valued (e.g. depth values), or even vector-valued (e.g. surface orientations). We first explain how to transfer discrete meta-data (Section III-A), and then extend the method to the real- or vector-valued case (Section III-B).

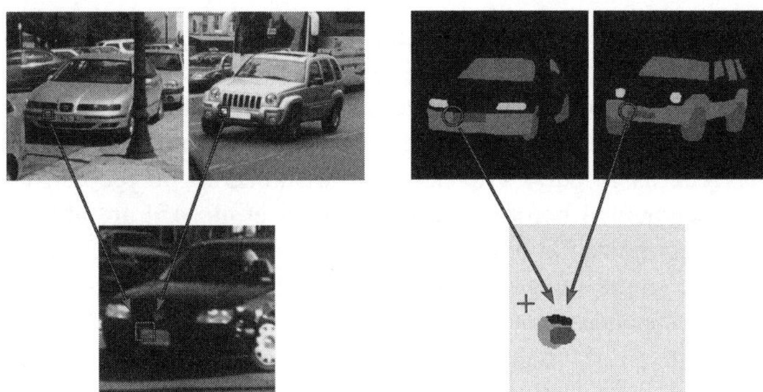

Fig. 3. Transferring (discrete) meta-data. Left: two training images and a test image. Right: the annotations for the training images, and the partial output annotation. The corner of the license plate matches with a codebook entry which has occurrences on similar locations in the training images. The annotation patches for those locations are combined and instantiated in the output annotation.

A. Transferring Discrete Meta-data

In case of discrete meta-data, the goal is to assign to each pixel of the detected object a label $a \in \{a_j\}_{j=1:N}$. We first compute the probability $p(\boldsymbol{p} = a_j)$ for each label a_j separately. This is achieved in a way analogous to what is done in eq. (2) for $p(\boldsymbol{p} = figure)$, but with some extensions necessary to adapt to the more general case of meta-data:

$$p(\boldsymbol{p} = a_j|o_n, \lambda) = \sum_{\boldsymbol{p} \in N(e)} \sum_i p(\boldsymbol{p} = a_j|c_i, o_n, \lambda) p(\hat{a}(\boldsymbol{p}) = a_e(\boldsymbol{p})|e) p(e, c_i|o_n, \lambda) \quad (3)$$

The components of this equation will be explained in detail next. The first and last factors are generalizations of their counterparts in eq. (2). They represent the annotations stored in the codebook, and the voting procedure respectively. One extension consists in transferring annotations also from image patches *near* the pixel $\boldsymbol{p}$, and not only from those *containing* it. With the original version, it is often difficult to obtain full coverage of the object, especially when the number of training images is limited. This is an important feature, because producing the training annotations can be labour-intensive (e.g. for the depth estimates of the cars in Section IV-B). Our notion of proximity is defined relative to the size of the image patch e, and parameterized by a scalefactor s_N. More precisely, let an image patch e be defined by the three-dimensional coordinates of its center and scale e_λ obtained from the interest point detector, i.e. $e = (e_x, e_y, e_\lambda)$. The neighbourhood $N(e)$ of e is defined as

$$N(e) = \{\boldsymbol{p}|\boldsymbol{p} \in (e_x, e_y, s_N \cdot e_\lambda)\} \quad (4)$$

A potential disadvantage of the above procedure is that for a pixel $\boldsymbol{p}$ outside the actual image patch, the transferred annotation gets less reliable. Indeed, the pixel may lie on an occluded image area, or small misalignment errors may get magnified. Moreover, some differences between the object instances shown in the training and test images that were not noticeable at the local scale can now affect the results. To compensate for this, we add the second factor to eq. (3), which indicates how probable it is that the transferred annotation $a_e(\mathbf{p})$ still corresponds to the 'true' annotation $\hat{a}(\mathbf{p})$. This probability is modeled by a Gaussian, decaying smoothly with the distance from the center of the image patch e, and with variance related to the size of e by a scalefactor s_G:

$$\begin{aligned} p(\hat{a}(\mathbf{p}) = a_e(\mathbf{p}) \,|\, e) &= \frac{1}{\sigma\sqrt{2\pi}} exp\left(-({d_x}^2 + {d_y}^2)/(2\sigma^2)\right) \\ \text{with} \qquad \sigma &= s_G \cdot e_\lambda \\ (d_x, d_y) &= (p_x - e_x, p_y - e_y) \end{aligned} \tag{5}$$

Once we have computed the probabilities $p(\mathbf{p} = a_j)$ for all possible labels $\{a_j\}_{j=1:N}$, we come to the actual assignment: we select the most likely label for each pixel. Note how for some applications, it might be better to keep the whole probability distribution $\{p(\mathbf{p} = a_j)\}_{j=1:N}$ rather than a hard assignment, e.g. when feeding back the information as prior probabilities to low-level image processing.

An interesting possible extension is to enforce spatial continuity between labels of neighboring pixels, e.g. by relaxation or by representing the image pixels as a Markov Random Field. In our experiments (Section IV), we achieved good results already without enforcing spatial continuity.

B. Transferring Real- or Vector-valued Meta-data

In many cases, the meta-data is not discrete, but rather real-valued (e.g. 3D depth) or vector-valued (e.g. surface orientation). We can approximate these cases by using a large number of quantization steps and interpolating the final estimate. This allows to re-use most of the discrete-case system.

First, we discretize the annotations into a fixed set of 'value labels' (e.g. 'depth 1', 'depth 2', etc.). Then we proceed in a way analogous to eq. (3) to infer for each pixel a probability for each discrete value. In the second step, we select for each pixel the discrete value label with the highest probability, as before. Next, we refine the estimated value by fitting a parabola (a $(D+1)$-dimensional paraboloid in the case of vector valued meta-data) to the probability scores for the maximum value label and the two immediate neighbouring value labels. We then select the value corresponding to the maximum of the parabola. This is a similar method as used in interest point detectors (e.g. [12, 1]) to determine continuous scale coordinates and orientations from discrete values. Thanks to this interpolation procedure, we obtain real-valued annotations. In our 3D depth estimation experiments this makes a significant difference in the quality of the results (Section IV-B).

C. Refining Hypotheses

When large areas of the object are insufficiently covered by interest points, no meta-data can be assigned to these areas. Using a large value for s_N will only partially solve this problem, because there is a limit as to how far information from neighboring points can be reliably extrapolated. A better solution is to actively search for additional codebook matches in these areas. The refinement procedure in early versions of the ISM system [10] achieved this by means of uniform sampling, which is untractable in the scale-invariant case. Therefore we implemented a more efficient refinement algorithm which only searches for matches in promising locations.

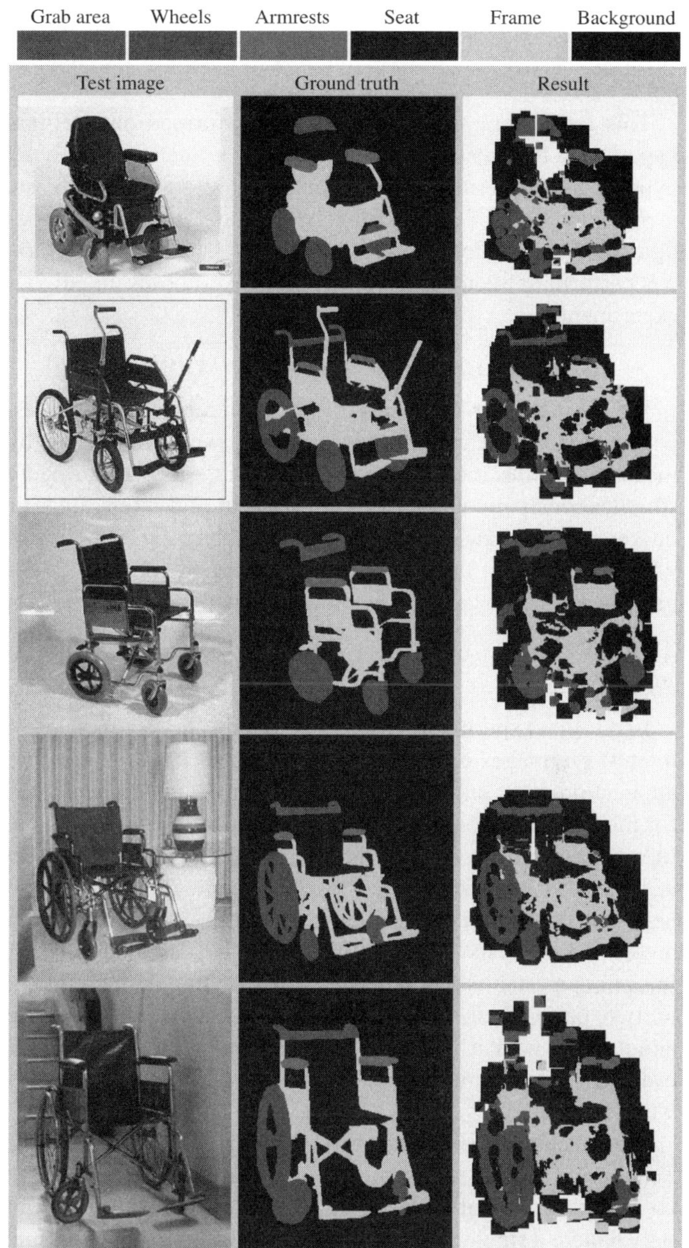

Fig. 4. Results for the annotation verification experiment on wheelchair images. From left to right: test image, ground-truth, and output of our system. White areas are unlabeled and can be considered background.

For each hypothesis, new candidate points are generated by backprojecting all occurrences in the codebook, excluding points nearby existing interest points. When the feature descriptor for a new point matches with the codebook cluster(s) that backprojected it, an additional hypothesis vote is cast. The confidence for this new vote is reduced by a penalty factor to reflect the fact that it was not generated by an actual interest point. The additional votes enable the meta-data transfer to cover those areas that were initially missed by the interest point detector.

This refinement step can either be performed on the final hypotheses that result from the MDL verification, or on all hypotheses that result from the initial voting. In the latter case, it will improve MDL verification by enabling it to obtain better figure area estimates of each hypothesis [10, 11]. Therefore, we perform refinement on the initial hypotheses in all our experiments.

IV. Experimental evaluation

We evaluate our approach on two different object classes, wheelchairs and cars. For both classes, we demonstrate by means of a discrete labeling experiment, how our system simultaneously recognizes object instances and infers areas of interest. For the cars, we additionally perform an experiment where a 3D depth map is recovered from a single image of a previously unseen car, which is a real-valued labeling problem.

A. Wheelchairs: Indicating Areas of Interest for an Assistive Robot

In our first experiment, the goal is to indicate certain areas of interest on images of various types of wheelchairs. A possible application is an assistive robot, for retrieving a wheelchair, for instance in a hospital or to help a disabled person at home. In order to retrieve the wheelchair, the robot must be able to both detect it, and determine where to grab it. Our method will help the robot to get close to the grabbing position, after which a detailed analysis of scene geometry in a small region of interest can establish the grasp [17]. We divide our experiment in two parts. First, we quantitatively evaluate the resulting annotations with a large set of controlled images. Next, we evaluate the recognition ability with a set of challenging real-world images.

We collected 141 images of wheelchairs from Google Image Search. We chose semi-profile views because they were the most widely available. Note that while the ISM system can only handle a single pose, it can be extended to handle multiple viewpoints [21]. All images were annotated with ground truth part segmentations for grab area, wheels, armrests, seat, and frame. The grab area is the most important for this experiment. A few representative images and their ground truth annotations can be seen in the left and middle columns of Fig. 4.

The images are randomly split into a training and test set. We train an ISM system using 80 images, using a Hessian-Laplace interest point detector [13] and Shape Context descriptors [2]. Next, we test the system on the remaining 61 images, using the method from Section III-A. Because each image only contains one object, we select the detection with highest score for meta-data transfer. Some of the resulting annotations can be seen in the third column of Fig. 4. The grab area is found quite precisely.

To evaluate this experiment quantitatively, we use the ground truth annotations to calculate the following error measures. We define *leakage* as the percentage of background

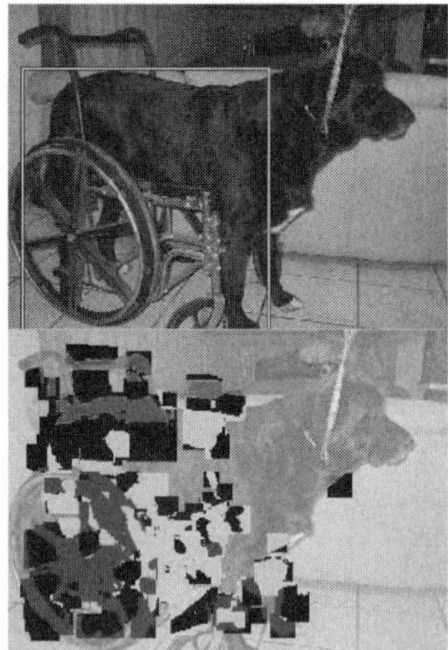
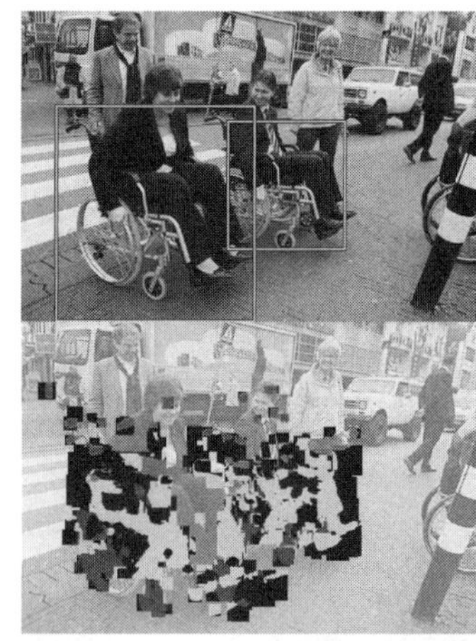

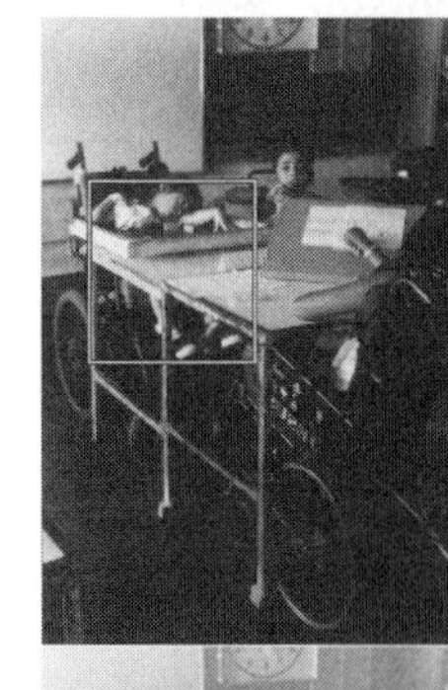
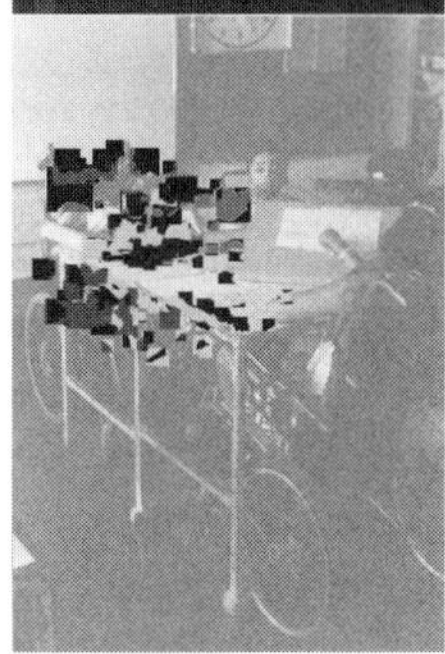

Fig. 5. Wheelchair detection and annotation results on challenging real-world test images (best viewed in color). Yellow and red rectangles indicate correct and false detections respectively. Note how one wheelchair in the middle right image was missed because it is not in the pose used for training.

	backgrnd	frame	seat	armrest	wheels	**grab-area**	unlabeled
backgrnd	**32.58**	1.90	0.24	0.14	1.10	0.37	**63.67**
frame	15.29	**66.68**	6.47	0.46	6.90	0.10	4.10
seat	2.17	15.95	**74.28**	0.97	0.33	1.55	4.75
armrest	11.22	5.62	29.64	**49.32**	1.25	0.63	2.32
wheels	13.06	9.45	0.36	0.07	**71.39**	0.00	5.67
grab-area	6.48	1.28	9.77	0.11	0.00	**76.75**	5.62

TABLE I

CONFUSION MATRIX FOR THE WHEELCHAIR PART ANNOTATION EXPERIMENT. THE ROWS REPRESENT THE ANNOTATION PARTS IN THE GROUND-TRUTH MAPS, THE COLUMNS THE OUTPUT OF OUR SYSTEM. THE LAST COLUMN SHOWS HOW MUCH OF EACH CLASS WAS LEFT UNLABELED. FOR MOST EVALUATIONS, THOSE AREAS CAN BE CONSIDERED 'BACKGROUND'.

pixels in the ground-truth annotation that were labeled as non-background by the system. The leakage for this experiment, averaged over all test images, is 3.75%. We also define a *coverage* measure, as the percentage of non-background pixels in the ground-truth images labeled non-background by the system. The coverage obtained by our algorithm is 95.1%. This means our method is able to reliably segment the chair from the background.

We evaluate the annotation quality of the separate parts with a confusion matrix. For each image, we count how many pixels of each part a_j in the ground-truth image are labeled by our system as each of the possible parts (grab, wheels, etc.), or remain unlabeled (which can be considered background in most cases). This score is normalized by the total number of pixels in the ground-truth $\hat{a}_j$. We average the confusion table entries over all images, resulting in Table I. The diagonal elements show how well each part was recovered in the test images. Not considering the armrests, the system performs well as it labels correctly between 67% and 77% of the pixels, with the latter score being for the part we are the most interested in, i.e. the grab area. The lower performance for the armrests is due to the fact that it is the smallest part in most of the images. Small parts have higher risk of being confused with the larger parts in their neighborhood.

To test the detection ability of our system, we collected a set of 34 challenging real-world images with considerable clutter and/or occlusion. We used the same ISM system as in the annotation experiment, to detect and annotate the chairs in these images. Some results are shown in Fig. 5. We consider a detection to be correct when its bounding box covers the chair. Out of the 39 wheelchairs present in the images, 30 were detected, and there were 7 false positives. This corresponds to a recall of 77% and a precision of 81%.

B. Cars: Optimizing an Automated Car Wash

In further experiments, we infer different types of meta-data for the object class 'car'. In the first experiment, we decompose recognized cars in their most important parts, similarly to the wheelchairs. In the second experiment, approximate 3D information is inferred. A possible application is an automated car wash. As illustrated in Fig. 1, the decomposition in parts can be used to apply different washing methods to the different parts. Moreover, even though such systems mostly have sensors to measure distances to the car, they are only used locally while the machine is already running. It could be useful to optimize the washing process beforehand, based on the car's global shape inferred by our system.

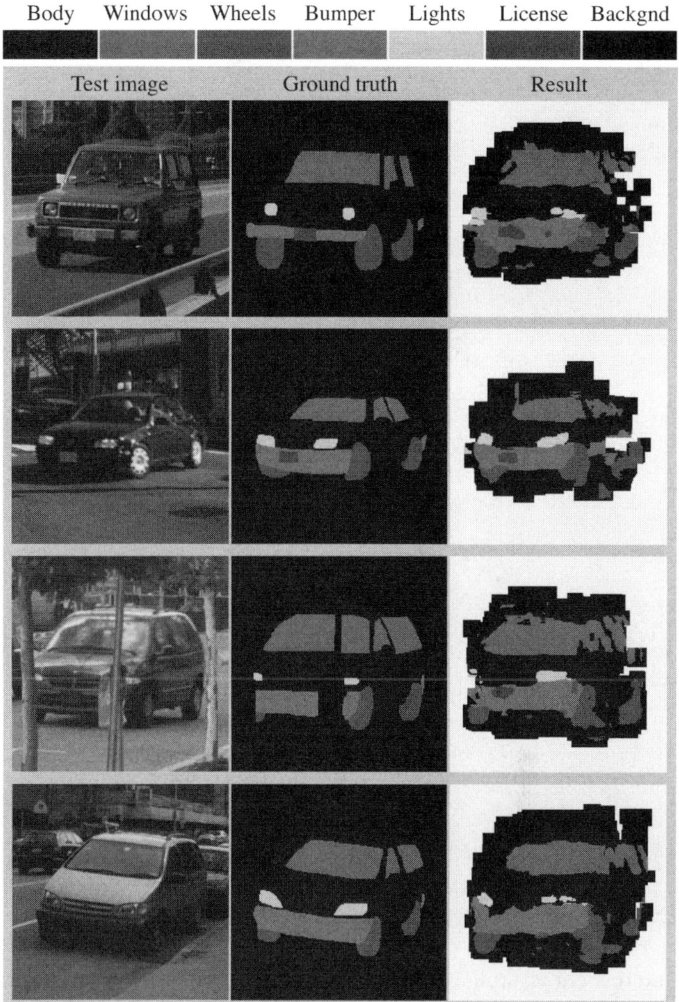

Fig. 6. Results for the car annotation experiment. From left to right: test image, ground-truth, and output of our system. White areas are unlabeled and can be considered background.

Our dataset is a subset of that used in [9]. It was obtained from the LabelMe website [16], by extracting images labeled as 'car' and sorting them according to their pose. For our experiments, we only use the 'az300deg' pose, which is a semi-profile view. In this pose both the front (windscreen, headlights, license plate) and side (wheels, windows) are visible. This allows for more interesting depth maps and part annotations compared to pure frontal or side views. The dataset contains a total of 139 images. We randomly picked 79 for training, and 60 for testing.

For parts annotation, the training and testing phase is analogous to the wheelchair experiment (section IV-A). Results are shown in Fig. 6. The leakage is 6.83% and coverage is 95.2%. The confusion matrix is shown in Table II. It again

	bkgnd	body	bumper	headlt	window	wheels	license	unlabeled
bkgnd	**23.56**	2.49	1.03	0.14	1.25	1.88	0.04	**69.61**
body	4.47	**72.15**	4.64	1.81	8.78	1.86	0.24	6.05
bumper	7.20	4.54	**73.76**	1.57	0.00	7.85	2.43	2.64
headlt	1.51	36.90	23.54	**34.75**	0.01	0.65	0.23	2.41
window	3.15	13.55	0.00	0.00	**80.47**	0.00	0.00	2.82
wheels	11.38	6.85	8.51	0.00	0.00	**63.59**	0.01	9.65
license	2.57	1.07	39.07	0.00	0.00	1.04	**56.25**	0.00

TABLE II

CONFUSION MATRIX FOR THE CAR PARTS ANNOTATION EXPERIMENT (CFR. TABLE I).

Fig. 7. Results for the car depthmap experiment. From left to right: test image, ground-truth, and output of our system. White areas are unlabeled and can be considered background.

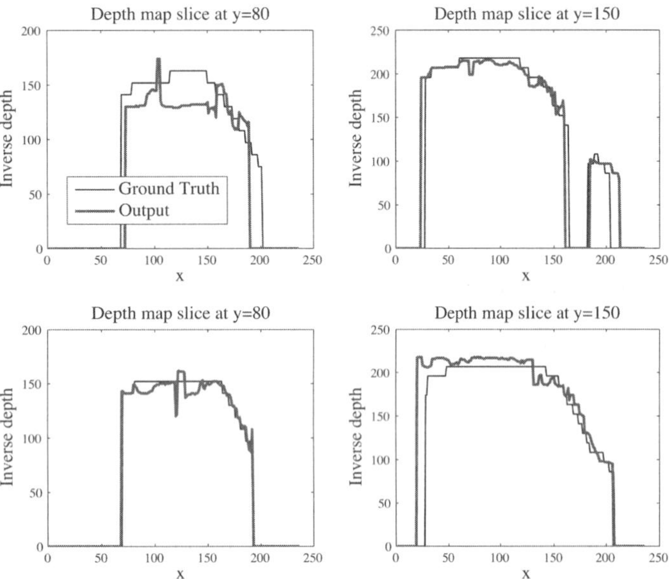

Fig. 8. Horizontal slices through the ground truth and output depthmaps of the second car (top row) and fourth car (bottom row) in Fig. 7.

shows good labeling performance, except for the headlights. Similarly to the armrests in the wheelchair experiments, this is as expected. The headlights are mostly very small, hence easily confused with the larger parts (body, bumper) in which they are embedded.

For the depth map experiment, we obtained ground-truth data by manually aligning the best matching 3D model from a freely available collection[2] to each image, and extracting the OpenGL Z-buffer. In general, any 3D scanner or active lighting setup could be used to automatically obtain depth maps. We normalize the depths based on the dimensions of the 3D models, by assuming that the width of a car is approximately constant. The depth maps are quantized to 20 discrete values. Using these maps as annotations, we use our method of section III-B to infer depths for the test images.

Results are shown in the rightmost column of Fig. 7. The leakage is 4.79% and the coverage 94.6%, hence the segmentation performance is again very good. It is possible to calculate a real-world depth error estimate, by scaling the normalized depth maps by a factor based on the average width of a real car, which we found to be approximately 1.8m. All depth maps are scaled to the interval $[0, 1]$ such that their depth range is 3.5 times the width of the car, and the average depth error is 0.042. This is only measured inside areas which are labeled non-background in both the ground-truth and result images, to eliminate bias from the background. A plausible real-world depth error can therefore be calculated by multiplying this figure by $3.5 \cdot 1.8$m, which yields a distance of 27cm. To better visualize how the output compares to the ground truth, Fig. 8 shows a few horizontal slices through two depth maps of Fig. 7.

To illustrate the combined recognition and annotation ability of our system for this object class, we again tested it on real-world images. We used the same system as in the annotation experiment on a few challenging images containing cars in a similar pose, including the car wash image from Fig. 1. Results are shown in Fig. 9.

V. CONCLUSIONS

We have developed a method to transfer meta-data annotations from training images to test images containing previously unseen objects, based on object class recognition. Instead of using extra processing for the inference of meta-data, it

[2]http://dmi.chez-alice.fr/models1.html

Fig. 9. Car detection and annotation results on real-world test images. Even though the car from Fig. 1 is in a near-frontal pose, it was still correctly detected and annotated by the system trained on semi-profile views.

is deeply intertwined with the actual recognition process. Low-level cues in an image can lead to the detection of an object, and the detection of the object itself causes a better understanding of the low-level cues from which it originated. The resulting meta-data inferred from the recognition can be used to initiate or refine robot actions.

Future research includes using the output from our system in a real-world application to guide a robot's actions, possibly in combination with other systems. We will extend the system to handle fully continuous and vector-valued meta-data. We will also investigate methods to improve the quality of the annotations by means of relaxation or Markov Random Fields, and ways to greatly reduce the amount of manual annotation work required for training.

ACKNOWLEDGMENT

The authors gratefully acknowledge support by IWT-Flanders, Fund for Scientific Research Flanders and European Project CLASS (3E060206).

REFERENCES

[1] H. Bay, T. Tuytelaars, and L. Van Gool. SURF: Speeded up robust features. *Proceedings ECCV, Springer LNCS*, 3951(1):404–417, 2006.

[2] Serge Belongie, Jitendra Malik, and Jan Puzicha. Shape context: A new descriptor for shape matching and object recognition. In *NIPS*, pages 831–837, 2000.

[3] F. Fraundorfer, C. Engels, and D. Nister. Topological mapping, localization and navigation using image collections. *IROS*, 2007.

[4] T. Goedemé, M. Nuttin, T. Tuytelaars, and L. Van Gool. Omnidirectional vision based topological navigation. *Int. Journal on Computer Vision and Int. Journal on Robotics Research*, 74(3):219–236, 2007.

[5] F. Han and S.-C. Zhu. Bayesian reconstruction of 3D shapes and scenes from a single image. *Workshop Higher-Level Knowledge in 3D Modeling Motion Analysis*, 2003.

[6] T. Hassner and R. Basri. Example based 3D reconstruction from single 2d images. In *Beyond Patches Workshop at IEEE CVPR06*, page 15, June 2006.

[7] D. Hoiem, A. Efros, and M. Hebert. Putting objects in perspective. *CVPR*, pages 2137–2144, 2006.

[8] D. Hoiem, A. A. Efros, and M. Hebert. Geometric context from a single image. *ICCV*, 2005.

[9] B. Leibe, N. Cornelis, K. Cornelis, and L. Van Gool. Dynamic 3D scene analysis from a moving vehicle. *CVPR*, 2007.

[10] B. Leibe and B. Schiele. Interleaved object categorization and segmentation. *BMVC*, 2003.

[11] B. Leibe and B. Schiele. Pedestrian detection in crowded scenes. *CVPR*, 2005.

[12] D. G. Lowe. Distinctive image features from scale-invariant keypoints. *IJCV*, 60(2):91–110, 2004.

[13] K. Mikolajczyk and C. Schmid. Scale & affine invariant interest point detectors. *IJCV*, 60(1):63–86, 2004.

[14] D. Mumford. Neuronal architectures for pattern-theoretic problems. *Large-Scale Neuronal Theories of the Brain*, pages 125–152, 1994.

[15] F. Rothganger, S. Lazebnik, C. Schmid, and J. Ponce. 3D object modeling and recognition using local affine-invariant image descriptors and multi-view spatial constraints. *IJCV*, 66(3):231–259, 2006.

[16] B.C. Russell, A. Torralba, K.P. Murphy, and W.T. Freeman. LabelMe: a database and web-based tool for image annotation. *MIT AI Lab Memo AIM-2005-025*, 2005.

[17] A. Saxena, J. Driemeyer, J. Kearns, and A. Y. Ng. Robotic grasping of novel objects. *NIPS 19*, 2006.

[18] A. Saxena, J. Schulte, and A. Y. Ng. Learning depth from single monocular images. *NIPS 18*, 2005.

[19] Sinisa Segvic, Anthony Remazeilles, Albert Diosi, and Francois Chaumette. Large scale vision-based navigation without an accurate global reconstruction. *CVPR*, 2007.

[20] E.B. Sudderth, A. Torralba, W.T. Freeman, and A.S. Willsky. Depth from familiar objects: A hierarchical model for 3D scenes. *CVPR*, 2006.

[21] A. Thomas, V. Ferrari, B. Leibe, T. Tuytelaars, B. Schiele, and L. Van Gool. Towards multi-view object class detection. *CVPR*, 2006.

Learning to Manipulate Articulated Objects in Unstructured Environments Using a Grounded Relational Representation

Dov Katz Yuri Pyuro Oliver Brock
Robotics and Biology Laboratory
University of Massachusetts Amherst

Abstract— **We introduce a learning-based approach to manipulation in unstructured environments. This approach permits autonomous acquisition of manipulation expertise from interactions with the environment. The resulting expertise enables a robot to perform effective manipulation based on partial state information. The manipulation expertise is represented in a relational state representation and learned using relational reinforcement learning. The relational representation renders learning tractable by collapsing a large number of states onto a single, relational state. The relational state representation is carefully grounded in the perceptual and interaction skills of the robot. This ensures that symbolically learned knowledge remains meaningful in the physical world. We experimentally validate the proposed learning approach on the task of manipulating an articulated object to obtain a model of its kinematic structure. Our experiments demonstrate that the manipulation expertise acquired by the robot leads to substantial performance improvements. These improvements are maintained when experience is applied to previously unseen objects.**

I. Introduction

Autonomous manipulation remains one of the great challenges in robotics. The successful endowment of autonomous robots with robust manipulation skills will have substantial impact in many important application areas, ranging from personal and professional service robotics to flexible manufacturing and planetary exploration.

We view autonomous manipulation as the purposeful and deliberate change of the configuration of an object. The object's configuration uniquely describes the object's pose by specifying every degree of freedom of the object. An object can have extrinsic and intrinsic degrees of freedom. Extrinsic degrees of freedom describe the spatial relationship between the object and its environment. Intrinsic degrees of freedom describe the relationship among the rigid bodies of an articulated object and are often relevant to the object's intended function. Examples of objects with intrinsic degrees of freedom include tools (scissors, pliers, etc.), doors, door handles, books, or drawers. Successful manipulation must be informed by knowledge of the extrinsic and intrinsic degrees of freedom of an object.

In unstructured environments, a robot cannot rely on a detailed and accurate *a priori* model of the environment. It must therefore be able to acquire task-relevant information to inform its interactions with the objects in the environment. Based on this information, the robot must adapt its manipulation behavior to ensure successful task execution. Manipulation becomes a continuous and interactive process of acquiring information about the environment and subsequently adapting the interaction with the environment in response to this information.

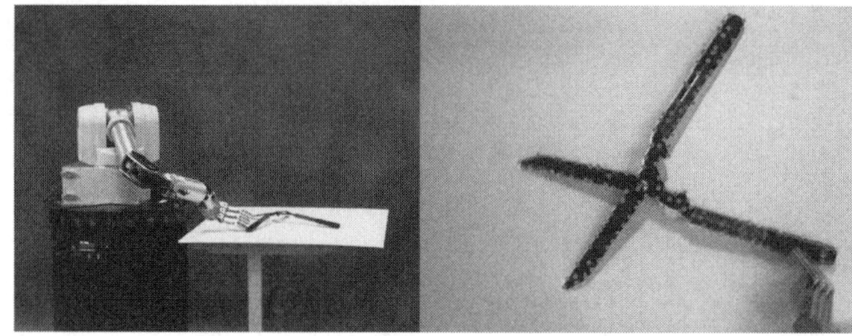

Fig. 1. UMan (UMass Mobile Manipulator) interacts with an articulated object to acquire information about the object's kinematic structure. The right image shows the scene as seen by the robot through an overhead camera; dots mark tracked visual features.

The contribution of this paper is a learning-based approach to manipulation in unstructured environments. This approach permits the robot to autonomously acquire manipulation expertise from its interactions with the environment. The resulting expertise enables the robot to select the most effective manipulation action based on partial state information. The manipulation expertise is learned in a relational state representation. This representation is essential, as it renders learning tractable by collapsing a large regions of the state space onto a single, task-relevant, relational state. The symbolic representation is carefully grounded in the perceptual and interaction skills of the robot to ensure that relationally learned knowledge remains applicable in the physical world.

Using this learning-based manipulation approach, we show how a robot can autonomously learn manipulation strategies to obtain a kinematic model of unknown articulated objects. The robot physically interacts with the object by pushing or pulling it and observes the object's motion (see Fig. 1). As these interactions create a change in the configuration of the object, the robot incrementally discovers the object's intrinsic and extrinsic degrees of freedom. The robot learns to select interactions that are most likely to reveal the maximum information about the kinematic structure. The acquired manipulation knowledge substantially reduces the number of interactions required to obtain an accurate kinematic model. Furthermore, the manipulation knowledge acquired with one object transfers to other objects, even if they have different kinematic structures.

In the following section we introduce the relational representations of kinematic structures that forms the basis of our learning-based approach to manipulation. We then describe how this representation can be grounded using the perceptual and interaction capabilities of the robot. We proceed to discuss the relational learning framework and how it can be grounded with respect to the relational representation. Finally, we demonstrate the effectiveness of our approach in manipulation experiments with articulated objects.

II. Related Work

We distinguish between three categories of approaches to manipulation [2, 15, 17]. The first category pertains to manipulation planning. Approaches in this category assume that an accurate geometric model of the manipulated object is available and devise manipulation plans based on this model. These manipulation planners address various flavors of manipulation, including grasping and in-hand manipulation [24], manipulation with sliding contacts [26], non-prehensile manipulation [1, 14], and gross motion planning for manipulation [21]. In contrast to these approaches, we focus on problems for which accurate models are not available.

The second category uses feedback control to achieve manipulation. Particularly relevant are approaches that use learning to design controllers [28]. These methods alleviate the difficulties that analytical methods for controller design encounter in the presence of modeling errors for systems with complex kinematics and dynamics. A different approach to learning-based controller design relies on memory-based learning [16]. Controllers can also be designed by searching in configuration space [22]. All methods in this category determine specific controllers that can serve to ground a relational state representation such as the one we use.

Whereas approaches in the previous category are concerned with individual controllers, approaches in the third category sequence [3, 18] or compose [20] multiple controllers to generate more complex manipulation behaviors. Composite controllers can be arranged into state transition diagrams to further increase robustness and versatility [19]. Most often, the necessary state transition diagrams are designed by the programmer, but they can also be learned using reinforcement learning [9]. Again, we view the resulting controllers as candidates to ground a relational state representation.

There are many other approaches that address aspects of manipulation but cannot easily be assigned to one of these three categories. We will discuss several with particular relevance to the work presented here. Christiansen et al. [4] learn manipulation strategies for a tray-tilting task in conjunction with a dynamic model of the domain. Edsinger and Kemp emphasize the importance of task-specific perceptual features that exploit common structural features of functionally related objects to facilitate manipulation in human environments [8]. Stoytchev presents an approach to learn tool affordances for robotic tool use [23]. He emphasizes the importance of grounding this representation in the robot's behavioral repertoire. This enables the immediate application of the robot's accumulated experience. In our work, we combine task-specificity for perception and grounding for action by requiring that an adequate grounding of our relational state representation has to rely on task-specific perception *and* task-specific behaviors.

III. Relational Representation

The relational representation is critical to the success of our learning-based approach to manipulation. Using a finite set of relations, we are able to describe an infinite number of states and actions. It thus becomes possible to represent and reason about situations that a propositional representation cannot handle. For example, a robot may encounter many types of scissors, varying in color, shape, and size. All scissors, however, have the same kinematic structure. A single relational formula can capture the kinematic structure of *all* scissors, irrespective of their other properties. Therefore, a single relational action can be applied to *all* objects. In contrast, a propositional representation would have to include a proposition for every encountered object and one for every action applicable to this object. The relational representation avoids this combinatorial explosion of actions and states, thus greatly reducing the state space and making learning possible.

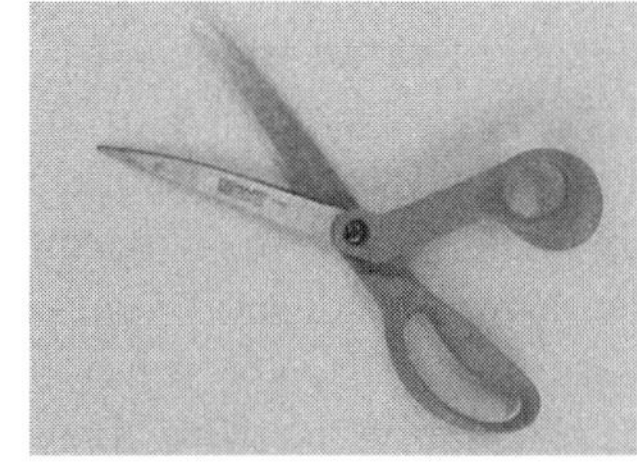

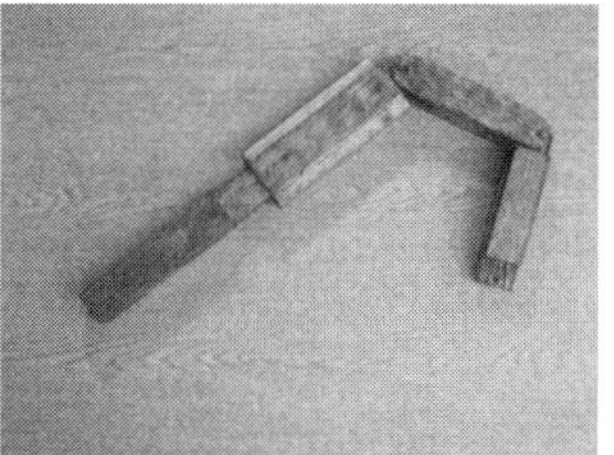

Fig. 2. Two examples of kinematic structures: scissors with a single revolute joint and a wooden toy with a prismatic joint and two revolute joints.

Our relational representation for kinematic models of articulated objects captures joint types, link properties, and kinematic relationships between links. Figure 2 shows two examples of planar kinematic structures. The scissors have a single revolute degree of freedom and the wooden toy is a serial kinematic chain with a prismatic joint (on the left of the figure) and two revolute joints. Our relational representation uses predicates $R(\cdot)$, $P(\cdot)$, and $D(\cdot)$ to describe that rigid bodies are connected by a revolute joint, a prismatic joint, or are disconnected, respectively. The predicates are n-ary, with $n \geq 2$, to capture branching kinematic structures. The rigid body passed as the first argument to the relation is the one in relationship with all other arguments. For example, $R(x, y, z)$ is equivalent to $R(x, y) \wedge R(x, z)$.

Using these relations, we can represent the kinematic structure of the scissors as

$$D(l_b, R(l_1, l_2)),$$

where l_1 and l_2 represent the two links of the scissors and l_b is a disconnected background link. The kinematic structure of the wooden toy can be represented as

$$D(l_b, R(l_4, R(l_3, P(l_1, l_2)))).$$

Note that this representation is not unique. The wooden toy could also be represented as

$$D(P(l_4, R(R(l_1, l_2), l_3)), l_b).$$

Which of these representations is used by the robot depends on the order of discovery of the links. The most deeply nested relation is discovered first.

Kinematic loops are represented by using the same link twice. A 5R kinematic loop is described by:

$$D(l_b, R(l_1, R(l_5, R(l_4, R(l_3, R(l_2, l_1)))))).$$

By extending our atomic representation of links to m-ary relations $L(\cdot)$, $m \geq 1$, we can include link properties in our description of kinematic chains. In this paper we will limit ourselves to a single property, namely the size of the link. The wooden toy can now be represented as

$$D(l_b, R(L(s, f_4), R(L(s, f_3), P(L(s, f_1), L(s, f_2))))),$$

where s stands for the property *small* and the f_i spatially identify links in the physical world (see section IV). The extension to an arbitrary number of link properties is straightforward.

With this relational representation of kinematic structures, it becomes possible to reason and learn about objects based on their kinematic structure. All experience acquired by manipulating scissors can be applied to all other scissors, as long as they have an identical kinematic structure. If specific properties of the links of an object affect the desired manipulation behavior, we can add these properties to the relational description of the links. Our representation is then able to differentiate between identical kinematic structures based on link properties. All properties irrelevant to manipulation are ignored during learning. This reduction in state space makes the learning problem considerably easier.

We also use a relational representation for the actions performed by the robot. Actions apply pushing or pulling forces to one of the links. The forces can be applied along the major axes of the link or along a forty-five degree angle to the major axes. An action is represented as $A(L(\cdot), \alpha)$, where $L(\cdot)$ represents a link and $alpha$ is an atom describing one of the possible six pushing/pulling directions relative to the link.

IV. Grounding the Relational Representation

The relational representation described in the previous section can only support the learning of manipulation knowledge if it is grounded in the physical capabilities of the robot. Grounding bridges between the symbols of our representation and the physical, continuous world [10]. Grounding ensures that we can symbolically interpret the observations made by the robot in regards to its interactions with the world. At the same time, grounding ensures that the resulting symbolic manipulation knowledge maintains its relevance and predictive power for the robot's real-world interactions.

To ground our relational representation, we bind the relations $R(\cdot,\cdot)$, $P(\cdot,\cdot)$, and $D(\cdot,\cdot)$ as well as the link properties to real-world perceptual capabilities of the robot.

In prior work we developed a skill for the robust perception of kinematic degrees of freedom and link properties of planar articulated objects [12]. Figure 1 shows a real-world interactive experiment with garden shears. The robot interacts with the shears to determine the location of the revolute joint and the spatial extent of the links. The image on the right shows the robot's view of its own interaction with the shears. Dots indicate tracked visual features.

This skill provides adequate grounding for our relational representation of links and their kinematic relationship. It extracts the degrees of freedom of an object by tracking the motion of the visual features in the scene. Tracked features are clustered into links using a graph representation in which the features correspond to vertices. Two vertices are connected by an edge if the relative distance of the corresponding visual features does not change throughout the interaction with the object. By clustering the features, it is possible to identify the spatial location and extent of the links. The features associated with a single link are grouped into the sets f_i (see previous section). The relationship between different clusters (links) in the graph can be analyzed to reveal their kinematic relationship.

The robustness of this skill has been proven in dozens of real-world experiments. The skill does not require prior knowledge of the object, is insensitive to lighting conditions and specularities, succeeds irrespective of the texture and color of the object's parts, works reliably even with low-quality cameras, and at the same time is computationally efficient. As a consequence, it is ideally suited for the grounding of our relational representations of kinematic structures for unstructured environments. For a detailed description of this interactive perception skill the reader is referred to reference [12].

V. Learning Manipulation with Grounded Relational Reinforcement Learning

The grounded relational description of states is the basis for our learning framework. To learn manipulation knowledge from interactions with the environment, we cast the incremental acquisition of kinematic representations of objects as a relational reinforcement learning [7, 25, 27] problem.

In reinforcement learning, an agent learns an optimal policy for solving a task. This policy tells the robot which action to perform in a particular state. The robot acquires the policy incrementally, by performing experiments. In our case, an experiment consists of the robot pushing an object. For every action, the robot receives a reward. In our experiments, the robot receives a reward for every degree of freedom it discovers. Over the course of multiple experiments, the robot incorporates new experiences into its policy. As a result, our robot learns an effective policy for acquiring kinematic models of articulated objects.

We formalize this problem as a Relational Markov Decision Process (RMDP) [27] and then apply Q-learning [29] to find an optimal policy. A Markov Decision Process (MDP) is a tuple $M = (S, A, T, R)$, where S designates the set of possible states, A is the set of actions available to the robot,

$T : S \times A \rightarrow \Pi(S)$ specifies a state transition function to determine a probability distribution $\Pi(S)$ over S, indicating the probability of attaining a successor state when an action is performed in an initial state, and $R : S \times A \rightarrow \mathbb{R}$ is a function to determine the reward obtained by taking a particular action in a particular state. In our case, the description of states and actions is relational and therefore we have a relational MPD.

The relational description of states and actions of the RMDP was presented in Section III. We now describe the remaining components of the RMDP and how Q-learning is employed to determine an optimal policy π for manipulating articulated objects in unstructured environments.

A. *Transition Function*

The transition function captures the state transitions that occur in the physical world when an action is applied. We never explicitly represent this function. Instead, we rely on the real world and on our perceptual capabilities to determine the new state after the application of an action has been completed.

B. *Reward Function*

The reward function $R : S \times A \rightarrow \mathbb{Z}$ returns the number of links and joints that were discovered by performing an action in a particular state.

C. *Q-Learning*

Q-learning [29] determines a policy $\pi : S \rightarrow A$ for selecting actions based on the current state. To determine this policy, our goal is to learn the Q-value function $Q : S \times A \rightarrow \mathbb{R}$ by performing a series of experiments, each of which reveals how much reward a particular action can obtain in a particular state. The Q-value function accumulates information about the total expected reward for an entire trial. The policy defined by the Q-value function is given by $\pi(s) = \arg\max_a Q(s, a)$.

As the robot performs actions in its environment and receives the resulting rewards, the Q function is updated according to the following rule:

$$Q(s_t, a_t) = (1-\alpha)\, Q(s_t, a_t) + \alpha \left(r_{r+1} + \gamma \max_a Q(s_{t+1}, a) \right),$$

where α is the learning rate, γ is the discount factor, and r_t is the reward received at time t.

D. *Representation of Q-Value Function*

Q-learning requires an adequate representation for the Q-value function. In our case, this representation is instance-based [6]. The robot stores each of its experiences as a tuple of state, action, and the Q-value obtained when performing the action in that state. Because states and actions are relational and stored uninstantiated, every stored experience is applicable to a possibly infinite number of situations.

Given the current state, the robot has to retrieve estimates of Q-values for actions from its experience. This is particularly important when the robot has not previously visited the current state. By doing so, the robot is able to leverage relevant prior experience in a new situation, thereby improving its learning performance.

To identify the experience most relevant to a state, we need a similarity measure for states. Similarity is affected by the state's kinematic structure and by the properties of the links in that structure. Neither of these aspects have to match perfectly for the robot to retrieve relevant experience. We first describe how unification is used to match properties of individual links between the kinematic structure of the current state and the state stored in the Q-value function. We then explain how similarity between kinematic structures can be identified.

Let us assume the robot at time t has uncovered the existence of three links (large, small, large), connected into a serial chain by revolute joints; the corresponding relational state description is

$$s_t = R(R(L(l, f_1), L(s, f_2)), L(l, f_3)),$$

ignoring the background for simplicity. Further assume that the Q-value function representation contains a single experience with a structure/action/reward tuple

$$(s, a, r) = (R(R(L(s, v_1), L(s, v_2)), L(s, v_3)), A(v_3, 45°), 1.6)\,.$$

The state s represents a serial chain with two revolute joints and three small links. Note that the Q-value function does not store the sets of features f_i for each link but instead includes a variable v_i. This variable can now be instantiated by unifying the memorized state s with the current state representation s_t. Due to the different instantiation of link size, however, unification fails in this case. We can still retrieve somewhat less relevant experience by ignoring the link size. The resulting unification leads to a binding of $v_3 \leftarrow f_3$. This instantiates the action to $A(f_3, 45°)$, telling the robot to push on the link described by the visual features in f_3 from $45°$ angle relative to the principal axes of the feature set.

This example illustrated how the unification process progressively ignores the least discriminative property of links until unification succeeds. We now explain how similar kinematic structures can be mapped onto each other to retrieve relevant experience.

We saw in Section III that the relational representation of kinematic structures is not unique. State s_t, for example, is equivalent to $R(L(l, f_3), R(L(s, f_2), L(l, f_3)))$. We would like to retrieve relevant experience in the presence of this ambiguity. Furthermore, for a state $s_t = R(L(l, f_1), L(l, f_2))$ we would like to be able to leverage our experience by realizing that $L(l, f_1)$ in s_t could represent $R(L(l, f_1), L(s, f_2))$ in s before the additional degree of freedom was discovered.

To identify closely related kinematic structures, we represent a relational state as an undirected, labeled graph $G = (V, E)$. A vertex $v \in V$ corresponds to a link. An edge $e \in E$ is labeled as either prismatic or revolute, corresponding to the kinematic relationship between two links.

This graph representation naturally supports the desired ability to retrieve relevant experience from the Q-value function, even for structure-preserving re-orderings of the relational representation as well as for super or sub-structures of the current state. Given two graphs G_t and G corresponding to s_t and s, we check for graph isomorphism to find exact structural matches and sub-matches, even when the relational descriptions of the underlying structure vary. To determine partial matches, we check for subgraph monomorphism between G_t and G. In contrast to subgraph isomorphism, which is a bijection, subgraph monomorphism is an injection, thus the match is one-to-one but not onto.

When one or multiple graph matches exist, the robot retrieves the experience associated with the closest match. When no graph match can be established or the action stored with the matching state cannot be instantiated based on the match, the robot is unable to retrieve relevant experience from the Q-value function.

Each time the robot performs an action and receives a reward, we store this experience in the instance-based Q-value function. If an exact graph match exists between the current state and a previously encountered state (graph isomorphism), we update the existing memory entry with the new experience. Otherwise, we add this experience as a new instance to the representation of the Q-value function.

Subgraph monomorphism is an NP-complete problem. However, efficient algorithms for small graphs exist [5]. Since most real-world articulated objects posses a small number of links, the theoretical computational complexity does not impose any practical limitations on our approach.

Similar to other memory-based approaches to learning, our approach may require large amounts of memory. Several methods to remedy this problem have been proposed, specifically in the context of relational reinforcement learning [6]. We believe that the consolidation of experiences based on domain-specific generalization is an important issue for future research. Ultimately, we expect to apply unification and graph matching to the obtained experience in order to generate general manipulation rules, greatly reducing the memory requirement of our instance-based representation for the Q-value function.

E. Action Selection: Balancing Exploration and Exploitation

To learn an optimal policy, the robot has to balance exploration and exploitation. Exploration refers to the execution of an action to improve the Q-value function's estimate of the associated reward. Exploitation, in contrast, refers to action selection based on maximizing reward. If the robot explores too much, it will learn slowly. If it exploits too early, it will perform poorly because it has not gathered enough experience. We complete the description of our approach by explaining how action selection during learning balances exploration and exploitation.

When selecting an action for the current state, the robot can either perform exploration by selecting a new action, or it can use its experience with previously performed actions. In the latter case, the robot again chooses between exploration and exploitation. It can either perform exploitation by choosing the most promising action based on its current estimates of Q-values, or perform exploration in an attempt to improve the current estimates of Q-values.

To decide if a new action should be executed (the first trade-off), we compute the fraction ϕ of actions for which the robot already has gathered experience. If a number drawn uniformly at random from the interval $[0, 1]$ is smaller than $e^{-\beta\phi}$, the robot performs exploration ($\beta = 2$ in our experiments). Otherwise it selects one of the actions associated with state s. The selection among those actions represents the second trade-off. We perform it using Interval Estimation (IE) [11]. Intuitively, IE picks the action that still has the highest potential to perform well. More advanced alternatives to IE guarantee polynomial bounds on the resources required to achieve near-optimal return [13].

VI. Grounding Relational Reinforcement Learning

The learning framework described in this paper is entirely symbolic. To ground this framework, we have to link updates to relational state descriptions to the perceptual capabilities of the robot and actions performed by the robot to the relational description of actions in the learning framework.

The state description is grounded using the perceptual skill described in Section IV. When the perceptual discovers a new link, observes internal motion of a link, or observes a different kinematic relationship than the one represented in the state, the relational state representation is updated. The state description is also updated when new properties of links are perceived.

An action is grounded using the set of visual features f_i in the robot's perceptual space. The relational action can be translated into a force-controlled physical action that establishes contact with the table on the appropriate side of the point cloud and then performs a compliant motion until contact with the object is made and the desired motion is observed.

The task-specific grounding of state updates and the action executions closes the loop between the physical world and the learning framework. It ensures that the learned manipulation experience is physically meaningful and can be translated back into a useful physical action.

VII. Experimental Validation

To demonstrate the effectiveness of our learning-based approach to manipulation in unstructured environments, we perform two types of experiments. First, we show that our approach permits the learning of manipulation knowledge from experience. Second, we show that the acquired experiences transfer to previously unseen objects.

A. Experimental Setup

Our experimental evaluation requires a large number of experiments. For practical reasons, we performed these experiments in a simulated environment. Due to the robustness of the perceptual skill described in Section IV and due to the simplicity of force guided pushing required for our experiments, we argue that our results remain valid in real-world experiments. Our simulation environment is based on the Open Dynamics Engine (ODE), a dynamics simulator. The simulation includes gravity, friction, and non-determinacy.

In each experiment, the robot interacts with an articulated object to extract its kinematic structure. Example objects are given in Figures 3 and 4. Revolute joints are shown as red cylinders, prismatic joints are represented by green boxes, and links are shown in blue. Currently, our approach is limited to planar objects. We also restrict our experimentation to serial chains, even though our implementation handles branching mechanisms and loops. Perceptual information about the manipulated objects is obtained from a simulation of the perceptual skill described in Section IV [12]. We do not use the simulator's internal object representation to obtain information of the object.

Each experiment consists of a sequence trials. For each trial we report the average over 10 independent experiments. A trial consists of a number of steps; in each step, the robot applies a pushing action to the the articulated object. The trial ends when an external observer signals that the obtained model accurately reflects the kinematic structure of the articulated object. The number of steps required to uncover the correct kinematic structure measures the effectiveness with which the robot accomplishes the task.

Each step of a trial can be divided into three phases. In the first phase, the robot selects an action and a link with which it wants to interact. The action is instantiated using the current state and the experience stored in the representation of the Q-value function. In the second phase, the selected action is applied to the link, and the ODE simulator generates the resulting object motion. The trajectories of the visual features tracked by the perception skill are reported to the robot. In the last phase, the robot analyzes the motion of visual features and determines the kinematic properties of the rigid bodies observed so far. These properties are then incorporated into the robot's current state representation. With each step, the robot accumulates manipulation experiences that improves its performance over time.

A trial ends when the kinematic model obtained by the robot corresponds to the structure of the articulated object. In our simulation experiments, an external supervisor issues a special reward signal to end the particular trial. Note that such a supervisor is not required for real-world experiments. The robot can decide to perform manipulation based on incomplete available information. If new kinematic information is discovered during manipulation, the robot simply updates its kinematic model and revises its manipulation strategy accordingly.

B. Learning Manipulation Knowledge

To demonstrate the ability of the proposed learning framework to acquire relevant manipulation knowledge, we observe the number of actions required to discover a kinematic structure. We compare the performance of the proposed grounded relational reinforcement learning approach to a random action selection strategy, using an object with seven degrees of freedom and eight links (Fig. 3(a)). The resulting learning curve is shown in Figure 3(b). Random action selection, as to be expected, does not improve its performance with additional trials. In contrast, action selection based on the proposed relational reinforcement learning approach results in a substantial reduction of the number of actions required to correctly identify the kinematic structure. This improvement already becomes apparent after about 20 trials. Using the learning-based strategy, an average of 8 pushing actions is required to extract the complete kinematic model, compared to the approximately 20 pushing actions required with random action selection. This corresponds to an improvement of 60%.

This first experiment demonstrates that our approach to manipulation enables robots to acquire manipulation knowledge and to apply this knowledge to improve manipulation performance.

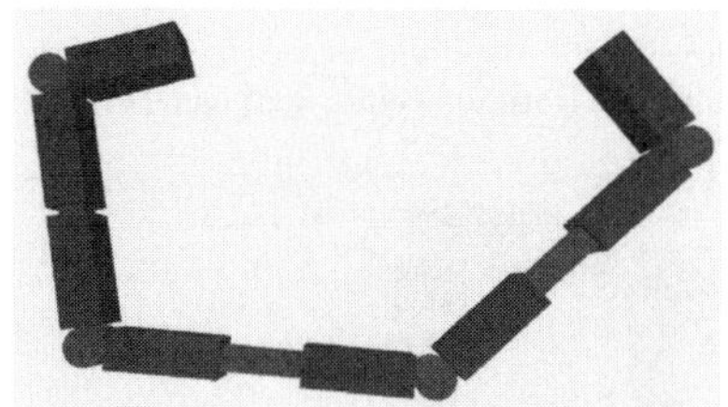

(a) Articulated object

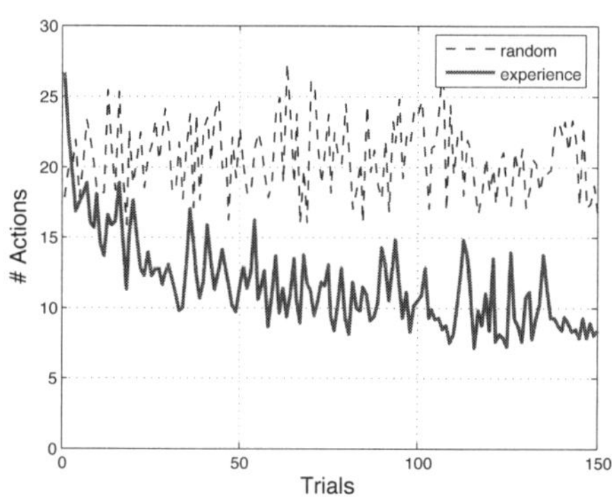

(b) Performance compared to random

Fig. 3. Experiments with a planar kinematic structure with seven degrees of freedom (RPRPRPR, R = revolute, P = prismatic). The learning curve for our learning-based approach to manipulation (green solid line) converges to eight required actions with a decreasing variance, representing an improvement of 60% over the random strategy (blue dashed line).

C. Transferring Manipulation Knowledge

To demonstrate that the manipulation experience acquired with one object transfers to other objects, we observe the number of actions required to discover a kinematic structure with and without prior experience.

In the first transfer experiment, the robot gathers experience with an articulated object with 5 degrees of freedom (see Fig. 4(a)). After 50 trials, the robot is given a more complex object with two additional degrees of freedom. The simple structure is a sub-structure of the more complex one. We compare the robot's performance with that of a robot without prior experience (see Fig. 4(a)). The robot with prior experience consistently outperforms the robot without experience. Over the first ten trials, this performance improvement is approximately 20%. In trials 10 to 40, the performance improvement is much smaller. Interestingly, as variance decreases (the robot decreases its exploration rate), the performance of the robot with experience again achieves a 20% performance

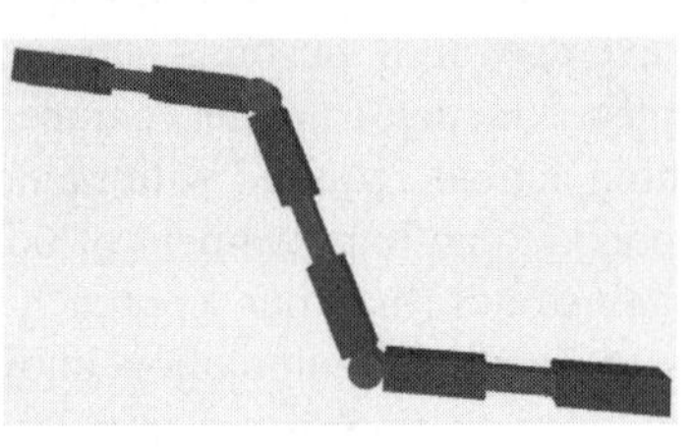

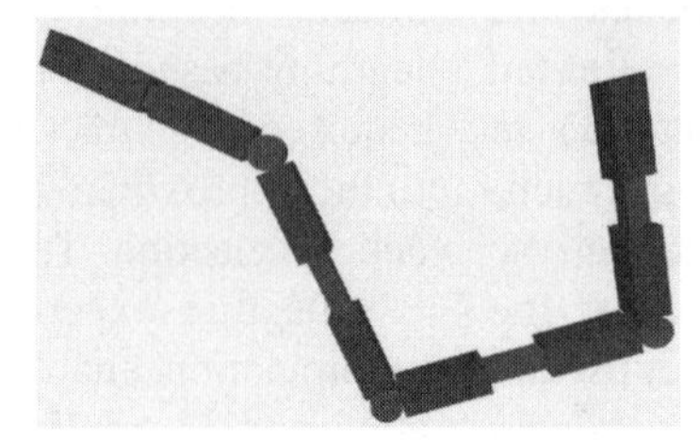

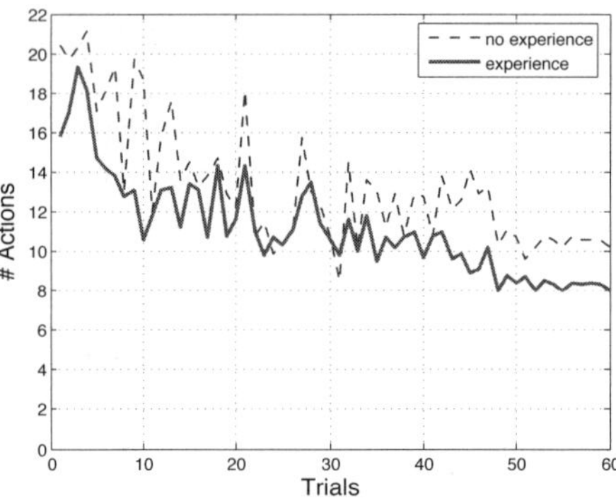

(a) Learning curves for a robot with experience manipulating the PRPRP object on the left (solid green line) compared to an inexperienced robot (dashed blue line). Both robots learn to acquire the kinematic structure of a more complex object (PRPRPRP, middle). Experience improves performance by 20%.

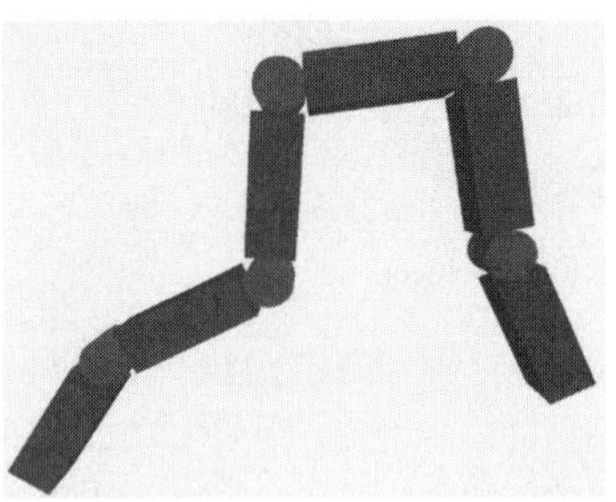

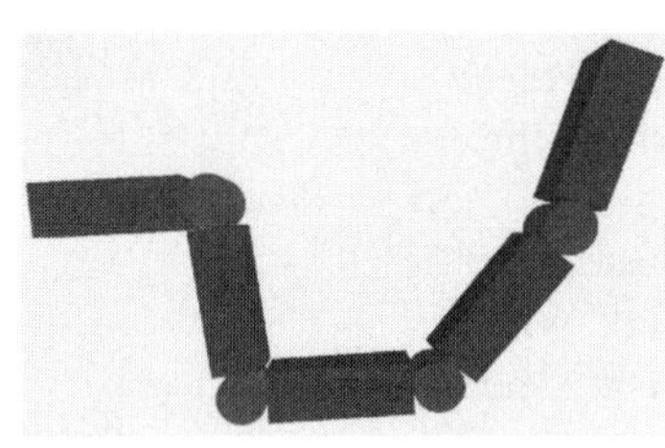

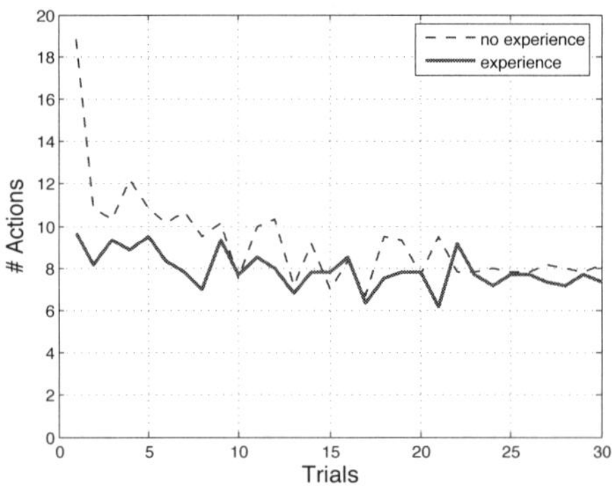

(b) Learning curves for a robot with experience manipulating the RRRRR object on the left (solid green line) compared to an inexperienced robot (dashed blue line). Both robots learn to acquire the kinematic structure of a simpler object (RRRR, middle). Experience leads to nearly immediate convergence.

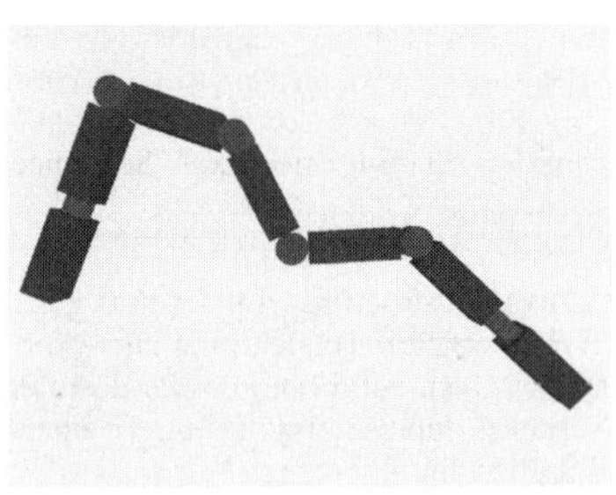

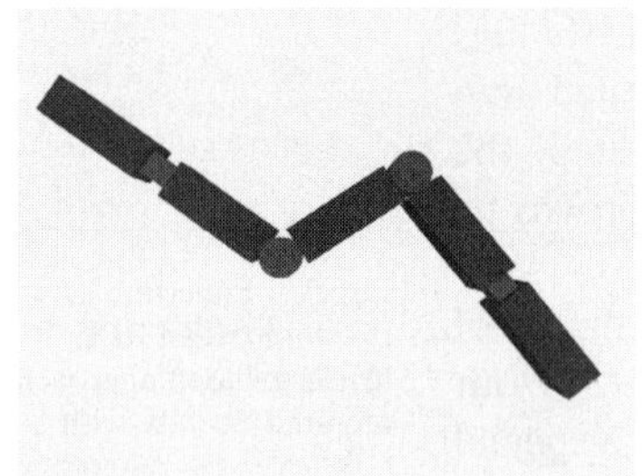

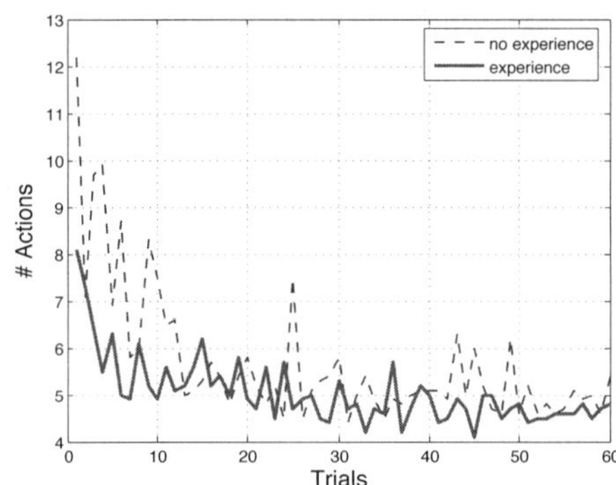

(c) Learning curves for a robot with experience manipulating the PRRRRP object on the left (solid green line) compared to an inexperienced robot (dashed blue line). Both robots learn to acquire the kinematic structure of a simpler object (PRRP, middle). The simpler object is **not** a sub-structure of the complex object. With experience, convergence is achieved in about five trials.

Fig. 4. Experimental validation of transfer of manipulation experience between different articulated objects.

improvement. We attribute this to the fact that some useful manipulation strategies can more easily be discovered in the smaller state space of the simpler structure.

In the second experiment, the robot learns to manipulate a complex articulated object with 5 revolute joints. After 50 trials, the robot is given a slightly simpler structure that only possesses four revolute joints. Again, the simpler structure is a sub-structure of the more complex one. We compare the robot's performance after these initial 50 trials to another robot's performance without prior experience (see Fig. 4(b)). Given prior experience, the robot achieves convergence almost immediately. This corresponds to a performance improvement of about 50% in the first trial, relative to the robot without experience. After about ten trials, both robots converge to approximately the same performance, which is to be expected for simple structures that exclusively consist of revolute joints.

In the third experiment, the robot learns to manipulate an articulated object with 6 degrees of freedom (see Fig. 4(c)). After 50 trials, the robot is given a different structure that is not a substructure of the other. We compare the robot's performance after these initial 50 trials to another robot's performance without prior experience (see Fig. 4(c)). Again, experience results in a much faster convergence (after only five trials) towards about five required interactions. In addition, the variance of successive trials is reduced. After about 15 trials, both robots converge towards the same number of interactions.

Our experimental results provide strong evidence that learning from past experience can significantly improve manipulation performance. We attribute the effectiveness of our approach leverages to the proper, task-specific grounding of our relational representation.

VIII. Conclusion

We proposed a learning-based approach for manipulation in unstructured environments. We provide experimental evidence that this approach enables robots to autonomously acquire manipulation expertise by interacting with the environment. This expertise transfers across different instances of the manipulation task and substantially improves manipulation performance.

Learning and generalization of manipulation knowledge becomes possible due to a relational representation of states and actions. This representation reduces the state space and renders relational reinforcement learning tractable, even in complex manipulation domains. The power of this symbolic representation is leveraged in the real world through careful grounding of the symbols in the robot's perceptual and interactive capabilities.

We validate the proposed approach in the context of extracting kinematic models of articulated objects. This is an important enabling skill for general manipulation in unstructured environments, as all manipulation tasks require a deliberate and purposeful change in the configuration of an object and therefore knowledge of the kinematic model of an object. We demonstrate that grounded relational reinforcement learning substantially improves the robot's performance in this task. Our experiments show that appropriately grounded relational reinforcement learning is a promising approach towards endowing robots with manipulation skills adequate for unstructured environments.

Acknowledgments

We gratefully acknowledge support by the National Science Foundation (NSF) under grants CNS-0454074, IIS-0545934, CNS-0552319, CNS-0647132, and CNS-0812986. We also thank Nick Roy for insightful suggestions and discussions.

References

[1] J. D. Bernheisel and K. M. Lynch. Stable transport of assemblies by pushing. *IEEE Transactions on Robotics*, 22(4):740–750, 2006.

[2] A. Bicchi and V. Kumar. Robotics grasping and contact: A review. In *Proceedings of the IEEE International Conference on Robotics and Automation (ICRA)*, San Francisco, CA, 2000.

[3] R. R. Burridge, A. A. Rizzi, and D. E. Koditschek. Sequential composition of dynamically dexterous robot behaviors. *International Journal of Robotics Research*, 18(6):534–555, 1999.

[4] A. D. Christiansen, M. T. Mason, and T. M. Mitchell. Learning reliable manipulation strategies without initial physical models. *Robotics and Autonomous Systems*, 8(1-2):7–18, 1991.

[5] L. P. Cordella, P. Foggia, C. Sandone, and M. Vento. Performance evaluation of the VF graph matching algorithm. In *Proceedings of the 10th ICIAP IEEE Computer Society Press*, volume 2, pages 1038–1041, 1999. http://amalfi.dis.unina.it/graph/db/vflib-2.0/doc/vflib.html.

[6] K. Driessens and J. Ramon. Relational instance based regression for relational reinforcement learning. In *Proceedings of the International Conference on Machine Learning (ICML)*, 2003.

[7] S. Džeroski, L. de Raedt, and K. Driessens. Relational reinforcement learning. *Machine Learning*, 43(1-3):7–52, 2001.

[8] A. Edsinger and C. C. Kemp. Manipulation in human environments. In *Proceedings of the IEEE International Conference on Humanoid Robots*, 2006.

[9] R. Grupen and M. Huber. A framework for the development of robot behavior. In *Proceedings of the 2005 AAAI Spring Symposium Series: Developmental Robotics*, Stanford, USA, 2005.

[10] S. Harnad. The symbol grounding problem. *Physica D*, 42:335–346, 1990.

[11] L. P. Kaelbling. *Learning in Embedded Systems*. MIT Press, 1993.

[12] D. Katz and O. Brock. Manipulating articulated objects with interactive perception. In *Proceedings of the IEEE International Conference on Robotics and Automation (ICRA)*, Pasadena, USA, 2008.

[13] M. Kearns and S. Singh. Near-optimal reinforcement learning in polynomial time. In *Proceedings of the International Conference on Machine Learning (ICML)*, pages 260–268. Morgan Kaufmann, San Francisco, CA, 1998.

[14] K. M. Lynch and M. T. Mason. Dynamic nonprehensile manipulation: Controllability, planning, and experiments. *International Journal of Robotics Research*, 18(1):64–92, 1999.

[15] M. T. Mason. *Mechanics of Robotic Manipulation*. MIT Press, 2001.

[16] A. W. Moore. Efficient memory-based learning for robot control. Technical Report UCAM-CL-TR-209, Computer Laboratory, University of Cambridge, 1990.

[17] A. M. Okamura, N. Smaby, and M. R. Cutkosky. An overview of dexterous manipulation. In *Proceedings of the IEEE International Conference on Robotics and Automation (ICRA)*, volume 1, pages 255–262, San Franscisco, USA, 2000.

[18] R. Platt, R. Burridge, M. Diftler, J. Graf, M. Goza, E. Huber, and O. Brock. Humanoid mobile manipulation using controller refinement. In *Proceedings of the IEEE International Conference on Humanoid Robots*, Genova, Italy, December 2006.

[19] R. Platt, A. H. Fagg, and R. A. Grupen. Manipulation gaits: Sequences of grasp control tasks. In *Proceedings of the IEEE International Conference on Robotics and Automation (ICRA)*, New Orleans, USA, 2004.

[20] L. Sentis and O. Khatib. Synthesis of whole-body behaviors through hierarchical control of behavioral primitives. *International Journal of Humanoid Robots*, 2(4):505–518, 2005.

[21] T. Siméon, J.-P. Laumonde, J. Cortés, and A. Sahbani. Manipulation planning with probabilistic roadmaps. *International Journal of Robotics Research*, 23(7-8):729–746, 2004.

[22] S. S. Srinivasan, M. A. Erdmann, and M. T. Mason. Control synthesis for dynamic contact manipulation. In *Proceedings of the IEEE International Conference on Robotics and Automation (ICRA)*, pages 2523–2528, 2005.

[23] A. Stoytchev. Behavior-grounded representation of tool affordances. In *Proceedings of the IEEE International Conference on Robotics and Automation (ICRA)*, pages 3071–3076, Barcelona, Spain, 2005.

[24] A. Sudsang and J. Ponce. In-hand manipulation: Geometry and algorithms. *Algorithmica*, 26(4):466–493, 2000.

[25] P. Tadepalli, R. Givan, and K. Driessens. Relational reinforcement learning: An overview. In *Proceedings of the Workshop on Relational Reinfocement Learning at ICML '04*, Banff, Canada, 2004.

[26] J. C. Trinkle and R. P. Paul. Planning for dexterous manipulation with sliding contacts. *International Journal of Robotics Research*, 9(3):24–48, 1990.

[27] M. van Otterlo. A survey of reinforcement learning in relational domains. Technical Report TR-CTIT-05-31, Department of Computer Science, University of Twente, The Netherlands, July 2005.

[28] S. Vijayakumar, A. D'Souza, T. Shibata, J. Conradt, and S. Schaal. Statistical learning for humaonid robots. *Autonomous Robots*, 12(1):59–72, 2002.

[29] C. J. C. H. Watkins and P. Dayan. *Q*-learning. *Machine Learning*, 8(3):279–292, 1992.

High Performance Outdoor Navigation from Overhead Data using Imitation Learning

David Silver, J. Andrew Bagnell, Anthony Stentz
Robotics Institute, Carnegie Mellon University
Pittsburgh, Pennsylvania USA

Abstract— **High performance, long-distance autonomous navigation is a central problem for field robotics. Efficient navigation relies not only upon intelligent onboard systems for perception and planning, but also the effective use of prior maps and knowledge. While the availability and quality of low cost, high resolution satellite and aerial terrain data continues to rapidly improve, automated interpretation appropriate for robot planning and navigation remains difficult. Recently, a class of machine learning techniques have been developed that rely upon expert human demonstration to develop a function mapping overhead data to traversal cost. These algorithms choose the cost function so that planner behavior mimics an expert's demonstration as closely as possible. In this work, we extend these methods to automate interpretation of overhead data. We address key challenges, including interpolation-based planners, non-linear approximation techniques, and imperfect expert demonstration, necessary to apply these methods for learning to search for effective terrain interpretations. We validate our approach on a large scale outdoor robot during over 300 kilometers of autonomous traversal through complex natural environments.**

I. Introduction

Recent competitions have served to demonstrate both the growing popularity and promise of mobile robotics. Although autonomous navigation has been successfully demonstrated over long distances through on-road environments, long distance off road navigation remains a challenge. The varying and complex nature of off-road terrain, along with different forms of natural and man made obstacles, contribute to the difficulty of this problem.

Although autonomous off-road navigation can be achieved solely through a vehicle's onboard perception system, both the safety and efficiency of a robotic system are greatly enhanced by prior knowledge of its environment. Such knowledge allows high risk sections of terrain to be avoided, and low risk sections to be more heavily utilized.

Overhead terrain data are a popular source of prior environmental knowledge, especially if the vehicle has not previously encountered a specific site. Overhead sources include aerial or satellite imagery, digital elevation maps, and even 3-D LiDAR scans. Much of this data is freely available at lower resolutions, and is commercially available at increasing resolution.

Techniques for processing such data have focused primarily on processing for human consumption. The challenge in using overhead data for autonomous navigation is to interpret the data in such a way as to be useful for a vehicle's planning and navigation system. This paper proposes the use of imitation learning to train a system to automatically interpret overhead data for use within an autonomous vehicle planning system. The proposed approach is based on the Maximum Margin Planning (MMP) [1] framework, and makes use of expert provided examples of how to navigate using the provided data.

The next section presents previous work in off-road navigation from prior data. Section III presents the MMP framework, and Section IV develops its application to this context. Results are presented in Section V, and conclusions in Section VI.

II. Outdoor Navigation from Overhead Data

At its most abstract level, the outdoor navigation problem in mobile robotics involves a robot driving itself from a start waypoint to a goal waypoint. The robot may or may not have prior knowledge of the environment, and may or may not be able to gather more environmental knowledge during its traverse. In either case, there exists some form of terrain model. For outdoor navigation, this is often a grid representation of the environment, with a set of environmental features at each grid cell.

The systems we consider in this work rely upon a *planning system* to make navigation decisions. This system is responsible for finding an optimal path that will lead the robot to its goal. The naturally raises the question of how "optimal" will be defined: is it the safest path, the fastest path, the minimum energy path, or the most stealthy path? Depending on the context, these are all valid answers; combinations of the above are even more likely to be desired. In practice, planners typically function by choosing the path with the minimum score according to some scalar function of the environment that approximates those metrics. Therefore, for a chosen metric, the robot's terrain model must be transformed into a single number for every planner state: that state's traversal cost. A planner can then determine the optimal path by minimizing the cumulative cost the robot expects to experience. Whether or not this path achieves the designer's desired behavior depends on how faithfully the cost function, mapping features of the environment to scalar traversal costs, reflects the designer's internal performance metric.

Previous work [2] [3] has demonstrated the effectiveness of overhead data for use in route planning. Overhead imagery, elevation maps, and point clouds are processed into features stored in 2-D grids. These feature maps can then be converted to traversal costs. Figure 1 demonstrates how this approach can lead to more efficient navigation by autonomous vehicles.

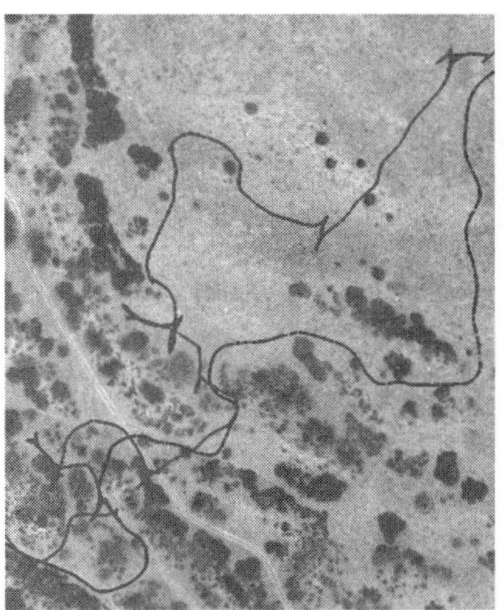

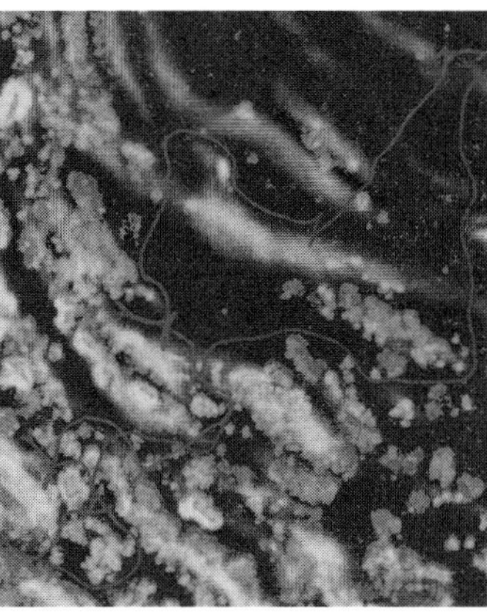

Fig. 1. The path traveled during two runs of an autonomous robot moving from top-right to bottom left. The **green** run had no prior map, and made several poor decisions that required backtracking. The **red** run had the prior costmap shown at right, and took a more efficient route. Brighter pixels indicate higher cost.

Despite good results, previous work has depended on hand tuned cost functions. That is, an engineer manually determined both the form and parameters of a function to map overhead features to a single cost value, in an attempt to express a desired planning metric. By far, human engineering is the most common approach to generating cost functions. This is somewhat of a black art, requiring a deep understanding of the features being used, the metric to be optimized, and the behavior of the planner. Making the situation worse, determining a cost function is not necessarily a one time operation. Each environment could have drastically different overhead input data available. One environment may have satellite imagery, another aerial imagery and an elevation map, another may have multispectral imagery, etc. Therefore, an entire family of cost functions is needed, and parameters must be continually retuned.

One approach to simplifying this problem is to transform the original "raw" feature space into a "cost" feature space, where the contribution of each feature to cost is more intuitive. Classifiers are a common example of this approach, where imagery and other features are transformed into classifications such as road, grass, trees, bushes, etc. However, the tradeoff that is faced is a loss of information. Also, while the cost functions themselves may be simpler, the mapping into this new feature space now must be recomputed or retrained when the raw features change, and the total complexity of the mapping to cost may have increased.

Regardless of any simplifications, human engineering of cost functions inevitably involves parameter tuning, usually until the results are correct by visual inspection. Unfortunately, this does not ensure the costs will actually reflect the chosen metric for planning. This can be validated by planning some sample start/goal waypoints to see if the resulting paths are reasonable. Unreasonable plans result in changes to the cost parameters. This process repeats, expert in the loop, until the expert is satisfied. Often times, continual modifications take the form of very specific rules or thresholds, for very specific shortcomings of the current cost function. This can result in a function with poor generalization.

Hidden in this process is a key insight. While experts have difficulty expressing a specific cost function, they are good at evaluating results. That is, an expert can look at the raw data in its human presentable form, and determine which paths are and are not reasonable. It is this ability of experts that will be exploited for learning better cost functions.

III. Maximum Margin Planning for Learning Cost Functions

Imitation Learning has been demonstrated as an effective technique for deriving suitable autonomous behavior from expert examples. Previous work specific to autonomous navigation [4, 5] has demonstrated how to learn mappings from features of a state to actions. However, these techniques do not generalize well to long range decisions, due to the dimensionality of the feature space that would be required to fully encode the such a problem.

A powerful recent idea for how to approach such long range problems is to structure the space of learned policies to be optimal solutions of search, planning or general Markov Decision Problems [1, 6, 7] . The goal of the learning procedure is then to identify a mapping from features of a state to costs such that the resulting minimum cost plans capture well the demonstrated behavior. We build on the approach of Maximum Margin Planning (MMP) [1, 7], which searches for a cost function that makes the human expert choices appear optimal.

In this method, an expert provided optimal example serves as a constraint on cost functions. If an example path is provided as the best path between two waypoints, then any acceptable cost function should cost the example as less then or equal to all other paths between those two waypoints. By providing multiple example paths, the space of possible cost functions can be further constrained.

Our interest in this work is in applying learning techniques to ease the development of cost functions appropriate for outdoor navigation. The approach to imitation learning described above directly connects the notion of learning a cost function to recovering demonstrated behavior. As we introduce new techniques that improve performance in our application, we sketch the derivation of the functional gradient version [8, 9] of MMP below. This approach allows us to adapt existing, off-the-shelf regression techniques to learn the potentially complicated cost function, leading to a modular and simple to implement technique.

A. Single Example

Consider a state space $\mathcal{S}$, and a feature space $\mathcal{F}$ defined over $\mathcal{S}$. That is, for every $x \in \mathcal{S}$, there exists a corresponding feature vector $F_x \in \mathcal{F}$. C is defined as a cost function over the feature space, $C : \mathcal{F} \rightarrow \mathbb{R}^+$. Therefore, the cost of a state x is $C(F_x)$.

A path P is defined as a sequence of states in $\mathcal{S}$ that lead from a start state s to a goal state g. The cost of P is simply the sum of the costs of all states along the path.

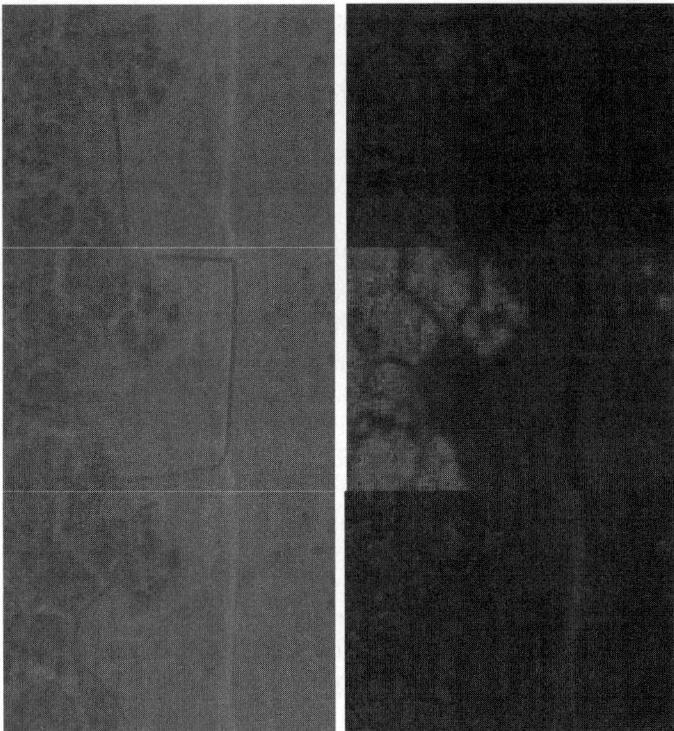

Fig. 2. **Left:** Example paths that imply different metrics (From top to bottom: minimize distance traveled, stay on roads, stay near trees) **Right:** Learned costmaps from the corresponding examples (brighter pixels indicate higher cost). Quickbird imagery courtesy of Digital Globe, Inc.

If an example path P_e is provided, then a constraint on cost functions can be defined such that the cost of P_e must be lower cost than any other path from s_e to g_e. The structured maximum margin approach [1] encourages good generalization and prevents trivial solutions (e.g. the everywhere 0 cost function) by augmenting the constraint to includes a margin: i.e. the demonstrated path must be BETTER than another path by some amount. The size of the margin is dependent on the similarity between the two paths. In this context, similarity is defined by how many states the two paths share, encoded in a function L_e. Finding the optimal cost function then involves constrained optimization of an objective functional over C

$$\min \mathcal{O}[C] = \textsc{Reg}(C) \tag{1}$$

subject to the constraints

$$\sum_{x \in \hat{P}} (C(F_x) - L_e(x)) \; - \; \sum_{x \in P_e} (C(F_x)) \geq 0$$

$$\forall \hat{P} \; s.t. \; \hat{P} \neq P_e, \;\; \hat{s} = s_e, \;\; \hat{g} = g_e$$

$$L_e(x) = \begin{cases} 1 & \text{if } x \in P_e \\ 0 & \text{otherwise} \end{cases}$$

where REG is a regularization operator that encourages generalization in the the cost function C.

There are typically an exponentially large (or even infinite) number of constraints, each corresponding to an alternate path. However, it is not necessarily to enumerate these constraints. For every candidate cost function, there is a minimum cost path between two waypoints and at each step it is only necessary to enforce the constraint on this path.

It may not always be possible to achieve all constraints and thus a "slack" penalty is added to account for this. Since the slack variable is tight, we may write an "unconstrained" problem that captures the constraints as penalties:

$$\min \mathcal{O}[C] = \textsc{Reg}(C) + \sum_{x \in P_e} (C(F_x)) - \min_{\hat{P}} \sum_{x \in \hat{P}} (C(F_x) - L_e(x)) \tag{2}$$

For linear cost functions (and convex regularization) $\mathcal{O}[C]$ is convex, and can be minimized using gradient descent. [1]

Linear cost functions may be insufficient, and using the argument in [8], it can be shown that the (sub-)gradient in the space of cost functions of the objective is given by [2]

$$\nabla \mathcal{O}_F[C] = \sum_{x \in P_e} \delta_F(F_x) \; - \; \sum_{x \in P_*} \delta_F(F_x) \tag{3}$$

Simply speaking, the functional gradient is positive at values of F that the example path pass through, and negative at values of F that the planned path pass through. The magnitude of the gradient is determined by the frequency of visits. If the example and planned paths agree in their visitation counts at F, the functional gradient is zero at that state. Applying gradient descent in this space of cost functions directly would involve an extreme form of overfitting: defining a (potentially unique) cost at every value of F encountered and involving no generalization. Instead, as in gradient boosting [9] we take a small step in a limited set of "directions" using a hypothesis space of functions mapping features to a real number. The result is that the cost function is of the form:

$$C(F) = \sum \eta_i R_i(F) \tag{4}$$

where R belongs to a chosen family of functions (linear, decision trees, neural networks, etc.) The choice of hypothesis space in turn controls the complexity of the resulting cost function. We must then find at each step the element R_* that maximizes the inner product $\langle - \nabla \mathcal{O}_F[C], R_* \rangle$ between the functional gradient and elements of the space of functions we are considering. As in [8], we note that maximizing the inner product between the functional gradient and the hypothesis space can be understand as a learning problem:

$$R_* = \arg\max_R \sum_{x \in P_e \cap P_*} \alpha_x y_x R(F_x) \tag{5}$$

$$\alpha_x = | \nabla \mathcal{O}_{F_x}[C] | \quad y_x = -\text{sgn}(\nabla \mathcal{O}_{F_x}[C])$$

In this form, it can be seen that finding the projection of the functional gradient essentially involves solving a weighted classification problem. Performing regression instead of classification adds regularization to each projection [8].

Intuitively, the regression targets are positive in places the planned path visits more than the example path, and negative

[1]Technically, sub-gradient descent as the function is non-differentiable

[2]Using δ to signify the dirac delta at the point of evaluation

```
C_0 = 1;
for i = 1...T do
    D = ∅;
    foreach P_e do
        M =
        buildCostmapWithMargin(C_{i-1}, s_e, g_e, F);
        P_* = planPath(s_e, g_e, M);
        D = D ⋃ {P_e, -1};
        D = D ⋃ {P_*, 1};
    end
    R_i = trainRegressor(F, D);
    C_i = C_{i-1} * e^{η_i R_i};
end
```

Algorithm 1: The proposed imitation learning algorithm

in places the example path visits more than the planned path. The weights on each regression target are the difference in visitation counts. In places where the visitation counts agree, both the target value and weight are 0. Pseudocode for the final algorithm is presented in Algorithm 1.

Gradient descent can be understood as encouraging functions that are "small" in the l_2 norm. If instead, we consider applying an *exponentiated functional gradient descent* update as described in [7, 10] we encourage functions that are "sparse" in the sense of having many small values and a few potentially large values. Instead of cost functions of the form in (4) this produces functions of the form

$$C(F) = e^{\sum \eta_i R_i(F)} \tag{6}$$

This naturally results in cost functions which map to $\mathbb{R}^+$, without any need for projecting the result of gradient descent into the space of valid cost functions. We believe, and our experimental work demonstrates, that this implicit prior is more natural and effective for problems of outdoor navigation.

B. Multiple Examples

Each example path defines a constraint on the cost function. However, it is rarely the case that a single constraint will sufficiently inform the learner to generalize and produce reasonable planner behavior for multiple situations. For this reason, it is often desireable to have multiple example paths.

Every example path will produce its own functional gradient as in (3). When learning with multiple paths, these individual gradients can either be approximated and followed one at a time, or summed up and approximated as a batch. To minimize training time, we have chosen the later approach.

One remaining issue is that of relative weighting between the individual gradients. If they are simply combined as is, then longer paths will have a greater contribution, and therefore a greater importance when balancing all constraints. Alternatively, each individual gradient can be normalized, giving every path equal weight. Again, we have chosen the latter approach. The intuition behind this is that every path is a single constraint, and all constraints should be equally important. However, this issue has not been thoroughly explored.

IV. Applied Imitation Learning on Overhead Data

At a high level, the application of this approach to overhead data is straightforward. $\mathcal{S}$ is $SE(2)$, with features extracted from overhead data at each 2-D cell. Example paths are drawn over this space by a domain expert. Paths can also be provided by driving over the actual terrain, when access to the terrain is possible before the learned map is required. However, there are several practical problems encountered during application of this algorithm that require solutions.

A. Adaptation to Different Planners

The algorithm suggested in the previous section requires determining the visitation counts for each cell along both the example and the current planned path. For some planners this is straightforward; a simple A* planner visits each cell on the path once. For other planners, it is not so simple. Since the goal of this work is to learn the proper cost function *for a specific planner*, planner details must be taken into account.

Many planners apply a configuration space expansion to an input cost map before planning, to account for the dimensions of the vehicle. When such an expansion does take place, it must also be taken into account by the learner. For example, if the planner takes the average cost of all cells within a window around the current cell, then the visitation count of all the cells within the same window must be incremented, not just the current cell. If the expansion applies different weightings to different cells within a window, then this weighting must also be applied when incrementing the visitation counts.

More complicated planners may also require non-unit changes to the visitation counts, regardless of an expansion. For example the interpolated A* algorithm [11] used in this work generates paths that are interpolated across cell boundaries. Therefore, the distance traveled through a cell may be anywhere in the interval $(0, \sqrt{2}]$ (in cells) As the cost of traveling through a cell under this scheme is proportional to the distance traveled through the cell, the visitation counts must also be incremented proportional to the distance traveled through the cell.

B. Regressor Choice

The choice of regressor for approximating the functional gradient can have a large impact on performance. Simple regressors may generalize better, and complex regressors may learn more complicated cost functions. This tradeoff must be effectively balanced. As in a classification task, one way to accomplish this is to observe performance on an independent validation set of example paths.

Linear regressors offer multiple advantages in this context. Aside from simplicity and good generalization, a weighted combination of linear functions simply produces another linear function. This provides a significant decrease in processing time for repeated cost evaluations.

The MMPBoost [8] algorithm can also be applied. This algorithm suggests using linear regressors, and then occasionally performing a nonlinear regression. However, instead of adding this nonlinear regressor directly to the cost function,

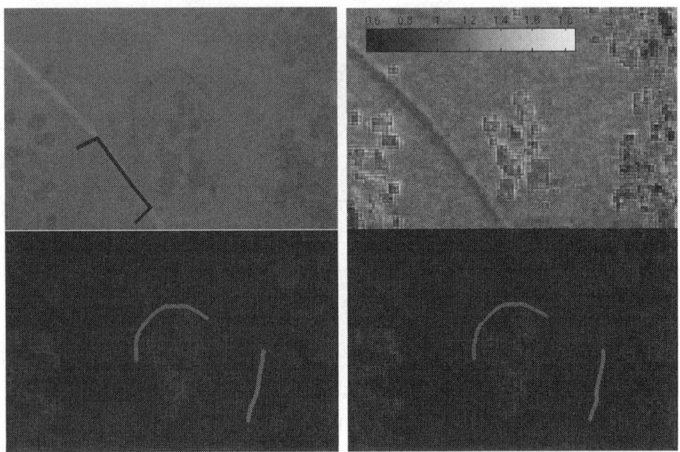

Fig. 3. **Top Left:**The red path is an unachievable example path, as it will be less expensive under any cost function to cut more directly accross the grass. **Bottom Left:** With standard weighting, the unachievable example forces down the cost of grass, and prevents the blue example from being achieved. **Bottom Right:** Balanced weighting prevents this bias, and the blue path is achieved **Top Right:** The ratio of the balanced to the standard costmap. The cost of grass is higher in the balanced map.

it is added as a new feature for future linear regressors. In this way, future iterations can tune the contribution of these occasional nonlinear steps, and the ammount of nonlinearity in the final cost function is tightly controlled.

C. Unachievable Example Paths

The derivations in Section III operate under the assumption that the example paths are either optimal or are just slightly sub-optimal, with the error being taken into the slack variables in the optimization problem. In practice, it is often possible to generate an example path such that no consistent cost metric will make the example optimal.

Figure 3 provides a simple example of such a case. The red example path takes a very wide berth around some trees. All the terrain the path travels over is essentially the same, and so will have the same cost. If the spacing the path gives the trees is greater than the expansion of the planner, then no cost function will ever consider that example optimal; it will always be better to cut more directly across the grass. In practice, such a case occurs in a small way on almost all human provided examples. People rarely draw or drive perfect straight lines or curves, and therefore generate examples that are a few percent longer than necessary. Further, a limited set of features and planner limitations mean that the path a human truly considers to be optimal may not be achievable using the planning system and features available to learn with.

It is interesting to note what happens to the functional gradient with an unachievable example. Imagine an example path that only travels through terrain with an identical feature vector F'. Under any cost function, the planned path will be the shortest path from start to goal. But what if the example path takes a longer route? The functional gradient will then be positive at F', as the example visitation count is higher than the planned. Therfore, the optimization will try to lower the cost of F'. At the next epoch, neither path will have changed. So unachievable paths result in components of the functional gradient that continually try to lower costs. These components can counteract the desires of other, achievable paths, resulting in worse cost functions (see Figure 3).

This effect can be counteracted with a small modification to the learning procedure of the functional gradient. Simplifying notation, define the negative gradient vector as

$$-\nabla \mathcal{O}_F[C] = U_* - U_e$$

Where U_* and U_e are the visitation counts of the planned and example paths, respectively.

Now consider a new gradient vector that looks at the feature counts normalized by the length of the corresponding path

$$G_{avg} = \frac{U_*}{|P_*|} - \frac{U_e}{|P_e|} \tag{7}$$

In the unachievable case described above, this new gradient would be zero instead of negative. That is, it would be satisfied with the function as is. It can be shown that this new gradient still correlates with the original gradient in other cases:

$$\begin{aligned} \langle -\nabla \mathcal{O}_F[C], G_{avg}\rangle &> 0 \\ \langle U_* - U_e, \frac{U_*}{|P_*|} - \frac{U_e}{|P_e|}\rangle &> 0 \\ \frac{\langle U_*, U_*\rangle - \langle U_*, U_e\rangle}{|P_*|} + \frac{\langle U_e, U_e\rangle - \langle U_*, U_e\rangle}{|P_e|} &> 0 \end{aligned}$$

The term $\langle U_*, U_e\rangle$ can be understood as related to the similarity in visitation counts between the two paths. If the two paths visit terrain with different features, this term will be small, and the inequality will hold. If the paths visit the exact same features ($U_* = U_e$) then the inequality does not hold. This is the situation discussed above, where the gradient is zero instead of negative.

In terms of the approximation to the gradient, this modification suggest performing a balanced classification or regression on the visitation counts as opposed to the standard, weighted one. As long as the classifier can still separate between the positive and negative classes, the balanced result will point in the same direction as the standard one. When the classifier can no longer distinguish between the classes, the gradient will tend more to towards zero, as opposed to moving in the direction of the more populous class (see Figure 3). This balancing also accomplishes the functional gradient normalization between paths described in section III-B.

D. Alternate Examples

When example paths are unachievable or inconsistent, it is often by a small amount, due to the inherent noise in a human provided example. One way to address this is to slightly redefine the constraints on cost functions. Instead of considering an example path as indicative of the exact optimal path, it can be considered instead as the approximately optimal path. That is, instead of trying to enforce the constraint that no path is cheaper than the example path, enforce the constraint that no path is cheaper than the cheapest path that exists completely within a corridor around the example path. With this constraint, the new objective functional becomes

$$\mathcal{O}[C] = \min_{\hat{P}_e} \sum_{x \in \hat{P}_e} C(F_x) \quad - \quad \min_{\hat{P}} \sum_{x \in \hat{P}} C(F_x) - L_e(x) \quad (8)$$

Where $\hat{P}_e$ is the set of all paths within the example corridor. The result of this new set of constraints is that the gradient is not affected by details smaller than the size of the corridor. In effect, this acts as a smoother that can reduce noise in the examples. However, if the chosen corridor size is too large, informative training information can be lost.

One downside of this new formulation is that the objective functional is no longer convex (due to the first $\min$ term). It is certainly possible to construct cases with a single path where this new formulation would converge to a very poor cost function. However, empirical observation has indicated that as long as more than a few example corridors are used, all the local minima achieved are quite satisfactory.

More complicated versions of this alternative formulation are also possible. For instance, instead of a fixed width corridor along the example path, a specified variable width corridor could be used. This would allow example paths with high importance at some sections (at pinch points) and low importance elsewhere. Another version of this formulation would involve multiple disjoint corridors. This could be used if an expert believed there were two different but equally desirable "best" paths from start to goal.

V. Experimental Results

A. Algorithm Verification

In order to understand the performance of this approach, a series of offline experiments was performed on real overhead data. The dataset consisted of Quickbird satellite imagery and a 3-D LiDAR scan with approximately 3 points per m^2. The metric used to evaluate the results of these experiments is the average *cost ratio* over all paths. The cost ratio is the cost of the example path over the cost of the planned path, and is proportional to the normalized value of the objective functional. As the cost function comes closer to meeting all constraints, this ratio approaches 1. All experiments were performed using linear regressors, unless otherwise stated.

1) Simulated Examples: In order to first verify the algorithm under ideal conditions, tests were run on simulated examples. A known cost function was used to generate a cost map, from which paths between random waypoints were planned. Different numbers of these paths were then used as input for imitation learning, and the performance measured on an independent test set of paths generated in the same manner.

Figure 4 shows the results using both the balanced and standard weighting schemes. As the number of training paths is increased, the test set performance continues to improve. Each input path further constrains the space of possible cost functions, bringing the learned function closer to the desired one. However, there are diminishing returns as additional

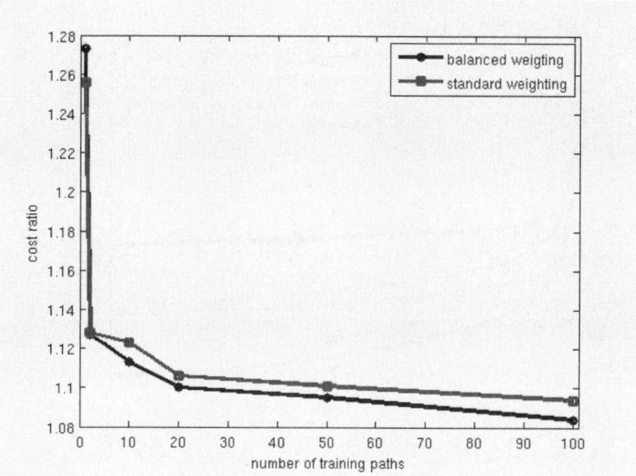

Fig. 4. Learning with simulated example paths. Test set performance is shown as a function of number of input paths.

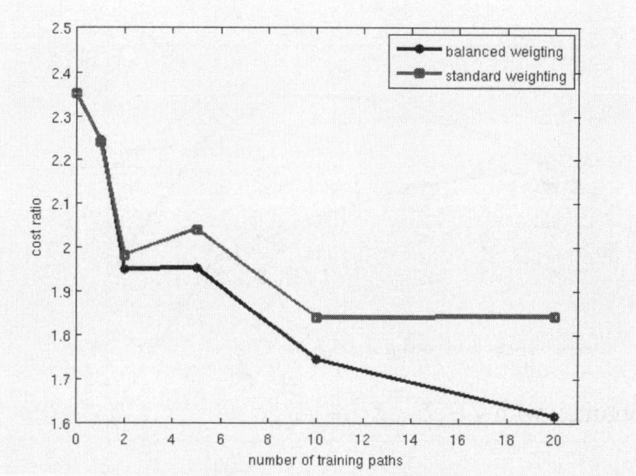

Fig. 5. Learning with expert drawn example paths. Test set performance is shown as a function of number of input paths.

paths overlap to some degree in their constraints. Finally, the performance of the balanced and standard weighting schemes is similar. Since all paths for this experiment were generated by a planner, they are by definition optimal under some metric.

2) Expert Drawn Examples: Next, experiments were performed with expert drawn examples. Figure 5 shows the results of an experiment of the same form as that performed with simulated examples. Again, the test set cost ratio decreases as the number of training examples increases. However, with real examples there is a significant difference between the two weighting schemes. The balanced weighting scheme achieved significantly better performance than the standard one. This difference in performance is further shown in Figure 6. Both the training and test performance are better with the balanced scheme. Further, with the standard weighting scheme, the distribution of costs is shifted lower, due to the negative contribution to the gradient of unachievable paths.

To demonstrate the differences in performance when using different regressors, cost functions were trained using linear

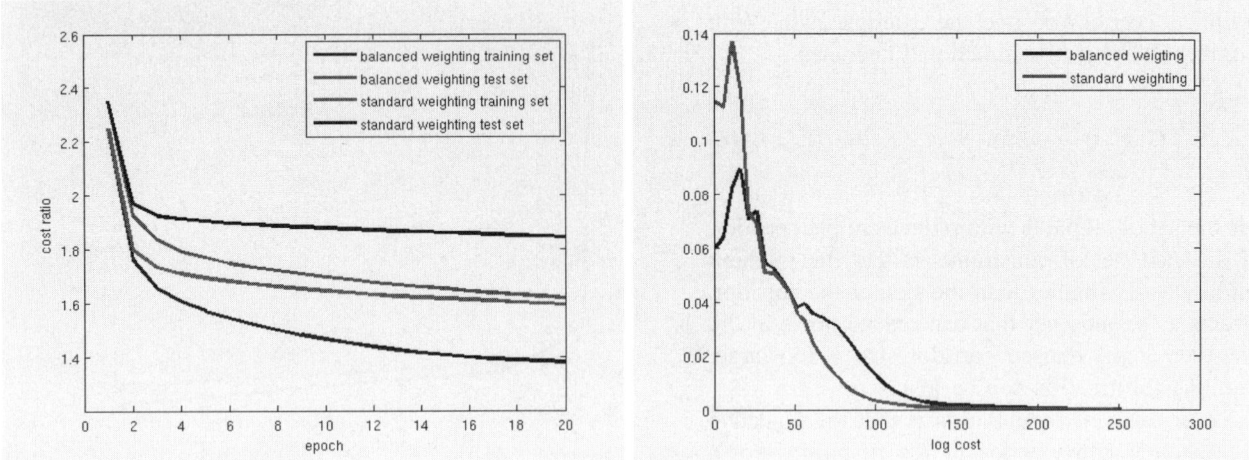

Fig. 6. **Left:** The training and test cost ratio, under both weighting schemes, as a function of the number of epochs of training. This test was run with 20 training and 20 test paths. **Right:** histogram of the costs produced by both weighting schemes. The standard weighting scheme produces lower costs.

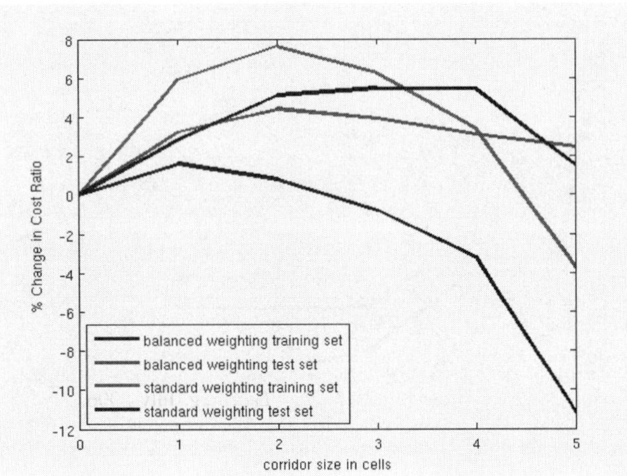

Fig. 7. Improvement in final cost ratio as a function of the corridor size

regressors, and simple regression trees (maximum 4 leaf nodes). The training/test cost ratio with linear regression was 1.23/1.35, and 1.13/1.61 with regression trees. This demonstrates the lack of generalization that can occur with even simple nonlinear regressors.

3) Corridor Constraints: A seres of experiments were performed to determine the effect of using corridor constraints. Figure 7 shows the results as a function of the corridor size in cells. Small corridors provide an improvement over no corridor. However, as the corridor gets too large, this improvement disappears; large corridors essentially over-smooth the examples. The improvement due to using corridor constraints is larger when using the standard weighting scheme, as the balanced scheme is more robust to noise in the examples.

B. Offline Validation

Next, experiments were performed in order to compare the performance of learned costmaps with engineered ones. A cost map was trained off of satellite imagery for an approximately 60 km^2 size area. An engineered costmap had been previously produced for this same area to support the DARPA UPI program (see section V-C). This map was produced by performing a supervised classification of the imagery, and then determining a cost for each class [3]. A subset of both maps is shown in Figure 8.

The two maps were compared using a validation set of paths generated by a UPI team member not directly involved in the development of overhead costing. The validation cost ratio was 2.23 with the engineered map, and 1.63 with the learned map.

C. Online Validation

The learning system's applicability to actual autonomous navigation was validated through the DARPA UPI program. These tests involved a six wheeled, skid steer autonomous vehicle. The vehicle is equipped with laser range finders and cameras for on-board perception. The perception system produces costs that are merged with prior costs into a single map. Due to the dynamic nature of this fused cost map, the Field D* [11] algorithm is used for real-time global planning. The full UPI Autonomy system is described in [12].

The entire UPI system was demonstrated during three large field tests during the past year. Each of these tests consisted of the vehicle autonomously traversing a series of courses, with each course defined as a set of widely spaced waypoints. The tests took place at different locations, each with highly varying local terrain characteristics.

Previous to these latest tests, prior maps for the vehicle were generated as described in [3]. Satellite Imagery and aerial fixed wing LiDAR scans were used as input to multiple feature extractors and classifiers. These features were then fed into a hand tuned cost function. During these most recent three tests, example paths were used to train prior cost maps from the available overhead features. The largest map covered an area of over 200 km^2. Overall, learned prior maps were used during over 300 km of sponsor monitored autonomous traverse.

In addition, two direct online comparisons were performed.

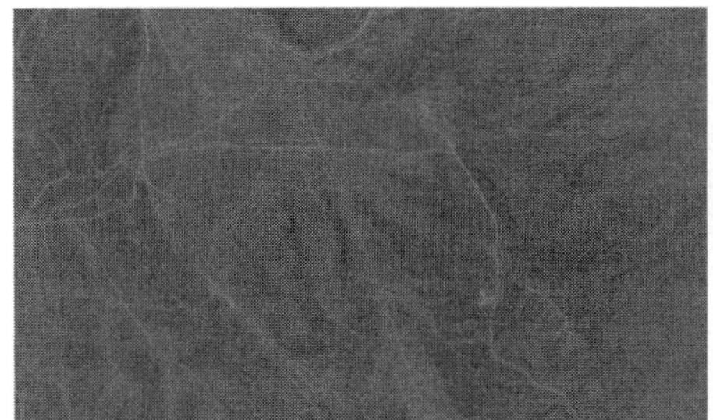  

Fig. 8. A 10 km^2 section of a UPI test site. From left to right: Quickbird imagery, MMP Cost, and Engineered Cost

Experiment	Total Net Distance(km)	Avg. Speed(m/s)	Total Cost Incurred	Max Cost Incurred
Experiment 1 Learned	6.63	2.59	11108	23.6
Experiment 1 Engineered	6.49	2.38	14385	264.5
Experiment 2 Learned	6.01	2.32	17942	100.2
Experiment 2 Engineered	5.81	2.23	21220	517.9
Experiment 2 No Prior	6.19	1.65	26693	224.9

TABLE I

These two tests were performed several months apart, at different test sites. During these experiments, the same course was run multiple times, varying only the prior cost map given to the vehicle. Each run was then scored according to the total cost incurred by the vehicle according to its onboard perception system.

The results of these experiments are shown in Table I. In both experiments, the vehicle traveled farther to complete the same course using MMP trained prior data, and yet incurred less total cost. Over both experiments, with each waypoint to waypoint section considered an independent trial, the improvement in average cost and average speed is statistically significant at the 5% and 10% levels, respectively. This indicates that the terrain the vehicle traversed was on average safer when using the learned prior, according to its own onboard perception system. This normalization by distance traveled is necessary because the learned prior and perception cost functions do not necessarily agree in their unit cost.

The course for Experiment 2 was also run without any prior data; the results are presented for comparison.

VI. Conclusion

This paper addresses the problem of interpreting overhead data for use in long range outdoor navigation. Once provided with examples of how a domain expert would navigate based on the data, the proposed imitation learning approach can learn mappings from raw data to cost that reproduce similar behavior. This approach produces cost functions with less human interaction than hand tuning, and with better performance in both offline and online settings.

Future research will focus more on interactive techniques for imitation learning. By specifically prompting an expert with candidate waypoint locations, it is hoped that more information can be derived from each path, thereby requiring fewer examples. The adaptation of these techniques to an onboard perception system will also be explored.

Acknowledgments

This work was sponsored by the Defense Advanced Research Projects Agency (DARPA) under contract "Unmanned Ground Combat Vehicle - PerceptOR Integration" (contract number MDA972-01-9-0005). The views and conclusions contained in this document are those of the authors and should not be interpreted as representing the official policies, either expressed or implied, of the U.S. Government.

References

[1] N. Ratliff, J. Bagnell, and M. Zinkevich, "Maximum margin planning," in *International Conference on Machine Learning*, July 2006.

[2] N. Vandapel, R. R. Donamukkala, and M. Hebert, "Quality assessment of traversability maps from aerial lidar data for an unmanned ground vehicle," in *Proceedings of the IEEE/JRS International Conference on Intelligent Robots and Systems*, October 2003.

[3] D. Silver, B. Sofman, N. Vandapel, J. A. Bagnell, and A. Stentz, "Experimental analysis of overhead data processing to support long range navigation," in *Proceedings of the IEEE/JRS International Conference on Intelligent Robots and Systems*, October 2006.

[4] D. Pomerleau, "Alvinn: an autonomous land vehicle in a neural network," *Advances in neural information processing systems 1*, pp. 305 – 313, 1989.

[5] C. Sammut, S. Hurst, D. Kedzier, and D. Michie, "Learning to fly," in *Proceedings of the Ninth International Conference on Machine Learning*, 1992, pp. 385–393.

[6] P. Abbeel and A. Ng, "Apprenticeship learning via inverse reinforcement learning," in *International Conference on Machine Learning*, 2004.

[7] J. A. Bagnell, J. Langford, Y. Low, N. Ratliff, and D. Silver, "Learning to search," 2008, manuscript in preparation.

[8] N. Ratliff, D. Bradley, J. Bagnell, and J. Chestnutt, "Boosting structured prediction for imitation learning," in *Advances in Neural Information Processing Systems 19*. Cambridge, MA: MIT Press, 2007.

[9] L. Mason, J. Baxter, P. Bartlett, and M. Frean, "Boosting algorithms as gradient descent," in *Advances in Neural Information Processing Systems 12*. Cambridge, MA: MIT Press, 2000.

[10] J. A. Bagnell, J. Langford, N. Ratliff, and D. Silver, "The exponentiated functional gradient algorithm for structured prediction problems," in *The Learning Workshop*, 2007.

[11] D. Ferguson and A. Stentz, "Using interpolation to improve path planning: The field d* algorithm," *Journal of Field Robotics*, vol. 23, no. 2, pp. 79–101, February 2006.

[12] A. Stentz, "Autonomous navigation for complex terrain," Carnegie Mellon Robotics Institute Technical Report, manuscript in preparation.

Classifying Dynamic Objects: An Unsupervised Learning Approach

Matthias Luber Kai O. Arras Christian Plagemann Wolfram Burgard

Albert-Ludwigs-University Freiburg, Department for Computer Science, 79110 Freiburg, Germany
{luber, arras, plagem, burgard}@informatik.uni-freiburg.de

***Abstract*— For robots operating in real-world environments, the ability to deal with dynamic entities such as humans, animals, vehicles, or other robots is of fundamental importance. The variability of dynamic objects, however, is large in general, which makes it hard to manually design suitable models for their appearance and dynamics. In this paper, we present an unsupervised learning approach to this model-building problem. We describe an exemplar-based model for representing the time-varying appearance of objects in planar laser scans as well as a clustering procedure that builds a set of object classes from given training sequences. Extensive experiments in real environments demonstrate that our system is able to autonomously learn useful models for, e.g., pedestrians, skaters, or cyclists without being provided with external class information.**

Fig. 1. Five examples of relevant object classes considered in this paper. Our proposed system learns probabilistic models of their appearance in planar range scans and the corresponding dynamics. The classes are denominated Pedestrian (PED), Buggy (BUG), Skater (SKA), Cyclist (CYC), and Kangaroo-shoes (KAN).

I. Introduction

The problem of tracking dynamic objects and modeling their time-varying appearance has been studied extensively in robotics, engineering, the computer vision community, and other areas. The problem is hard as the appearance of objects is ambiguous, partly occluded, may vary quickly over time, and is perceived via a high-dimensional measurement space. On the other hand, the problem is highly relevant in practice, especially in future applications for mobile robots and intelligent cars. Consider, for example, a service robot deployed in a populated environment, e.g., a pedestrian precinct. A number of tasks such as collision-free navigation or interaction require the ability to recognize, distinguish, and track moving objects including reliable estimates of object classes like 'adult', 'infant', 'car', 'dog', etc.

In this paper, we consider the problem of detecting, tracking, and classifying moving objects in sequences of planar range scans acquired by a laser sensor. We present an exemplar-based model for representing the time-varying appearance of moving objects as well as a clustering procedure that builds a set of object classes from given training sequences in conjunction with a Bayes filtering scheme for classification. The proposed system, which has been implemented and tested on a real robot, does not require labeled object trajectories, but rather uses an unsupervised clustering scheme to automatically build appropriate class assignments. By pre-processing the sensor stream using state-of-the-art feature detection and tracking algorithms, we achieve a system that is able to learn and re-use object models on-the-fly without human intervention. The resulting set of object models can then be used to (1) recognize previously seen object classes and (2) improve data segmentation and association in ambiguous multi-target tracking situations. We furthermore believe that the object models may be used in various applications to associate semantics with recognized objects depending on their classes.

II. Related Work

Exemplar-based models are frequently applied in computer vision systems for dealing with the high dimensionality of visual input. Toyama and Blake [1], for instance, used probabilistic exemplar models for representing and tracking human motion. Their approach is similar to ours in that they also learn probabilistic transition models. As the major differences, the range-bearing observations used in this work are substantially more sparse than visual input and we also address the problem of learning different object classes in an unsupervised way. Plagemann *et al.* [2] used exemplars to represent the visual appearance of 3D objects in the context of an object localization framework. Krüger *et al.* [3] learned exemplar models to realize a face recognition system for video streams. Exemplar-based approaches have also been used in other areas such as action recognition [4] or word sense disambiguation [5].

There exists a large body of work on laser-based object and people tracking in the robotics literature [6, 7, 8, 9, 10]. People tracking typically requires carefully engineered or learned features for track identification and data association and often a-priori information about motion models. This has been shown to be the case also for geometrically simpler and rigid object such as vehicles in traffic scenarios [11]. Cui *et al.* [12] describe a system for tracking single persons within a larger set of persons, given the relevant motion models are known.

The work most closely related to ours has recently been presented by Schulz [13], who combined vision- and laser-based exemplar models to realize a people tracking system. In contrast to his work, our main contribution is the unsupervised learning of *multiple* object classes that can be used for tracking as well as for classifying dynamic objects.

Periodicity and self-similarity have been studied by Cutler and Davis [14], who developed a classification system based on the autocorrelation of appearances, which is able to distinguish, for example, walking humans from dogs.

A central component of our approach detailed in the following section is an unsupervised clustering algorithm to produce a suitable set of exemplars. Most approaches to cluster analysis [15] assume that all data is available from the beginning and that the number of clusters is given. Recent work in this area also deals with sequential data and incremental model updates [16, 17]. Ghahramani [18] gives an easily accessible overview of the state-of-the-art in unsupervised learning.

As an alternative to the exemplar-based approach, researchers have applied generic dimensionality reduction techniques in order to deal with high-dimensional and/or dynamic appearance distributions. PCA and ICA have, for example, been used to recognize people from iris images [19] or their faces [20]. Recent advances in this area include Isomap [21] and latent variable models, such as GP-LVM [22]. For more details on dimensionality reduction, we refer to standard text books like [23].

III. Modeling Object Appearance and Dynamics Using Exemplars

Exemplar models are non-parametric representations for both, appearance and appearance dynamics. They are a choice consistent with the motivation for an unsupervised learning approach avoiding manual feature selection, parameterized physical models (e.g., human gait models) and hand-tuned classifier creation.

This section describes how the exemplar-based models of dynamic objects are learned. Based on a segmentation and tracking system presented in Section VI, we assume to have a discrete track for each dynamic object in the current scene. Over time, these tracks describe trajectories that we analyze regarding the object's appearance and dynamics.

A. Problem Description

The problem we address in this work can be formally stated as follows. Let $T = \langle Z_1, \ldots, Z_m \rangle$ be a *track*, i.e., a time-indexed observation sequence of appearances Z_t, $t = 1, \ldots, m$, of an object belonging to an *object class* C. Then we face the following two problems:

1) **Unsupervised learning:** Given a set of observed tracks $\mathcal{T} = \{T_1, T_2, \ldots\}$, learn object classes $\{C_1, \ldots, C_n\}$ in an unsupervised manner. This amounts to setting an appropriate number n of classes and to learn for each class C_j a probabilistic model $p(T \mid C_j)$ that characterizes the time-varying appearance of tracks T associated with that class.
2) **Classification:** Given a newly observed track T and a set of known object classes $\mathcal{C} = \{C_1, \ldots, C_n\}$, estimate the class probabilities $p(C_j \mid T)$ for all classes.

Note that 'unsupervised' in this context shall not mean that *all* model parameters are learned from scratch, but rather that just the important class information (e.g. 'pedestrian') is not supplied to the system. The underlying segmentation, tracking, and feature extraction subsystems are designed to capture a wide variety of possible object appearances and the unsupervised learning task is to build a *compact* representation of object appearance that generalizes well across instances.

B. The Exemplar Model

Exemplar models [1] aim at approximating the typically high-dimensional and dynamic appearance distribution of objects using a sparse set $\mathcal{E} = \{E_1, \ldots, E_r\}$ of significant observations E_i, termed *exemplars*. Similarities between concrete observations and exemplars as well as between two exemplars are specified by a distance function $\rho(E_i, E_j)$ in exemplar space. Furthermore, each exemplar is given a prior probability $\pi_i = p(E_i)$, which reflects the prior probability of a new observation being associated with this exemplar. Changes in appearance over time are dealt with by introducing transition probabilities $p(E_i \mid E_j)$ between exemplars w.r.t. a predefined iteration frequency. Formally, this renders the exemplar model a first-order Markov chain, specified by the four elements $\mathcal{M} = (\mathcal{E}, B, \pi, \rho)$, which are the exemplar set $\mathcal{E}$, the transition probability matrix B with elements $b_{i,j} = p(E_i \mid E_j)$, the priors π, and the distance function ρ. All these components can be learned from data, which is a central topic of this paper.

C. Exemplars for Range-Bearing Observations

In a laser-based object tracking scenario, the raw laser measurements associated with each track constitute the objects' appearance $Z = \{(\alpha_i, r_i)\}_{i=1}^{l}$, where α_i is the bearing, r_i is the range measurement, and l is the number of laser end points in the respective laser segment.

To cluster the laser segments into exemplars, the individual laser segments need to be normalized with respect

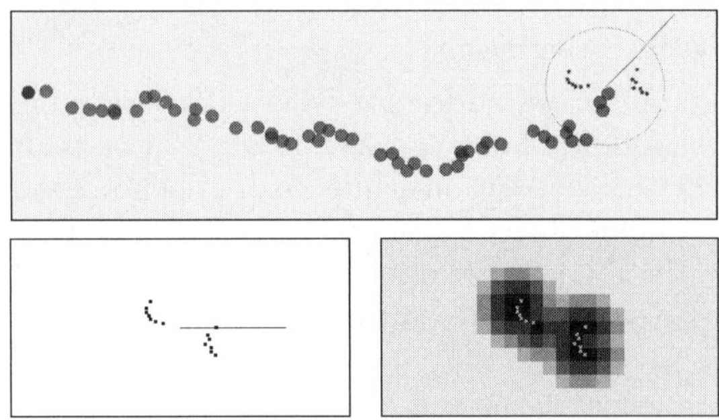

Fig. 2. Pre-processing steps illustrated with a pedestrian observed via a laser range finder. First, the segmentation and tracking system yields estimates of the objects' location, orientation and velocity (top). Second, the raw range readings are normalized such that the estimated direction of motion is zeroed (bottom left). Third, a grid-based representation is generated from the set of normalized laser end points (bottom right).

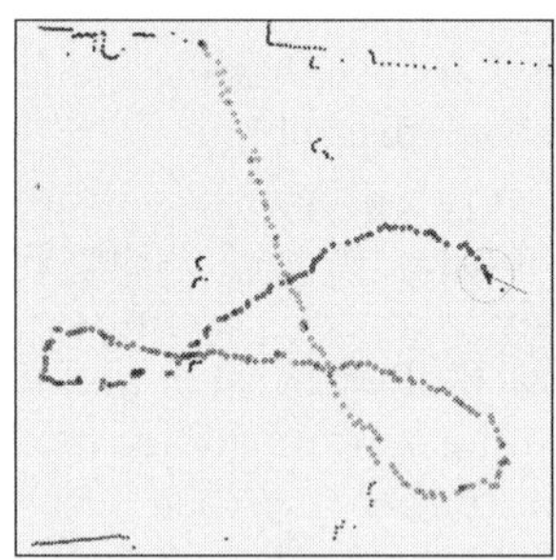

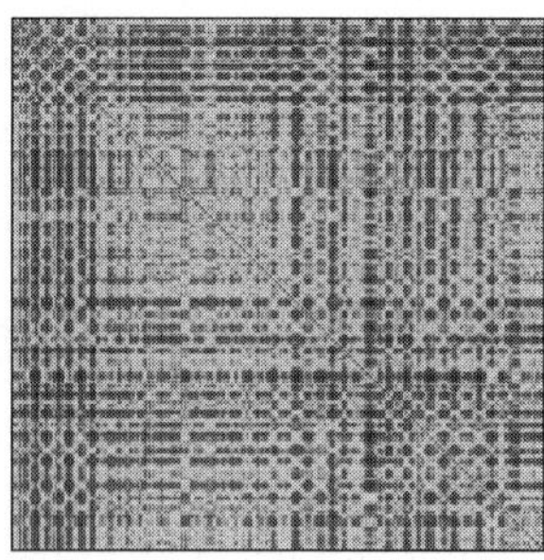

Fig. 3. Trajectory (left) and self-similarity matrix (right) of a pedestrian walking in a large hallway. The track consist of 387 observations.

to rotation and translation. This is achieved using the object's state information estimated by the underlying tracker. Here, the state of a track $\mathbf{x} = (x, y, v_x, v_y)^T$ is composed of the position (x, y) and the velocities (v_x, v_y). The velocity vector can then be used to obtain the object's heading. Translational invariance is achieved by shifting the segment's center of gravity to $(0, 0)$, rotational invariance is gained from zeroing the orientation in the same way. After normalization, all segments appear in a fixed position and orientation.

Rather than using the raw laser end points of the normalized segments as exemplars (see Schulz [13]), we integrate the points into a regular metric grid. This is done by adding l Gaussian density functions centered at the beam end points to the grid. The main advantage of this approach is that the distance function for exemplars can be defined independently of the number of laser end points in the segment and that likelihood estimation for new observations can be performed easily and efficiently. We will henceforth denote the grid representation of an appearance Z_i as G_i. Figure 2 shows an example of a track, a laser segment corresponding to a walking pedestrian, the normalized segment, and the corresponding grid.

D. Validation of the Exemplar Approach

Obviously, the exemplar representation has a strong impact on both the creation of the exemplar set from a sequence of appearances and the unsupervised creation of new object classes. This motivates a careful analysis of the choices made. To identify the general usefulness of the exemplar model described above, we analyzed the self-similarity of exemplars for tracks of objects from relevant object classes. For this purpose, we define the similarity S_{t_1,t_2} of two observations obtained at times t_1 and t_2 as the absolute correlation

$$S_{t_1,t_2} := \sum_{(x,y)\in\mathcal{B}} \mid G_{t_1}(x,y) - G_{t_2}(x,y) \mid , \qquad (1)$$

where $\mathcal{B}$ is the bounding box of the grid-based representations of the observations Z_{t_1} and Z_{t_2}.

Figure 3 visualizes the self-similarity of a pedestrian over a sequence of 387 observations. Both axes of the self-similarity matrix (Fig. 3, right) show time with t_1 horizontally and t_2 along the vertical axis. The colors that encode self-similarity range from green to black where green stands for maximal, black for minimal correlation. The diagonal is maximal by definition as the distance of an observation to itself is zero.

We clearly recognize a periodicity across the entire matrix that is caused by the strong self-similarity of the pedestrian's appearance along the trajectory. This is not self-evident as the appearance of the walking person in laser data changes with the heading of the person relative to the sensor. Poor normalization (e.g., by inaccurate heading estimates of the underlying tracker) or a poor exemplar representation (e.g., too sensitive to measurement noise) could have failed to produce a good self-similarity. We conclude from this analysis that the normalization and the grid-based exemplar representation have good invariance properties, such that a compact representation of trajectories can be achieved.

E. Learning the Exemplar Model

This section describes how the exemplar model is learned from observation sequences. This involves the exemplar set $\mathcal{E}$, the prior probabilities π_i and the transition probabilities $p(E_i \mid E_j)$.

1) Exemplar Set: Exemplars are representations that generalize typical object appearances. To this aim, similar appearances are associated and merged into clusters. We used k-means clustering [15] to partition the full data set into r clusters $\mathcal{P}_1, \mathcal{P}_2, \ldots, \mathcal{P}_r$.

Strong outliers in the training set—which cannot be merged with other observations—are retained by the clustering process as additional, non-representative exemplars. Such observations may occur for several reasons, e.g., when a tracked object performs atypical movements, when the underlying segmentation method fails to produce a proper foreground segment, or due to sensor noise. To achieve robustness with respect to such outliers, we accept an exemplar only if it was created from a minimum number of observations. This assures that the resulting exemplars

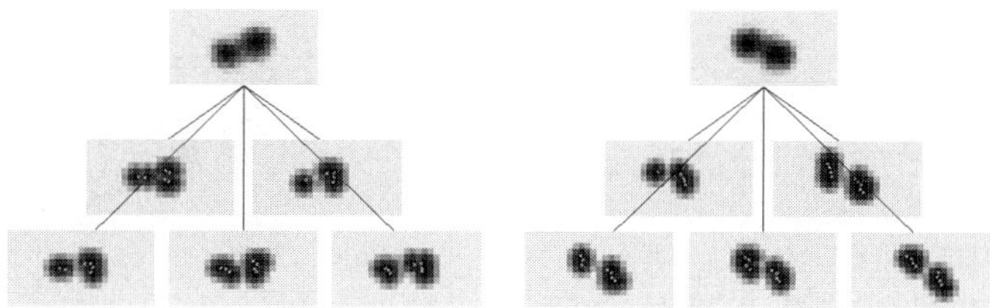

Fig. 4. Example clusters of a pedestrian. The diagram shows the centroids of two clusters (exemplars) each created from a set of 5 observations.

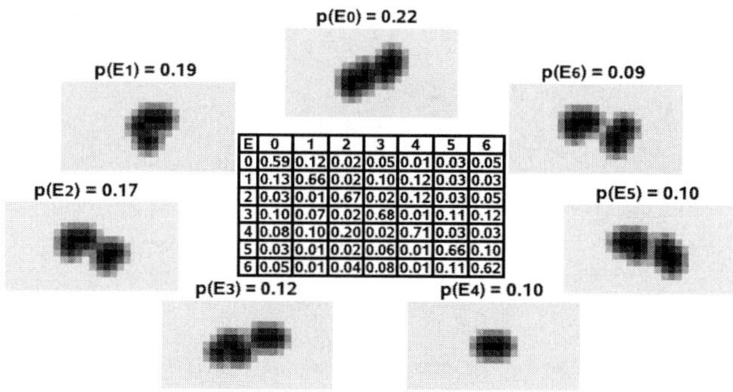

E	0	1	2	3	4	5	6
0	0.59	0.12	0.02	0.05	0.01	0.03	0.05
1	0.13	0.66	0.02	0.10	0.12	0.03	0.03
2	0.03	0.01	0.67	0.02	0.12	0.03	0.05
3	0.10	0.07	0.02	0.68	0.01	0.11	0.12
4	0.08	0.10	0.20	0.02	0.71	0.03	0.03
5	0.03	0.01	0.02	0.06	0.01	0.66	0.10
6	0.05	0.01	0.04	0.08	0.01	0.11	0.62

Fig. 5. Laser-based exemplar model of a pedestrian. The transition matrix is shown in the center with the exemplars sorted counterclockwise according to their prior probability.

characterize only states of the appearance dynamics that occur often and are representative.

2) Transition Probabilities: Once the clustered exemplar set has been generated from the training set, the transition probabilities between exemplars can be learned. As defined in Sec. III-B, we model the dynamics of an object's appearance using a Hidden Markov model (HMM). The transition probabilities are obtained by pairwise counting. A transition between two exemplars E_i and E_j is counted each time when an observation that has minimal distance to E_i is followed by an observation with minimal distance to E_j. As there is a non-zero probability that some transitions are never observed although they exist, the transition probabilities are initialized with a small value to moderately smooth the resulting model.

3) Distance Function: We assess the similarity of two observations Z_i and Z_j based on a distance function applied to the corresponding grid-based representations G_i and G_j. Interpreting the grids as histograms we employ the Euclidean distance for this purpose:

$$\rho_e(G_i, G_j) = \sqrt{\sum_{(x,y)} (G_i(x,y) - G_j(x,y))^2} \qquad (2)$$

IV. Classification

Having learned the exemplar set and transition probabilities as described in the previous section, they can be used to classify tracks of different objects in a Baysian filtering framework. More formally, given the grid representations $\langle G_1, \ldots, G_m \rangle$ of the observations of a track T and a set of learned classes $\mathcal{C} = \{C_1, \ldots C_n\}$, we want to estimate the class probabilities $p_t(C_k \mid T)_{k=1}^n$ for every time step t. The estimates for the last time step m then reflect the consistency of the whole track with the different exemplar models. These quantities can thus be used to make classification decisions.

A. Estimating Class Probabilities over Time

Each exemplar model $\mathcal{M}^i$ represents the distribution of track appearances for its corresponding object class C_i. Thus, a combination of *all* known exemplar models $\mathcal{M}^{comb} = \{\mathcal{M}^1, \ldots \mathcal{M}^n\}$ covers the whole space of possible appearances – or, more precisely, of all appearances that the robot has seen in the training phase. We construct the exemplar set $\mathcal{E}^{comb}$ of $\mathcal{M}^{comb}$ by simply building the union set of the individual exemplar sets $\mathcal{E}^k$ of all models $\mathcal{M}^k$. The transition probability matrix B^{comb} as well as the exemplar priors π^{comb} can be obtained from the B^k matrices and the π^k in a straight forward way since the corresponding exemplar sets do not intersect.

Given this combined exemplar model, a belief function Bel_t for the class probabilities $p_t(C_k \mid T)_{k=1}^n$ can be updated recursively over time using the well-known Bayes filtering scheme. For better readability, we introduce the notation E_i^k to refer to the ith exemplar of model $\mathcal{M}^k$. According to the Bayes filter, the belief about object classes is initialized as,

$$Bel_0(E_i^k) = p(\mathcal{M}^k) \cdot \pi_i^k \ , \qquad (3)$$

where π_i^k denotes the prior probability of E_i^k and $p(\mathcal{M}^k)$ stands for the model prior, which can be estimated from the training set. Starting with G_1, we now perform the following recursive update of the belief function for every G_t:

$$\begin{aligned} Bel_t(E_i^k) &= \eta_t \cdot p(G_t \mid E_i^k) \\ &\quad \cdot \sum_k \sum_j p(E_i^k \mid E_j^l) \cdot \ Bel_{t-1}(E_j^l) \end{aligned} \qquad (4)$$

Here, the normalizing factor η_t is calculated such that

$$\sum_{i,k} \ Bel_t(E_i^k) = 1. \qquad (5)$$

The estimates $Bel_t(E_i^k)$ of exemplar probabilities at time t can be summed up to yield the individual class probabilities

$$p_t(\mathcal{M}^k \mid T) = \sum_i Bel_t(E_i^k) \ . \qquad (6)$$

At time $t = m$, that is, when the whole observation sequence has been processed, $p_m(\mathcal{M}^k \mid T)$ constitute the resulting estimates of the class probabilities of our proposed model. In particular, we define

$$\mathcal{M}^{best}(T) := \text{argmax}_k \ p_m(\mathcal{M}^k \mid T) \qquad (7)$$

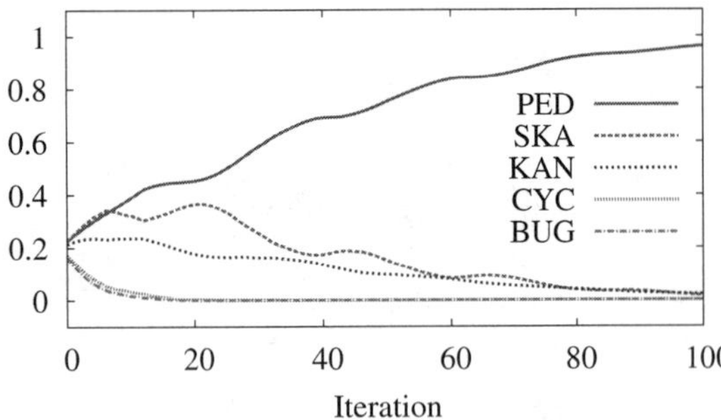

Fig. 6. The figure shows a typical probability evolution of a successfully classified pedestrian. The x-axis refers to the time t. The graphs show the probabilities of different classes. The red one belongs to the pedestrian class.

as the most likely class assignment for track T. To visualize the filtering process described above, we give an example run for a pedestrian track T in Fig. 6 and plot the class probabilities for five alternative object classes over time.

V. Unsupervised Learning Of Object Classes

As the variety of dynamic objects in the world is hard to predict a-priori, we seek to learn such objects without external class information. In this section we explain how the creation of new classes is handled in the unsupervised case.

Objects of a previously unknown type will always be assigned to some class by the Bayes filter. The class with the highest resulting probability estimate provides the current best, yet suboptimal description of the object at the time. A better fit would always be achieved by creating a new, specifically trained model for this particular object instance. Thus, we are faced with the classic model selection problem, i.e., choosing between a more compact vs. a more precise model for explaining the observed data. As a selection criterion, we employ the *Bayes factor* [24] which considers the amount of evidence in favor of a model relative to an alternative one.

More formally, given a set of known classes $\mathcal{C} = \{C_1, \ldots, C_n\}$ and their respective models $\{\mathcal{M}^1, \ldots, \mathcal{M}^n\}$, let T be the track of an object to be classified. We determine the best matching model $\mathcal{M}^{best}(T)$ and learn a new, fitted model $\mathcal{M}^{new}(T)$. To decide whether T should be added to $\mathcal{M}^{best}(T)$ or rather to $\mathcal{M}^{new}(T)$ by adding a new object class C^{new} to the existing set of classes, we calculate the model probabilities $p(\mathcal{M}^{best}(T) \mid T)$ and $p(\mathcal{M}^{new}(T) \mid T)$ using the Bayes filter. The ratio of these probabilities yields the factor

$$K = \frac{p(\mathcal{M}^{new}(T) \mid T)}{p(\mathcal{M}^{best}(T) \mid T)}, \tag{8}$$

that quantifies how much better the fitted model describes this object instance relative to the existing, best matching model. While large values for a threshold on K favor more compact models (less classes and lower data-fit), lower values lead to more precise models (more classes, in the extreme case overfitting the training set). As alternative model selection criteria, one could use, e.g., the Bayesian Information Criterion (BIC), or Akaike Information Criterion (AIC), or various others. The comparison of these criteria to K [see Eq. (8)], which worked well in our experiments, is not part of this work.

	$K \geq 1$	$K \geq 2$	$K \geq 4$	$K \geq 8$	$K \geq 20$
PED/PED	41%	2%	0%	0%	0%
SKA/SKA	58%	7%	0%	0%	0%
CYC/CYC	79%	32%	14%	10%	8%
BUG/BUG	78%	47%	21%	9%	1%
KAN/KAN	60%	40%	21%	11%	3%
PED/KAN	46%	3%	0%	0%	0%
PED/SKA	100%	83%	40%	10%	0%
CYC/BUG	100%	100%	100%	99%	50%
BUG/KAN	100%	100%	100%	100%	82%
CYC/KAN	100%	100%	100%	98%	92%

TABLE I

Percentages of incorrectly (top five rows) and correctly (bottom five rows) separated track pairs. A Bayes-Factor is sought that trades off separation of tracks from different classes and association of tracks from the same class.

We now describe how the threshold for K can be learned, such that the system achieves similar classification results as a human. Interestingly, our learned threshold for K coincides with the interpretation of "substantial evidence against the alternative model" of Kass and Raftery [24]. Note that fitting the threshold K to a labeled data-set does not render our approach a supervised one, since no specific class labels are supplied to the system. This step can rather be compared to learning regularization parameters in alternative models to balance data-fit against model complexity.

Concretely, for determining a suitable threshold on K, we collected a training set of *pedestrian*, *skater*, *cyclist*, *buggy*, and *kangaroo* instances. We first compared the best models and the fitted models of objects of the same class and calculated the factors K according to Eq. (8). Then we made the same comparison with objects of different classes with randomly selected tracks. Table I gives the relative amount of compared pairs for which different values of K—ranging from 1 to 20—were exceeded. It can be seen that, e.g., for $K \geq 4$, all pedestrians are merged to the same class (PED/PED), but also that there is a poor separation (40%) between pedestrians and skaters (PED/SKA). Given this set of tested thresholds K, the best trade-off between precision and recall is achieved between $K \geq 2$ and $K \geq 4$. We therefore chose $K \geq 3$.

VI. Segmentation and Tracking

The segmentation and tracking system takes the raw laser scans as input and produces the tracks with asso-

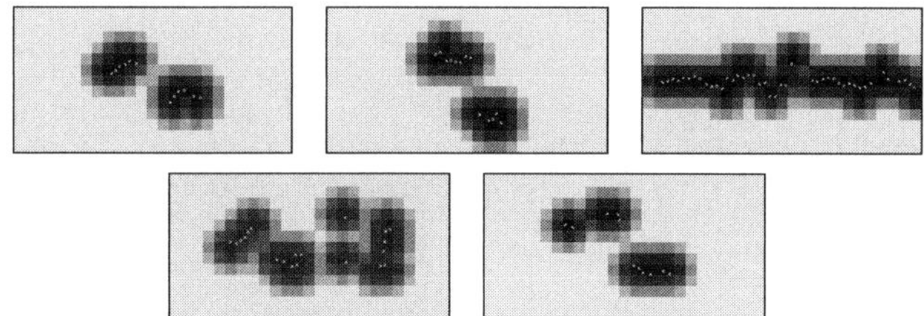

Fig. 7. Top left to bottom right: Typical exemplars of the classes *pedestrian*, *skater*, *cyclist*, *buggy* and *kangaroo*. Direction of motion is from left to right. Pedestrians and skaters have very similar appearance but differ in their dynamics. Pedestrians and subjects on kangaroo-shoes have a similar dynamics but different appearances (mainly due to metal springs attached at the backside of the shoes). We use both information to classify these objects.

ciated laser segments for the exemplar generation step. To this end, we employ a Kalman filter-based multi-target tracker with a constant velocity motion model. The observation step in the filter amounts to the problem of partitioning the laser range image into segments that consist in measurements on the same dynamic objects and to estimate their center. This is done by subtracting successive laser scans to extract beams that belong to dynamic objects. If the beam-wise difference is above the sensor noise level, the measurement is marked and grouped into a segment with other moving points in a pre-defined radius.

We compared four different techniques to calculate the segment center: mean, median, average of extrema, and the center of a circle fitted through the segments points (for the latter the closed-form solutions from [10] were taken). The last approach leads to very good results when tracking pedestrians, skaters, and people on kangaroo shoes but fails to produce good estimates with person pushing a buggy and cyclists. The mean turned out to be the smoothest estimator of the segment center.

Data association is realized with a modified Nearest Neighbor filter. It was adapted so as to associate multiple observations to a single track. This is necessary to correctly associate the two legs of pedestrians, skaters, and kangaroo shoes that appear as nearby blobs in the laser range image. Although more advanced data association strategies, motion models or segmentation techniques have been described in the related literature, the system was useful enough for the purposes of this paper.

VII. Experiments

We experimentally evaluated our approach with five object classes: *pedestrian* (PED), *skater* (SKA), *cyclist* (CYC), *person pushing a buggy* (BUG), and *people on kangaroo-shoes* (KAN), see Fig. I. We recorded a total of 436 tracks of subjects belonging to one of the five classes. The sensor was a SICK LMS291 laser range finder mounted at a height of 15 cm above ground. The tracks include walking and running pedestrians, skaters with small, wide, or no pace (just rolling), cyclists at slow and medium speeds, people pushing a buggy, and subjects

Classes	PED	SKA	CYC	BUG	KAN
Pedestrian	92.8%	7.2%	0%	0%	0%
Skater	5.4%	94.6%	0%	0%	0%
Cyclist	0%	0%	90.8%	1.5%	7.7%
Buggy	0%	0%	0%	97.9%	2.1%
Kangaroo	12.5%	0%	0%	0%	87.5%

TABLE II

Classification rates in the supervised case. Rows denote ground truth and columns the classification results.

on kangaroo shoes that walk slowly and fast. Note that pedestrians, skaters, and partly also kangaroo shoes have very similar appearance in the laser data but differ in their dynamics. See Fig. 7 for typical exemplars of each class.

A. *Supervised Learning Experiments*

In the first group of experiments, we test the classification performance in the supervised case. The training set was composed of a single, typical track for each class including their labels *PED*, *SKA*, *CYC*, *BUG*, or *KAN*. The exemplar models were then learned from these tracks. Based on the resulting prototype models, we classified the remaining 431 tracks. The results are shown in Tab. II.

Pedestrians are classified correctly in 92.8% of the cases whereas 7.2% are found to be skaters. A manual analysis of these 7.2% revealed that the misclassification occurred typically with running pedestrians whose appearance *and* dynamics resemble those of skaters. We obtain a rate of 94.6% for skaters with 5.4% falsely classified as pedestrians. The latter group was found to skate slower than usual with a small pace, thereby resembling pedestrians. Cyclists are classified correctly in 90.8% of the cases. None of them was wrongly recognized as pedestrians or skaters. It appeared that the bicycle wheels produced measurements that resemble subjects on kangaroo shoes taking big steps. This lead to a rate of 7.7% of cyclists falsely classified as kangaroo shoes. There was one cyclist (1.5%), that was classified to belong to the buggy class. A percentage of 97.9% of the buggy tracks were classified correctly. Only one (2.1%) was found to be a subject on kangaroo-shoes. In this particular case, the track contained measurements with the buggy's front partially outside the sensor's field of view with two legs of the person still visible. Subjects on kangaroo shoes were correctly recognized at a rate of 87.5% with 12.5% of the tracks wrongly classified as pedestrians. A manual analysis revealed that the latter group consisted mainly of kangaroo shoe novices taking small steps thereby appearing like pedestrians.

Given the limited information in the laser data and the high level of self-occlusion of the objects, the results demonstrate that our exemplar models are expressive enough to yield high classification rates. Misclassifications typically occur at boundaries where objects of different classes appear or move similarly.

Classes	PED	SKA	CYC	BUG	KAN	
class 1 (209)	187	5	0	0	17	"PED"
class 2 (114)	7	107	0	0	0	"SKA"
class 3 (41)	0	0	41	0	0	"CYC"
class 4 (23)	0	0	23	0	0	"CYC"
class 5 (26)	0	0	1	25	0	"BUG"
class 6 (23)	0	0	0	23	0	"BUG"
total (436)	194	112	65	48	17	

TABLE III

UNSUPERVISED LEARNING RESULTS. ROWS CONTAIN THE LEARNED CLASSES, COLUMNS SHOW THE NUMBER OF CLASSIFIED OBJECTS. THE LAST COLUMN SHOWS THE MANUALLY ADDED LABELS, THE LAST ROW HOLDS THE TOTAL NUMBER OF TRACKS OF EACH CLASS.

B. Unsupervised Learning Experiments

In the second experiment the classes were learned in an unsupervised manner. The entire set of 436 tracks from all five classes was presented to the system in random order.

Each track was either assigned to an existing class or was taken as basis for a new class according to the learning procedure described above. As can be seen in Tab. III, six classes have been generated for our data set: one class for *pedestrians* (PED), one for *skaters* (SKA), two for *cyclists* (CYC), two for *buggies* (BUG), and none for *kangaroo shoes* (KAN).

Class number one (labeled PED) contains 187 pedestrian tracks (out of 194), 5 skater tracks and 17 kangaroo tracks resulting in a true positive rate of 89.5%. Class number two (labeled SKA) holds 107 skater tracks (out of 112) and 7 pedestrian tracks yielding a true positive rate of 93.9%. Given the resemblance of pedestrians and skaters, the total number of tracks and the extent of intra-class variety, this is an encouraging result that shows the ability of the system to discriminate objects that vary predominantly in their dynamics.

Classes number three and four (labeled CYC) contain 41 and 23 cyclist tracks respectively. No misclassifications occurred. The last two classes, number five and six (labeled BUG), hold 25 and 23 buggy tracks with a bicycle track as the single false negative in class number five. The representation of cyclists and buggies by two classes is due to the larger variability in their appearance and more complex dynamics. The discrimination from the other three classes is exact—no pedestrians, skaters, or subjects on kangaroo shoes were classified to be a cyclist or a buggy.

The system failed to produce a class for subjects on kangaroo shoes as all instances of the latter class were summarized in the pedestrian class. The best known model for all 17 kangaroo tracks was always class number one which has previously been created from a pedestrian track. This results in a false negative rate of 8.1% from the view point of the pedestrian class. The result confirms the outcome in the supervised experiment where the highest

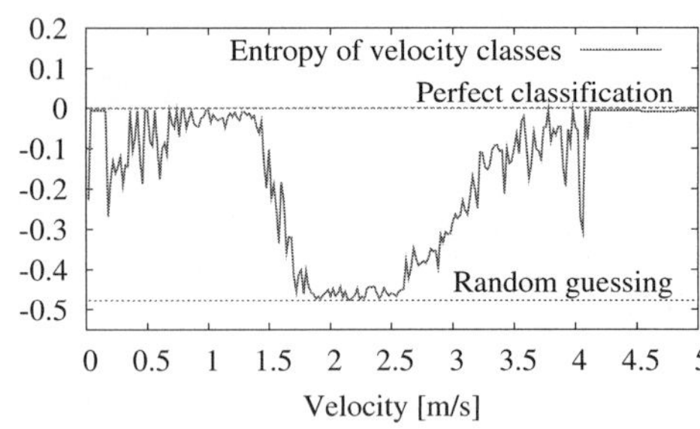

Fig. 8. Analysis of the track velocities as alternative features for classification. While high and low velocities are strong indicators for certain classes, there is a high level of confusion in the medium range.

misclassification rate (12.5%) was found to be between pedestrians and subjects on kangaroo shoes (see Tab. II).

C. Analysis of Track Velocities

The data set of test trajectories that was used in our experiments contains a high level of intra-class variation, like for example skaters moving significantly slower than average pedestrians or even pedestrians running at double their typical velocity. To visualize this diversity and to show that simple velocity-based classification would fail, we calculated a velocity histogram for the classes PED, SKA, and CYC. For every velocity bin, we calculated the entropy $H(v_i) = \sum_{j=1}^{3}(p(c_j|v_i) \cdot \log p(c_j|v_i))$ and visualized the result in Fig. 8. Note that the uniform distribution over three classes, which corresponds to random guessing, has an entropy of $3 \cdot (1/3 \cdot \log(1/3)) \approx -0.477$, which is visualized by a straight, dashed line. As can be seen from the diagram, high and low velocities are strong indicators for certain classes while there is a high level of confusion in the medium range.

D. Classification with a Mobile Robot

An additional supervised and an unsupervised experiment was carried out with a moving platform. A total of 12 tracks has been collected: 3 pedestrian tracks, 5 skater tracks and 4 cyclist tracks (kangaroo shoes and buggies were unavailable for this experiment). The robot moved with a maximal velocity of 0.75 m/s and an average velocity of 0.35 m/s. A typical robot trajectory is depicted in Fig. 9.

For the supervised experiment, the trained models from the supervised experiment in Sec. VII-A have been reused to classify the tracks collected from the moving platform.

TABLE IV

AVERAGED CLASSIFICATION PROBABILITIES FOR THE SUPERVISED EXPERIMENT WITH THE MOVING PLATFORM. ALL OBJECTS HAVE BEEN CLASSIFIED CORRECTLY.

Classes	PED	SKA	CYC	BUG	KAN
Pedestrian	0.99	0	0	0	0.01
Skater	0.12	0,87	0	0	0.01
Cyclist	0.01	0	0.90	0.07	0.02

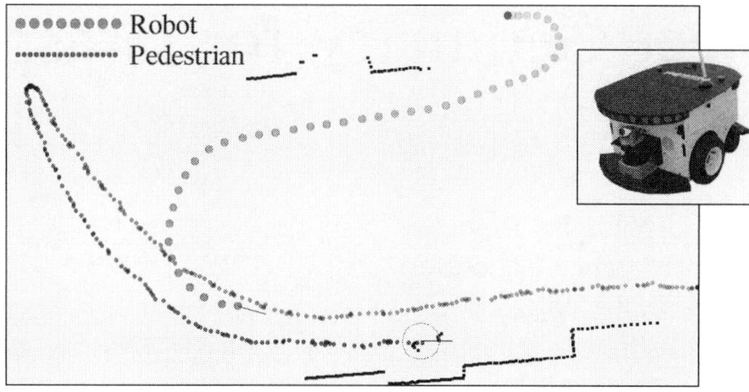

Fig. 9. Trajectory of the robot (an ActivMedia PowerBot) and a pedestrian over a sequence of 450 observations.

All objects were classified correctly by the moving robot. Table IV contains the classification probabilities of Eq. (6) (t being the track length), averaged over all tracks in the respective class. The last two columns contain the probabilities for the classes BUG and KAN, all being close to zero. The lowest classification probability in this experiment was a skater track which still had the probability 0.76 of being a skater.

In the unsupervised experiment, the tracks have been presented to the system in random order without prior class information. The result was exact: three classes have been created that each contain the tracks of the same object category.

VIII. Conclusions and Outlook

We have presented an unsupervised learning approach to the problem of tracking and classifying dynamic objects. In our framework, the appearance of objects in planar range scans is represented using a probabilistic exemplar model in conjunction with a hidden Markov model for dealing with the dynamically changing appearance over time. Extensive real-world experiments including more than 400 recorded trajectories show that (a) the model is expressive enough to yield high classification rates in the supervised learning case and that (b) the unsupervised learning algorithm produces meaningful object classes consistent with the true underlying class assignments. Additionally, our system does not require any manual class labeling and runs in real-time.

In future research, we first plan to strengthen the interconnection between the tracking process and the classification module, i.e., to improve segmentation and data association given the estimated posterior over future object appearances.

Acknowledgments

This work has partly been supported by the EC under contract numbers FP6-004250-CoSy, FP6-IST-045388 and by the German Federal Ministry of Education and Research (BMBF) within the research project DESIRE under grant no. 01IME01F.

References

[1] K. Toyama and A. Blake, "Probabilistic tracking with exemplars in a metric space," *Int. Journal of Computer Vision*, vol. 48, no. 1, pp. 9–19, June 2002.

[2] C. Plagemann, T. Müller, and W. Burgard, "Vision-based 3d object localization using probabilistic models of appearance." in *Pattern Recognition, 27th DAGM Symposium, Vienna, Austria*, vol. 3663. Springer, 2005, pp. 184–191.

[3] V. Kruger, S. Zhou, and R. Chellappa, "Integrating video information over time. example: Face recognition from video," in *Cognitive Vision Systems*. Springer, 2006, pp. 127–144.

[4] E. Drumwright, O. C. Jenkins, and M. J. Mataric, "Exemplar-based primitives for humanoid movement classification and control," in *Proc. of the Int. Conf. on Robotics & Automation (ICRA)*, 2004.

[5] H. T. Ng and H. B. Lee, "Integrating multiple knowledge sources to disambiguate word sense: An exemplar-based approach," in *Proc. of the 34th annual meeting on Association for Computational Linguistics*, 1996, pp. 40–47.

[6] D. Schulz, W. Burgard, D. Fox, and A. Cremers, "Tracking multiple moving targets with a mobile robot using particle filters and statistical data association," in *Proc. of the Int. Conf. on Robotics & Automation (ICRA)*, 2001.

[7] A. Fod, A. Howard, and M. Mataríc, "Fast and robust tracking of multiple moving objects with a laser range finder," in *Proc. of the Int. Conf. on Robotics & Automation (ICRA)*, 2001.

[8] M. Montemerlo and S. Thrun, "Conditional particle filters for simultaneous mobile robot localization and people tracking," in *Proc. of the Int. Conf. on Robotics & Automation (ICRA)*, 2002.

[9] A. Fod, A. Howard, and M. Mataríc, "Laser-based people tracking," in *Proc. of the Int. Conf. on Robotics & Automation (ICRA)*, 2002.

[10] K. O. Arras, O. Martínez Mozos, and W. Burgard, "Using boosted features for the detection of people in 2d range data," in *Proc. of the Int. Conf. on Robotics & Automation*, 2007.

[11] R. MacLachlan and C. Mertz, "Tracking of moving objects from a moving vehicle using a scanning laser rangefinder," in *Intelligent Transportation Systems 2006*. IEEE, September 2006, pp. 301–306.

[12] J. Cui, H. Zha, H. Zhao, and R. Shibasaki, "Robust tracking of multiple people in crowds using laser range scanners," in *18th Int. Conf. on Pattern Recognition (ICPR)*, Washington, DC, USA, 2006.

[13] D. Schulz, "A probabilistic exemplar approach to combine laser and vision for person tracking." in *Robotics: Science and Systems*. The MIT Press, 2006.

[14] R. Cutler and L. Davis, "Robust real-time periodic motion detection, analysis, and applications," *PAMI*, vol. 22, no. 8, pp. 781–796, August 2000.

[15] J. A. Hartigan, *Clustering Algorithms*. John Wiley&Sons, 1975.

[16] D. K. Tasoulis, N. M. Adams, and D. J. Hand, "Unsupervised clustering in streaming data," in *6th Int. Conf. on Data Mining - Workshops(ICDMW)*, Washington, DC, 2006, pp. 638–642.

[17] M. Chis and C. Grosan, "Evolutionary hierarchical time series clustering," in *6th Int. Conf. on Intelligent Systems Design and Applications (ISDA)*, Washington, DC, USA, 2006, pp. 451–455.

[18] Z. Ghahramani, *Unsupervised Learning*. Springer, 2004.

[19] Y. Wang and J.-Q. Han, "Iris recognition using independent component analysis," *Machine Learning and Cybernetics, 2005.*, vol. 7, pp. 4487–4492, 2005.

[20] J. Fortuna and D. Capson, "Ica filters for lighting invariant face recognition," in *17th Int. Conf. on Pattern Recognition (ICPR)*, Washington, DC, USA, 2004, pp. 334–337.

[21] J. B. Tenenbaum, V. de Silva, and J. C. Langford, "A global geometric framework for nonlinear dimensionality reduction." *Science*, vol. 290, no. 5500, pp. 2319–2323, December 2000.

[22] N. Lawrence, "Probabilistic non-linear principal component analysis with gaussian process latent variable models," *J. Mach. Learn. Res.*, vol. 6, pp. 1783–1816, 2005.

[23] C. M. Bishop, *Pattern Recognition and Machine Learning (Information Science and Statistics)*. Springer, August 2006.

[24] R. E. Kass and A. E. Raftery, "Bayes factors," *Journal of the American Statistical Association*, vol. 90, pp. 733–795, 1995.

Probabilistic Models of Object Geometry for Grasp Planning

Jared Glover, Daniela Rus, and Nicholas Roy
Computer Science and Artificial Intelligence Laboratory
Massachusetts Institute Of Technology
Cambridge, MA 02139
{jglov,rus,nickroy}@mit.edu

***Abstract*— Robot manipulators generally rely on complete knowledge of object geometry in order to plan motions and compute successful grasps. However, manipulating real-world objects poses a substantial modelling challenge. New instances of known object classes may vary from learned models. Objects that are not perfectly rigid may appear in new configurations that do not match any of the known geometries.**

In this paper we describe an algorithm for learning generative probabilistic models of object geometry for the purposes of manipulation; these models capture both non-rigid deformations of known objects and variability of objects within a known class. Given a single image of partially occluded objects, the model can be used to recognize objects based on the visible portion of each object contour, and then estimate the complete geometry of the object to allow grasp planning.

We provide two main contributions: a probabilistic model of shape geometry and a graphical model for performing correspondence between shape descriptions. We show examples of learned models from image data and demonstrate how the learned models can be used by a manipulation planner to grasp objects in cluttered visual scenes.

I. Introduction

Robot manipulators largely rely on complete knowledge of object geometry in order to plan their motion and compute successful grasps. If an object is fully in view, the object geometry can be inferred from sensor data and a grasp computed directly. If the object is occluded by other entities in the scene, manipulations based on the visible part of the object may fail; to compensate, object recognition is often used to identify the location of the object and compute the grasp from a prior model. However, new instances of a known class of objects may vary from the prior model, and known objects may appear in novel configurations if they are not perfectly rigid. As a result, manipulation can pose a substantial modelling challenge when objects are not fully in view.

Consider the camera image[1] of four toys in a box in figure 1(a). Having a prior model of the objects is extremely useful in that visible segments (such as the three visible parts of the stuffed bear) can be aggregated into a single object, and a grasp can be planned appropriately as in figure 1(b). However, having a prior model of the geometry of every object in the world is not only infeasible but unnecessary. Although an object such as the stuffed bear may change shape as it is handled and placed in different configurations, the

[1]Note that for the purposes of reproduction, the images have been cropped and modified from the original in brightness and contrast. They are otherwise unchanged.

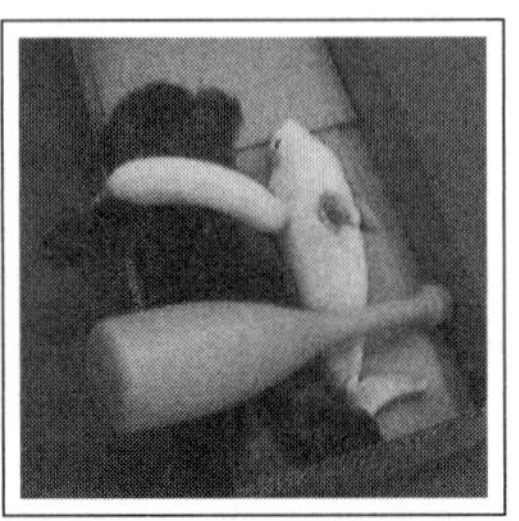

(a) Original Image

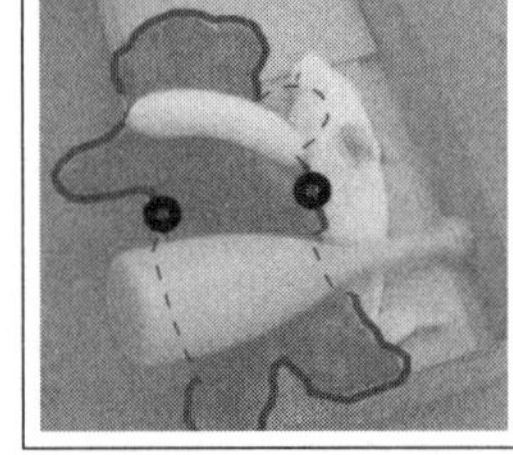

(b) Recovered Geometries

Figure 1. (a) A collection of toys in a box. The toys partially occlude each other, making object identification and grasp planning difficult. (b) By using learned models of the bear, we can identify the bear from the three visible segments and predict its complete geometry (shown by the red line; the dashed lines are the predicted outline of the hidden shape). This prediction of the complete shape can then be used in planning a grasp of the bear (planned grasp points shown by the blue circles).

general shape in terms of a head, limbs, etc. are roughly constant. Regardless of configuration, a single robust model which accounts for deformations in shape should be sufficient for recognition and grasp planning for most object types.

In this paper we describe an algorithm for learning a probabilistic model of visual object geometry. Although statistical models of shape geometry have received attention in a number of domains including computer vision [9, 7] and robotics, existing techniques have largely been coupled to tasks such as shape localization [7], recognition and retrieval [18, 1]. Many effective recognition and retrieval algorithms are discriminative in nature and create representations of the shape that make it difficult to perform additional inference such as recovering hidden object geometry. Our primary contribution is an algorithm for learning generative models of object shapes as dense 2-D contours, as we are specifically interested in object geometry for manipulation planning. We use a model of object shape, known as *Procrustean shape* [6, 13], that provides model invariance to translation, scale and rotation; we generalize this technique to learn object models that are robust to object variation and deformations.

One of the challenges in inferring dense models of shape is that in order to compute the likelihood of a particular shape given a model, we must *a priori* know which points in the measured shape correspond to which points in the model. Thus, our second contribution is to provide a graphical model for computing correspondences between shapes as a pre-processing step to the model learning. We conclude with experimental demonstrations of object detection in cluttered

Algorithm 1 The Manipulation Process.

Require: An image of a scene, and learned models of objects

1: Segment the image into object components
2: Extract contours of components
3: Determine maximum-likelihood correspondence between observed contours and known models
4: Infer complete geometry of each object from matched contours
5: Return planned grasp strategy based on inferred geometries

scenes, geometry prediction and grasp planning.

II. The Manipulation Process

Our goal is to manipulate an object in a cluttered scene–for example to grasp the bear in figure 1(a). Our proposed manipulation process is given in algorithm 1. The input to the algorithm is a single image which is first segmented into perceptually similar regions. (Although image segmentation is a challenging research problem, it is outside the scope of this paper and we rely on existing segmentation algorithms such as [23].) The boundaries or contours of the image segments are extracted, and it is these representations of object geometry that we use throughout this paper.

We first describe how to learn a generative probabilistic model of a class of objects, given a set of object contours of the same class. Using the learned models of class geometry, we next describe how different instances of an object class can be recognized and localized in a single image of partially occluded objects. We use the generative model to infer the hidden parts of each object in order to complete the model of each object. Finally, we describe how the inferred complete geometry can be used to compute a grasp.

III. Probabilistic Models of 2-D Shape

Formally, we represent an object $\mathbf{Z}$ in an image as a set of n ordered points on the contour of the shape, $\{\mathbf{z}_1, \mathbf{z}_2, \ldots \mathbf{z}_n\}$, in a two-dimensional Euclidean space, so that $\mathbf{z}_i = (x_i, y_i)$). Our goal is to learn a probabilistic, generative model of $\mathbf{Z}$. We begin by making the contour invariant with respect to position and scale, normalizing $\mathbf{Z}$ so as to have unit length with centroid at the origin, that is,

$$\mathbf{Z}' = \{\mathbf{z}'_i = (x_i - \bar{x}, y_i - \bar{y})\} \quad (1)$$

$$\tau = \frac{\mathbf{Z}'}{|\mathbf{Z}'|}, \quad (2)$$

where τ is the *pre-shape* of the contour $\mathbf{Z}$. Since τ is a unit vector, the space of all possible pre-shapes of n points is the unit hyper-sphere, $\mathbb{S}_*^{2n-3}$, called *pre-shape space*. Since we can rotate any pre-shape through a great circle orbit $\mathcal{O}(\tau)$ of maximal length of the hypersphere without changing the geometry of z, we define the "shape" of $\mathbf{Z}$ as an equivalence class of pre-shapes over rotations.

If we can define a distance metric between shapes, then we can infer a parametric distribution over the shape space. The spherical geometry of the pre-shape space requires a geodesic distance rather than Euclidean distance. The distance between

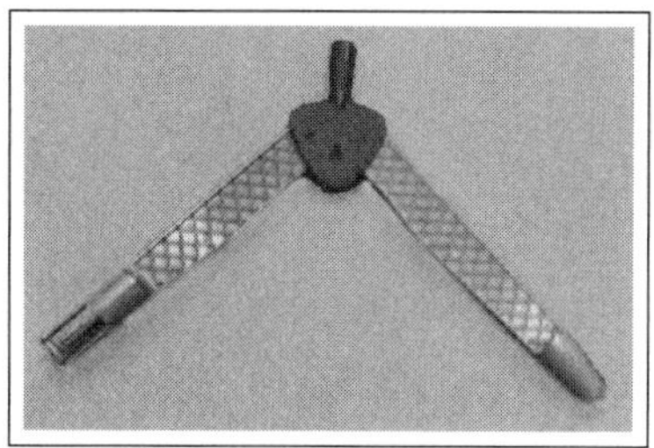

(a) Example Object

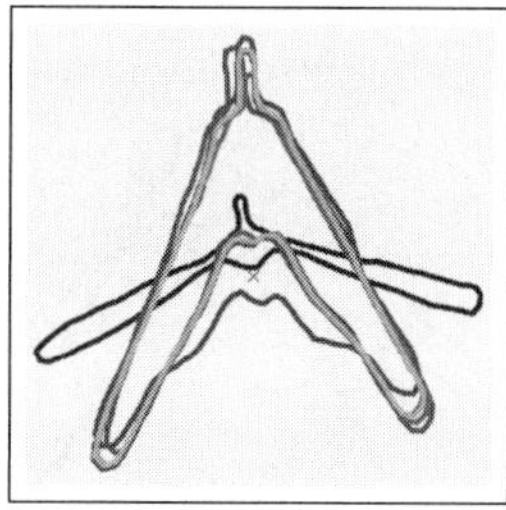

(b) Class Distribution

Figure 2. (a) An example image of a chalk compass. The compass can deform by opening and closing. (b) Sample shapes from the learned distribution along different eigenvalues of the distribution.

τ_1 and τ_2 is defined as the smallest distance between their orbits,

$$d_p[\tau_1, \tau_2] = \inf[d(\varphi, \psi) : \varphi \in \mathcal{O}(\tau_1), \psi \in \mathcal{O}(\tau_2)] \quad (3)$$

$$d(\varphi, \psi) = \cos^{-1}(\varphi \cdot \psi). \quad (4)$$

Kendall [13] defined d_p as the *Procrustean metric* where $d(\varphi, \psi)$ is the geodesic distance between φ and ψ. We can solve for the minimization of equation (3) in closed form by representing the points of τ_1 and τ_2 in complex coordinates, which naturally encode rotation in the plane by scalar complex multiplication. This gives d_p as

$$d_p[\tau_1, \tau_2] = \cos^{-1} |\tau_2^H \tau_1| \quad (5)$$

where τ_2^H is the *Hermetian*, or complex conjugate transpose of the complex vector τ_2.

A. Learning Shape Models

In order to complete our probabilistic model of object geometry, we compute a distribution for each object class from training images. We choose a Gaussian approximation to the distribution over shapes, which only requires us to compute the mean and covariance of the training data. This Gaussian lies in the tangent space to the hypersphere at the mean shape vector. For each object class i, we compute a mean shape μ_i, from a set of pre-shapes $\{\tau_1, \ldots, \tau_n\}$ by minimizing the sum of Procrustean distances from each pre-shape to the mean,

$$\mu_i = \underset{\mu}{\operatorname{arginf}} \sum_j [d_p(\tau_j, \mu)]^2, \quad (6)$$

subject to the constraint that $\|\mu_i\| = 1$. In two dimensions, this minimization can be done in closed form; iterative algorithms exist for computing μ_i in higher dimensions [2, 10].

In order to estimate the covariance of the shape distribution from the sample pre-shapes $\{\tau_1, \ldots, \tau_n\}$, we rotate each τ_j to fit the mean shape μ_i (i.e. to minimize Procrustean distance), and then project the rotated pre-shapes into the tangent space of the pre-shape hypersphere at the mean shape. We then use Principle Components Analysis (PCA) in tangent space to model the principle axes of the Gaussian shape distribution of $\{\tau_1, \ldots, \tau_n\}$. Figure 2(a) shows one example out of a training set of images of a deformable object. Figure 2(b) shows sample objects drawn from the learned distribution. The red contour is the mean, and the green and blue samples are taken along the first two principal components of the distribution.

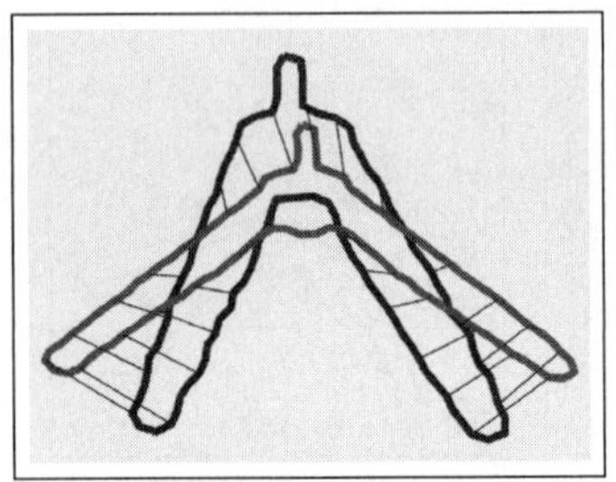
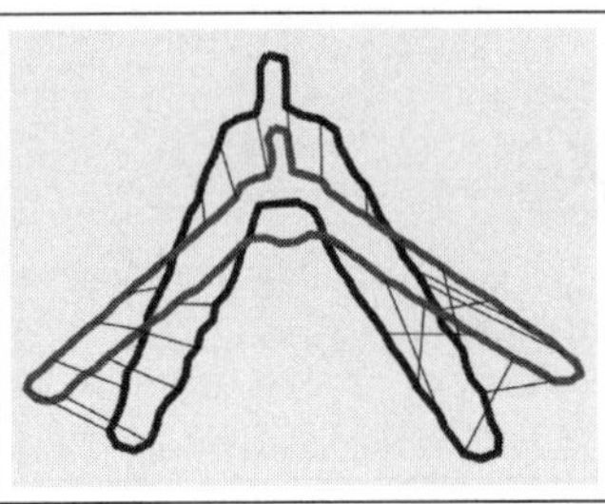

Figure 3. Order-preserving matching (left) vs. Non-order-preserving matching (right). The thin black lines depict the correspondences between points in the red and blue contour. Notice the violation of the cyclic-ordering constraint between the right arms of the two contours in the right image.

B. Shape Classification

Given k previously learned shape classes $C_1, \ldots, C_k$ with shape means $\mu_1, \ldots, \mu_k$ and covariance matrices $\Sigma_1, \ldots, \Sigma_k$, and given a measurement $\mathbf{m}$ of an unknown object shape, we can now compute the likelihood of a shape class given a measured object: $\{P(C_i|\mathbf{m}) : i = 1 \ldots k\}$. The shape classification problem is to find the maximum likelihood class, $\hat{C}$, which we can compute as

$$\hat{C} = \arg\max_{C_i} P(C_i|\mathbf{m}) \tag{7}$$

$$= \arg\max_{C_i} P(\mathbf{m}|C_i)P(C_i). \tag{8}$$

Given the mean and covariance of a shape class, we can compute the likelihood of a measured object given a class as $p(\mathbf{m}|C_i) = \mathcal{N}(\mathbf{m}; \mu_i, \Sigma_i)$. Assuming a uniform prior on C_i, we can compute the maximum likelihood class as

$$\hat{C} = \underset{C_i}{\text{argmax}} \mathcal{N}(\mathbf{m}; \mu_i, \Sigma_i). \tag{9}$$

IV. Data Association and Shape Correspondences

Evaluating the likelihood given by equation 9 requires calculating the Procrustes distance d_p between the observed contour $\mathbf{m}$ and the mean μ_i. The distance between any two contours τ_1 and τ_2 implicitly assumes that there is a known correspondence between a point $\mathbf{x}_i$ in τ_1 and some point $\mathbf{y}_j$ in τ_2. (There is also an assumption that the lengths of τ_1 and τ_2 are the same.) Before we can compute the probability of a contour, or even learn the mean and covariance of a set of pre-shapes, we must therefore be able to compute the correspondences between contours, matching each point in τ_1 to a corresponding point on τ_2.

Solving for the most likely correspondences between sets of data is an open problem in a number of fields, including computer vision and robot mapping. As object geometries vary due to projection distortions, sensor error, or even natural object dynamics, determining *which* part of an object image corresponds to *which* part of a previous image is non-trivial.

Furthermore, by the nature of object contours, our specific shape correspondence problem contains a *cyclic order-preserving* constraint, that is, correspondences between the two contours cannot "cross" each other. Scott and Nowak [22] define the Cyclic Order-Preserving Assignment Problem (COPAP) as the problem of finding an optimal one-to-one matching such that the assignment of corresponding points preserves the cyclic ordering inherited from the contours. Figure 3 shows an example set of correspondences (the thin black lines) that preserve the cyclic order-preserving constraint on the left, whereas the correspondences on the right of figure 3 violate the constraint at the right of the shape (notice that the association lines cross.) In the following sections, we show how the original COPAP algorithm can be written as a linear graphical model with the introduction of additional book-keeping variables.

Our goal is to match the points of one contour, $\mathbf{x}_1, \ldots, \mathbf{x}_n$ to the points on another, $\mathbf{y}_1, \ldots, \mathbf{y}_m$. Let Φ denote a correspondence vector, where ϕ_i is the index of $\mathbf{y}$ to which $\mathbf{x}_i$ corresponds; that is: $\mathbf{x}_i \rightarrow \mathbf{y}_{\phi_i}$. We wish to find the most likely Φ given $\mathbf{x}$ and $\mathbf{y}$, that is, $\Phi^* = \text{argmax}_\Phi\, p(\Phi|\mathbf{x}, \mathbf{y})$. If we assume that the likelihood of individual points $\{\mathbf{x}_i\}$ and $\{\mathbf{y}_j\}$ are conditionally independent given Φ, then

$$\Phi^* = \underset{\Phi}{\text{argmax}} \frac{1}{Z} p(\mathbf{x}, \mathbf{y}|\Phi) p(\Phi) \tag{10}$$

$$= \underset{\Phi}{\text{argmax}} \frac{1}{Z} \prod_{i=1}^{n} p(x_i, y_{\phi_i}) p(\Phi) \tag{11}$$

where Z is a normalizing constant.

A. Priors over Correspondences

There are two main terms to equation (10), the prior over correspondences, $p(\Phi)$, and the likelihood of object points given the correspondences, $p(\mathbf{x}_i, \mathbf{y}_{\phi_i})$. We model the prior over correspondences, $p(\Phi)$, as an exponential distribution subject to the cyclic-ordering constraint. We encode this constraint in the prior by allowing $p(\Phi) > 0$ if and only if

$$\exists \omega \ s.t. \ \phi_\omega < \phi_{\omega+1} < \cdots < \phi_n < \phi_1 < \cdots < \phi_{\omega-1}. \tag{12}$$

We call ω the *wrapping point* of the assignment vector Φ. Each assignment vector, Φ, which obeys the cyclic-ordering constraint must have a unique wrapping point, ω.

Due to variations in object geometry, the model must allow for the possibility that some sequence of points of $\{\mathbf{x}_i, \ldots, \mathbf{x}_j\}$ do not correspond to any points in $\mathbf{y}$, for example, if sensor noise has introduced spurious points along an object edge or if the shapes vary in some significant way, such as an animal contour with three legs where another has four. We "skip" individual correspondences in $\mathbf{x}$ by allowing $\phi_i = 0$. (Points $\mathbf{y}_j$ are skipped when $\nexists i \ s.t. \ \phi_i = j$). We would like to minimize the number of such skipped assignments, so we give diminishing likelihood to ϕ as the number of skipped points increases. Therefore, for Φ with k skipped assignments (in $\mathbf{x}$ and $\mathbf{y}$),

$$p(\Phi) = \begin{cases} \frac{1}{Z_\Phi} \exp\{-k(\Phi) \cdot \lambda\} & \text{if } \Phi \text{ is cyclic ordered} \\ 0 & \text{otherwise,} \end{cases} \tag{13}$$

where Z_Φ is a normalizing constant and λ is a likelihood penalty for skipped assignments.

B. Correspondence Likelihoods

Given an expression for the correspondence prior, we also need an expression for the likelihood that two points $\mathbf{x}_i$ and $\mathbf{y}_{\phi_i}$ correspond to each other, $p(\mathbf{x}_i, \mathbf{y}_{\phi_i})$, which we model as the likelihood that the local geometry of the contours match. Section III described a probabilistic model for global geometric similarity using the Procrustes metric, and we specialize this model to computing the likelihood of local geometries, which we call the *Procrustean Local Shape Distance* (PLSD).

We first need a description of the local shape about $\mathbf{x}_i$. In order to be robust to the local spacing of $\mathbf{x}$'s points, we sample points evenly spaced about $\mathbf{x}_i$. We define the *local neighborhood* of size k about $\mathbf{x}_i$ as:

$$\eta_k(x_i) = \langle \delta_x^i(-2^k\Delta), ..., \delta_x^i(0), ..., \delta_x^i(2^k\Delta) \rangle \quad (14)$$

where $\delta_x^i(d)$ returns the point from x's contour interpolated a distance of d starting from $\mathbf{x}_i$ and continuing clockwise for d positive or counter-clockwise for d negative. (Also, $\delta_x^i(0) = \mathbf{x}_i$.) The parameter Δ determines the step-size between interpolated neighborhood points, and thus the resolution of the local neighborhood shape. We have found that setting Δ such that the largest neighborhood is 20% of the total shape circumference yields good results on most datasets.

The Procrustean Local Shape Distance, d_{PLS}, between two points, x_i and y_j is the mean Procrustean shape distance over neighborhood sizes k:

$$d_{PLS}(x_i, y_j) = \int_k \xi_k \cdot d_P[\eta_k(x_i), \eta_k(y_j)] \quad (15)$$

with neighborhood size prior ξ. No closed form exists for this integral so we approximate it using a sum over a discrete set of neighborhood sizes.

C. A Graphical Model for Shape Correspondences

Although we assume independence between local features $\mathbf{x}_i$ and $\mathbf{y}_j$, the cyclic-ordering constraint leads to dependencies between the assignment variables ϕ_i in a non-trivial way—in fact, the sub-graph of Φ is fully connected since each ϕ_i must know the values of all the other assignments, ϕ_j, in order to determine whether the matching is order-preserving or not. Computing the maximum likelihood Φ is therefore a non-trivial loopy graphical inference problem.

We can avoid this problem and break most of these dependencies by introducing variables α_i and ω, where α_i corresponds to the last non-zero assignment before ϕ_i and ω corresponds to the wrapping point from section IV-A. With these additional variables, each ϕ_i depends only on the wrapping point, which is stored in ω as well as the last non-zero assignment, α_i; the cyclic ordering-constraint is thus encoded by $p_{co}(\phi_i)$, such that

$$p_{co}(\phi_i) = \begin{cases} \frac{1}{Z_{co}} & : \text{if } \phi_i > \alpha_i \text{ or} \\ & \phi_i < \alpha_i \text{ and } \omega_i = i \text{ or} \\ & \phi_i = 0 \\ 0 & : \text{otherwise,} \end{cases} \quad (16)$$

which gives (17)

$$p(\Phi) = \frac{1}{Z_\Phi}(\exp\{-k(\Phi)\cdot\lambda)\prod_i p_{co}(\phi_i). \quad (18)$$

If we initially assign the wrapping point ω, the state vector $\{\alpha_i, \phi_i\}$ then yields a cyclic Markov chain. The standard approach to solving this cyclic Markov chain is to try setting the wrapping point, ω, to each possible value from 1 to n. Given $\omega = k$, the cycle is broken into a linear chain (according to equation 12), which can be solved by dynamic programming. It is this introduction of the α_i and ω variables that is the key to the efficient inference procedure by converting the loopy graphical model into a linear chain.

In this approach, the point-assignment likelihoods are converted into a cost function $C(i, \phi_i)$ by taking a log likelihood, and ϕ is optimized using

$$\Phi^* = \underset{\Phi}{\text{argmax}} \ \log \prod_i p(\mathbf{x}_i, \mathbf{y}_{\phi_i}) p(\Phi) \prod_i p_{co}(\phi_i) \quad (19)$$

$$= \underset{\Phi}{\text{argmin}} \left(\sum_i C(i, \phi_i) \right) + \lambda \cdot k(\phi) \quad (20)$$

$$\text{s.t. } \forall \phi_i \ p_co(\phi_i) > 0$$

where $k(\Phi)$ is the number of points skipped in the assignment Φ. Solving for Φ using equation (20) takes $O(n^2m)$ running time; however a bisection strategy exists in the dynamic programming search graph which reduces the complexity to $O(nm \log n)$ [22].

V. Shape Completion

We now turn to the problem of estimating the complete geometry of an object from an observation of part of its contour. We phrase this as a maximum likelihood estimation problem, estimating the missing points of a shape with respect to the Gaussian tangent space shape distribution.

Let us represent a shape as:

$$\mathbf{z} = [\mathbf{z}_1 \ \mathbf{z}_2]^T \quad (21)$$

where $\mathbf{z}_1 = \mathbf{m}$ contains the p points of our partial observation of the shape, and $\mathbf{z}_2$ contains the $n - p$ unknown points that complete the shape. Given a shape distribution D on n points with mean μ and covariance matrix Σ, and given $\mathbf{z}_1$ containing p measurements ($p < n$) of our shape, our task is to compute the last $n - p$ points which maximize the joint likelihood, $P_D(\mathbf{z})$. (We implicitly assume that correspondences from the partial shape $\mathbf{z}$ to the model D are known–we later show how to compute partial shape correspondences in order to relax this assumption.)

In order for us to transform our completed vector, $\mathbf{z} = (\mathbf{z}_1, \mathbf{z}_2)^T$, into a pre-shape, we must first normalize translation and scale. However, this cannot be done without knowing the last $n - p$ points. Furthermore, the Procrustes minimizing rotation from $\mathbf{z}$'s pre-shape to μ depends on the missing points, so any projection into the tangent space (and corresponding likelihood) will depend in a highly non-linear way on the location of the missing points. We can, however, compute the missing points $\mathbf{z}_2$ given an orientation and scale. This leads to an iterative algorithm that holds the orientation and scale fixed, computes $\mathbf{z}_2$ and then computes a new orientation and scale given the new $\mathbf{z}_2$. The translation term can then be computed from the completed contour $\mathbf{z}$.

We derive $\mathbf{z}_2$ given a fixed orientation θ and scale α in the following manner. For a complete contour $\mathbf{z}$, we normalize for orientation and scale using

$$\mathbf{z}' = \frac{1}{\alpha} R_\theta \mathbf{z} \quad (22)$$

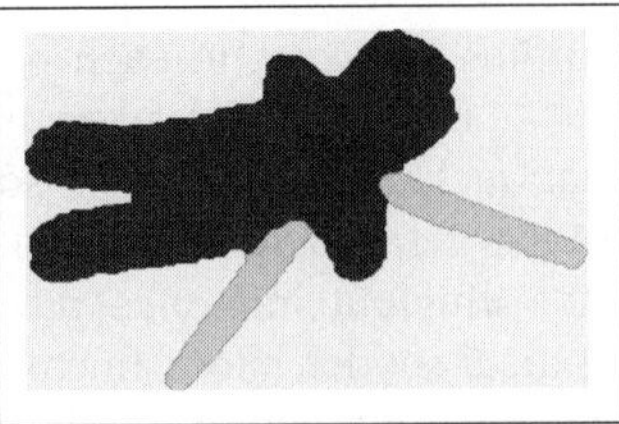

Figure 4. An example of occluded objects, where the bear occludes the compass. (a) The original image and (b) the image segmented into (unknown) objects. The contour of each segment must be matched against a known model.

where R_θ is the rotation matrix of θ. To center $\mathbf{z}'$, we then subtract off the centroid:

$$\mathbf{w} = \mathbf{z}' - \frac{1}{n}C\mathbf{z}' \tag{23}$$

where C is the $2n \times 2n$ checkerboard matrix,

$$C = \begin{bmatrix} 1 & 0 & \cdots & 1 & 0 \\ 0 & 1 & \cdots & 0 & 1 \\ \vdots & \vdots & \ddots & \vdots & \vdots \\ 1 & 0 & \cdots & 1 & 0 \\ 0 & 1 & \cdots & 0 & 1 \end{bmatrix}. \tag{24}$$

Thus $\mathbf{w}$ is the centered pre-shape. Now let M be the matrix that projects into the tangent space defined by the Gaussian distribution (μ, Σ):

$$M = I - \mu\mu^T \tag{25}$$

The Mahalanobis distance with respect to D from $M\mathbf{w}$ to the origin in the tangent space is:

$$d_\Sigma = (M\mathbf{w})^T \Sigma^{-1} M\mathbf{w} \tag{26}$$

Minimizing d_Σ is equivalent to maximizing $P_D(\cdot)$, so we continue by setting $\frac{\partial d_\Sigma}{\partial \mathbf{z}_2}$ equal to zero, and letting

$$W_1 = M_1(I_1 - \frac{1}{n}C_1)\frac{1}{\alpha}R_\theta^1 \tag{27}$$

$$W_2 = M_2(I_2 - \frac{1}{n}C_2)\frac{1}{\alpha}R_\theta^2 \tag{28}$$

where the subscripts "1" and "2" indicate the left and right sub-matrices of M, I, and C that match the dimensions of $\mathbf{z}_1$ and $\mathbf{z}_2$. This yields the following system of linear equations which can be solved for the missing data, $\mathbf{z}_2$:

$$(W_1\mathbf{z}_1 + W_2\mathbf{z}_2)^T \Sigma^{-1} W_2 = 0 \tag{29}$$

As described above, equation (29) holds for a specific orientation and scale. We can then use the estimate of $\mathbf{z}_2$ to re-optimize θ and α and iterate. Alternatively, we can simply sample a number of candidate orientations and scales, complete the shape of each sample, and take the completion with highest likelihood (lowest d_Σ).

To design such a sampling algorithm, we must choose a distribution from which to sample orientations and scales. One idea is to match the partial shape, $\mathbf{z}_1$, to the partial mean shape, μ_1, by computing the pre-shapes of $\mathbf{z}_1$ and μ_1 and finding the Procrustes fitting rotation, θ^*, from the pre-shape of $\mathbf{z}_1$ onto the pre-shape of μ_1. This angle can then be used as a mean for a von Mises distribution (the circular analog of a Gaussian) from which to sample orientations. Similarly, we can sample scales from a Gaussian with mean α_0–the ratio of scales of the partial shapes $\mathbf{z}_1$ and μ_1 as in

$$\alpha_0 = \frac{\|\mathbf{z}_1 - \frac{1}{p}C_1\mathbf{z}_1\|}{\|\mu_1 - \frac{1}{p}C_1\mu_1\|}. \tag{30}$$

Any sampling method for shape completion will have a *scale bias*–completed shapes with smaller scales project to a point closer to the origin in tangent space, and thus have higher likelihood. One way to fix this problem is to solve for $\mathbf{z}_2$ by performing a constrained optimization on d_Σ where the scale of the centered, completed shape vector is constrained to have unit length:

$$\|x' - \frac{1}{n}Cx'\| = 1. \tag{31}$$

This constrained optimization problem can be attacked with the method of Lagrange multipliers, and reduces to the problem of finding the zeros of a $(n-p)$th order polynomial in one variable, for which numerical techniques are well-known.

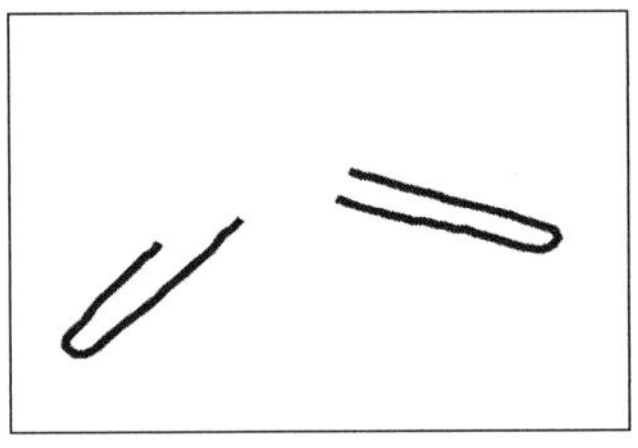

(a) Partial contour to be completed

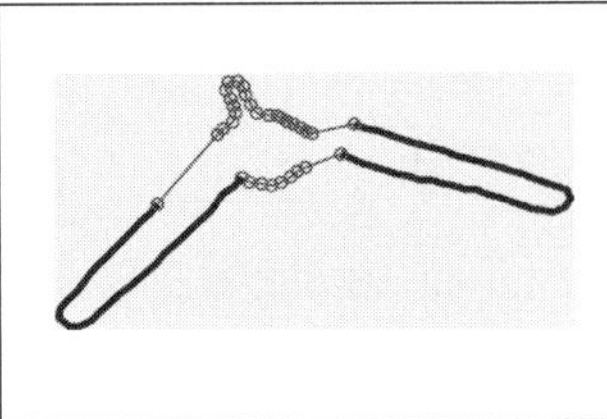

(b) Completed as compass

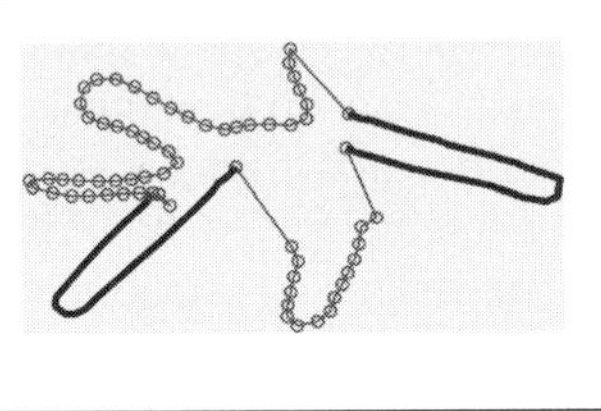

(c) Completed as stuffed animal

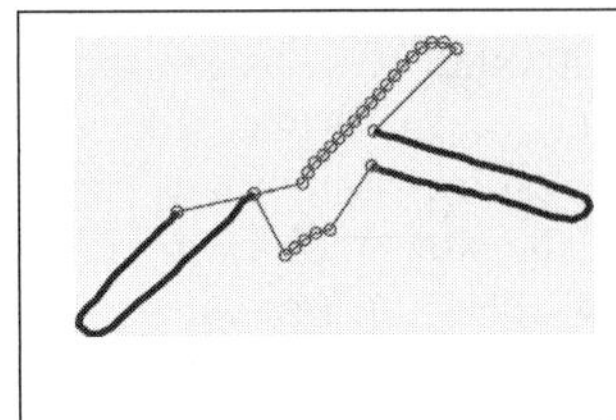

(d) Completed as jump rope

Figure 5. Shape completion of the partial contour of the compass in figure 4. Note that the correct completion (b) captures the knob in the top of the compass. The hypothesized completions in (c) and (d) lead to very unlikely shapes.

A. Partial Shape Class Likelihood

Let $\mathbf{z} = \{\mathbf{z}_1, \mathbf{z}_2\}$ be the completed shape, where $\mathbf{z}_1$ is the partial shape corresponding to measurement $\mathbf{m}$, and $\mathbf{z}_2$ is unknown. The probability of the class given the observed part of the contour $\mathbf{z}_1$ is then

$$P(C_i|\mathbf{z}_1) = \frac{P(C_i, \mathbf{z}_1)}{P(\mathbf{z}_1)} \propto \int P(C_i, \mathbf{z}_1, \mathbf{z}_2) d\mathbf{z}_2 \tag{32}$$

Rather than marginalize over the hidden data, $\mathbf{z}_2$, we can approximate this marginal with an estimate $\hat{\mathbf{z}}_2$, the output of our shape completion algorithm, yielding:

$$P(C_i|\mathbf{z}_1) \approx \eta \cdot P(\mathbf{z}_1, \hat{\mathbf{z}}_2|C_i) \tag{33}$$

where η is a normalizing constant (and can be ignored during classification), and $P(\mathbf{z}_1, \hat{\mathbf{z}}_2|C_i)$ is the complete shape class likelihood of the completed shape.

B. Partial Shape Correspondences

In order to calculate the maximum likelihood shape completion $\hat{\mathbf{z}}_2$ with respect to a shape model D, we must know which points in D the observed points $\mathbf{z}_1$ correspond to. In practice, $\mathbf{z}_1$ may contain multiple disconnected contour segments which must be associated with hidden contour segments to form a complete contour–take for example, the two compass handles in figure 5. Before the hidden contours can be inferred between the handles, observable contours must be ordered. We can constrain the connection ordering by noting that the interiors of all the observed object segments must remain on the interior of any completed shape. For most real-world cases, this topological constraint is enough to identify a unique connection ordering; in cases where the ordering of components is still ambiguous, a search process through the orderings can be used to identify the most likely correspondences.

Given a specific ordering of observed contour segments, we can adapt our graphical model from section IV to compute the correspondence between an ordered set of partial contour segments and a model mean shape, μ. First, we add a set of hidden, or "wildcard" points connecting the partial contour segments. This forms a complete contour, $\mathbf{z}_c$, where some of the points are hidden and some are observed. We then run a modified COPAP algorithm, where the only modification is that all "wildcard" points on $\mathbf{z}_c$ may be assigned to any of μ's points with no cost. (We must still pay a penalty of λ for skipping hidden points, however.)

In order to identify how large the hidden contour is (and therefore, how many hidden points should be added to connect the observed contour segments), we use the insight that objects of the same type generally have a similar scale. We can therefore use the ratio of the observed object segment areas to the expected full shape area to (inversely) determine the ratio of hidden points to observed points. If no size priors are available, one may also perform multiple completions with varying hidden points ratios, and select the best completion using a generic prior such as the minimum description length (MDL) criterion.

Using this partial shape correspondence algorithm, we employ an iterative procedure to complete the hidden parts of an object contour–(1) compute the partial shape correspondences, (2) complete the shape given the partial correspondences, (3) compute the *full* shape correspondences from the completed shape to the model, (4) re-complete the shape using the new correspondences, and repeat (3) and (4) until convergence.

VI. Grasp Planning

Recall from Section II that our manipulation strategy is a pipelined process–first, we estimate the complete geometric structure of the scene; then, we plan a grasp. But before we can get into the details about how an individual object is grasped, we must first decide *which* object to grasp. The problem domains which we are primarily interested in–such as the "box-of-toys" world of Figure 1–are domains in which there is a single "desired" object or object type; for example, a teddy bear. Thus, our ultimate goal is to retrieve a specific object or class of object from the scene. Sometimes, the desired object will be at the top of the pile, fully in view. In this case, after analyzing the image and recognizing the object, we will be able to plan a grasp to retrieve the object, irrespective of the placement of other objects in the scene. However, if the desired object is occluded, before attempting to pick it up, we must determine the probability that the sensed object is actually the desired object, and the probability that a planned grasp on the accessible part of the object will be successful. If either of these probabilities are below a pre-determined threshold, we first remove one or more occluding objects and then re-analyze the scene before planning a grasp of the desired object. We implement the first test as a threshold on the class likelihood of the sensed object, $p(C_i|\mathbf{m}) > 0.7$; the second test is a function of our strategy for grasping a single object, described below.

A. Grasping a Single Object

We have developed a grasp planning system for our mobile manipulator (shown in figure 6), a two-link arm on a mobile base with an in-house-designed gripper with two opposable fingers. Each finger is a structure capable of edge and surface contact with the object to be grasped.

Figure 6. Our mobile manipulator with a two link arm and gripper.

The input to the grasp planning system is the object geometry with the partial contours completed as described in Section V. The output of the system is two regions, one for each finger of the gripper, that can provide an equilibrium grasp for the object following the algorithms for stable grasping described in [19]. Intuitively, the fingers are placed on opposing edges so that the forces exerted by the fingers can cancel each other out. Friction is modeled as Coulomb friction with empirically estimated parameters.

The grasp planner is implemented as search for a pair of grasping edges that yield maximal regions for the two grasping fingers using the geometric conditions derived by Nguyen [19]. Two edges can be paired if their friction cones are overlapping. Given two edges that can be paired we identify maximal regions for placing the fingers so that we can tolerate maximal uncertainty in the finger placement using Nguyen's criterion [19].

If the desired object is fully observed, we can use the above grasping algorithm unchanged. If it is partially occluded, we must filter out finger placements which lie on hidden (inferred) portions of the object's boundary. If, after filtering out infeasible grasps, there is still an accessible grasp of sufficient quality according to Nguyen's criterion, we can attempt a grasp of the object.

VII. Results

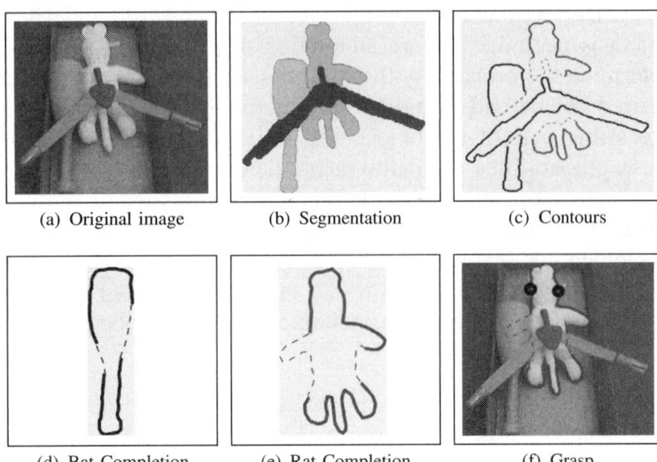

(a) Original image (b) Segmentation (c) Contours (d) Bat Completion (e) Rat Completion (f) Grasp

Figure 7. An example of a very simple planning problem involving three objects. The chalk compass is fully observed, but the stuffed rat and green bat are partially occluded by the compass. After segmentation (b), the image decomposes into five separate segments shown in (c). The learned models of the bat and the rat can be completed (d) and (e), and the complete contour of the stuffed rat is correctly positioned in the image (f). The two blue circles correspond to the planned grasp that results from the computed geometry.

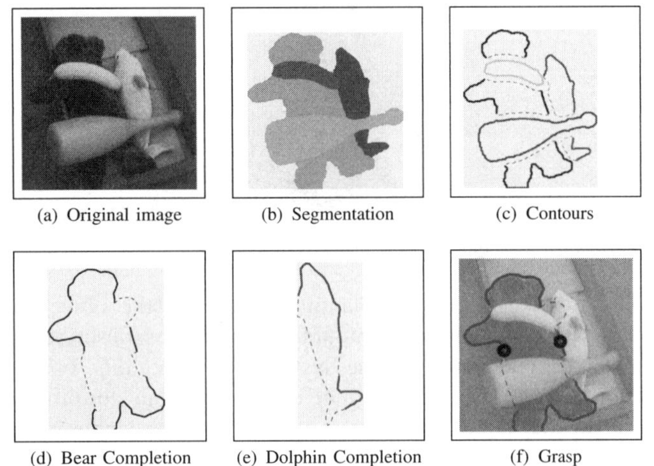

(a) Original image (b) Segmentation (c) Contours (d) Bear Completion (e) Dolphin Completion (f) Grasp

Figure 8. A more complex example involving four objects. The blue bat and the yellow banana are fully observed, but the stuffed bear and dolphin are significantly occluded. After segmentation (b), the image decomposes into five separate segments shown in (c). The learned models of the bear and the dolphin can be completed (d) and (e), and the complete contour of the stuffed bear is is correctly positioned in the image (f). The two blue circles correspond to the planned grasp given the geometry.

We built a shape dataset containing 11 shape classes (6 of which are seen in figures 7 and 8). We collected 10 images of each object type, segmented the object contours from the background, and used the correspondence and shape distribution learning algorithms of sections III and IV to build probabilistic shape models for each class, using contours of 100 points each. We reduced the dimensionality of the covariance using Principal Components Analysis (PCA). Reducing the covariance to three principal components led to 100% prediction accuracy of the training set, and 98% cross-validated ($k = 5$) prediction accuracy.

Object	Partial	Complete
ring	3/8	15/15
bat	7/10	8/10
rat	9/13	4/4
bear	7/7	7/7
fish	9/9	6/6
banana	-	1/2
dolphin	1/2	-
compass	1/3	5/5
totals	37/52	46/49
	71.15%	93.88%
detect $> 5\%$	42/52	48/49
	80.77%	97.96%

Table I

Classification rates on test set.

In figures 7 and 8 we show the results of two manipulation experiments, where in each case we seek to retrieve a single type of object from a box of toys, and we must locate and grasp this object while using the minimum number of object grasps possible. In both cases, the object we wish to retrieve is occluded by other objects in the scene, and so a naive grasping strategy would first remove the objects on top of the desired object until the full object geometry is observed, and only then would it attempt to retrieve the object. Using the inferred geometry of the occluded object boundaries to classify and plan a grasp for the desired object, we find in both cases that we are able to grasp the object immediately, reducing the number of grasps required from 3 to 1. In addition, we were able to successfully complete and classify the other objects in each scene, even when a substantial portion of their boundaries was occluded. The classification of this test set of 7 object contours (from 6 objects classes) was 100% (note the correct completions in figures 7 and 8 of the occluded objects).

For a more thorough evaluation, we repeated the same type of experiment on 20 different piles of toys. In each test, we again sought to retrieve a single type of object from the box of toys, and in some cases, the manipulation algorithm required several grasps in order to successfully retrieve an object, due to either not being able to find the object right away, or because the occluding objects were blocking access to a stable grasp of the desired object.

In total, 52 partial and 49 complete contours were classified, 33/35 grasps were successfully executed (with 3 failures due to a hardware malfunction which were discounted). In table I, we show classification rates for each class of object present in the images. Partially-observed shapes were correctly classified 71.15% of the time, while fully-observed shapes were correctly classified 93.88% of the time. Several of the errors were simply a result of ambiguity–when we examine the $> 5\%$ detection rates (i.e. the percentage of objects for which the algorithm gave at least 5% likelihood to the correct class), we see an improvement to 80.77% for partial shapes, and 97.96% for full shapes. While a few of the detection errors were from poor or noisy image segmentations, most were from failed correspondences from the observed contour to the correct shape model. The most common reason for these failed correspondences was a lack of local features for the COPAP algorithm to latch onto with the PLSD point assignment cost. These failures would seem to argue for a combination of local and global match likelihoods in the correspondence algorithm, which is a direction we hope to explore in future work.

VIII. Related Work

Statistical shape modeling began with the work on landmark data by Kendall [13] and Bookstein [4] in the 1980s. In recent years, more complex statistical shape models have arisen, for example, in the active contours literature [3]. We believe ours is one of the first works to perform probabilistic inference of deformable objects from partially occluded views. In terms of shape classification, shape contexts [1] and spin images [12] provide robust frameworks for estimating correspondences between shape features for recognition and modelling problems; our work is very related but our initial experiments with these descriptors motivated our work for a better shape model for partial views of objects. Classical statistical shape models require a large amount of human intervention (e.g. hand-labelled landmarks) in order to learn accurate models of shape [6]; only recently have algorithms emerged that require little human intervention [9, 7].

We also build on classical and recent results on motion planning and grasping, manipulation, uncertainty for modeling in robot manipulation, POMDPs applied to mobile robots, kinematics, and control. The initial formulation of the problem of planning robot motions under uncertainty was the preimage backchaining paper [16]. It was followed up with further analysis and implementation [5, 8], analysis of the mechanics and geometry of grasping [17], and grasping algorithm that guarantees geometrically closure properties [19]. Lavalle and Hutchinson [15] formulated both probabilistic and nondeterministic versions of the planning problem through information space. Our manipulation planner currently does not take advantage of the probabilistic representation of the object, but we plan to extend our work to this domain.

More recently, Grupen and Coelho [11] have constructed a system that learns optimal control policies in an information space that is derived from the changes in the observable modes of interaction between the robot and the object it is manipulating. Ng et al. [21] have used statistical inference techniques to learn manipulation strategies directly from monocular images; while these techniques show promise, the focus has been generalizing as much as possible from as simple a data source as possible. It is likely that the most robust manipulation strategies will result from including geometric information such as used by Pollard and Zordan [20].

IX. Conclusions

In future work, we hope to demonstrate improved performance on recognition tasks by incorporating additional priors into the correspondence and completion models, in order to bias the inference procedure towards smoother, more natural correspondences and completions. The shape classes that we have found to cause the most problems for our model contain multiple articulations and self-occlusions, which suggests that it may be useful to combine a skeleton or parts-based models with our global parametric models in order to achieve robustness to these highly variable shapes.

X. Acknowledgements

Jared Glover and Nicholas Roy were supported by the National Science Foundation Division of Information and Intelligent Systems under grant # 0546467 and the Air Force Office of Scientific Research under STTR Contract FA9550-06-C-0088. Nicholas Roy and Daniela Rus were supported by the National Science Foundation Division of Computer and Network Systems under grant # 0707601. Daniela Rus was supported by the National Science Foundation under grants # 0426838 and 0735953, and Boeing.

References

[1] Serge Belongie, Jitendra Malik, and Jan Puzicha. Shape matching and object recognition using shape contexts. *IEEE Trans. Pattern Analysis and Machine Intelligence*, 24(24):509–522, April 2002.

[2] Jos M. F. Ten Berge. Orthogonal procrustes rotation for two or more matrices. *Psychometrika*, 42(2):267–276, June 1977.

[3] A. Blake and M. Isard. *Active Contours*. Springer-Verlag, 1998.

[4] F.L. Bookstein. A statistical method for biological shape comparisons. *Theoretical Biology*, 107:475–520, 1984.

[5] Bruce Donald. A geometric approach to error detection and recovery for robot motion planning with uncertainty. *Artificial Intelligence*, 37:223–271, 1988.

[6] I. Dryden and K. Mardia. *Statistical Shape Analysis*. John Wiley and Sons, 1998.

[7] G. Elidan, G. Heitz, and D. Koller. Learning object shape: From drawings to images. In *Proceedings of the Conference on Computer Vision and Pattern Recognition (CVPR)*, 2006.

[8] Michael Erdmann. Using backprojection for fine motion planning with uncertainty. *IJRR*, 5(1):240–271, 1994.

[9] P. Felzenszwalb. Representation and detection of deformable shapes. *IEEE Trans. Pattern Analysis and Machine Intelligence*, 27(2), 2005.

[10] J. C. Gower. Generalized procrustes analysis. *Psychometrika*, 40(1):33–51, March 1975.

[11] Roderic A. Grupen and Jefferson A. Coelho. Acquiring state from control dynamics to learn grasping policies for robot hands. *Advanced Robotics*, 16(5):427–443, 2002.

[12] Andrew Johnson and Martial Hebert. Using spin images for efficient object recognition in cluttered 3-d scenes. *IEEE Transactions on Pattern Analysis and Machine Intelligence*, 21(5):433 – 449, May 1999.

[13] D.G. Kendall, D. Barden, T.K. Carne, and H. Le. *Shape and Shape Theory*. John Wiley and Sons, 1999.

[14] Walter Kristof and Bary Wingersky. Generalization of the orthogonal procrustes rotation procedure to more than two matrices. In *Proceedings, 79th Annual Convention, APA*, pages 89–90, 1971.

[15] S. M. LaValle and S. A. Hutchinson. An objective-based stochastic framework for manipulation planning. In *Proc. IEEE/RSJ/GI Int'l Conf. on Intelligent Robots and Systems*, pages 1772–1779, September 1994.

[16] Tomás Lozano-Pérez, Matthew Mason, and Russell H. Taylor. Automatic synthesis of fine-motion strategies for robots. *International Journal of Robotics Research*, 3(1), 1984.

[17] Matthew T. Mason and J. Kenneth Salisbury Jr. *Robot Hands and the Mechanics of Manipulation*. MIT Press, Cambridge, Mass., 1985.

[18] F. Mokhtarian and A. K. Mackworth. A theory of multiscale curvature-based shape representation for planar curves. In *IEEE Trans. Pattern Analysis and Machine Intelligence*, volume 14, 1992.

[19] Van-Duc Nguyen. Constructing stable grasps. *I. J. Robotic Res.*, 8(1):26–37, 1989.

[20] N. S. Pollard and Victor B. Zordan. Physically based grasping control from example. In *Proceedings of the ACM SIGGRAPH / Eurographics Symposium on Computer Animation*, Los Angeles, CA, 2005.

[21] Ashutosh Saxena, Justin Driemeyer, Justin Kearns, Chioma Osondu, and Andrew Y. Ng. Learning to grasp novel objects using vision. In *Proc. International Symposium on Experimental Robotics (ISER)*, 2006.

[22] C. Scott and R. Nowak. Robust contour matching via the order preserving assignment problem. *IEEE Transactions on Image Processing*, 15(7):1831–1838, July 2006.

[23] J. Shi and J. Malik. Normalized cuts and image segmentation. *IEEE Trans. Pattern Analysis and Machine Intelligence*, 22(8):888–905, 2000.

Dynamic Modeling of Stick Slip Motion in an Untethered Magnetic Micro-Robot

Chytra Pawashe*, Steven Floyd*, and Metin Sitti

***Abstract*— This work presents the dynamic modeling of an untethered electromagnetically actuated magnetic micro-robot, and compares computer simulations to experimental results. The micro-robot, which is composed of neodymium-iron-boron with dimensions 250 μm x 130 μm x 100 μm, is actuated by a system of 5 macro-scale electromagnets. Periodic magnetic fields are created using two different control methods, which induce stick-slip motion in the micro-robot. The effects of model parameter variations on micro-robot velocity are explored and discussed. Micro-robot stick-slip motion is accurately captured in simulation. Velocity trends of the micro-robot on a silicon surface as a function of magnetic field oscillation frequency and magnetic field strength are also captured. Mismatch between simulation and reality is discussed.**

I. Introduction

The fundamental challenge with decreasing robot size below the centimeter scale is providing power and actuation to the robot. Most current micro-robots rely on external actuation and/or power to function, and usually have further limitations as well. These limitations include restrictions such as requiring tethers [1]–[3], a fluid environment [4]–[7], or a specialized operating surface [8]–[11]. Further, though many miniature robots exist on the centimeter or millimeter scale, true micron-scale robots, with all characteristic lengths on the order of tens to hundreds of microns, are still quite rare [4], [5], [9], [11], [12].

Thc micro-robot presented in this and in earlier work [13] utilizes magnetic torque provided by large-scale electromagnets to achieve its motion and does not require any physical tethers or an on-board power source. Controlled by dynamically adjusting magnetic fields, it does not need a patterned work surface or electrostatic coupling to a sub-surface. Additionally, while the robot does not require a fluid medium to operate, it can move and perform tasks in a fluid environment. Somc limitations of this design are that the working surface cannot be composed of a ferromagnetic or strongly diamagnetic material, nor can the surface itself be magnetized, and the robot must remain within the working volume of the electromagnets.

In this paper, we present a comprehensive dynamic model of the magnetic micro-robot and its interactions with the magnetic field and the silicon surface on which it operates. Such a model is necessary to understand the nature of the robot's motion, and can be used as a tool for future optimization and control. With an accurate model, the number of necessary experiments can be greatly reduced, and time consuming or difficult tests can be performed in computer simulation, allowing for quick parameter optimization.

Micro-robots that do not rely on specialized surfaces for power delivery and control are a vital step toward advancing the field of micro-robotics, which is filled with many potential applications. Examples include micro-manipulation of micro-components, and micro-assembly and fabrication of hybrid micro-systems [8], [10]. Tetherless micro-robots with appropriate tools can be used for micro-scale measurement and surface inspection [1]. More advanced micro-robots could even be used for undetectable surveillance as micro-unmanned vehicles, or as micro-surgeons in medical applications inside living bodies [2].

II. Experimental Setup

Five independent electromagnetic coils were constructed large enough to enclose a cube 10 cm on a side, which contains the working volume. Of these coils, four were placed upright to control the direction and gradient of the horizontal magnetic field, and one is placed below the work plane to control electromagnetic clamping, as seen in Fig. 1. Within the tolerances of machining, the coils were constructed to be identical, with the same dimensions, wire gauge, and number of turns of the wire. Imaging of the the magnetic micro-robot is accomplished with a CCD camera connected to a variable magnification microscope lens. For high-framerate video, a high speed camera (Phantom V7.0) and an additional microscope lens was placed horizontally inside one of the four upright magnets to achieve a side view of the micro-robot during actuation. Parameters for each of the electromagnets are provided in Table I. Control of the electromagnetic coils is performed by a PC with a data acquisition system at a control bandwidth of 1 kHz. The coils are powered by custom-made electronic amplifiers, controlled by the PC.

III. Robot Fabrication

The micro-robot used in these experiments was made of neodymium-iron-boron (NdFeB), a hard magnetic material. To create the robot, a magnetized piece of NdFeB was cut using a laser machining system (NewWave LaserMill). First, the NdFeB was cut parallel to the direction of magnetization, making planar slices approximately 100 μm thick. These slices were then laid flat so that the magnetization of the slice was in a horizontal plane. Robots were then cut from

* Equally contributing co-first authors

C. Pawashe and S. Floyd are with the Department of Mechanical Engineering, Carnegie Mellon University, Pittsburgh, PA 15213, USA [csp, srfloyd]@andrew.cmu.edu

M. Sitti is with the Department of Mechanical Engineering and Robotics Institute, Carnegie Mellon University, Pittsburgh, PA 15213, USA sitti@cmu.edu

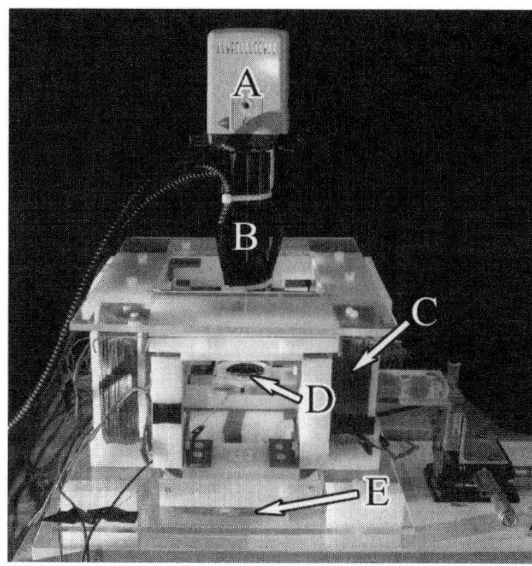

Fig. 1. Photograph of the electromagnetic coil setup, where (A) is the camera, (B) is the microscope lens, (C) is one of four horizontal coils that move the micro-robot within the plane, (D) is the wafer where the micro-robot resides, and (E) is the clamping coil beneath the wafer that holds the micro-robot to the surface.

these slices such that the magnetization vector was pointing towards the front of the robot. High translational speeds, small laser spot size, and low cutting depths per pass were employed to minimize local demagnetization due to heating by the laser.

IV. Modeling

A computer simulation was created to model the behavior of the micro-robot as it interacts with the silicon surface it moves upon, and how it is affected by the magnetic fields created by the five macro-scale electromagnets over time. Being made of hard magnetic material, the magnetic field does not affect the robot's internal magnetization. Furthermore, the small size of the robot implies that the magnetic field it creates does not significantly affect the electromagnets.

A. Magnetic Field

The magnetic field produced is determined by the current, I, through the electromagnetic coil. It is a function of the voltage across the coil, V, which is a control input in the simulation, and the resistance and inductance of the coil. Resistance and inductance were measured experimentally, and their values are presented in Table I. Using these values, the differential equation for the current through an inductor is incorporated in the dynamic simulation:

$$\frac{dI}{dt} = \frac{-R}{L_i} I + \frac{1}{L_i} V \tag{1}$$

This current is used to determine the magnetic field produced by each of the coils.

Inside of the control volume, the principle of superposition is valid for determining the magnetic field at a point in space. Hence, the contributions from all five electromagnets can be determined separately and then added together. To determine the contribution of each electromagnet, one must apply the Biot-Savart law for each square turn coil:

$$\vec{B}(\vec{X}) = \frac{\mu_0 N_t I}{4\pi} \oint_S \frac{\vec{dl'} \times \vec{a}_R}{R^2} \tag{2}$$

where $\vec{B}(\vec{X})$ is the magnetic field at the robot's position $\vec{X} = x\vec{e}_x + y\vec{e}_y + z\vec{e}_z$, μ_0 is the permeability of free space $(4\pi \times 10^{-7})$, $\vec{dl'}$ is an infinitesimal line segment along the direction of integration, $\vec{a}_R$ is the unit vector from the line segment to the point in space of interest, and R is the distance from the line segment to the space of interest.

For a square turn coil, this contour integral simplifies into four line integrals, the definite integral of which exists and can be evaluated at the end points [14]. For the x-directed coils, the primary field in x and the fringe fields in y and z can be determined at any point in space by evaluating the following [15]:

$$R = \sqrt{(x-c)^2 + (y-y_j)^2 + (z-z_i)^2} \tag{3}$$

$$\begin{aligned} B_x = \; & \tfrac{\mu_0 N_t I}{4\pi} \textstyle\sum_{i=1}^{2} \sum_{j=1}^{2} (-1)^{i+j} * \tfrac{(z-z_i)(y-y_j)}{R} \\ & * \left[\tfrac{1}{(x-c)^2+(y-y_j)^2} + \tfrac{1}{(x-c)^2+(z-z_i)^2} \right] \end{aligned} \tag{4}$$

$$\begin{aligned} B_y = \; & \tfrac{\mu_0 N_t I}{4\pi} \textstyle\sum_{i=1}^{2} \sum_{j=1}^{2} (-1)^{i+j+1} \\ & * \left(\tfrac{(z-z_i)(x-c)}{R} * \left[\tfrac{1}{(x-c)^2+(y-y_j)^2} \right] \right) \end{aligned} \tag{5}$$

$$\begin{aligned} B_z = \; & \tfrac{\mu_0 N_t I}{4\pi} \textstyle\sum_{i=1}^{2} \sum_{j=1}^{2} (-1)^{i+j+1} \\ & * \left(\tfrac{(x-c)(y-y_j)}{R} * \left[\tfrac{1}{(x-c)^2+(z-z_i)^2} \right] \right) \end{aligned} \tag{6}$$

$$z = [a, -a], y = [a, -a] \tag{7}$$

where B_i is the magnetic field in the i direction, a is half the effective length of the electromagnet, and c is $\pm$ the distance from the center, where the $\pm$ is evaluated based upon the location of the coil of interest (i.e. plus for the coil in the positive x or positive y-directions). Similar equations are used for the y-directed and clamping coils.

This derivation is for a concentrated electromagnet, i.e. all the current carrying wires can be described by a single line with zero thickness. This assumption is found to be accurate to within 2.2% because the travel distances are small in comparison to the magnetic coils, and the distribution of wires within each electromagnet is small compared to the

Description	Value	Units
Coil Resistance (R)	10.0	Ω
Inductance (L_i)	70	mH
Inner length	0.120	m
Outer length	0.157	m
Number of Turns (N_t)	510	–
Effective Length	0.1385	m
Distance From Center	.099	m
Maximum Field at Center	6.5	mT
Maximum Gradient at Center	149	mT/m

TABLE I
Electromagnet Properties

size of the magnet. As a result, an effective length must be used instead of the actual inner or outer length; these values are listed in Table I.

B. Magnetic Forces

Within a magnetic field any magnetized object, in this case the micro-robot, will experience both a torque and a force. This magnetic torque is proportional to the magnetic field strength, and acts in a direction to bring the internal magnetization of the object into alignment with the field. The magnetic force is proportional to the gradient of the magnetic field, and acts to move the object to a local maximum. The equations that govern these interactions are:

$$\vec{T_m} = V_m \vec{M} \times \vec{B}(\vec{X}) \tag{8}$$

$$\vec{F_m} = V_m (\vec{M} \bullet \nabla) \vec{B}(\vec{X}) \tag{9}$$

where $\vec{T_m}$ is the torque the robot experiences, V_m is the volume of the robot, $\vec{M}$ is the magnetization of the robot (assumed to be uniform) [7], [16].

The gradient is determined analytically by taking the derivatives of (4-6) in each direction, yielding nine terms. Torque is determined by finding the magnetic field at the center of the robot's body. It was found that for a cube 2.5 cm on a side in the center of the working volume, fringing fields were less than 2.5% of the primary field for each coil individually, and the maximum force and torque that can be exerted on the micro-robot are 240 nN and 10.5 $\mu N \cdot mm$, respectively.

C. Magnetic Micro-Robot

In modeling the micro-robot, we assume it is an isotropic rectilinear solid with the properties tabulated in Table II.

Description	Value	Units
Length (L)	250	μm
Width (W)	130	μm
Height (H)	100	μm
Density (ρ)	7400	kg/m^3
Mass (m)	25.6	ng
Weight (mg)	251	nN
Magnetization (M)	5×10^5	A/m

TABLE II

MAGNETIC ROBOT PROPERTIES

From experimental high-speed video of the micro-robot, it appears to exhibit a stick-slip motion when actuated by a pulsed magnetic field. During motion, the robot moves by rocking forward and backward around a steady state angle. Motion occurs when the contact point between the robot and the silicon surface slips. This occurs either when the robot "falls" while rocking down, or "jumps" while rocking up. Several images from a high speed video at 200 frames per second (fps) are presented in Fig. 2(a-d) to illustrate this motion. A video can be found online at [17]. We attempt to model this behavior in a computer simulation.

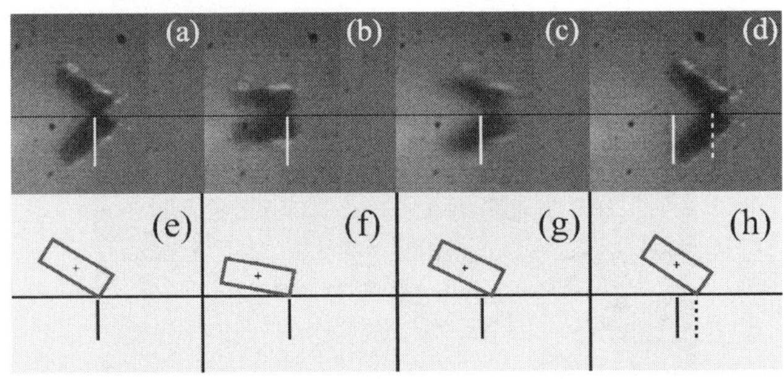

Fig. 2. Stick-slip motion of the micro-robot observed with a high speed camera, compared to simulated results. (a) The robot is initially at its steady state angle, and the point of contact is highlighted by a solid white line. (b) When the magnetic field changes, the robot rocks downward, changing its angle relative to the silicon wafer. (c) During the next upswing, the robot slides forward. The former contact point is highlighted with a solid white line. (d) After slipping forward, the robot assumes its steady state angle again at a new position, shown by a dashed white line. Analogous steps are performed in simulation in images (e) through (h).

To simulate the dynamics of the magnetic micro-robot, we restrict modeling to a side-view of the robot in the x-z plane, shown in Fig. 3. The robot has a center of mass (COM) at $\vec{X}$, an orientation angle θ from the ground, a distance r from its COM to a corner, and an angle ϕ determined from geometry. The robot experiences external forces, including its weight, mg, a normal force from the surface, N, an adhesive force to the surface, F_{adh}, an x-directed externally applied magnetic force, F_x, a z-directed externally applied magnetic force, F_z, a linear damping force in the x-direction, L_x, a linear damping force in the z-direction, L_z, an externally applied magnetic torque, T_y, a rotational damping torque, D_y, and a Coulomb sliding friction force F_f. F_f depends on N, the sliding friction coefficient μ, and the velocity of the contact point, $\frac{dP_x}{dt}$, where (P_x, P_z) is the bottom-most point on the micro-robot (nominally in contact with the surface). Using these forces, we develop the dynamic relations:

$$m\ddot{x} = F_x - F_f - L_x \tag{10}$$

$$m\ddot{z} = F_z - mg + N - F_{adh} - L_z \tag{11}$$

$$\begin{aligned} J\ddot{\theta} &= T_y + (F_f) r sin(\theta + \phi) \\ &\quad -(N - F_{adh}) r cos(\theta + \phi) - D_y \end{aligned} \tag{12}$$

where J is the polar moment of inertia of the robot, calculated as $J = m(H^2 + L^2)/12$.

The robot is first assumed pinned to the surface at (P_x, P_y), where $0 < \theta < \frac{\pi}{2}$. This gives the following additional equations:

$$x = P_x - r cos(\theta + \phi)$$

$$\ddot{x} = \ddot{P}_x + r\ddot{\theta} sin(\theta + \phi) - r\dot{\theta}^2 cos(\theta + \phi) \tag{13}$$

$$z = P_z + r sin(\theta + \phi)$$

$$\ddot{z} = \ddot{P}_z + r\ddot{\theta} cos(\theta + \phi) - r\dot{\theta}^2 sin(\theta + \phi) \tag{14}$$

To solve equations (10-14), we realize that there are 7 unknown quantities (N, $\ddot{\theta}$, $\ddot{x}$, $\ddot{z}$, $\ddot{P}_x$, $\ddot{P}_z$, F_f) for 5 equations, indicating an under-defined system. As the stick-slip motion in this system is similar to the case outlined by Painlevé's paradox, we resolve the paradox by taking the friction force,

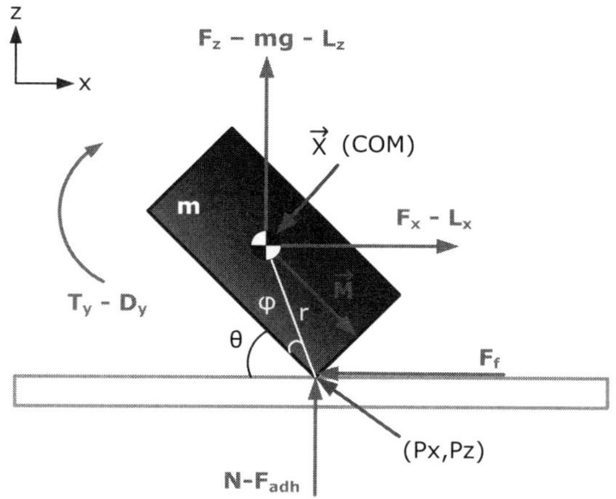

Fig. 3. Schematic of a rectangular magnetic micro-robot with applied external forces and torques. Magnetization vector denoted by $\overrightarrow{M}$.

F_f, as an unknown value (as opposed to setting $F_f = \mu N$) [18]. Using the pinned assumption, we can set $\ddot{P}_x = \ddot{P}_z = 0$; then, equations (10-14) are solved directly. There are three possible types of solution that can occur during each time step:

1. The solution results in $N < 0$ (an impossible case). This implies that the pinned assumption was false, and the micro-robot has broken contact with the surface. Equations (10-12) are resolved using $N = 0$ and $F_f = 0$.

2. The solution results in $F_f > F_{fmax}$, where $F_{fmax} = \mu \cdot N$. This also implies that the pinned assumption was false, and the point of contact is slipping; thus the robot is translating in addition to rocking. Equations (10-14) are resolved using $F_f = F_{fmax}$ and $\ddot{P}_x$ left as an unknown.

3. All of the variables being solved for are within physically reasonable bounds. The robot is in contact with the surface at the pinned location and is rocking in place.

When a satisfactory solution is reached for each time step, the solutions for orientation and location are used as initial conditions in the next solution step.

V. Simulation Results

To simulate the micro-robot, a 5th ordered Runge-Kutta solver is used to solve the time-dependent system. A magnetic pulsing signal is given as a voltage waveform, and equation (1) is solved for the currents. With given initial conditions, equations (1-9) are used to determine the magnetic field forces, and equations (10-14) are solved for the three position states of the micro-robot: x, z, and θ. The results of the simulation are displayed in Figure 2(e-h), where the simulated micro-robot exhibits stick-slip motion, agreeing with experiment.

From previous work [13], two different types of magnetic actuation can be used to move the magnetic micro-robot in a reliable fashion. The first is In Plane Pulsing (IPP), where the magnetic field within the plane of motion is varied, while the clamping magnetic field is held constant. The second method is Out of Plane Pulsing (OPP), where the magnetic field within the plane of motion is held constant, but the clamping magnetic field is varied. Since a DC magnetic field will not translate a micro-robot due to high static friction, these periodic excitation methods are necessary to induce stick-slip translation.

A. In Plane Pulsing

For IPP control, the voltages across the clamping electromagnet and one in-plane electromagnet (in the direction of robot motion) are slowly ramped up. This causes the robot to orient in the desired direction while remaining locked down on the silicon surface. Next, the in-plane electromagnet is pulsed using a sawtooth waveform at a higher voltage, varying the horizontal, or x-directed magnetic field. The magnetic fields generated in the simulation from IPP control are shown in Fig. 4.

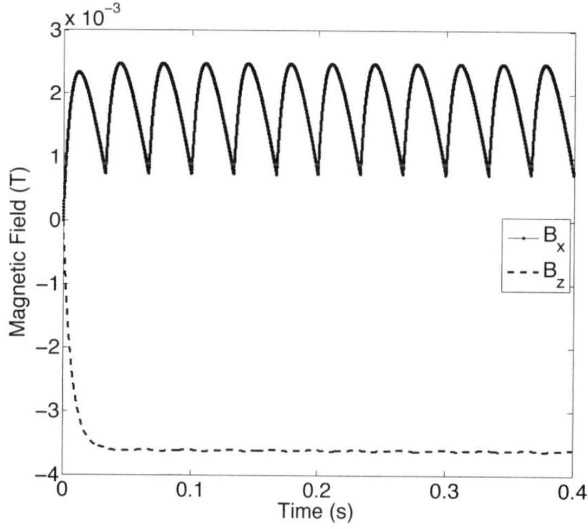

Fig. 4. Simulation of the x and z-directed magnetic fields generated by IPP control at a maximum pulsing magnetic field strength of 2.5 mT with a pulsing frequency of 30 Hz in simulation.

This alternating in-plane magnetic field causes the robot to rock downward (θ decreases) as the field is increased, and rock upward when the field is decreased. The robot translates by slipping each time it rocks downward, which agrees with experimentally observed behavior. After reaching its target destination, the voltage across the in-plane electromagnet is ramped down, leaving only the clamping electromagnet active. A simulation of this motion is presented in Fig. 5.

B. Out of Plane Pulsing

In OPP control, the voltages across the clamping electromagnet and one in-plane electromagnet (in the direction of robot motion) are slowly ramped up, like in IPP control. After orienting the robot, however, the voltage across the clamping electromagnet is varied using a sawtooth waveform, while the horizontal magnetic field is held constant. The magnetic fields generated in the simulation from OPP control are shown in Fig. 6.

By alternating the out-of-plane magnetic field, the robot tends to rock upward (θ increases) as the clamping field is increased, and rock downward when the field is decreased;

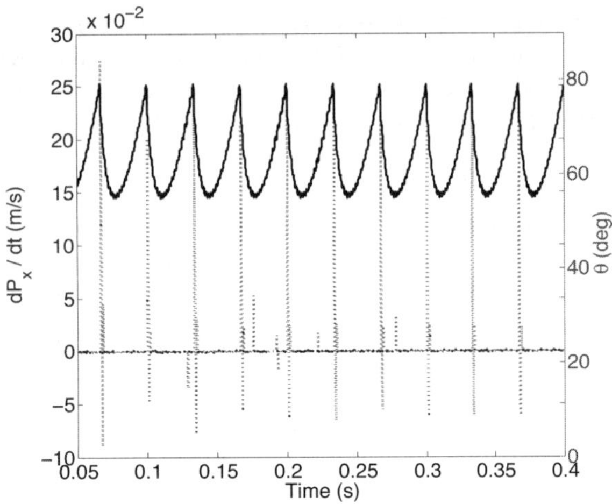

Fig. 5. Simulation of the robot's angle θ (solid line) and the x-direction velocity of the contact point, $\frac{dP_x}{dt}$ (dotted line) in a simulation of IPP control. The robot slides over the surface during the downstroke, when θ is decreasing.

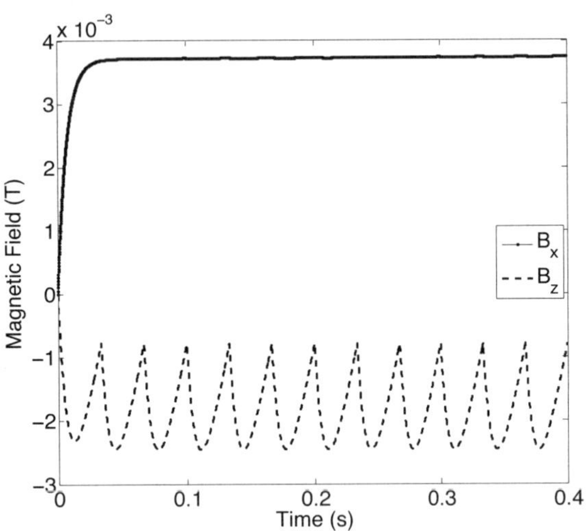

Fig. 6. Simulation of the x and z-directed magnetic fields generated by OPP control at a maximum pulsing magnetic field strength of 2.5 mT with a pulsing frequency of 30 Hz in simulation.

this is an opposite effect when compared to IPP control. In this case, the robot translates by slipping each time it rocks upward, which agrees with experimentally observed behavior, and is explicitly shown in Fig. 2. After reaching its target destination, the voltage across the in-plane electromagnet is ramped down, leaving only the clamping electromagnet active. A simulation of this motion is presented in Fig. 7.

VI. Matching Model Parameters

In order to accurately model the physical system, several parameters had to be determined empirically. As with any empirical determination, there exists the possibility of error.

A. Friction

The Coulomb sliding friction coefficient between the robot and the surface, μ, was determined by using a load-cell to

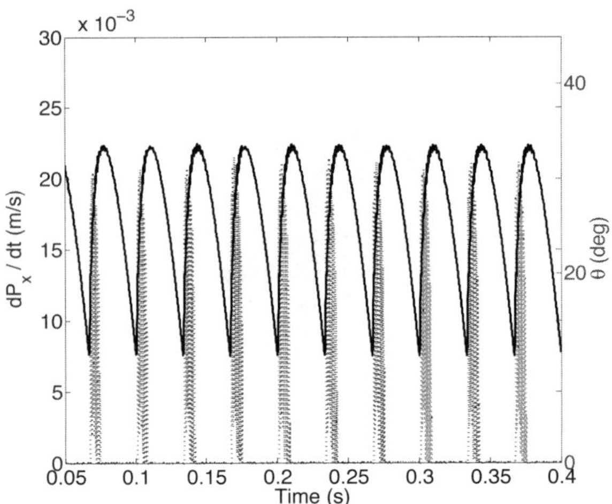

Fig. 7. Simulation of the robot's angle θ (solid line) and the x-direction velocity of the contact point, $\frac{dP_x}{dt}$ (dotted line) in a simulation of OPP control. The robot slides over the surface during the upstroke, when θ is increasing.

measure the force required to slide a bulk NdFeB micro-magnet with a known downward load across a Si surface. From this, the friction coefficients between NdFeB and Si were found to be $\mu_k = 0.145$ for kinetic, and $\mu_s = 0.2$ for static. If $|\frac{dP_x}{dt}| > 0$, $\mu = \mu_k$, otherwise $\mu = \mu_s$.

Noting that the kinetic friction coefficient is approximately 70% the static friction coefficient, the effect of varying μ was explored in simulation while maintaining this ratio, as shown in Fig. 8. At low friction coefficients, the micro-robot dominantly slides; motion due to stick-slip behavior is less dominant as stick-slip motion requires friction. Speeds are higher for OPP control, because the in-plane coil is always on. As the friction coefficient increases, stick-slip behavior becomes more dominant and results in very similar velocity profiles for the two control methods between $\mu_s = 0.2$ and $\mu_s = 1.0$. At friction coefficients $\mu_s > 1.0$, IPP results in higher velocities. This is due to the reduced normal force which arises during IPP downstroke motion, shown in Fig. 5, as opposed to the increased normal force which arises in upstroke motion for OPP, shown in Fig. 7.

B. Adhesion

Adhesive forces between the surface and the micro-robot are present due to stiction effects that become significant at the micro-scale, such as capillary and van der Waals forces [19]; these are lumped together into one force F_{adh}. We estimate this force to be $F_{adh} = 0.22\ \mu N$, or about 90% of the micro-robot's weight. To determine this, a micro-robot was placed on a silicon wafer, and the wafer was slowly rotated. Using the previously determined friction coefficient, μ, a measured angle at which the robot begins to slide on the wafer, α, and a simple free body diagram, the magnitude of the adhesion force can be derived:

$$F_{adh} = mg\left(\frac{sin(\alpha)}{\mu} - cos(\alpha)\right) \tag{15}$$

From Fig. 9, OPP velocities are higher at low adhesion because the in-plane coil is always on, allowing sliding

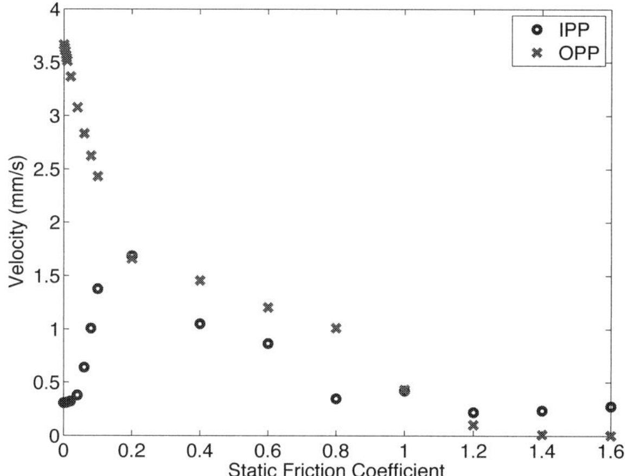

Fig. 8. Simulated effect of changing the static and kinetic friction coefficients on the steady state velocity of the robot. The kinetic friction coefficient is taken to be 70% of the static. Pulsing frequency is 100 Hz, with a 2.5 mT maximum pulsing magnetic field.

motion. As adhesion increases, OPP motion is influenced more than IPP for reasons similar to increasing friction: OPP translates the robot during upswing, when normal forces (which increase with adhesion) and friction forces are at their highest, whereas IPP translates the robot during the downswing, when the importance of these forces is minimized.

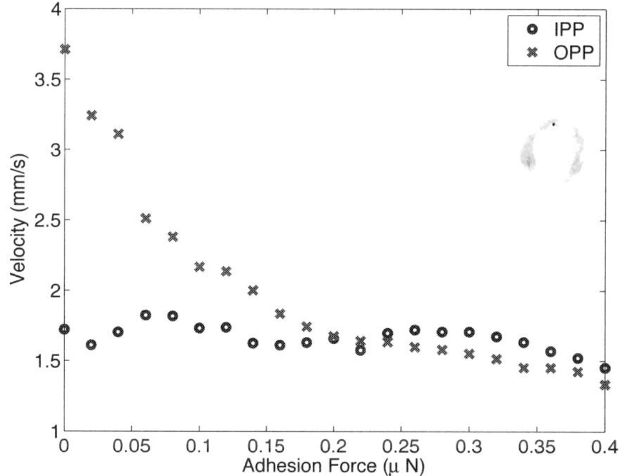

Fig. 9. Simulated effect of changing the adhesive force, F_{adh}, on the steady state velocity of the robot. Pulsing frequency is 100 Hz, with a 2.5 mT maximum pulsing magnetic field.

C. Damping

Linear damping forces, L_x, L_z, and rotational damping torque D_y, are a result of fluid drag effects through a liquid layer that is assumed to form between the robot and the silicon surface. For modeling, linear damping forces are first estimated using a Couette flow fluid drag model:

$$L_{x,z} \approx \frac{A_b \mu_d}{g_0} \times \{\dot{x}, \dot{z}\} \; [N] \tag{16}$$

where A_b is the area of the bottom of the robot, μ_d is the dynamic viscosity of the liquid layer, and g_0 is the separation from the surface, assumed to be 1.5 μm from measurements of the surface roughness. This results in a linear drag $L_{x,z} \approx 2.4 \times 10^{-5} \times \{\dot{x}, \dot{z}\} \; [N]$. In a similar fashion, torque damping forces are estimated using the integral of a viscous drag force equation:

$$D_y = C_D \frac{1}{2} \rho_w \int_0^{L/2} W s^3 ds \times \dot{\theta}^2 \; [N \cdot m] \tag{17}$$

where $C_D \approx \frac{19}{Re}$ is the constant drag coefficient at $Re \approx \frac{L\dot{x}}{\nu} = 0.22$ [20], ν is the kinematic viscosity of water, ρ_w is the density of water, and s is a variable of integration. This results in a rotational damping torque $D_y \approx 6.7 \times 10^{-16} \times \dot{\theta}^2 \; [N \cdot m]$.

Both damping coefficients are later adjusted to match the simulation to experimental results. In reality, the damping forces will depend on the micro-robot's orientation angle, θ. For purposes of simulation, estimated average damping coefficients are used to capture the overall behavior trends; in the future, angular-dependent drag coefficients may be used in simulation. The estimated constant damping forces used are:

$$L_{x,z} = 1.0 \times 10^{-5} \times \{\dot{x}, \dot{z}\} \; [N] \tag{18}$$

$$D_y = 9.0 \times 10^{-17} \times \dot{\theta}^2 \; [N \cdot m] \tag{19}$$

These damping terms are necessary to keep the simulated robot stable in both rotational and translational motion. With increasing linear damping, average micro-robot velocity decreases for both IPP and OPP control. Changes in the linear damping coefficient cause different behavior regimes to emerge. At very low values of linear damping, the velocity is controlled almost exclusives by friction forces. At higher values of linear damping, all motion is suppressed. Only in a very small range is micro-robot velocity controlled by linear damping. For lower values of rotational damping, there is little effect on micro-robot velocity. This is likely the case when rotational drag is much smaller than magnetic torque. As rotational damping increases, both OPP and IPP velocities decrease at about the same rate, supporting this theory.

VII. Results and Discussion

All testing was performed on the back side of a silicon wafer in open air. No special polishing or preparation was performed on the wafer. For both control methods, two different parameters were examined: (1) Maintaining a constant waveform pulsing frequency while varying the maximum voltage across the coils (as a result, varying the maximum magnetic field), and (2) maintaining a maximum voltage across the coils while varying the frequency of waveform pulsing.

For each experiment at each pulsing frequency, three trials were performed to attain an error estimate of the velocity. During the experiment, a video of the robot motion was recorded and post-processed. Two frames of the video, one near the beginning and one near the end of the robot's journey, were taken. In each position, the robot's central position was determined, and the total travel distance was

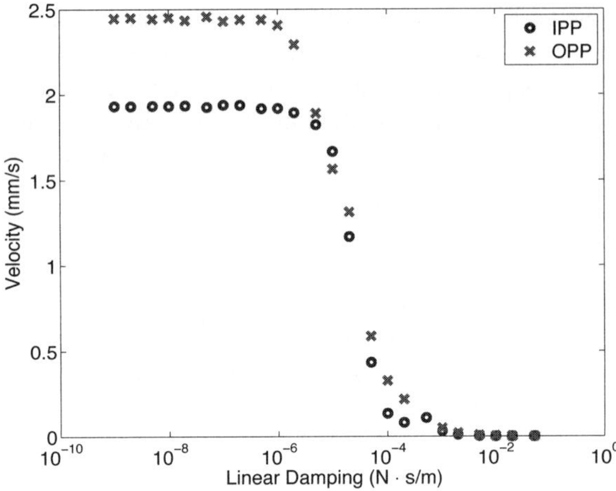

Fig. 10. Simulated effect of changing the linear damping on the steady state velocity of the robot. Pulsing frequency is 100 Hz, with a 2.5 mT maximum pulsing magnetic field.

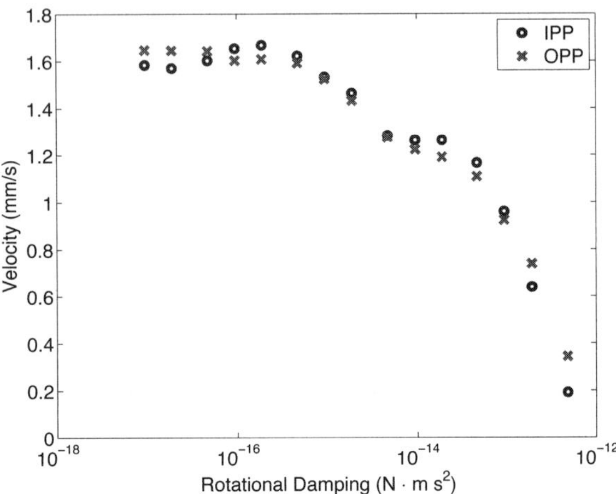

Fig. 11. Simulated effect of changing the rotational damping coefficient on the steady state velocity of the robot. Pulsing frequency is 100 Hz, with a 2.5 mT maximum pulsing magnetic field.

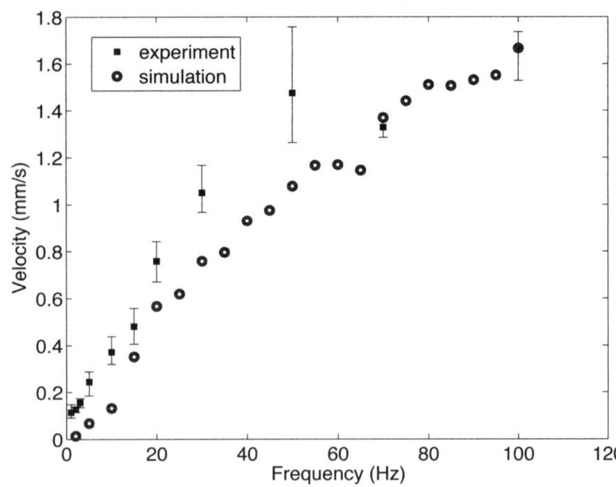

Fig. 12. Simulated and experimental robot velocity at varying frequencies, at a maximum pulsing magnetic field strength of 2.5 mT under IPP control.

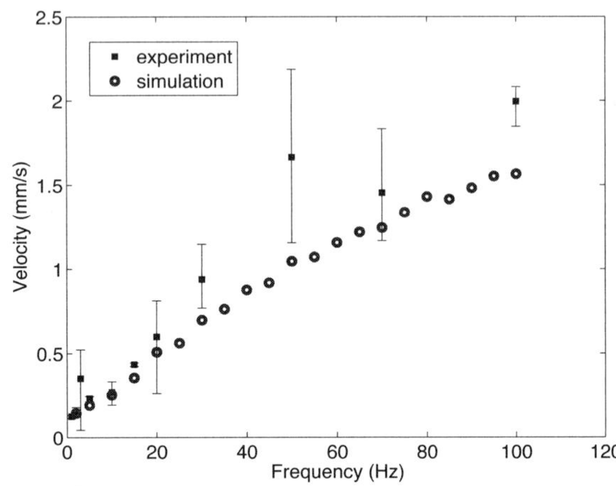

Fig. 13. Simulated and experimental robot velocity at varying frequencies, at a maximum pulsing magnetic field strength of 2.5 mT under OPP control.

measured in pixels. A conversion ratio from the image to real-world distances in microns/pixel was empirically determined by counting the pixels across a known length. The total time for travel was also recorded to determine the velocity. Across a travel distance of about 5 mm with a positioning error of 1-2 pixels (about 50 μm), results in a 1% error in measured distance.

For simulations, the average velocity was determined in a similar manner. The x-position and time was reported shortly after steady state motion was reached (after the third cyclic pulse), and also at a determined time signifying the end of the simulation. These two values were used to determine the average velocity of the simulated micro-robot.

Micro-robot velocity as a function of frequency for both IPP and OPP control appears to be linear for low frequencies in both experiment and simulation, as seen in Figs. 12 and 13. At higher frequencies, velocities appear to exhibit a slight roll-off in experiment. This roll-off may be due to the linear and rotational damping effects experienced by the micro-robot.

Simulated results seem to underestimate micro-robot velocity for both IPP and OPP translation modes as a function of maximum field strength, shown in Figs. 14 and 15. For both IPP and OPP, the simulation suggests a linearly increasing dependence of velocity on field strength with some roll off for higher fields in IPP. This dependence is apparent in the IPP case, but is not as clear for OPP, which may be linearly increasing or may be relatively constant.

VIII. Conclusion

A detailed computer simulation that modeled the dynamics of a magnetically controlled micro-robot on a flat surface is proposed in this study. The parameters of the simulations were adjusted until an approximate match with reality was achieved. Both in simulation and in experiment, the magnetic micro-robot was subjected to alternating magnetic fields, which induced a stick-slip motion over the surface. The velocity of this motion as a function of both excitation

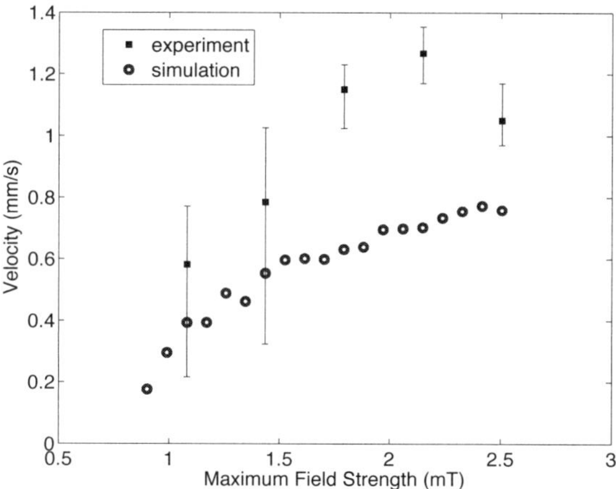

Fig. 14. Experimental robot velocity at varying maximum coil voltages (displayed as maximum field strength) at a constant pulse frequency of 30 Hz for IPP control.

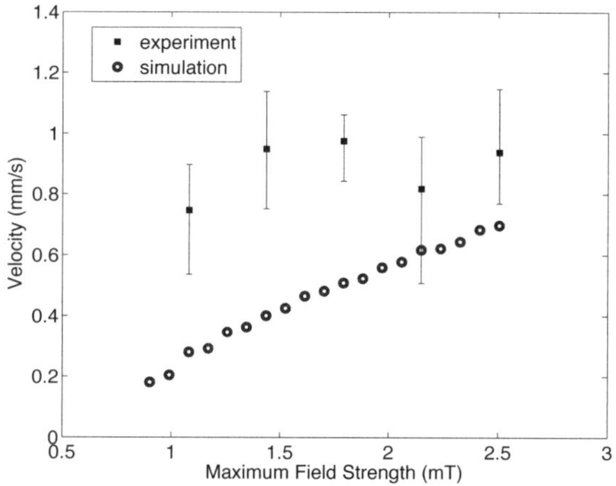

Fig. 15. Experimental robot velocity at varying maximum coil voltages (displayed as maximum field strength) at a constant pulse frequency of 30 Hz for OPP control.

frequency and maximum magnetic field for two different control methods were recorded and analyzed.

It was found that in both control methods, IPP and OPP control, the robot first attains a steady state angle with respect to the surface. This angle is much larger for IPP than for OPP. When one of the magnetic fields begins to oscillate, the robot will rock upward and downward. For IPP control, the robot will slide over the surface each time it rocks downward. Alternatively, in OPP control, the robot will slide each time it rocks upward. These behaviors were observed experimentally and accurately reproduced in simulation.

The dependence of robot velocity on friction, adhesion, linear damping, and rotational damping was explored in simulation. In addition, both frequency and peak voltage of OPP and IPP control were varied in simulation and experiment to determine their effects on robot velocity. The results obtained in simulation show general agreement with experiments.

Future work will include adapting the system for both coarse and fine motion control by using the simulation to refine the control signals and determine appropriate control laws in each case. In addition, vision algorithms are being developed for closed-loop computer control of the micro-robot. Possible applications in micro-object manipulation, underwater control, and cooperative robotics are currently being explored.

ACKNOWLEDGMENT

The authors would like to thank L. Weiss and the Pittsburgh Infrastructure Technology Alliance for the use of the high-speed camera, and the NanoRobotics Laboratory members for all of their support and suggestions.

REFERENCES

[1] T. Ebefors et. al., "A Walking Silicon Micro-Robot," The 10th Int Conference on Solid-State Sensors and Actuators (Transducers99), Sendai, Japan, pp 1202-1205. 1999.

[2] P. E. Kladitis, V. M. Bright, "Prototype Microrobots for Micro-positioning and Micro-unmanned Vehicles," Sensors and Actuators, Vol. 80, No. 2, pp. 132-137, 2000.

[3] M. Mohebbi et. al., "Omnidirectional Walking Microrobot Realized by Thermal Microactuator Arrays," Proc. ASME Int. Mechanical Engineering Congress and Exposition, pp. 1-7, 2001.

[4] S. Martel, C. Tremblay, S. Ngakeng, and G. Langlois, "Controlled manipulation and actuation of micro-objects with magnetotactic bacteria," Applied Physics Letters, Vol. 89, No. 233904, 2006.

[5] A. Yamazaki et. al., "Wireless micro swimming machine with magnetic thin film," Journal of Magnetism and Magnetic Materials, Vol. 272276, pp. e1741-e1742, 2004.

[6] R. Dreyfus et. al., "Microscopic artificial swimmers," Nature, Vol. 437, pp. 862-865, 2005.

[7] K. Yesin, K. Vollmers, and B. Nelson, "Modeling and Control of Untethered Biomicrorobots in a Fluidic Environment Using Electromagnetic Fields," International Journal of Robotics Research, Vol. 25, pp. 527-536, 2006.

[8] S. Martel et. al., "Three-Legged Wireless Miniature Robots for Mass-Scale Operations at the Sub-Atomic Scale," Proceedings of the 2001 IEEE International Conference on Robotics and Automation, Seoul, Korea, pp. 3423- 3428, 2001.

[9] B. Donald et. al., "An Untethered, Electrostatic, Globally Controllable MEMS Micro-Robot," Journal of Microelectromechanical Systems, Vol. 15, No. 1, pp. 1-15, 2006.

[10] S. Martel, "Special surface for power delivery to wireless micro-electro-mechanical systems," J. Micromech Microeng., 15, pp. S251-258, 2005.

[11] K. Vollmers et. al., "Wireless Resonant Magnetic Microactuator for Untethered Mobile Microrobots," Applied Physics Letters, Vol. 92, No. 144103, 2008.

[12] O. Sul et. al., "Thermally Actuated Untethered Impact-driven Locomotive Microdevices," Applied Physics Letters, Vol. 89, No. 203512, 2006.

[13] S. Floyd, C. Pawashe, and M. Sitti, "An Untethered Magnetically Actuated Micro-Robot Capable of Motion on Arbitrary Surfaces," Proceedings of the 2008 IEEE International Conference on Robotics and Automation, Pasadena, CA, *In Press*.

[14] D. K. Cheng, "Field and Wave Electromagnetics, 2nd Edition," Addison-Wesley Publishing Company, Inc., 1992.

[15] W. Frix, G. Karady, and B. Venetz, "Comparison of Calibration Systems for Magnetic Field Measurement Equipment," IEEE Transactions on Power Delivery, Vol. 9, pp. 100-108, 1994.

[16] T. Boyer, "The Force on a Magnetic Dipole," Am. J. Phys. Vol. 56, Issue 8, pp. 688-692, 1987.

[17] Available Online: http://nanolab.me.cmu.edu/

[18] D. Stewart, "Finite-Dimensional Contact Mechanics," Philosophical Transactions: Mathematical, Physical, and Engineering Sciences, 359 (1789), Non-Smooth Mechanics, pp. 2467-2482, 2001.

[19] J. Israelachvili, "Intermolecular and Surface Forces," Academic Press, London, 1992.

[20] B. Munson, D. Young, and T. Okiishi, "Fundamentals of Fluid Mechanics, 4th Edition," John Wiley and Sons, Inc., 2002.

Super-Flexible Skin Sensors Embedded on the Whole Body, Self-Organizing Based on Haptic Interactions

Tomoyuki Noda*, Takahiro Miyashita†, Hiroshi Ishiguro*†‡, and Norihiro Hagita†
* AMS, Graduate school of Engineering, Osaka Univ., 2–1 Yamada–oka, Suita, Osaka, 565–0871, Japan.
Email: noda@ed.ams.eng.osaka-u.ac.jp
†ATR Intelligent Robotics and Communication Laboratories, Kyoto, 619–0288, Japan
‡Asada Synergistic Intelligence Project, ERATO, JST, Osaka Univ. Osaka, 565–0871, Japan

Abstract— **As robots become more ubiquitous in our daily lives, humans and robots are working in ever-closer physical proximity to each other. These close physical distances change the nature of human robot interaction considerably. First, it becomes more important to consider safety, in case robots accidentally touch (or hit) the humans. Second, touch (or *haptic*) feedback from humans can be a useful additional channel for communication, and is a particularly natural one for humans to utilize. Covering the whole robot body with malleable tactile sensors can help to address the safety issues while providing a new communication interface. First, soft, compliant surfaces are less dangerous in the event of accidental human contact. Second, flexible sensors are capable of distinguishing many different types of touch (e.g., hard v.s. gentle stroking). Since soft skin on a robot tends to invite humans to engage in even more touch interactions, it is doubly important that the robot can process haptic feedback from humans. In this paper, we discuss attempts to solve some of the difficult new technical and information processing challenges presented by flexible touch sensitive skin. Our approach is based on a method for sensors to self-organize into sensor banks for classification of touch interactions. This is useful for distributed processing and helps to reduce the maintenance problems of manually configuring large numbers of sensors. We found that using sparse sensor banks containing as little as 15% of the full sensor set it is possible to classify interaction scenarios with accuracy up to 80% in a 15-way forced choice task. Visualization of the learned subspaces shows that, for many categories of touch interactions, the learned sensor banks are composed mainly of physically local sensor groups. These results are promising and suggest that our proposed method can be effectively used for automatic analysis of touch behaviors in more complex tasks.**

I. Introduction

Robots are becoming more ubiquitous in our daily lives [1][2][3][4], and humans and robots are working in ever-closer physical proximity to each other. Due to this proximity, there is increased potential for robots to inadvertently harm users. Physical nearness also increases the need for robots to be able to interpret the meaning of touch (or *haptic*) feedback from humans. Covering the whole robot body with malleable tactile sensors allows us to address both of these concerns. First, soft, compliant surfaces are less dangerous in the event of accidental human contact[5]. Second, flexible sensors are capable of distinguishing many different types of touch (e.g., hard v.s. gentle stroking). This is important, as soft skin actually invites more natural types of touch interaction from humans, so it is critical that the soft surfaces of robots be touch sensitive.

To extract information about humans' physical contact with robots, the distribution density of tactile sensor elements, sampling rate, and resolution of kinesthetic sense all must be high [3], resulting in a high volume of tactile information that must be processed. To do so, the following three problems must be solved. First, there is the problem of reduced system robustness due to an increased number of possible failing components. The second is the high cost of data processing. The third is the administration of the sensors' configuration.

The previous study of [6] proposed highly dense distributed skin sensor processing based on interconnecting a self-organized sensor network. Spatiotemporal calculation in each node with spatially seamless tactile information gathered from adjacent nodes enabled haptic interaction features to be extracted, solving the first and second challenges. For instance, an edge detection method is applied to extract features of haptic interaction within the local sensor, yielding an efficient data compression. This type of distributed processing requires that the configuration of tactile sensor position is described in the distributed programs of network nodes, which is the remaining third challenge. Manually describing 3-dimensional tactile sensor positions, changing with robot's postures, is very labor intensive and error prone. Moreover, distributed processing typically requires predefined sensor banks, defining which tactile sensors are used in distributed processing for each network node, also a labor intensive task.

In [7], we found that interaction scenarios could be successfully classified simple k-nearest neighbors (KNN) using a novel feature space based on cross-correlation between tactile sensors, achieving performance of 60% in a 13-way forced choice task. We also found that many categories of touch interactions can be easily visualized by arranging sensors into a "Somatosensory Map" using MultiDimensional Scaling (MDS)[8] applied to this feature space as a similarity measure. These promising results suggest that this feature space can be effectively used for automatic analysis of touch behaviors in more complex tasks.

In this paper, we propose a method for learning "self-organizing tactile sensors" using the feature space from [7] to solve the remaining third challenge. In the proposed method, a classifier is constructed using CLAss-Featuring Information Compression (CLAFIC) [9], a type of a subspace method, applied to a data set consisting of the full cross-correlation based feature space of [7]. Instead of directly using a learned subspace as the input for a classifier, we select the sensor pairs that are most "useful", i.e., have the highest relevance for the classifier output, to form a more compact sensor bank to be used as input to the classifier. Since now different sensor nodes are involved in different classifiers, it is possible to distribute processing around the body, which can be implemented as in-network processing on the self-organizing sensor network [6]. We call each learned sensor bank a "self-organizing tactile sensor". We found that classifiers based on self-organizing tactile sensors could classify interaction scenarios with an accuracy of up to 80% in a 15-way forced choice task, a significant improvement over prior work. The learned subspaces can also be visualized in a "Somatosensory Map", showing sensor point distribution in a 2D plane.

The rest of this paper is organized as follows. In Section we II describe related work, and contrast our study with others. Section III describes the basic idea, and then details the proposed method for self-organizing tactile sensors. Section IV describes experiments dealing with human-robot haptic interaction used to construct a haptic interaction database. In section V, using the database, the performance of the proposed method is shown. Also, the learned subspaces are visualized in the Somatosensory 2D Map. Section VI discusses the results, and Section VII concludes.

II. Background and Comparisons

Prior work on studies of robots with tactile sensors has tended to focus on the development of the physical sensors and transmitting sensor data. For instance, [10] proposes a Large-Scale Integration (LSI) technique for processing data from tactile sensors. Iwata et al.[4] demonstrated physical interaction with users via a skin equipped with 6-axis-kinesthetic sensors. Pan et al. [11] and Inaba et al.[2] described tactile sensors using electrically conductive fabric and strings as a whole-body distributed tactile sensor for humanoid robots. And Shinoda et al.[12] proposed a wireless system for transmitting tactile information by burying wireless sensors under the robot "skin." Thus so far, this research is mainly limited to the problems of collecting tactile information, and solving the necessary wiring and physical implementation problems.

Compared to other sensory modalities such as vision and audio, relatively little prior work has been done on processing haptic interaction from incoming tactile sensor signals. Miyashita et al. [1] estimated user position and posture in interaction using whole-body distributed tactile sensors. Naya et al. [3] classified haptic user interaction based on output from tactile sensors covering a robot pet. Francois et al. [13] also classify different two touch styles, namely "strong" and "gentle". Though the above research classifies human-robot interaction using tactile sensors, these and other prior studies have not to our knowledge been successful in classifying several haptic interactions while robots are interacting with users.

Pierce and Kuipers [14] proposed self-organizing techniques for building a "cognitive map", which represents knowledge of the body corresponding to physical position of sensors. This map shows the position of each sensor installed on the surface of a robot. However, this method will not work out for a robot having high degree of freedom and soft skin because the positions of the tactile sensors in 3-d "world" coordinates dynamically change during an interaction. Rather than construct spatial maps to acquire physical sensor positions our objective is to use, interpret and visualize underlying haptic interaction features.

Kuniyoshi et al. [15] proposed a method for learning a "Somatosensory Map," showing the topographic relationship of correlations between incoming signals from tactile sensors distributed on the whole body surface of a simulated baby. In their somatosensory map highly correlated sensor points are plotted on a 2D plane close to each other. As the result, the map showed the structure of robot body parts rather than the physical sensor positions as in Pierce and Kuipers[14].

In this paper, our goal is similar to that of [15], so that highly correlated sensor points will be located close to each other, and thus we keep the name "Somatosensory Map[15]." However, we use a different technique to acquire the map, and use real-world human-robot interaction rather than a simulated baby. Moreover, we attempt to classify haptic interactions using correlations between incoming signals from tactile sensors distributed on the whole body surface.

III. Self-organizing tactile sensor method to decide sensor boundaries

A. Basic idea

Suppose that N tactile sensors are implemented on a robot, and that i-th tactile sensor stream during one human-robot interaction is called S_i $(i = 1, \ldots, N)$ where S_i is a vector of the n time-step sampling result of the sensor outputs ($S_i = \{(S_i)_1, \cdots, (S_i)_n\}$, $(S_i)_t \in \mathbf{R}$). Features need to be extracted from this time series of the data stream. However, less work has been done on processing tactile features than on audio or vision. In conventional works[3][13], since the data are high dimensional, summary statistics, such as mean, standard error, minimum, max, and coefficients of fast Fourier transformation, are computed from one sensor or all of the sensors to be used as features. A feature space defined from one sensor will not be enough when several sensor are activated by touches, e.g., distinguishing a finger tap from a hand tap or a tickle. On the other hand, the feature space computed from combining all the sensors is less robust, since the features could be drastically changed if e.g. one sensor is broken. Feature space defined from several sensors, at least from two sensors, could be an effective happy medium.

We proposed a feature space using cross-correlations computed from sensor pairs, satisfying the above condition, in [7]. The cross-correlation is one important statistics in human tactile system – Dince et al.[16] reported that discrimination ability of the two point stimulus is improved when correlated stimulus is added continually to two close separated point of a human finger. In fact, the visualization results of our feature space, i.e., the Somatosensory Map, show characteristics of many categories of touch interactions[7] by arranging sensor point. Fig. 7 is the Somatosensory Map made from a 2 minute interaction between human and a robot. This presentation simplifies the haptic interaction; e.g., distinctive sensor point cluster of both arms' sensors are to the result of subject touching the robot's arms at the same time.

The cross-correlation based feature space has another advantage that the choice of sensors to be included in a sensor bank for a certain computation can be decided via the elements used in the classifier's selected subspace. Often the useful subspace is composed of combination of sensors located in close spatial proximity, since distant sensors usually have low correlated signals and thus have less mutual information than adjacent sensors.

B. Feature space

A feature vector $\mathbf{a}$ is computed from the cross-correlation matrix defined by equation (2) as follows:

$$\mathbf{a} = \big(\underbrace{R_{(1,2)}, \cdots, R_{(1,N)}}_{N-1,}, \underbrace{R_{(2,3)}, \cdots, R_{(2,N)}}_{N-2,}, \cdots, \underbrace{R_{(N-1,N)}}_{1}\big)^t \tag{1}$$

where $R_{(i,j)}$ is cross-correlation matrix element at (i, j) between N sensors, i.e.,

$$R_{ij}(S_i, S_j) = \frac{C_{ij}}{\sqrt{C_{ii} C_{jj}}} \qquad (-1 \leq R_{ij} \leq 1) \tag{2}$$

where

$$C_{ij} = \sum_{t=1}^{n} \left((S_i)_t - \bar{S}_i\right)\left((S_j)_t - \bar{S}_j\right) \tag{3}$$

is cross variation of (i, j), and $\bar{S}_i$ is average of time series data S_i.

C. Overview

We construct a classifier for detecting haptic interaction between robot and human uses the CLAFIC method[9], a type of subspace method. This method represents each class as eigenpairs computed from a training data set. The subspace method starts from an idea of Watanabe et al.[9] that, as the feature space grows, the data set will converge to a limited small subspace. The CLAFIC method approximates this subspace with eigenpairs. Since the feature space defined in III-B is also a high dimensional space of $O(N^2)$, feature vectors of a data set should also be restricted mostly to a limited feature subspace.

In our proposed method, a dimension reduction of the extracted subspace is additionally applied by selecting base vectors having large inner product values. The classifier output is then calculated from a selected subspace consisting of only a few base vectors chosen from cross-correlation elements of the coefficient R_{ij} computed from all sensor pairs. Hence, this subspace and feature selection results in useful subsets of sensors that can be used to distribute processing. Since these subsets are found automatically, we call this method finding "self-organizing tactile sensors".

D. CLAFIC method

Fig. 1 shows the overview of classification. At first, a data set $\mathbf{X}_k$ is prepared from p_k feature vectors ($\mathbf{a}_k$) computed from a time series of sensor streams labeled as a class k ($\stackrel{def}{\equiv} \omega_k$), where p_k is the number of training data set. Thus

$$\mathbf{X}_k = \{\mathbf{a}_{k1}, \ldots, \mathbf{a}_{kp_k}\} \tag{4}$$

The approximated subspace of CLAFIC method begins by performing singular value decomposition (SVD):

$$\mathbf{X}_k = \mathbf{U} D_\lambda \mathbf{V}^t \tag{5}$$

where the columns of $\mathbf{U}$ is left singular vectors; D_λ has singular values and is diagonal; and $\mathbf{V}^t$ has rows that are the right singular vectors. We perform a forward feature selection of singular vectors to perform classification using a low dimensional subspace. First, we arrange the vectors obtained from the SVD performed only on data from class k in decreasing order of the $((d_k)_1, (d_k)_2, \cdots, (d_k)_{smallest})$ singular values, and compute a discrimination function DF_k of class k, which takes input an unknown vector $\mathbf{x}$, and computes

$$DF_k(\mathbf{x}) = \sum_{j=1}^{m} \left(\mathbf{x}^t \mathbf{u}_{kj}\right)^2 \tag{6}$$

using only the first m vectors, where $\mathbf{u}_{kj}$ is a left singular vector derived only from data in class k. We choose m by starting at $m = 1$ and increasing m until the cumulative contribution ratio in eq. 6 exceeds a threshold value C_1. The output of the discrimination function corresponds to the square of the length of an unknown vector orthographically projected onto the low dimensional subspace. The classifier outputs the class name which has maximum DF output.

$$\max_{k=1,\cdots,c} \{DF_k(\mathbf{x})\} = DF_l(\mathbf{x}) \quad \Rightarrow \quad \mathbf{x} \in \omega_l \tag{7}$$

Using the fidelity value τ as Watanabe et al. proposes[9], the unknown is vector classified as "unknown" class if the maximum DF is not significantly difference from the second maximum DF, i.e., if

$$\frac{DF_l(\mathbf{x})}{\max\limits_{k \neq l}\left(DF_k(\mathbf{x})\right)} > {}^{1}\!/_{\tau} \tag{8}$$

evaluates to "false", the classification is "unknown".

E. Learning Sensor Banks

A sensor bank for a classification task is decided by selecting a useful subspace that has high relevance for the classifier. As equation (6) shows, the output of the classifier is composed of inner products. Considering the $\mathbf{u}_{ij}$ elements are weights for the unknown vector elements, if a p-th element $\{\mathbf{u}_{ij}\}_p$ is close to 0, the element $\{\mathbf{x}\}_p$ could be ignored. In computer vision, Ishiguro et al.[17] has proposed this type of idea, describing it as a form of "attention control".

Let $\tilde{\mathbf{u}}_{ij}$ be an approximated subspace ignoring the all elements close to 0, where all $\{\mathbf{u}_{ij}\}_p$ smaller than C_2 are simply replaced by 0. Now we construct approximated discrimination function $DF_k^*(\mathbf{x})$ as follows.

$$DF_k^*(\mathbf{x}) = \sum_{j=1}^{m} \left(\mathbf{x}^t \mathbf{u}_{kj}^*\right)^2 \quad (9)$$

where $\mathbf{u}_{kj}^*$ are assumed to be nearly orthogonal, and are normalized ($\mathbf{u}_{kj}^* = \tilde{\mathbf{u}}_{kj} / \|\tilde{\mathbf{u}}_{kj}\|$).

The approximated discrimination function $DF_k^*(\mathbf{x})$ is also composed of inner products of $\mathbf{u}_{kj}^*$ and $\mathbf{x}$, however now we need to know only the q-th elements ($\{\mathbf{x}\}_q$), where

$$\left\{q : \{\mathbf{u}_{kj}^*\}_q \neq 0\right\} \quad (10)$$

From the definition of feature space in equation (1), only elements $\mathbf{R}_{(r_q, s_q)}$ need to be computed, where $(r_q,\ s_q)$ is a sensor pair needed to compute q-th element $\{\mathbf{x}\}_q = \mathbf{R}_{(r_q, s_q)}$. These sensor pairs define whether the sensor is used or not used in the sensor bank, facilitating distributed processing, since each classifier only needs a subset of sensors.

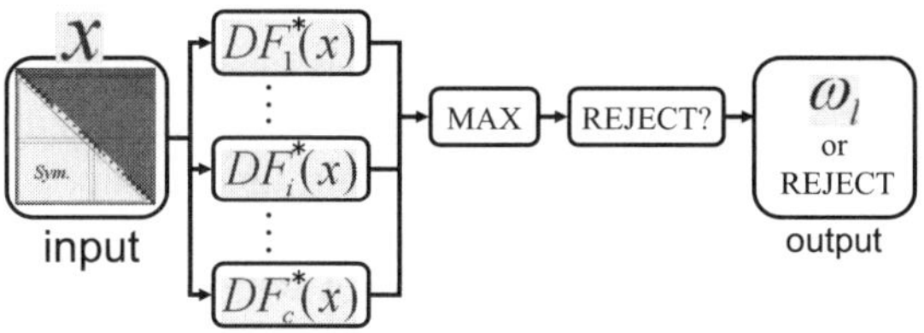

Fig. 1. Overview of the classification

TABLE I

SCENARIO BASED DEFENITION OF THE INTERACTION CLASSES

Class name (step)	Approaching a person	m ($C_1 = 0.15$)
$class1$	"Hello."	4
$class2$	"Let's shake hands."	4
$class3$	"Nice to meet you."	5
$class4$	"What's your name?"	5
$class5$	"Where are you from?"	5
$class6$	"let's play!"	5
$class7$	"Do you think I'm cute?"	4
$class8$	"I wish you'd pat me on the head"	3
$class9$	"Whee!"	6
$class10$	"I want to play more."	5
$class11$	"Tickle me."	7
$class12$	"That tickles!"	3
$class13$	"Thanks"	4
$class14$	"Give me a hug."	7
$class15$	"Bye-bye!"	4

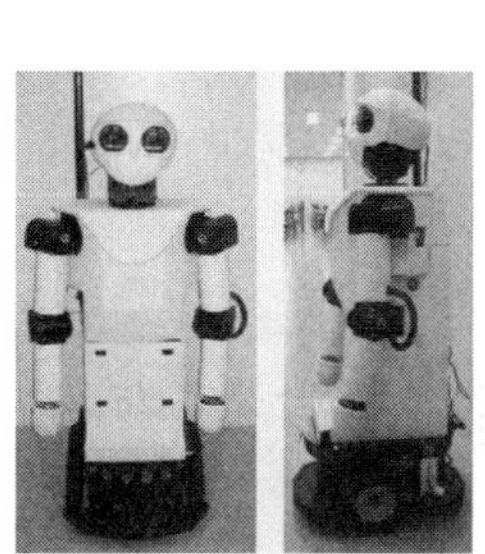

(a) Robovie-IIF (ATR)

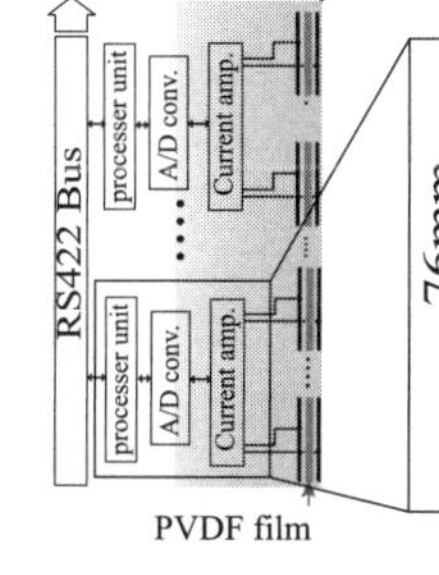

(b) Tactile sensor network consisting of an RS422 bus network via which nodes are connected to a host PC

Fig. 2. Overview of our tactile sensor system

F. Somatosensory Map

To visualize feature vectors and sensor banks, we define dissimilarity as d_{ij} converted from the coefficient of the cross-correlation matrix R_{ij} with the following equation (11),

$$d_{ij}(R_{ij}) = -\log\left(|R_{ij}|\right). \qquad (0 \leq d_{ij} \leq \infty) \quad (11)$$

This dissimilarity definition defines a "distance" between (i, j) sensors, i.e., higher correlated (or negatively correlated) sensor pairs have smaller dissimilarity. (Note that it does not satisfy all properties of a true distance notion.) Since the self-correlation coefficient is always 1, the dissimilarity with itself is always 0, i.e.,

$$d_{ii} = 0. \qquad (S_i \neq constant) \quad (12)$$

In the Somatosensory Map, the N sensor points are arranged into a 2D map using MDS[8] based on the dissimilarity definition of the equation (11). This 2D map can be used to visualize a vector of the cross-correlation feature space, e.g., fig. 7 is the Somatosensory Map plotted using a feature vector during a human-robot interaction[7]. In section V we apply this method to $\mathbf{u}_{ij}$ to interpret experimental results.

IV. Experiments

A. Hardware

The hardware on which we are testing our proposed technique is detailed below. Fig. 2 shows an outline of the hardware for the experiments described in this section. Fig. 2 (a) shows the communication robot Robovie-IIF[20], provided with high-density soft tactile sensors and a sensor network consisting of a RS422 bus network via which nodes connected to a host PC (Fig. 2(b)). Fig. 3 shows the structure and materials of the skin sensors installed on the Robovie-IIF surface, and Fig. 4 is the location of embedded piezofilms.

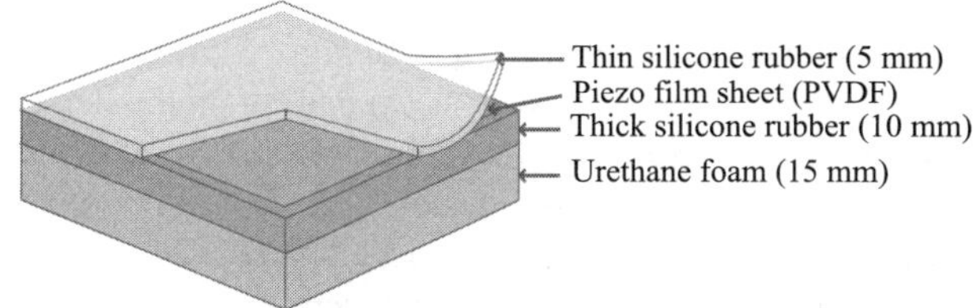

Fig. 3. Architecture of skin sensor devices

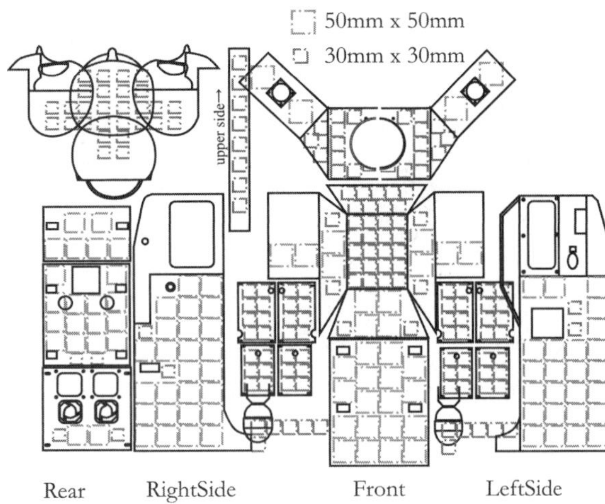

Fig. 4. Position of the skin sensors (PVDF films) on the deployed surface of the Robovie-IIF

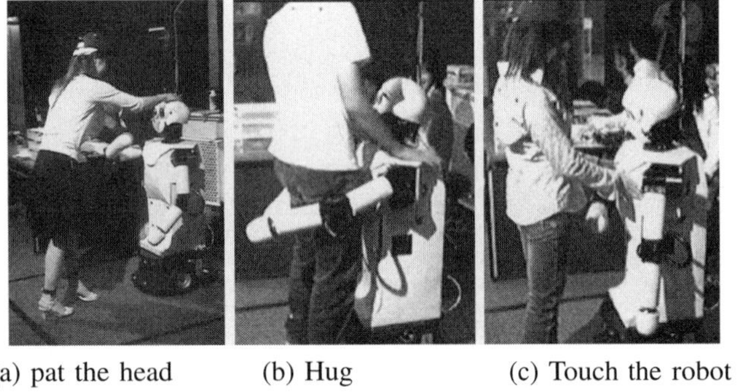

(a) pat the head (b) Hug (c) Touch the robot

Fig. 5. Observed haptic interactions between the robot and the visitors at Osaka Science Museum May. 2005

274 Piezofilms (3 cm × 3 cm, or 5 cm × 5 cm) are embedded in soft silicone rubber. Sampling time for the 16 bit A/D converter is set to 100 Hz, and tactile sensor outputs are read to the host PC of Robovie-IIF at every sampling. Using this hardware, we conducted two experiments detailed below and shown in Fig. 5 and Fig. 6.

B. Field experiment

Our research group conducted several events in the Osaka Science Museum with socially interactive robots[19]. The first experiment is a field experiment during the event named "Let's play with Robovie". In the event, the Robovie-IIF was displayed in the Osaka Science Museum during May 2005. We asked visitors to play with the Robovie-IIF and with the goal of investigating what kind of haptic interaction can be realized between humans and the Robovie-IIF. Figs. 5 (a) to (c) show the three haptic interactions such as (a) Patting on the head, (b) Hug and (c) Touch the robot body, observed in this experiment. From these observations we designed a "haptic interaction scenario" for a second experiment to encourage human-robot interaction in which we expected subjects to touch Robovie-IIF, detailed below. Table I shows the interaction scenario stages. Each stage in the scenario consists of the control rule that the Robovie-IIF tries to sustain interest of a subject to keep interaction going and proceeds to the next stage after finishing each interaction. (See "approaching a person" of Table I.)

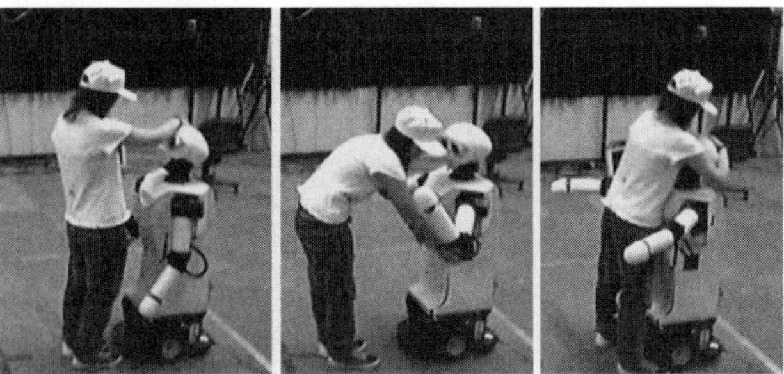

(a) "I wish you'd pat me on the head." (b)"Tickle me." (c) "Give me a hug."

Fig. 6. Observed subject's behaviors in each step of the scenario during the experiment for database construction

C. Construction of the haptic interaction database

In a second experiment, using a "Wizard of OZ" method [18] based on the scenario-based rules described in Section IV-B, Robovie-IIF is controlled by an experimenter with several monitoring displays. We expect the robot to be touched by subjects during these interactions. We constructed a scenario based interaction database which includes the monitoring videos, all of the tactile sensor signals, the command signals sent to control the robot, with all data time stamped using a common clock.

Each subject was asked to interact with Robovie-IIF in 3 trials separated by 10 minutes each. There were a total of 48 subjects, 24 males and 24 females, all of them college students. Each trial took around 5 minutes and was set up with the same condition except for the subject's position at the start of the scenario. These positions were each $2m$ away from Robovie-IIF, at 45 degrees to the right, 45 degrees to the left, and 0 degrees (where 0 degree is defined as in front of the robot). We asked the subjects to simply play with Robovie-IIF (which has a child-like voice and uses other cues to encourage humans to treat it as a child), and explained to them before each trial the following rules: (1) The subjects can touch the whole body of Robovie-IIF, (2) The subjects are required to listen carefully to what Robovie-IIF is saying, and (3) The subjects are required to be close to the robot in order to turn on the robot by touching it at the start of each trial.

Fig. 6(a) to (c) show the observed haptic interactions in the experiment, such as (a) "I wish you'd pat me on the head." of $class8$, (b) "Tickle me." of $class11$ and (c) "Give me a hug." of $class14$. Excluding approximately 24 cases in which there were technical difficulties, approximately 120 cases of data were acquired to form a "haptic interaction database" of data collected from real interaction scenarios. Segments of the tactile sensor data are automatically clipped and labeled using the time stamps for when each scenario stage (as defined in Table I) begins and ends. Thus, unlike previous work in which interaction segments were hand-labeled by an experimenter, we do not perform any manual coding of the data.

V. RESULTS

To emphasize the tactile sensor data when subjects touch the robot, we prepared $\hat{S}_i$ from the output of tactile sensors as shown in the equation (13) since our database includes lots of information caused by the robot movements.

$$\hat{S}_i = \left\{ \begin{array}{l} S_i, \left(\left|S_i - \bar{S}_i\right| > \sigma_i\right) \\ \bar{S}_i, \left(\left|S_i - \bar{S}_i\right| \leq \sigma_i\right) \end{array} \right\} \quad (13)$$

where S_i is the i-th sensor output, average of time series data of S_i is $\bar{S}_i$, and standard deviation of S_i is σ_i. If the absolute difference between S_i and $\bar{S}_i$ is smaller than standard variation σ_i, S_i is replaced by the average $\bar{S}_i$. In this section, all of the results are from $\hat{S}_i$.

Fig. 7 shows the 2D Somatosensory Map obtained from a cross-correlation matrix of each tactile sensor during an interaction in a field experiment between the Robovie-IIF and a subject (Fig. 5 (c)). Figs. 8 and 9 show the results of Leave-One-Out cross validation tests for evaluation of the classifier using the K-nearest neighbor (KNN) method ($k = 3$) and the currently proposed method, respectively, on the whole dataset. Figs. 9 through 12 show the results of choosing different values of C_1 and C_2 using the proposed method. In these figures, $class\ k$ ($k = 1, \ldots, 15$) corresponds to the classes defined in Table I. For example, the data set labeled $class2$ consists of tactile sensor data collected between the time when the start command of $class2$ ("Let's shake hands") was sent to the Robovie-IIF and the time when the start command of $class3$ ("Nice to meet you") was sent. (The data sets include some cases in which subjects did not deliver expected interaction.) As can be seen in Fig 8, the KNN method achieved classification of over 60% for many haptic interactions such as $class2$, $class6$, $class7$, $class8$, $class11$, and $class15$, using only the correlation patterns of all tactile sensors.

Fig. 9 shows the correct recognition rates for the proposed method, while figure Fig. 10 shows the "false alarm" rate, computed for each class as the number of times an example was incorrectly classified as belonging to that class, divided by the number of examples that actually belong to that class (thus these numbers can be greater than 1). For these experiments, the fidelity value τ was experimentally fixed to be 0.95, which did not change the recognition rate but improved the false alarm rate. Each figure has 6 conditions that are in the set $\{(C_1, C_2) : C_1 = 1, 0.15\ C_2 = 0, 1, 2\}$. The number of orthogonal base vectors (that are left singular vectors, $\mathbf{u}_{ij}$,) is decided by the parameter C_1, shown in Table. I. The parameter C_2 determines the reduction of size of the feature space, e.g., the reduced feature space in case $C_2 = 0, 1, 2$ were 0%, around $80 - 85\%$, and around $94 - 96\%$, respectively.

The proposed method improved classification to 80% for most haptic interactions, including $class1$, $class2$, $class6$, $class7$, $class9$, $class11$, $class14$, and $class15$ when using $(C_1, C_2) = (0.15, 0)$. This performance was almost the same as for $(C_1, C_2) = (0.15, 1)$, which used only 15% of the feature space. When the feature space size is reduced to 5% in the condition $(C_1, C_2) = (0.15, 2)$, performance is still as high as 60% for haptic interactions of $class1$, $class8$, $class11$, and $class15$ (note that random, "by chance" performance is less than 7%).

Fig. 11 shows the result of the feature space reduction. Elements of vector $\mathbf{u}_{81}$ are arrayed onto a matrix of the same size as a cross-correlation matrix, and large weighted cross-correlation elements are visualized with darker (more black) colors. Adjacent tactile sensors usually have closer numbers, and are displayed as square line boxes corresponding to their part names. As expected, the useful feature spaces are composed mostly of adjacent sensor pairs. The boxes shown, which include several highly weighted elements, are the result of self-organizing results corresponding to the boundaries of tactile sensors. These self-organizing results are also shown in Fig. 12 (a) and (b), which shows the arrangement of sensors in a 2D Somatosensory Map using $\mathbf{u}_{(8)1}$ and $\mathbf{u}_{(11)1}$. Closer sensor pairs have larger weight in their cross-correlation element. Fig. 12 (a) shows that head sensors, probably touched in the "pat me" interaction, are clustered apart from other sensors. Fig. 12 (b) also clustering of the front side of the body (F-body) and of the left and right side of the body (LSide-body and RSide-body), which are often touched together in the "tickle" interaction. Note that these results do not make use of any knowledge about the spatial position of sensors but only using sensor streams from the whole robot body.

VI. DISCUSSION

Comparing our Somatosensory Map with previous work [14][15], we found that haptic interactions form clusters in the map that often can be grouped by body part. Using this representation for human-robot interaction we achieved good classification results for those interaction categories in which there was some human touching. In previous work for classification of haptic interaction [3][13], the data sets for learning classifiers were hand labeled by the experimenter. In our case, the database is self-labeled during scenario based interactions. The label of each data point is based on what the current designed scenario is, rather than given *post hoc* by the experimenter asking subjects to touch the robot. We assume this is a more natural and practical database construction.

The Somatosensory Map shows large weighted elements mainly between spatially-localized sensors. This is consistent with the idea of Watanabe et al. that, as the feature space grows, data sets converge to limited subspaces. Additionally, the learned subspace was composed mainly of adjacent sensor pairs in the tactile system, as seen in Fig. 11. Thus, the CLAFIC method is able to achieve higher classification performance even when using smaller subspaces of 15% size. In fact, performance of 80% classification was achieved, improving over the KNN method of 60%, despite a much more challenging evaluation than in previous work. Instead of using static objects consisting of tactile sensors, we constructed the database from real human robot interactions. Since the robot has malleable tactile sensors embedded under soft skin, and the robot is moving during the experiment, it is possible that the results come from the classification of self-sensations provided by self-movements. However, it seems like more classification is provided from subject's touches, because the recognition rate of the classes that we don't expect to see subject's touches had low recognition rates. Nevertheless, this problem is unavoidable while the robot moving during the touch from other, so in the future we probably will also need to use proprioception in the tactile system.

VII. Conclusions

In conclusion, the proposed method was found to be efficient with the classification of real human-robot interactions, and was able to be implemented as distributed in-network processing.

In this paper, we describe a haptic interaction classification method using cross-correlation matrix features, and propose a self-organizing technique to define a bank of sensors to be used in distributed processing of each class. The cross validation rests results in recognition of 80% for those interactions in which we expect subjects to touch the robot, using only 15% of the feature subspace. The Somatosensory Map visualization shows that the selected feature space was composed mainly of spatially-adjacent sensor pairs. These promising results suggest that our proposed method may be useful for automatic analysis of touch behaviors in more complex future tasks.

Acknowledgment

Valuable comments from Dr. Ian Fasel improved the presentation quality of this paper. The research in Section IV was supported in part by Japan's Ministry of Internal Affairs and Communications. The analysis of the experimental data in Section V was conducted as part of the JST ERATO Asada Project.

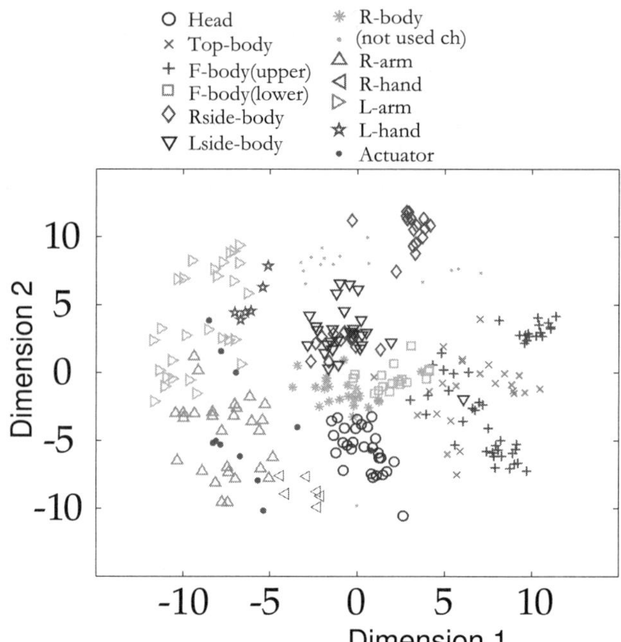

Fig. 7. This map[7] is the 2D Somatosensory Map obtained from the haptic interaction shown in fig. 5(c) in the field experiment. The time window used to compute the distance matrix, $\{d_{ij}\}$, is 2 minute of the whole interaction between the robot and the subject. The two cluster of the "F-body (upper)" sensors' distribution, around the locations of $(Dimension1, Dimension2) = (7, -6)$ and $(11, 4)$, can be interpreted as a result of that she touched the two area of the upper body with her left/right hands. This result suggests that cross-correlations contain touch features.

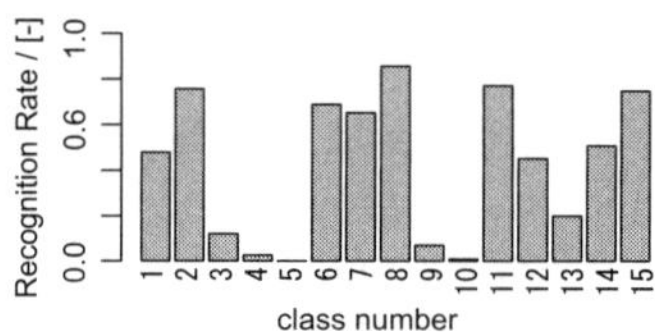

Fig. 8. Recognition Rate of the classifier using the K-Nearest Neighbor method in 15-way forced choice task ($k = 3$ is the same condition as [7], but 2 classes added to the choice. [7] was in 13-way forced chice task.)

References

[1] Takahiro Miyashita, Taichi Tajika, Kazuhiko Shinozawa, Hiroshi Ishiguro, Kiyoshi Kogure and Norihiro Hagita, *"Human Position and Posture Detection based on Tactile Information of the Whole Body"*, In Proc. of IEEE/RSJ 2004 International Conference on Intelligent Robots and Systems Workshop (IROS'04 WS), Sep. 2004.

[2] M. Inaba, Y. Hoshino, K. Nagasaka, T. Ninomiya, S. Kagami and H. Inoue, *"A Full-Body Tactile Sensor Suit Using Electrically Conductive Fabric and Strings"*, Proc. 1996 IEEE/RSJ International Conference on Intelligent Robots and Systems (IROS 96), Vol. 2, pp. 450.457, 1996.

[3] Naya, F., Yamato, J. and Shinozawa, K.,*"Recognizing Human Touching Behaviors using a Haptic Interface for a Pet-robot"*, Proc. 1999 IEEE International Conference on Systems, Man, and Cybernetics (SMC'99), pp. II-1030–1034.

[4] H. Iwata, H. Hoshino, T. Morita, and S. Sugano, *"Force Detectable Surface Covers for Humanoid Robots*, Proc. 2001 IEEE/ASME International Conference on Advanced Intelligent Mechatronics (AIM'01), pp. 1205–1210, 2001.

[5] H. Iwata, S. Sugano, *"Whole-body covering tactile interface for human robot coordination"* Proceedings - IEEE International Conference on Robotics and Automation. Vol. 4, pp. 3818-3824. 2002.

[6] Tomoyuki Noda, Takahiro Miyashita, Hiroshi Ishiguro, Kiyoshi Kogure, Norihiro Hagita, *"Detecting Feature of Haptic Interaction Based on Distributed Tactile Sensor Network on Whole Body"*, Journal of Robotics and Mechatronics, Vol.19, No. 1, pp. 42-51, 2007.

[7] Tomoyuki Noda, Takahiro Miyashita, Hiroshi Ishiguro, Norihiro Hagita, *"Map Acquisition and Classification of Haptic Interaction Using Cross Correlation between Distributed Tactile Sensors on the Whole Body Surface"*, Proceedings of IEEE/RSJ International Conference on Intelligent Robots and Systems (IROS 07), 2007.

[8] Torgerson, W., *"Multidimensional scaling : I. Theory and method."*, Psychometrika, 17, pp.401-419, 1952.

[9] S. Watanabe and N. Pakvasa, *"Subspace method in pattern recognition,"* Proc. 1st Int. J. Conf on Pattern Recognition, Washington DC, pp. 2-32, Feb. 1973.

[10] Atsushi Iwashita, Makoto Shimojo,*"Development of a Mixed Signal LSI for Tactile Data Processing"*, in Proc. of IEEE Int. Conf. on Systems, Man and Cybernetics 2004 (SMC2004), 2004.

[11] Z. Pan, H. Cui, and Z. Zhu, *"A Flexible Full-body Tactile Sensor of Low Cost and Minimal Connections"*, Proc. 2003 IEEE International Conference on Systems, Man, and Cybernetics (SMC'03), Vol. 3, pp. 2368–2373, 2003.

[12] H. Shinoda, N. Asamura, T. Yuasa, M. Hakozaki, X. Wang, H. Itai, Y. Makino, and A. Okada, *"Two-Dimensional Communication Technology Inspired by Robot Skin"*, Proc. IEEE TExCRA 2004 (Technical Exhibition Based Conf. on Robotics and Automation), pp.99-100, 2004.

[13] D. Francois, D. Polani, K. Dautenhahn *"On-line Behaviour Classifica-*

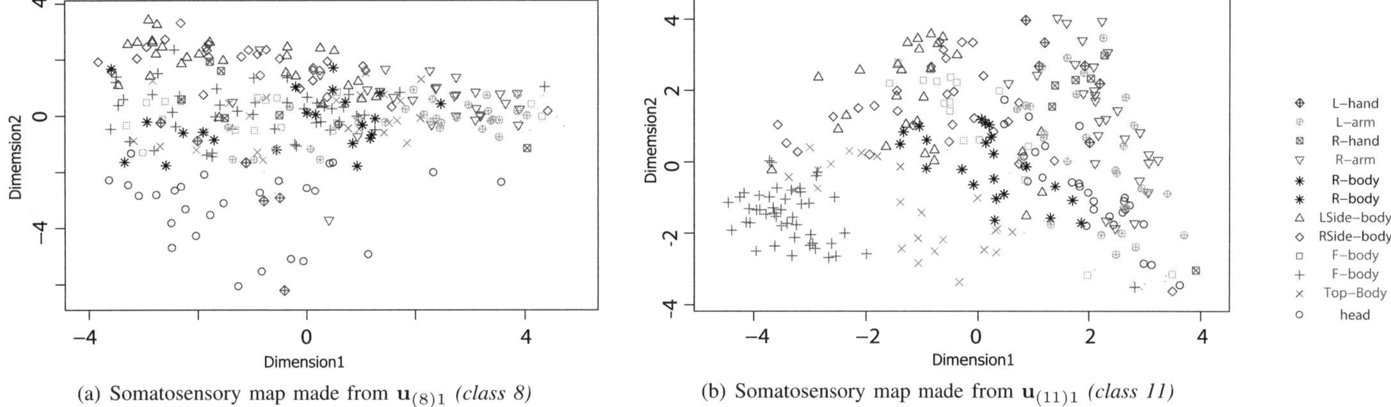

(a) Somatosensory map made from $\mathbf{u}_{(8)1}$ *(class 8)*

(b) Somatosensory map made from $\mathbf{u}_{(11)1}$ *(class 11)*

Fig. 12. Somatosensory map can also visualize a $\mathbf{u}_{kj}$ as sensor point distribution in a 2D plane, since $\mathbf{u}_{kj}$ has the same dimension as the feature vector $\mathbf{a}$ has. In these two maps, closer sensor pairs have larger weight in their cross-correlation element.

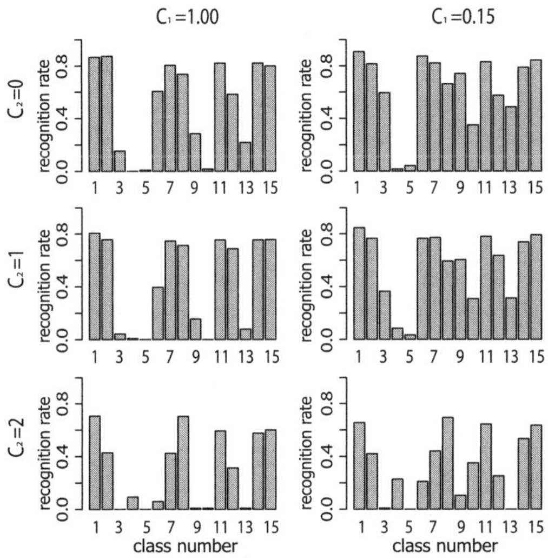

Fig. 9. Recognition rate

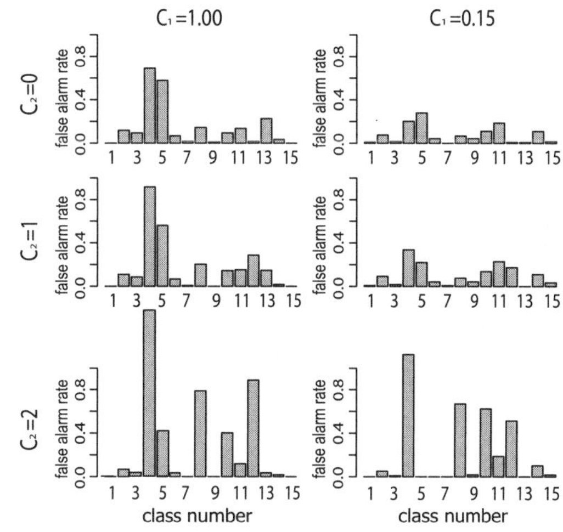

Fig. 10. False alarm rate

tion and Adaptation to Human-Robot Interaction Styles" , poster session of HRI(2007), HRI, 2007

[14] D. Pierce and B. Kuipers, *"Map learning with uninterpreted sensors and effectors"*, Artificial Intelligence, vol. 92, 1997.

[15] Yasuo Kuniyoshi, Yasuaki Yorozu, Yoshiyuki Ohmura, Koji Terada, Takuya Otani, Akihiko Nagakubo, Tomoyuki Yamamoto,*"From Humanoid Embodiment to Theory of Mind"*, July 7-11, 2003, Revised Papers, pp.202–218.

[16] Hubert R. Dince et. al., *"Improving Human Haptic Performance in Normal and Impaired Human Populations through Unattended Activation-Based Learning"*, ACM Transactions on Applied Perception, Vol. 2, No. 2, pp.71-88., April 2005,

[17] Hiroshi Ishiguro, Masatoshi Kamiharako and Toru Ishida, *"State Space Construction by Attention Control"*, International Joint Conference on Artificial Intelligence (IJCAI-99), pp. 1131-1137, 1999.

[18] Dahlback, N., Jonsson, A. and Ahrenberg, L., *"Wizard of Oz studies - Why and How"*, Proc. of the international workshop on Intelligent user interfaces, pp.193-200, 1993.

[19] Masahiro Shiomi, Takayuki Kanda, Hiroshi Ishiguro, and Norihiro Hagita, *"Interactive Humanoid Robots for a Science Museum"*, IEEE Intelligent Systems, vol. 22, no. 2, pp. 25-32, Mar/Apr, 2007

[20] Takahiro Miyashita, Taichi Tajika, Hiroshi Ishiguro, Kiyoshi Kogure, and Norihiro Hagita, *"Haptic Communication between Humans and Robots"*, in Proc. of 12th International Symposium of Robotics Research, CD-ROM, San Francisco, CA, USA, Oct. 2005.

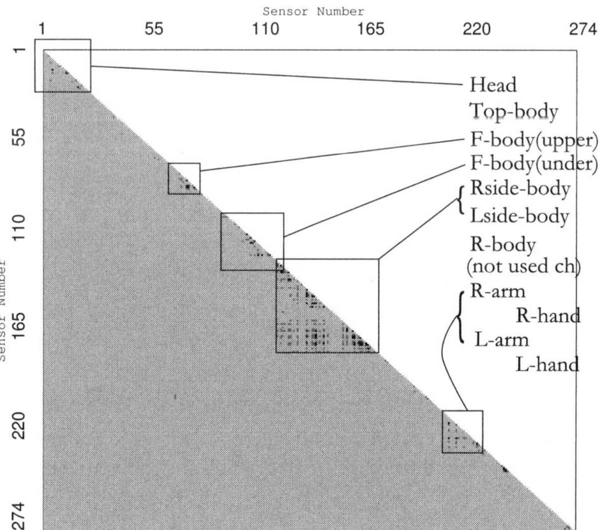

Fig. 11. Color image matrix reconstructed from orthogonal vector $\mathbf{u}_{81}$, highlighting the large weighted elements of the cross-correlation

NanoNewton Force Sensing and Control in Microrobotic Cell Manipulation

Xinyu Liu, Keekyoung Kim, Yong Zhang, and Yu Sun*
Advanced Micro and Nanosystems Laboratory
University of Toronto, Canada
*sun@mie.utoronto.ca

Abstract— **Cellular force sensing and control techniques are capable of enhancing the dexterity and reliability of microrobotic cell manipulation systems. This paper presents a vision-based cellular force sensing technique using a microfabricated elastic cell holding device and a sub-pixel visual tracking algorithm for resolving forces down to 3.7nN during microrobotic mouse embryo injection. The technique also experimentally proves useful for *in situ* differentiation of healthy mouse embryos from those with compromised developmental competence without the requirement of a separate mechanical characterization process. Concerning force-controlled microrobotic cell manipulation (pick-transport-place), this paper presents the first demonstration of nanoNewton force-controlled cell micrograsping using a MEMS-based microgripper with integrated two-axis force feedback. On-chip force sensors are used for detecting contact between the microgripper and cells to be manipulated (resolution: 38.5nN) and sensing gripping forces (resolution: 19.9nN) during force-controlled grasping. The experimental results demonstrate that the microgripper and the control system are capable of rapid contact detection and reliable force-controlled micrograsping to accommodate variations in size and stiffness of cells with a high reproducibility.**

I. Introduction

Manipulation of single living cells represents an enabling technology that is important for a range of biological disciplines (e.g., genetics [1][2], *in vitro* fertilization [3], cell mechanical characterization [4], and single cell-based sensing [5]). The past decade has witnessed significant progress in the development of robotic systems and tools for conducting complex cell manipulation tasks, such as probing, characterizing, grasping, and injecting single cells.

Robotic cell manipulation is universally conducted under an optical microscope; thus, visual feedback is the main sensing modality in all existing microrobotic cell manipulation systems. Meanwhile, due to the fact that biological cells are delicate and highly deformable, quantification of interaction forces between the end-effector and cells can enhance the capability of a robotic cell manipulation system. For example, cellular force feedback was demonstrated to be useful for the alignment between a probe and a cell [4]. The measurement of cellular forces also enables the prediction of cell membrane penetration in the injection of zebrafish embryos [6][7][8].

In order to obtain cellular force feedback during microrobotic cell manipulation, the development of force sensing devices has been a focus, resulting in capacitive force sensors [4] and piezoelectric force sensors [6][9], to name just a few. Inherent limitations prevent their use in practical cell manipulation tasks: (1) these force sensors are typically limited to resolving forces at the microNewton level while the manipulation of most cell lines requires a resolution of nanoNewton or sub-nanoNewton; (2) the integration of an end-effector (e.g., glass micropipette) and the force sensors is via epoxy glue, complicating the task of end-effector exchange.

Overcoming limitations of existing cellular force sensing approaches, this paper presents a vision-based cellular force measurement technique with a nanoNewton force resolution employing a microfabricated elastic cell holding device and a sub-pixel visual tracking algorithm. The technique allows for accurately resolving cellular forces during microrobotic cell manipulation without disturbing the manipulation process or imposing difficulties in end-effector exchange. The effectiveness of the technique is demonstrated in microrobotic mouse embryo injection. Furthermore, the force sensing technique proves useful for *in situ* distinguishing normal embryos from those with compromised developmental competence, without requiring a separate cell characterization process.

On the front of cellular force sensing and control, the paper also presents the first demonstration of force-controlled micrograsping of biological cells at the nanoNewton force level. As mechanical end-effectors, microgrippers enable pick-transport-place of biological cells in an aqueous environment. The microrobotic system employs a novel microgripper that integrates two-axis force sensors for resolving both gripping forces and contact forces between the gripping arm tips and a sample/substrate. The force-controlled microrobotic system experimentally demonstrated the capability of rapid contact detection and reliable force-controlled micrograsping of interstitial cells to accommodate variations in sizes and mechanical properties of cells with a high reproducibility.

II. Vision-Based Cellular Force Measurement During Cell Injection

Vision-based force measurement techniques are capable of retrieving both vision and force information from a single vision sensor (CCD/CMOS camera) under microscopic environments [10][11]. For cellular force sensing during microrobotic cell manipulation, this concept is realized by visually tracking flexible structural deformations, and subsequently, transforming material deformations into forces.

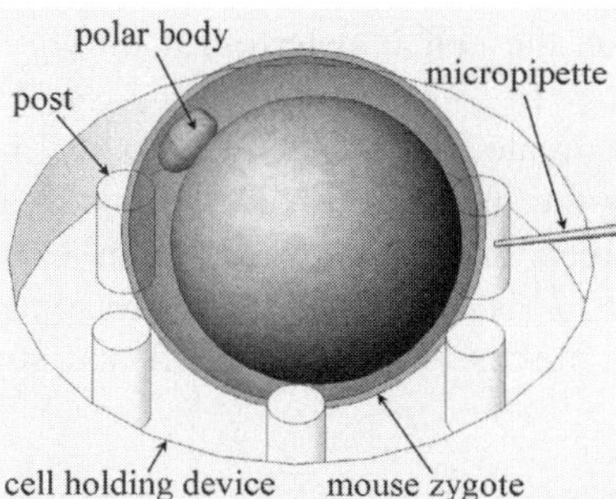

Fig. 1. Cellular force measurement using low-stiffness elastic posts during microrobotic cell injection.

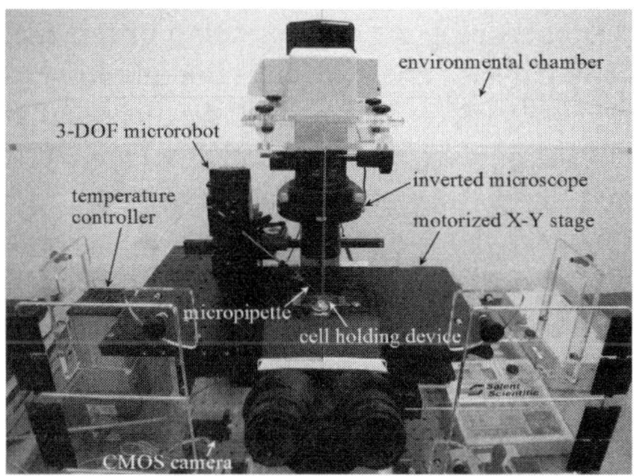

Fig. 2. Microrobotic mouse embryo injection system.

Fig. 1 schematically illustrates the principle of vision-based cellular force measurement using an elastic cell holding device during microrobotic cell injection. While the micropipette injects individual cells inside cavities on a cell holding device, applied forces are transmitted to low-stiffness, supporting posts. In real time, a sub-pixel visual tracking algorithm measures post deflections that are fitted into an analytical mechanics model to calculate the force exerted on the cell.

This technique was previously demonstrated on zebrafish embryos [12]. The study presented in this paper focuses on investigating the feasibility of further miniaturizing the cell holding devices to accommodate mouse embryos (100μm in diameter vs. 1.3mm zebrafish embryos) for measuring nanoNewton cellular forces during microinjection; and the possibility of using cellular force information to distinguish normal mouse embryos from those with compromised developmental competence for better selecting healthy embryos in genetics and reproductive research.

A. Microrobotic Mouse Embryo Injection System

The microrobotic mouse embryo injection system (Fig. 2) consists of a polydimethylsiloxane (PDMS) cell holding device, an inverted microscope (TE2000, Nikon) with a CMOS digital camera (A601f, Basler), a 3-DOF microrobot (MP-285, Sutter) for controlling the micropipette motion, a motorized X-Y stage (ProScan II, Prior Scientific) for positioning cell samples, and a temperature-controlled chamber (Solent Scientific) to maintain cells at 37°C.

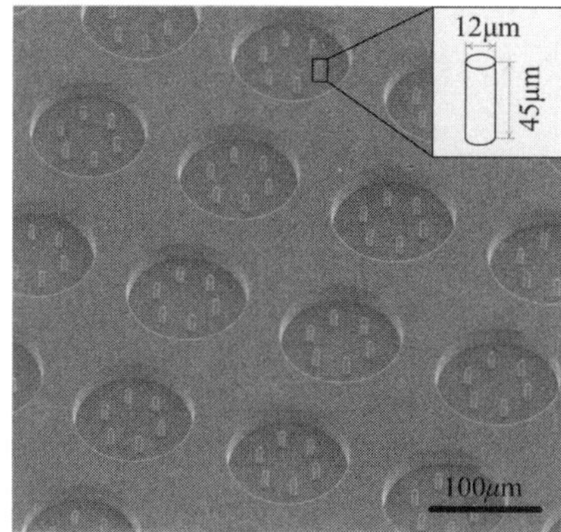

Fig. 3. SEM image of a PDMS cell holding device.

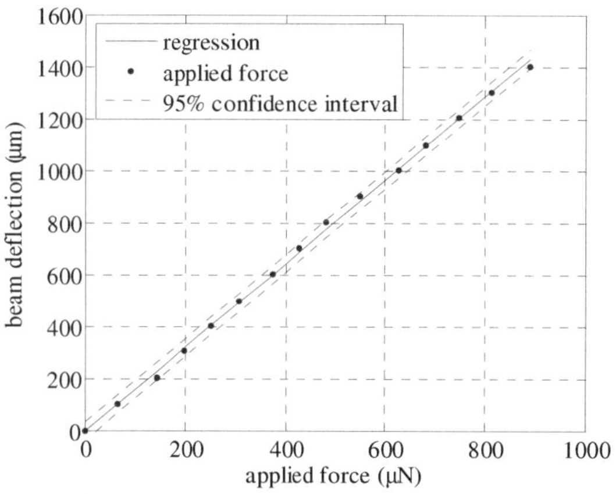

Fig. 4. Young's modulus calibration on a bulkier PDMS beam.

B. Fabrication and Characterization of Cell Holding Devices

The cell holding device shown in Fig. 3 was constructed with PDMS via soft lithography [12]. Briefly, PDMS prepolymer prepared by mixing Sylgard 184 (Dow Corning) and its curing agent with a weight ratio of 15:1, was poured over a SU-8 mold (SU-8 50, MicroChem) made on a silicon wafter using standard photolithography. After curing at 80°C for 8hr, the PDMS devices were peeled off the SU-8 mold. The depth of the cavity and protruding posts is 45μm, and the diameter of the posts is 12μm (Fig. 3). In order to make the PDMS surface hydrophilic, the devices were oxygen plasma treated for 10sec before use.

To determine the Young's modulus of the cell holding device, a bulkier PDMS beam produced under exactly the same processing conditions was calibrated with a piezoresistive force sensor (AE801, SensorOne) as described in [12]. It has been demonstrated that the Young's modulus values characterized from bulk PDMS and a micro PDMS structure, both constructed with the same microfabrication parameters, differ within 5% [13]. Fig. 4 shows the calibration data of applied force vs. beam deflection. The determined Young's modulus value is 422.4kPa.

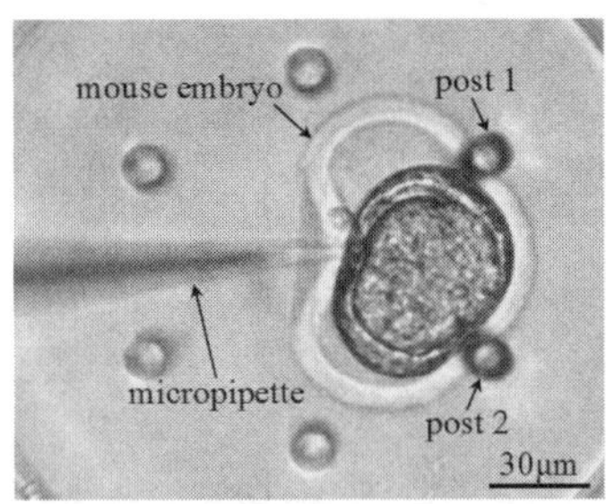

Fig. 5. Indentation forces deform the mouse embryo and deflect two supporting posts.

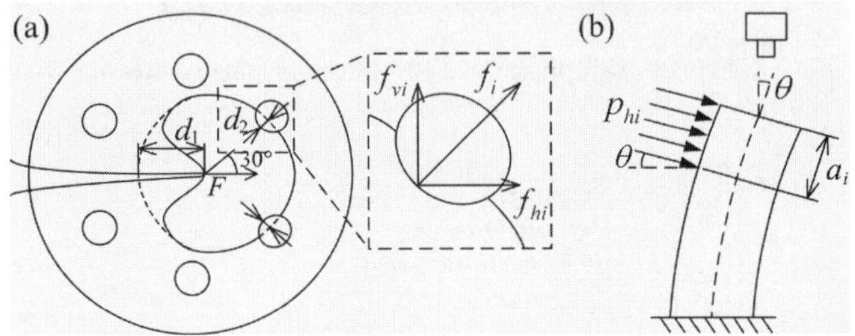

Fig. 6. Injection force analysis. (a) Force balance on the cell under indentation. (b) Post deflection model.

C. Mouse Embryo Preparation

As a model organism, mouse is a primary animal for genetics and reproductive research. Besides the importance in *in vitro* fertilization, microinjection of mouse oocytes and embryos is important for screening molecular targets linked to the study of basic biology of embryo development, such as mitochondrial-associated recombinant proteins, neutralizing antibody, morpholinos, and expression vectors for siRNA.

The mouse embryos used in this research were collected according to standard protocols approved by the Mount Sinai Hospital Animal Care Committee in Toronto. Young (8-12 weeks old) and older (40 weeks old) ICR female mice were used for obtaining normal embryos and those with blastomere fragmentation. ICR females with different ages were superovulated by injecting equine pregnant mare's serum gonadotropin (PMSG) and human chorionic gonadotropin (hCG) 48hr later. The mice were subsequently mated with ICR males of proven fertility, and plugs were verified the next morning. *In vivo* fertilized embryos were collected from the mated female mice at 24hr post-hCG and cultured in human tubal fluid (HTF) to two-cell stage (at 48hr post-hCG). The average diameter of the mouse embryos is 98μm.

D. Force Analysis

Fig. 5 shows a snapshot captured in the cell injection process. The microrobot controls an injection micropipette to exert an indentation force to a mouse embryo, deflecting the two supporting posts on the opposite side. Post deflections, measured by a visual tracking algorithm that will be discussed in Section II-E, are fitted to an analytical mechanics model to obtain contact forces between the cell and posts. Based on the contact forces, the indentation force applied by the micropipette on the cell is determined through the following force analysis.

The cell is treated as elastic due to the fact that quick indentation by the micropipette does not leave sufficient time for cellular creep or relaxation to occur. Consequently, the injection force, F is balanced by the horizontal components, f_{hi} of contact forces between the cell and supporting posts (Fig. 6(a)),

$$F = \sum_{i=1}^{2} f_{hi} \tag{1}$$

Much higher deformability of mouse embryos than that of zebrafish embryos results in different contact behavior between a cell and supporting posts, necessitating different treatments of forces in analysis. In the device configuration, the radius of the cell ($\sim$49μm) is larger than the depth of the cavity and posts (45μm), resulting in an initial point contact between the cell and supporting posts before post deflections occur. However, the high deformability of mouse embryos makes cell membrane conform to the posts when an injection force is applied to the cell. It is assumed that the contact forces are evenly distributed over the contact areas. Thus, the horizontal components, f_{hi} are expressed by a constant force intensity, p_{hi} and a contact length, a_i (Fig. 6(b))

$$f_{hi} = p_{hi} a_i \tag{2}$$

Slope θ of the posts' free ends shown in Fig. 6(b) was measured to verify the validity of linear elasticity that requires small structural deflections. The maximum slope was determined to be $11.1°$, which satisfies $sin\theta \approx \theta$; thus, the small deflection assumption of linear elasticity holds [14]. Therefore, the relationship of the horizontal force intensity, p_{hi} and post deflections can be established [14].

$$p_{hi} = \frac{\delta_i}{\frac{40a_i(1+\gamma)(2H-a_i)}{9\pi ED^2} + \frac{8(a_i^4+8H^3a_i-6H^2a_i^2)}{3\pi ED^4}} \tag{3}$$

where $i = 1, 2$; δ_i is the horizontal deflection; H and D are post height and diameter; E and γ are Young's modulus and Poisson's ratio ($\gamma = 0.5$ for PDMS [12]).

Combining (1)-(3) yields the injection force applied by the micropipette to the cell.

$$F = \sum_{i=1}^{2} \frac{\delta_i a_i}{\frac{40a_i(1+\gamma)(2H-a_i)}{9\pi ED^2} + \frac{8(a_i^4+8H^3a_i-6H^2a_i^2)}{3\pi ED^4}} \tag{4}$$

In (4), the unknown parameters are post horizontal deflections, δ_i and the contact length, a_i. Experimentally, imaging with a side-view microscope confirmed that the contact length, a_i increases at a constant speed, v for a given indentation speed. Hence, $a_i = vt$, where t denotes time.

Note that for a constant indentation speed of the micropipette, the variation speed of a_i, v varies for different cells. At 20μm/sec used throughout the experiments, v of the six tested mouse embryos were measured to be 0.8-1.2μm/sec. Interestingly, the sensitivity of the mechanics model (4) to variations in v is low. The injection force varies only by

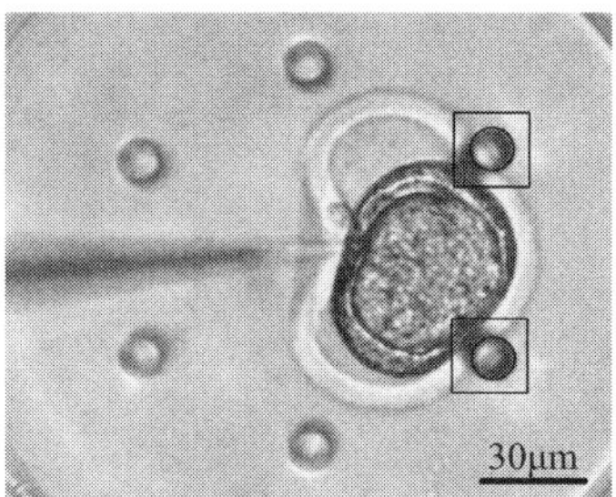

Fig. 7. Image patches tracked by template matching and LSCD detected post top circles.

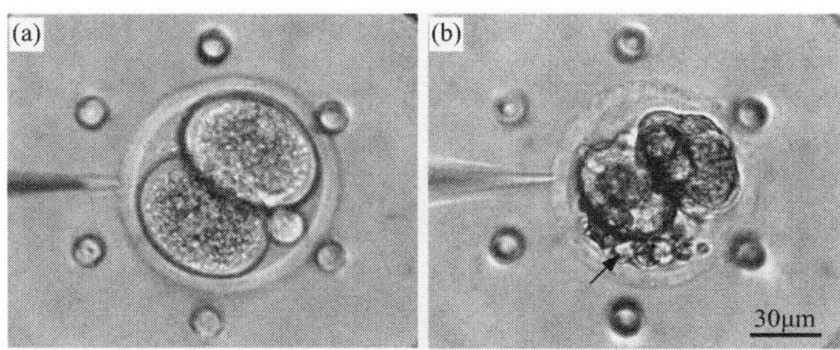

Fig. 8. Mouse embryos for cellular force measurement. (a) Normal embryo. (b) Embryo with blastomere fragmentation (arrow labeled).

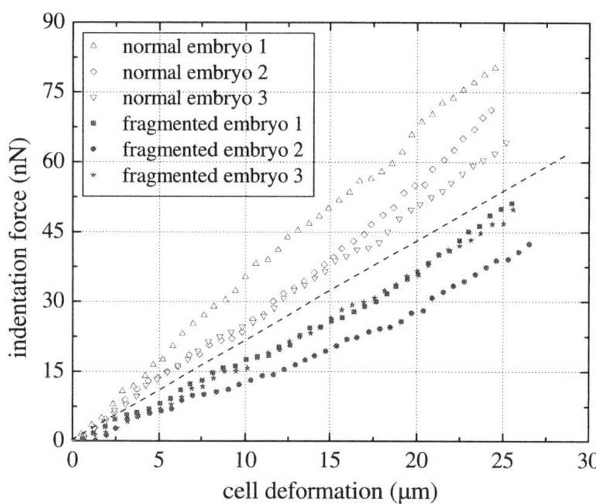

Fig. 9. Force-deformation curves of normal embryos (blue) and fragmented embryos (red). They separate themselves into two distinct regions.

TABLE I
SLOPES OF CELLULAR FORCE-DEFORMATION CURVES.

mouse embryo	slope (nN/μm)	slope average (nN/μm)
normal embryo 1	3.35	
normal embryo 2	2.74	2.87±0.43
normal embryo 3	2.52	
fragmented embryo 1	1.84	
fragmented embryo 2	1.45	1.70±0.22
fragmented embryo 3	1.81	

1% when v changes from 0.8μm/sec to 1.2μm/sec. Thus, the average value of the measured speeds, 1μm/sec was used to calculate injection forces for all the embryos.

E. Visual Tracking of Post Deflections

In order to accurately track post deflections, a visual tracking algorithm with a resolution of 0.5 pixel was developed and described in detail in [12]. A template matching algorithm with constant template updates first tracks the motion of the supporting posts, providing processing areas for a least squares circle detection (LSCD) algorithm to determine posts' center positions. The LSCD algorithm utilizes Canny edge detector to obtain an edge image and then extracts a portion of the post top surface for circle fitting. Fig. 7 shows the tracked image patches and LSCD detected post top circles.

F. Experimental Results and Discussion

As mouse embryos are exquisitely sensitive to slight temperature variations, experiments were conducted at 37°C inside the temperature-controlled chamber. With a 40× objective (NA 0.55), the pixel size of the imaging system was calibrated to be 0.36μm×0.36μm. Micropipette tips used for indenting mouse embryos was 1.8μm in diameter.

Three normal ICR embryos and three ICR embryos with blastomere fragmentation at the two-cell stage were used for cellular force measurements. Blastomere fragmentation is often indicative of future programmed cell death [15]. Although the blastomere fragmented embryo shown in Fig. 8 can be distinguished morphologically from normal embryos, using morphological differences alone is not always effective to distinguish diseased embryos from normal embryos due to the fact that one fourth of the fragmented embryos are able to develop normally [15]. We hypothesized that subtle changes in the cytoskeleton structure could lead to stiffness changes between abnormal and normal embryos. Thus, cellular force-deformation measurements were expected to provide additional information for detecting embryonic dysfunctions that require assisted hatching and for helping better select healthy embryos for implantation after microinjection.

The six embryos were manually delivered onto the cell holding device using a transfer pipette and then indented via microrobotic teleoperation. The micropipette was controlled to indent each embryo by 30μm at the speed of 20μm/sec. During the indentation process, force data were collected in real time (30 data points per sec).

Fig. 9 shows force-deformation curves of both normal and fragmented embryos. The horizontal axis represents cell deformation, $d = d_1 + d_2 cos30°$, where d_1 and d_2 were defined in Fig. 6. The vertical axis shows vision-based cellular force data. With the current cell holding devices and imaging system, the force measurement resolution was determined to be 3.7nN.

From Fig. 9, it can be seen that the force-deformation curves of normal and fragmented embryos separate themselves into two distinct regions. Table I summarizes the curve slopes calculated by linear regression. The slopes for normal embryos range from 2.52nN/μm to 3.35nN/μm while the slopes for fragmented embryos are between 1.45nN/μm and 1.84nN/μm, quantitatively demonstrating that the normal embryos and fragmented embryos are mechanically different.

All the indented embryos were subsequently cultured in an incubator at 37°C with 5% CO_2. The three normal embryos successfully developed into the four-cell stage; however, the

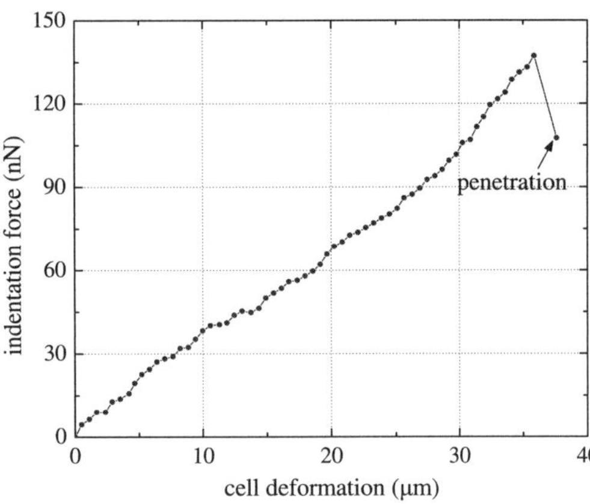

Fig. 10. Abrupt force change indicates cell membrane penetration during microrobotic cell injection.

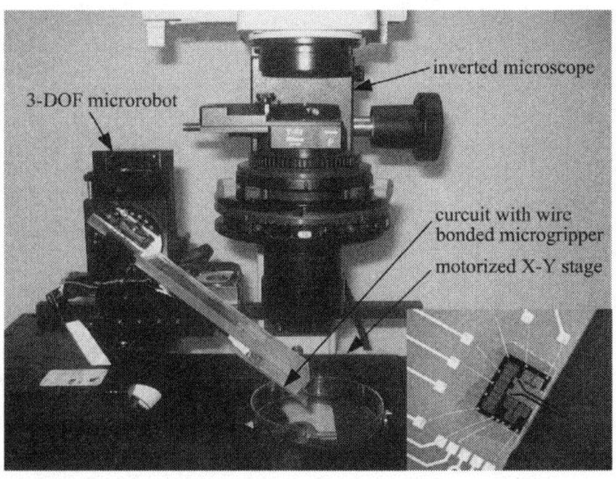

Fig. 11. Microrobotic system setup for Force-controlled micrograsping. Inlet picture shows the wire-bonded microgripper.

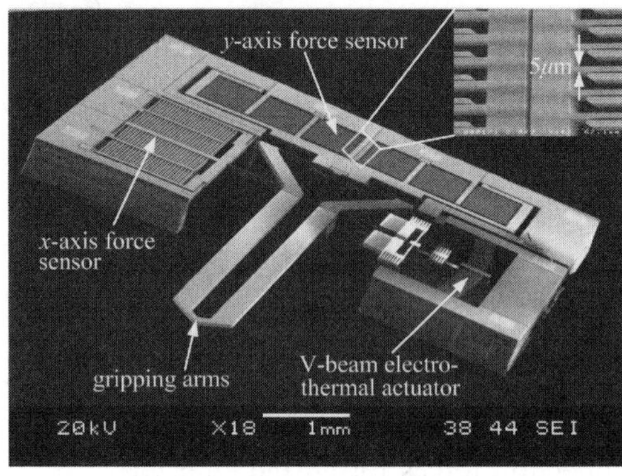

Fig. 12. MEMS-based microgripper with integrated two-axis capacitive force sensors.

three fragmented embryos were arrested at the two-cell stage, proving that the cellular force measurement results could be useful for distinguishing normal embryos from those with embryonic defects during microrobotic cell injection without a separate cell characterization process.

In addition, the cellular force sensing technique can also be used for detecting the penetration of cell membrane in microrobotic cell injection. An abrupt change of cellular forces (Fig. 10) indicates cell membrane penetration for subsequent material deposition. The cellular force does not return to the zero level immediately after penetration since the indented cell does not have sufficient time for recovery during injection. The force required to penetrate the outside membrane (*zona pellucida*) of a healthy ICR embryo was 137.3nN.

III. NanoNewton Force-Controlled Micrograsping of Biological Cells

Compared with end-effectors with a single tip such as a probe or a micropipette for microrobotic cell manipulation, a microgripper with two gripping arms is a more powerful tool for reliable pick-transport-place tasks. Concerning force sensing and control in microgripper-based microrobotic cell manipulation, this section presents the first demonstration of force-controlled micrograsping of biological cells at the nanoNewton force level, which was conducted with a novel, monolithic MEMS-based microgripper with integrated two-axis force sensors.

A. Microrobotic System for Force-Controlled Micrograsping

1) System Setup: The microrobotic system shown in Fig. 11 includes a 3-DOF microrobot (MP-285, Sutter) for positioning the microgripper, a motorized X-Y stage (ProScan II, Prior Scientific) for positioning cell samples, an inverted microscope (TE2000, Nikon) with a CMOS camera (A601f, Basler), a microgripper wire bonded on a circuit board, and a motion control board (6259, National Instruments) mounted on a host computer. The microgripper was tilted at an angle of 40° to enable the gripping arm tips to reach samples on the substrate without immersing the actuator or force sensors into the culture medium.

2) MEMS Microgripper: Over the past two decades, many MEMS microgrippers were developed using different mechanical structures and actuation principles. For example, electrothermal microgrippers without force feedback were developed for cell manipulation [16] and carbon nanotube grasping [17]. Electrostatic microgrippers with a single-axis force sensor was reported for open-loop micrograsping [18] and for investigating charge transport through DNA [19]. Force-controlled micro and nanomanipulation requires microgrippers ideally capable of providing multi-axis force feedback: to protect the fragile microgripper by detecting contact between the microgripper and object to be manipulated; and to provide gripping force feedback for achieving secured grasping without applying excessive forces.

The MEMS microgripper with integrated two-axis force sensors, shown in Fig. 12 was constructed through a modified DRIE-SOI process. The device employs a V-beam electrothermal actuator that is connected to the lower part of a long gripping arm to generate large gripping displacements at gripping arm tips with low driving voltages. As shown in Fig. 13, the gripping arm tip moves by 32μm at 6V; and due to the many heat sink beams, the measured temperature at the gripping arm tip is 29°C in air, demonstrating a low temperature suitable for biomaterial manipulation.

Integrated two-axis capacitive force sensors are imple-

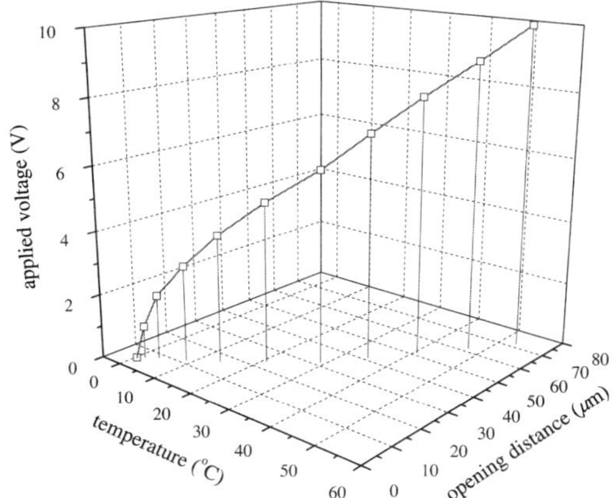

Fig. 13. Measured gripping arm tip displacement and temperature at actuation voltages of 1-10V.

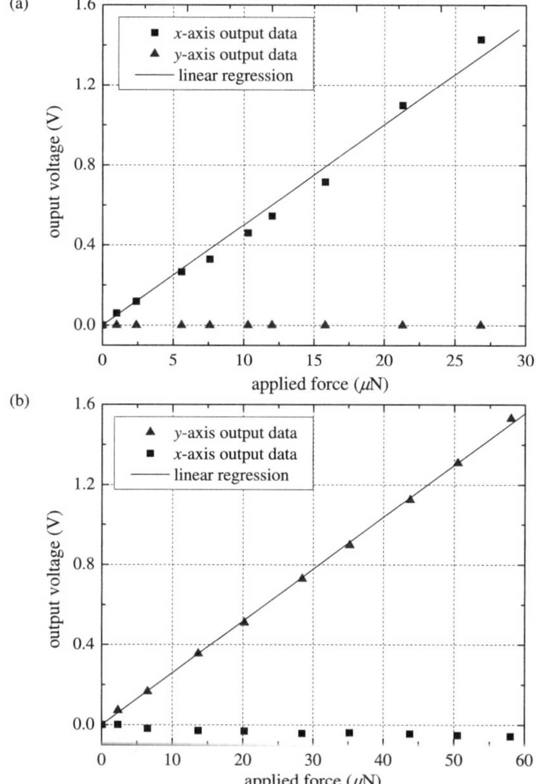

Fig. 14. Force sensor calibration results. Forces applied only (a) along the x direction; (b) only along the y direction. Also shown are coupled responses.

mented with transverse differential comb drives and are orthogonally configured. The contact force feedback (y-directional) enables contact detection and protection of the microgripper from breakage. The gripping force feedback (x-directional) permits force-controlled micrograsping with a force controller to accommodate size and stiffness variations of objects to achieve secured grasping with no excessive forces applied. Fig. 14 shows the force sensor calibration results, demonstrating a high input-output linearity and minimized cross-axis coupling. The integrated force sensors are capable of resolving gripping forces up to 30μN (resolution: 19.9nN) and contact forces up to 58μN (resolution: 38.5nN).

B. Interstitial Cell Preparation

Porcine aortic valve interstitial cells (PAVICs) were manipulated to demonstrate force-controlled micrograsping. Manipulation of single PAVICs with cellular force feedback is required for cell transfer and mechanical characterization in pharmacological studies, such as heart aortic valve calcification.

Aortic valve leaflets were harvested from healthy pig hearts obtained at a local abattoir. After rinsing with antibiotics, each leaflet was treated with collagenase (150U/mL, 37°C, 20min) and the leaflet surfaces were scraped to remove endothelial cells. The leaflets were then minced, and digested with collagenase (150U/mL, 37°C, 2hr). The interstitial cells were enzymatically isolated, grown on tissue culture flasks, and kept in an incubator in standard tissue culture medium (DMEM supplemented with 10% FBS and 1% antibiotics). The medium was changed every 2 days, and the cells were passaged when confluent. P2 cells were trypsinized and re-suspended in standard tissue culture medium at 10^5cells/mL for use in experiments.

C. Experimental Results and Discussion

The experiments were conducted at room temperature (23°C). In order to reduce adhesion of cells to the gripping arm tips and thus, facilitate cell release, the microgripper tips were dip coated with 10% SurfaSil siliconizing fluid (Pierce Chemicals) and 90% histological-grade xylenes (Sigma-Aldrich) for 10sec before use.

1) Contact Detection: A droplet of cell culture medium containing suspended PAVICs (10-20μm in diameter) was dispensed through pipetting on a polystyrene petri dish. After PAVICs settle down on the substrate, a microrobot controls the microgripper to immerse gripping arm tips into the liquid droplet and conducts contact detection.

Contact detection is important to protect the microgripper from damage. After the tips of gripping arms are immersed into the medium, the microrobot controls the microgripper at a constant speed of 20μm/sec to approach the substrate while force data along the y direction are sampled. The contact detection process completes within 5sec. Without the integrated contact force sensor, this process would be extremely time consuming and operator skill dependent.

When the monitored contact force level reaches a pre-set threshold value, it indicates that contact between the gripping arm tips and the substrate is established. Subsequently, the microrobot stops lowering the microgripper further and moves the microgripper upwards until the contact force returns to zero (Fig. 15). After the initial contact position is detected, the microgripper is positioned a few micrometers above the detected contact position. The pre-set threshold force value used in the experiments was 150nN, which was effective for reliably determining the initial contact between the gripping arm tips and the substrate.

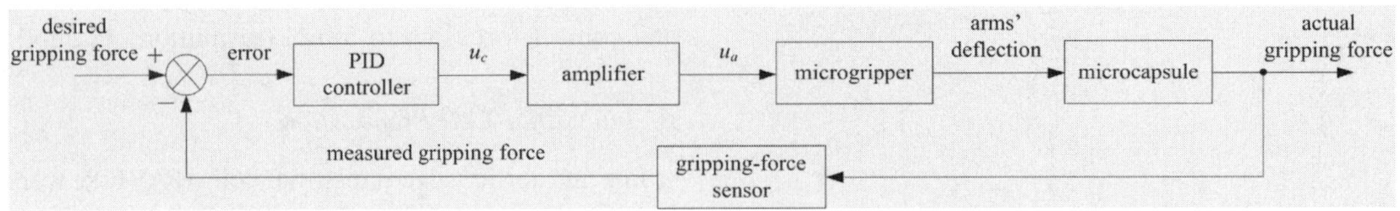

Fig. 17. Block diagram of force-controlled micrograsping.

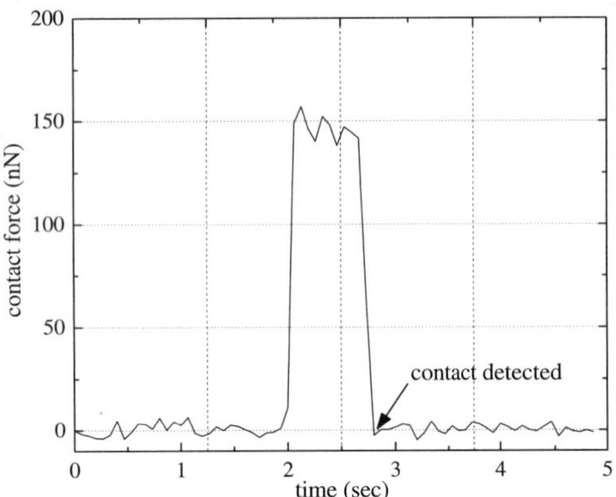

Fig. 15. Contact force monitoring for reliable contact detection.

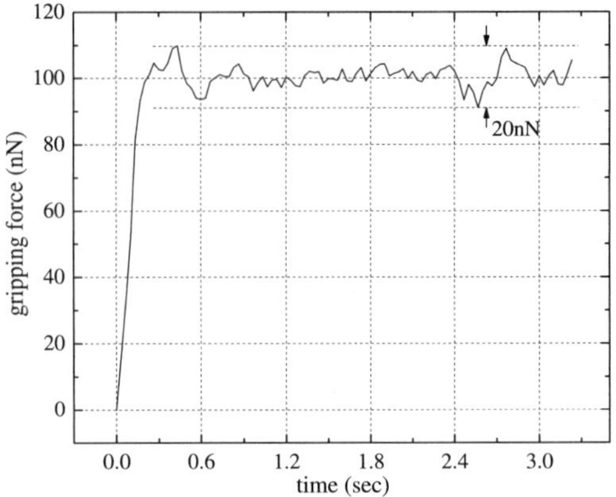

Fig. 18. Step response of force-controlled micrograsping.

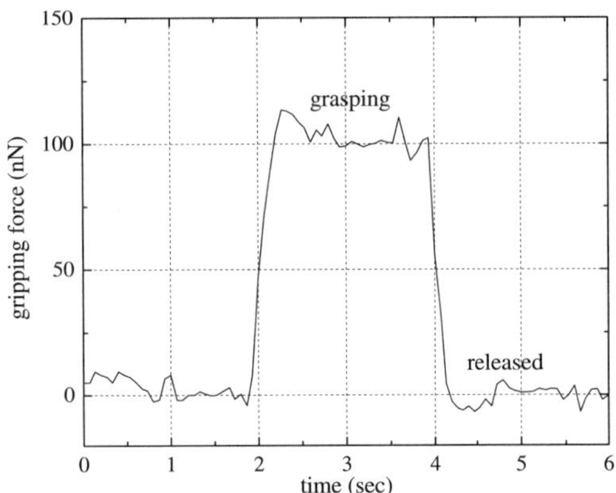

Fig. 16. Gripping force profile during micrograsping and releasing of a cell.

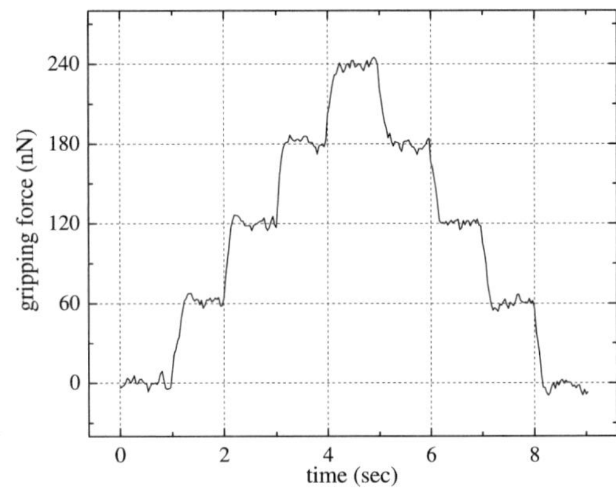

Fig. 19. Tracking force steps with an increment of 60nN.

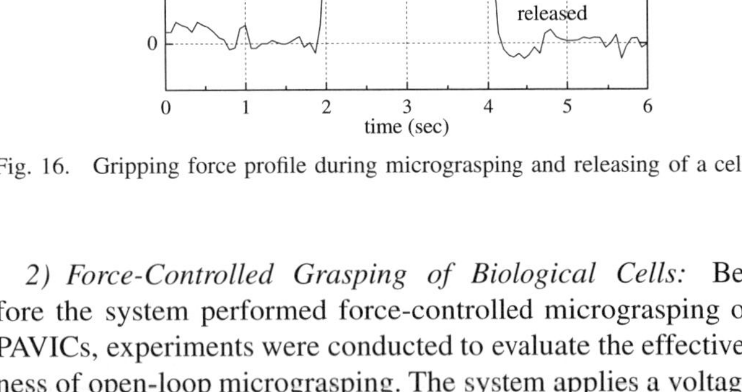

2) Force-Controlled Grasping of Biological Cells: Before the system performed force-controlled micrograsping of PAVICs, experiments were conducted to evaluate the effectiveness of open-loop micrograsping. The system applies a voltage to the V-beam electrothermal actuator to produce an opening larger than the size of a PAVIC between the two gripping arms. When grasping a target PAVIC, the system reduces the applied voltage level, which decreases the arm opening and realizes grasping.

Fig. 16 shows the force profile during cell grasping and releasing, where a sequence of actuation voltages was applied (5V opening voltage, 1.5V grasping voltage, and 5V releasing voltage) to grasp and release a 15μm PAVIC. At 1.5V grasping voltage, the PAVIC was experiencing a gripping force of 100nN that produced 15% cell deformation of its diameter. Due to different sizes of PAVICs and their stiffness variations, a single fixed grasping voltage can often cause either unsecured grasping or cell rupturing from excessively applied forces, necessitating closed-loop force-controlled micrograsping.

To achieve reliable micrograsping, a closed-loop control system was implemented by using gripping force signals as feedback to form a closed loop. Fig. 17 shows the block diagram of the force control system that accepts a pre-set force level as reference input and employs proportional-integral-derivative (PID) control for force-controlled micrograsping. Fig. 18 shows the step response of the force-controlled micrograsping system to track a reference input of 100nN. The settling time is approximately 200ms. Corresponding to reference input force steps with an increment of 60nN, tracking results are shown in Fig. 19.

Enabled by the monolithic microgripper with two-axis force feedback, the microrobotic system demonstrates the capability of rapidly detecting contact, accurately tracking nanoNew-

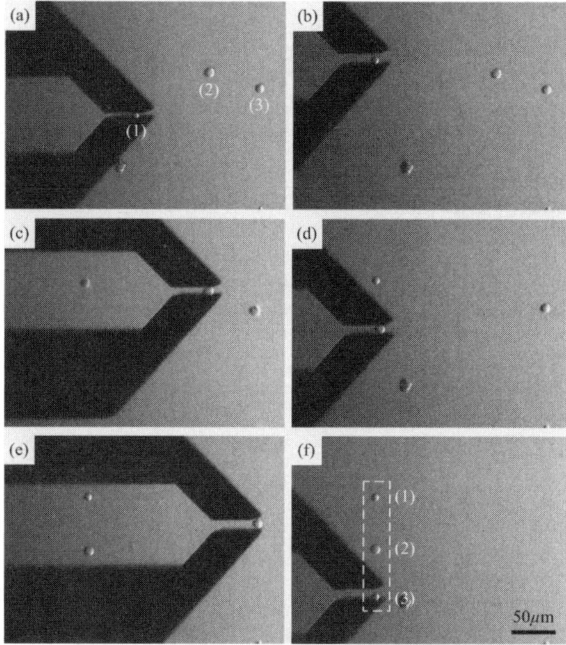

Fig. 20. Cell manipulation and alignment with force-controlled micrograsping. (a) After contact detection, the microgripper grasps a first cell. (b) The microgripper transfers the cell to a new position and releases the cell. (c) The microgripper grasps a second cell. (d) Transferring and releasing the second cell. (e) The microgripper approaches a third cell. (f) Transferring and releasing the third cell. Three cells of different sizes are transferred to desired positions and aligned.

ton gripping forces, and performing reliable force-controlled micrograsping to accommodate size and mechanical property variations of objects. Fig. 20 shows three PAVICs of different sizes that were picked, placed, and aligned. Force-controlled micrograsping of the aligned PAVICs was conducted at a force level of 65nN.

IV. Conclusion

This paper presented nanoNewton cellular force sensing and control in microrobotic cell manipulation. A vision-based cellular force sensing technique was demonstrated for resolving forces with a resolution of 3.7nN during microrobotic mouse embryo injection, based on a microfabricated PDMS cell holding device and a sub-pixel computer vision algorithm with a 0.5 pixel resolution. The acquired cellular force-deformation curves provided important information for differentiating normally developed mouse embryos from those with compromised developmental competence. Additionally, nanoNewton force-controlled micrograsping of cells was demonstrated. The microrobotic system used a close-loop force controller to control a MEMS-based microgripper with integrated two-axis force sensors. The contact force sensor enables rapid detection of the contact between the substrate and gripping arm tips. A PID force controller was used to regulate gripping forces for force-controlled micrograsping. Experimental results on force-controlled grasping of interstitial cells demonstrated that the microrobotic system is capable of performing reliable force-controlled manipulation at a force level of 20nN.

Acknowledgment

The authors thank Jurisicova and Casper groups for the assistance with mouse embryo preparation, and Simmons group for assistance with interstitial cell preparation. This work was supported by the Natural Sciences and Engineering Research Council of Canada, Ontario Centers of Excellence, and the Ontario Ministry of Research and Innovation.

References

[1] Y. Sun and B.J. Nelson, "Biological cell injection using an autonomous microrobotic system," *Int. J. Robot. Res.*, Vol. 21, No. 10-11, pp. 861-868, 2002.

[2] W.H. Wang, X.Y. Liu, D. Gelinas, B. Ciruna, and Y. Sun, "A fully automated robotic system for microinjection of zebrafish embryos," *PLoS ONE*, Vol. 2, No. 9, e862. doi:10.1371/journal.pone.0000862, 2007.

[3] Y. Kimura and R. Yanagimachi, "Intracytoplasmic sperm injection in the mouse," *Biol. Reprod.*, Vol. 52, pp. 709-720, 1995.

[4] Y. Sun, K.T. Wan, B.J. Nelson, J. Bischof, and K. Roberts, "Mechanical property characterization of the mouse zona pellucida," *IEEE Trans. NanoBioSci.*, Vol. 2, No. 4, pp. 279-286, 2003.

[5] R.J. Whelan and R.N. Zare, "Single-cell immunosensors for protein detection," *Biosens. Bioelectron.*, Vol. 19, No. 4, pp. 331-336, 2003.

[6] A. Pillarisetti, M. Pekarev, A.D. Brooks, and J.P. Desai, "Evaluating the effect of force feedback in cell injection," *IEEE Trans. Autom. Sci. Eng.*, Vol. 4, No. 3, pp. 322-331, 2007.

[7] Z. Lu, P.C.Y. Chen, J. Nam, R. Ge, and W. Lin, "A micromanipulation system with dynamic force-feedback for automatic batch microinjection," *J. Micromech. Microeng.*, Vol. 17, pp. 314-321, 2007.

[8] H.B. Huang, D. Sun, J.K. Mills, and W.J. Li, "Visual-based Impedance Force Control of Three-dimensional Cell Injection System," IEEE Conf. ICRA, Rome, Italy, April 2007.

[9] D.H. Kim, C.N. Hwang, Y. Sun, S.H. Lee, B.K. Kim, and B.J. Nelson, "Mechanical analysis of chorion softening in prehatching stages of zebrafish embryos," *IEEE Trans. NanoBioSci.*, Vol. 5, No. 2, pp. 89-94, 2006.

[10] M.A. Greminger and B.J. Nelson, "Vision-based force measurement," *IEEE Trans. Pattern Anal. Mach. Intell.*, Vol. 26, No. 3, pp. 290-298, 2004.

[11] V.V. Dobrokhotov, M.M. Yazdanpanah, S. Pabba, A. Safir, and R.W. Cohn, "Visual force sensing with flexible nanowire buckling springs," *Nanotechnology*, Vol. 19, No. 3, 035502, 2008.

[12] X.Y. Liu, Y. Sun, W.H. Wang, and B.M. Lansdorp, "Vision-based cellular force measurement using an elastic microfabricated device," *J. Micromech. Microeng.*, Vol. 17, pp. 1281-1288, 2007.

[13] Y. Zhao, C.C. Lim, D.B. Sawyer, R. Liao, and X. Zhang, "Microchip for subcelluar mechanics study in living cells," *Sens. Actuat. B - Chem.*, Vol. 114, pp.1108-1115, 2006.

[14] A.C. Ugural and S.K. Fenster, *Advanced Strength and Applied Elasticity*, Prentice Hall, 2003.

[15] A. Jurisicova, S. Varmuza, R.F. Casper, "Involvement of programmed cell death in preimplantation embryo demise," *Hum. Reprod. Update*, Vol. 1, pp. 558-566, 1995.

[16] N. Chronis and L.P. Lee, "Electrothermally activated SU-8 microgripper for single cell manipulation in solution," *J. Microelectromech. Syst.*, Vol. 14, pp. 857-863, 2005.

[17] K. Carlson, K.N. Andersen, V. Eichhorn, D.H. Petersen, K. Mølhave, I.Y.Y. Bu, K.B.K. Teo, W.I. Milne, S. Fatikow, and P. Bøggild, "A carbon nanofibre scanning probe assembled using an electrothermal microgripper,", *Nanotechnology*, Vol. 18, 375501, 2007.

[18] F. Beyeler, A. Nield, S. Oberti, D.J. Bell, Y. Sun, J. Dual, and B.J. Nelson, "Monolithically fabricated microgripper with integratred force sensor for manipulating micro-objects and biological cells aligned in an ultrasonic field," *J. Microelectromech. Syst.*, Vol. 16, No. 1, pp. 7-15, 2007.

[19] C. Yamahata, D. Collard, T. Takekawa, M. Kumemura, G. Hashiguchi, and H. Fujita, "Humidity dependence of charge transport through DNA revealed by silicon-based nanotweezers manipulation," *Biophys. J.*, Vol. 94, pp. 63-70, 2008.

Gas Distribution Modeling using Sparse Gaussian Process Mixture Models

Cyrill Stachniss[1] Christian Plagemann[1] Achim Lilienthal[2] Wolfram Burgard[1]

[1]Albert-Ludwigs-University Freiburg, Dept. for Computer Science, D-79110 Freiburg, Germany
[2]Örebro University, AASS, Dept. of Technology, S-70182 Örebro, Sweden
{stachnis, plagem, burgard}@informatik.uni-freiburg.de, achim.lilienthal@tech.oru.se

***Abstract*— In this paper, we consider the problem of learning a two dimensional spatial model of a gas distribution with a mobile robot. Building maps that can be used to accurately predict the gas concentration at query locations is a challenging task due to the chaotic nature of gas dispersal. We present an approach that formulates this task as a regression problem. To deal with the specific properties of typical gas distributions, we propose a sparse Gaussian process mixture model. This allows us to accurately represent the smooth background signal as well as areas of high concentration. We integrate the sparsification of the training data into an EM procedure used for learning the mixture components and the gating function. Our approach has been implemented and tested using datasets recorded with a real mobile robot equipped with an electronic nose. We demonstrate that our models are well suited for predicting gas concentrations at new query locations and that they outperform alternative methods used in robotics to carry out in this task.**

***Index Terms*— Gas distribution modeling, gas sensing, Gaussian processes, mixture models**

I. Introduction

Gas distribution modeling has important applications in industry, science, and every-day life. Mobile robots equipped with gas sensors are deployed, for example, for pollution monitoring in public areas [1], surveillance of industrial facilities producing harmful gases, or inspection of contaminated areas within rescue missions.

Although humans have a natural odor sensor, it is hard for us to build a spatial representation of a sensed gas distribution. Building gas distribution maps is actually a challenging task due to the chaotic nature of gas dispersal. The complex interaction of gas with its surroundings is dominated by two physical effects. First, on a comparably large timescale, diffusion mixes the gas with the surrounding atmosphere to achieve a homogeneous mixture of both in the long run. Second, turbulent air flow fragments the gas emanating from a source into intermittent *patches* of high concentration with steep gradients at their edges [16]. Especially this chaotic system of localized patches of gas makes the modeling problem a hard one. In addition to that, gas sensors provide information about a small spatial region only since gas sensor measurements require direct interaction between the sensor surface and the analyze molecules. This makes gas sensing different to perceiving the environment with laser range finders or other popular robotic sensors.

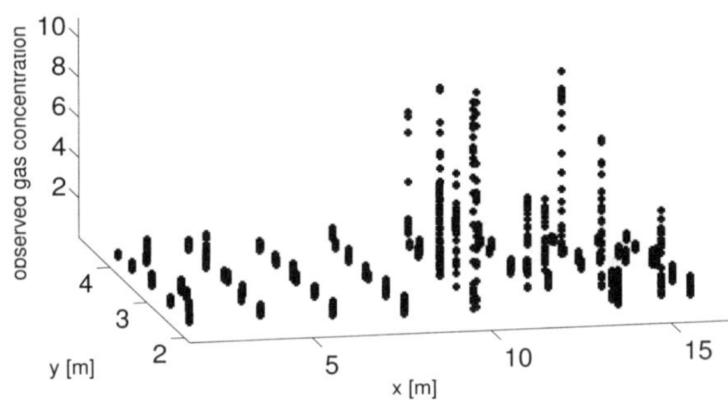

Fig. 1. Gas concentration measurements acquired by a mobile robot in a corridor. The distribution consists of a rather smooth "background" signal and several peaks indicating high gas concentrations.

Fig. 1 illustrates actual gas concentration measurements recorded with a mobile robot along a corridor containing a single gas source. The distribution consists of a rather smooth "background" signal and several peaks indicating high gas concentrations. The challenge in gas distribution mapping is to model this background signal while being able to cover also the areas of high concentration and their sharp boundaries. Since performing measurements is a comparably costly operation, one is also interested in reducing the number of samples needed to build a representation. It is important to note that the noise is dominated by the large fluctuations of the instantaneous gas distribution and not by the electronic noise of the gas sensors. From a probabilistic point of view, the task of modeling a gas distribution can be described as finding a model that best explains the observations and that is able to accurately predict new ones. Thus, the data likelihood in combination with cross validation is the standard criterion to evaluate such a model.

Simple spatial averaging, which represents a straightforward approach to the modeling problem, disregards the different nature of the background noise and the peaks resulting from areas of high gas concentrations and, thus, achieves only limited prediction accuracy. On the other hand, precise physical simulation of the gas dynamics in the environment would require immense computational resources as well as precise knowledge about the physical conditions, which is not known in most practical scenarios.

To achieve a balance between model accuracy and tractability, we treat gas distribution mapping as a two-dimensional regression problem. We derive a solution by means of a sparse mixture model of Gaussian process experts [21] that is able to handle both physical phenomena highlighted above.

Formally, we interpret gas sensor measurements obtained from static sensors or from a mobile robot at locations as noisy samples from a time-constant distribution. This implies that the gas distribution in fact exhibits a time-constant structure, an assumption that is often made in unventilated and un-populated indoor environments [22].

While existing approaches to gas distribution mapping such as local averaging [6, 11], kernel extrapolation techniques [7], or standard GP models represent the average concentration per location only, our mixture model explicitly distinguishes different components of the distribution, i.e., concentration layers varying smoothly due to dispersion processes versus those containing localized *patches* of gas. This leads to a more accurate prediction of the gas concentration. Our model actually allows us to do both, computing the average gas concentration per location (as existing models supply) as well as the multi-modal predictive densities.

The contribution of this paper is a novel approach that learns gas distribution models from sensor data using a sparse Gaussian process mixture model. As a by-product, we present an algorithm that learns a GP mixture model and simultaneously reduces the model complexity in order to achieve an efficient representation even for large data sets. Our technique provides gas concentration estimates for each location in space and also the corresponding predictive uncertainties. The mixture model allows us to improve the gas concentration estimate close to the boundaries and in areas with high gas concentration compared to standard models. As we will demonstrate in experiments carried out with a real robot, our model has a lower mean squared error and a higher data likelihood than other methods and thus allows to more accurately predict gas concentration at query locations.

This paper is organized as follows. After a discussion of related work, we introduce in Sec. III Gaussian processes for regression. Then, Sec. IV explains our approach to learn a sparse GP mixture to model gas distributions from observations. Finally, we present the experimental evaluation of our work with a real mobile robot.

II. Related Work

A straightforward method to create a representation of the time-averaged concentration field is to perform measurements over a prolonged time with a grid of gas sensors. Equidistant gas sensor locations can be used to represent the average concentration values directly on a grid map. This method, though with partially simultaneous measurements, was applied by Ishida *et al.* [6]. A similar method was used in [11] but instead of the average concentration, the peak concentration observed during a sampling period of 20 s was considered to create the map.

Consecutive measurements with a single sensor and time-averaging over 2 minutes for each sensor location were used by Pyk *et al.* [12] to create a map of the distribution of ethanol. Methods, which aim at determining a map of the instantaneous gas distribution from successive concentration measurements, rely on the assumption of a time-constant distribution profile, i.e., uniform delivery and removal of the analyze gas and stable environmental conditions. Thus, the experiments of Pyk *et al.* were performed in a wind tunnel with a constant airflow and a uniform gas source. To make predictions at locations different from the measurement points, they apply bi-cubic interpolation in the case of equidistant measurements and triangle-based cubic filtering in the case where the measurement points are not equally distributed [12]. A problem with these interpolation methods is that there is no means of "averaging out" instantaneous response fluctuations at measurement locations. Even if response values were measured very close to each other, they will appear independently in the gas distribution map with interpolated values in-between. Consequently, interpolation maps tend to get more and more jagged while new measurements are added [8].

Histogram methods take the spatial correlation of concentration measurements into account because of the implicit extrapolation on the measurements by the quantization into histogram bins. Hayes *et al.* [5] suggest a two-dimensional histogram where the bins contain the accumulated number of "odor hits" received in the corresponding area. Odor hits are counted whenever the response level of a gas sensor exceeds a defined threshold. In addition to the dependency of the gas distribution map on the selected threshold, a problem with using only binary information from the gas sensors is that much useful information about fine gradations in the average concentration is discarded. A further disadvantage of histogram methods for gas distribution modeling is their dependency on the bin size and that they require perfectly even coverage of the inspected area.

Kernel extrapolation gas distribution mapping, which can be seen as an extension of histogram methods, was introduced by Lilienthal and Duckett [7]. Spatial integration is carried out by convolving sensor readings and modeling the information content of the point measurements with a Gaussian kernel. As discussed in [8], this method has also an analogy with non-parametric estimation of density functions using a Parzen window method.

Model-based approaches as in Ishida *et al.* [6] infer the parameters of an analytical gas distribution model from the measurements. They depend crucially on the underlying model. Complex numerical models based on fluid dynamics simulations are computationally expensive and depend sensitively on accurate knowledge of the state of the environment (boundary conditions) which is not available in practical situations. Simpler analytical models, on the other hand, often rest on rather unrealistic assumptions and are of course only applicable for situations in which the model assumptions hold. Model-based approaches also rely on well-calibrated gas sensors and an established understanding of the sensor-environment interaction.

The majority of approaches proposed in the literature create a two-dimensional representation and represent time-constant structures in the gas distribution. Also the effort (either in terms of time consumption or the number of sensors) of the model-free approaches to converge to a stable representation, scales quadratically with the size of the environment. None of the approaches suggested so far models the variance together with the time-average of the concentration field.

In contrast to those approaches, we apply Gaussian processes in a mixture model setting to learn probabilistic gas distribution maps. GPs allow us to model the dependency between nearby locations by means of a covariance function. They enable us to make predictions at locations not observed so far and do not only provide the mean gas distribution but also a predictive variance. Our mixture model can furthermore model sharp boundaries of areas with high gas concentration.

Gaussian processes (GPs) are a non-parametric method frequently used to solve regression and classification problems [13]. A drawback of the standard GP approach is its computational complexity. However, several methods for learning sparse GP models [18, 19] have been presented that overcome this limitation and lead to a near-linear complexity [19]. Tresp [21] introduced a mixture model of GP experts to better deal with spatially varying properties in the data. Extensions of this technique using infinite mixtures have been proposed by Rasmussen and Ghahramani [15] and Meeds and Osindero [9].

GPs have already received considerable attention within the robotics community. Schwaighofer *et al.* [17] introduced a positioning system for cellular networks based on Gaussian processes. Brooks *et al.* [2] proposed a Gaussian process model in the context of appearance-based localization with an omni-directional camera. Ferris *et al.* [3] applied Gaussian processes to locate a mobile robot from wireless signal strength. Related Bayesian regression approaches have been also followed for example by Ting *et al.* [20] to identify rigid body dynamics and Grimes *et al.* [4] to learn imitative whole-body motions.

III. Gaussian Processes for Regression

The general gas distribution mapping problem, given a set of gas concentration measurements $y_{1:n}$ acquired at locations $\mathbf{x}_{1:n}$, is to learn a predictive model $p(y_* \mid \mathbf{x}_*, \mathbf{x}_{1:n}, y_{1:n})$ for gas concentrations y_* at a query location $\mathbf{x}_*$. We address this estimation problem as a regression problem. Gaussian processes (GPs) offer a flexible way of solving such regression problems [13]. GPs are a "non-parametric" method, since no parametric form of the underlying function $\mathbf{x} \mapsto y$ is assumed. The model is represented directly using the given training data. GPs can be seen as a generalization of the Gaussian probability distribution to a distribution over functions. A GP for real-valued functions f is defined by a mean function $m(\cdot)$ and a covariance function $k(\cdot,\cdot)$

$$\begin{aligned} m(\mathbf{x}) &= \mathbb{E}[f(\mathbf{x})] & (1)\\ k(\mathbf{x}_p,\mathbf{x}_q) &= \mathbb{E}[(f(\mathbf{x}_p)-m(\mathbf{x}_p))(f(\mathbf{x}_q)-m(\mathbf{x}_q))]. & (2)\end{aligned}$$

In the following, we set $m(\mathbf{x}) = 0$ for simplicity of notation and apply the squared exponential covariance function

$$k(\mathbf{x}_p,\mathbf{x}_q) = \sigma_f^2 \cdot \exp\left(-\frac{1}{2}\frac{|\mathbf{x}_p-\mathbf{x}_q|^2}{l^2}\right). \quad (3)$$

Observations y obtained from the process are assumed to be affected by Gaussian noise, $y \sim \mathcal{N}(m(\mathbf{x}), \sigma_n^2)$. The variables $\Phi = \{\sigma_f, l, \sigma_n\}$ are the so-called hyperparameters of the process which have to be learned from data.

Given a set $\mathcal{D} = \{(\mathbf{x}_i, y_i)\}_{i=1}^n$ of training data where $\mathbf{x}_i \in \mathbb{R}^d$ are the inputs and $y_i \in \mathbb{R}$ the targets, the goal in regression is to predict target values $y_* \in \mathbb{R}$ at a new input point $\mathbf{x}_*$. Let $X = [\mathbf{x}_1; \ldots; \mathbf{x}_n]$ be the $n \times d$ matrix of the inputs and X_* be defined analogously for multiple test data points. In the GP model, any finite set of samples is jointly Gaussian distributed

$$\begin{bmatrix} \mathbf{y} \\ f(X_*) \end{bmatrix} \sim \mathcal{N}\left(\mathbf{0}, \begin{bmatrix} K(X,X)+\sigma_n^2 I & K(X,X_*) \\ K(X_*,X) & K(X_*,X_*) \end{bmatrix}\right), \quad (4)$$

where $K(\cdot,\cdot)$ refers to the matrix with the entries given by the covariance function $k(\cdot,\cdot)$ and $\mathbf{y}$ the vector of the (observed) targets y_i. To actually make predictions at X_*, we obtain for the predictive mean

$$\bar{f}(X_*) := \mathbb{E}[f(X_*)] = K(X_*,X)[K(X,X)+\sigma_n^2 I]^{-1}\mathbf{y} \quad (5)$$

and for the (noise-free) predictive variance

$$\begin{aligned}\mathbb{V}[f(X_*)] = {} & K(X_*,X_*) \\ & -K(X_*,X)[K(X,X)+\sigma_n^2 I]^{-1}K(X,X_*), \quad (6)\end{aligned}$$

where I is the identity matrix. The corresponding (noisy) predictive variance for an observation $\mathbf{y}_*$ can be obtained by adding the noise term σ_n^2 to the individual components of $\mathbb{V}[f(X_*)]$.

GPs are a sound mathematical framework with many practical applications. The standard GP model as described above, however, has two major limitations in our problem domain. First, the computational complexity is high, since to compute the predictive variance given in Eq. (6), one needs to invert the matrix $K(X,X) + \sigma_n^2 I$, which introduces a complexity of $\mathcal{O}(n^3)$ where n is the number of training examples. As a result, an important issue for GP-based solutions to practical problems is the reduction of this complexity. This can, as we will show in Sec. IV, be achieved by artificially limiting the training data set in a way that introduces small loss in the data likelihood of the whole training set while at the same time minimizing the runtime. As a second limitation, the standard GP model generates a *uni-modal* distribution per input location $\mathbf{x}$. This assumption hardly fits our application domain in which a relatively smooth "background" signal is typically *mixed* with high-concentration "packets" of gas. In the following, we address this issue by deriving a *mixture model* of Gaussian processes.

A. Mixtures of Gaussian Process Models

The GP mixture model [21] constitutes a locally weighted sum of several Gaussian process models. For simplicity of notation, we consider without loss of generality the case of single predictions only ($\mathbf{x}_*$ instead of X_*). Let $\{\mathcal{GP}_1, \ldots, \mathcal{GP}_m\}$ be a set of m Gaussian processes representing the individual mixture components. Let $P(z(\mathbf{x}_*) = i)$ be the probability that $\mathbf{x}_*$ is associated with the i-th component of the mixture. Let $\bar{f}_i(\mathbf{x}_*)$ be the mean prediction of the $\mathcal{GP}_i$ at $\mathbf{x}_*$. The likelihood of observing y_* in such as model is thus given by

$$h(\mathbf{x}_*) := p(y_* \mid \mathbf{x}_*) = \sum_{i=1}^{m} P(z(\mathbf{x}_*) = i) \cdot \mathcal{N}_i(y_*; \mathbf{x}_*), \quad (7)$$

where we define $\mathcal{N}_i(y; \mathbf{x})$ as the Gaussian density function with mean $\bar{f}_i(\mathbf{x})$ and variance $\mathbb{V}[f_i(\mathbf{x})] + \sigma_n^2$ evaluated at

y. One can sample from such a mixture by first sampling the mixture component according to $P(z(\mathbf{x}_*) = i)$ and then sampling from the corresponding Gaussian. For some applications such as information-driven exploration missions, it is practical to estimate the mean and variance for this multi-modal model. The mean $\mathbb{E}[h(\mathbf{x}_*)]$ of the mixture model is given by

$$\bar{h}(\mathbf{x}_*) := \mathbb{E}[h(\mathbf{x}_*)] = \sum_{i=1}^{m} P(z(\mathbf{x}_*) = i) \cdot \bar{f}_i(\mathbf{x}_*) \tag{8}$$

and the corresponding variance is computed as

$$\begin{aligned}\mathbb{V}[h(\mathbf{x}_*)] &= \sum_{i=1}^{m}(\mathbb{V}[f_i(\mathbf{x}_*)] + (\bar{f}_i(\mathbf{x}_*) - \bar{h}(\mathbf{x}_*))^2) \\ &\quad \cdot P(z(\mathbf{x}_*) = i).\end{aligned} \tag{9}$$

IV. Learning the Mixture Model from Data

Given a training set $\mathcal{D} = \{(\mathbf{x}_j, y_j)\}_{j=1}^{n}$ of gas concentration measurements y_j and the corresponding sensing locations $\mathbf{x}_j$, the task is to jointly learn the assignment $z(\mathbf{x}_j)$ of data points to mixture components and, given this assignment, the individual regression models $\mathcal{GP}_i$. Tresp [21] describes an approach based on Expectation Maximization (EM) for solving this task. We take his approach, but also seek to minimize the model complexity to achieve a computationally tractable model even for large training data sets $\mathcal{D}$. This is of major importance in our application, since typical gas concentration data sets easily exceed $n = 1000$ data points and the standard GP model (see Sec. III) is of cubic complexity $\mathcal{O}(n^3)$. Different solutions have been proposed for lowering this upper bound, such as dividing the input space into different regions and solving these problems individually or the usage of the so called *sparse GPs*. Sparse GPs [18, 19] use a reduced set of inputs to approximate the full space. This new set can be either a subset of the original inputs [18] or a set of new *pseudo-inputs* [19] which are obtained using an optimization procedure. This reduces the complexity from $\mathcal{O}(n^3)$ to $\mathcal{O}(nm^2)$ with $m \ll n$, which in practice results in a nearly linear complexity. In this section, we describe a greedy forward-selection algorithm integrated into the EM-learning procedure which achieves a sparse mixture model while also maximizing the data likelihood of the whole training set $\mathcal{D}$.

A. Initializing the Mixture Components

In a first step, we subsample n_1 data points and learn a standard GP for this set. This model $\mathcal{GP}_1$ constitutes the first mixture component. To cover areas of gas concentrations that are poorly modeled by this initial model, we learn an "error GP" which models the absolute differences between a set of target values and the predictions of $\mathcal{GP}_1$. We then sample points according to the error GP and use them as the initialization for the next mixture component. In this way, the new mixture is initialized with the data points that are poorly approximated by the first one. This process is continued until the desired number m of model components is reached. For typical gas modeling scenarios, we found that two mixture components are often sufficient to achieve good results. In our experiments, the converged mixture models nicely reflect the bi-modal nature of gas distributions, having one smooth "background" component and a layer of locally concentrated structures as outlined in the introduction of this paper.

B. Iterative Learning via Expectation-Maximization

The Expectation Maximization (EM) algorithm can be used to obtain a maximum likelihood estimate when hidden and observable variables need to be estimated. It consists of two steps, the so-called estimation (E) step and the maximization (M) step which are executed alternately.

In the E-step, we estimate the probability $P(z(\mathbf{x}_j) = i)$ that the data point j corresponds to the model i. This is done by computing the marginal likelihood of each data point for all models individually. Thus, the new $P(z(\mathbf{x}_j) = i)$ is computed given the previous one as

$$P(z(\mathbf{x}_j) = i) \;\leftarrow\; \frac{P(z(\mathbf{x}_j) = i) \cdot \mathcal{N}_i(y_j; \mathbf{x}_j)}{\sum_{k=1}^{m} P(z(\mathbf{x}_j) = k) \cdot \mathcal{N}_k(y_j; \mathbf{x}_j)} \tag{10}$$

In the M-step, we update the components of our mixture model. This is achieved by integrating the probability that a data point belongs to a model component into the individual GP learning steps (see also [21]). This is achieved by modifying Eq. (5) to

$$\bar{f}_i(X_*) \;=\; K(X_*, X)[K(X, X) + \Psi^i]^{-1}\mathbf{y}, \tag{11}$$

where Ψ^i is a matrix with

$$\Psi^i_{jj} \;=\; \frac{\sigma_n^2}{P(z(\mathbf{x}_j) = i)} \tag{12}$$

and zeros in the off-diagonal elements. Eq. (6) is updated respectively. The matrix Ψ^i allows us to consider the probabilities that the individual inputs belong to the corresponding components. The contribution of an unlikely data point to a model is reduced by increasing the data point specific noise term. If the probability, however, is one, only σ_n^2 remains as in the standard GP model.

Learning a GP model also involves the estimation of its hyperparameters $\Phi = \{\sigma_f, l, \sigma_n\}$. To estimate them for $\mathcal{GP}_i$, we first apply a variant of the hyperparameter heuristic used by Snelson and Ghahramani [19] in their open-source implementation. We extended it to incorporate the correspondence probability $P(z(\mathbf{x}_k) = i)$ into this initial guess

$$l \;\leftarrow\; \max_{\mathbf{x}_j} \; P(z(\mathbf{x}_j) = i) \; ||\mathbf{x}_j - \bar{\mathbf{x}}|| \tag{13}$$

$$\sigma_f^2 \;\leftarrow\; \frac{\sum_{j=1}^{n} P(z(\mathbf{x}_j) = i) \; (y_j - \mathbb{E}[y])^2}{\sum_{j=1}^{n} P(z(\mathbf{x}_j) = i)} \tag{14}$$

$$\sigma_n^2 \;\leftarrow\; 0.25 \cdot \sigma_f^2 \,, \tag{15}$$

where $\bar{\mathbf{x}}$ refers to the weighted mean of the inputs—each $\mathbf{x}_j$ having a weight of $P(z(\mathbf{x}_j) = i)$.

To optimize the hyperparameters further given this initial estimate, one could apply, for example, Rasmussen's conjugate-gradient–based optimization technique [14] to minimize the negative log marginal likelihood. In our experiments, however, this approach lead to serious overfitting and we therefore resorted to cross validation-based optimization. Concretely, we randomly sample the hyperparameters and evaluate the model accuracy according to Sec. IV-B on a separate validation set. As a sampling strategy, we draw in each even iteration new parameters from an uninformed prior and in each odd iteration, we improve the current best parameters Θ' by sampling from a Gaussian with mean Θ'. The standard deviation of that Gaussian decreases with the iteration. In our experiments, this strategy found appropriate hyperparameters quickly while significantly reducing the risk of overfitting.

C. Learning the Gating Function

In our mixture model, the gating function defines for each data point the likelihood that it belongs to the individual mixture components. The EM algorithm learns these assignment probabilities for all inputs $\mathbf{x}_j$, maximizing the overall data likelihood. These learned hidden variables are then used to estimate the assignment at an unknown location $\mathbf{x}_*$ by means of regression. Concretely, we learn a gating GP for each component i that uses the $\mathbf{x}_j$ as inputs and the $z(\mathbf{x}_j)$ obtained from the EM algorithm as targets. Let $\bar{f}_i^z(\mathbf{x})$ be the prediction of z for $\mathcal{GP}_i$. Given this set of m GPs, we can compute the correspondence probability for a new test point $\mathbf{x}_*$ as

$$P(z(\mathbf{x}_*) = i) \quad = \quad \frac{\exp(\bar{f}_i^z(\mathbf{x}_*))}{\sum_{j=1}^{m} \exp(\bar{f}_j^z(\mathbf{x}_*))}. \tag{16}$$

D. Illustrating Example

We have specified all quantities that are needed to model gas distributions with sparse Gaussian process mixture models. To summarize the approach, we use a a simple, simulated, one-dimensional example.

The first part of the data points where uniformly distributed around a y value of 2 while the second part was generated with higher noise at two distinct locations. The left image of Fig. 2 depicts the standard GP learned from the input data and the right one the resulting error GP. Based on the error GP, a second mixture component is initialized and used as the input to the EM algorithm.

The individual images in Fig. 3 illustrate the iterations of the EM algorithm. They depict the two components of the mixture model. After convergence, the gating function is learned using the hidden variables reported by the EM algorithm. The learned gating function is depicted in the left image of Fig. 4 and the final GP mixture model is shown in the right image. It is obvious that this model is a better representation of the distribution than the standard GP model shown in the left image of Fig. 2 (averaged negative log likelihood of -1.70 vs. -0.24).

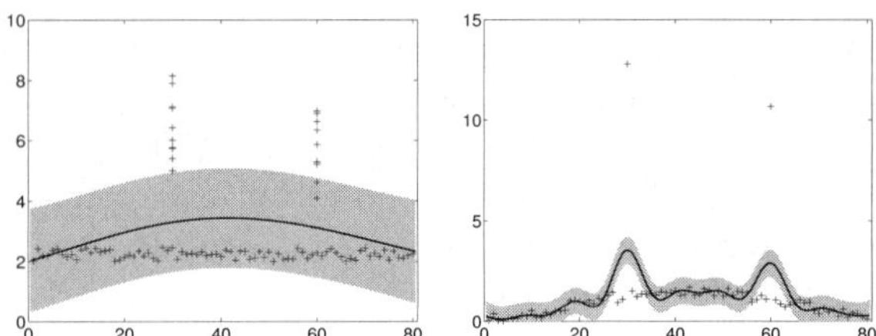

Fig. 2. Left: The standard GP used to initialize the first mixture component. Right: The error GP used to initialize the next mixture component.

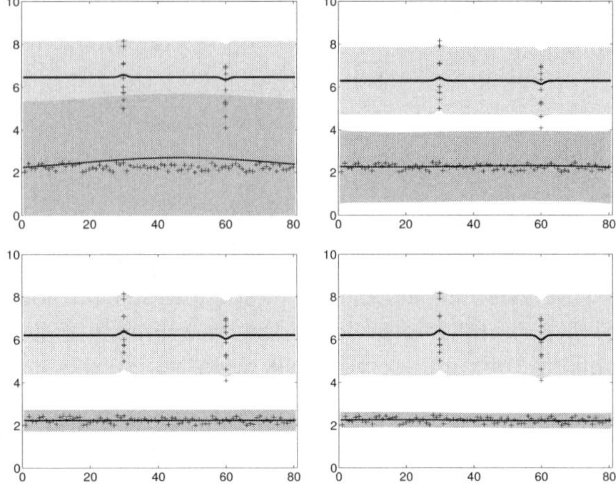

Fig. 3. Components during different iterations of the EM algorithm.

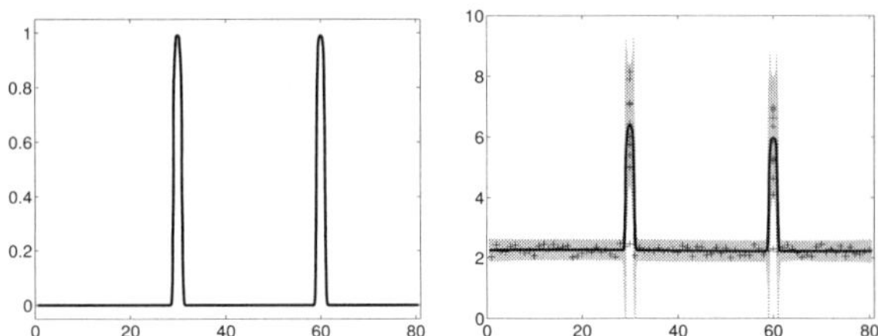

Fig. 4. Left: The learned gating function. Right: Resulting distribution of the GP mixture model.

V. EXPERIMENTS

We carried out pollution monitoring experiments in which the robot followed a predefined sweeping trajectory covering the area of interest. Along its path, the robot was stopped at a pre-defined set of grid points to carry out measurements on the spot between 10 s (outdoors) and 30 s (indoors). The spacing between the grid points was set to values between 0.5 m to 2.0 m depending on the topology of the available space. The sweeping motion was performed twice in opposite directions and the robot was driven at a maximum speed of 5 cm/s in between the stops (to reduce the risk of turbulent air flow due to the motion of the robot). The gas source was a small cup filled with ethanol.

Apart from a SICK laser range scanner used for pose correction, the robot was equipped with an electronic nose and an anemometer. The electronic nose comprises six Figaro gas sensors (2 $\times$ TGS 2600, TGS 2602, TGS 2611, TGS 2620, TGS 4161) enclosed in an aluminum tube. This tube is horizontally mounted at the front side of the robot (see also Fig. 5). The electronic nose is actively ventilated through a fan that creates a constant airflow towards the gas sensors. This lowers the effect of external airflow and the movement of the robot on the sensor response.

Note that in this work, we concentrate only on the gas concentration measurements and do not consider the pose uncertainty of the vehicle. One can apply one of the various SLAM systems available to account for the uncertainty in the robot's pose.

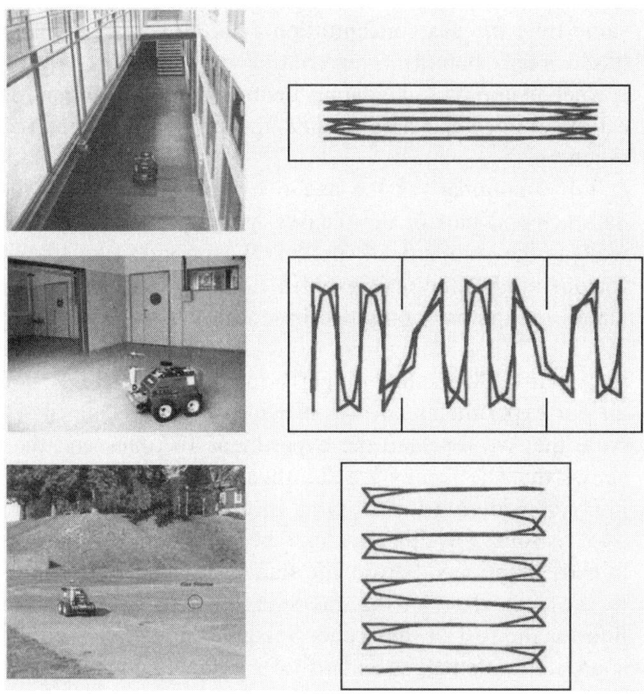

Fig. 5. Pictures of the robot inspecting three different environments as well as the corresponding sweeping trajectories.

TABLE I

AVERAGED NEGATIVE LOG LIKELIHOODS OF TEST DATA POINTS GIVEN THE DIFFERENT MODELS

Dataset	GP	GPM	GPM avg
3-rooms	-1.22	-1.54	-1.50
corridor	-0.98	-1.60	-1.58
outdoor	-1.01	-1.77	-1.69

Three environments with different properties have been selected for the pollution monitoring experiments. The first experiment (*3-rooms*) was carried out in an enclosed indoor area that consists of three rooms which are separated by slightly protruding walls in between them. The area covered by the path of the robot is approximately $14\times6\,\mathrm{m}^2$. There is very little exchange of air with the "outer world" in this environment. The gas source was placed in the central room and all three rooms were monitored by the robot. The second location was a part of a *corridor* with open ends and a high ceiling. The area covered by the trajectory of the robot is approximately $14\times2\,\mathrm{m}^2$. The gas source was placed on the floor in the middle of the investigated corridor segment. Finally, an *outdoor* scenario was considered. Here, the experiments were carried out in an $8\times8\,\mathrm{m}^2$ region that is part of a much bigger open area.

We used the raw sensor readings in all three environments and applied our approach to learn gas distribution models. In the experiments shown here, the robot moved through the environment twice. Therefore, we used the first run for learning the model and the second one for evaluating it. For a comparison with our technique, we also computed a gas distribution model using a standard GP. We furthermore compared our mean estimates to the one of the grid-based method with interpolation and the kernel extrapolation technique.

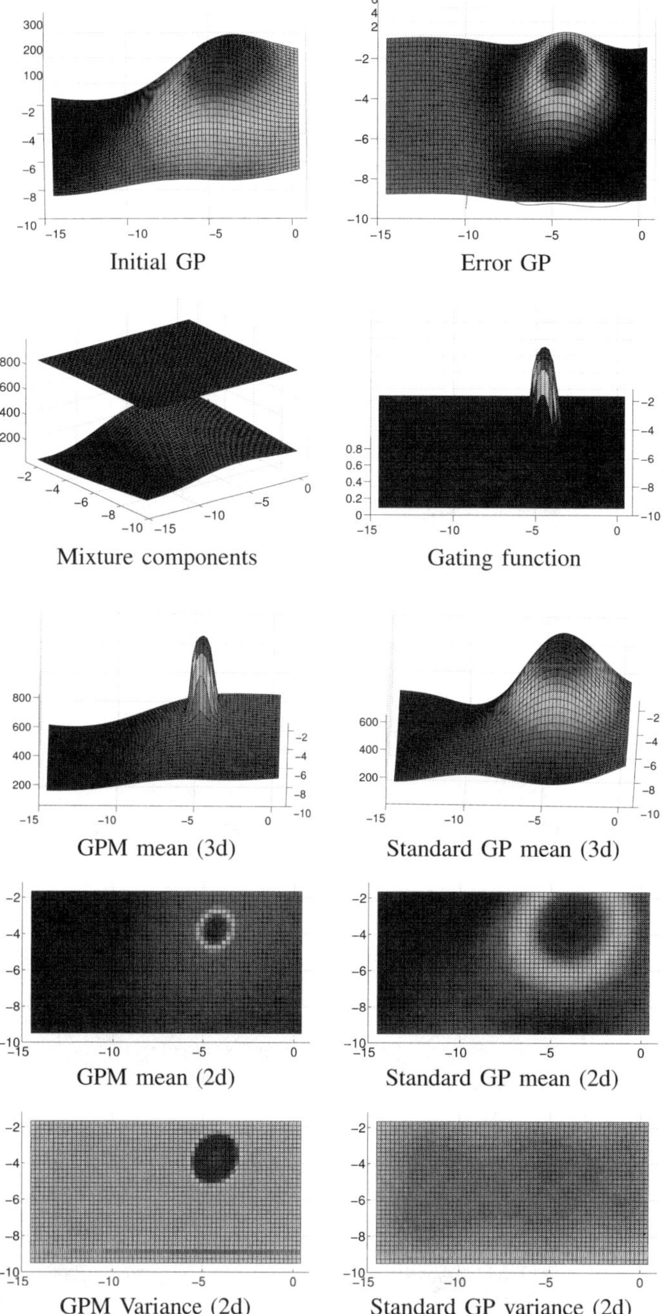

Fig. 6. The 3-rooms dataset with one ethanol gas source in the central room. The room structure itself is not visualized here. In all plots, blue represents low, yellow reflect medium, and red refers to high values.

Fig. 6 depicts the learned models for the 3-room dataset. The left plot in the first row illustrates the mean prediction for the standard GP on the sub-sampled training set which serves as the first mixture component. The right image depicts the error GP representing the differences between the initial prediction and a set of observations. Based on the error GP, a new mixture component is initialized and the EM algorithm is carried out. After convergence, the gating function is learned based on the hidden variables reported by the EM (right image, second row). The left image in the third row shows the final mean prediction of our mixture model. As can be seen, the "background" distribution is smoothly modeled while at the

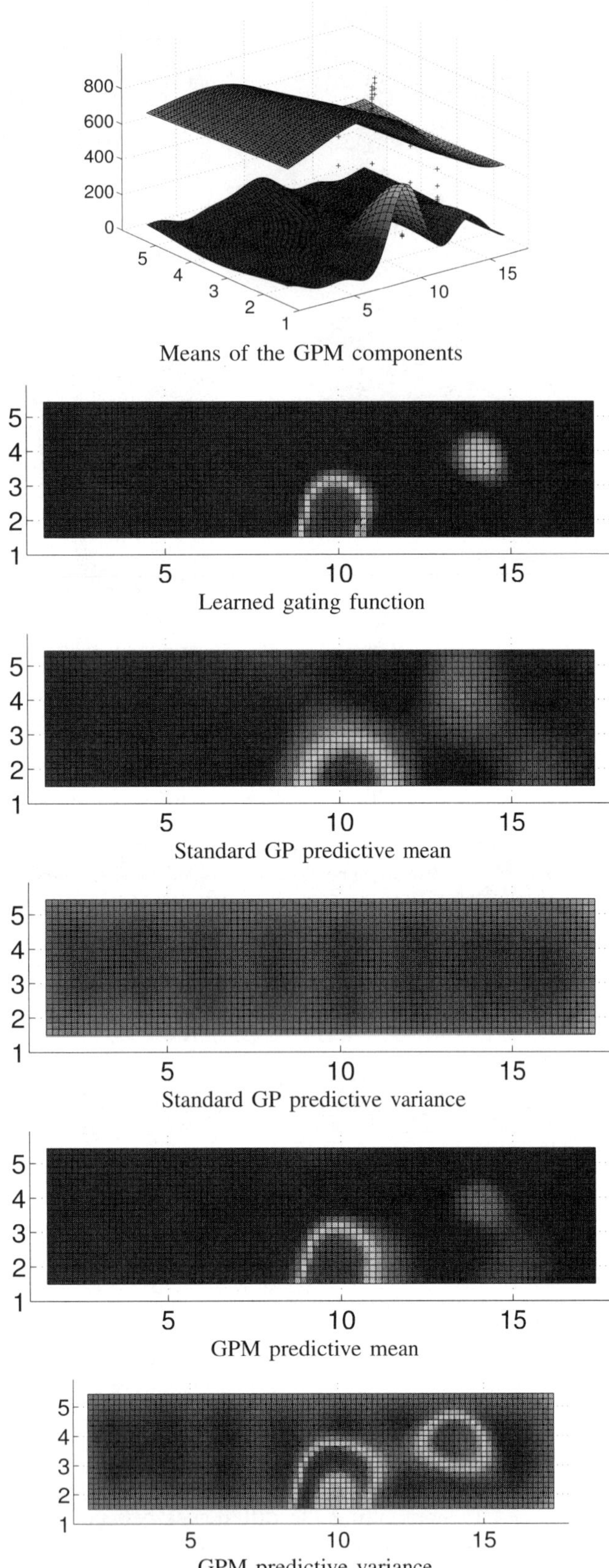

Fig. 7. Models learned from concentration data recorded in the corridor environment (see Fig. 1 for the raw data). The gas source was placed at the location 10, 3. The standard GP and our GPM model provide similar mean estimates. Our approach, however, provides a better predictive uncertainty and thus a higher likelihood given the test data (see Tab. I).

same time the gas concentration peak close to the gas source has a sharp boundary. In contrast to this, the standard GP learned using the same data is unable to provide an appropriate estimate since the area around the peak is to smoothed too much.

Tab. I summarizes the negative log likelihoods of the test data (second part of the dataset) given our mixture model as well as the standard GP model. We provide two likelihoods for our model, the one given in Eq. (7) (called 'GPM' in the table) and the one computed based on the averaged prediction specified in Eq. (8) and Eq. (9) (called 'GPM avg'). As can be seen, our GPM method outperforms the standard GP model in all our experiments since it provides the best data likelihood. Note that we repeated the experiment 10 times and the t-test shows that the results are significant.

By considering the 2d plots in the last two rows of Fig. 6, two reasons for this fact can be observed easily. First, as already mentioned before, the standard GP smoothes too much in the area close to the gas source while this smoothing is fine for the rest of the scene. Second, the variance around the source is too small (standard GPs assumes constant noise for all inputs).

In the corridor experiment, the area of high gas concentration was mapped appropriately also by the standard GP, but again the variance was too small close to the area of high gas concentration. This can be observed by considering Fig. 7. In contrast to this, our GPM model provides a high variance in this area – which actually models the observations in a more precise way. Similar results are obtained in the outdoor dataset. Mean and variance predictions of the standard GP and our model are provided in Fig. 9.

In all our experiments, we limited the number of data points in the reduced input set to $n_1 = 100$ (taken from the first part of the datasets). The datasets itself contained between 2,500 and 3,500 measurements so our model was able to make accurate predictions with less than 5% of the data. Matrices of that size can be easily inverted and as a result the overall computation time to learn our model including cross validation using unoptimized Matlab code on a notebook computer takes around 1 minute for all datasets shown above.

Finally, we compared the mean estimates of our mixture model to the results obtained with the method of Lilienthal and Duckett [7] as well as with the standard approach of using a grid in combination with interpolation. The results of this comparison is shown in Fig. 8. As can be seen, our method outperforms both alternative methods.

VI. Conclusions

In this paper, we considered the problem of modeling gas distributions from sensor measurements by means of sparse Gaussian process mixture models. Gaussian processes are an attractive modeling technique in this context since they do not only provide a gas concentration estimate for each point in the space but also the predictive uncertainty. Our approach learns a GP mixture model and simultaneously decreases the model complexity by reducing the training set in order to achieve an efficient representation even for a large number of observations. This overcomes the major drawback of GPs, their high

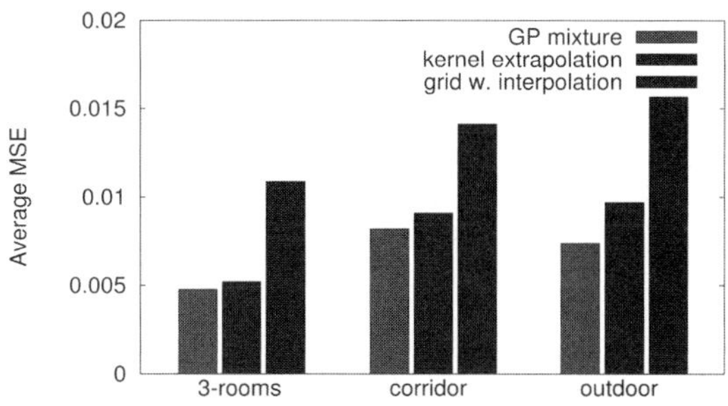

Fig. 8. Mean squared error of the GP mixture model mean and the kernel extrapolation technique and the grid approximation with interpolation.

computational complexity. The mixture model allows us to explicitly distinguish the different components of the spatial gas distribution, namely areas of high gas concentration from the smoothly varying background signal. This improves the accuracy of the gas concentration prediction.

Our method has been implemented and tested using gas sensors mounted on a real robot. With our method, we obtain gas distribution models that better explain the sensor data compared to techniques such as the standard GP regression for gas distribution mapping. Our approach and the one of Lilienthal and Duckett [7] provide similar mean gas concentration estimates, their approach as well as the majority of techniques in the field, however, lack the ability of estimating their predictive uncertainties.

Despite this encouraging results, there is space for further optimizations. Considering non-stationary kernels [10] might further improve the estimates or might serve as an alternative to explictly modeling mixtures. In addition, we are currently exploring the possibility to model the diffusion in high concentration areas by smoothing the gating function over time.

ACKNOWLEDGMENT

This work has partly been supported by the DFG under contract number SFB/TR-8 as well as by the EC under contract number FP6-IST-34120-muFly and FP6-2005-IST-6-RAWSEEDS.

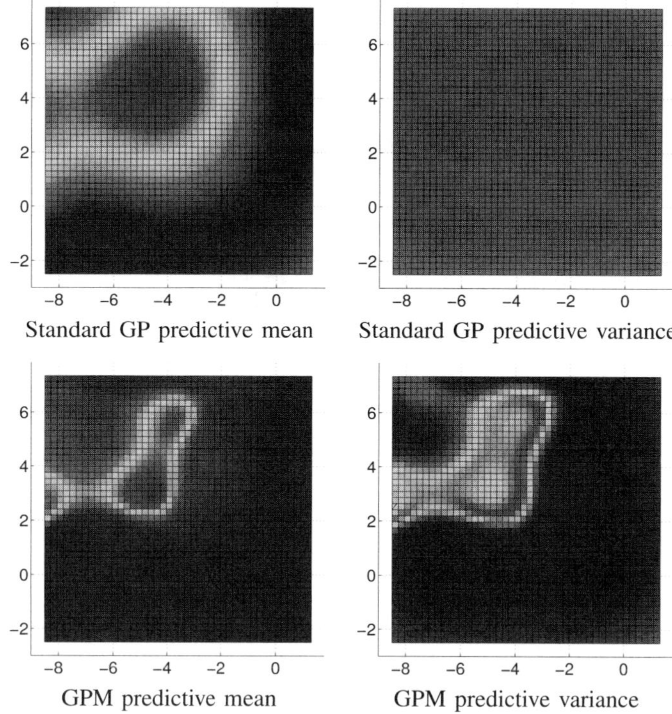

Fig. 9. Outdoor dataset of a 8 m by 8 m area with an ethanol source in the center and airflow. approximatively from south-east to north-west.

REFERENCES

[1] DustBot - Networked and Cooperating Robots for Urban Hygiene. http://www.dustbot.org.

[2] A. Brooks, A. Makarenko, and B. Upcroft. Gaussian process models for sensor-centric robot localisation. In *Proc. of the IEEE Int. Conf. on Robotics & Automation (ICRA)*, 2006.

[3] B. Ferris, D. Haehnel, and D. Fox. Gaussian processes for signal strength-based location estimation. In *Proceedings of Robotics: Science and Systems*, 2006.

[4] D. Grimes, R. Chalodhorn, and R. Rao. Dynamic imitation in a humanoid robot through nonparametric probabilistic inference. In *Proceedings of Robotics: Science and Systems*, 2006.

[5] A.T. Hayes, A. Martinoli, and R.M. Goodman. Distributed Odor Source Localization. *IEEE Sensors Journal, Special Issue on Electronic Nose Technologies*, 2(3):260–273, 2002.

[6] H. Ishida, T. Nakamoto, and T. Moriizumi. Remote Sensing of Gas/Odor Source Location and Concentration Distribution Using Mobile System. *Sensors and Actuators B*, 49:52–57, 1998.

[7] A. Lilienthal and T. Duckett. Building Gas Concentration Gridmaps with a Mobile Robot. *Robotics and Autonomous Systems*, 48(1):3–16, 2004.

[8] A. Lilienthal, A. Loutfi, and T. Duckett. Airborne Chemical Sensing with Mobile Robots. *Sensors*, 6:1616–1678, 2006.

[9] E. Meeds and S. Osindero. An alternative infinite mixture of gaussian process experts. In *Advances in Neural Information Processing Systems*, 2006.

[10] C. Plagemann, K. Kersting, and W. Burgard. Nonstationary gaussian process regression using point estimates of local smoothness. In *Proc. of the European Conference on Machine Learning (ECML)*, Antwerp, Belgium, 2008.

[11] A.H. Purnamadjaja and R.A. Russell. Congregation Behaviour in a Robot Swarm Using Pheromone Communication. In *Proc. of the Australian Conf. on Robotics and Automation*, 2005.

[12] P. Pyk et al. An Artificial Moth: Chemical Source Localization Using a Robot Based Neuronal Model of Moth Optomotor Anemotactic Search. *Autonomous Robots*, 20:197–213, 2006.

[13] C. E. Rasmussen and C. K.I. Williams. *Gaussian Processes for Machine Learning*. The MIT Press, 2006.

[14] C.E. Rasmussen. Minimize. http://www.kyb.tuebingen.mpg.de/bs/people/carl/code/minimize, 2006.

[15] C.E. Rasmussen and Z. Ghahramani. Infinite mixtures of gaussian process experts. In *Advances in Neural Information Processing Systems 14*, 2002.

[16] P.J.W. Roberts and D.R. Webster. Turbulent Diffusion. In H. Shen, A. Cheng, K.-H. Wang, M.H. Teng, and C. Liu, editors, *Environmental Fluid Mechanics - Theories and Application*. ASCE Press, Reston, Virginia, 2002.

[17] A. Schwaighofer, M. Grigoras, V. Tresp, and C. Hoffmann. Gpps: A gaussian process positioning system for cellular networks. In *Proc. of the Conf. on Neural Information Processing Systems (NIPS)*, 2003.

[18] A.J. Smola and P.L. Bartlett. Sparse greedy gaussian process regression. In *NIPS*, pages 619–625, 2000.

[19] E. Snelson and Z. Ghahramani. Sparse gaussian processes using pseudo-inputs. In *Advances in Neural Information Processing Systems 18*, pages 1259–1266, 2006.

[20] J. Ting, M. Mistry, J. Peters, S. Schaal, and J. Nakanishi. A bayesian approach to nonlinear parameter identification for rigid body dynamics. In *Proceedings of Robotics: Science and Systems*, 2006.

[21] V. Tresp. Mixtures of gaussian processes. In *Proc. of the Conf. on Neural Information Processing Systems (NIPS)*, 2000.

[22] M. Wandel, A. Lilienthal, T. Duckett, U. Weimar, and A. Zell. Gas distribution in unventilated indoor environments inspected by a mobile robot. In *Proc. of the Int. Conf. on Advanced Robotics (ICAR)*, pages 507–512, 2003.